Principles of Free Electron Lasers

Henry P. Freund • T. M. Antonsen, Jr.

Principles of Free Electron Lasers

Fourth Edition

Springer

Henry P. Freund
University of New Mexico
University of Maryland
Vienna, VA, USA

T.M. Antonsen, Jr.
University of Maryland
Potomac, MD, USA

ISBN 978-3-031-40944-8 ISBN 978-3-031-40945-5 (eBook)
https://doi.org/10.1007/978-3-031-40945-5

This Springer imprint is published by the registered company Springer Nature Switzerland AG
The registered company address is: Gewerbestrasse 11, 6330 Cham, Switzerland

Paper in this product is recyclable.

To Lena Marion, Anna Jane, Thea, Kal, Annika, Tracy, Thomas Alexander, Margaret Elise, and Christine Marie

Preface

Five years have now passed since the publication of the third edition of *Principles of Free-Electron Lasers* and, while the field has matured considerably, we felt that there have been sufficient new developments to warrant the publication of a new edition. There were four developments that we felt were of particular interest.

The first of these relates to the limitations of the slowly varying envelope approximation (SVEA) which forms the basis for many of the free-electron laser simulation codes that are in widespread use. Because the SVEA is based upon an average of the wave equation over the fast time scale of the optical field, it cannot treat arbitrarily short pulse durations. In order to ascertain and quantify this limitation [1], an SVEA has been compared with a Particle-in-Cell code [2] for ultra-short pulse durations. The results indicate that the SVEA is able to treat pulses as short as a cooperation length. Which corresponds to the spike separation in self-amplified spontaneous emission (SASE) free-electron lasers.

Self-amplified spontaneous emission (SASE) in which the optical field grows from shot noise on the electron beam in a single pass through the wiggler(s) forms the basis for the majority of short wavelength (x-ray) free-electron lasers. However, the SASE interaction is notoriously characterized by relatively large fluctuations from shot to shot in both the power and the spectrum. In order to stabilize the interaction at x ray wavelengths, various oscillator configurations are under study. In particular, the regenerative amplifier (RAFEL) which is a high-gain/low-Q oscillator [3, 4] is receiving increased attention as an alternative to SASE x-ray free electron, and this merits some discussion in this new edition.

The recent development of superconducting wigglers is driven by the desire to go to shorter wiggler periods without sacrificing field strength in order to generate short wavelengths without the need for excessively high beam energies. However, this has also opened up the possibility of generating terawatt x-ray pulses [5, 6], and this is a development of great interest which also merits inclusion in the present edition.

As the number of x-ray free-electron lasers increases worldwide, new applications and uses also grow. While the most common wigglers in use in these facilities are plane-polarized and produce plane-polarized photons, there is also interest in generating a wider range of polarized x-rays. Of course, this can be accomplished by using wigglers which are circularly or elliptically polarized; it may be more convenient to use configurations where successive wigglers are oriented in different directions or in which a plane-polarized wiggler line is capped with an afterburner wiggler with a different polarization state. Most SVEA codes employ a fixed-phasor representation for the optical field and cannot model such varied wiggler configurations. Hence, it was felt to be important to develop an SVEA formalism in which the x- and y-components of the optical field are treated independently [7], and this too merits inclusion in this new edition.

Finally, we felt that the decision made by one of us (HPF) and Dr. P.J.M. van der Slot for the release of the MINERVA SVEA code as an open source to the community would receive more widespread attention if the announcement was made in this new edition.

Vienna, VA, USA

Henry P. Freund
T.M. Antonsen, Jr.

References

1. L.T. Campbell, H.P. Freund, J.R. Henderson, B.W.J. McNeil, P. Traczykowski, P.J.M. van der Slot, Analysis of ultra-short bunches in free-electron lasers. New. J. Phys. **22**, 073031 (2020)
2. L.T. Campbell, B.W.J. McNeil, PUFFIN: a three-dimensional, unaveraged free-electron laser simulation code. Phys. Plasmas **19**, 093119 (2012)
3. H.P. Freund, P.J.M. van der Slot, Yu. Shvyd'ko, An x-ray regenerative amplifier free-electron laser using diamond pinhole mirrors. New J. Phys. **21**, 093028 (2019)
4. G. Marcus, A. Halavanau, Z. Huang, J. Krzywinski, J. MacArthur, R. Margraf, T. Raubenheimer, D. Zhu, Refractive guide switching a regenerative amplifier free-electron laser for high peak and average power hard x-rays. Phys. Rev. Lett. **125**, 254801 (2020)
5. C. Emma, K. Fang, J. Wu, C. Pellegrini, High efficiency, multiterawatt x-ray free-electron lasers. Phys. Rev. Accel. Beams, **19**, 020705 (2016)
6. H.P. Freund, P.J.M. van der Slot, Studies of a terawatt x-ray free-electron laser. New. J. Phys. **20**, 073017 (2018)
7. H.P. Freund, P.J.M. van der Slot, Variable polarization states in free-electron lasers. J. Phys. Commun. **5**, 085011 (2021)

Preface to the Third Edition

It has been more than two decades since the publication of the second edition of *Principles of Free-Electron Lasers*, and it has become increasingly clear that both experimental and theoretical developments in the field have progressed far beyond the content of that prior edition. As a result, we judged it to be important to prepare a third edition to bring the material up to date. It is our intention in this regard to provide a comprehensive description of the present understanding of the principles, theory, and simulation techniques of the free-electron laser that can be used both as a reference work and a handbook. To that end, extensive derivations are given for the spontaneous emission, the single-particle orbits, and both the linear and nonlinear formulations of the interaction as a resource for the student or the experienced researcher in the field. However, we also provide useful formulae that can be used as a starting point in the design and analysis of specific free-electron laser configurations.

To that end, the work builds upon the presentation of the second edition. At that time, free-electron laser research and development encompassed both long wavelength free-electron masers using pulse line accelerators, modulators, and induction linacs and short wavelength infrared through ultraviolet free-electron lasers based upon radio-frequency (rf) linacs and storage rings. Since that time, however, the research into long wavelength free electron lasers has withered, although it has not completely disappeared, while the development of short wavelength free-electron lasers has flowered. While we retain the description(s) relevant the long wavelength free-electron lasers in the book, the bulk of the new material in this edition is devoted to the analysis and simulation of short wavelength free-electron lasers.

The flowering of short wavelength free-electron lasers has its genesis in the development of laser driven photo-cathodes [1–3] and in the application of photo-cathodes to the electron guns in the injectors of rf linacs [4], which has enabled the production of high quality electron beams with low emittances. There are two principal thrusts of this development. One is the quest for high average power, and this is exemplified in the application of an energy recovery rf linac for use in the high-average power infrared oscillator at the Thomas Jefferson National Accelerator Facility which produced an average power of 14-kW at a wavelength of 1.6 μm [5, 6]. The second thrust is toward ever shorter wavelengths with high peak powers, and this is epitomized by the Linac Coherent Light Source (LCLS) at the Stanford Linear Accelerator Center [7]. The LCLS is the first of its kind fourth generation x-ray light source user facility that came online in 2009 and produces intense, short pulses of x-rays at wavelengths as short as 1.5 Å. At the present time, there is a great deal of activity worldwide in the design and construction of fourth generation free-electron lasers.

The LCLS and many of the other fourth generation light sources rely on the amplification of shot noise on the electron beam to high peak power levels on a single pass through a long wiggler. Because of this, there are relatively large shot to shot fluctuations in the output spectra and power levels. Since this is undesirable for many applications, research into techniques to produce coherent short wavelength free-electron lasers is ongoing. One possible approach is through what is termed high-gain harmonic generation (HGHG). In HGHG, a two-segment wiggler is used in which a magnetic dispersive element is placed in the gap between the wigglers. In this configuration, the electrons are injected into the first wiggler (termed the *modulator*) in synchronism with a high-power seed pulse which acts to impose a modulation on the longitudinal velocity and density profiles of the electrons. In most cases, the dispersive element is a chicane composed of dipole magnets in which higher energy electrons in the tail of the electron bunch overtake lower energy electrons in the head of the bunch. As a result, the modulation imposed on the electrons in the *modulator* is enhanced by the chicane, and this acts to precondition the electrons for rapid radiation in the second wiggler (called the *radiator*). In the HGHG configuration, the *radiator* is tuned to a harmonic of the resonant wavelength in the *modulator* so that coherent radiation at wavelengths shorter than the seed laser used for the *modulator* can be produced. The first HGHG user facility operating at extreme ultraviolet wavelengths is presently in existence at Fermi-Elletra in Trieste, Italy [8].

In order to provide an analysis of many of these new developments in short wavelength free-electron lasers, one of the principal additions included in this edition is a derivation and justification of the application of the slowly varying envelope approximation to the time-dependent simulation of short pulse free-electron lasers. This is one of the most important developments over the last two decades. In so doing, we have made an effort in this book to compare theory and simulations with as many actual experiments as possible in order to demonstrate the validity of our present simulation capabilities.

Although the principal focus of the book is on the free-electron laser interaction in the wiggler, the current thrust toward ultra-short wavelength light sources utilizing very long multi-segment wigglers with strong focusing systems and phase shifters necessitates some discussion of the elements of electron beam optics needed to understand these systems for readers that may not be familiar with these concepts. As a result, we have included an Appendix to explain the basic elements needed to understand the complex transport and focusing of electron beams, and we provide references to many of the fundamental works in the field. The number of reference works dealing solely with electron optics is extensive and complete, and we hope that the student of the field will use this Appendix, not only as a description of the relevant electron optics as used in free-electron lasers, but also as a gateway to more complete treatises.

Vienna, VA, USA

Henry P. Freund
T.M. Antonsen, Jr.

References

1. E. Garwin, F. Meier, T. Pierce, K. Sattler, H.-C. Siegmann, A pulsed source of spin-polarized electrons by photoemission from EuO. Nucl. Instrum. Meth. **120**, 483 (1974)
2. D.T. Pierce, F. Meier, Photoemisison of spin-polarized electrons from GaS. Phys. Rev. B **13**, 5484 (1976)
3. C.K. Sinclair, R.H. Miller, A high current, short pulse, rf synchronized electron gun for the Stanford linear accelerator. IEEE Trans. Nuclear Sci. **NS-28**, 2649 (1981)
4. R.L. Sheffield, E.R. Gray, J.S. Fraser, The Los Alamos photoinjector program. Nucl. Instrum. Meth. **A272**, 222 (1988)
5. G.R. Neil, C. Behre, S.V. Benson, M. Bevins, G. Biallas, J. Boyce, J. Coleman, L.A. Dillon-Townes, D. Douglas, H.F. Dylla, R. Evans, A. Grippo, D. Gruber, J. Gubeli, D. Hardy, C. Hernandez-Garcia, K. Jordan, M.J. Kelley, L.Merminga, J. Mammosser, W. Moore, N. Nishimori, E. Pozdeyev, J. Preble, R. Rimmer, M. Shinn, T. Siggins, C. Tennant, R. Walker, G.P. Williams, S. Zhang, The JLab high power ERL light source. Nucl. Instrum. Methods Phys. Res. **A557**, 9 (2006)
6. P.J.M. van der Slot, H.P. Freund, W.H. Miner, Jr., S.V. Benson, M. Shinn, K.-J. Boller, Time-dependent, three-dimensional simulation of free-electron laser oscillators. Phys. Rev. Lett. **102**, 244802 (2009)
7. P. Emma et al., First lasing and operation of an Ångstrom-wavelength free-electron laser. Nature Phot. **4**, 641 (2009)
8. E. Allaria et al., Highly coherent and stable pulses from the FERMI seeded free-electron laser in the extreme ultraviolet. Nature Phot. **6**, 699 (2012)

Preface to the Second Edition

The primary consideration involved in contemplating the utility of a second edition of any book is whether or not the weight of new developments in the field warrants the effort. In the case of free-electron lasers, the field has been growing so rapidly that we judged this to be the case. This rapid growth has occurred both in the number of active free-electron laser experiments and user facilities worldwide, and in the theoretical understanding of the various operating regimes. As such, we felt that the first edition was becoming out-dated, and that a second edition was necessary to give readers a complete description of the current state of the theory of free-electron lasers.

In organizing the material to be included in the second edition, we did not feel it practicable to rewrite the entire volume. If we were to start with a clean slate, then much of the new material would be incorporated directly into the existing chapters. However, in order to minimize the composition costs of the second edition, we chose in most cases to add new chapters rather than rewrite the existing material.

There were, however, compelling reasons for modifying several chapters. Firstly, while our purpose is not that of a review in which a history of the experimental development of free-electron lasers is important, we do provide (1) a brief description of selected experiments in order to illustrate and validate the theory, and (2) an overview of the primary applications of free-electron lasers. This has necessitated some revision in Chap. 1 as well as a substantial updating of the references contained therein. Secondly, the material dealing with spontaneous undulator radiation in Chap. 3 deals exclusively with an infinite and uniform transverse configuration. This is insufficient in the treatment of superradiance in Chap. 15 (new to this edition); hence, we have added some description of the spontaneous radiation in a waveguide in the new chapter. Finally, we have corrected an omission in Chap. 5 dealing with the nonlinear formulation of free-electron laser amplifiers. The first edition was written largely with separate discussions of planar and helical wiggler geometries, and Chap. 5 included discussions of the nonlinear analysis of helical wigglers in both one and three dimensions, but discussed only the three-dimensional analysis of planar wigglers. The lack of a one-dimensional analysis of planar wigglers is not a serious omission, but one which we felt is important to correct. As a result, a new section dealing with this subject has been incorporated into Chap. 5.

With these exceptions, all new material has been organized into four new chapters dealing with (1) wiggler imperfections, (2) the reversed-field configuration, (3) collective effects, and (4) amplification of spontaneous emission and superradiance. These are, in our view, the primary fields in which advances have been made since the publication of the first edition.

The issue of the effects of wiggler imperfections has important practical implications in the design of free-electron lasers. A great deal of effort has been expended in the design of wigglers which minimize field imperfections, as well as in the incorporation of external steering magnets to correct for known imperfections. As a result, we felt that the inclusion of a chapter on the effects and importance of wiggler imperfections would be an important addition.

The reversed-field configuration refers to a recent experiment in which a helical wiggler was used in conjunction with an axial solenoidal field directed anti-parallel to the wiggler. This configuration had not been studied previously since the combined effects of the two fields in this orientation would result in a reduction in the transverse wiggler-induced electron velocity and, in turn, a reduction in the gain. Indeed, this has proven to be the case; however, the reduction in gain occurred along with a relatively high efficiency. The maximum efficiency found in the experiment (which used a uniform helical wiggler) was in the neighborhood of 27% at a frequency of 35 GHz, which compared favorably with the previous record high efficiency of 35% at the same frequency using a tapered wiggler. As a result, no second edition would be complete without the inclusion of a description of this important experiment.

The treatment of collective effects in free-electron lasers has been addressed in the first edition. It became clear in the three years since the initial publication, however, that there were still misunderstood aspects of both the importance of and subtleties in the theoretical analysis of collective effects. Hence, a chapter discussing these points was felt to be important. This includes both a discussion of the Raman regime in which the beam space-charge wave is important and of the analysis of self-electric and self-magnetic fields due to the DC charge and current densities of the beam. Both cases involve an analysis of several experiments in order to illustrate criteria for evaluation of the importance of collective effects.

The last new chapter deals with superradiance in free-electron lasers. There is some ambiguity in what this term means. In early work on free-electron lasers the term *superradiant amplifier* was used to denote an experiment in which no drive signal was imposed and the radiation grew from noise in a single pass through the wiggler. However, this type of radiation is now referred to as Self-Amplified Spontaneous Emission (SASE), and the term superradiance is also used to refer to cases in which the radiation pulse breaks up into large-amplitude spikes. The nature of this process was still controversial at the time the first edition was published, and we chose to omit it from the discussion. Since this is no longer the case, we felt it important to include a discussion of these effects in the second edition.

Of course, it is also important to correct the inevitable typographical errors which creep into the text during the typesetting process, and we apologize to any readers of the first edition for the confusion they may have caused. Finally, we refer interested readers to the more recent proceedings of the annual free-electron laser conferences for more complete summaries of the experimental progress in the field [1–3].

Vienna, VA, USA

Henry P. Freund
T.M. Antonsen, Jr.

References

1. Free-electron Lasers, in *Nuclear Instruments and Methods in Physics Research*, Vol. A318, ed. by J.C. Goldstein, B.E. Newnam (North-Holland, Amsterdam, 1992)
2. Free-electron Lasers, in *Nuclear Instruments and Methods in Physics Research*, Vol. A331, ed. by C. Yamanaka, K. Mima (North-Holland, Amsterdam, 1993)
3. Free-electron Lasers, in *Nuclear Instruments and Methods in Physics Research*, Vol. A341, ed. by P.W. van Amersfoort, P.J.M. van der Slot (North-Holland, Amsterdam, 1994)

Preface to the First Edition

At the time that we decided to begin work on this book, several other volumes on the free-electron laser had either been published or were in press. The earliest work of which we were aware was published in 1985 by Dr. T.C. Marshall of Columbia University [1]. This book dealt with the full range of research on free-electron lasers, including an overview of the extant experiments. However, the field has matured a great deal since that time and, in our judgement, the time was ripe for a more extensive work which includes the most recent advances in the field.

The fundamental work in this field has largely been approached from two distinct and, unfortunately, separate viewpoints. On the one hand, free-electron lasers at sub-millimeter and longer wavelengths driven by low-energy and high-current electron beams have been pursued by the plasma physics and microwave tube communities. This work has largely confined itself to the high-gain regimes in which collective effects may play an important role. On the other hand, short wavelength free-electron lasers in the infrared and optical regimes have been pursued by the accelerator and laser physics community. Due to the high-energy and low-current electron beams appropriate to this spectral range, these experiments have largely operated in the low-gain single-particle regimes. The most recent books published on the free-electron laser by Dr. C.A. Brau [2] and Drs. P. Luchini and H. Motz [3] are excellent descriptions of the free-electron laser in this low-gain single-particle regime. In contrast, it is our intention in this book to present a coherent description of the linear and nonlinear aspects of both the high- and low-gain regimes. In this way, we hope to illustrate the essential unity of the interaction mechanism across the entire spectral range. However, the reader should bear in mind that our own principal research interests derive from the high-gain millimeter and sub-millimeter regime, and the specific examples of experiments we describe are largely confined to this regime. In most cases, however, these cases are adequate to demonstrate the essential physics of the free-electron laser. Indeed, many of the first laboratory demonstrations of the physical principals of the free-electron laser were conducted in the microwave/millimeter-wave regime.

The organization of the book was chosen to appeal to the reader on a variety of levels. It was our intention to write a book which can be approached both by the novice seeking to begin study of the free-electron laser as well as the expert who wishes to approach the subject at great depth. This has forced us to adopt a multi-level approach to the subject. At the lowest level, the reader unfamiliar with free-electron lasers can find an extensive description of the fundamental physics of the free-electron laser in Chap. 1. This chapter is largely taken from an article on free-electron lasers written by Drs. H.P. Freund and R.K. Parker which appeared in the 1991 yearbook of the Academic Press Encyclopedia on Physical Science and Technology [4]. It includes (1) a brief history and summary of the experimental research on the free-electron laser, (2) a description of the essential operating regimes of the free-electron laser, including formulae for the linear gain, nonlinear saturation level, electron beam quality requirements, and efficiency enhancement by means of a tapered wiggler, and (3) a survey of proposed applications. In this regard, an extensive bibliography of the experimental literature is included for the reader who desires to conduct a more complete survey. At the next level, each subsequent chapter includes an introductory section which describes the essential physics to be discussed in that chapter. The highest level is contained in the bulk of each chapter which is devoted to an extensive and in-depth presentation of the appropriate subject.

At the highest level, we have not shrunk from the task of presenting a detailed derivation of each topic of interest, and have included several different methods of derivation of some important quantities. For example, we have employed the Vlasov-Maxwell equations in the study of the linear stability of the free-electron laser in Chap. 4 in the idealized one-dimensional analysis of both the low-gain and high-gain regimes. The purpose of this is twofold. In the first place, it serves to illustrate the relationship between these two operating regimes. In the second place, it allows the effect of a beam thermal spread on the linear gain to be analyzed. In contrast, the Vlasov-Maxwell formalism is retained in the three-dimensional stability analysis only in the high-gain regime. The small-signal gain in the three-dimensional analysis is treated by the more conventional approach based upon a phase average of electron motion in the ponderomotive wave formed by the beating of the wiggler and radiation field. The result obtained by this method, however, is a straightforward extension of that found in the idealized one-dimensional approach, and includes a filling factor which describes the overlap of the electron beam and the radiation field.

The bulk of free-electron laser designs have employed wigglers with either helical or planar symmetry. Heretofore, texts dealing with free-electron lasers have typically concentrated on a discussion of the physics of the interaction for one or the other wiggler geometry, with a brief discussion of the generalization required to obtain the results for the other. However, there are essential differences in the character of the interaction for each of these wiggler designs. Hence, we have chosen to present the linear and nonlinear analyses of each of these configurations. While this approach adds to the length of the presentation, we feel that this is necessary in order to treat the field in adequate depth.

The essential focus of the book is on the theory of free-electron lasers. While the meaning of this term has sometimes been extended to alternate concepts such as cyclotron and Cerenkov masers which also make use of free electrons, we use the term here to refer solely to devices which rely upon a periodic magnetic field to mediate the interaction of the electron beam with the electromagnetic wave. In this regard, we do not include a detailed description of the wide range of experiments conducted in the field. The only experiments which are described in the text are those which illustrate some essential point of the discussion. We have chosen this course both because of constraints on the length of the manuscript and because this aspect of the field is subject to rapid changes due to advances in the technology of electron beams, wiggler designs, and optics. Similarly, we have not included examples of the application of the theory of free-electron lasers to speculative designs of coherent ultraviolet and X-ray sources, although the theoretical tools to analyze such designs are included. Such devices are under study at the present time, but are limited by the technology related to the production of high brightness electron beams and mirrors with high reflectivities at these wavelengths.

A more appropriate place for a broad discussion of the experimental base and speculative designs are review papers on the field as well as the proceedings of the annual free-electron laser conferences. We recommend the interested reader to the recent article by C.W. Roberson and P. Sprangle [5] for an extensive summary of the experimental base. In addition, excellent year-by-year summaries of the experimental and speculative literature are to be found in the proceedings of the annual free-electron laser conferences dating back to 1977 [6–17].

Vienna, VA, USA
January, 1992

Henry P. Freund
T.M. Antonsen, Jr.

References

1. T.C. Marshall, *Free-Electron Lasers* (McMillan, New York, 1985)
2. C.A. Brau, *Free-Electron Lasers* (Academic, Boston, 1990)
3. P. Luchini, H. Motz, *Undulators and Free-Electron Lasers* (Clarendon Press, Oxford, 1990)
4. H.P. Freund, R.K. Parker, *Free-Electron Lasers*, in the 1991 *Yearbook of the Encyclopedia of Physical Science and Technology* (Academic Press, Inc., San Diego, California), pp. 49–71
5. C.W. Roberson, P. Sprangle, A review of free-electron lasers. Phys. Fluids B **1**, 3 (1989)
6. *The Physics of Quantum Electronics: Novel Sources of Coherent Radiation*, eds. by S.F. Jacobs, M. Sargent, M.O. Scully, Vol. 5 (Addison-Wesley, Reading, 1978)
7. *The Physics of Quantum Electronics: Free-Electron Generators of Coherent Radiation*, eds. by S.F. Jacobs, H.S. Pilloff, M. Sargent, M.O. Scully, R. Spitzer, Vol. 7 (Addison-Wesley, Reading, Massachusetts, 1980)
8. *The Physics of Quantum Electronics: Free-Electron Generators of Coherent Radiation*, eds. by S.F. Jacobs, G.T. Moore, H.S. Pilloff, M. Sargent, M.O. Scully, R. Spitzer, Vols. 8 and 9 (Addison-Wesley, Reading, 1982)

9. *Bendor Free-Electron Laser Conference*, eds. by D.A.G. Deacon, M. Billardon, Journal de Physique Colloque **C1-44** (1983)
10. *Free-Electron Generators of Coherent Radiation*, eds. by C.A. Brau, S.F. Jacobs, M.O. Scully (Proc. SPIE 453, Bellingham, Washington, 1984)
11. Free-Electron Lasers, eds. J.M.J. Madey, A. Renieri, in *Nuclear Instruments and Methods in Physics Research*, Vol. A237 (North-Holland, Amsterdam, 1985)
12. Free-Electron Lasers, eds. E.T. Scharlemann, D. Prosnitz, in *Nuclear Instruments and Methods in Physics Research*, Vol. A250 (North-Holland, Amsterdam, 1986)
13. Free-Electron Lasers, ed. by M.W. Poole, in *Nuclear Instruments and Methods in Physics Research*, Vol. A259 (North-Holland, Amsterdam, 1987)
14. Free-Electron Lasers, eds. by P. Sprangle, C.M. Tang, J. Walsh, in *Nuclear Instruments and Methods in Physics Research*, Vol. A272 (North-Holland, Amsterdam, 1988)
15. Free-Electron Lasers, eds. by A. Gover, V.L. Granatstein, in *Nuclear Instruments and Methods in Physics Research*, Vol. A285 (North-Holland, Amsterdam, 1989)
16. Free-Electron Lasers, eds. by L.R. Elias, I. Kimel, in *Nuclear Instruments and Methods in Physics Research*, Vol. A296 (North-Holland, Amsterdam, 1990)
17. Free-Electron Lasers, eds. by J.M. Buzzi, J.M. Ortega, in *Nuclear Instruments and Methods in Physics Research*, Vol. A304 (North-Holland, Amsterdam, 1991)

Acknowledgments

The contents of this book are based upon the expertise developed by the authors over several years of research in the field of coherent radiation sources in general, and of free-electron lasers in particular. As such, we would like to express our appreciation to our many collaborators who, in a very real sense, made this book possible. This includes Drs. Steven V. Benson, Sandra G. Biedron, Charles A. Brau, Joseph Blau, William B. Colson, Mark Curtin, David Douglas, Adam T. Drobot, David J. Dunning, Pietro Falgari, L. Giannessi, Steven H. Gold, Victor L. Granatstein, Dennis L.A.G. Grimminck, Robert H. Jackson, Roger McGinnis, Stephen V. Milton, George R. Neil, Patrick G. O'Shea, Robert K. Parker, Joseph R. Peñano, Dean E. Pershing, Charles W. Roberson, Irwan D. Setya, Michelle Shinn, Todd Smith, Phillip Sprangle, Cha-Mei Tang, Neil R. Thompson, Edward Stanford, and Raymond Gilbert. Our special appreciation is for Drs. Achintya K. Ganguly, and Peter J.M. van der Slot for their essential collaborations in the development of nonlinear simulation codes.

Contents

Chapter 1
Introduction

In its fundamental concept, the free-electron laser is an extremely adaptable light source that can produce high-power coherent radiation across virtually the entire electromagnetic spectrum. In contrast, gas and solid-state lasers generate light at well-defined wavelengths corresponding to discrete energy transitions within atoms or molecules in the lasing media. Dye lasers are tunable over a narrow spectral range but require a gas laser for optical pumping and operate at relatively low-power levels. Further, while conventional lasers are typically characterized by energy conversion efficiencies of only a few percent, theoretical calculations indicate that the free-electron laser is capable of efficiencies as high as 65%, while efficiencies of 40% have been demonstrated in the laboratory at a wavelength of 8 mm.

The free-electron laser was first conceived almost seven decades ago and has since operated over a spectrum ranging from microwaves to X-rays. In a free-electron laser, high-energy electrons emit coherent radiation, as in a conventional laser, but the electrons travel in a beam through a vacuum instead of remaining in bound atomic states within the lasing medium. Because the electrons are free streaming, the radiation wavelength is not constrained by a particular transition between two discrete energy levels. In quantum mechanical terms, the electrons radiate by transitions between energy levels in the continuum, and, therefore, radiation is possible over a much larger range of frequencies than is found in a conventional laser. However, the process can be described by classical electromagnetic theory alone for all presently operational free-electron lasers.

The radiation is produced by an interaction among three elements: the electron beam, an electromagnetic wave traveling in the same direction as the electrons, and an undulatory magnetic field produced by an assembly of magnets known as a *wiggler* or *undulator*. The distinction in use between these terms is arbitrary; however, *wiggler* is generally used to describe the periodic magnets in free-electron lasers, while *undulator* is used for incoherent synchrotron light sources. The wiggler magnetic field acts on the electrons in such a way that they acquire an undulatory motion. The acceleration associated with this curvilinear trajectory is what makes radiation possible. In this process, the electrons lose energy to the electromagnetic

H. P. Freund, T. M. Antonsen, Jr., *Principles of Free Electron Lasers*,
https://doi.org/10.1007/978-3-031-40945-5_1

wave that is amplified and emitted by the laser. The tunability of the free-electron laser arises because the wavelength of light required for the interaction between these three elements is determined by both the periodicity of the wiggler field and the energy of the electron beam.

Although the basic principle underlying the free-electron laser is relatively simple, the practical application of the concept can be difficult. In 1951, Hans Motz [1–3] of Stanford University first calculated the emission spectrum from an electron beam in an undulatory magnetic field. At the time, coherent optical emission was not expected due to the difficulty of bunching the electron beam at short wavelengths; however, it was recognized that maser (microwave amplification through stimulated emission of radiation) operation was possible. Experiments performed by Motz and coworkers shortly thereafter produced both incoherent radiation in the blue-green part of the spectrum and coherent emission at millimeter wavelengths. The use of undulatory magnetic fields in a maser was independently invented by Robert Phillips [4, 5] in 1957 in search of higher power than was currently available from microwave tubes. The term *ubitron* was coined at this time as an acronym for undulating beam interaction. Over the succeeding 7 years, Phillips performed an extensive study of the interaction and pioneered many innovative design concepts in use today. Whereas the original microwave experiment at Stanford observed an output power of 1–10 W, Phillips achieved 150 kW at a 5-mm wavelength. However, the full potential of the free-electron laser was unrecognized, and the ubitron program was terminated in 1964 due to a general shift in interest from vacuum electronics to solid-state physics and quantum electronics.

A resurgence of interest in the concept began in the mid-1970s when the term free-electron laser was coined in 1975 by John Madey [6, 7] to describe an experiment at Stanford University. This experiment [8–10] produced stimulated emission in the infrared spectrum at a wavelength of 10.6 μm using an electron beam from a radiofrequency linear accelerator (rf linac). The first optical free-electron laser [11–14] was built using the ACO storage ring at the Université de Paris Sud and has been tuned over a broad spectrum. Stimulated emission at visible and ultraviolet wavelengths [15] was also reported using the VEPP storage ring at Novosibirsk in the Soviet Union. Visible wavelength free-electron lasers were also built at both Stanford University [16–18] and by a Boeing Aerospace/Los Alamos National Laboratory collaboration [19–26] based upon rf linacs. The rf linac formed the basis for a free-electron laser program at Los Alamos [27–48]. At the present time, the rf linac forms the basis for virtually all the cutting-edge short-wavelength free-electron laser development.

In parallel with the work at Stanford, experimenters at several laboratories began work on microwave free-electron lasers, successors to the ubitron. Those projects, at the Naval Research Laboratory [49–64], Columbia University [65–72], the Massachusetts Institute of Technology [73–84], Lawrence Livermore National Laboratory [85–91], TRW [92–94], the Ecole Polytechnique [95, 96] in France, and Hughes [97, 98]. These programs differed from the original work by Phillips by using intense relativistic electron beams with currents of the order of a kiloamperes and voltages in excess of a megavolt. The principal goal of this effort was the production of high absolute powers, and the results ranged from a peak power of the order of 2 MW at a wavelength of 2.5 mm at Columbia, through 70 MW at a 4 mm wavelength at the Naval Research Laboratory, to a maximum power figure of 1 GW obtained by Livermore at an 8 mm wavelength. This latter result represents an efficiency (defined as the ratio of the output radiation power to the initial electron beam power) of 35% and was made possible by the use of a nonuniform wiggler field.

Free-electron lasers have been constructed over the entire electromagnetic spectrum. At wavelengths above 0.1 mm, free-electron lasers already either match or exceed power levels obtainable from conventional technology. At shorter wavelengths, the comparison is mixed, and conventional lasers can be found at specific wavelengths with higher powers than is currently available from free-electron lasers. However, free-electron laser technology is rapidly maturing, and this situation is likely to change in the future. In particular, the advent of high peak power X-ray free-electron laser light sources is having an important impact on free-electron laser development and research in various fields of science.

1.1 Principles of Operation

An electron beam that traverses an undulatory magnetic field emits incoherent radiation. Indeed, this is the mechanism employed in synchrotron light sources. In conventional terminology, the periodic magnetic field used in free-electron lasers and synchrotron light sources is referred to either as a *wiggler* or an *undulator*, although there is no fundamental difference between them. It is necessary for the electron beam to form coherent bunches in order to give rise to the stimulated emission required for a free-electron laser. This can occur when a light wave traverses an undulatory magnetic field such as a wiggler because the spatial variations of the wiggler and the electromagnetic wave combine to produce a beat wave, which is essentially an interference pattern. It is the interaction between the electrons and this beat wave which gives rise to the stimulated emission in free-electron lasers.

This beat wave has the same frequency as the light wave, but its wavenumber is the sum of the wavenumbers of the electromagnetic and wiggler fields. With the same frequency, but a larger wavenumber (and thus a shorter wavelength), the beat wave travels more slowly than the light wave; for this reason, it is called a *ponderomotive* wave. Since the ponderomotive wave is the combination of the light wave and the stationary (or magnetostatic) field of the wiggler, it is the effective field experienced by an electron as it passes through the free-electron laser. In addition, since the ponderomotive wave propagates at less than the speed of light in vacuo, it can be in synchronism with the electrons which are limited by that velocity. Electrons moving in synchronism with the wave are said to be in *resonance* with it and will experience a constant field—that of the portion of the wave with which it is traveling. In such cases, the interaction between the electrons and the ponderomotive wave can be extremely strong.

A good analogy to the interaction between the electrons and the ponderomotive wave is that of a group of surfers and a wave approaching a beach. If the surfers remain stationary in the water, the velocity difference between the wave and the surfers is large, and an incoming wave will merely lift them up and down briefly and then return them to their previous level. There is no bulk, or average, translational motion or exchange of energy between the surfers and the wave. But if the surfers "catch the wave" by paddling so as to match the speed of the wave, then they can gain significant momentum from the wave and be carried inshore. This is the physical basis underlying the resonant interaction in a free-electron laser. However, in a free-electron laser, the electrons amplify the wave, so the situation is more analogous to the surfers "pushing" on the wave and increasing its amplitude.

The frequency of the electromagnetic wave required for this resonant interaction can be determined by matching the velocities of the ponderomotive wave and the electron beam. This is referred to as the phase matching condition. The interaction is one in which an electromagnetic wave is characterized by an angular frequency ω and wavenumber k and the magnetostatic wiggler with a wavenumber k_w produce a beat wave with the same frequency as the electromagnetic wave but a wavenumber equal to the sum of the wavenumbers of the wiggler and electromagnetic waves (i.e., $k + k_w$). The velocity of the ponderomotive wave is given by the ratio of the frequency of the wave to its wavenumber. As a result, matching this velocity to that of the electron beam gives the resonance condition in a free-electron laser

$$\frac{\omega}{k+k_w} = \upsilon_z, \tag{1.1}$$

for a beam with a bulk streaming velocity υ_z in the z-direction. The z-direction is used throughout to denote both the bulk streaming direction of the electron beam and the symmetry axis of the wiggler field. The dispersion relation between the frequency and wavenumber for waves propagating in free space is $\omega \cong ck$, where c denotes the speed of light in vacuo. The combination of the free-space dispersion relation and the free-electron laser resonance condition gives the standard relation for the wavelength as a function of both the electron beam energy and the wiggler period

$$\lambda \cong \frac{\lambda_w}{2\gamma_z^2}, \tag{1.2}$$

where $\gamma_z = (1 - \upsilon_z^2/c^2)^{-1/2}$ is the relativistic time dilation factor which is related to the electron streaming energy and $\lambda_w = 2\pi/k_w$ is the wiggler wavelength. The wavelength, therefore, is directly proportional to the wiggler period and inversely proportional to the square of the streaming energy. This results in a broad tunability that permits the free-electron laser to operate across virtually the entire electromagnetic spectrum.

How does a magnetostatic wiggler and a forward-propagating electromagnetic wave, both of whose electric and magnetic fields are directed transversely to the direction of propagation, give rise to an axial ponderomotive force which can extract energy from the electron beam? The wiggler is the predominant influence on the electron's motion. In order to understand the dynamical relationships between the electrons and the fields, consider the motion of an electron into a helically symmetric wiggler field. An electron propagating through a magnetic field experiences a force that acts at right angles to both the direction of the field and to its own velocity. The wiggler field is directed transversely to the direction of bulk motion of the electron beam and rotates through 360° in one wiggler period. An electron streaming in the axial direction, therefore, experiences a transverse force and acquires a transverse velocity component upon entry into the wiggler. The resulting trajectory is helical and describes a bulk streaming along the axis of symmetry as well as a transverse circular rotation that lags 180° behind the phase of the wiggler field. The transverse wiggle velocity, denoted by v_w, is proportional to the product of the wiggler amplitude and period. This relationship may be expressed in the form

$$\frac{v_w}{c} \cong -\frac{K}{\gamma_b}, \tag{1.3}$$

where $K \cong 0.9337 B_w \lambda_w$ is referred to as the wiggler strength parameter (where the wiggler period is expressed in units of centimeters and the wiggler amplitude, B_w, is in Tesla) and $\gamma_b = 1 + E_b/m_e c^2$ denotes the relativistic time dilation factor associated with the total kinetic energy E_b of the electron beam (where m_e denotes the rest mass of the electron and $m_e c^2$ denotes the electron rest energy).

Since the motion is circular in a helical wiggler, both axial and transverse velocities have a constant magnitude. This is important because the resonant interaction depends upon the axial velocity of the beam. In addition, since the wiggler induces a constant-magnitude transverse velocity, the relation between the total electron energy and the streaming energy can be expressed in terms of the time dilation factors in the form

$$\gamma_z \cong \frac{\gamma_b}{\sqrt{1+K^2}}. \tag{1.4}$$

As a result, the resonant wavelength depends upon the total beam energy, and the wiggler amplitude and period through

$$\lambda_{res} \cong \frac{\lambda_w}{2\gamma_b^2}\left(1+K^2\right). \tag{1.5}$$

It is the interaction between the transverse wiggler-induced velocity with the transverse magnetic field of an electromagnetic wave that induces a force normal to both in the axial direction. This is the ponderomotive force. The transverse velocity and the radiation magnetic field are directed at right angles to each other and undergo a simple rotation about the axis of symmetry. A resonant wave must be circularly polarized with a polarization vector that is normal to both the transverse velocity and the wiggler field and which rotates in synchronism with the electrons. This

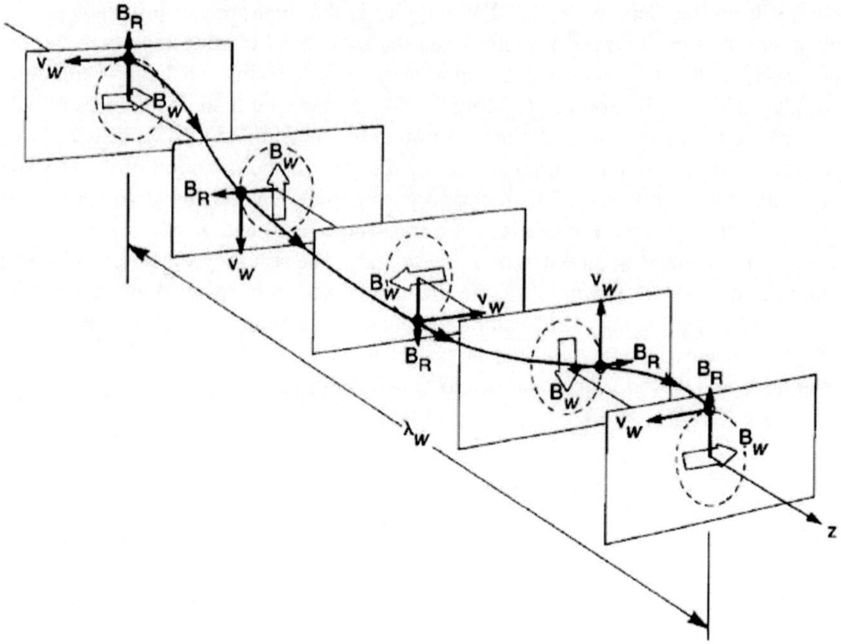

Fig. 1.1 The electron trajectory in a helical wiggler includes bulk streaming parallel to the axis of symmetry as well as a helical gyration. The vector relationships between the wiggler field $\mathbf{B}_w$, the transverse velocity $\mathbf{v}_w$, and the radiation field $\mathbf{B}_R$ of a resonant wave are shown in the figure projected onto planes transverse to the symmetry axis at intervals of one quarter of a wiggler period. This projection is circular, and the transverse velocity is directed opposite to that of the wiggler. A resonant wave must be circularly polarized with a polarization vector that is normal to both the transverse velocity and the wiggler field and which rotates in synchronism with the electrons. The electrons then experience a slowly varying wave amplitude. The transverse velocity and the radiation field are directed at right angles to each other and undergo a simple rotation. The interaction between the transverse velocity and the radiation field induces a force in the direction normal to both which coincides with the symmetry axis

synchronism is illustrated in Fig. 1.1 and is maintained by the aforementioned resonance condition.

In order to understand the energy transfer, we return to the surfer analogy and consider a group of surfers attempting to catch a series of waves. In the attempt to match velocities with the waves, some will catch a wave ahead of the crest and slide forward, while others will catch a wave behind the crest and slide backward. As a result, clumps of surfers will collect in the troughs of the waves. The surfers that slide forward ahead of the wave are accelerated and gain energy at the expense of the wave, while those that slide backward are decelerated and lose energy to the wave. The wave grows if more surfers are decelerated than accelerated, and there is a net transfer of energy to the wave. The free-electron laser operates by an analogous process. Electrons in near resonance with the ponderomotive wave lose energy to the wave if their velocity is slightly greater than the phase velocity of the wave and gain energy at the expense of the wave in the opposite case. As a result, wave amplification occurs if the wave lags behind the electron beam.

This process in a free-electron laser is described by a nonlinear pendulum equation. The *ponderomotive phase* ψ [$= (k + k_w)z - \omega t$] is a measure of the position of an electron in both space and time with respect to the ponderomotive wave. The ponderomotive phase satisfies the circular pendulum equation

$$\frac{d^2}{dz^2}\psi = -\kappa^2 \sin\psi, \tag{1.6}$$

where the pendulum constant is proportional to the square root of the product of the wiggler and radiation fields

$$\kappa \cong 8.29\frac{\sqrt{B_w B_R}}{\gamma_b}. \tag{1.7}$$

Here, κ is expressed in units of inverse centimeters and the magnetic fields are expressed in Tesla. A detailed derivation of the pendulum equation for the free-electron laser is given in the introduction in Chap. 4 (see Eq. 4.13). There are two classes of trajectory: trapped and untrapped. The untrapped, or free-streaming, orbits correspond to the case in which the pendulum swings through the full 360° cycle. The electrons pass over the crests of many waves traveling fastest at the bottom of the troughs and slowest at the crests of the ponderomotive wave. In contrast, the electrons are confined within the trough of a single wave in the trapped orbits. This corresponds to the motion of a pendulum that does not rotate full circle but is confined to oscillate about the lower equilibrium point. The dynamical process is one in which the pendulum constant evolves during the course of the interaction. Electrons lose energy as the wave is amplified; hence, the electrons decelerate and both the pendulum constant and separatrix grow. Ultimately, the electrons cross the growing separatrix from untrapped to trapped orbits.

Wigglers in free-electron lasers are typically either helical or linearly polarized. The interaction in a linear (or planar) wiggler operates in the same manner as in a helical wiggler except that since the electron orbits oscillate in a plane, the transverse velocity oscillates in both magnitude and direction. As a result, the effective wiggler field is given by the rms magnitude (i.e., $B_w/\sqrt{2}$) of the peak on-axis field strength, and the resonant wavelength is given by

$$\lambda \cong \frac{\lambda_w}{2\gamma_b^2}\left(1 + K_{rms}^2\right). \tag{1.8}$$

Usage in the literature varies, but the rms wiggler strength (*i.e.*, $K/\sqrt{2}$) *is sometimes referred to as* a_w; *however, we use* a_w *and* K *interchangeably in this work and refer to the rms wiggler strength parameter as* K_{rms} *herein.* It should be understood in the remainder of this chapter that, unless otherwise noted, the formulae are valid for both helical and planar wigglers as long as the rms magnitude for the on-axis field is used.

Free-electron lasers have been operated in a variety of configurations. In *amplifiers* (sometimes referred to as master oscillator power amplifiers or MOPAs for short), a seed pulse is injected into the wiggler in synchronism with the electron beam and then grows exponentially until the interaction saturates over the length of the wiggler. Depending upon the strength of the interaction, the wigglers used in amplifiers can be quite long. In contrast, *oscillators* use a relatively short wiggler in conjunction with a resonant optical cavity. The radiation starts from shot noise on the electron beam and is amplified to high-power levels over the course of many round trips of the optical field through the wiggler and resonator. The gain per pass through the wiggler is, typically, typically small (i.e., less than or of the order of 100%) in which case the out-coupling from the resonator must be less than the gain or the oscillator will not lase. As a result, the Q-factor of the resonator in these free-electron lasers is high reflecting high levels of stored power in the resonator. However, when a sufficiently long undulator is used that the gain per pass is high, then the out-coupling can be large denoting a high-gain/low-Q free-electron laser. These systems are sometimes referred to as *regenerative amplifiers* or *RAFELs* for short. A third class of free-electron lasers is used when neither high-power seed lasers nor highly reflective mirrors are unavailable. These cases are typically found for ultra-short wavelengths (i.e., ultraviolet or X-ray). In these cases, the radiation grows from shot noise on the electron beam as in oscillators but is then amplified to high-power levels in a single pass through an extremely long wiggler. The wigglers in X-ray free-electron lasers can be 60–100 m in length. This process is referred to as *self-amplified spontaneous emission* (*SASE*). In the early literature, this was often called *superradiant amplifiers*, although this term has gone out of usage. Examples of each of these configurations will be discussed in later chapters.

A variation on the amplifier is the *optical klystron*. The configuration used in an optical klystron consists of two wigglers separated by a drift space that contains a magnetic dispersive element. The magnetic dispersive element is usually formed by a series of dipoles that act as a chicane in which path length through the chicane decreases with increasing energy. A high-power seed pulse injected into the first wiggler (called the modulator) in synchronism with the electrons induces a modulation on the electron beam velocity and density. Since high-energy electrons in the tail of the bunch will overtake lower-energy electrons near the head of the bunch, this modulation is enhanced in the chicane. The enhanced modulation preconditions the electron beam for radiation in the second wiggler (called the radiator) leading to more rapid amplification of the optical field than would be obtained by simple exponential growth in a single long wiggler. The radiator can be configured to be resonant at the same wavelength as the modulator or at a harmonic of the modulator wavelength. In the latter case, the process is referred to as *high-gain harmonic generation* (*HGHG*). The advantage of HGHG is that a relatively long-wavelength seed laser can be used in the modulator, which then gives rise to coherent radiation at a much shorter wavelength.

In typical operation, electrons entering the free-electron laser are free streaming on untrapped trajectories and begin interacting with the radiation field immediately upon entry. Ultimately, in sufficiently long wigglers, the field exponentiates; however, there is some lethargy after the wiggler entrance during which initial transients in the field decay before the exponentiation begins. This initial transient phase is often referred to as the low-gain regime and applies to most oscillator configurations. High-gain configurations include most amplifiers and SASE free-electron lasers.

The power in the *low-gain regime* increases as the cube of the distance z along the wiggler. This case is often referred to as the low-gain Compton regime and is relevant to the single-pass gain in low-gain/high-Q oscillators. Here, the peak power gain over N_w wiggler periods is given by

$$G \approx 0.54(2\pi\rho N_w)^3, \tag{1.9}$$

where ρ is the Pierce parameter. The Pierce parameter was originally formulated to describe the interaction in traveling wave tubes [99] and is defined as

$$\rho^3 = \frac{JJ^2 K_{rms}^2}{4\gamma_b^3 k_w^2 R_b^2} \frac{I_b}{I_A}, \tag{1.10}$$

for free-electron laser applications, where I_b and R_b are the electron beam current and radius, respectively, and $I_A = m_e c^3/e \approx 17{,}045$ A is the Alfvén current. The JJ-factor is common usage for the effect of the wiggler on the electron motion. In a helical wiggler, the orbits are characterized by constant axial and transverse velocities, and $JJ = 1$. However, the dynamics are more complex in a planar wiggler. In the first place, the magnitude of the transverse velocity is oscillatory, and we must use the rms wiggler magnitude. In the second place, the oscillations in the axial velocity are only approximately sinusoidal. In general, the axial velocity is governed by elliptic functions. In the third place, while the lower beat wave $[\omega = (k - k_w)v_z]$ is suppressed by the symmetry in a helical wiggler, it is present in a planar wiggler and affects the overall strength of the interaction. As a result,

$$JJ = \left[J_0\left(\frac{K_{rms}^2}{1+K_{rms}^2}\right) - J_1\left(\frac{K_{rms}^2}{1+K_{rms}^2}\right)\right], \tag{1.11}$$

in a planar wiggler. This will be rigorously derived in Chap. 4. The Pierce parameter typically ranges over $10^{-4} < \rho < 10^{-2}$ for a variety of operational free-electron lasers. A detailed derivation of this expression for the gain is given in Chap. 4 (see Eq. 4.59). The wavelength corresponding to this peak gain is shifted from the resonant wavelength, λ_{res}, by

$$\frac{\lambda}{\lambda_{res}} \cong \frac{N_w}{N_w - 0.414} \tag{1.12}$$

and is characterized by a spectral linewidth of

$$\frac{\Delta\lambda}{\lambda} \cong \frac{1}{N_w}. \tag{1.13}$$

This low-gain regime is relevant to operation of oscillators at short wavelengths in the infrared and optical spectra. These free-electron lasers typically employ electron beams generated by radiofrequency linear accelerators, microtrons, storage rings, and electrostatic accelerators in which the total current is small. The extraction efficiency, η, in this regime is given approximately by (see Chap. 10)

$$\eta \cong \frac{1}{2.4N_w}, \tag{1.14}$$

so that the spectral linewidth $\Delta\lambda/\lambda \approx 2\eta$. As a result, while the gain increases and the spectral linewidth decreases with increasing wiggler length, the efficiency decreases. Hence, oscillator design requires a balance between achieving the highest gain possible consistent with acceptable extraction efficiencies. In practice, it is necessary to also balance the gain against the losses in the resonator. As the oscillator starts up, the interaction is linear, and the power grows by an amount given by the gain in Eq. (1.9) on each pass through the wiggler. As the power nears saturation, the gain per pass decreases. Saturation is found when the residual gain per pass is decreasing to the point where it is balanced by the losses in the resonator. If L denotes the losses per pass, then the residual gain per pass at saturation is given by $G_{sat} \approx L/(1 - L)$. The initial gain must exceed this value so that $0.54\ (2\pi\rho N_w)^3 > L/(1 - L)$. At wavelengths where low-loss mirrors are not available, these conditions may not be achievable.

The *high-gain regime* is found when the wiggler is long enough to overcome the lethargy and the growth is exponential. A comprehensive derivation of the growth rates in this high-gain regime appears in Chap. 4. Two distinct exponential regimes exist. The high-gain Compton (sometimes called the *strong-pump*) regime is found when the ponderomotive potential is much stronger than the space-charge potential and space-charge waves can be neglected in the analysis. This occurs when

$$\frac{\omega_b}{ck_w} \ll \frac{1}{16}\gamma_b^3\left(\frac{v_w}{c}\right)^2, \tag{1.15}$$

where ω_b is the beam plasma frequency. In this regime, the minimum exponentiation length (which corresponds to the maximum growth rate) in the power is given by (see Eq. 4.94)

$$L_G \cong \frac{\lambda_w}{4\pi\sqrt{3}\rho} \tag{1.16}$$

and is found at the resonant wavelength. This regime is relevant to most modern free-electron laser amplifier and SASE configurations driven by radiofrequency linear accelerators.

The opposite limit, referred to as the collective Raman regime, is fundamentally different from either the high- or low-gain Compton regimes. It occurs when the current density of the beam is high enough that the space-charge force exceeds that exerted by the ponderomotive wave. The exponentiation length in this regime varies as (see Eq. 4.97)

$$L_G \cong \frac{\lambda_w}{4\pi\sqrt{2}\rho^{3/4}}. \tag{1.17}$$

In this regime, the space-charge forces result in electrostatic waves that co-propagate with the beam and are characterized by the dispersion relations

$$\omega = k_{sc}v_z \pm \frac{\omega_b}{\gamma_b^{1/2}\gamma_z}, \tag{1.18}$$

which describe the relation between the frequency ω and wavenumber k_{sc}. These dispersion relations describe positive- and negative-energy waves corresponding to the "+" and "−" signs, respectively. The interaction results from a stimulated three-wave scattering process. This is best visualized from the perspective of the electrons, in which the wiggler field appears to be a backward-propagating electromagnetic wave called a *pump* wave. This pump wave can scatter off the negative-energy electrostatic wave (the *idler*) to produce a forward-propagating electromagnetic wave (the *signal*). The interaction occurs when the wavenumbers of the pump, idler, and signal satisfy the condition $k_{sc} = k + k_w$, which causes a shift in the wavelength of the signal to

$$\lambda \approx \lambda_{res}\left(1 - \frac{\omega_b\sqrt{1+K^2}}{\gamma_b^{3/2}ck_w}\right)^{-1}. \tag{1.19}$$

Observe that the interaction in the Raman regime is shifted to a somewhat longer wavelength than occurs in the high-gain Compton regime.

Wave amplification can saturate by several processes. The highest efficiency occurs when the electrons are trapped in the ever-deepening ponderomotive wave and undergo oscillations within the troughs. In essence, the electrons are initially free streaming over the crests of the ponderomotive wave. Since they are traveling at a velocity faster than the wave speed, they come upon the wave crests from behind. However, the ponderomotive wave grows together with the radiation field, and the electrons ultimately will come upon a wave that is too high to cross. When this happens, they rebound and become trapped within the trough of the wave. In analogy to the oscillation of a pendulum, the trapped electrons lose energy as they rise and gain energy as they fall toward the bottom of the trough. As a result, the energy transfer between the wave and the electrons is cyclic, and the wave amplitude ceases to grow and oscillates with the electron motion in the trough. The ultimate saturation efficiency for this mechanism can be estimated from the requirement that the net change in electron velocity at saturation is equal to twice the velocity difference between the electron beam and the ponderomotive wave. This results in a saturation efficiency of

$$\eta \approx \frac{\gamma_b \rho}{\gamma_b - 1}, \tag{1.20}$$

in the high-gain Compton regime, and

$$\eta \approx \frac{\gamma_b}{\gamma_b - 1} \frac{\omega_b}{\gamma_b^{3/2} c k_w} \left(1 + K^2\right), \tag{1.21}$$

in the collective Raman regime.

The spectral bandwidth in the high-gain regimes depends upon the interaction strength and generally increases with the Pierce parameter. Since the Raman regime is important at longer wavelengths for free-electron masers while most free-electron lasers today operate at short wavelengths in the Compton regime, we now focus on this regime. It is clear from the resonance condition that $\Delta\lambda/\lambda = -2\Delta\gamma/\gamma$. Since $|\Delta\gamma/\gamma| \approx \rho$ at saturation, the spectral linewidth produced in the high-gain Compton regime is

$$\frac{\Delta\lambda}{\lambda} \cong 2\rho \approx 2\eta, \tag{1.22}$$

in the highly relativistic limt ($\gamma_b \gg 1$). Observe that the spectral linewidth in this regime is the same as that for the low-gain Compton regime when expressed in terms of the extraction efficiency.

The free-electron laser interaction places stringent requirements on the quality of the electron beam. The preceding formulae apply to the idealized case of a monoenergetic (or cold) beam. This represents a theoretical maximum for the gain and efficiency since each electron has the same axial velocity and interacts with the wave in an identical manner. A monoenergetic beam is physically unrealizable, however, and all beams exhibit a velocity spread that determines a characteristic temperature. Electrons with axial velocities different from the optimal resonant velocity are unable to participate fully in the interaction. If this axial velocity spread is sufficiently large that the entire beam cannot be in simultaneous resonance with the wave, then the fraction of the electron beam that becomes trapped must fall. Ultimately, the trapping fraction falls to the point where the trapping mechanism becomes ineffective, and saturation occurs through the thermalization of the beam. Thus, there are two distinct operating regimes: the *cold* beam limit characterized by a narrow bandwidth and relatively high efficiencies and the *thermal* regime characterized by a broader bandwidth and sharply lower efficiencies.

The question of electron beam quality is the most important single issue facing the development of the free-electron laser [100]. In order to operate in the cold beam regime, the axial velocity spread of the beam must be small. It is convenient to relate the axial velocity spread to an energy spread to obtain an invariant measure of the beam quality suitable for a wide range of electron beams. In the case of the low-gain limit, this constraint on the beam thermal spread is

$$\frac{\Delta E_b}{E_b} \ll \frac{1}{N_w} \cong 2\eta, \tag{1.23}$$

where ΔE_b represents the beam thermal spread. In the high-gain regimes, the maximum permissible energy spread for saturation by particle trapping is determined by the depth of the ponderomotive or space-charge waves, which is measured by twice the difference between the streaming velocity of the beam and the wave speed. The maximum permissible thermal spread corresponds to this velocity difference and is one-half the saturation efficiency for either the high-gain Compton or collective Raman regimes, that is,

$$\frac{\Delta E_b}{E_b} \ll 2\rho \cong 2\eta. \tag{1.24}$$

Typically, this energy spread must be approximately 1% or less of the total beam energy for the trapping mechanism to operate at millimeter wavelengths and decreases approximately with the radiation wavelength. Hence, the requirement on beam quality becomes more restrictive at shorter wavelengths and places greater emphasis on accelerator design.

Two related quantities often used as measures of beam quality are the *emittance* and *brightness*. The emittance measures the collimation of the electron beam and may be defined in terms of the product of the beam radius and the average pitch angle (i.e., the angle between the velocity and the symmetry axis). It describes a random pitch angle distribution of the beam which, when the velocities are projected onto the symmetry axis, is equivalent to an axial velocity spread. In general, therefore, even a monoenergetic beam with a nonvanishing emittance displays an axial velocity spread. The electron beam brightness is an analog of the brightness of optical beams and is directly proportional to the current and inversely proportional to the square of the emittance. As such, it describes the average current density per unit pitch angle and measures both the beam intensity and the degree of collimation of the electron trajectories. Since the gain and efficiency increase with increasing beam current for fixed emittance, the brightness is a complementary measure of the beam quality. While it is important to minimize the emittance and maximize the brightness in order to optimize performance, both of these measures relate to the free-electron laser only insofar as they describe the axial velocity spread of the beam.

Typical free-electron laser efficiencies range up to approximately 12% at the longer wavelengths and decrease with the wavelength; however, significant enhancements are possible when either the wiggler amplitude or period is systematically tapered. The free-electron laser amplifier operating at 35 GHz at Lawrence Livermore National Laboratory which achieved an extraction efficiency of 35% employed a wiggler with an amplitude that decreased along the axis of symmetry and contrasts with an observed efficiency of about 6% in the case of a uniform wiggler. The use of tapered wigglers was pioneered by Phillips in 1960. The technique has received intensive study, and tapered wiggler designs have also been shown to be effective at infrared wavelengths in experiments at Los Alamos National Laboratory and using the superconducting rf linac at Stanford University [16–18]. More recently, a tapered wiggler amplifier has been demonstrated at a wavelength of 0.8 μm at the Source Development Laboratory at Brookhaven National Laboratory [101].

The effect of a tapered wiggler is to alter both the transverse and axial velocities. Since the transverse velocity is directly proportional to the product of the amplitude and period, the effect of gradually decreasing either of these quantities is to decrease the transverse velocity and, in turn, increase the axial velocity. The energy extracted during the interaction results in an axial deceleration that drives the beam out of resonance with the wave; hence, efficiency enhancement occurs because the tapered wiggler maintains a relatively constant axial velocity (and phase relationship between the electrons and the wave) over an extended interaction length. The physical basis for this process is discussed in more detail in the introduction in Chap. 5 (see Eq. 5.12), which also includes detailed discussions of nonlinear simulations of the tapered wiggler interaction. The enhancement in the tapered wiggler interaction efficiency is proportional to the decrement in the wiggler field of ΔB_w and satisfies

$$\Delta\eta \approx \frac{0.872 B_w^2 \lambda_w^2}{1 + 0.872 B_w^2 \lambda_w^2} \frac{\Delta B_w}{B_w}. \tag{1.25}$$

In practice, a tapered wiggler is effective only after the bulk of the beam has become trapped in the ponderomotive wave. In single-pass amplifier configurations, therefore, the taper is not begun until the signal has reached saturation in a section of uniform wiggler, and the total extraction efficiency is the sum of the uniform wiggler efficiency and the tapered wiggler increment. Numerical simulations indicate that total efficiencies as high as 65% are possible under the right conditions, but this has only been shown at longer wavelengths in the submillimeter regime.

Once particles have been trapped in the ponderomotive wave and begin executing a bounce motion between the troughs of the wave, then the potential exists for exciting secondary emission referred to as sideband waves. These sidebands arise from the beating of the primary signal with the ponderomotive bounce motion. The bounce period is, typically, much longer than the radiation wavelength, and these sidebands are found at wavelengths close to that of the primary signal. The difficulties imposed by the presence of sidebands are that they may compete with and drain energy from the primary signal. This is particularly crucial in long tapered wiggler systems that are designed to trap the beam at an early stage of the wiggler and then extract a great deal more energy from the beam over an extended interaction length. In these systems, unrestrained sideband growth can be an important limiting factor. As a result, a great deal of effort has been expended on techniques of sideband suppression. One method of sideband suppression was employed in a free-electron laser oscillator at Columbia University [102]. This experiment operated at a 2 mm wavelength in which the dispersion due to the waveguide significantly affected the resonance condition. As a consequence, it was found to be possible by proper choice of the size of the waveguide to shift the sideband frequencies out of resonance with the beam. Experiments on an infrared free-electron laser oscillator at Los Alamos National Laboratory [41, 42] indicate that it is also possible to suppress sidebands by (1) using a Littrow grating to deflect the sidebands out of the optical cavity or (2) changing the cavity length.

The preceding description of the principles and theory of the free-electron laser is, necessarily, restricted to the idealized case in which the transverse inhomogeneities of both the electron beam and wiggler field are unimportant. This is sufficient for an exposition of the fundamental physics of the free-electron laser. In practice, however, these gradients can have important consequences on the performance of the free-electron laser. The most important effect is found if the wiggler field varies substantially across the diameter of the electron beam, since the electron response to the wiggler will vary across the beam as well. In practice, this means that an electron at the center of the beam will experience a different field than an electron at the edge of the beam, and the two electrons will follow different trajectories with different velocities. As a result of this, the wave-particle resonance that drives the interaction will be broadened and the gain and efficiency will decline. In essence, therefore, the transverse wiggler inhomogeneity is manifested as an effective beam thermal spread. The bounded nature of the electron beam also affects the interaction since wave growth will occur only in the presence of the beam. Because of this, it is important in amplifier configurations to ensure good overlap between the injected signal and the electron beam. Once such overlap has been accomplished, however, the dielectric response of the electron beam in the presence of the wiggler can act in much the same way as an optical fiber to refractively guide the light through the wiggler.

1.2 Quantum Mechanical Effects

The early analyses of free-electron lasers relied upon classical treatments of the interaction physics. The free-electron laser gain formula was rederived by Madey [6] using quantum mechanical principles, and a great deal of work has since been published on the quantum mechanics of the free-electron laser [102–114]. However,

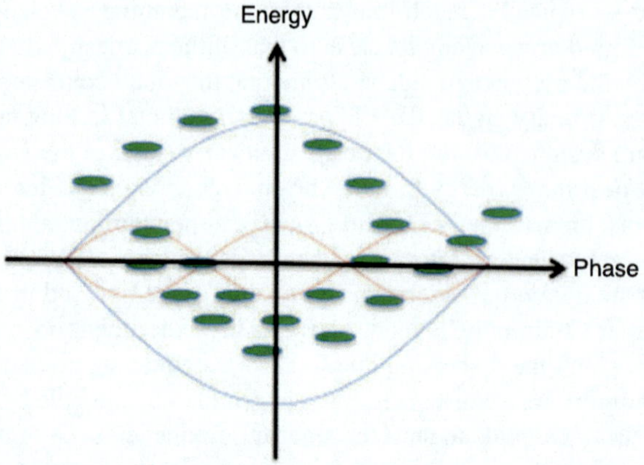

Fig. 1.2 The schematic illustration of the electron phase space at saturation showing the separatrices of the fundamental (blue lines) and of the third harmonic (red lines) and the electron wave packets (green ovals). Quantum mechanical effects become important when the wave packets become sufficiently large that the electron locations within the separatrix becomes uncertain. Observe that since the extent of the separatrix at harmonics is smaller than that of the fundamental, quantum mechanical effects on the generation of harmonics become important before that may be the case for the fundamental

for most applications of practical interest, quantum mechanical effects are negligible, and the classical limit of the quantum mechanical treatments is valid.

In general quantum mechanical effects can be neglected when the spreading of the electron wave packet is less than one wave period over the length of the wiggler L. What this means is that the electrons can be accurately located within the trapped phase space. A free-electron laser saturates by particle trapping in the ponderomotive wave formed by the beating of the undulator and radiation fields (see Sect. 4.1). The efficiency is limited since not all the electrons are trapped, and it is important to determine the *trapping fraction* accurately in order to simulate FEL performance. Electron dynamics are governed by a nonlinear pendulum equation, which exhibits a *separatrix* between the trapped and untrapped orbits as shown schematically in Fig. 1.2, where the blue lines denote the separatrix of the fundamental, the red line is the separatrix for the third harmonic, and the green ovals denote electron wave packets. As mentioned above, quantum effects are negligible when the spreading of the wave packet is much less than the bucket length. Evidently, when the wave packets become comparable to the bucket length, the uncertainty in electron locations can significantly impact whether the electron is trapped.

In order to estimate the magnitude of the quantum mechanical effect, consider the spreading of a one-dimensional wave packet. If the electron wave function is represented by a Gaussian wave packet

$$\psi(z,0) \approx \frac{1}{\left(\sqrt{2\pi}\Delta z_0\right)^{1/2}} \exp(ik_0 z)\exp(-z^2/4\Delta z_0^2), \tag{1.26}$$

upon entry to the wiggler with an initial spread or width Δz_0. This may be decomposed into Fourier components

$$\psi(z,0) \approx \frac{1}{\sqrt{2\pi}} \int_{-\infty}^{\infty} dk\,\psi(k)\exp(ikz), \tag{1.27}$$

where

$$\psi(k) = \left(\frac{2\Delta z_0}{\sqrt{2\pi}}\right)^{1/2} \exp\left[-(k-k_0)^2\Delta z_0^2\right]. \tag{1.28}$$

The time-dependent solution for the wave function may be constructed using these Fourier amplitudes

$$\psi(z,t) = \frac{1}{\sqrt{2\pi}} \int_{-\infty}^{\infty} dk\,\psi(z)\exp[ikz - i\omega(k)t], \tag{1.29}$$

where

$$\omega(k) = 2\pi\frac{m_e c^2}{h}\left[\sqrt{1+\frac{h^2k^2}{4\pi^2 m_e^2 c^2}} - 1\right]. \tag{1.30}$$

For a wave packet with a narrow spread, this frequency can be expanded about $k = k_0$ as

$$\omega(k) \cong \omega(k_0) + \upsilon_0(k-k_0) + \frac{\upsilon_0}{\gamma_0^2 k_0}(k-k_0)^2, \tag{1.31}$$

where $\upsilon_0 = hk_0/2\pi\gamma_0 m_e$ denotes the bulk electron velocity and $\gamma_0 = (1 + h^2k_0^2/4\pi^2m_e^2c^2)^{1/2}$ is the corresponding relativistic factor. Hence, the time-dependent wave function

$$\psi(z,t) = \frac{1}{\left(\sqrt{2\pi}\Delta z_0\right)^{1/2}} \frac{\exp[ik_0 z - i\omega(k_0)t]}{\sqrt{1 + iv_0 t/2\gamma_0^2 k_0 \Delta z_0^2}} \exp\left(-\frac{z^2}{4\Delta z_0^2} \frac{1}{1 + iv_0 t/2\gamma_0^2 k_0 \Delta z_0^2}\right). \tag{1.32}$$

Evaluation of $|\psi(z,t)|^2$ shows that the width of the wave packet increases in time via

$$\Delta z^2(t) = \Delta z_0^2 + \frac{v_0^2 t^2}{4\gamma_0^4 k_0^2 \Delta z_0^2}. \tag{1.33}$$

The width of the wave packet at any given time is minimized by a choice of $\Delta z_0^2 = v_0 t/2\gamma_0^2 k_0$. Over the length of the wiggler, L $(= ct)$, the wave packet extent is

$$\Delta z^2(L) \cong \frac{\lambda_c L}{2\pi\gamma_0^3}, \tag{1.34}$$

where $\lambda_c = h/m_e c$ is the Compton wavelength. The requirement that the spreading of the wave packet must be much less than the wavelength, therefore, can be expressed as

$$\frac{\lambda_c L}{\pi\gamma_0\lambda_w} \ll \lambda. \tag{1.35}$$

This is well satisfied for virtually all cases of practical interest to date and would become important only for extremely short-wavelength operation.

For example, consider a free-electron laser at a 1.0 Å wavelength which employs a 14 GeV electron beam and has a wiggler with a 3 cm period that is 100 m in length. In this case, $\lambda_c L/\pi\gamma_0\lambda_w \approx 9.4 \times 10^{-4}$ Å, and the inequality is satisfied by more than three orders of magnitude.

Another requirement for the neglect of quantum mechanical effects is that the electron recoil upon the emission of a photon be small. This criterion may be stated in the form that the downshift in the frequency of the emitted photon due to the electron recoil must be much smaller than the gain linewidth. However, as shown in both Brau [112] and Luchini and Motz [113], this requirement results in a criterion identical to (1.35).

As shown above, quantum mechanical effects are expected to be negligible for most free-electron lasers of practical interest. One aspect of free-electron laser operation in which quantum mechanical effects may play a role, however, is in the oscillator and SASE start-up regime since they typically start from the spontaneous emission of the electrons propagating through the wiggler. It has been speculated [115] that if the spontaneous emission is weak (i.e., the number of photons emitted is small), then quantum noise may prevent the start-up even if the wiggler is long enough that the classical gain formula predicts substantial gain. This is most likely to occur for short period wigglers in which the wiggler-induced transverse velocity is low, however, and we shall henceforth ignore quantum mechanical effects and deal with a classical treatment of the interaction in the free-electron laser.

1.3 Experiments and Applications

There are three basic experimental configurations for free-electron lasers: amplifiers, oscillators, and SASE free-electron lasers. In an amplifier, the electron beam is injected into the wiggler in synchronism with the signal to be amplified. The external radiation source that drives the amplifier is referred to as the master oscillator and can be any convenient radiation source such as a conventional laser or microwave tube. As a consequence, this configuration is often referred to as a master oscillator power amplifier (MOPA). Because amplification occurs during one pass through the wiggler, MOPAs require intense electron beam sources that can operate in the high-gain regime. Oscillators differ from amplifiers in that some degree of reflection is introduced at the ends of the wiggler so that a signal will make multiple passes through the system. The signal is amplified during that part of each pass in which the radiation co-propagates with the electrons and allows for a large cumulative amplification over many passes even in the event of a low gain per pass. Oscillators are typically constructed to amplify the spontaneous (i.e., shot) noise within the beam, and no outside signal is necessary for their operation. However, a long-pulse accelerator is required because a relatively long time may be required to build up to saturation. SASE free-electron lasers are devices in which the shot noise in the beam is amplified over the course of a single pass through the wiggler and, like amplifiers, require high-current accelerators to drive them. Since the shot noise present in the beam is generally broadband, the radiation from a SASE free-electron laser is typically characterized by a broader bandwidth than a MOPA.

The optimal configuration and type of accelerator used in a free-electron laser design depend upon the specific application, and issues such as the electron beam quality, energy, and current are important considerations in determining both the wavelength and power. In general, however, each accelerator type is suited to the production of a limited range of wavelengths, as shown in Fig. 1.3. In addition, the temporal structure of the output light from a free-electron laser corresponds to that of the electron beam. Thus, either a pulsed or continuous electron beam source will give rise to a pulsed or continuous free-electron laser. Free-electron lasers have been constructed using virtually every type of electron source including storage rings, radiofrequency linear accelerators, microtrons, induction linacs, electrostatic accelerators, pulse line accelerators, and modulators. Since the gain in a free-electron laser increases with current but decreases with energy, accelerators producing low-current/high-energy beams are generally restricted to the low-gain regime. Accelerator types that fall into this category include radiofrequency (rf) linacs, microtrons, storage rings, and electrostatic accelerators. In contrast, intense beam

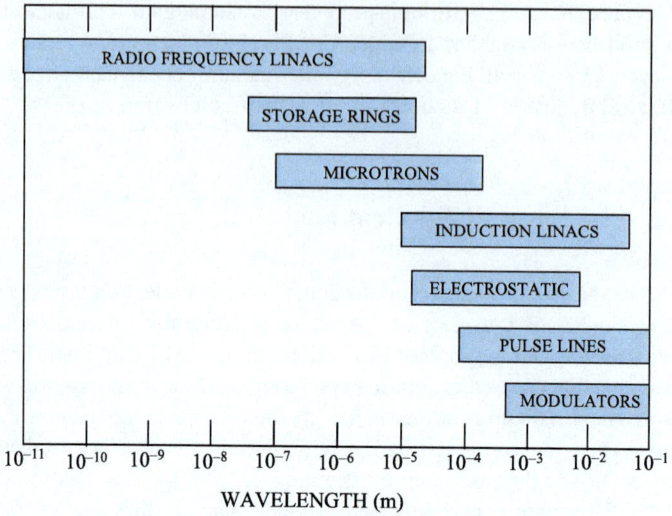

Fig. 1.3 The limits of the wavelengths possible with different accelerators depend both on the electron beam energies and on the state of wiggler technology. The approximate wavelength ranges that may be addressed with current accelerator and wiggler technology are indicated

accelerators such as induction linacs, pulse line accelerators, and modulators are suitable electron beam sources for high-gain systems.

Storage rings are typically characterized by multiple electron pulses continuously circulating through the ring. Each pulse is several nanoseconds in duration, and the output light from a free-electron laser driven by a storage ring is a continuous stream of picosecond bursts. In addition, while storage rings produce high-quality and high-energy beams of low to moderate currents, the electron pulses are recirculated through both the ring and the wiggler, and the stability of the ring is disrupted by the extraction of too much energy. Hence, storage rings are feasible for applications that require uniform and continuous short-wavelength radiation sources but do not demand high output powers.

The first successful operation of a storage ring free-electron laser was at the Université de Paris Sud at Orsay [11], and the performance of this system was extended to the VUV by harmonic operation [116]. This experimental configuration was that of an oscillator and made use of the ACO storage ring that operates at energies and average currents in the range of 160–224 MeV and 16–100 mA. The laser was tuned across a broad band of the visible spectrum but was first operated at wavelengths in the neighborhood of approximately 0.65 μm. The peak output power from the oscillator was 60 mW over the 1 ns duration of the micropulses, which corresponds to an intracavity power level of 2 kW. The average power extracted from the system was typically of the order of 75 μW. Higher harmonic emission was also detected in the ultraviolet, however, which posed a problem since radiation at these wavelengths resulted in the ultimate degradation of the optical system. An ultraviolet free-electron laser oscillator has also been achieved using the VEPP storage ring at Novosibirsk [15, 117, 118]. This experiment employed an optical klystron configuration in which the wiggler was composed of two distinct sections separated by a dispersive drift space. In this configuration, the first section operates as a prebuncher for the electron beam that subsequently enhances the gain in the second wiggler section. Operating the storage ring at 350 MeV and a peak current of 6 A, experimenters were able to obtain coherent emission at wavelengths as short as 0.3 μm and average output powers as high as 6 mW.

Radiofrequency linacs employ a series of cavities that contain rapidly varying electromagnetic (rf) fields to accelerate streams of electrons. The beams they produce are composed of a sequence of macropulses (typically of microseconds in duration) each of which consists of a train of shorter picosecond pulses. Microtrons produce beams with a temporal structure similar to that of rf linacs but, unlike the rf linac, are composed of a single accelerating cavity coupled to a magnet that causes the electron beam to recirculate through the cavity many times. The output light from a free-electron laser built with these accelerators, therefore, is similar to that from a storage ring. Radiofrequency linacs have developed into the most versatile accelerator for free-electron laser applications since they can be configured to produce beams with energies ranging from a few tens of MeV to a few tens of GeV and with bunch charges of up to several nano-Coulombs. They have been used in amplifiers, oscillators, and SASE free-electron lasers. They have been configured as energy recovery linacs (ERLs) where the electron beam makes a single pass through a wiggler and is then redirected back into the linac where the beam is decelerated and the energy recovered to support the rf fields in the accelerating cavities [119, 120]. In energy recovery mode, rf linacs are suitable drivers for high-average power free-electron lasers and can compete with storage rings by producing pulses with very high repetition frequencies over long periods of operation. In contrast with storage rings, however, the electron beam is used only once and so can achieve very high brightness.

Free-electron lasers based upon rf linacs have demonstrated operation over a broad spectrum extending from the infrared to X-rays. Experiments at Stanford University [16–18] and Boeing Aerospace [24, 121, 122] first demonstrated the feasibility of the rf linac to produce visible light. The initial experiments conducted by Madey and coworkers at Stanford University resulted in (1) an amplifier which operated at a wavelength of 10.6 μm with an overall gain of 7% and (2) a 3.4 μm oscillator which produced peak and average output powers of 7 kW and 0.1 mW, respectively. In collaboration with TRW [123–125], the superconducting rf linac (SCA) at Stanford University has been used to drive a free-electron laser oscillator which has demonstrated efficiency enhancement with a tapered wiggler, operation at visible wavelengths, and electron beam recirculation. The tapered wiggler experiment is operated in the infrared at a wavelength of 1.6 μm and peak power levels of 1.3 MW. This yields a peak extraction efficiency of approximately 1.2%, which constitutes an enhancement by a factor of three over the efficiency in the case of an untapered wiggler. Operation at visible wavelengths was also found at 0.52 μm and peak power levels of 21 kW. The superconducting technology embodied in the SCA can enable the rf linac to further compete with storage rings by operating in a near steady-state mode. Both energy recovery and enhancement of the extraction efficiency by means of tapered wiggler were also demonstrated at Los Alamos National Laboratory. Starting in 1981 with a tapered wiggler free-electron laser amplifier which obtained an extraction efficiency of 4% at a wavelength of 10.6 μm, researchers went on to (1) extend that to a 5% extraction efficiency in an oscillator configuration and (2) demonstrate a 70% energy recovery rate with beam recirculation. Using an energy recovery linac, a series of high-average power oscillators at infrared wavelengths have been demonstrated at the Thomas Jefferson National Accelerator Facility in Newport News, Virginia [126, 127]. These free-electron lasers currently hold the record for the highest-average power (14 kW) at any wavelength. Free-electron laser oscillators based on ERLs have also been built at the Japan Atomic Energy Research Institute [128] and at Daresbury Laboratory in the United Kingdom [129].

More recently, rf linacs have provided the basis for the design and development of short-wavelength EUV and X-ray light sources that will supplant synchrotrons for biological and material research. The Linac Coherent Light Source (LCLS) at the Stanford Linear Accelerator Center (SLAC) was the first X-ray free-electron laser to become operational [130] using 1/3 of the SLAC linac. This facility was commissioned in 2009 and produces 2 mJ pulses of photons at a wavelength 1.5 Å at a repetition rate of 120 Hz. The second operational X-ray free-electron is at Spring-8 in Japan and produces photons at a wavelength of 0.63 Å and peak power levels of 10 GW over a duration of 10 fs [131]. Another X-ray free-electron laser facility became operational at the Pohang Accelerator Laboratory in South Korea producing wavelengths ranging from 1 Å to 10 nm at a 60 Hz repetition rate [132]. These X-ray free-electron lasers will be supplemented by new facilities coming online in the near future at the Deutsches Elektronen-Synchrotron (DESY) in Hamburg, Germany [133], and at the Paul Scherrer Institute in Switzerland [134]. These X-ray free-electron lasers are complemented by the existence of EUV free-electron laser user facilities at DESY [135] and an HGHG free-electron laser at Fermi-Elettra in Trieste, Italy [136].

The limitations which storage rings, rf linacs, and microtrons impose on free-electron laser design stem from restrictions on the peak (or instantaneous) currents which made be obtained and which limits the peak power from a free-electron laser. High peak powers may be obtained by using induction linacs, pulse line accelerators, or modulators which produce electron beams with currents ranging from several amperes to several thousands of amperes and with pulse times ranging from several tens of nanoseconds to several microseconds. Induction linacs operate by inducing an electromotive force in a cavity through a rapid change in the magnetic field strength. In effect, the electron beam acts as the secondary winding in a transformer. For example, the Advanced Test Accelerator (ATA) built at Lawrence Livermore National Laboratory (LLNL) was an induction linac that demonstrated energies and currents as high as 50 MeV and 10 kA with a pulse duration of 50 ns. At lower energies, pulse line accelerators and conventional microwave tube modulators are available. Pulse line accelerators produce beams with energies up to several tens of MeV, currents of several tens of kiloamperes, and pulse times up to 50–100 ns. As a result, pulse line accelerators and modulators have been applied exclusively to microwave generation.

Amplifier experiments employing induction linacs have been performed at the Naval Research Laboratory, Lawrence Livermore National Laboratory, and at the Institute for Laser Engineering at Osaka University in Japan [137–141] and have demonstrated operation from the microwave to the infrared spectra. A superradiant amplifier at the Naval Research Laboratory employed a 650 keV/200 A electron beam and produced 4 MW at a wavelength of 1 cm [55–57]. The LLNL experiments employed both the Experimental Test Accelerator (ETA) and the ATA. The free-electron laser amplifier experiment at an 8 mm wavelength was conducted with the ETA [85, 86] operating at approximately 3.5 MeV and a current of 10 kA. However, due to beam quality requirements, only about 10% of the beam was found to be usable in the free-electron laser. This ETA-based MOPA was operated in both uniform and tapered wiggler configurations. In the case of a uniform wiggler, the measured output power was in the neighborhood of 180 MW, which corresponds to an extraction efficiency of about 6%. A dramatic improvement in the efficiency was achieved, however, using a tapered wiggler. In this case, the total output power rose to 1 GW for an efficiency of 35%. The ATA was used for a high power MOPA design at a wavelength of 10.6 μm.

An important consideration in the construction of free-electron lasers with the intense beams generated by these accelerators is that an additional source of focusing is required to confine the electrons against the self-repulsive forces generated by the beam itself. This can be accomplished by the use of additional magnetic fields generated by either solenoid or quadrupole current windings. Quadrupole field windings were employed in the 8 mm amplifier experiment at the Lawrence Livermore National Laboratory; and the interaction mechanism in a free-electron laser is largely, though not entirely, transparent to the effect of the quadrupole field. In contrast, a solenoidal field has a deep and subtle effect on the interaction mechanism. This arises because a solenoidal field results in a precession, called Larmor rotation, about the magnetic field lines that can resonantly enhance the helical motion induced by the wiggler. This enhancement in the transverse velocity associated with the helical trajectory occurs when the Larmor period is comparable to the wiggler period. Since the Larmor period varies with beam energy, this relation can be expressed as

$$B_0 \approx 1.07 \frac{\gamma_b}{\lambda_w}, \tag{1.36}$$

where the solenoidal field B_0 is expressed in Tesla and the wiggler period is in centimeters. For fixed beam energies, therefore, this resonant enhancement in the wiggler-induced velocity requires progressively higher solenoidal fields as the wiggler period is reduced.

The effect of a resonant solenoidal magnetic field is to enhance both the gain and saturation efficiency of the interaction. This was demonstrated in a superradiant amplifier experiment using the VEBA pulse line accelerator at the Naval Research

Laboratory [51–54]. In this experiment, the output power from the free-electron laser was measured as a function of the solenoidal field as it varied over the range of 0.6–1.6 T. The beam energy and current in the experiment were 1.35 MeV and 1.5 kA, respectively, and the wiggler period was 3.0 cm. It should be remarked that due to the high current in the experiment, the beam was unable to propagate through the wiggler for solenoidal fields below 0.6 T. The magnetic resonance was expected for a solenoidal field in the neighborhood of 1.3 T, and the experiment showed a dramatic increase in the output power for fields in this range. Other experiments that have demonstrated the effect of a solenoidal field have been performed at the Massachusetts Institute of Technology [73–82], Columbia University [66–70], and the Ecole Polytechnique in France [95, 96].

The maximum enhancement in the output power in the experiment at the Naval Research Laboratory was observed for solenoidal fields slightly above the resonance. In this case, the nature of the interaction mechanism undergoes a fundamental change. In the absence of a solenoidal field (or for fields below the magnetic resonance), the axial velocity of the electrons decreases as energy is lost to the wave, while the transverse velocity remains relatively constant. In contrast, the result of the strong solenoidal field is to cause a negative-mass effect in which the electrons accelerate in the axial direction as they lose energy to the wave. The bulk of the energy used to amplify the wave is extracted from the transverse motion of the electrons. Computer simulations of free-electron lasers operating in this strong solenoidal regime indicate that extremely high extraction efficiencies (in the neighborhood of 50%) are possible without recourse to a tapered wiggler field. However, operation in this regime is precluded below submillimeter wavelengths. The reason for this is that the solenoidal field required to achieve this magnetic resonance varies directly with the beam energy and inversely with the wiggler period. Because of this, impractically, high fields are required for wavelengths in the far infrared and below.

The preceding discussion of the effect of a solenoidal magnetic field dealt with solenoidal fields directed parallel to the wiggler field. However, a recent experiment at the Massachusetts Institute of Technology [142, 143] has demonstrated the utility of a solenoidal field directed antiparallel to the wiggler field. In contrast to the parallel orientation of the wiggler and solenoidal fields, this orientation results in a decrease in the transverse electron velocity and, hence, the linear gain. However, high efficiencies are still possible, and this experiment measured a maximum extraction efficiency of 27% at a frequency of 35 GHz without any tapering of the wiggler field. This compares favorably with the efficiency found in the tapered wiggler experiment using the ETA accelerator at LLNL [85, 86]. A detailed analysis of this experiment is discussed in Chap. 5.

It is important to bear in mind that the high peak versus average power and oscillator versus MOPA distinctions between the different aforementioned accelerator technologies are becoming blurred by advances in the design of both rf and induction linacs. On the one hand, the application of laser-driven photocathodes to free-electron lasers at Los Alamos National Laboratory [43] has dramatically increased both the peak (of the order of 400 A) and average currents achievable with rf linacs. It is significant in this regard that a collaborative effort between

Stanford University and Rocketdyne Inc. and TRW [144–147] has already achieved MOPA operation using the Mark III rf linac. Successful completion of these development programs will enable high-average power free-electron lasers to be constructed using this technology.

Although their average power is lower than that of linacs, electrostatic accelerators can produce continuous electron beams using charge recovery techniques. In such a process, the electron beam is recirculated through the wiggler and back into the accelerator in a continuous stream. Using this technology, the electrostatic accelerator holds promise as an electron beam source for a continuous or long-pulse free-electron laser. However, given practical restrictions on the size of such accelerators, which limit energies and currents, electrostatic accelerators have been restricted to the construction of free-electron laser oscillators that operate from the microwave regime to the infrared spectrum. In particular, a high-average power free-electron laser development program was conducted at FOM in the Netherlands for the heating of magnetic fusion reactors that is based upon an electrostatic accelerator with beam recovery [148].

The principal design issue for accelerators at the present time is that of beam quality, since it is crucial to have a low axial velocity spread for efficient operation of the free-electron laser. This process is accomplished in a number of different ways for the various accelerator types. In the low-energy cases relevant to modulators and pulse line accelerators, this is accomplished by careful design of the cathodes and focusing systems in the electron guns, as well as by attention to the transport system which brings the beam into the wiggler. In rf linacs and microtrons, the development of laser-driven photocathodes permitted the production of exceptionally high brightness electron beams [37, 43, 149, 150].

The breadth of free-electron laser experiments includes many different wiggler configurations and virtually every type of accelerator in use today. The wiggler has been produced in planar, helical, and cylindrical forms by means of permanent magnets, current carrying coils, and hybrid electromagnets with ferrite cores. Helical wiggler fields can be produced by a current carrying bifilar helical coil in which field increases radially outward from the symmetry axis and provides magnetic focusing to confine the beam against the mutually repulsive forces between the electrons. In a planar wiggler, both the transverse and axial components of the velocity oscillate in synchronism with the wiggler. As such, the interaction is determined by the average, or root-mean-square, wiggler field. Because of this, planar wigglers require a stronger field to produce the same effect as a helical wiggler. This is compensated for, however, by the ease of adjustment allowed by a planar design, in which the strengths or positions of the individual magnets can be altered to provide either a uniform or tapered field. In contrast, the only adjustment possible for a bifilar helix is the strength of the field.

One practical constraint on the development of free-electron laser oscillators at the present time is mirror technology for infrared and shorter wavelengths and relates to both reflectivity and durability. The reflectivity is important since the net gain of an oscillator decreases as the mirror losses increase, and oscillation is possible only if the amplification due to the free-electron laser interaction exceeds the losses at the mirrors. The reflectivity is a measure of this loss rate and must be kept sufficiently high that the energy losses at the mirrors do not overwhelm the gain. The issue of durability relates to the power level that any specific mirror material can endure without suffering optical damage. In this sense, optical damage refers to a decrease in the reflectivity. Note that the extreme case of the complete burning out of the mirrors might be described as a catastrophic drop in the reflectivity. Problems exist in finding materials with a high enough reflectivity and durability to operate in the infrared and ultraviolet spectra, and even the visible presents problems. For example, the visible free-electron laser oscillator at the Université de Paris Sud experienced mirror degradation due to harmonic emission in the ultraviolet. At extremely high-power levels, solutions can be found through such techniques as the grazing incidence mirrors used by Boeing Aerospace in which the optical beam is allowed to expand to the point where the power density on the mirrors is low. In the infrared, oscillator experiments at Los Alamos National Laboratory originally employed a dielectric mirror material with a high reflectivity at low-power levels. However, recent observations indicate that nonlinear phenomena occur in this material at high-power levels which effectively reduces the reflectivity and that the use of copper mirrors substantially improves performance. An additional problem occurs at high-power levels due to thermal distortion of the optical surface. In order to combat this problem, actively cooled mirrors were developed for the 10-kW upgrade experiment at the Thomas Jefferson National Accelerator Facility [126, 127].

The principal biomedical applications of the free-electron laser are surgery, photocoagulation and cauterization, photodynamic therapy, and the in vivo thermal destruction of tissue through a process called photothermolysis. The most common surgical technique is the thermal ablation of tissue that requires a laser producing powers of 10–100 W at a wavelength of approximately 3 μm. This corresponds to a strong absorption resonance of the water molecule characterized by relatively little scattering of the light by the tissues. In contrast, photocoagulation requires a shorter wavelength of approximately 1–1.5 μm that is also strongly absorbed by the water molecule but exhibits a higher degree of scattering throughout the surrounding tissue. It is important to observe that the tunability of the free-electron laser holds the potential for a surgical laser that may be tuned in a single sequential process from 3 μm down to 1 μm to give both clean surgical incisions and cauterization. Another advantage is that the optical pulse can be tailored to meet specific requirements by control of the temporal structure of the electron beam. For example, short pulses are useful in opthalmic therapy for the surgical disruption of pigmented tissue, while longer pulses are required for retinal photocoagulation. Photodynamic therapies rely on the injection of photosensitive dyes that are preferentially concentrated in malignant tissue. Subsequent irradiation excites a photooxidation process that is toxic to the tumorous tissue. The principal dyes are photosensitive at wavelengths between 0.6 and 1.7 μm and are used to treat tumors of the lung, bladder, and gastrointestinal tract at early stages in their development as well as a palliative treatment at later stages of growth. At relatively high-average powers (up to 100 W), a free-electron laser makes possible the simultaneous treatment of relatively large masses of tissue. However, the high-energy electron beams needed to produce these wavelengths also produce relatively large X-ray fluxes, and the entire facility including power supply, accelerator, wiggler, optical system, and X-ray shielding is likely to be rather bulky and complex. Hence, in consideration of the rapid development of conventional laser sources at these wavelengths, the long-term biomedical applications are envisioned to be in (1) the initial refinement of these therapeutic techniques, (2) as large centralized facilities for tumor treatment, and (3) experimental research tools.

Applications to research are unimpaired by considerations of the bulk and complexity of a free-electron laser facility, and the first user facility was established by Luis Elias at the University of California at Santa Barbara in 1984 [151–158] and employs a long-pulse 3 MeV electrostatic accelerator. The free-electron laser produces a peak power of as much as 10 kW over a range in wavelengths of 390–1000 μm and is suitable for a wide range of experiments in the biomedical, solid-state, and surface sciences. In the field of photobiology, since the DNA molecule is sensitive to infrared wavelengths, the free-electron laser can study such behavior as the variation in the DNA mutation rate with wavelength. In addition, experiments have been conceived in the linear and nonlinear excitations of phonons and magnons, ground and excited state Stark splitting, the generation of coherent phonons, phonon amplification by stimulated emission, induced phase transitions, and semiconductor bandgap structure. The latter application may prove relevant to the study of high critical temperature superconductors. This facility is capable of producing kilowatts of tunable radiation in three frequency bands spanning the submillimeter through mid-infrared spectra: 120 GHz–1 THz, 1–5 THz, and 5–10 THz.

Two user facilities were constructed at Stanford University based upon both the Mark III rf linac [159–167] and the SCA [16–18, 168]. The Mark III facility, established by John Madey and coworkers, is tunable over the range 0.5–10 μm and produces 60 kW over a pulse time of several microseconds. Due to the electron energies available from the Mark III, however, the fundamental operation occurs at wavelengths in the neighborhood of 1–3 μm. Shorter wavelengths are achieved through the use of frequency doubling techniques common to laser engineering. In this case, the tunability, high power, and temporal structure offer a unique opportunity to study surgical applications. In particular, experiments have been conducted in the cutting of both bone and soft tissue with encouraging results. Since spot sizes of the order of 100–1000 μm and power densities as high as several megawatts per square centimeter are possible, the cutting mechanism is not thermal ablation but direct plasma formation of the irradiated tissue with extremely clean and localized incisions. In contrast with conventional lasers (which produce lower powers over longer pulse times), the combination of high power and short pulses results in less scar formation and more rapid healing. Indeed, the power densities available with this free-electron laser have raised concern that current optical fiber technology may ultimately prove inadequate to the task of directing the radiation, and research has begun in the development of optical fibers capable of handling higher intensities. In addition, experiments have been conducted to study semiconductor band gap structure as well as the multiphoton spectroscopy of germanium and poly-acetylene.

The SCA free-electron laser facility is operated in a continuous mode in which the macropulses were generated at a frequency of 20 Hz (i.e., 20 macropulses per second) over a timescale of several hours. A typical macropulse is of 5 ms duration and is composed of a train of 3 ps micropulses separated by 84.6 ns. Average electron beam currents over a macropulse can reach 200 μA with a peak current of 6 A over a micropulse. The peak voltage of the superconducting linac is 75 MeV, but the recirculation system allows this to rise as high as 150 MeV. The free-electron laser has operated over wavelengths ranging from 0.5 to 3.5 μm and is capable of operation in the ultraviolet. This range was extended to 15 μm using a wiggler supplied by Spectra Technologies, Inc. [21]. Experiments were concerned with dynamical processes on picosecond timescales in materials such as, photon echoes in dye molecule/glass systems. To this end, the picosecond micropulses and the relatively large macropulse separation (which is long enough for conventional optical pulse selection techniques to be used) produced by the SCA were crucial. In one experiment, the free-electron laser was operated at a wavelength of 1.54 μm with a linewidth of 0.08% and was stable over a timescale of several hours. The output power was approximately 300 kW over a micropulse, 12 W averaged over a macropulse, or 1 W over the longer timescale.

The operation of the SCA was upgraded by the implementation of a novel electron injector/acceleration scheme. In a conventional rf linac, the electron beam is accelerated by a single-frequency rf signal which varies sinusoidally over time. Since the amount of electron acceleration depends upon the rf power, the electron pulse must be synchronized to the peak of the rf signal and its duration kept short. In general, the longer the electron pulse, the larger the variation in rf power over the pulse, and the greater the energy spread of the electron beam. In order to minimize the beam energy spread, which degrades the performance of the free-electron laser, researchers at the SCA implemented a plan to use a composite rf signal composed of waves at multiple frequencies. In a manner analogous to the way in which a *square-wave* signal can be built up from a large composite spectrum of waves, this process extended the duration of the peak rf signal. As a consequence, both the duration and power in the electron beam increased at little or no cost to the beam quality. It was estimated that the peak and average beam currents increased to approximately 50 A and 1 mA, respectively, with corresponding increases in the output of the free-electron laser.

Experiments using the SCA/FEL spanned a wide range of fields including solid-state and surface science, molecular chemistry, biophysics, and medical science. In solid-state and surface science, the applications have dealt studies of picosecond physics of spectroscopy, nonlinear optics, free-carrier lifetimes and dynamics, inter-subband transitions, harmonic generation, and hole relaxation. In molecular chemistry, research has been conducted in photon echoes, non-photochemical hole burning, optical dephasing, ultrafast temperature jump spectroscopy, laser ablation, infrared vibrational photon echoes, ultrafast multiphoton up-pumping, ultrafast spectroscopy, and vibrational dynamics. Biophysical applications include photon photochemistry, spontaneous vibrational Stark effect spectroscopy, and the manipulation of single DNA molecules using optical tweezers. Research in the field of medical science includes microscopy of transient local heating in single living cells, optical tomography, and the detection of pathogens in living hosts.

As a measure of the impact of advances in accelerator technology on the free-electron laser, it should be noted that the Mark III linac was the accelerator used by Hans Motz in 1951 and that the range of free-electron laser experiments conducted with this accelerator was made possible by the continual improvement of the original design. Indeed, a user facility was built at Vanderbilt University [169, 170] based upon a further improved Mark III linac. This free-electron laser, which began operating in the summer of 1991, was intended to operate over a spectral range of 0.2–10 μm and has produced an average power of 11 W. Experiments have been conducted in biology and materials science and include matrix-assisted laser desorption-ionization mass spectroscopy, studies of the electronic structure of semiconductor heterojunctions, and laser surgery of hard and soft tissues. In the latter case, experiments using radiation at 3–4 μm wavelengths have demonstrated the ability to cut the bone with heretofore unrealized precision. Soft tissue surgery has been conducted using radiation at 6.45 μm wavelengths corresponding to a protein band that eliminates tissue by ablation.

In general, the principal advantages of the free-electron laser as a research tool are high intensity (relative to currently available sources), tunability, and a temporal structure that is controlled by the characteristics of the accelerator. Tunability permits the selection of specific energy states for study, while an appropriate tailoring of the temporal structure of the pulses allows the time evolution and decay of excited states to be investigated. In addition, it should be remarked that there are no good infrared sources available with wavelengths ranging from 2 to 5 μm. Important areas of investigation are the bulk and surface properties of semiconductors: in particular, the bandgap structure. Some of the most commercially important semiconductors exhibit bandgaps in the range of 0.25–2.5 μm, and it is important to extend the spectral range of study to within 0.1 μm about this range. A high-intensity, tunable free-electron laser producing pulses of approximately 10–100 ps duration at a repetition rate of several MHz would permit the study of the dynamic excitation and subsequent decay of electrons into unoccupied energy states. Other applications include laser photochemistry and photophysics that require sources in the visible and ultraviolet spectra. The development of an ultraviolet free-electron laser was carried out at Los Alamos National Laboratory using rf linac technology [171]. Since the electron beam is composed of short bursts at high repetition rates, rf linacs produce radiation with the desired temporal structure. The principal competition in the ultraviolet comes from incoherent synchrotron light sources that also make use of undulator magnets. The advantage of a free-electron laser over synchrotron sources is the increased counting rates resulting from a larger photon flux that would make practical a large number of currently marginal experiments. A nonexhaustive list of experiments possible with a coherent visible through ultraviolet source is the multiphoton ionization of liquids, chemistry of combustion and of molecular ions, high-resolution polyatomic and flourescence spectroscopy, time-resolved resonance Raman spectroscopy, spin-polarized photoemission and photoemission microscopy, magneto-optical studies of rare-earth elements, and studies of optical damage from high-intensity ultraviolet radiation.

Industrial applications are envisioned in materials production and photolithography. The requirements for materials production are sources in the near infrared (2.5–100 μm) and the ultraviolet through soft X-ray spectra and involve the pyrolytic production of powders for catalysts, the near-stoichiometric production of high-value chemicals and pharmaceuticals, and the pyrolytic and photolytic deposition of thin films on substrates. In photolithography, the surface of a wafer is coated with a layer of a photoresisting substance of which a part is illuminated. The wafer is then processed by the removal of either the exposed or unexposed portions of the resist. There are three principal lithographic techniques either in use or under consideration. Contact printing is performed by bringing a mask and wafer into contact, while in proximity printing, a small space separates these two components. The preferred technique is that of projection printing in which the image of the mask is projected onto the wafer from a greater distance than in contact or proximity printing, which allows the use of larger masks and improves usable mask lifetime. Each of these techniques requires uniform and stable sources of illumination. The sources may be either continuous or pulsed; however, possible difficulties with pulsed sources (i.e., rf linac-driven free-electron lasers) are that (1) the instantaneous pulse power necessary to give a sufficiently high-average power may also be high enough to damage the wafer and (2) that extremely high uniformity from pulse to pulse is required. In addition, short-wavelength sources may render the non-lithographic direct printing of wafers possible and eliminate the need for photoresists.

Applications also exist for high-power microwaves and submillimeter waves in the fields of communications, radar, and plasma heating. We confine the discussion to the heating of a magnetically confined plasma for controlled thermonuclear fusion. The reactor design of greatest interest is the Tokamak, which confines a high temperature plasma within a toroidal magnetic *bottle*. A thermonuclear reactor must confine a plasma at high density and temperature for a sufficiently long time to ignite a sustained fusion reaction. In its original conception, the Tokamak was to be heated to ignition by an Ohmic heating technique whereby a current is induced by means of a coil threading the torus. As such, the Tokamak acts like the secondary winding in a transformer. However, recent developments indicate the need for auxiliary heating, and the resonant absorption of submillimeter radiation has been proposed for this purpose. The frequencies of interest are the harmonics of the electron cyclotron frequency and range from about 280 to 560 GHz. The free-electron laser has operated in this spectral region at power levels of this order but over a much shorter pulse time. As such, it represents one of several competing concepts [172–174].

A user facility in the Netherlands referred to as the Free Electron Laser for Infrared eXperiments (FELIX) facility employs a 21 MeV rf linac which produces a 3 μs macropulse train of 1–10 ps micropulses with an overall repetition frequency of 5 Hz. Temporal overlap of these pulses is possible and a pulse of approximately 7 ns is achievable. Two spectral bands have been covered thus far: 6–20 μm and 40–60 μm. In addition to medical and materials research, this facility has been used to study many aspects of the device physics of free-electron laser oscillators including phase locking [175], single-mode operation [176], limit cycle oscillations [177], hole coupling and dynamic resonator desynchronization [178, 179], coherent start-up [180], and short pulse effects [181]. Research into medical and materials applications include induced two photon absorption, photon drag effects, inter-subband absorption, sum-frequency generation at surfaces, photoablation of corneal tissue, and phosphorescence detection of triplet-triplet transitions.

User facilities were constructed in France at the Université de Paris Sud using the Super-ACO storage ring [182–189] and an rf linac [190–193] system (CLIO). Experiments in time-resolved fluorescence have been conducted using the Super-ACO facility and experiments dealing with solid-state and surface interactions (sum frequency generation), near-field infrared microscopy, and vibrational energy transfer in molecules.

A more controversial application of high-power, long-pulse free-electron lasers was to the strategic defense against intercontinental ballistic missiles. In this regard, planners envisioned a large-scale ground-based laser that would direct light toward a target by means of both ground-based and orbiting mirrors. Designs based on both amplifier and oscillator free-electron lasers were being pursued. In experiments, a free-electron laser amplifier at LLNL amplified a 14 kW input signal from a carbon dioxide laser at 10.6 μm to a level of approximately 7 MW, a gain of 500 times. Boosting the input beam to five megawatts yielded a saturated power of 50 MW. Boeing Aerospace built an experimental free-electron laser oscillator in collaboration with Los Alamos National Laboratory based on a 5-m-long planar wiggler and an advanced radiofrequency linac. In its initial design, the linac produced electron beams with energies as high as 120 MeV. The oscillator has lased in the red region of the visible spectrum at a wavelength of 0.62 μm and at power levels a billion times that of the normal spontaneous emission within the cavity. The average power over the course of a 100 μs pulse was about 2 kW. The corresponding conversion efficiency is about 1%, but the peak power is a more respectable 40 MW. Even though oscillators typically generate short-wavelength harmonics that can damage the cavity mirrors, no degradation has been observed.

Estimates indicated that pulses of visible or near-infrared light at an average power of about 10–100 MW over a duration of approximately 1 s are required to destroy a missile during its boost phase. This means lengthening the pulses or increasing the peak power levels of existing free-electron lasers by a factor of a million or more. Depending on laser efficiency and target hardness, a collection of ground-based free-electron lasers would require somewhere between 400 MW and 20 GW of power for several minutes during an attack. (For comparison, a large power plant generates about 1 GW.) For these and other reasons, it became clear that it was impractical to scale up free-electron lasers to the power levels required.

While the strategic defense application was ultimately canceled, the effort did result in an advance that forms the basis for much of the more recent activities. Specifically, the development of laser-driven photocathode electron guns at Los Alamos [42, 194] which built on previous work [195, 196]. The high brightness electron beams that this technology enables form the basis for the development for all the more modern free-electron lasers based on radiofrequency linear accelerators.

One important line of research is aimed at the development of high-average power oscillators driven by superconducting energy recovery linear accelerators. This work was centered at the Thomas Jefferson National Accelerator Facility in Newport News [120, 126, 127], Virginia, and at the Japan Atomic Energy Research Institute [128, 197]. The free-electron laser in Newport News ultimately achieved an average output power of 14.3 kW at a wavelength of 1.6 μm.

In addition to the high-average power/long-wavelength free-electron laser oscillators, activity in the generation of high peak power/short-wavelength X-ray free-electron lasers is increasing in intensity worldwide [198–207]. Because of the dearth of sources to provide a seed at X-ray wavelength, most of these facilities rely on self-amplified spontaneous emission (SASE) generated as the electron beam propagates through the wiggler line in a single pass. However, since SASE is subject to shot-to-shot fluctuations, there is interest in alternative approaches including (1) self-seeding in which the interaction is broken at a stage prior to saturation and then the optical output is passed through a monochromator before being reinjected into the wiggler line in synchronism with the electron beam and (2) employing a high-gain harmonic generation (HGHG) scheme (see Chap. 13) in which the interaction is seeded at a longer wavelength which preconditions the electron beam for emission at harmonics in successive wigglers [136, 208]. A summary of existing X-ray free-electron lasers and synchrotron light sources is shown in Fig. 1.4. Finally, progress is being made in using laser wakefield accelerators in driving short-wavelength free-electron lasers [210].

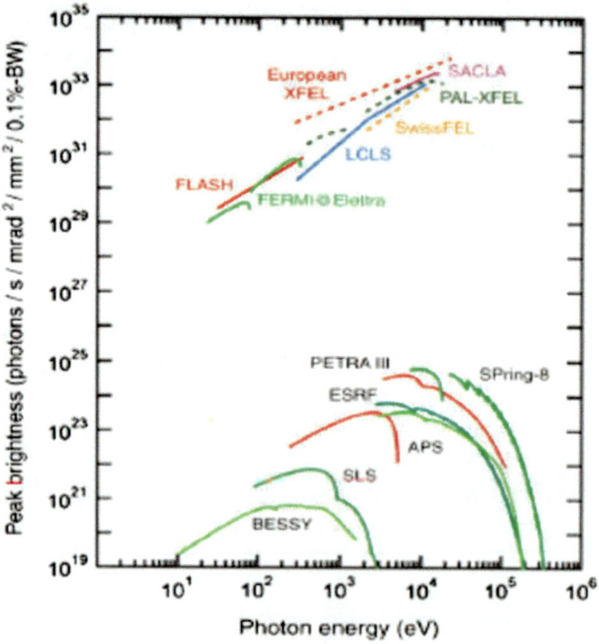

Fig. 1.4 Comparison of XFEL and synchrotron light sources [209]

1.4 Discussion

This chapter includes a necessarily abridged list of experiments and recently conceived applications of the free-electron laser. The fundamental principles of the free-electron laser are understood at the present time, and the future direction of research is toward evolutionary improvements in electron beam sources (in terms of beam quality and reliability) and wiggler designs. The issues, therefore, are technological rather than physical, and the free-electron laser can be expected ultimately to cover the entire spectral range shown. In this regard, it is important to recognize that the bulk of the early experiments were performed with accelerators not originally designed for use in a free-electron laser and issues of the beam quality important to free-electron laser operation were not adequately addressed in those early designs. As a consequence, the results often did not represent the full potential of the free-electron laser although many of the experiments have produced record power levels. However, advances in the design of photocathode injectors and high-energy beamlines have given birth to the current range of X-ray free-electron laser light sources, the development of which is still in its infancy and which a plethora of innovative new designs in undoubtedly to be seen.

References

1. H. Motz, Applications of radiation from fast electron beams. J. Appl. Phys. **22**, 527 (1951)
2. H. Motz, W. Thon, R.N. Whitehurst, Experiments on radiation by fast electron beams. J. Appl. Phys. **24**, 826 (1953)
3. H. Motz, M. Nakamura, Radiation of an electron in an infinitely long waveguide. Ann. Phys. **7**, 84 (1959)
4. R.M. Phillips, The ubitron, a high power traveling wave tube based on a periodic beam interaction in unloaded waveguide. IRE Trans. Electron Dev. **ED-7**, 231 (1960)
5. R.M. Phillips, History of the ubitron. Nucl. Instrum. Methods Phys. Res. **A272**, 1 (1988)
6. J.M.J. Madey, Stimulated emission of bremsstrahlung in a periodic magnetic field. J. Appl. Phys. **42**, 1906 (1971)
7. J.M.J. Madey, H.A. Schwettman, W.M. Fairbank, A free-electron laser. IEEE Trans. Nucl. Sci. **NS-20**, 980 (1973)
8. L.R. Elias, W.M. Fairbank, J.M.J. Madey, H.A. Schwettman, T.I. Smith, Observation of stimulated emission of radiation by relativistic electrons in a spatially periodic transverse magnetic field. Phys. Rev. Lett. **36**, 717 (1976)
9. D.A.G. Deacon, L.R. Elias, J.M.J. Madey, G.J. Ramian, H.A. Schwettman, T.I. Smith, First operation of a free-electron laser. Phys. Rev. Lett. **38**, 892 (1977)
10. S.V. Benson, D.A.G. Deacon, J.N. Eckstein, J.M.J. Madey, K. Robinson, T.I. Smith, R. Taber, Optical autocorrelation of a 3.2 μm free-electron laser. Phys. Rev. Lett. **48**, 235 (1982)
11. M. Billardon, P. Ellaume, J.M. Ortega, C. Bazin, M. Bergher, M. Velghe, Y. Petroff, First operation of a storage ring free-electron laser. Phys. Rev. Lett. **51**, 1652 (1983)
12. M. Billardon, P. Ellaume, J.M. Ortega, C. Bazin, M. Bergher, M. Velghe, D.A.G. Deacon, Y. Petroff, Free-electron laser experiment at Orsay: A review. IEEE J. Quantum Electron. **QE-21**, 805 (1985)

13. J.M. Ortega, Y. Lapierre, B. Girard, M. Billardon, P. Ellaume, C. Bazin, M. Bergher, M. Velghe, Y. Petroff, Ultraviolet coherent generation from an optical klystron. IEEE J. Quantum Electron. **QE-21**, 909 (1985)
14. M. Billardon, P. Ellaume, Y. Lapierre, J.M. Ortega, C. Bazin, M. Bergher, J. Marilleau, Y. Petroff, The Orsay storage ring free-electron laser: New results. Nucl. Instrum. Methods Phys. Res. **A250**, 26 (1986)
15. G.N. Kuliapanov, V.N. Litvinenko, I.V. Panaev, V.M. Popik, A.N. Skrinsky, A.S. Sokolov, N.A. Vinokurov, The VEPP-3 storage-ring optical klystron: Lasing in the visible and ultraviolet regions. Nucl. Instrum. Methods Phys. Res. **A296**, 1 (1990)
16. T.I. Smith, H.A. Schwettman, R. Rohatgi, Y. Lapierre, J. Edighoffer, Development of the SCA free-electron laser for use in biomedical and materials science experiments. Nucl. Instrum. Methods Phys. Res. **A259**, 1 (1987)
17. R. Rohatgi, H.A. Schwettman, T.I. Smith, R.L. Swent, The SCA free-electron laser program: Operation in the infrared, visible, and ultraviolet. Nucl. Instrum. Methods Phys. Res. **A272**, 32 (1988)
18. T.I. Smith, J.C. Frisch, R. Rohatgi, H.A. Schwettman, R.L. Swent, Status of the SCA free-electron laser. Nucl. Instrum. Methods Phys. Res. **A296**, 33 (1990)
19. J.M. Slater, J.L. Adamski, D.C. Quimby, T.L. Churchill, L.Y. Nelson, R.E. Center, Electron spectrum measurements for a tapered-wiggler free-electron laser. IEEE J. Quantum Electron. **QE-19**, 374 (1983)
20. J.M. Slater, T. Churchill, D.C. Quimby, K.E. Robinson, D. Shemwell, A. Valla, A.A. Vetter, J. Adamski, W. Gallagher, R. Kennedy, B. Robinson, D. Shoffstall, E. Tyson, A. Vetter, A. Yeremian, Visible wavelength free-electron laser oscillator. Nucl. Instrum. Methods Phys. Res. **A250**, 228 (1986)
21. K.E. Robinson, D.C. Quimby, J.M. Slater, The tapered hybrid undulator of the visible free-electron laser oscillator experiment. IEEE J. Quantum Electron. **QE-23**, 1497 (1987)
22. D.M. Shemwell, K.E. Robinson, R.I. Gellert, D.C. Quimby, J.M. Ross, J.M. Slater, A.A. Vetter, D. Trost, J. Zumdieck, Optical cavities for visible free-electron laser experiments. IEEE J. Quantum Electron. **QE-23**, 1522 (1987)
23. J.M. Eggleston, J.M. Slater, Baseline conceptual design for high-power free-electron laser ring cavities. IEEE J. Quantum Electron. **QE-23**, 1527 (1987)
24. K.E. Robinson, T.L. Churchill, D.C. Quimby, D.M. Shemwell, J.M. Slater, A.S. Valla, A.A. Vetter, J. Adamski, T. Doering, W. Gallagher, R. Kennedy, B. Robinson, D. Shoffstall, E. Tyson, A. Vetter, A. Yeremian, Panorama of the visible wavelength free-electron laser oscillator. Nucl. Instrum. Methods Phys. Res. **A259**, 49 (1987)
25. A.H. Lumpkin, N.S.P. King, M.D. Wilke, S.P. Wei, K.J. Davis, Time-resolved spectral measurements for the Boeing free-electron laser experiments. Nucl. Instrum. Methods Phys. Res. **A285**, 17 (1989)
26. A.H. Lumpkin, R.L. Tokar, D.H. Dowell, A.R. Lowrey, A.D. Yeremian, R.E. Justice, Improved performance of the Boeing/LANL free-electron laser experiment: Extraction efficiency and cavity-length detuning effects. Nucl. Instrum. Methods Phys. Res. **A296**, 169 (1990)
27. R.W. Warren, B.E. Newnam, J.G. Winston, W.E. Stein, L.M. Young, C.A. Brau, Results of the Los Alamos free-electron laser experiment. IEEE J. Quantum Electron. **QE-19**, 391 (1983)
28. B.E. Newnam, R.W. Warren, R.L. Sheffield, W.E. Stein, M.T. Lynch, J.S. Fraser, J.C. Goldstein, J.E. Sollid, T.A. Swann, J.M. Watson, C.A. Brau, Optical performance of the Los Alamos free-electron laser. IEEE J. Quantum Electron. **QE-21**, 867 (1985)
29. R.W. Warren, B.E. Newnam, J.C. Goldstein, Raman spectra and the Los Alamos free-electron laser. IEEE J. Quantum Electron. **QE-21**, 882 (1985)
30. R.L. Sheffield, W.E. Stein, R.W. Warren, J.S. Fraser, A.H. Lumpkin, Electron beam diagnostics and results for the Los Alamos free-electron laser. IEEE J. Quantum Electron. **QE-21**, 895 (1985)

31. M.T. Lynch, R.W. Warren, P.J. Tellerico, The effects of linear accelerator noise on the Los Alamos free-electron laser. IEEE J. Quantum Electron. **QE-21**, 904 (1985)
32. J.M. Watson, Status of the Los Alamos free-electron laser. Nucl. Instrum. Methods Phys. Res. **A250**, 1 (1986)
33. J.C. Goldstein, B.E. Newnam, R.W. Warren, R.L. Sheffield, Comparison of the results of theoretical calculations with experimental measurements from the Los Alamos free-electron laser oscillator experiment. Nucl. Instrum. Methods Phys. Res. **A250**, 4 (1986)
34. W.E. Stein, R.L. Sheffield, Electron micropulse diagnostics and results for the Los Alamos free-electon laser. Nucl. Instrum. Methods Phys. Res. **A250**, 12 (1986)
35. R.W. Warren, J.C. Goldstein, B.E. Newnam, Spiking mode operation for a uniform-period wiggler. Nucl. Instrum. Methods Phys. Res. **A250**, 19 (1986)
36. J.S. Fraser, R.L. Sheffield, E.R. Gray, A new high brightness electron injector for free-electron lasers driven by rf linacs. Nucl. Instrum. Methods Phys. Res. **A250**, 71 (1986)
37. R.W. Warren, D.W. Feldman, B.E. Newnam, S.C. Bender, W.E. Stein, A.H. Lumpkin, R.A. Lohsen, J.C. Goldstein, B.D. McVey, K.C.D. Chan, Recent results from the Los Alamos free-electron laser. Nucl. Instrum. Methods Phys. Res. **A259**, 8 (1987)
38. D.W. Feldman, R.W. Warren, W.E. Stein, J.S. Fraser, G. Spalek, A.H. Lumpkin, J.M. Watson, B.F. Carlsten, H. Takeda, T.S. Wang, Energy recovery in the Los Alamos free-electron laser. Nucl. Instrum. Methods Phys. Res. **A259**, 26 (1987)
39. D.W. Feldman, R.W. Warren, B.E. Carlsten, W.E. Stein, A.H. Lumpkin, S.C. Bender, G. Spalek, J.M. Watson, L.M. Young, J.S. Fraser, J.C. Goldstein, H. Takeda, T.S. Wang, K.C.D. Chan, B.D. McVey, B.E. Newnam, R.A. Lohsen, R.B. Feldman, R.K. Cooper, W.J. Johnson, C.A. Brau, Recent results from the Los Alamos free-electron laser. IEEE J. Quantum Electron. **QE-23**, 1476 (1987)
40. J.C. Goldstein, B.E. Newnam, R.W. Warren, Sideband suppression by an intracavity optical filter in the Los Alamos free-electron laser oscillator. Nucl. Instrum. Methods Phys. Res. **A272**, 150 (1988)
41. R.W. Warren, J.C. Goldstein, The generation and suppression of synchrotron sidebands. Nucl. Instrum. Methods Phys. Res. **A272**, 155 (1988)
42. R.L. Sheffield, E.R. Gray, J.S. Fraser, The Los Alamos photoinjector program. Nucl. Instrum. Methods Phys. Res. **A272**, 222 (1988)
43. B.E. Carlsten, D.W. Feldman, A.H. Lumpkin, J.E. Sollid, W.E. Stein, R.W. Warren, Emittance studies at the Los Alamos national laboratory free-electron laser. Nucl. Instrum. Methods Phys. Res. **A272**, 247 (1988)
44. R.W. Warren, J.E. Sollid, D.W. Feldman, W.E. Stein, W.J. Johnston, A.H. Lumpkin, J.C. Goldstein, Near-ideal lasing with a uniform wiggler. Nucl. Instrum. Methods Phys. Res. **A285**, 1 (1989)
45. D.W. Feldman, H. Takeda, R.W. Warren, J.E. Sollid, W.E. Stein, W.J. Johnson, A.H. Lumpkin, R.B. Feldman, High extraction efficiency experiments with the Los Alamos free-electron laser. Nucl. Instrum. Methods Phys. Res. **A285**, 11 (1989)
46. R.W. Warren, L.C. Haynes, D.W. Feldman, W.E. Stein, S.J. Gitomer, Lasing on the third harmonic. Nucl. Instrum. Methods Phys. Res. **A296**, 84 (1990)
47. A.H. Lumpkin, D.W. Feldman, J.E. Sollid, R.W. Warren, W.E. Stein, W.J. Johnson, J.M. Watson, B.E. Newnam, J.C. Goldstein, First direct observation of free electron lasing fron $\lambda = 20$ to 45 μm. Nucl. Instrum. Methods Phys. Res. **A296**, 181 (1990)
48. D.W. Feldman, S.C. Bender, B.E. Carlsten, J. Early, R.B. Feldman, W.J.D. Johnson, A.H. Lumpkin, P.G. O'Shea, W.E. Stein, R.L. Sheffield, K. McKenna, Performance of the Los Alamos HIBAF accelerator at 17 MeV. Nucl. Instrum. Methods Phys. Res. **A304**, 224 (1991)
49. V.L. Granatstein, S.P. Schlesinger, M. Herndon, R.K. Parker, J.A. Pasour, Production of megawatt submillimeter pulses by stimulated magneto-Raman scattering. Appl. Phys. Lett. **30**, 384 (1977)

50. D.B. McDermott, T.C. Marshall, S.P. Schlesinger, R.K. Parker, V.L. Granatstein, High-power free-electron laser based on stimulated Raman backscattering. Phys. Rev. Lett. **41**, 1368 (1978)
51. R.K. Parker, R.H. Jackson, S.H. Gold, H.P. Freund, V.L. Granatstein, P.C. Efthimion, M. Herndon, A.K. Kinkead, Axial magnetic field effects in a collective-interaction free-electronlaser at millimeter wavelengths. Phys. Rev. Lett. **48**, 238 (1982)
52. R.H. Jackson, S.H. Gold, R.K. Parker, H.P. Freund, P.C. Efthimion, V.L. Granatstein, M. Herndon, A.K. Kinkead, J.E. Kosakowski, T.J.T. Kwan, Design and operation of a collective millimeter-wave free-electron laser. IEEE J. Quantum Electron. **QE-19**, 346 (1983)
53. S.H. Gold, W.M. Black, H.P. Freund, V.L. Granatstein, R.H. Jackson, P.C. Efthimion, A.K. Kinkead, Study of gain, bandwidth, and tunability of a millimeter-wave free-electron laser operating in the collective regime. Phys. Fluids **26**, 2683 (1983)
54. S.H. Gold, W.M. Black, H.P. Freund, V.L. Granatstein, A.K. Kinkead, Radiation growth in a millimeter-wave free-electron laser operating in the collective regime. Phys. Fluids **27**, 746 (1984)
55. J.A. Pasour, R.F. Lucey, C.A. Kapetanakos, Long-pulse, high-power free-electron laser with no external beam focusing. Phys. Rev. Lett. **53**, 1728 (1984)
56. S.H. Gold, D.L. Hardesty, A.K. Kinkead, L.R. Barnett, V.L. Granatstein, High-gain 35 GHz free-electron laser amplifier experiment. Phys. Rev. Lett. **52**, 1218 (1984)
57. J.A. Pasour, R.F. Lucey, C.W. Roberson, Long pulse free-electron laser driven by a linear induction accelerator, in *Free-Electron Generators of Coherent Radiation*, Proceedings of SPIE, ed. by C.A. Brau, S.F. Jacobs, M.O. Scully, vol. 453, (SPIE, Bellingham, 1984), p. 328
58. J.A. Pasour, S.H. Gold, Free-electron laser experiments with and without a guide magnetic field: A review of the millimeter-wave free-electron laser research at the Naval Research Laboratory. IEEE J. Quantum Electron. **QE-21**, 845 (1985)
59. S.H. Gold, A.K. Ganguly, H.P. Freund, A.W. Fliflet, V.L. Granatstein, D.L. Hardesty, A.K. Kinkead, Parametric behavior of a high-gain 35 GHz free-electron laser amplifier with guide magnetic field. Nucl. Instrum. Methods Phys. Res. **A250**, 366 (1986)
60. J. Mathew, J.A. Pasour, High-gain, long-pulse free-electron laser oscillator. Phys. Rev. Lett. **56**, 1805 (1986)
61. J.A. Pasour, J. Mathew, C. Kapetanakos, Recent results from the Naval Research Laboratory experimental free-electron laser program. Nucl. Instrum. Methods Phys. Res. **A259**, 94 (1987)
62. R.H. Jackson, D.E. Pershing, F. Wood, Naval Research Laboratory ubitron amplifier cold test results. Nucl. Instrum. Methods Phys. Res. **A259**, 99 (1987)
63. D.E. Pershing, R.H. Jackson, H.P. Freund, H. Bluem, The K_u-band ubitron experiment at the Naval Research Laboratory. Nucl. Instrum. Methods Phys. Res. **A285**, 56 (1989)
64. D.E. Pershing, R.H. Jackson, H. Bluem, H.P. Freund, Fundamental mode amplifier performance of the Naval Research Laboratory ubitron. Nucl. Instrum. Methods Phys. Res. **A296**, 199 (1990)
65. D.S. Birkett, T.C. Marshall, S.P. Schlesinger, D.B. McDermott, A submillimeter free-electron laser experiment. IEEE J. Quantum Electron. **QE-17**, 1348 (1981)
66. J. Masud, T.C. Marshall, S.P. Schlesinger, F.G. Yee, Gain measurements from start-up and spectrum of a Raman free-electron laser oscillator. Phys. Rev. Lett. **56**, 1567 (1986)
67. J. Masud, T.C. Marshall, S.P. Schlesinger, F.G. Yee, W.M. Fawley, E.T. Scharlemann, S.S. Yu, A.M. Sessler, E.J. Sternbach, Sideband control in a millimeter-wave free-electron laser. Phys. Rev. Lett. **58**, 763 (1987)
68. J. Masud, T.C. Marshall, S.P. Schlesinger, F.G. Yee, Regenerative gain in a Raman free-electron laser oscillator. IEEE J. Quantum Electron. **QE-23**, 1594 (1987)
69. F.G. Yee, J. Masud, T.C. Marshall, S.P. Schlesinger, Power and sideband studies of a Raman free-electron laser. Nucl. Instrum. Methods Phys. Res. **A259**, 104 (1987)
70. F.G. Yee, T.C. Marshall, S.P. Schlesinger, Efficiency and sideband observations of a Raman free-electron laser oscillator with a tapered undulator. IEEE Trans. Plasma Sci. **PS-16**, 162 (1988)

71. S.Y. Cai, S.P. Chang, J.W. Dodd, T.C. Marshall, H. Tang, Optical guiding in a Raman free-electron laser: Computation and experiment. Nucl. Instrum. Methods Phys. Res. **A272**, 136 (1988)
72. J.W. Dodd, T.C. Marshall, Spiking radiation in the Columbia free-electron laser. Nucl. Instrum. Methods Phys. Res. **A296**, 4 (1990)
73. J. Fajans, G. Bekefi, Y.Z. Yin, B. Lax, Spectral measurements from a tunable, Raman free-electron laser. Phys. Rev. Lett. **53**, 246 (1984)
74. J. Fajans, G. Bekefi, Y.Z. Yin, B. Lax, Microwave studies of a tunable free-electron laser in combined axial and wiggler magnetic fields. Phys. Fluids **28**, 1995 (1985)
75. J. Fajans, J.S. Wurtele, G. Bekefi, D.S. Knowles, K. Xu, Nonlinear power saturation and phase in a collective free-electron laser amplifier. Phys. Rev. Lett. **57**, 579 (1986)
76. J. Fajans, G. Bekefi, Measurements of amplification and phase shift in a free-electron laser. Phys. Fluids **29**, 3461 (1986)
77. J. Fajans, G. Bekefi, Effect of electron beam temperature on the gain of a collective free-electron laser. IEEE J. Quantum Electron. **QE-23**, 1617 (1987)
78. F. Hartemann, K. Xu, G. Bekefi, J.S. Wurtele, J. Fajans, Wave-profile modification induced by the free-electron laser interaction. Phys. Rev. Lett. **59**, 1177 (1987)
79. F. Hartemann, K. Xu, G. Bekefi, Pulse compression in a free-electron laser amplifier. Nucl. Instrum. Methods Phys. Res. **A272**, 125 (1988)
80. F. Hartemann, K. Xu, G. Bekefi, Generation of short pulses of coherent electromagnetic radiation in a free-elecron laser amplifier. IEEE J. Quantum Electron. **QE-24**, 105 (1988)
81. C. Leibovitch, K. Xu, G. Bekefi, Effects of electron prebunching on the radiation growth rate of a collective free-electron laser amplifier. IEEE J. Quantum Electron. **QE-24**, 1825 (1988)
82. K. Xu, G. Bekefi, C. Leibovitch, Observations of field profile modifications in a Raman free-electron laser amplifier. Phys. Fluids B **1**, 2066 (1989)
83. D.A. Kirkpatrick, G. Bekefi, A.C. DiRienzo, H.P. Freund, A.K. Ganguly, A millimeter and submillimeter wavelength free-electron laser. Phys. Fluids B **1**, 1511 (1989)
84. K. Xu, G. Bekefi, Experimental study of multiple frequency effects in a free-electron laser amplifier. Phys. Fluids B **2**, 678 (1990)
85. T.J. Orzechowski, B.R. Anderson, W.M. Fawley, D. Prosnitz, E.T. Scharlemann, S.M. Yarema, D.B. Hopkins, A.C. Paul, A.M. Sessler, J.S. Wurtele, Microwave radiation from a high-gain free-electron laser amplifier. Phys. Rev. Lett. **54**, 889 (1985)
86. T.J. Orzechowski, B.R. Anderson, J.C. Clark, W.M. Fawley, A.C. Paul, D. Prosnitz, E.T. Scharlemann, S.M. Yarema, D.B. Hopkins, A.M. Sessler, J.S. Wurtele, High-efficiency extraction of microwave radiation from a tapered-wiggler free-electron laser. Phys. Rev. Lett. **57**, 2172 (1986)
87. T.J. Orzechowski, E.T. Scharlemann, D.B. Hopkins, Measurement of the phase of the electromagnetic wave in a free-electron laser amplifier. Phys. Rev. A **35**, 2184 (1987)
88. R.A. Jong, E.T. Scharlemann, High gain free-electron laser for heating and current drive in the ALCATOR-C tokamak. Nucl. Instrum. Methods Phys. Res. **A259**, 254 (1987)
89. A.L. Throop, W.M. Fawley, R.A. Jong, T.J. Orzechowski, D. Prosnitz, E.T. Scharlemann, R.D. Stever, G.A. Westenskow, D.B. Hopkins, A.M. Sessler, S.G. Evangelides, K.E. Kreischer, Experimental results of a high gain microwave free-electron laser operating at 140 GHz. Nucl. Instrum. Methods Phys. Res. **A272**, 15 (1988)
90. R.A. Jong, R.R. Stone, Induction-linac based free-electron laser amplifier for fusion applications. Nucl. Instrum. Methods Phys. Res. **A285**, 387 (1989)
91. A.L. Throop, R.A. Jong, S.L. Allen, D.P. Atkinson, J.C. Clark, B. Felker, S.W. Ferguson, M.A. Makowski, W.E. Nexsen, B.W. Rice, B.W. Stallard, W.C. Turner, 140 GHz microwave free-electron laser experiments using ELF-II. Nucl. Instrum. Methods Phys. Res. **A296**, 41 (1990)
92. H. Boehmer, M.Z. Caponi, J. Edighoffer, S. Fornaca, J. Munch, G. Neil, B. Saur, C.C. Shih, An experiment on free-electron laser efficiency enhancement with a variable wiggler. IEEE Trans. Nucl. Sci. **NS-28**, 3156 (1981)

93. H. Boehmer, M.Z. Caponi, J. Edighoffer, S. Fornaca, J. Munch, G.R. Neil, B. Saur, C.C. Shih, Variable wiggler free-electron laser experiment. Phys. Rev. Lett. **48**, 141 (1982)
94. D. Arnush, H. Boehmer, M.Z. Caponi, C.C. Shih, Design of a high power CW free-electron maser. Int. J. Electron. **53**, 605 (1982)
95. K.L. Felch, L. Vallier, J.M. Buzzi, P. Drossart, H. Boehmer, H.J. Doucet, B. Etlicher, H. Lamain, C. Rouillé, Collective free-electron laser studies. IEEE J. Quantum Electron. **QE-17**, 1354 (1981)
96. J.M. Buzzi, Collective free-electron lasers, in *Fifth International Conference on High-Power Particle Beams*, ed. by R. Briggs, A.J. Toepfer, (LLNL, Livermore, 1983), p. 546
97. F.A. Dolezal, R.J. Harvey, A.J. Palmer, R.W. Schumacher, Mode structure of a low-gain low-voltage free-electron laser, in *Free-Electron Generators of Coherent Radiation*, Proceedings of SPIE, ed. by C.A. Brau, S.F. Jacobs, M.O. Scully, vol. 453, (SPIE, Bellingham, 1984), p. 356
98. R.J. Harvey, F.A. Dolezal, The Hughes low-voltage free-electron laser program. Nucl. Instrum. Methods Phys. Res. **A250**, 274 (1986)
99. J.R. Pierce, *Traveling Wave Tubes* (Van Nostrand & Co., New York, 1950)
100. C.W. Roberson, Free-electron laser beam quality. IEEE J. Quantum Electron. **QE-21**, 860 (1985)
101. X.J. Wang, H.P. Freund, D. Harder, W.H. Miner Jr., J.B. Murphy, H. Qian, Y. Shen, X. Yang, Efficiency and spectrum enhancement in a tapered free-electron laser amplifier. Phys. Rev. Lett. **103**, 154801 (2009)
102. W.B. Colson, Theory of a free-electron laser. Phys. Lett. **A59**, 187 (1976)
103. W. Becker, Multiphoton analysis of the free-electron laser. Opt. Commun. **33**, 69 (1980)
104. W. Becker, M.O. Scully, M.S. Zubairy, Generation of squeezed coherent states via a free-electron laser. Phys. Rev. Lett. **48**, 475 (1982)
105. W. Becker, M.S. Zubairy, Photon statistics of a free-electron laser. Phys. Rev. A **25**, 2200 (1982)
106. W. Becker, J.K. McIver, Fully quantized many-particle theory of a free-electron laser. Phys. Rev. A **27**, 1030 (1983)
107. G. Dattoli, A. Renieri, Free-electron laser quantum aspects. J. Phys. **44**, C1–C125 (1983)
108. P. Bosco, W.B. Colson, R.A. Freedman, Quantum/classical mode evolution in free-electron laser oscillators. IEEE J. Quantum Electron. **QE-19**, 272 (1983)
109. G. Dattoli, M. Richetta, Free-electon laser quantum theory: Comments on Glauber coherence, antibunching and squeezing. Opt. Commun. **50**, 165 (1984)
110. F. Ciocci, G. Dattoli, M. Richetta, Analysis of the harmonic Raman-Nath equation. J. Phys. A **17**, 1333 (1984)
111. G. Dattoli, J. Gallardo, A. Renieri, M. Richetta, Quantum statistical properties of a free-electron laser amplifier. IEEE J. Quantum Electron. **QE-21**, 1069 (1985)
112. J. Gea-Banacloche, G.T. Moore, R.R. Schlicher, M.O. Scully, H. Walther, Soft X-ray free-electron laser with a laser undulator. IEEE J. Quantum Electron. **QE-23**, 1558 (1987)
113. C.A. Brau, *Free-Electron Lasers* (Academic Press, Boston, 1990)
114. P. Luchini, H. Motz, *Undulators and Free-Electron Lasers* (Clarendon Press, Oxford, 1990)
115. W.B. Colson, A limitation on the start-up of compact free-electron lasers. Nucl. Instrum. Methods Phys. Res. **A296**, 348 (1990)
116. R. Prazeres, P. Guyot-Sionnest, J.M. Ortega, D. Jaroszynski, M. Billardon, M.E. Couprie, M. Velghe, Y. Petroff, Coherent harmonic generation in VUV with the optical klystron on the storage ring Super-ACO. Nucl. Instrum. Methods Phys. Res. **A304**, 72 (1991)
117. M.E. Couprie, N.G. Gavrilov, G.N. Kulipanov, V.N. Litvinenko, I.V. Pinaev, V.M. Popik, A.N. Skrinsky, N.A. Vinokurov, The results of lasing linewidth narrowing on the VEPP-3 storage ring optical klystron. Nucl. Instrum. Methods Phys. Res. **A304**, 47 (1991)
118. V.N. Litvinenko, V.N. Vinokurov, Lasing spectrum and temporal structure in storage ring free-electron lasers: Theory and experiment. Nucl. Instrum. Methods Phys. Res. **A304**, 66 (1991)

119. M. Tigner, A possible apparatus for electron clashing-beam experiments. Nuovo Cimento **37**, 1228 (1965)
120. G.R. Neil, C.L. Bohn, S.V. Benson, G. Biallas, D. Douglas, H.F. Dylla, R. Evans, J. Fugitt, A. Grippo, J. Gubeli, R. Hill, K. Jordan, R. Li, L. Merminga, P. Piot, J. Preble, M. Shinn, T. Siggins, R. Walker, B. Yunn, Sustained kilowatt lasing in a free-electron laser with same-cell energy recovery. Phys. Rev. Lett. **84**, 662 (2000)
121. D.H. Dowell, M.L. Laucks, A.R. Lowrey, M. Bemes, A. Currie, P. Johnson, K. McCrary, L. Milliman, C. Lancaster, J. Adamski, D. Pistoresi, D.R. Shoffstall, M. Bentz, R. Burns, J. Guha, R. Hudyma, K. Sun, W. Tomita, W. Mower, S. Bender, J.C. Goldstein, A.H. Lumpkin, B.D. McVey, R.L. Tokar, D. Shemwell, First operation of a free-electron laser using a ring resonator. Nucl. Instrum. Methods Phys. Res. **A304**, 1 (1991)
122. M.L. Laucks, D.H. Dowell, A.R. Lowrey, M. Bemes, A. Currie, P. Johnson, K. McCrary, J. Adamski, D. Pistoresi, D.R. Shoffstall, M. Bentz, R. Burns, R. Hudyma, K. Sun, W. Mower, S. Bender, J.C. Goldstein, A.H. Lumpkin, B.D. McVey, R.L. Tokar, D. Shemwell, Optical measurements on the Boeing free-electron laser ring resonator experiment. Nucl. Instrum. Methods Phys. Res. **A304**, 25 (1991)
123. J.C. Frisch, J.E. Edighoffer, Time-dependent measurements on the SCA free-electron laser. Nucl. Instrum. Methods Phys. Res. **A296**, 9 (1990)
124. J.A. Edighoffer, H. Boehmer, M.Z. Caponi, S. Fornaca, J. Munch, G.R. Neil, B. Saur, C.C. Shih, Free-electron laser small signal gain measurement at 10.6 μm. IEEE J. Quantum Electron. **QE-19**, 316 (1983)
125. J.A. Edighoffer, G.R. Neil, C.E. Hess, T.I. Smith, S.W. Fornaca, H.A. Schwettman, Variable-wiggler free-electron laser oscillation. Phys. Rev. Lett. **52**, 344 (1984)
126. S. Benson, G. Biallas, C. Bohn, D. Douglas, H.F. Dylla, R. Evans, J. Fuggit, R. Hill, K. Jordan, G. Krafft, R. Legg, R. Li, L. Merminga, G.R. Neil, D. Oepts, P. Piot, J. Preble, M. Shinn, T. Siggins, R. Walker, B. Yunn, First lasing of the Jefferson Lab IR demo FEL. Nucl. Instrum. Methods Phys. Res. **A429**, 27 (1999)
127. G. Neil, C. Behre, S.V. Benson, M. Bevins, G. Biallas, J. Boyce, J. Coleman, L.A. Dillon-Townes, D. Douglas, H.F. Dylla, R. Evans, A. Grippo, D. Gruber, J. Gubeli, D. Hardy, C. Hernandez-Garcia, K. Jordan, M.J. Kelley, L. Merminga, J. Mammosser, W. Moore, N. Nishimori, E. Pozdeyev, J. Preble, R. Rimmer, M. Shinn, T. Siggins, C. Tennant, R. Walker, G.P. Williams, S. Zhang, The JLab high power ERL light source. Nucl. Instrum. Methods Phys. Res. **A557**, 9 (2006)
128. N. Nishimori, R. Hajima, R. Nagai, E.J. Minehara, Sustained saturation in a free-electron laser oscillator at perfect synchronism of an optical cavity. Phys. Rev. Lett. **86**, 5707 (2001)
129. D.J. Dunning, N.R. Thompson, J.A. Clarke, B.W.J. McNeil, 3D modelling of the ERLP IR-FEL, in Proceedings of the 2007 Free-Electron Laser Conference, Paper MOPPH069, Novosibirsk, Russia, 2007
130. P. Emma et al., First lasing and operation of an Ångstrom-wavelength free-electron laser. Nat. Photonics **4**, 641 (2009)
131. T. Tanaka, S. Goto, T. Hara, T. Hatsui, H. Ohashi, K. Togawa, M. Yabashi, H. Tanaka, Undulator commissioning by characterization of radiation in x-ray free-electron lasers. Phys. Rev. Spec. Top. Accel. Beams **15**, 110701 (2012)
132. J.-H. Han et al., Status of the PAL-XFEL project, in Proceedings of the 2012 International Particle Accelerator Conference, New Orleans, Louisiana, 2012
133. http://www.desy.de/research/facilities__projects/european_xfel/index_eng.html
134. C. Erny, C.P. Hauri, The SwissFEL expermental laser facility. J. Synchrotron Radiat. **23**, 1143 (2016)
135. V. Ayvazyan et al., First operation of a free-electron laser generating GW power radiation at 32 nm wavelength. Eur. Phys. J. D **37**, 297 (2006)
136. E. Allaria et al., Highly coherent and stable pulses from the FERMI seeded free-electron laser in the extreme ultraviolet. Nat. Photonics **6**, 699 (2012)

137. N. Ohigashi, K. Mima, S. Miyamoto, K. Imasaki, Y. Kitagawa, H. Fujita, S.I. Kuruma, S. Nakai, C. Yamanaka, Free-electron laser experiments with the pulse power beam at the Institute for Laser Engineering at Osaka. Nucl. Instrum. Methods Phys. Res. **A259**, 111 (1987)
138. T. Akiba, M. Fukuda, M. Noma, N. Inoue, K. Tateishi, K. Imasaki, S. Miyanoto, N. Ohigashi, K. Mima, S. Nakai, C. Yamanaka, Develoment of an inductive voltage accumulating system fro a free-electron laser. Nucl. Instrum. Methods Phys. Res. **A259**, 115 (1987)
139. K. Mima, Y. Kitagawa, T. Akiba, K. Imasaki, S. Kuruma, N. Ohigashi, S. Miyamoto, S. Fujita, S. Nakayama, Y. Tsunawaki, H. Motz, T. Taguchi, S. Nakai, C. Yamanaka, Experiment and theory on CO_2 laser powered wiggler and induction linac free-electron laser. Nucl. Instrum. Methods Phys. Res. **A272**, 106 (1988)
140. K. Mima, K. Imasaki, S. Kuruma, T. Akiba, N. Ohigashi, Y. Tsunawaki, K. Tanaka, C. Yamanaka, S. Nakai, Theory and experiment for the induction linac free-electron laser. Nucl. Instrum. Methods Phys. Res. **A285**, 47 (1989)
141. S. Hiramatsu, K. Ebihara, Y. Kimura, J. Kishiro, M. Kumada, H. Kurino, Y. Mizumaki, T. Ozaki, K. Takayama, Proposal for an X-band single-stage free-electron laser. Nucl. Instrum. Methods Phys. Res. **A285**, 83 (1989)
142. M.E. Conde, G. Bekefi, Experimental study of a 33.3 GHz free-electron laser amplifier with a reversed axial guide magnetic field. Phys. Rev. Lett. **67**, 3082 (1991)
143. M.E. Conde, G. Bekefi, Amplification and superradiant emission from a 33.3 GHz free-electron laser with a reversed axila guide magnetic field. IEEE Trans. Plasma Sci. **20**, 240 (1992)
144. A. Bhowmik, M.S. Curtin, W.A. McMullin, S.V. Benson, J.M.J. Madey, B.A. Richman, L. Vintro, First operation of the Rocketdyne/Stanford free-electron laser. Nucl. Instrum. Methods Phys. Res. **A272**, 10 (1988)
145. M. Curtin, G. Bennett, R. Burke, A. Bhowmik, P. Metty, S.V. Benson, J.M.J. Madey, First demonstration of a free-electron laser driven by electrons from a laser-irradiated photocathode. Nucl. Instrum. Methods Phys. Res. **A296**, 127 (1990)
146. A. Bhowmik, M.S. Curtin, W.A. McMullin, S.V. Benson, J.M.J. Madey, B.A. Richman, L. Vintro, Initial results from the free-electron laser master oscillator power amplifier experiment. Nucl. Instrum. Methods Phys. Res. **A296**, 20 (1990)
147. N.H. Lazar, H. Boehmer, S. Fornaca, B. Hauss, B. Quon, N. Roy, S. Talmadge, H. Thompson Jr., G. Neil, High brightness injector for a high power superconducting rf free-electron laser. Nucl. Instrum. Methods Phys. Res. **A304**, 243 (1991)
148. W.H. Urbanus, C.A.J. van der Geer, W.A. Bongers, G. van Dijk, B.S.Q. Elzendoorn, J.P. van Ieperen, A. van der Linden, P. Manintveld, A.B. Sterk, A.V. Tulupov, A.G.A. Verhoeven, M.J. van der Wiel, A.A. Varfolomeev, S.N. Ivanchenkov, A.S. Khlebnikov, V.L. Bratman, G.G. Denisov, A free-electron maser for fusion. Nucl. Instrum. Methods Phys. Res. **A358**, 155 (1995)
149. P. Balleyguier, J.P. Bléśés, A. Bloquet, C. Bonetti, J.P. De Brion, J. Brisset, R. Dei-Cas, J. Di Crescenzo, P. Fourdin, G. Haouat, R. Joly, S. Joly, C. Laspalles, J.P. Laget, H. Leboutet, F. Schumann, S. Striby, J. Vouillarmet, Production of high-brightness electron beams for the ELSA free-electron laser. Nucl. Instrum. Methods Phys. Res. **A304**, 308 (1991)
150. H. Takeda, K. Davis, L. Delo, Beam quality simulation of the Boeing photoinjector accelerator for the MCTD project. Nucl. Instrum. Methods Phys. Res. **A304**, 396 (1991)
151. L.R. Elias, G. Ramian, R.J. Hu, A. Amir, Observation of single-mode operation in a free-electron laser. Phys. Rev. Lett. **57**, 424 (1986)
152. A. Amir, L.R. Elias, D.J. Gregoire, J. Kotthaus, G.J. Ramian, A. Stern, Spectral characteristics of the University of California at Santa Barbara free-electron laser 400 μm experiment. Nucl. Instrum. Methods Phys. Res. **A250**, 35 (1986)
153. L.R. Elias, Free-electron laser research at the University of California at Santa Barbara. IEEE J. Quantum Electron. **QE-23**, 1470 (1987)
154. G. Ramian, L.R. Elias, The new University of California at Santa Barbara far-infrared free-electron laser. Nucl. Instrum. Methods Phys. Res. **A272**, 81 (1988)

155. A. Amir, R.J. Hu, F. Kielmann, J. Mertz, L.R. Elias, Injection locking experiment at the University of California at Santa Barbara free-electron laser. Nucl. Instrum. Methods Phys. Res. **A272**, 174 (1988)
156. A. Amir, J.F. Knox-Seith, M. Warden, Narrow-bandwidth operation of a free-electron laser enforced by seeding. Phys. Rev. Lett. **66**, 29 (1991)
157. L.R. Elias, I. Kimmel, Confirmation of single-mode free-electron laser operation. Nucl. Instrum. Methods Phys. Res. **A296**, 144 (1990)
158. J. Kaminski, Current applications of the University of California at Santa Barbara free-electron laser. Nucl. Instrum. Methods Phys. Res. **A296**, 784 (1990)
159. S.V. Benson, J.M.J. Madey, Demonstration of harmonic lasing in a free-electron laser. Phys. Rev. A **39**, 1579 (1989)
160. S.V. Benson, J.M.J. Madey, J. Schultz, M. Marc, W. Wadensweiler, G.A. Westenskow, M. Velghe, The Stanford Mark III infrared free-electron laser. Nucl. Instrum. Methods Phys. Res. **A250**, 39 (1986)
161. S.V. Benson, J. Schultz, B.A. Hooper, R. Crane, J.M.J. Madey, Status report on the Stanford Mark III infrared free-electron laser. Nucl. Instrum. Methods Phys. Res. **A272**, 22 (1988)
162. K. Carlson, W. Fann, J.M.J. Madey, Spatial distribution of visible coherent harmonics generated by the Mark III free-electron laser. Nucl. Instrum. Methods Phys. Res. **A272**, 92 (1988)
163. B.A. Hooper, S.V. Benson, A. Cutolo, J.M.J. Madey, Experimental results of two stage harmonic generation with picosecond pulses on the Stanford Mark III free-electron laser. Nucl. Instrum. Methods Phys. Res. **A272**, 96 (1988)
164. J.E. La Sala, D.A.G. Deacon, J.M.J. Madey, Optical guiding measurements on the Mark III free-electron laser oscillator. Nucl. Instrum. Methods Phys. Res. **A272**, 141 (1988)
165. D.J. Bamford, D.A.G. Deacon, Measurement of the coherent harmonics emitted in the Mark III free-electron laser. Nucl. Instrum. Methods Phys. Res. **A285**, 23 (1989)
166. D.J. Bamford, D.A.G. Deacon, Harmonic generation experiments on the Mark III free-electron laser. Nucl. Instrum. Methods Phys. Res. **A296**, 89 (1990)
167. S.V. Benson, W.S. Fann, B.A. Hooper, J.M.J. Madey, E.B. Szarmes, B. Richman, L. Vintro, A review of the Stanford Mark III infrared free-electron laser program. Nucl. Instrum. Methods Phys. Res. **A296**, 110 (1990)
168. H.A. Schwettman, T.I. Smith, R.L. Swent, The Stanford free-electron laser center. Nucl. Instrum. Methods Phys. Res. **A341**, ABS 19 (1994)
169. G.S. Edwards, N.H. Tolk, Vanderbilt University Free-Electron Laser Research Center for biomedical and materials research. Nucl. Instrum. Methods Phys. Res. **A272**, 37 (1988)
170. C.A. Brau, M.H. Mendenhall, Medical and materials research at the Vanderbilt University free-electron laser center. Nucl. Instrum. Methods Phys. Res. **A341**, ABS 21 (1994)
171. J.C. Goldstein, B.D. McVey, B.E. Newnam, Optical design and performance of an XUV free-electron laser oscillator. Nucl. Instrum. Methods Phys. Res. **A296**, 288 (1990)
172. J.H. Booske, D.J. Radack, T.M. Antonsen, S.W. Bidwell, Y. Carmel, W.W. Destler, H.P. Freund, V.L. Granatstein, P.E. Latham, B. Levush, I.D. Mayergoyz, A. Serbeto, Design of high average-power near-millimeter free-electron laser oscillators using short-period wigglers and sheet electron beams. IEEE Trans. Plasma Sci. **PS-18**, 399 (1990)
173. S.W. Bidwell, D.J. Radack, T.M. Antonsen Jr., J.H. Booske, Y. Carmel, W.W. Destler, V.L. Granatstein, B. Levush, P.E. Latham, I.D. Mayergoyz, Z.X. Zhang, A high-average-power tapered free-electron laser amplifier at submillimeter frequencies using sheet electron beams and short-period wigglers. Nucl. Instrum. Methods Phys. Res. **A304**, 187 (1991)
174. P.W. van Amersfoort, W.H. Urbanus, A.G.A. Verhoeven, A. Verheul, A.B. Sterk, A.M. van Ingen, M.J. van der Wiel, An electrostatic free-electron maser for fusion: Design consideration. Nucl. Instrum. Methods Phys. Res. **A304**, 168 (1991)
175. D. Oepts, R.J. Bakker, D.A. Jaroszynski, A.F.G. van der Meer, P.W. van Amersfoort, Induced and spontaneous interpulse phase locking in a free-electron laser. Phys. Rev. Lett. **68**, 3543 (1992)

176. D. Oepts, A.F.G. van der Meer, R.J. Bakker, P.W. van Amersfoort, Selection of single-mode radiation from a short pulse free-electron laser. Phys. Rev. Lett. **70**, 3255 (1993)
177. D.A. Jaroszynski, R.J. Bakker, A.F.G. van der Meer, D. Oepts, P.W. van Amersfoort, Experimental observation of limit-cycle oscillations in a short pulse free-electron laser. Phys. Rev. Lett. **70**, 3412 (1993)
178. B. Faatz, R.W.B. Best, D. Oepts, P.W. van Amersfoort, Hole coupling in free-electron lasers. IEEE J. Quantum Electron. **QE-29**, 2229 (1993)
179. R.J. Bakker, G.M.H. Knippels, A.F.G. van der Meer, D. Oepts, D.A. Jaroszynski, P.W. van Amersfoort, Dynamic desynchronization of a free-electron laser resonator. Phys. Rev. E **48**, R3256 (1993)
180. D.A. Jaroszynski, R.J. Bakker, A.F.G. van der Meer, D. Oepts, P.W. van Amersfoort, Coherent start-up of an infrared free-electron laser. Phys. Rev. Lett. **71**, 3798 (1993)
181. R.J. Bakker, D.A. Jaroszynski, A.F.G. van der Meer, D. Oepts, P.W. van Amersfoort, Short-pulse effects in a free-electron laser. IEEE J. Quantum Electron. **QE-30**, 1635 (1994)
182. K.E. Robinson, D.A.G. Deacon, M.F. Velghe, J.M.J. Madey, Laser induced bunch lengthening on the ACO storage ring free-electron laser. IEEE J. Quantum Electron. **QE-19**, 365 (1983)
183. J.M. Ortega, Harmonic generation in the VUV on a storage ring and prospects for Super-ACO at Orsay. Nucl. Instrum. Methods Phys. Res. **A250**, 203 (1986)
184. M. Billardon, P. Ellaume, J.M. Ortega, C. Bazin, M. Bergher, M.E. Couprie, Y. Lapierre, Y. Petroff, R. Prazeres, M. Velghe, Status of the Orsay free-electron laser experiment. Nucl. Instrum. Methods Phys. Res. **A259**, 72 (1987)
185. R. Prazeres, J.M. Ortega, C. Bazin, M. Bergher, M. Billardon, M.E. Couprie, M. Velghe, Y. Petroff, Coherent harmonic generation in the vacuum ultraviolet spectral range on the storage ring ACO. Nucl. Instrum. Methods Phys. Res. **A272**, 68 (1988)
186. M.E. Couprie, M. Billardon, M. Velghe, C. Bazin, M. Bergher, H. Fang, J.M. Ortega, Y. Petroff, R. Prazeres, Optical properties of multilayer mirrors exposed to synchrotron radiation. Nucl. Instrum. Methods Phys. Res. **A272**, 166 (1988)
187. M.E. Couprie, C. Bazin, M. Billardon, M. Velghe, Spontaneous emission of the Super-ACO free-electron laser optical klystron DOMINO. Nucl. Instrum. Methods Phys. Res. **A285**, 31 (1989)
188. M.E. Couprie, M. Billardon, M. Velghe, C. Bazin, J.M. Ortega, R. Prazeres, Y. Petroff, Free-electron laser oscillation on the Super-ACO storage ring at Orsay. Nucl. Instrum. Methods Phys. Res. **A296**, 13 (1990)
189. D. Garzella, M.E. Couprie, A. Delboulbé, T. Hara, M. Billardon, Temporal behavior and longitudinal instabilities on the Super-ACO free-electron laser. Nucl. Instrum. Methods Phys. Res. **A341**, 24 (1994)
190. J.M. Ortega, M. Bergher, R. Chaput, A. Dael, M. Velghe, Y. Petroff, J.C. Bourdon, R. Belbeoch, P. Brunet, Y. Dabin, B. Mouton, J.P. Perrine, E. Plouvier, R. Pointal, M. Renard, M. Roch, J. Rodier, P. Roudier, Y. Thiery, P. Bourgeois, P. Carlos, C. Hezard, J. Fagot, J.L. Fallou, P. Garganne, J.C. Malglaive, D.T. Tran, CLIO: Collaboration for an infrared laser at Orsay. Nucl. Instrum. Methods Phys. Res. **A285**, 97 (1989)
191. F. Glotin, J.M. Berset, R. Chaput, D.A. Jaroszynski, J.M. Ortega, R. Prazéres, Bunch length measurements on CLIO. Nucl. Instrum. Methods Phys. Res. **A341**, 49 (1994)
192. R. Prazéres, F. Glotin, J.M. Ortega, Optical mode analysis on the CLIO infrared free-electron laser. Nucl. Instrum. Methods Phys. Res. **A341**, 54 (1994)
193. J.M. Ortega, Operation of the CLIO infrared user facility. Nucl. Instrum. Methods Phys. Res. **A341**, 138 (1994)
194. J. Fraser, R.L. Sheffield, High-brightness injectors for RF-driven free-electron lasers. IEEE J. Quantum Electron. **23**, 1489 (1987)
195. D.T. Pierce, F. Meier, Photoemission of spin-polarized electrons from GaAs. Phys. Rev. B **13**, 5484 (1976)

196. C.K. Sinclair, R.E. Miller, A high current, short pulse, RF synchronized electron gun for the Stanford linear accelerator. IEEE Trans. Nucl. Sci. **NS-28**, 2649 (1981)
197. R. Hajima, T. Shizuma, M. Sawamura, R. Nagai, N. Nishimori, N. Kikuzawa, E.J. Minehara, First demonstration of energy-recovery operation in the JAERI superconducting linac for a high-power free-electron laser. Nucl. Instrum. Methods Phys. Res. **A507**, 115 (2003)
198. P. Emma et al., First lasing and operation of an Angstrom-wavelength free-electron laser. Nat. Photonics **4**, 641 (2009)
199. W. Ackermann et al., Operation of a free-electron laser from the extreme ultraviolet to the water window. Nat. Photonics **1**, 336 (2007)
200. T. Ishikawa et al., A compact, x-ray free-electron laser emitting in the sub-Angstrom region. Nat. Photonics **6**, 540 (2012)
201. B. Faatz et al., Simultaneous operation of two soft x-ray free electron lasers driven by one linear accelerator. New J. Phys. **18**, 062002 (2016)
202. H.-S. Kang et al., Hard x-ray free-electron laser with femtosecond-scale timing jitter. Nat. Photonics **11**, 708 (2017).; I.S. Ko, Construction and commissioning of the PAL-XFEL facility. Appl. Sci. **7**, 479 (2017)
203. A. Madsden, H. Sinn, Europe enters the extreme x-ray era. CERN Cour. **57**(July/August), 18 (2017)
204. C.J. Milne et al., SwissFEL: The Swiss x-ray free-electron laser. Appl. Sci. **7**, 720 (2017)
205. Z. Zhao et al., SXFEL: A soft x-ray free electron laser in China. Synchrotron Radiat. News **30**, 29 (2017)
206. J. Grünert et al., First photon diagnostics commission at the European XFEL. AIP Conf. Proc. **2054**, 030014 (2019)
207. W. Decking et al., A MHz-repetition-rate hard x-ray free-electron laser driven by a superconducting linear accelerator. Nat. Photonics **14**, 391 (2020)
208. E. Allaria et al., Two-stage seeded soft x-ray free-electron laser. Nat. Photonics **7**, 913 (2013)
209. N. Huang, H. Deng, B. Liu, D. Wang, Z. Zhao, Features and futures of X-ray free-electron lasers. Innovation (Camb) **2**(2), 100097 (2021). https://doi.org/10.1016/j.xinn.2021.100097
210. W. Wang et al., Free-electron lasing at 27 nanometres based on a laser wakefield accelerator. Nature **595**, 516 (2021)

Chapter 2
The Wiggler Field and Electron Dynamics

The electron trajectories in the external magnetostatic fields in free-electron lasers are fundamental to any understanding of the operational principles and have been the object of study for a considerable time [1–9]. The basic concept relies upon a spatially periodic magnetic field to induce an oscillatory motion in the electron beam, and the emission of radiation is derived from the corresponding acceleration. The specific character of the wiggler field can take on a variety of forms exhibiting both helical and planar symmetries. The most common wiggler configurations that have been employed to date include helically symmetric fields generated by bifilar current windings and linearly symmetric fields generated by alternating stacks of permanent magnets. However, wiggler fields generated by rotating quadrupole fields (helical symmetry) and pinched solenoidal fields (cylindrical symmetry) have also been considered.

The helical wiggler field generated by a bifilar helical current winding results in electron trajectories which exhibit circular motion in the plane transverse to the axis of symmetry of the wiggler. As a result, the parallel component of the velocity is relatively constant. This is an important consideration from the standpoint of the interaction between the beam and the radiation fields. In cylindrical coordinates, the field generated by a bifilar helix can be expressed in the form

$$\mathbf{B}_w(\mathbf{x}) = 2B_w\left[I_1'(\lambda)\hat{e}_r\cos\chi - \frac{1}{\lambda}I_1(\lambda)\hat{e}_\theta\sin\chi + I_1(\lambda)\hat{e}_z\sin\chi\right], \tag{2.1}$$

where B_w denotes the wiggler amplitude, $\lambda = k_w r$, $\chi = \theta - k_w z$, k_w ($= 2\pi/\lambda_w$, where λ_w is the wiggler period) denotes the wiggler wavenumber, and I_n and I_n' denote the modified Bessel function of the first kind of order n and its derivative, respectively. This field exhibits a local minimum on axis (i.e., at $r = 0$) that acts to focus and to confine the beam against the effects of self-generated electric and magnetic fields. However, a uniform axial solenoidal field $\mathbf{B}_0$ [$= B_0\hat{\mathbf{e}}_z$] is often employed in conjunction with a helical wiggler in order to provide enhanced focusing for extremely

H. P. Freund, T. M. Antonsen, Jr., *Principles of Free Electron Lasers*,
https://doi.org/10.1007/978-3-031-40945-5_2

intense electron beams. In such cases, the combined influence of the wiggler and axial magnetic fields has important consequences on the character of both the electron trajectories and the interaction.

Linearly symmetric wiggler fields induce an oscillatory electron motion in the plane transverse to that of the wiggler field itself. As a result, all components of the velocity oscillate about some mean value. While this is not an advantageous quality in comparison with the helical wiggler from the standpoint of the resonant interaction in the free-electron laser, the planar wiggler has the advantage of greater ease of adjustment than is possible with a helical wiggler. Planar wigglers may be constructed of alternating stacks of either permanent magnets or electromagnets. In its simplest form, in which the pole faces of the individual magnets are flat, the wiggler field may be expressed as

$$\mathbf{B}_w(\mathbf{x}) = B_w\left[\hat{\mathbf{e}}_y \cosh k_w y \sin k_w z + \hat{\mathbf{e}}_z \sinh k_w y \cos k_w z\right]. \tag{2.2}$$

In this case, the transverse component of the wiggler field is directed in along the y-axis and the electron motion is predominantly in the x–z plane. This form of planar wiggler has a local minimum at the symmetry plane for $y = 0$ and, hence, provides a focusing force for the electron beam in the direction normal to the wiggler-induced motion of the electron beam. If additional focusing is required, then the pole faces may be tapered to provide for an inhomogeneity in the plane of the wiggler-induced motion [10, 11]. In the case in which the pole faces are parabolically shaped, the wiggler field is of the form [11]

$$\begin{aligned}\mathbf{B}_w(\mathbf{x}) = B_w\Bigg\{ & \cos k_w z\left[\hat{\mathbf{e}}_x \sinh\left(\frac{k_w x}{\sqrt{2}}\right) \sinh\left(\frac{k_w y}{\sqrt{2}}\right) + \hat{\mathbf{e}}_y \cosh\left(\frac{k_w x}{\sqrt{2}}\right) \cosh\left(\frac{k_w y}{\sqrt{2}}\right)\right] \\ & - \sqrt{2}\,\hat{\mathbf{e}}_z \cosh\left(\frac{k_w x}{\sqrt{2}}\right) \sinh\left(\frac{k_w y}{\sqrt{2}}\right) \sin k_w z\Bigg\}.\end{aligned} \tag{2.3}$$

An axial solenoidal field is not employed to provide enhanced focusing in conjunction with a planar wiggler because the combined fields result in a $\mathbf{B}_0 \times \nabla \mathbf{B}_w$ drift that causes the beam to diverge rapidly [5].

While alternate wiggler concepts have been studied analytically, there has been little effort to perfect them in the laboratory. In the case of a rotating quadrupole configuration [12–15], the wiggler field vanishes along the axis of symmetry. For this reason, the radial profile of the electron beam employed with such a wiggler must be hollow. This requirement presents difficulties in generating beams of sufficiently low emittance and energy spread to ensure high-efficiency operation. Configurations employing pinched solenoidal fields are often referred to as *lowbitrons* in the literature [16, 17]. These devices operate by a different principle than the typical free-electron laser since the magnetic field itself does not impart a coherent transverse velocity to the beam, and the beam must be created with a large transverse velocity component. For this reason, lowbitrons can be more properly considered as a form of axially modulated cyclotron maser than as a free-electron laser. For these reasons, we shall focus our attention on both planar and bifilar helical wiggler configurations.

2.1 Helical Wiggler Configurations

For the sake of generality, we shall consider the case of the electron dynamics in a configuration that consists both a helical wiggler and a uniform axial solenoidal field. The orbits in such a case are determined by the Lorentz force equation

$$\frac{d}{dt}\mathbf{p} = -\frac{e}{c}\mathbf{v}\times(B_0\hat{\boldsymbol{e}}_z + \mathbf{B}_w), \tag{2.4}$$

where e denotes the charge of the electron and $\mathbf{v}$ and $\mathbf{p}$ denote the velocity and momentum, respectively. This set of equations displays a wide variation of properties that depend upon the amplitudes of the wiggler and solenoidal fields, the wiggler period, and the electron energy. For the sake of clarity of presentation, it is often useful to first consider the idealized one-dimensional approximation in which the electron displacement from the axis of symmetry remains much less than a wiggler period (i.e., $r \ll \lambda_w$). In this limit, the helical wiggler field assumes the particularly simple form

$$\mathbf{B}_w = B_w\left(\hat{\boldsymbol{e}}_x \cos k_w z + \hat{\boldsymbol{e}}_y \sin k_w z\right) \tag{2.5}$$

that exhibits no transverse inhomogeneities to within terms of order $(k_w r)^2$. This wiggler field model describes an idealized picture of the interaction in a free-electron laser, since the lack of transverse inhomogeneities also implies that there will be no variation in the either the electron trajectories or the interaction with the electromagnetic field across the beam profile.

2.1.1 *Idealized One-Dimensional Trajectories*

The trajectories in the idealized one-dimensional limit may be obtained very simply in the absence of an axial solenoidal field by noting that the vector potential of the wiggler is

$$\mathbf{A}_w = -\frac{B_w}{k_w}\left(\hat{\boldsymbol{e}}_x \cos k_w z + \hat{\boldsymbol{e}}_y \sin k_w z\right). \tag{2.6}$$

Therefore, both x and y are ignorable coordinates, and the corresponding canonical momenta, which we denote by P_x and P_y, are constants of the motion. Since the total energy is conserved, the relativistic factor γ [$= (1 - \upsilon^2/c^2)^{-1/2}$] is also a constant of the motion. As a result, the trajectories are given by

$$\upsilon_x = -\frac{P_x}{\gamma m_e} + \upsilon_w \cos k_w z, \tag{2.7}$$

$$\upsilon_y = -\frac{P_y}{\gamma m_e} + \upsilon_w \sin k_w z, \tag{2.8}$$

$$\upsilon_z = c\sqrt{1 - \frac{1}{\gamma^2} - \frac{\upsilon_w^2 + V_\perp^2}{c^2} + 2\frac{\upsilon_w V_\perp}{c^2}\cos(k_w z - \varphi)}, \tag{2.9}$$

where $\upsilon_w = -cK/\gamma$ is the wiggler-induced velocity, $K = eB_w/m_e c^2 k_w$ is the wiggler strength parameter, m_e is the electron mass, c is the speed of light in vacuo, $V_\perp = (P_x^2 + P_y^2)^{1/2}/\gamma m_e$, and $\varphi = \tan^{-1}(P_y/P_x)$. Observe that the magnitude of both the transverse and axial components of the velocity is constant in the limit in which both P_x and P_y vanish, and the orbit describes a helix that is in phase with the wiggler field.

In the presence of an axial solenoidal field, the orbits are more complicated, since x and y are no longer ignorable coordinates and the associated canonical momenta are not constants of the motion. For convenience, we shall work in the coordinate frame which rotates with the wiggler field and is defined by the basis vectors: $\hat{e}_1 = \hat{e}_x \cos k_w z + \hat{e}_y \sin k_w z$, $\hat{e}_2 = -\hat{e}_x \sin k_w z + \hat{e}_y \cos k_w z$, and $\hat{e}_3 = \hat{e}_z$. We shall henceforth refer to this as the *wiggler frame*. In this coordinate frame, substitution of the idealized wiggler model into the Lorentz force equation yields

$$\frac{d}{dt}\upsilon_1 = -(\Omega_0 - k_w\upsilon_3)\upsilon_2, \tag{2.10}$$

$$\frac{d}{dt}\upsilon_2 = (\Omega_0 - k_w\upsilon_3)\upsilon_1 - \Omega_w\upsilon_3, \tag{2.11}$$

$$\frac{d}{dt}\upsilon_3 = \Omega_w\upsilon_2, \tag{2.12}$$

where $\Omega_{0,w} = eB_{0,w}/\gamma m_e c$. Two constants of the motion are readily obtained. The first, as in the absence of the axial field, is the total energy or γ. The second may be obtained by elimination of υ_2 from Eqs. (2.10) and (2.12), which yields

$$\frac{d}{dt}\upsilon_1 = -\frac{(\Omega_0 - k_w\upsilon_3)}{\Omega_w}\frac{d}{dt}\upsilon_3. \tag{2.13}$$

As a result,

$$V_1 = v_1 - \frac{(\Omega_0 - k_w v_3)^2}{2k_w \Omega_w}, \tag{2.14}$$

is a second constant of the motion. The solutions to the orbit equations (2.10)–(2.12) may be reduced to quadrature by means of these constants.

2.1.1.1 Steady-State Trajectories

Before we obtain closed form solutions for the orbits, it is instructive to consider the *steady-state* solutions [2, 4] for which v_1, v_2, and v_3 are constants. If the derivatives in Eqs. (2.10)–(2.12) vanish, then the solutions are $v_1 = v_w$, $v_2 = 0$, and $v_3 = v_\parallel$, where

$$v_w = \frac{\Omega_w v_\parallel}{\Omega_0 - k_w v_\parallel} \tag{2.15}$$

describes the wiggler-induced transverse velocity. In contrast to the orbits in the absence of an axial magnetic field, the presence of the axial field establishes a preferred direction of propagation through the wiggler in which propagation parallel to the axial field results in an enhancement in the transverse velocity. The constants which determine the transverse and axial velocities (v_w, $v_\parallel$) are related to the requirement that the energy is conserved; hence,

$$v_w^2 + v_\parallel^2 = \left(1 - \frac{1}{\gamma^2}\right)c^2. \tag{2.16}$$

One effect of the axial solenoidal field is to resonantly increase the magnitude of the transverse wiggler-induced velocity when the Larmor period associated with the axial field is comparable to the wiggler period (i.e., $\Omega_0 \approx k_w v_\parallel$).

2.1.1.2 Stability of the Steady-State Trajectories

The stability of these steady-state solutions was first examined by Friedland [2] by means of an expansion $v_1 = v_w + \delta v_1$, $v_2 = \delta v_2$, and $v_3 = v_\parallel + \delta v_3$, where the velocity perturbations are assumed to be small. Expansion of the orbit equations (2.10)–(2.12) to first order in the perturbed velocity yields

$$\frac{d}{dt}\delta v_1 = -\left(\Omega_0 - k_w v_\parallel\right)\delta v_2, \tag{2.17}$$

$$\frac{d}{dt}\delta v_2 = \left(\Omega_0 - k_w v_\parallel\right)\delta v_1 - \Omega_0 \beta_w \delta v_3, \tag{2.18}$$

$$\frac{d}{dt}\delta v_3 = \Omega_w \delta v_2, \tag{2.19}$$

where $\beta_w = v_w/v_\parallel$. Differentiation of Eq. (2.18) with respect to time and substitution of the derivatives of δv_1 and δv_3 from Eqs. (2.17) and (2.19) gives

$$\left(\frac{d^2}{dt^2} + \Omega_r^2\right)\delta v_2 = 0, \tag{2.20}$$

where the natural response frequency of the perturbation Ω_r is defined by

$$\Omega_r^2 = \left(\Omega_0 - k_w v_\parallel\right)\left[\left(1 + \beta_w^2\right)\Omega_0 - k_w v_\parallel\right]. \tag{2.21}$$

As a result, the steady-state trajectories are unstable whenever $\Omega_r^2 < 0$. It should be remarked that in the absence of the wiggler $\Omega_r^2 = (\Omega_0 - k_w v_\parallel)^2$. The character of the perturbations in this limit, therefore, describe stable Larmor rotation in the axial magnetic field.

Substitution of (2.15) into (2.16) yields a quartic polynomial in the axial velocity that may be solved for $v_\parallel$ as a function of B_0, B_w, k_w, and γ. There are in general four solutions for $v_\parallel$ for any given sets of the parameters. Of these solutions, one corresponds to backward propagation (i.e., $v_\parallel < 0$), which will be ignored. The remaining steady-state solutions may be divided into two classes corresponding to the cases in which $\Omega_0 < k_w v_\parallel$ referred to as Group I and $\Omega_0 > k_w v_\parallel$ referred to as Group II. A representative solution for the axial velocity is shown in Fig. 2.1 as a function of the axial magnetic field for a beam energy of 1 MeV ($\gamma = 2.957$) and $\Omega_w/ck_w = 0.05$. The question of the stability of these orbits is illustrated in Fig. 2.2 in

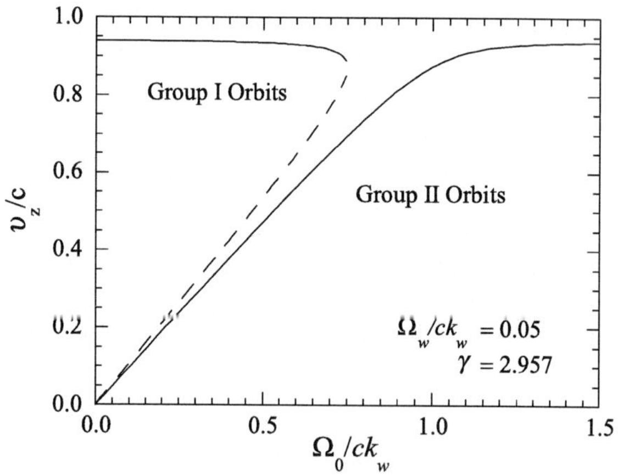

Fig. 2.1 Graph of the axial velocity of the steady-state orbits as a function of the solenoidal magnetic field showing the Group I and Group II orbits. The dashed line indicates unstable Group I trajectories

Fig. 2.2 Graph of Ω_r^2 corresponding to the Group I and Group II orbits shown in Fig. 2.1

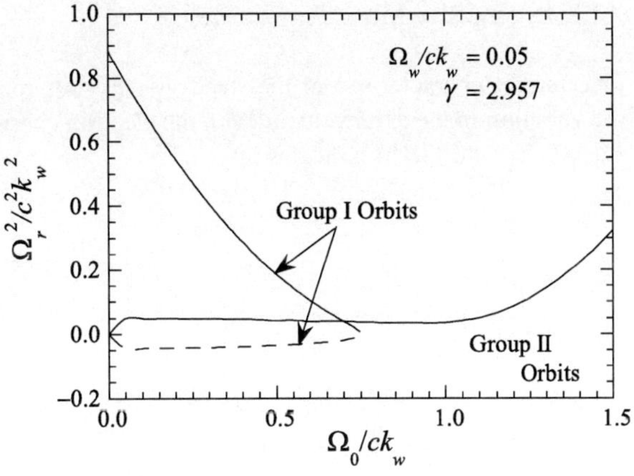

which Ω_r^2 is plotted as a function of the axial magnetic field. The Group I trajectories are, in general, multivalued functions of the axial field, and unstable orbits occur whenever

$$v_\parallel < \frac{\Omega_0}{k_w}\left[1 + \left(\frac{\Omega_w}{\Omega_0}\right)^{2/3}\right]. \tag{2.22}$$

The unstable Group I trajectories are indicated by a dashed line in the figure. This limits the Group I trajectories to axial magnetic fields below some critical value (given by the criterion $\Omega_0/ck_w < 0.75$ for the present choice of parameters). The stable trajectories decrease monotonically with increasing B_0 up to this point corresponding to the increase in the transverse velocity as the resonance (at $\Omega_0 \simeq k_w v_\parallel$) is approached. In contrast, Group II orbits are always stable, and the axial velocity increases monotonically with the axial magnetic field.

The steady-state trajectories describe helical orbits in the combined wiggler and axial magnetic fields. Observe that for the Group I trajectories, $v_w = -cK/\gamma$ when the axial magnetic field vanishes and that this corresponds with the results given in Eqs. (2.7) and (2.8) in the case for which $P_x = P_y = 0$. It should be remarked that the transverse velocity given in (2.15) undergoes a change in sign between the Group I and Group II orbits. It will be shown in the succeeding section that this does not indicate a change in the direction of rotation; rather, it indicates a phase shift of the trajectory in which the phase of the Group II orbits is shifted by 180° relative to both the wiggler and the Group I orbits.

2.1.1.3 Negative-Mass Trajectories

One further characteristic of the steady-state orbits merits discussion: specifically, the variation in the axial velocity with the electron energy. Implicit differentiation of Eqs. (2.15) and (2.16) indicates that

$$\frac{d}{d\gamma}v_{\|} = \frac{c^2}{\gamma\gamma_{\|}^2 v_{\|}}\Phi, \tag{2.23}$$

where $\gamma_{\|} = (1 - v_{\|}^2/c^2)^{-1/2}$ and

$$\Phi = 1 - \frac{\gamma_{\|}^2\beta_w^2\Omega_0}{(1+\beta_w^2)\Omega_0 - k_w v_{\|}}. \tag{2.24}$$

The function Φ is shown in Fig. 2.3 as a function of the axial magnetic field for both the stable Group I and Group II orbits. Observe that for Group I orbits, Φ increases monotonically from unity (at $B_0 = 0$) and exhibits a singularity at the transition to orbital instability (at which point both $\Omega_r = 0$ and $(1 + \beta_w^2)\Omega_0 = k_w v_{\|}$). The behavior of Φ for the Group II orbits is more interesting, however, since it is negative whenever

$$\frac{\gamma_{\|}^2\Omega_0}{\gamma^2} < k_w v_{\|}. \tag{2.25}$$

This implies the existence of a *negative-mass* regime in which the axial velocity will increase with decreasing energy.

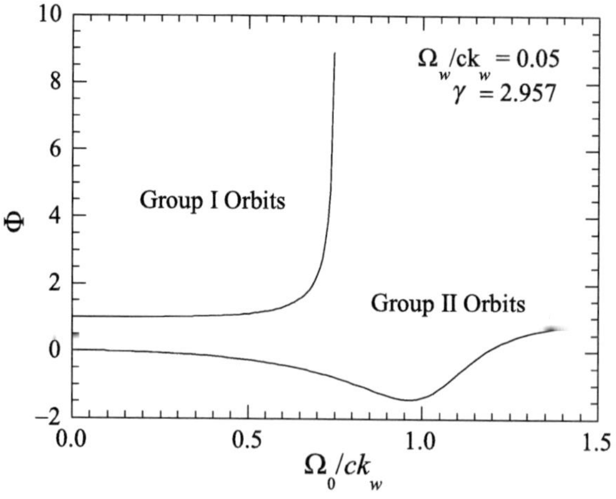

Fig. 2.3 Graph of Φ as a function of the axial magnetic field corresponding to the Group I (stable) and Group II orbits shown in Fig. 2.1

2.1.1.4 General Integration of the Orbit Equations

The orbit equations (2.10)–(2.12) can be reduced to a single integrable differential equation [4] using the constants of the motion. Since the total energy is a conserved quantity, $v_2^2 = v_0^2 - v_1^2 - v_3^2$, where v_0 [$= (1 - 1/\gamma^2)^{1/2}c$] denotes the magnitude of the velocity. Elimination of v_1 using Eq. (2.14) and substitution of the result into Eq. (2.12) yields an equation for u_3 ($= v_3/c - \Omega_0/ck_w$)

$$\left(\frac{d}{dt}u_3\right)^2 + c^2k_w^2 U(u_3) = 0, \tag{2.26}$$

where the pseudopotential $U(u_3)$ is a quartic polynomial in u_3 defined by

$$U(u_3) = \frac{1}{4}u_3^4 + \frac{\Omega_w}{ck_w}\left(\frac{V_1}{c} + \frac{\Omega_w}{ck_w}\right)u_3^2 + \frac{2\Omega_0}{ck_w}\frac{\Omega_w^2}{c^2k_w^2}u_3 + \frac{\Omega_w^2}{c^2k_w^2}\left[\frac{V_1^2}{c^2} + \frac{\Omega_0^2}{c^2k_w^2} - \frac{v_0^2}{c^2}\right]. \tag{2.27}$$

The form of the pseudopotential determines the character of the solutions to Eq. (2.26). We observe that real solutions are possible only when $U(u_3) \le 0$ and, as a result, the roots of the pseudopotential define the bounds on the oscillations in u_3. An example is shown in Fig. 2.4 in which a family of curves of the pseudopotential is plotted versus u_3 for $\Omega_0/ck_w = 0.10$, $\Omega_w/ck_w = 0.01$, an electron energy of 1 MeV, and $V_1/c = -0.50$, -0.94, and -1.50. For the case in which $V_1/c = -1.50$, the pseudopotential has four real roots, and the orbit is characterized by oscillations in u_3 over the approximate ranges of ± 0.21 through ± 0.11. Two of these roots tend to coalesce as V_1 increases until, at $V_1/c = -0.94$, they become degenerate and the pseudopotential at the central maximum vanishes. An orbit exists at this point with a uniform axial velocity, but it corresponds to one of the unstable

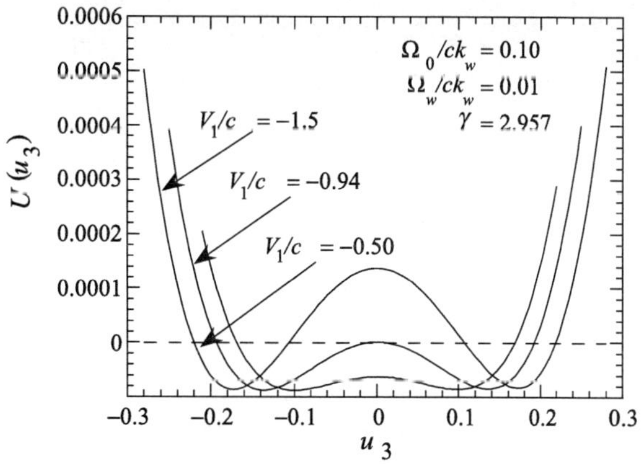

Fig. 2.4 Plots of the pseudopotential for characteristic values of $\Omega_0/ck_w = 0.10$, $\Omega_w/ck_w = 0.01$, an electron energy of 1 MeV, and $V_1/c = -0.50$, -0.94, and -1.50

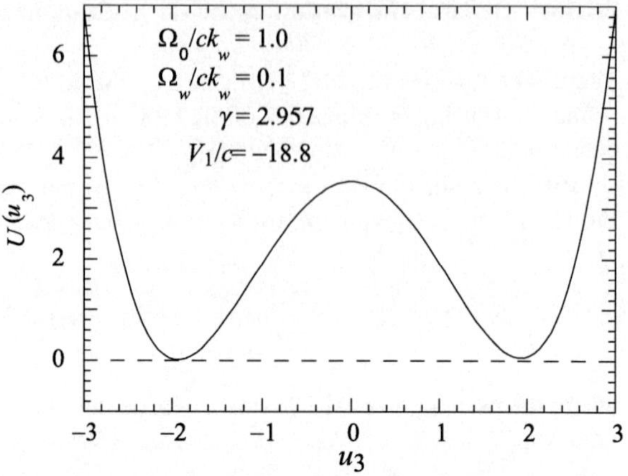

Fig. 2.5 Plot of the pseudopotential corresponding to a stable steady-state trajectory for $\Omega_0/ck_w = 1.0$, $\Omega_w/ck_w = 0.1$, $V_1/c = -18.8$, and an electron energy of 1 MeV

steady-state trajectories. As V_1 increases further, the degenerate roots become a complex conjugate pair, and there are only two real roots for the pseudopotential which define a large-amplitude oscillation which for $V_1/c = -0.50$ extends over the approximate range of from -0.16 to 0.16.

An example of the form of the pseudopotential in the vicinity of a stable steady-state orbit for $u_3 \approx -1.94$ is shown in Fig. 2.5 for $\Omega_0/ck_w = 1.0$, $\Omega_w/ck_w = 0.1$, $V_1/c = -18.8$, and an energy of 1 MeV. For this choice of parameters, the root of the pseudopotential corresponds to electron streaming antiparallel to the solenoidal magnetic field with a velocity of $v_\parallel/c \approx -0.94$. For $V_1/c > -18.8$ there are no real roots to the quartic and no trajectories are possible. In the opposite case in which $V_1/c < -18.8$, the orbits are oscillatory.

Equation (2.26) can be integrated in closed form in terms of the Jacobi elliptic functions, which describe anharmonic oscillations. We first consider the case in which only two real roots exist and the pseudopotential is of the form $U(u_3) = (u_3 - \alpha_+)(u_3 - \alpha_-)[(u_3 - \rho)^2 + \zeta^2]$, where it is assumed that α_+, α_-, ρ, and ζ are real and that $\alpha_+ > \alpha_-$. The oscillation is confined to within the range $\alpha_- \le u_3 \le \alpha_+$. and formal integration of (2.26) gives

$$\int_{\alpha_-}^{u_3(t)} \frac{dx}{\sqrt{(x-\alpha_-)(\alpha_+ - x)\left[(x-\rho)^2 + \zeta^2\right]}} = \pm ck_w\tau, \tag{2.28}$$

where $\tau = t - t_0$ and t_0 is the time at which the trajectory passes through $u_3 = \alpha_-$. Integration of (2.28) yields

$$\pm ck_w = \frac{1}{\sqrt{pq}} F\left(2\cot^{-1}\sqrt{\frac{q(\alpha_+ - u_3)}{p(u_3 - \alpha_-)}}, k\right), \tag{2.29}$$

where $F(x,k)$ is the incomplete elliptic integral of argument x and modulus k, $p^2 = (\rho - \alpha_+)^2 + \zeta^2$, $q^2 = (\rho - \alpha_-)^2 + \zeta^2$, and

$$k = \frac{1}{2}\sqrt{\frac{(\alpha_+ - \alpha_-)^2 - (p-q)^2}{pq}}. \tag{2.30}$$

Inversion of Eq. (2.29) yields an expression for the axial velocity in terms of the Jacobi elliptic functions

$$\frac{v_\parallel}{c} = \frac{\Omega_0}{ck_w} + \frac{\alpha_+ q\,\mathrm{sn}^2\left(ck_w\sqrt{pq}\tau, k\right) + \alpha_- p\left[1 \pm \mathrm{cn}\left(ck_w\sqrt{pq}\tau, k\right)\right]^2}{q\,\mathrm{sn}^2\left(ck_w\sqrt{pq}\tau, k\right) + p\left[1 \pm \mathrm{cn}\left(ck_w\sqrt{pq}\tau, k\right)\right]^2}, \tag{2.31}$$

where sn and cn denote the Jacobi elliptic sine and cosine functions, respectively. Observe that the steady-state trajectories are recovered for the case in which the roots are degenerate (i.e., $\alpha = \alpha_+ = \alpha_-$), since $v_\parallel/c = \Omega_0/ck_w + \alpha$ in this limit.

In the case in which there are four real roots, the pseudopotential is of the form $U(u_3) = (u_3 - \alpha_+)(u_3 - \alpha_-)(u_3 - \beta_+)(u_3 - \beta_-)$, where $\alpha_- < \alpha_+ < \beta_- < \beta_+$. Oscillatory solutions occur within the ranges $\alpha_- \le u_3 \le \alpha_+$ and $\beta_- \le u_3 \le \beta_+$. Therefore, we formally integrate

$$\int_{\alpha_-}^{u_3(t)} \frac{dx}{\sqrt{(x-\alpha_-)(\alpha_+ - x)(x-\beta_+)(x-\beta_-)}} = \pm ck_w\tau, \tag{2.32}$$

for orbits within the lower range, and

$$\int_{\beta_-}^{u_3(t)} \frac{dx}{\sqrt{(x-\alpha_-)(\alpha_+ - x)(\beta_+ - x)(x-\beta_-)}} = \pm ck_w\tau, \tag{2.33}$$

for orbits within the upper range. Integration yields, respectively,

$$F(x,k) = \pm \mu ck_w\tau, \tag{2.34}$$

where $2\mu = [(\beta_+ - \alpha_+)(\beta_- - \alpha_-)]^{1/2}$, the modulus

$$\kappa = \sqrt{\frac{(\beta_+ - \beta_-)(\alpha_+ - \alpha_-)}{(\beta_+ - \alpha_+)(\beta_- - \alpha_-)}}, \tag{2.35}$$

and

$$x = \begin{cases} \sin^{-1}\sqrt{\dfrac{(\beta_+ - \alpha_+)(u_3 - \alpha_-)}{(\alpha_+ - \alpha_-)(\beta_+ - u_3)}}; & \alpha_- \le u_3 \le \alpha_+ \\ \sin^{-1}\sqrt{\dfrac{(\beta_+ - \alpha_+)(u_3 - \beta_-)}{(\beta_+ - \beta_-)(u_3 - \alpha_+)}}; & \beta_- \le u_3 \le \beta_+ \end{cases}. \tag{2.36}$$

The axial velocity can be found by inversion of Eq. (2.34)

$$\frac{v_\parallel}{c} = \frac{\Omega_0}{ck_w} + \frac{\alpha_-(\beta_+ - \alpha_+) - \beta_+(\alpha_+ - \alpha_-)\mathrm{sn}^2(\mu ck_w\tau, \kappa)}{(\beta_+ - \alpha_+) - (\alpha_+ - \alpha_-)\mathrm{sn}^2(\mu ck_w\tau, \kappa)}, \tag{2.37}$$

in the lower region, and

$$\frac{v_\parallel}{c} = \frac{\Omega_0}{ck_w} + \frac{\beta_-(\beta_+ - \alpha_+) - \alpha_+(\beta_+ - \beta_-)\mathrm{sn}^2(\mu ck_w\tau, \kappa)}{(\beta_+ - \alpha_+) - (\beta_+ - \beta_-)\mathrm{sn}^2(\mu ck_w\tau, \kappa)}, \tag{2.38}$$

in the upper region. As in the case in which there are only two real roots, these orbits reduce to the steady-state trajectories in the limit in which the roots become degenerate.

The general solutions for the trajectories can be expanded to obtain orbits that are arbitrarily close to the steady-state trajectories; however, it is more instructive to expand the orbit equations directly. In order to simplify the comparison with the trajectories in the absence of the axial magnetic field, we treat the orbit equations in Cartesian coordinates and separate the x- and y-components to obtain

$$\left(\frac{d^2}{dt^2} + \Omega_0^2\right)\left(v_x + \frac{\Omega_w}{k_w}\cos k_w z\right) = \frac{\Omega_w\Omega_0}{k_w}(\Omega_0 + k_w v_z)\cos k_w z, \tag{2.39}$$

$$\left(\frac{d^2}{dt^2} + \Omega_0^2\right)\left(v_y + \frac{\Omega_w}{k_w}\sin k_w z\right) = \frac{\Omega_w\Omega_0}{k_w}(\Omega_0 + k_w v_z)\sin k_w z, \tag{2.40}$$

$$\frac{d}{dt}v_z = -\Omega_w\left(v_x \sin k_w z - v_y \cos k_w z\right). \tag{2.41}$$

Under the assumption that $v_z = v_\parallel + \delta v_z(t)$ where $|\delta v_z(t)| \ll v_\parallel$, we find that the transverse velocity is given by

$$v_x \cong v_w \cos k_w z + V_\perp \cos(\Omega_0 t + \varphi), \tag{2.42}$$

$$v_y \cong v_w \sin k_w z + V_\perp \sin(\Omega_0 t + \varphi), \tag{2.43}$$

where $V_\perp$ and φ are constants and we require that $V_\perp^2 \ll v_w^2$. Substitution of (2.42) and (2.43) into (2.41) shows that the axial velocity satisfies

$$v_z \cong v_\parallel - \frac{v_w V_\perp}{v_\parallel} \cos(k_w z - \Omega_0 t - \varphi), \tag{2.44}$$

where the average axial velocity satisfies the equation

$$\frac{v_\parallel}{c} = \sqrt{1 - \frac{1}{\gamma^2} - \frac{v_w^2 + V_\perp^2}{c^2}}. \tag{2.45}$$

The solution given by Eqs. (2.42)–(2.44) describes the combined effect of the wiggler-induced oscillation and Larmor gyromotion due to the axial solenoidal field and is valid as long as the magnitude of the Larmor motion is small. The solution also reduces to that shown in Eqs. (2.7)–(2.9) in the absence of the axial field.

2.1.2 *Trajectories in a Realizable Helical Wiggler*

The trajectories in the one-dimensional idealized wiggler are valid under the assumption that the electron displacement from the axis of symmetry is much less than a wiggler period (i.e., $k_w r \ll 1$). Since the radius of curvature of an orbit is proportional to the transverse velocity, this implies that the idealized analysis is valid as long as $v_w/v_\parallel \ll 1$. As a consequence, a three-dimensional analysis [3, 7, 8] of the trajectories is required for axial magnetic fields in the neighborhood of the resonance at $\Omega_0 \approx k_w v_\parallel$.

The fundamental equations governing the electron trajectories in a realizable helical wiggler (2.1) and an axial magnetic field are

$$\frac{d}{dt} v_1 = -\left(\Omega_0 - k_w v_\parallel + 2\Omega_w I_1(\lambda) \sin\chi\right) v_2 + \Omega_w v_3 I_1(\lambda) \sin 2\chi, \tag{2.46}$$

$$\frac{d}{dt} v_2 = \left(\Omega_0 - k_w v_\parallel + 2\Omega_w I_1(\lambda) \sin\chi\right) v_1 - \Omega_w v_3 \left[I_0(\lambda) + I_2(\lambda) \cos 2\chi\right], \tag{2.47}$$

$$\frac{d}{dt} v_3 = \Omega_w v_2 \left[I_0(\lambda) + I_2(\lambda) \cos 2\chi\right] - \Omega_w v_1 I_2(\lambda) \sin 2\chi, \tag{2.48}$$

$$\frac{d}{dt} \lambda = k_w (v_1 \cos\chi + v_2 \sin\chi), \tag{2.49}$$

and

$$\frac{d}{dt} \chi = \frac{k_w}{\lambda} (v_2 \cos\chi - v_1 \sin\chi - \lambda v_3), \tag{2.50}$$

in the wiggler frame, where $\lambda = k_w r$ and $\chi = \theta - k_w z$.

In addition to the total energy, a second constant of the motion can be obtained from the symmetry properties of the vector potential that, in cylindrical coordinates, can be expressed in the form

$$\mathbf{A}_w(\mathbf{x}) = -2\frac{B_w}{k_w}\left(\frac{1}{\lambda}I_1(\lambda)\hat{\mathbf{e}}_r\cos\chi - I_1'(\lambda)\hat{\mathbf{e}}_\theta\sin\chi\right) \pm \frac{1}{2}B_0\hat{\mathbf{e}}_\theta. \tag{2.51}$$

The Lagrangian associated with this vector potential has the functional form $\mathcal{L} = \mathcal{L}(r,\chi,\dot{r},\dot{\chi},\dot{z})$ which is independent of z. As a consequence, the z-component of the canonical momentum [7]

$$P_z = \frac{\gamma m_e}{k_w}\left[\lambda^2\left(\frac{d\theta}{dt} - \frac{1}{2}\Omega_0\right) + k_w v_3 - 2\Omega_w \lambda I_1(\lambda)\sin\chi\right] \tag{2.52}$$

is also a conserved quantity. This is sometimes referred to as the *helical invariant.*

2.1.2.1 Steady-State Trajectories

The helical orbits were first obtained by Diament [3] by the requirement of steady-state solutions in which v_1, v_2, v_3, λ, and χ are constant. This implies that $v_1 = v_w$, $v_2 = 0$, $v_3 = v_\|$, $\chi = \pm\pi/2$, and $\lambda = \lambda_0$, where $v_\|$ is constant, $\lambda_0 = \mp v_w/v_\|$, and

$$v_w = \frac{2\Omega_w v_\| I_1(\lambda_0)/\lambda_0}{\Omega_0 - k_w v_\| \pm 2\Omega_w I_1(\lambda_0)}. \tag{2.53}$$

Observe that this result for the transverse wiggler-induced velocity (2.53) reduces to that found for the idealized wiggler (2.15) in the limit as $\lambda_0 \to 0$. The complete determination of the orbit requires knowledge of either v_w, $v_\|$, or λ_0. Specification of any one of these is sufficient to determine the other two. This is accomplished by means of the energy conservation requirement that can be expressed in the form

$$\sqrt{\frac{\gamma^2 - 1}{1 + \lambda_0^2}} = \frac{\Omega_0}{ck_w} \pm 2\frac{\Omega_w}{ck_w}\left(\frac{1 + \lambda_0^2}{\lambda_0^2}\right)I_1(\lambda_0). \tag{2.54}$$

Solution of these equations produces the general form of the Group I and Group II orbits discussed for the case of the idealized wiggler corresponding to $\chi = \pm\pi/2$, respectively. It should be remarked that while v_w changes sign between the Group I and II orbits, the direction of rotation of these steady-state helical trajectories does not change. The sign change indicates, rather, a relative phase shift of 180° in the orbits.

2.1.2.2 Stability of the Steady-State Trajectories

The stability of these orbits is determined by perturbation of the orbit equations about the steady-state trajectories: $v_1 = v_w + \delta v_1$, $v_2 = \delta v_2$, $v_3 = v_\parallel + \delta v_3$, $\chi = \pm\pi/2 + \delta\chi$, and $\lambda = \lambda_0 + \delta\lambda$. The orbit equations, therefore, can be written as

$$\frac{d}{dt}\delta v_1 = -\frac{\Omega_0 - k_w v_\parallel}{1+\lambda_0^2}\delta v_2 - 2\Omega_w I_2(\lambda_0)\delta\chi, \tag{2.55}$$

$$\begin{aligned}\frac{d}{dt}\delta v_2 = &\frac{\Omega_0 - k_w v_\parallel}{1+\lambda_0^2}\delta v_1 - \frac{v_w}{v_\parallel}\frac{\Omega_0 + \lambda_0^2 k_w v_\parallel}{1+\lambda_0^2}\delta v_3 \\ &- 2\Omega_w v_\parallel\left[\frac{1+\lambda_0^2}{\lambda_0}I_2(\lambda_0) + I_1(\lambda_0)\right]\delta\lambda,\end{aligned} \tag{2.56}$$

$$\frac{d}{dt}\delta v_3 = 2\Omega_w\left[v_w I_2(\lambda_0)\delta\chi + \frac{1}{\lambda_0}I_1(\lambda_0)\delta v_2\right], \tag{2.57}$$

$$\frac{d}{dt}\delta\chi = -k_w\left[\delta v_3 \pm \frac{1}{\lambda_0}\delta v_1 + \frac{1}{\lambda_0}v_\parallel\delta\lambda\right], \tag{2.58}$$

$$\frac{d}{dt}\delta\lambda = \pm k_w(\delta v_2 - v_w\delta\chi), \tag{2.59}$$

to lowest order in the perturbations.

The perturbed quantities can be isolated from this system of coupled first-order differential equations, and the result can be written as a set of homogeneous higher-order differential equations. Thus, [7]

$$\left(\frac{d^2}{dt^2} + \Omega_+^2\right)\left(\frac{d^2}{dt^2} + \Omega_-^2\right)\begin{bmatrix}\delta v_2 \\ \delta\chi\end{bmatrix} = 0, \tag{2.60}$$

and

$$\frac{d}{dt}\left(\frac{d^2}{dt^2} + \Omega_+^2\right)\left(\frac{d^2}{dt^2} + \Omega_-^2\right)\begin{bmatrix}\delta v_1 \\ \delta v_3 \\ \delta\lambda\end{bmatrix} = 0, \tag{2.61}$$

where

$$\Omega_\pm^2 = \frac{1}{2}\left(\omega_1^2 + \omega_2^2\right) \pm \frac{1}{2}\sqrt{\left(\omega_1^2 - \omega_2^2\right)^2 + k_w v_\parallel \Omega_w \omega_3^2}, \tag{2.62}$$

$$\omega_1^2 = k_w^2 v_w^2 + 2\Omega_w k_w v_w \left(\frac{1+\lambda_0^2}{\lambda_0^2} \right) I_2(\lambda_0), \tag{2.63}$$

$$\omega_2^2 = \frac{\left(\Omega_0 - k_w v_\parallel\right)^2}{1+\lambda_0^2} - 2\Omega_w k_w v_w \left(\frac{1+\lambda_0^2}{\lambda_0^2} \right) I_2(\lambda_0), \tag{2.64}$$

and

$$\omega_3^2 = 8\frac{v_\parallel}{v_w}\left(\Omega_0 - 2k_w v_\parallel\right)\left\{\Omega_0 I_2(\lambda_0) + \lambda_0 k_w v_\parallel \left[I_1(\lambda_0) + 2\lambda_0 I_2(\lambda_0)\right]\right\}. \tag{2.65}$$

It is evident that $\Omega_+^2 \geq 0$, so that orbital instability occurs only when $\Omega_-^2 < 0$. In order to obtain an orbital instability criterion, therefore, it is more convenient to deal with the product $\Omega_-^2 \ \Omega_+^2$. This is negative, and orbital instability occurs, whenever

$$\left[(1+\lambda_0^2)\Omega_0 - k_w v_\parallel\right] Z(\lambda_0) - \lambda_0^2 k_w v_\parallel W(\lambda_0) < 0, \tag{2.66}$$

where

$$Z(\lambda_0) = (1+\lambda_0^2) I_1'(\lambda_0) - \frac{2}{\lambda_0} I_1(\lambda_0) \tag{2.67}$$

and

$$W(\lambda_0) = (1+\lambda_0^2) I_1'(\lambda_0) - \frac{1-\lambda_0^2}{\lambda_0} I_1(\lambda_0). \tag{2.68}$$

Observe that this instability criterion (2.66) reduces to that found in the idealized wiggler limit (2.21) when $\lambda_0 \ll 1$.

Solutions for these steady-state trajectories as a function of the axial magnetic field that corresponds to those found in the idealized limit are shown in Fig. 2.6, in which the dashed lines indicate unstable orbits. These orbits correspond to the idealized steady-state trajectories shown in Fig. 2.1, which are reproduced here for comparison. The steady-state orbits are close to those obtained in the idealized limit whenever $\lambda_0 \ll 1$ which, in general, requires that $\Omega_w/ck_w \ll 1$. However, λ_0 can approach and even exceed unity near the resonance at $\Omega_0 \approx k_w v_\parallel$ due to the enhancement in both the transverse velocity and λ_0. It is in this regime, as shown in the figure, in which the effects of a realizable wiggler are most pronounced. The discrepancy with the idealized trajectories is most in evidence for the Group II trajectories, which exhibit an orbital instability not found in the idealized wiggler limit.

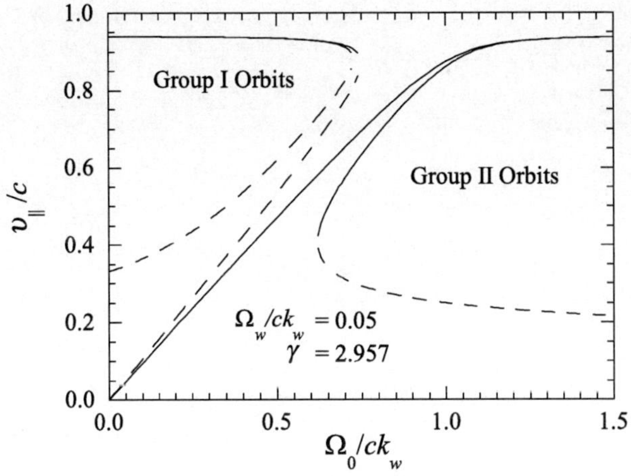

Fig. 2.6 Graph of the axial velocity of the steady-state orbits as a function of the axial magnetic field for both the idealized and realizable wiggler models. The dashed lines indicate unstable trajectories

2.1.2.3 Negative-Mass Trajectories

The negative-mass regime found for Group II orbits in the idealized limit also occurs for the realizable wiggler, in particular [7]

$$\frac{d}{d\gamma}v_{\|} = \frac{c^2}{\gamma\gamma_{\|}^2 v_{\|}}\Phi(\lambda_0), \tag{2.69}$$

where

$$\Phi(\lambda_0) = 1 - \frac{\gamma_{\|}^2\lambda_0^2\left[\Omega_0 Z(\lambda_0) - \lambda_0^2 k_w v_{\|} W(\lambda_0)\right]}{\left[\left(1+\lambda_0^2\right)\Omega_0 - k_w v_{\|}\right]Z(\lambda_0) - \lambda_0^2 k_w v_{\|} W(\lambda_0)}. \tag{2.70}$$

This is similar to Eq. (2.23) found in the idealized wiggler limit, and we note that the generalized $\Phi(\lambda_0)$ reduces to Eq. (2.24) in the idealized wiggler limit in which $\lambda_0 \ll 1$. The form of $\Phi(\lambda_0)$ as a function of the axial field is shown in Fig. 2.7 for parameters which correspond to the steady-state orbits shown in Fig. 2.6. Observe that, as in the idealized limit, $\Phi(\lambda_0)$ exhibits singularities at the transitions to orbital instability which now occur for both Group I and Group II orbits.

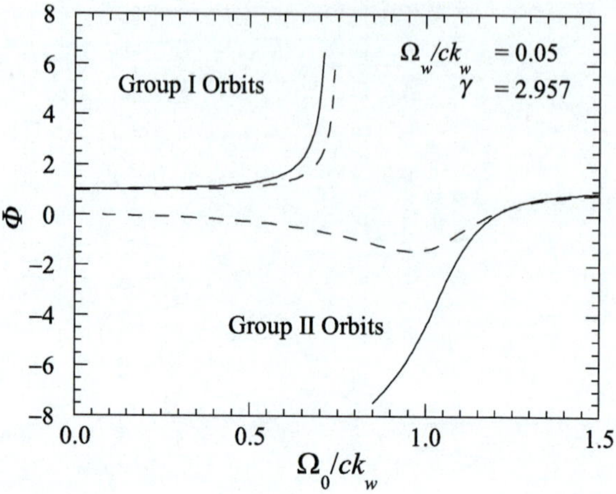

Fig. 2.7 Graphs of the Φ as a function of the axial magnetic field for both the idealized (dashed) and realizable (solid) wiggler representations

2.1.2.4 Generalized Trajectories: Larmor and Betatron Oscillations

The homogeneous equations for the perturbed quantities (Eqs. 2.60 and 2.61) describe orbits that, while close to the steady-state orbits, describe non-helical, axis-encircling trajectories. The solutions to Eq. (2.60) are of the form [7]

$$\delta \upsilon_2 = V_+ \sin(\Omega_+ t - \phi_+) + V_- \sin(\Omega_- t - \phi_-), \tag{2.71}$$

and

$$\delta \chi = X_+ \sin(\Omega_+ t - \phi_+) + X_- \sin(\Omega_- t - \phi_-), \tag{2.72}$$

where $V_\pm$, $X_\pm$, $\phi_\pm$, and $\theta_\pm$ are the integration constants. Substitution of these solutions into the first-order equations indicates that $\delta \upsilon_2$ and $\delta \chi$ are related by $\theta_\pm = \phi_\pm$ and

$$X_\pm = \frac{V_\pm}{\upsilon_w} \frac{k_w \upsilon_\| (\Omega_0 - 2k_w \upsilon_\|)}{\Omega_\pm^2 - \omega_1^2}. \tag{2.73}$$

As a consequence,

$$\delta \upsilon_1 = \Delta_+ V_+ \sin(\Omega_+ t - \phi_+) + \Delta_- V_- \sin(\Omega_- t - \phi_-), \tag{2.74}$$

$$\delta \upsilon_3 = \pm \lambda_0 \delta \upsilon_1, \tag{2.75}$$

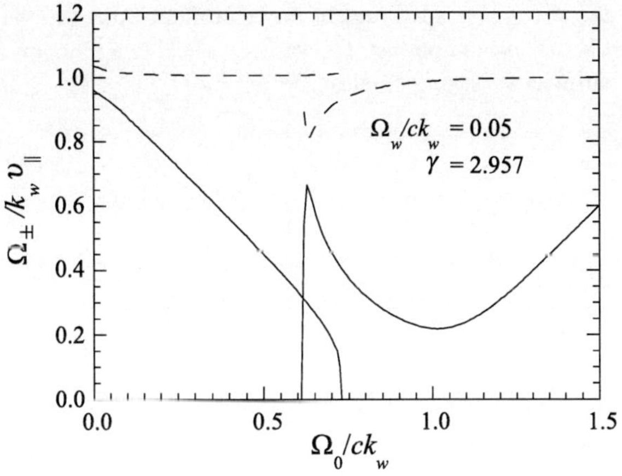

Fig. 2.8 Graph of $\Omega_+/k_w v_\|$ (dashed line) and $\Omega_-/k_w v_\|$ (solid line) versus the axial magnetic field for parameters corresponding to the stable steady-state orbits shown in Fig. 2.6

$$\frac{\delta\lambda}{\lambda_0} = \rho_+ \frac{V_+}{v_w} \frac{k_w v_\|}{\Omega_+} \cos\left(\Omega_+ t - \phi_+\right) + \rho_- \frac{V_-}{v_w} \frac{k_w v_\|}{\Omega_-} \cos(\Omega_- t - \phi_-), \tag{2.76}$$

where

$$\Delta_\pm = 2\frac{\Omega_w}{\Omega_\pm} \frac{v_\|}{v_w} \left[\frac{1}{\lambda_0} I_1(\lambda_0) + \frac{k_w v_\| \left(\Omega_0 - 2k_w v_\|\right)}{\Omega_\pm^2 - \omega_1^2} I_2(\lambda_0) \right], \tag{2.77}$$

$$\rho_\pm = 1 - \frac{k_w v_\| \left(\Omega_0 - 2k_w v_\|\right)}{\Omega_\pm^2 - \omega_1^2}. \tag{2.78}$$

These non-helical trajectories exhibit oscillations at the frequencies $\Omega_\pm$ in addition to the large-scale, wiggler induced oscillation. In order to determine the physical basis for these oscillations, we consider the characteristics of these frequencies in more detail. A plot of the variation of Ω_+ and Ω_- versus the axial magnetic field is shown in Fig. 2.8 for parameters corresponding to the stable steady-state orbits shown in Fig. 2.6. Expansion of $\Omega_\pm$ in powers of λ_0 in the absence of an axial magnetic field indicates that to lowest nontrivial order

$$\Omega_\pm \approx k_w v_\| - \Omega_\beta, \tag{2.79}$$

where $\Omega_\beta = \Omega_w/\sqrt{2}$ defines the betatron frequency. As a result, transformation from the wiggler frame to Cartesian coordinates indicates that these oscillations are degenerate in the absence of an axial magnetic and correspond to betatron oscillations due to the inhomogeneous magnetic field of the wiggler. Betatron oscillations arise because the wiggler field is a minimum on axis and, therefore, exerts a restoring force on the electrons. In the absence of an axial magnetic field, this restoring force gives rise to uncoupled oscillations in x and y. It is evident from Fig. 2.8 that

$\Omega_+ \approx k_w v_\parallel$ for all values of the axial field except in the vicinity of the resonance at $\Omega_0 \approx k_w v_\parallel$. In contrast, Ω_- varies widely as a function of the axial field. Indeed, for sufficiently strong axial fields that $2\lambda_0 \Omega_w \Omega_0 < \Omega_0^2 < 4k_w^2 v_\parallel^2$

$$\Omega_+ = k_w v_\parallel - \frac{\Omega_\beta^2}{\Omega_0} \frac{k_w v_\parallel}{\Omega_0 - k_w v_\parallel}, \tag{2.80}$$

and

$$\Omega_- \approx \left| k_w v_\parallel - \Omega_0 \right| \tag{2.81}$$

correct to within terms of order λ_0^2. Here, Ω_- describes the rapid Larmor oscillations, while Ω_+ describes a modified betatron oscillation. The presence of the axial magnetic field modifies both the frequency and character of the betatron oscillation. The modification of the frequency is evident by comparison of Eqs. (2.79) and (2.80). The character of the oscillation is altered because the presence of an axial field results in a coupling of the x and y motion which results in an elliptic oscillation in the transverse plane.

The aforementioned orbits refer to motion that departs only marginally from the steady-state helical trajectories and is axis encircling. Indeed, the steady-state trajectories are axis centered. However, not all motion is axis encircling. The opposite limit in which the electron displacement from the symmetry axis is large in comparison with all rapid oscillatory motion (i.e., either wiggler or Larmor) is referred to as the *guiding-center* approximation. In order to treat this limit [7], we decompose the electron position and velocity into $\mathbf{x} = \mathbf{x}_0 + \mathbf{x}_c$ and $\mathbf{v} = \mathbf{v}_0 + \mathbf{v}_c$, where $(\mathbf{x}_0, \mathbf{v}_0)$ describes the rapid oscillatory motion about the slowly moving guiding center $(\mathbf{x}_c, \mathbf{v}_c)$ and it is assumed that $|\mathbf{r}_0| \ll |\mathbf{r}_c|$ (where $\mathbf{r}_c$ and $\mathbf{r}_0$ denote the transverse components of $\mathbf{r}_0$ and $\mathbf{r}_c$) while $|\mathbf{v}_c| \ll |\mathbf{v}_0|$. A schematic illustration of this decomposition is illustrated in Fig. 2.9. In this regime, the electron reacts to the local magnetic field at the guiding center, and we may expand

$$\mathbf{B}(x) = \mathbf{B}(\mathbf{x}_c) + \mathbf{x}_c \cdot \nabla_c \mathbf{B}(\mathbf{x}_c), \tag{2.82}$$

where ∇_c denotes the divergence with respect to the guiding-center coordinates. The motion of the guiding center is obtained by means of an average of the orbit equations over a wiggler period, which yields

$$\frac{d}{dt}\mathbf{v}_c \cong -\Omega_0 \mathbf{v}_c \times \hat{\mathbf{e}}_z - \frac{e}{\gamma m_e c} \langle \mathbf{v}_0 \times [\mathbf{B}(\mathbf{x}_c) + \mathbf{x}_0 \cdot \nabla_c \mathbf{B}(\mathbf{x}_c)] \rangle - \frac{e}{\gamma m_e c} \mathbf{v}_c \times \langle \mathbf{x}_0 \cdot \nabla_c \mathbf{B}(\mathbf{x}_c) \rangle, \tag{2.83}$$

where $<\mathbf{B}(\mathbf{x}_c)> = B_0 \hat{\mathbf{e}}_z$. In addition, it is assumed that $<\mathbf{r}_0> = 0$ and $<\mathbf{v}_0> = v_\parallel \hat{\mathbf{e}}_z$. The equation governing the rapid oscillatory motion

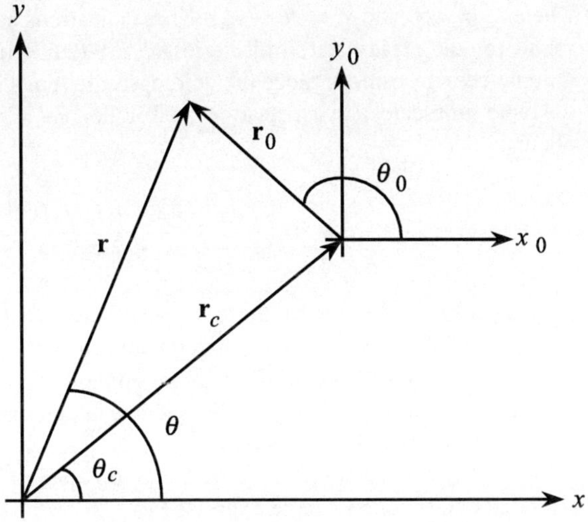

Fig. 2.9 Schematic representation of the vector relationships between the electron position **r**, the guiding center location $\mathbf{r}_c$, and the wiggler-induced component of the trajectory $\mathbf{r}_0$

$$\frac{d}{dt}\mathbf{v}_0 \cong -\frac{e}{\gamma m_e c}\mathbf{v}_0 \times [\mathbf{B}(\mathbf{x}_c) + \mathbf{x}_0 \cdot \nabla_c \mathbf{B}(\mathbf{x}_c)] + \frac{e}{\gamma m_e c}\mathbf{v}_0 \times \langle \mathbf{B}(\mathbf{x}_c) + \mathbf{x}_0 \cdot \nabla_c \mathbf{B}(\mathbf{x}_c)\rangle$$
$$- \frac{e}{\gamma m_e c}\mathbf{v}_c \times [\mathbf{B}_w(\mathbf{x}_c) + \mathbf{x}_0 \cdot \nabla_c \mathbf{B}(\mathbf{x}_c) - \langle \mathbf{x}_0 \cdot \nabla_c \mathbf{B}(\mathbf{x}_c)\rangle], \tag{2.84}$$

is obtained by subtraction of the guiding-center motion from the complete orbit equation.

If we solve the equation for the rapid oscillatory motion to lowest order in $k_w|\mathbf{r}_0|$, the only the first term survives, and we write

$$\frac{d}{dt}\mathbf{v}_c \cong -\frac{e}{\gamma m_e c}\mathbf{v}_0 \times \mathbf{B}(\mathbf{x}_c). \tag{2.85}$$

This equation essentially describes the effect of the local magnetic field at the electron guiding center upon the trajectory. The orbit equations in the wiggler frame, therefore, take the form

$$\frac{d}{dt}v_{01} \cong -\left[\Omega_0 - k_w v_\parallel + 2\Omega_w I_1(\lambda_c)\sin\chi_c\right]v_{02} + \Omega_w v_{03} I_2(\lambda_c)\sin 2\chi_c, \tag{2.86}$$

$$\frac{d}{dt}v_{02} \cong \left[\Omega_0 - k_w v_\parallel + 2\Omega_w I_1(\lambda_c)\sin\chi_c\right]v_{01} - \Omega_w v_{03}[I_0(\lambda_c) + 2I_2(\lambda_c)\cos 2\chi_c], \tag{2.87}$$

$$\frac{d}{dt}v_{03} \cong \Omega_w v_{02}[I_0(\lambda_c) + 2I_2(\lambda_c)\cos 2\chi_c] - \Omega_w v_{01} I_2(\lambda_c)\sin 2\chi_c, \tag{2.88}$$

where $\lambda_c = k_w r_c$ and $\chi_c = \theta_c - k_w z$. These equations are formally identical to those found for the steady-state trajectories (2.55)–(2.57) under the substitution of the guiding-center position. Since the guiding-center position is assumed to be fixed on the rapid timescale, a quasi-steady-state solution $v_{01} = v_{0w}$, $v_{02} = 0$, and $v_{03} = v_{0\|}$ is obtained where

$$v_{0w} = \frac{\Omega_w v_{0\|}\left[I_0(\lambda_c) + I_2(\lambda_c)\cos 2\chi_c\right]}{\Omega_0 - k_w v_{0\|} + 2\Omega_w I_1(\lambda_c)\sin\chi_c}. \tag{2.89}$$

This solution is similar to the previously discussed helical steady-state orbits. However, while the values of v_{0w} and $v_{0\|}$ are constant on the rapid wiggler timescale, they vary slowly with the position of the guiding center.

The slow timescale motion of the guiding center can be determined by substitution of the lowest-order solution for $(\mathbf{x}_0,\mathbf{v}_0)$ into Eq. (2.81). Since $<\mathbf{v}_0 \times \mathbf{x}_0 \cdot \nabla_c \mathbf{B}(\mathbf{x}_c)> = -B_w(v_{0w}/v_{0\|})\hat{\mathbf{e}}_z$ and $<\mathbf{v}_0 \times \mathbf{B}(\mathbf{x}_c)> = -B_w k_w v_{0w} r_c/2$, the equation governing the guiding-center motion can be expressed as

$$\frac{d}{dt}\mathbf{v}_c = \omega_c \hat{\mathbf{e}}_z \times \mathbf{v}_c + \frac{1}{2}\Omega_w k_w v_{0w} \Upsilon_c, \tag{2.90}$$

where

$$\omega_c \equiv \Omega_0 - \Omega_w \frac{v_{0w}}{v_{0\|}}. \tag{2.91}$$

In general, the guiding-center motion is a coupled oscillation in (x_c, y_c) which can be separated into the following two fourth-order differential equations:

$$\left(\frac{d^2}{dt^2} + \omega_+^2\right)\left(\frac{d^2}{dt^2} + \omega_-^2\right)\begin{bmatrix} x_c(t) \\ y_c(t) \end{bmatrix} = 0, \tag{2.92}$$

where

$$\omega_\pm^2 \equiv \frac{1}{2}\left(\omega_c^2 - \Omega_w k_w v_{0w}\right) \pm \frac{\omega_c}{2}\sqrt{\omega_c^2 - \Omega_w k_w v_{0w}}, \tag{2.93}$$

subject to the requirement that $\left|\omega_\pm^2\right| \ll k_w^2 v_{|0\|}^2$.

In the limit in which the axial magnetic field disappears, the frequencies $\omega_\pm \cong \Omega_\beta$ and the (x_c,y_c) motion of the guiding center reduce to well-known uncoupled betatron oscillations in the inhomogeneous wiggler

$$\begin{bmatrix} x_c(t) \\ v_{xc}(t) \end{bmatrix} = \begin{bmatrix} \cos\Omega_\beta t & \frac{1}{\Omega_\beta}\sin\Omega_\beta t \\ -\Omega_\beta \sin\Omega_\beta t & \cos\Omega_\beta t \end{bmatrix}\begin{bmatrix} x_c(0) \\ v_{xc}(0) \end{bmatrix}, \tag{2.94}$$

and

$$\begin{bmatrix} y_c(t) \\ v_{yc}(t) \end{bmatrix} = \begin{bmatrix} \cos \Omega_\beta t \frac{1}{\Omega_\beta} \sin \Omega_\beta t - \Omega_\beta \sin \Omega_\beta t \cos \Omega_\beta t \end{bmatrix} \begin{bmatrix} y_c(0) \\ v_{yc}(0) \end{bmatrix}. \tag{2.95}$$

This motion describes an oscillation in which the electron trajectories pass through the symmetry axis at $x_c = y_c = 0$. It arises because the wiggler field strength increases with the displacement from the symmetry axis and, therefore, exerts a restoring force on the electron trajectories.

The presence of the axial magnetic field couples the (x_c,y_c) motion and results in a precession of the guiding center around the symmetry axis. If the axial magnetic field is sufficiently strong that $\Omega_0^2 \gg |\Omega_w k_w v_{0w}|$, then $\omega_c \cong \Omega_0$, $\omega_+^2 \cong \Omega_0^2 - \Omega_w k_w v_{0w}$, and

$$\omega_- \cong \frac{\Omega_\beta^2}{\Omega_0} \frac{k_w v_{0\|}}{\Omega_0 - k_w v_{0\|}}. \tag{2.96}$$

Observe that this is precisely the modified betatron frequency found previously by perturbation about the steady-state trajectories (Eq. 2.80). As a result, $\omega_\pm$ characterizes modified Larmor and betatron motion, respectively, which now describe ellipses. In particular, electrons in the combined magnetic fields execute betatron motion characterized by

$$\begin{bmatrix} x_c(t) \\ y_c(t) \end{bmatrix} = \begin{bmatrix} \cos \omega_- t - \alpha \sin \omega_- t \frac{1}{\alpha} \sin \omega_- t \cos \omega_- t \end{bmatrix} \begin{bmatrix} x_c(0) \\ y_c(0) \end{bmatrix}, \tag{2.97}$$

where

$$\alpha \equiv \frac{2\Omega_0 \omega_-}{2\Omega_0^2 + \Omega_w k_w v_{0w}}. \tag{2.98}$$

A similar expression can be found for the modified Larmor motion.

2.2 Planar Wiggler Configurations

Because the wiggler magnitude is itself oscillatory in the case of planar wiggler fields, both the transverse and axial components of the electron trajectories are periodic as well. This is in marked contrast to a helical wiggler for which steady-state orbits characterized by constant-magnitude transverse and axial velocities are possible. The absence of steady-state solutions for the electron trajectories in planar wiggler geometries implies that the wave-particle resonance condition can be satisfied only in an average sense and that electrons will drift into and out of resonance with the wave over the course of a wiggler period. In practice, this means that the interaction will be governed by the root-mean-square value of the wiggler field and that the effective wiggler field is reduced. In comparison to the effect of a helical wiggler, therefore, a planar wiggler must be approximately 71% higher in order to have a comparable effect. In view of this disadvantage, the widespread use of planar wigglers is attributed to the ease with which they may be adjusted. In particular, the ability to adjust the field strength of the wiggler in the direction of the axis of symmetry is important for efficiency enhancement schemes that employ nonuniform (i.e., tapered) wiggler fields. The effect of axial wiggler tapers on the electron trajectories will be discussed in the next section.

2.2.1 *Idealized One-Dimensional Trajectories*

In the limit in which the electron displacements from the plane of symmetry are small in comparison with a wiggler period (i.e., $k_w|x| \ll 1$ and $k_w|y| \ll 1$), both planar wiggler configurations given by Eq. (2.2) may be represented in the form

$$\mathbf{B}_w = B_w \hat{\mathbf{e}}_y \sin k_w z, \tag{2.99}$$

with the corresponding vector potential

$$\mathbf{A}_w = -\frac{B_w}{k_w} \hat{\mathbf{e}}_x \cos k_w z. \tag{2.100}$$

As in the case of the idealized one-dimensional limit of the helical wiggler, both x and y are ignorable coordinates, and the transverse canonical momenta are constants of the motion. The third constant of the motion is the total energy. As a consequence, υ_y is also a constant of the motion and the x- and z-components of the velocity are given by

$$\upsilon_x = \frac{P_x}{\gamma m_e} + \upsilon_w \cos k_w z, \tag{2.101}$$

$$\frac{\upsilon_z}{c} = \sqrt{1 - \frac{1}{\gamma^2} - \frac{V_\perp^2}{c^2} - \frac{\upsilon_w^2}{c^2} \cos^2 k_w z - 2\frac{\upsilon_w V_\perp}{c^2} \cos k_w z}, \tag{2.102}$$

where $P_{x,y}$ denote the canonical momenta, $\upsilon_w = -cK/\gamma$ is the wiggler-induced transverse velocity where K is the wiggler strength parameter, $V_\perp = (P_x^2 + P_y^2)^{1/2}/\gamma m_e$, and $V_{x,y} = P_{x,y}/\gamma m_e$. Observe that the assumption of small displacements from the symmetry plane is equivalent to the requirement that $|\upsilon_w/\upsilon_\parallel| \ll 1$.

In contrast with the helical wiggler where the orbits describe steady-state helices in the limit in which the canonical momenta vanish, there are no steady-state orbits for a planar wiggler. In this case, the transverse and axial velocity components can be written in the form

$$v_x = v_w \cos k_w z \tag{2.103}$$

that describes the wiggler-induced oscillation in the transverse direction and

$$v_z = v_{\|} \sqrt{1 - \frac{v_w^2}{2c^2} \cos 2k_w z}, \tag{2.104}$$

where

$$\frac{v_{\|}}{c} \equiv \sqrt{1 - \frac{1}{\gamma^2} - \frac{v_w^2}{2c^2}} \tag{2.105}$$

denotes the bulk axial velocity. Observe that this bulk axial velocity corresponds to an average transverse velocity that corresponds to the root-mean square value of the wiggler field. In addition, it should be remarked that the oscillatory component of the axial velocity is of second order in the wiggler magnitude and is, therefore, negligible under most circumstances.

2.2.1.1 Quasi-Steady-State Trajectories

In the presence of an axial magnetic field, the canonical momenta are no longer conserved, and the orbits take on an even more complex nature. Using the Lorentz force equation (Eq. 2.4), we write the orbit equations in the form

$$\frac{d}{dt} v_x = -\Omega_0 v_y + \Omega_w v_z \sin k_w z, \tag{2.106}$$

$$\frac{d}{dt} v_y = \Omega_0 v_x, \tag{2.107}$$

$$\frac{d}{dt} v_z = -\Omega_w v_x \sin k_w z. \tag{2.108}$$

In order to obtain quasi-steady-state solutions, we assume that the orbits are characterized by a large bulk axial motion with a small amplitude oscillation superimposed upon it. As in the case shown in Eqs. (2.103)–(2.106), this oscillation in the axial velocity is of second order in the wiggler amplitude. Therefore, $v_z = v_{\|}$ (constant) to within first order in the wiggler amplitude, and we obtain

$$\left(\frac{d^2}{dt^2} + \Omega_0^2 \right) v_x = k_w v_{\|}^2 \Omega_w \cos k_w z, \tag{2.109}$$

and

$$\left(\frac{d^2}{dt^2}+\Omega_0^2\right)v_y=\Omega_w\Omega_0 v_\parallel \sin k_w z. \tag{2.110}$$

The homogeneous solutions to these equations describe the Larmor rotation due to the solenoidal magnetic field. This Larmor motion is dependent upon the initial conditions of the orbit and can, in principle, be made arbitrarily small. In contrast, the wiggler-induced motion corresponds to the particular solution and describes an ellipse in the *xy*-plane characterized by

$$\mathbf{v}_\perp=\alpha_x\hat{\mathbf{e}}_x\cos k_w z+\alpha_y\hat{\mathbf{e}}_y\sin k_w z, \tag{2.111}$$

where

$$\alpha_x=\frac{\Omega_w k_w v_\parallel^2}{\Omega_0^2-k_w^2 v_\parallel^2}, \tag{2.112}$$

and

$$\alpha_y=\frac{\Omega_w\Omega_0 v_\parallel}{\Omega_0^2-k_w^2 v_\parallel^2}. \tag{2.113}$$

Observe that in the limit in which the solenoidal magnetic field vanishes $\alpha_x=-cK/\gamma$, $\alpha_y=0$, and the result given in Eq. (2.103) is recovered. These orbits show a resonant enhancement in the magnitude of the transverse velocity when $\Omega_0\approx k_w v_\parallel$ which is similar to that found for helical wigglers. The average axial velocity over a wiggler period for this trajectory is obtained from the conservation of energy using the root-mean-square values of the transverse velocity components, and we find that

$$\frac{v_\parallel^2}{c^2}\left[1+\frac{\Omega_w^2\left(\Omega_0^2+k_w^2 v_\parallel^2\right)}{2\left(\Omega_0^2-k_w^2 v_\parallel^2\right)^2}\right]=1-\frac{1}{\gamma^2}. \tag{2.114}$$

It should be noted that this equation is symmetric in $v_\parallel$ and yields trajectories that are independent of the direction of propagation. This contrasts with the helical wiggler that establishes a preferred direction of propagation with respect to the orientation of the axial magnetic field. The second-order oscillation in the axial velocity is found by substitution of Eq. (2.111) into Eq. (2.108) to be

$$v_z=v_\parallel+\frac{1}{4}\frac{\Omega_w^2 v_\parallel}{\Omega_0^2-k_w^2 v_\parallel^2}\cos 2k_w z. \tag{2.115}$$

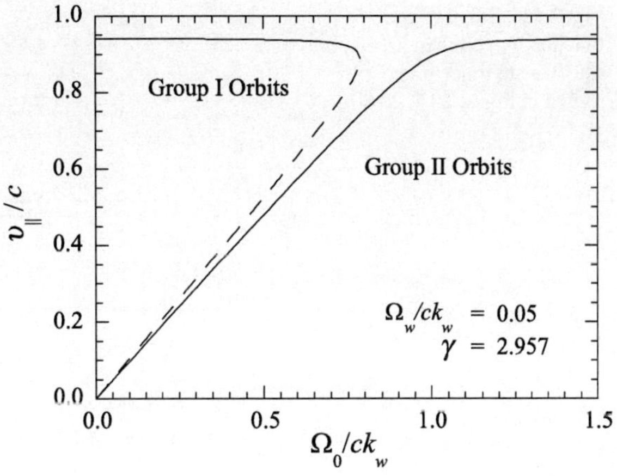

Fig. 2.10 Graph of the axial velocity of the quasi-steady-state orbits as a function of the solenoidal magnetic field showing the Group I and Group II orbits

The bulk axial velocity can be obtained by solution of Eq. (2.114) as a function of the wiggler field strength and period, the axial magnetic field, and the electron energy. The characteristic solutions are very similar to those found in the case of a helical wiggler field and can be divided into Group I ($\Omega_0 < k_w v_\parallel$) and Group II ($\Omega_0 > k_w v_\parallel$) orbits. These solutions are illustrated in Fig. 2.10 as a function of the axial magnetic field for parameters consistent with the steady-state solutions shown for the helical wiggler in Fig. 2.1. Observe that only those solutions for $v_\parallel > 0$ are shown as the curves are symmetric about the direction of propagation.

2.2.1.2 Negative-Mass Trajectories

These quasi-steady state trajectories exhibit similar positive- and negative-mass regimes as found in the case of a helical wiggler. This is shown by the implicit differentiation of Eq. (2.114) with respect to γ, which gives

$$\frac{d}{d\gamma} v_\parallel = \frac{c^2}{\gamma \gamma_\parallel^2 v_\parallel} \Phi_p, \tag{2.116}$$

where

$$\Phi_p \equiv 1 - \frac{\gamma_\parallel^2 \Omega_w^2 \Omega_0^2 \left(\Omega_0^2 + 3k_w^2 v_\parallel^2 \right)}{2\left(\Omega_0^2 - k_w^2 v_\parallel^2 \right)^3 + \Omega_w^2 \Omega_0^2 \left(\Omega_0^2 + 3k_w^2 v_\parallel^2 \right)} \tag{2.117}$$

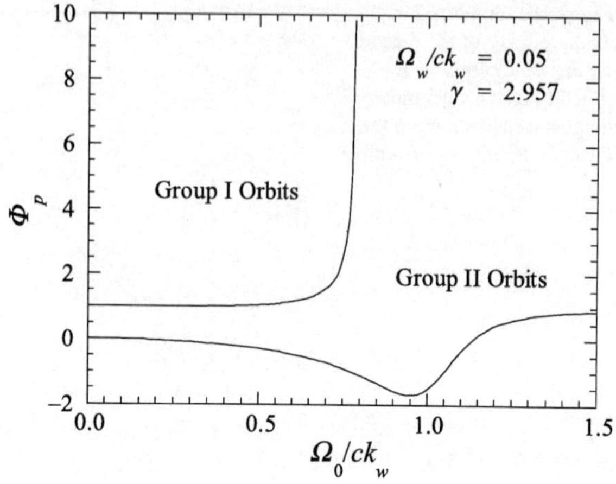

Fig. 2.11 Graph of the function Φ_p to the quasi-steady-state trajectories shown in Fig. 2.10

is the analog of the function Φ (Eq. 2.24) for a planar wiggler. A plot of the function Φ_p is given in Fig. 2.11 corresponding to the quasi-steady-state trajectories shown in Fig. 2.10. As in the case of helical wiggler fields, these quasi-steady-state orbits exhibit a negative-mass regime for Group II orbits close to the resonance at $\Omega_0 \approx k_w v_{\parallel}$.

The general character of the quasi-steady-state trajectories in a planar wiggler is similar to that of the steady-state trajectories in a helical wiggler. The major distinctions, however, are that (1) there is no preferred direction of propagation with respect to the axial solenoidal field, (2) the projection of the orbits in the *xy*-plane in a planar wiggler is elliptic rather than circular, and (3) that the magnitudes of the transverse components of the velocity are determined by the root-mean-square magnitude of the wiggler. This latter distinction implies that a planar wiggler must be approximately 41% larger in magnitude than a helical wiggler in order to have a comparable effect.

There are two effects that we have ignored in the idealized treatment of the quasi-steady-state trajectories. The first is that we have neglected the oscillations in the axial velocity. However, these oscillations are of second order in the wiggler magnitude and become important for large-amplitude wigglers and for the generation of harmonic radiation. The second effect arises within the context of a three-dimensional analysis that employs a self-consistent wiggler field. In this case, a $\mathbf{B}_0 \times \nabla \mathbf{B}_w$ drift appears which drives the electron beam away from the symmetry plane and will ultimately result in the loss of the beam to the drift tube wall. It is for this reason that alternate methods of beam focusing are usually employed in conjunction with planar wigglers.

2.2.2 *Trajectories in Realizable Planar Wigglers*

When the electron trajectories diverge substantially from the plane of symmetry (i.e., $k_w|x| \approx 1$ and $k_w|y| \approx 1$), then the wiggler inhomogeneities become important, and the idealized wiggler model breaks down. We first treat the case of the wiggler model given in Eq. (2.2) for flat pole faces, which can be described by a vector potential of the form

$$\mathbf{A}_w(y,z) = -\frac{B_w}{k_w}\hat{\mathbf{e}}_x \cosh k_w y \cos k_w z. \tag{2.118}$$

Since the vector potential is independent of x, this is an ignorable coordinate and the canonical momentum in the x-direction is a conserved quantity. This allows us to eliminate one of the equations of motion. However, before we treat this case, it is instructive to consider the effect of an axial solenoidal magnetic field upon the trajectories, particularly in regard to the aforementioned $\mathbf{B}_0 \times \nabla \mathbf{B}_w$ drift.

The orbit equations in a realizable planar wiggler with flat pole faces take the form

$$\frac{d}{dt}v_x = -\Omega_0 v_y - \Omega_w \left(v_y \sinh k_w y \cos k_w z - v_z \cosh k_w y \sin k_w z\right), \tag{2.119}$$

$$\frac{d}{dt}v_y = \Omega_0 v_x + \Omega_w v_x \sinh k_w y \cos k_w z, \tag{2.120}$$

$$\frac{d}{dt}v_z = -\Omega_w v_x \cosh k_w y \sin k_w z, \tag{2.121}$$

where we observe that the total energy is a conserved quantity. A second constant of the motion corresponds to the x-component of the canonical momentum. Equation (2.118) can be integrated to obtain v_x as a function of the canonical momentum P_x and the (y,z) coordinates, and we obtain yields

$$v_x = \frac{P_x}{\gamma m_e} - \Omega_0 y + v_w \cosh k_w y \cos k_w z, \tag{2.122}$$

where $v_w = -\Omega_w/k_w = -cK/\gamma$.

2.2.2.1 Gradient Drifts Due to an Axial Magnetic Field

Under the assumption that the orbit does not diverge significantly from the symmetry plane, we may expand in powers of $k_w y$ about the quasi-steady-state solutions obtained previously (Eqs. 2.111–2.113) and write that

$$\mathbf{v}_\perp = \alpha_x \hat{\mathbf{e}}_x \cos k_w z + \alpha_y \hat{\mathbf{e}}_y \sin k_w z + \delta \mathbf{v}_\perp, \tag{2.123}$$

and

$$v_z = v_\parallel + \delta v_z, \tag{2.124}$$

where we have neglected contributions of second order in the wiggler amplitude to the quasi-steady-state trajectories. We now expand Eqs. (2.119)–(2.121) to first order in $k_w y$ and average the resulting equations over a wiggler period. As a result, the perturbations in the transverse velocities are governed by

$$\left(\frac{d^2}{dt^2} + \Omega_0^2 + \Omega_\beta^2\right)\delta v_x \cong -y\frac{\Omega_0\Omega_\beta^2 k_w^2 v_\parallel^2}{\Omega_0^2 - k_w^2 v_\parallel^2}, \tag{2.125}$$

and

$$\left(\frac{d^2}{dt^2} + \Omega_0^2 - \frac{\Omega_\beta^2 k_w^2 v_\parallel^2}{\Omega_0^2 - k_w^2 v_\parallel^2}\right)\delta v_y \cong 0, \tag{2.126}$$

where Ω_β denotes the betatron frequency. The latter Eq. (2.126) describes a modified betatron oscillation in the y-direction that arises from the wiggler inhomogeneity. Observe that the frequency of this oscillation reduces to the Ω_β in the limit in which the solenoidal magnetic field vanishes. The homogeneous solution to Eq. (2.125) represents a modified Larmor oscillation in the combined wiggler and solenoidal fields. However, the particular solution describes the $\mathbf{B}_0 \times \nabla\mathbf{B}_w$ drift due to the combined fields. This bulk drift in the x-direction varies as

$$\delta v_x \cong -y\frac{B_w^2}{2B_0^2 + B_w^2}\frac{\Omega_0 k_w^2 v_\parallel^2}{\Omega_0^2 - k_w^2 v_\parallel^2}, \tag{2.127}$$

so that

$$x(y,z) \cong -yz\frac{B_w^2}{2B_0^2 + B_w^2}\frac{\Omega_0 k_w^2 v_\parallel^2}{\Omega_0^2 - k_w^2 v_\parallel^2}. \tag{2.128}$$

We observe that the magnitude of this drift is (1) proportional to the displacement from the plane of symmetry in the y-direction and (2) resonant at $\Omega_0 \approx k_w v_\parallel$.

As a consequence of this drift, the electron beam can readily be driven to the wall of the drift tube. In order to minimize particle losses, this imposes stringent conditions on the parameters of free-electron lasers that employ this configuration. Since the drift is in the x-direction and is proportional to the y-displacement from the plane of symmetry, we must require that the maximum thickness of the beam in the y-direction, Δy_b, must be sufficiently thin that

$$\Delta y_b < \frac{L_x}{k_w L_z} \frac{\Omega_0^2 - k_w^2 v_\parallel^2}{\Omega_0 k_w v_\parallel} \frac{B_w^2}{2B_0^2 + B_w^2}, \tag{2.129}$$

where L_x defines the x-dimension of the drift tube and L_z is the axial length of the wiggler. The maximum permissible thickness of the beam in the y-direction, therefore, decreases as the wiggler length increases and as the resonance at $\Omega_0 \approx k_w v_\parallel$ is approached. As a result, few attempts have been made to use a planar wiggler in conjunction with an axial magnetic field, and alternate means of enhancing beam confinement in planar wiggler configurations have been employed.

2.2.2.2 Betatron Oscillations

Few experiments have employed both a planar wiggler and a solenoidal magnetic field due to the $\mathbf{B}_0 \times \nabla \mathbf{B}_w$ drift. In the absence of the solenoidal magnetic field, the x-component of the transverse velocity is determined by

$$v_x = \frac{P_x}{\gamma m_e} + v_w \cosh k_w y \cos k_w z. \tag{2.130}$$

For simplicity, we consider the limit in which $P_x = 0$. Substitution of Eq. (2.130) into the equation for v_y (2.119) yields

$$\frac{d}{dt} v_y = -\frac{\Omega_w}{k_w} \sinh k_w y \cosh k_w y \cos^2 k_w z. \tag{2.131}$$

This equation describes both rapid oscillations with a period half that of the wiggler and slow betatron oscillations. If we average this equation over a wiggler period to remove the rapid oscillations, then we find that the slow betatron oscillations are described by

$$\frac{d^2}{dt^2} y = -\frac{\Omega_w^2}{2k_w} \sinh k_w y \cosh k_w y. \tag{2.132}$$

This nonlinear second-order differential equation reduces to simple harmonic oscillations when the electron displacement from the symmetry axis is small. In this limit, $k_w y \ll 1$ and to lowest order we obtain

$$\left(\frac{d^2}{dt^2} + \Omega_\beta^2\right) y \cong 0. \tag{2.133}$$

Solution of this equation results in the typical form for betatron oscillations given in Eqs. (2.95) and (2.126). It should be remarked that the magnitude of the betatron oscillations is independent of the wiggler amplitude and period and is dependent on the initial conditions on the electrons, specifically, on the magnitudes of the initial velocity and displacement from the plane of symmetry.

In the case of large-amplitude excursions from the plane of symmetry, Eq. (2.132) describes a nonlinear oscillation that can be described in terms of the elliptic functions. Equation (2.132) can be integrated once to obtain

$$\left(\frac{du}{d\tau}\right)^2 \cong H^2 - \sinh^2 u, \tag{2.134}$$

where H is an integration constant, $u = k_w y$, and $\tau = \Omega_\beta t$. This describes oscillatory motion in which $\sinh^2 u < H^2$. Equation (2.134) takes on the usual form of an elliptic differential equation under the transformation $w = \sinh u$, for which we obtain

$$\left(\frac{dw}{d\tau}\right)^2 \cong \left(1 + w^2\right)\left(H^2 - w^2\right). \tag{2.135}$$

This has the solution

$$w = \frac{\kappa H \operatorname{sn}\left(\sqrt{1 + H^2}\tau, \kappa\right)}{\sqrt{1 - \kappa^2 \operatorname{sn}^2\left(\sqrt{1 + H^2}\tau, \kappa\right)}}, \tag{2.136}$$

subject to the initial condition $w(\tau = 0) = 0$, where sn denotes the Jacobi elliptic sine function, and

$$\kappa \equiv \frac{H}{\sqrt{1 + H^2}}. \tag{2.137}$$

Hence, the large-amplitude betatron oscillation is described by

$$y(t) = \frac{1}{k_w} \sinh^{-1}\left[\frac{\kappa H \operatorname{sn}\left(\Omega_\beta \sqrt{1 + H^2}\tau, \kappa\right)}{\sqrt{1 - \kappa^2 \operatorname{sn}^2\left(\Omega_\beta \sqrt{1 + H^2}\tau, \kappa\right)}}\right], \tag{2.138}$$

subject to the initial condition $y(t = 0) = 0$. Equation (2.138) reduces to the small amplitude result in the limit in which $H^2 \ll 1$.

2.2.2.3 The Effect of Parabolic Pole Faces

Betatron oscillations describe the electron response to the inhomogeneity of the wiggler that provides a restoring force that confines the electron beam in the neighborhood of the symmetry plane. However, the inhomogeneity is manifested only in the y-direction in the case of a planar wiggler with flat pole faces (Eq. 2.2), and there is no restoring force to enhance the confinement of the electron beam in the x-direction. Both solenoidal and quadrupole magnetic fields have been employed to confine the electron beam against spreading in the x-direction. However, these techniques are not entirely satisfactory since orbital resonances between these fields and the wiggler can deleteriously affect the interaction strength. An alternate technique is to use a wiggler with tapered pole faces such as that given in Eq. (2.3), which describes the effect of parabolic pole faces on the wiggler field. This technique for improving the confinement of the electron beam has the advantage that the resulting betatron oscillations occur at low frequencies and couple only weakly to the electromagnetic waves. Using Eq. (2.3), we find that the detailed equations that describe the electron trajectories in this wiggler are

$$\frac{d}{dt}v_x = \Omega_w \cosh\left(\frac{k_w x}{\sqrt{2}}\right)\left[\sqrt{2}v_y \sinh\left(\frac{k_w y}{\sqrt{2}}\right)\sin k_w z + v_z \cosh\left(\frac{k_w y}{\sqrt{2}}\right)\cos k_w z\right], \tag{2.139}$$

$$\frac{d}{dt}v_y = -\Omega_w \sinh\left(\frac{k_w x}{\sqrt{2}}\right)\left[\sqrt{2}v_x \cosh\left(\frac{k_w y}{\sqrt{2}}\right)\sin k_w z + v_z \sinh\left(\frac{k_w y}{\sqrt{2}}\right)\cos k_w z\right], \tag{2.140}$$

and

$$\frac{d}{dt}v_z = -\Omega_w \cos k_w z\left[v_x \cosh\left(\frac{k_w x}{\sqrt{2}}\right)\cosh\left(\frac{k_w y}{\sqrt{2}}\right) - v_y \sinh\left(\frac{k_w x}{\sqrt{2}}\right)\sinh\left(\frac{k_w y}{\sqrt{2}}\right)\right]. \tag{2.141}$$

Observe that there are no ignorable coordinates for this form of the wiggler and the only constant of the motion is the total energy.

In order to obtain solutions for the equations of motion, we assume that displacements from the symmetry plane are much less than the wiggler period and expand in powers of $k_w x$ and $k_w y$. The principal component of the wiggler-induced oscillation remains in the x-direction for the planar wiggler with parabolic pole faces (Eq. 2.3). As a consequence, there will be two components of the motion in the x-direction: a rapid oscillation at the wiggler period and a slow variation due to the wiggler inhomogeneity. For simplicity, we shall assume that both components of the motion are small and expand the orbit equations about the idealized trajectories (Eqs. 2.103 and 2.104)

$$\mathbf{v}^{(0)} \cong -v_w \hat{\mathbf{e}}_x \sin k_w z + v_\parallel \hat{\mathbf{e}}_z \left(1 + \frac{v_w^2}{4v_\parallel^2}\cos 2k_w z\right) \tag{2.142}$$

that is correct to within terms of order $(v_w/v_\parallel)^2$. Observe that the form we employ for the planar wiggler with parabolic pole faces introduces a phase shift in the idealized trajectories relative to the planar wiggler with flat pole faces (2.2). Expanding about this solution, we write $x = (v_w/k_w v_\parallel)\cos k_w z + \delta x$, $y = \delta y$, $z = v_\parallel t + \delta z$, $v_x = -v_w \sin k_w z + \delta v_x$, $v_y = \delta v_y$, and $v_z = v_\parallel + \delta v_z$. To lowest order in the perturbed quantities, therefore, the orbit equations governing the transverse velocities are

$$\frac{d}{dt}\delta v_x \cong -\frac{1}{2}\Omega_\beta^2 \delta x, \tag{2.143}$$

and

$$\frac{d}{dt}\delta v_y \cong -\frac{1}{2}\Omega_\beta^2 \delta y, \tag{2.144}$$

where we have performed an average over the wiggler period. The inhomogeneity introduced by the parabolic pole faces results in a modified form of the betatron oscillation at a frequency of $\Omega_\beta/\sqrt{2}$. The detailed solution for this modified betatron oscillation in the x-direction is

$$\begin{bmatrix} \delta x(t) \\ \delta v_x(t) \end{bmatrix} = \left[\cos\left(\frac{\Omega_\beta t}{\sqrt{2}}\right) \frac{\Omega_\beta}{\sqrt{2}} \sin\left(\frac{\Omega_\beta t}{\sqrt{2}}\right) - \frac{\Omega_\beta}{\sqrt{2}} \sin\left(\frac{\Omega_\beta t}{\sqrt{2}}\right) \cos\left(\frac{\Omega_\beta t}{\sqrt{2}}\right)\right] \begin{bmatrix} \delta x(0) \\ \delta v_x(0) \end{bmatrix}. \tag{2.145}$$

A similar form is found for the betatron oscillation in the y-direction.

The effect of the betatron oscillations in the x- and y-directions on the axial component of the velocity is not identical. Expansion of Eq. (2.141) to first order in the perturbed quantities shows that the axial velocity is dependent upon δv_x in the following manner:

$$\frac{d}{dt}\delta v_z \cong -\Omega_w \delta v_x \cos k_w v_\parallel t. \tag{2.146}$$

As a result, substitution of the form for the betatron oscillation from Eq. (2.145) gives a perturbed axial velocity that varies as

$$\begin{aligned} \delta v_z \cong \frac{\Omega_w}{2}\Bigg[\frac{1}{\Omega_+}\left(\frac{\Omega_\beta}{\sqrt{2}}\delta x(0)\cos\Omega_+ t + \delta v_x(0)\sin\Omega_+ t\right) \\ + \frac{1}{\Omega_-}\left(\frac{\Omega_\beta}{\sqrt{2}}\delta x(0)\cos\Omega_- t - \delta v_x(0)\sin\Omega_- t\right)\Bigg], \end{aligned} \tag{2.147}$$

where $\Omega_\pm = k_w v_\parallel \pm \Omega_\beta/\sqrt{2}$. The result of a planar wiggler with parabolic pole faces, therefore, is to introduce sideband oscillations in the axial velocity about the wiggler period. In the event that the displacement from the plane of symmetry in the x-direction is large (i.e., when $|\delta x| \approx \Omega_w/k_w^2 v_\parallel$, the sideband perturbations to the axial velocity are comparable in magnitude to the oscillation with a wavenumber of $2k_w$), then these sideband oscillations can have a significant impact on the operation of the free-electron laser.

2.3 Tapered Wiggler Configurations

Axially nonuniform (i.e., tapered) magnetostatic fields are important for the purpose of enhancing the extraction efficiency in free-electron lasers. The electron beam loses energy to the wave during the course of the interaction and decelerates in the axial direction. The interaction saturates when the electron beam loses sufficient energy to drop out of resonance with the wave. The purpose of a tapered magnetic field is to transfer energy from the transverse motion of the beam to accelerate the beam and maintain the wave-particle resonance over an extended interaction region.

This reacceleration of the electron beam can be accomplished by means of a wiggler field in which either the amplitude or period varies in the axial direction. In addition, an axially tapered solenoidal magnetic field can also be used for this purpose. This process can be most easily described under the assumption of a slow variation in the magnetostatic fields in which the scale length of the field inhomogeneity is much longer than the wiggler period, specifically

$$\frac{1}{B_{w,0}}\left|\frac{d}{dz}B_{w,0}\right| \ll k_w, \tag{2.148}$$

and

$$\left|\frac{d}{dz}k_w\right| \ll k_w^2. \tag{2.149}$$

The tapered fields, therefore, can be treated as a small perturbation, and the orbits can be obtained by expansion about the trajectories obtained in the uniform field limit.

2.3.1 *The Idealized One-Dimensional Limit*

In the case of a configuration that consists of a helical wiggler field and a solenoidal magnetic field, we expand about the steady state trajectories. The magnitude of the axial velocity is dependent upon the magnitudes of the wiggler and solenoidal magnetic fields and upon the period of the wiggler field and varies with these parameters as

$$\frac{d}{dz}v_\parallel = \frac{\partial v_\parallel}{\partial B_w}\frac{d}{dz}B_w + \frac{\partial v_\parallel}{\partial B_0}\frac{d}{dz}B_0 + \frac{\partial v_\parallel}{\partial k_w}\frac{d}{dz}k_w. \tag{2.150}$$

It is implicitly assumed within the context of this formulation that the trajectory varies continuously from one steady-state orbit to another. Because the total energy is conserved, the derivatives of $v_\parallel$ can be obtained by implicit differentiation of

$v_w^2 + v_\parallel^2 = (1 - \gamma^{-2})c^2$, where v_w is given by Eqs. (2.15) and (2.53) for the idealized and realistic wigglers, respectively. Therefore, the axial acceleration occurs at the expense of a transverse deceleration. In the idealized limit, we find that

$$\frac{\partial v_\parallel}{\partial B_w} = -\frac{1}{B_w}\frac{\Omega_w v_w}{(1+\beta_w^2)\Omega_0 - k_w v_\parallel}, \tag{2.151}$$

$$\frac{\partial v_\parallel}{\partial k_w} = -\frac{v_w^2}{(1+\beta_w^2)\Omega_0 - k_w v_\parallel}, \tag{2.152}$$

and

$$\frac{\partial v_\parallel}{\partial B_0} = \frac{1}{B_0}\frac{\Omega_w v_\parallel \beta_w^2}{(1+\beta_w^2)\Omega_0 - k_w v_\parallel}. \tag{2.153}$$

Substitution of these derivatives into Eq. (2.150) gives the rate of change of the axial velocity

$$\frac{d}{dz}v_\parallel = -\frac{v_\parallel \beta_w^2}{(1+\beta_w^2)\Omega_0 - k_w v_\parallel}\left(\frac{\Omega_0 - k_w v_\parallel}{B_w}\frac{d}{dz}B_w - \frac{k_w v_\parallel}{\lambda_w}\frac{d}{dz}\lambda_w - \frac{\Omega_0}{B_0}\frac{d}{dz}B_0\right). \tag{2.154}$$

It is important to observe that in order to accelerate the electron beam, both the wiggler amplitude and period must be tapered downward (i.e., $dB_w/dz < 0$ and $d\lambda_w/dz < 0$). However, the situation is more complex in the case of the solenoidal field. Observe that $(1 + \beta_w^2)\Omega_0 < k_w v_\parallel$ for the stable Group I orbits; hence, the solenoidal field amplitude must be tapered downward to achieve an axial acceleration of the electrons. In contrast, $\Omega_0 > k_w v_\parallel$ for the Group II trajectories, and the solenoidal field must be tapered upward (i.e., increased) in order to accelerate the beam. Observe that in the limit in which the solenoidal magnetic field vanishes, the variation in the axial velocity is given by

$$\frac{d}{dz}v_\parallel = -\frac{\Omega_w^2}{k_w^2 v_\parallel}\left(\frac{1}{B_w}\frac{d}{dz}B_w + \frac{1}{\lambda_w}\frac{d}{dz}\lambda_w\right). \tag{2.155}$$

2.3.2 *The Realizable Three-Dimensional Formulation*

These results may be generalized to treat the realistic helical wiggler given by Eq. (2.1), for which

$$\frac{\partial v_\parallel}{\partial B_w} = \frac{1}{B_w} \frac{2\Omega_w v_w (1+\lambda_0^2)^2 I_1^2(\lambda_0)/\lambda_0^2}{\left[(1+\lambda_0^2)\Omega_0 - k_w v_\parallel\right] Z(\lambda_0) - \lambda_0^2 k_w v_\parallel W(\lambda_0)}, \tag{2.156}$$

$$\frac{\partial v_\parallel}{\partial k_w} = \frac{\lambda_0 v_\parallel^2 (1+\lambda_0^2) I_1(\lambda_0)}{\left[(1+\lambda_0^2)\Omega_0 - k_w v_\parallel\right] Z(\lambda_0) - \lambda_0^2 k_w v_\parallel W(\lambda_0)}, \tag{2.157}$$

and

$$\frac{\partial v_\parallel}{\partial B_0} = -\frac{1}{B_0} \frac{\lambda_0 v_\parallel \Omega_0 (1+\lambda_0^2) I_1(\lambda_0)}{\left[(1+\lambda_0^2)\Omega_0 - k_w v_\parallel\right] Z(\lambda_0) - \lambda_0^2 k_w v_\parallel W(\lambda_0)}. \tag{2.158}$$

As a consequence, the axial velocity varies as

$$\begin{aligned} \frac{d}{dz} v_\parallel = &\frac{\lambda_0 v_\parallel (1+\lambda_0^2) I_1(\lambda_0)}{\left[(1+\lambda_0^2)\Omega_0 - k_w v_\parallel\right] Z(\lambda_0) - \lambda_0^2 k_w v_\parallel W(\lambda_0)} \\ &\times \left(\frac{\Omega_0}{B_w} \frac{k_w v_\parallel}{dz} \frac{d}{dz} B_w - \frac{k_w v_\parallel}{\lambda_w} \frac{d}{dz} \lambda_w - \frac{\Omega_0}{B_0} \frac{d}{dz} B_0 \right), \end{aligned} \tag{2.159}$$

This differs from the result obtained in the idealized limit only in terms of an overall factor that describes the effect of large displacements from the axis of symmetry and reduces to that result in the limit as $\lambda_0 \to 0$.

2.3.3 Planar Wiggler Geometries

The corresponding axial acceleration due to a tapered planar wiggler magnetic field can be determined in an analogous manner by perturbation about the quasi-steady-state orbits. In this case, however, we neglect the solenoidal field. This is because the presence of an axial magnetic field results in a transverse drift which, over the course of the tapered wiggler, will cause a loss of the beam to the walls of the drift tube. As a result, the acceleration of the bulk axial velocity (Eq. 2.105) due to gradients in the amplitude and period is given by

$$\frac{d}{dz} v_\parallel = -\frac{\Omega_w^2}{2k_w^2 v_\parallel} \left(\frac{1}{B_w} \frac{d}{dz} B_w + \frac{1}{\lambda_w} \frac{d}{dz} \lambda_w \right). \tag{2.160}$$

Observe that this is identical to the result obtained for the helical wiggler in the absence of the solenoidal magnetic field under the substitution of the root-mean-square value of the wiggler field amplitude.

References

1. J.P. Blewett, R. Chasman, Orbits and fields in the helical wiggler. J. Appl. Phys. **48**, 2692 (1977)
2. L. Friedland, Electron beam dynamics in combined guide and pump magnetic fields for free-electron laser applications. Phys. Fluids **23**, 2376 (1980)
3. P. Diament, Electron orbits and stability in realizable and unrealizable wigglers of free-electron lasers. Phys. Rev. A **23**, 2537 (1981)
4. R.D. Jones, Constants of the motion in a helical magnetic field. Phys. Fluids **24**, 564 (1981)
5. J.A. Pasour, F. Mako, C.W. Roberson, Electron drift in a linear magnetic wiggler with an axial guide field. J. Appl. Phys. **53**, 7174 (1982)
6. H.P. Freund, A.T. Drobot, Relativistic electron trajectories in free-electron lasers with an axial guide field. Phys. Fluids **25**, 736 (1982)
7. H.P. Freund, A.K. Ganguly, Electron orbits in free-electron lasers with helical wiggler and axial guide magnetic fields. IEEE J. Quantum Electron. **QE-21**, 1073 (1985)
8. J. Fajans, D.A. Kirkpatrick, G. Bekefi, Off-axis electron orbits in realistic helical wigglers for free-electron laser applications. Phys. Rev. A **32**, 3448 (1985)
9. R.G. Littlejohn, A.N. Kaufman, Hamiltonian structure of particle motion in an ideal helical wiggler with guide field. Phys. Lett. A **120**, 291 (1987)
10. R.M. Phillips, History of the ubitron. Nucl. Instrum. Methods Phys. Res. **A272**, 1 (1988)
11. E.T. Scharlemann, Wiggler plane focusing in linear wigglers. J. Appl. Phys. **58**, 2154 (1985)
12. B. Levush, T.M. Antonsen Jr., W.M. Manheimer, P. Sprangle, A free-electron laser with a rotating quadrupole wiggler. Phys. Fluids **28**, 2273 (1985)
13. B. Levush, T.M. Antonsen Jr., W.M. Manheimer, Spontaneous radiation of an electron beam in a free-electron laser with a quadrupole wiggler. J. Appl. Phys. **60**, 1584 (1986)
14. T.M. Antonsen Jr., B. Levush, Nonlinear theory of a quadrupole free-electron laser. IEEE J. Quantum Electron. **QE-23**, 1621 (1987)
15. S.F. Chang, O.C. Eldridge, J.E. Sharer, Analysis and nonlinear simulation of a quadrupole wiggler free-electron laser at millimeter wavelengths. IEEE J. Quantum Electron. **QE-24**, 2309 (1988)
16. R.E. Shefer, G. Bekefi, Cyclotron emission from intense relativistic electron beams in uniform and rippled magnetic fields. Int. J. Electron. **51**, 569 (1981)
17. W.A. McMullin, G. Bekefi, Stimulated emission from relativistic electrons passing through a spatially periodic longitudinal magnetic field. Phys. Rev. A **25**, 1826 (1982)

Chapter 3
Incoherent Undulator Radiation

The spontaneous synchrotron radiation produced by individual electrons executing undulatory trajectories in a magnetostatic field is incoherent and is the radiation mechanism used in synchrotron light sources. The magnetostatic fields in these devices are formally identical to those employed in free-electron lasers but are commonly referred to as *undulators* rather than *wigglers*. The reason for this is that electron synchrotrons produce high-energy electron beams which permit the use of extremely long-period undulations. The use of long-period undulations makes possible the production of relatively large amplitude magnetostatic fields which are required to ensure the production of a relatively high radiation intensity from this incoherent mechanism. In contrast, since the free-electron laser relies on a coherent emission process, the wiggler magnets employed can be of shorter periods and lower amplitudes. However, incoherent synchrotron radiation is produced in free-electron lasers as well [1–5].

In this chapter, we present a derivation of the spontaneous undulator radiation emitted as individual electrons propagate through the wiggler. However, it should be remarked that this is only one part of the process. The spontaneously emitted photons can stimulate the emission of more photons or be reabsorbed. The complete physics must include both the spontaneous and stimulated emission mechanisms. Aspects of the stimulated emission, including both the linear instability and nonlinear saturation of that instability, are presented in the succeeding chapters.

3.1 Test Particle Formulation

The time-averaged radiated power by this incoherent process can be calculated from Poynting's Theorem by means of the equation

H. P. Freund, T. M. Antonsen, Jr., *Principles of Free Electron Lasers*,
https://doi.org/10.1007/978-3-031-40945-5_3

$$P = -\lim_{T\to\infty}\frac{1}{T}\int_{-T/2}^{T/2} dt \int d^3x\, \mathrm{E}(\mathrm{x},t)\cdot \mathrm{J}(\mathrm{x},t), \tag{3.1}$$

where $\mathbf{E}(\mathbf{x},t)$ denotes the microscopic radiation field, and

$$\mathrm{J}(\mathrm{x},t) = -e\sum_{j=1}^{N_e} \mathrm{v}_j(t)\delta[\mathrm{x}-\mathrm{x}_j(t)], \tag{3.2}$$

is the microscopic source current consisting of the sum of all the N_e electrons in the beam during the time $-T/2 < t < T/2$. The radiated power can be expressed in terms of the Fourier amplitudes of the microscopic fields and source currents by noting that for frequency ω and wavenumber $\mathbf{k}$

$$P = -2(2\pi)^4\lim_{T\to\infty}\frac{1}{T}\int_{-T/2}^{T/2} d\omega \int d^3k\, \mathrm{Re}\left[\mathrm{E}_{\mathrm{k},\omega}\cdot \mathrm{J}^*_{\mathrm{k},\omega}\right], \tag{3.3}$$

where the asterisk (*) denotes the complex conjugate and the Fourier transforms are defined as

$$f_{\mathrm{k},\omega} = \int_{-\infty}^{\infty} d\omega \int d^3k\, \exp(i\omega t - i\mathrm{k}\cdot\mathrm{x}) f(\mathrm{x},t), \tag{3.4}$$

A self-consistent relation between the fields and the source current depends upon the dielectric properties of the beam and can be expressed in the form

$$\Lambda_{\mathrm{k},\omega}\cdot \mathrm{E}_{\mathrm{k},\omega} = \frac{4\pi i}{\omega}\mathrm{J}_{\mathrm{k},\omega}, \tag{3.5}$$

where $\boldsymbol{\Lambda}_{\mathbf{k},\omega}$ denotes the dispersion tensor

$$\Lambda_{\mathrm{k},\omega} = \frac{c^2}{\omega^2}\left(\mathrm{kk} - k^2\mathrm{I}\right) + {}_{\mathrm{k},\omega}, \tag{3.6}$$

$\mathbf{I}$ denotes the unit dyadic, and $\boldsymbol{\varepsilon}_{\mathbf{k},\omega}$ is the dielectric tensor of the beam. For the case of a diffuse beam in which the wave frequency greatly exceeds the electron plasma frequency, $\boldsymbol{\varepsilon}_{\mathbf{k},\omega} \cong \mathbf{I}$ and Eq. (3.5) can be inverted to give

$$\mathrm{E}_{\mathrm{k},\omega} = -\frac{4\pi i\omega}{\omega^2 - c^2k^2}\left[\mathrm{J}_{\mathrm{k},\omega} - \frac{c^2}{\omega^2}\mathrm{kk}\cdot \mathrm{J}_{\mathrm{k},\omega}\right]. \tag{3.7}$$

The power radiated per unit frequency, volume V, and solid angle subtended by **k** is referred to as the *emissivity* and can be expressed as

$$\eta(\omega,\Omega_{\mathrm{k}})\equiv\frac{1}{V}\frac{dP}{d\omega d\Omega_{\mathrm{k}}}=(2\pi)^6\frac{\omega^2}{Vc^3}\lim_{T\to\infty}\frac{1}{T}\left[\left|\mathrm{J}_{\mathrm{k},\omega}\right|^2-\frac{1}{k^2}\left|\mathrm{k}\cdot\mathrm{J}_{\mathrm{k},\omega}\right|^2\right], \tag{3.8}$$

where V is the total volume of the interaction region and $\Omega_{\mathbf{k}}$ is the solid angle subtended by **k**. The radiation spectrum, therefore, can be determined from (3.8) based upon a knowledge of the single-particle orbits which is necessary to specify the source current. The Fourier transform of the source current is

$$\mathrm{J}_{\mathrm{k},\omega}=-\frac{e}{2\pi^4}\sum_{j=1}^{N_e}\lim_{T\to\infty}\int_{-T/2}^{T/2}dt\,\mathrm{v}_j(t)\exp\left[i\omega t-i\mathrm{k}\cdot\mathrm{x}_j(t)\right], \tag{3.9}$$

and the electron trajectories must now be specified.

Let us now consider the case of a helical wiggler configuration in the one-dimensional limit. In this limit, the electron trajectories are given approximately by the steady-state orbits,

$$\mathrm{v}=v_w\left[\hat{\boldsymbol{e}}_x\cos k_wz+\hat{\boldsymbol{e}}_y\sin k_wz\right]+v_\parallel\hat{\boldsymbol{e}}_z, \tag{3.10}$$

and

$$\mathrm{x}=\mathrm{x}_0+\frac{v_w}{k_wv_\parallel}\left[\hat{\boldsymbol{e}}_x(\sin k_wz-\sin k_wz_0)-\hat{\boldsymbol{e}}_y(\cos k_wz-\cos k_wz_0)\right]+v_\parallel t\hat{\boldsymbol{e}}_z, \tag{3.11}$$

where v_w and $v_\parallel$ are the transverse and axial velocities and $\mathbf{x}_0$ denotes the initial position. In view of the Bessel function identity

$$\exp(ib\sin\theta)=\sum_{n=-\infty}^{\infty}J_n(b)\exp(in\theta), \tag{3.12}$$

where J_n is the regular Bessel function of the first kind, it is apparent that

$$\mathrm{v}_j(t)\exp\left(i\omega t-i\mathrm{k}\cdot\mathrm{x}_j\right)=\mathrm{ex}\left(-i\mathrm{k}\cdot\mathrm{x}_{0j}\right)\sum_{n=-\infty}^{\infty}\mathrm{V}_n(b_j)\exp\left[i\omega t-i\left(k_\parallel+nk_w\right)v_\parallel t\right]$$
$$\times\exp\left[ib_j\sin(k_wz_0-\theta)-in(k_wz_0-\theta)\right], \tag{3.13}$$

where $b_j=k_\perp v_{jw}/k_wv_{j\parallel}$, the wavenumber is $\mathbf{k}=k_\perp[\hat{\boldsymbol{e}}_x\cos\varphi+\hat{\boldsymbol{e}}_y\sin\varphi]+k_\parallel\hat{\boldsymbol{e}}_z$, and

$$\mathrm{V}_n \equiv \frac{nk_w v_{\|}}{k_{\perp}} J_n(b)\hat{e}_{\perp} + iv_w J_n{}'(b)\hat{e}_{\varphi} + v_{\|} J_n(b)\hat{e}_z, \tag{3.14}$$

in the basis $\hat{\boldsymbol{e}}_{\perp} = \hat{\boldsymbol{e}}_x \cos\varphi + \hat{\boldsymbol{e}}_y \sin\varphi$ and $\hat{\boldsymbol{e}}_{\varphi} = -\,\hat{\boldsymbol{e}}_x \sin\varphi + \hat{\boldsymbol{e}}_y \cos\varphi$. As a result, the source current is

$$\begin{aligned} \mathrm{J}_{\mathrm{k},\omega} = & -\frac{e}{(2\pi)^4} \lim_{T\to\infty} \sum_{j=1}^{N_e} \exp\left(-i\mathrm{k}\cdot \mathrm{x}_{0j}\right) \sum_{n=-\infty}^{\infty} \mathrm{V}_n(b_j) \frac{\sin\left(\Delta\omega_{jn}T/2\right)}{\Delta\omega_{jn}T/2} \\ & \times \exp\left[ib_j \sin(k_w z_0 - \theta) - in(k_w z_0 - \theta)\right], \end{aligned} \tag{3.15}$$

where $\Delta\omega_{jn} = \omega - (k_{\|} + nk_w)v_{j\|}$ defines the frequency mismatch parameter.

In the computation of the quadratic forms which appear in the emissivity (3.8), we impose a random phase approximation and obtain

$$\begin{aligned} \eta(\omega, \Omega_{\mathrm{k}}) = & \frac{e^2\omega^2}{2\pi^2 V c^3} \lim_{T\to\infty} \sum_{j=1}^{N_e} \sum_{n=-\infty}^{\infty} \left[\left|\mathrm{V}_n(b_j)\right|^2 - \frac{1}{k^2}\left|\mathrm{k}\cdot \mathrm{V}_n(b_j)\right|^2\right] \\ & \times \left.\frac{\sin^2\left(\Delta\omega_{jn}T/2\right)}{\Delta\omega_{jn}^2 T/2}\right|_{k=\omega/c}. \end{aligned} \tag{3.16}$$

Observe that this expression describes emission at each harmonic of the wiggler resonance.

We now convert the discrete sum over individual electrons into a continuous integral over a six-dimensional beam distribution function $F_b(\mathbf{x},\mathbf{p})$ by making the replacement

$$\frac{1}{V}\sum_{j=1}^{N_e} \to n_b \iiint d^3x \iiint d^3p F_b(\mathrm{x}, \mathrm{p}), \tag{3.17}$$

subject to the normalization condition that

$$\iiint d^3x \iiint d^3p F_b(\mathrm{x}, \mathrm{p}) = 1, \tag{3.18}$$

and where n_b is the electron beam density. The above expression is perfectly general; however, in what follows, we take a one-dimensional limit of a uniform spatial distribution. In this case, we observe that specifying the total energy for our choice of wiggler field also specifies $v_{\|}$ and v_w so that the distribution collapses to a one-dimensional integral over p. Therefore, the emissivity can be written as

$$\eta(\omega,\theta) = \frac{e^2 n_b \omega^2}{2\pi^2 c^3} \lim_{T\longrightarrow\infty} \sum_{n=-\infty}^{\infty} \int_0^{\infty} dp v_w^2 F_b(p) \left[\left(\cos\theta - \frac{k}{nk_w} \sin^2\theta \right)^2 \frac{n^2}{b^2} J_n^2(b) + J_n'^2(b) \right] \frac{\sin^2(\Delta\omega_n T/2)}{(\Delta\omega_n^2 T/2)} \bigg|_{k=\omega/c}, \tag{3.19}$$

upon substitution for $\mathbf{V}_n(b)$. This describes the total emissivity due to contributions from all harmonics. Observe that the emissivity is independent of the angle φ [$= \tan^{-1}(k_y/k_x)$], and the emission is azimuthally symmetric. It should be remarked that in the conversion of the discrete sum over electrons to the continuous integral over the beam distribution, we have made use of the fact that the electron trajectories have been constrained to follow the steady-state orbits. In this case, apart from the field information, only one parameter [i.e., the total momentum or energy] is required to specify v_w and $v_\parallel$ for the orbit. In addition, we have not restricted steady-state trajectories to the idealized one-dimensional form. Equations (3.10) and (3.11) are valid for either the one-dimensional or three-dimensional solutions subject to the inclusion of the guiding-center motion in $\mathbf{x}_0$ in the realizable case. This inclusion does not affect the emissivity (3.19) since the guiding-center oscillation occurs over a much longer period than the emitted radiation.

The situation is somewhat different for the case of a planar wiggler field since the azimuthal symmetry present for the helical wiggler geometry does not exist. Using the quasi-steady-state orbits described in Sect. 2.2.1 to treat the trajectories in the presence of both a planar wiggler and a solenoidal magnetic field, we find that the emissivity is of the form

$$\begin{aligned}\eta(\omega,\theta,\varphi) = {} & \frac{e^2 n_b \omega^2}{2\pi^2 c^3} \lim_{T\to\infty} \sum_{l,n=-\infty}^{\infty} \int_0^{\infty} dp_\parallel \alpha_x^2 F_b(p_\parallel) J_l^2(b_y) \\ & \times \left\{ \left[\sin^2\varphi + \cos^2\varphi \left(\cos\theta - \frac{k}{nk_w} \sin^2\theta \right)^2 \right] \frac{n^2}{b_x^2} J_n^2(b_x) \right. \\ & \left. + \left[\sin^2\varphi \cos^2\theta + \cos^2\varphi \right] \frac{\alpha_y^2}{\alpha_x^2} J_n'^2(b_x) \right\} \frac{\sin^2(\Delta\omega_{n-l} T/2)}{(\Delta\omega_{n-l}^2 T/2)} \bigg|_{k=\omega/c},\end{aligned} \tag{3.20}$$

where $b_x = k_x \alpha_x / k_w v_\parallel$, $b_y = k_y \alpha_y / k_w v_\parallel$, and we have employed the quasi-steady-state trajectories given in Eqs. (2.111)

$$\mathrm{v} = \alpha_x \hat{\mathbf{e}}_x \cos k_w z + \alpha_y \hat{\mathbf{e}}_y \sin k_w z + v_\parallel \hat{\mathbf{e}}_z. \tag{3.21}$$

It should be remarked that the oscillatory component of the axial velocity has been neglected in the derivation of Eq. (3.20). This is valid under the assumption of the quasi-steady-state trajectories subject to the requirement that $\alpha_x^2 << 1$ and $\alpha_y^2 << 1$. In this case, the emissivity in the presence of an axial solenoidal magnetic field does depend upon the angle φ. This means that the spontaneous emission from an electron beam propagating through a planar wiggler is not azimuthally symmetric as in the case of a helical wiggler. This polarization in the direction of propagation reaches its most extreme limit when the solenoidal magnetic field vanishes, at which point $\alpha_y = 0$ and only the $l = 0$ term in the summation survives.

3.2 The Cold-Beam Regime

The cold (i.e., monoenergetic)-beam limit is obtained in the regime in which the width of the frequency spectrum is determined primarily by the bounded nature of the system (i.e., the axial length $L = v_\| T$ of the interaction region) rather than the thermal spread of the beam. This condition can be expressed as

$$\frac{\Delta v_\|}{v_\|} << \frac{\lambda}{L}, \tag{3.22}$$

where $\Delta v_\|$ denotes the axial velocity spread, and $\gamma_\| = (1 - v_\|^2/c^2)^{-1/2}$. In this regime, the distribution can be written as $F_b(p) = \delta(p - p_0)$, where p_0 is the bulk momentum of the beam, and the emissivity in the case of a helical wiggler becomes

$$\eta(\omega,\theta) = \frac{e^2 n_b \omega^2 v_w^2 L}{4\pi^2 c^3 v_\|} \sum_{n=-\infty}^{\infty} \left[\left(\cos\theta - \frac{k}{nk_w} \sin^2\theta \right)^2 \frac{n^2}{b^2} J_n^2(b) + J_n'^2(b) \right] \frac{\sin^2\Theta_n}{\Theta_n^2}, \tag{3.23}$$

where θ [$= \tan^{-1}(k_\perp/k_\|)$] denotes the polar angle between the wavevector and the symmetry axis, and $\Theta_n = \Delta\omega_n L/2v_\|$. Observe that Eq. (3.23) is to be evaluated for the steady-state solutions of v_w and $v_\|$ corresponding to a total energy given by $\gamma_0 = (1 + p_0^2/m_e^2 c^2)^{1/2}$.

The emission band at each harmonic is determined by the spectral function $\sin^2 \Theta_n/\Theta_n^2$, which is plotted in Fig. 3.1 for general n. The spontaneous emission spectrum peaks in the limit of a vanishing frequency mismatch for which $\omega = (k_\| + nk_w)v_\|$. The bandwidth is given by the FWHM points of the spectral function which occur at $\Theta \cong \pm 1.4$; hence

$$\frac{\Delta\omega}{\omega} < \frac{\lambda_w}{L}, \tag{3.24}$$

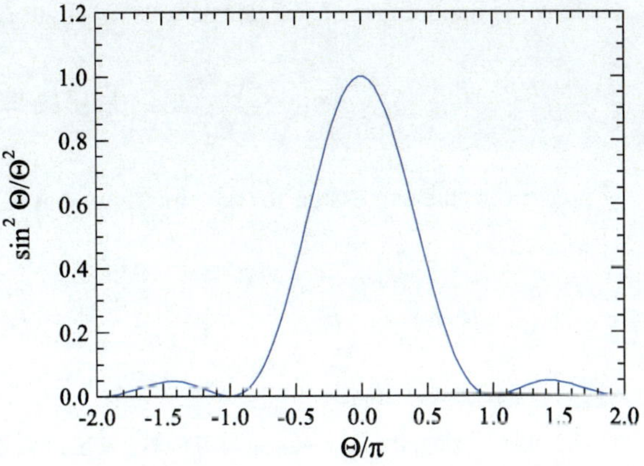

Fig. 3.1 Graph of the spectral function for the spontaneous radiation from a cold beam

where the center frequency

$$\omega = \frac{nk_w v_\parallel}{1-\beta_\parallel \cos\theta}, \tag{3.25}$$

and $\beta_\parallel = v_\parallel/c$. This reduces to the more familiar form $\omega \cong 2{\gamma_\parallel}^2 k_w v_\parallel$ for the fundamental resonance frequency for parallel propagation in the highly relativistic limit (i.e., $v_\parallel \cong c$).

The emission is symmetric about the axis of symmetry and is emitted in the forward direction where, typically, $k_\perp \ll k_\parallel$. Since it is also usually the case that $v_w \ll v_\parallel$, we expect that $b \ll 1$ so that we can apply a small argument expansion of the Bessel functions. Therefore, the emissivity in the neighborhood of the fundamental resonance ($n = 1$) is given approximately by

$$\eta(\omega,\theta) \approx \frac{e^2 n_b \omega^2 L}{16\pi^2 c v_\parallel}\frac{v_w^2}{c^2}\sum_{n=-\infty}^{\infty}\left[1+\left(\cos\theta - \frac{k_{\rm res}}{k_w}\sin^2\theta\right)^2\right]\frac{\sin^2\Theta_1}{\Theta_1^2}, \tag{3.26}$$

where $L\ (= Tv_\parallel)$ is the undulator length,

$$\Theta_1 = \frac{L}{2v_\parallel}\left[\omega - (k_\parallel + k_w)v_\parallel\right] \approx \frac{k_w L}{2\omega_{\rm res}}(\omega - \omega_{res}), \tag{3.27}$$

and where $\omega_{res} \approx 2{\gamma_\parallel}^2 k_w c$ and $k_{\rm res} \approx 2{\gamma_\parallel}^2 k_w$ are the resonant frequency and wavenumber. We can now define

$$\Delta\omega = \frac{2\pi\omega_{\rm res}}{k_w L} = \frac{\omega_{\rm res}}{N_w}, \tag{3.28}$$

where N_w is the number of wiggler periods in length L, so that

$$\sin c^2\Theta_1 \equiv \frac{\sin^2\Theta_1}{\Theta_1^2} = \sin c^2\left(\pi\frac{\omega-\omega_{\rm res}}{\omega_{\rm res}}\right). \tag{3.29}$$

As a result, the expression for the emissivity can be rewritten as

$$\eta(\omega,\theta) \approx \frac{1}{16\pi^2}\frac{e^2 n_b v_\parallel}{\gamma m_e c^3}\frac{v_w^2}{c^2}(\gamma m_e c^2)\left[1+\left(\cos\theta-\frac{k_{\rm res}}{k_w}\sin^2\theta\right)^2\right]\frac{\omega^2 L}{c^2}\sin c^2\Theta_1. \tag{3.30}$$

The current density $J_b = en_b v_\parallel = I_b/\pi R_b^2$, where I_b denotes the beam current and R_b is the beam radius. As a result

$$\frac{e^2 n_b v_\parallel}{m_e c^3} = \frac{1}{\pi R_b^2}\frac{I_b}{I_A}, \tag{3.31}$$

where $I_A = m_e c^3/e \approx 17{,}045$ amperes is the Alfvén current. Defining the beam energy $E_b = \gamma m_e c^2$, we write the emissivity as

$$\eta(\omega,\theta) \approx \rho^3 E_b \frac{c}{\lambda_{\rm res}}\frac{\omega_{\rm res}^2}{\lambda_w c^2}\left[1+\left(\cos\theta-\frac{k_{\rm res}}{k_w}\sin^2\theta\right)^2\right]\frac{\omega^2}{\omega_{res}^2}\frac{2}{\Delta\omega}\sin c^2\Theta_1, \tag{3.32}$$

where

$$\rho = \frac{1}{2\gamma}\left(\frac{I_b}{I_A}\frac{1}{k_w^2 R_b^2}\frac{v_w^2}{c^2}\right)^{1/3} \tag{3.33}$$

is the Pierce parameter.

The spectral width of the spontaneous emission is determined by the *sinc* function which varies as Θ_1 as shown in Fig. 3.1. Observe that the principal peak is found over the range $-\pi \le \Theta_1 \le \pi$ that corresponds to $\omega_{\rm res} - \Delta\omega \le \omega \le \omega_{\rm res} + \Delta\omega$. Integrating the emissivity over this range of frequencies, we note that

$$\int_0^\infty d\omega\omega^2\frac{\sin^2\Theta_1}{\Theta_1^2} \cong \int_{\omega_{res}-\Delta\omega}^{\omega_{\rm res}+\Delta\omega} d\omega\omega^2\frac{\sin^2\Theta_1}{\Theta_1^2} \cong \frac{\Delta\omega}{\pi}\omega_{\rm res}^2\int_{-\pi}^{\pi} d\varpi\frac{\sin^2\varpi}{\varpi^2}$$
$$\approx 2.836\frac{\Delta\omega}{\pi}\omega_{\rm res}^2. \tag{3.34}$$

Using this expression, we can write the emissivity in the form

$$\int_0^\infty d\omega\omega^2\eta(\omega,\theta) \approx 1.8\rho^3 E_b \frac{c}{\lambda_{\rm res}}\frac{\omega_{\rm res}^2}{\lambda_w c^2}\left[1+\left(\cos\theta-\frac{k_{\rm res}}{k_w}\sin^2\theta\right)^2\right]. \tag{3.35}$$

The radiation is focused within small angles about the forward direction. Expanding for small angles θ, $\cos\theta \cong 1-\theta^2/2$ and $\sin\theta \cong \theta-\theta^3/6$, so that

$$1+\left(\cos\theta-\frac{k_{\rm res}}{k_w}\sin^2\theta\right)^2 \cong 2-\frac{\theta^2}{\theta_{\max}^2}. \tag{3.36}$$

where

$$\theta_{\max}^2 \equiv \frac{1}{1+\frac{2k_{\rm res}}{k_w}} \approx \frac{k_w}{2k_{\rm res}} \approx \frac{1}{4\gamma_\parallel^2}. \tag{3.37}$$

This angle decreases as $\theta_{\max}$ increases and reaches unity when $\theta = \theta_{\max}$, so that forward emission is within a cone with an opening half-angle of $\theta_{\max} = \gamma_\parallel/2$. As a result, integration of the emissivity over frequency and the polar and azimuthal angles (forward emission) yields the power per unit volume

$$\frac{P}{V} = \iint d\Omega_{\rm k}\int_0^\infty d\omega\eta(\omega,\Omega_{\rm k}) \cong 1.8\rho^3 E_b \frac{c}{\lambda_{\rm res}}\frac{\omega_{\rm res}^2}{\lambda_w c^2}\int_0^{2\pi} d\varphi\int_0^{\theta_{\max}} d\theta\theta\left(2-\frac{\theta^2}{\theta_{\max}^2}\right), \tag{3.38}$$

so that

$$P \cong 0.68\pi\rho^3 E_b V \frac{c}{\lambda_{\rm res}}\frac{\omega_{\rm res}^2}{\lambda_w \gamma_\parallel^2 c^2}, \tag{3.39}$$

which describes the power radiated from a volume V into the forward cone ($\theta \le \theta_{\max}$). Observe that $\rho^3 V$ is proportional to the number of radiating electrons, N_e, as expected. The emission is over a frequency band of

$$\frac{\delta\omega}{\omega_{res}} = \frac{2}{N_w}. \tag{3.40}$$

3.3 The Temperature-Dominated Regime

The spectral band of the incoherent radiation is dominated by the thermal spread of the beam in the limit in which

$$\frac{\Delta v_{\parallel}}{v_{\parallel}} > \frac{\lambda}{L}. \tag{3.41}$$

In this regime, we observe the δ-function identity

$$\lim_{T\to\infty} \frac{\sin^2(xT/2)}{x^2T/2} = \pi\delta(x), \tag{3.42}$$

and the emissivity for a helical wiggler configuration is of the form,

$$\begin{aligned}\eta(\omega,\theta) = \frac{e^2 n_b\omega^2}{2\pi c^3} \sum_{n=-\infty}^{\infty} \int_0^{\infty} dp_{\parallel} v_{\parallel}^2 F_b\left(p_{\parallel}\right)\delta\left[\omega - \left(k_{\parallel} + nk_w\right)v_{\parallel}\right] \\ \left[\left(\cos\theta - \frac{k}{nk_w}\sin^2\theta\right)^2 \frac{n^2}{b^2} J_n^2(b) + J_n'^2(b)\right]_{k=\omega/c},\end{aligned} \tag{3.43}$$

As in the cold-beam limit, only the fundamental contribution ($n = 1$) survives for forward-directed emission and the emissivity reduces to

$$\eta(\omega,\theta=0) = \frac{e^2 n_b\omega^2}{4\pi c^3} \sum_{n=-\infty}^{\infty} \int_0^{\infty} dp_{\parallel} v_{\parallel}^2 F_b\left(p_{\parallel}\right)\delta\left[\omega - \left(k_{\parallel} + nk_w\right)v_{\parallel}\right], \tag{3.44}$$

which varies as the square of the wiggler-induced transverse velocity.

The spectral bandwidth of the emission is determined by the thermal spread of the electron beam. For illustrative purposes, we restrict the analysis to the case of forward propagation and assume an electron beam which has a Gaussian temperature spread

$$F_b\left(p_{\parallel}\right) = \frac{2\exp\left[-\left(p_{\parallel} - p_0\right)^2/\Delta p_{\parallel}^2\right]}{\sqrt{\pi}\Delta p_{\parallel}\left[1 + \operatorname{erf}\left(p_0/\Delta p_{\parallel}\right)\right]}, \tag{3.45}$$

where p_0 and $\Delta p_{\parallel}$ describe the bulk streaming momentum and the momentum spread, respectively, and

$$\mathrm{erf}(x) = \frac{2}{\sqrt{\pi}} \int_0^x dt \exp(-t^2), \tag{3.46}$$

denotes the error function. The emission occurs at a resonant axial phase velocity

$$\frac{v_{ph}}{c} = \frac{\omega}{\omega + ck_w}, \tag{3.47}$$

which is associated with a transverse velocity

$$v_w(v_{ph}) = \frac{\Omega_w v_{ph}}{\Omega_0 - k_w v_{ph}}, \tag{3.48}$$

where $\Omega_{0,w} = eB_{0,w}/\gamma m_e c$. This variation in the axial and transverse velocities with frequency corresponds to a variation in the energy which satisfies the equation

$$v_{ph}^2 + v_w^2(v_{ph}) = 1 - \frac{c^2}{\gamma^2}. \tag{3.49}$$

This defines a quartic polynomial equation for γ as a function of frequency which specifies the Group I and Group II trajectories. As a consequence, the emissivity for this choice of distribution is

$$\eta(\omega, \theta = 0) = \frac{e^2 n_b \omega}{2\pi c} \frac{p_{ph}}{\Delta p_\parallel} \frac{v_{ph}^2(v_{ph})}{c^2} \frac{\left[\gamma_{ph}^2 \beta_{ph}^2 + \Phi(v_{ph})\right]}{\Phi(v_{ph})} \frac{\exp\left[-(p_{ph} - p_0)^2/\Delta p_\parallel^2\right]}{\sqrt{\pi}\left[1 + \mathrm{erf}\left(p_0/\Delta p_\parallel\right)\right]}, \tag{3.50}$$

where $\beta_{ph} = v_{ph}/c$, $p_{ph} = \gamma_{ph} m_e v_{ph}$ is the associated resonant momentum, $\gamma_{ph} = (1 - \beta_{ph}{}^2)^{-1/2}$, and Φ is defined in Eq. (2.24).

This expression for the emissivity simplifies considerably in the limit in which the solenoidal magnetic field vanishes. In this case, $\Phi = 1$,

$$v_w = -\frac{\Omega_w}{\gamma k_w}, \tag{3.51}$$

and the energy is given by

$$\gamma = \gamma_{ph}\sqrt{1 + K^2}. \tag{3.52}$$

where $K = eB_w/m_ec^2k_w$ is the wiggler strength parameter. The emissivity, therefore, becomes

$$\eta(\omega,\theta=0) = \frac{e^2 n_b \omega}{2\pi c}\frac{p_{ph}}{\Delta p_\parallel}\frac{K^2}{1+K^2}\frac{\exp\left[-\left(p_{ph}-p_0\right)^2/\Delta p_\parallel^2\right]}{\sqrt{\pi}\left[1+\text{erf}\left(p_0/\Delta p_\parallel\right)\right]}. \tag{3.53}$$

The frequency spectrum is governed by the Gaussian, and peak emission occurs for $p_{ph} = p_0$, which corresponds to the frequency

$$\omega_0 = \frac{k_w v_0}{1 - v_0/c}, \tag{3.54}$$

where the streaming velocity is $v_0 = p_0/\gamma_0 m_e$, and

$$\gamma = \sqrt{1 + K^2 + \frac{p_0^2}{m_e^2 c^2}}. \tag{3.55}$$

The width of the frequency spectrum, under the assumption that $\Delta p_\parallel << p_0$, yields

$$\frac{\Delta\omega}{\omega_0} = 2\gamma_\parallel^2 \frac{\Delta v_\parallel}{v_0}, \tag{3.56}$$

where $\gamma_\parallel = (1 - v_0^2/c^2)^{-1/2}$, and $\Delta v_\parallel = \gamma_0 \gamma_\parallel^2 m_e \Delta p_\parallel$. Observe that the criteria which define the cold-beam versus temperature-dominated regimes are obtained by comparison of the respective frequency bandwidths.

References

1. H. Motz, M. Nakamura, Radiation of an electron in an infinitely long waveguide. Ann. Phys. **7**, 84 (1959)
2. N.M. Kroll, Relativistic Synchrotron radiation in a medium and its implications for stimulated electromagnetic shock radiation, in *Physics of Quantum Electronics: Free-Electron Generators of Coherent Radiation*, ed. by S.F. Jacobs, H.S. Pilloff, M. Sargent, M.O. Scully, R. Spitzer, vol. 7, (Addison-Wesley, Reading, 1980), p. 355
3. J.M.J. Madey, Relationship between mean radiated energy, mean squared radiated energy and spontaneous power spectrum in a power series expansion of the equations of motion in a free-electron laser. Nuovo Cimento **50B**, 64 (1979)
4. H.P. Freund, P. Sprangle, D. Dillenburg, E.H. da Jornada, B. Liberman, R.S. Schneider, Coherent and incoherent radiation from free-electron lasers with an axial guide field. Phys. Rev. A **24**, 1965 (1981)
5. K.J. Kim, Characteristics of synchrotron radiation, in *The Physics of Particle Accelerators*, AIP Conf. Proceedings No. 184, ed. by M. Month, M. Dienes, (American Institute of Physics, New York, 1989), p. 565

Chapter 4
Coherent Emission: Linear Theory

In order to give rise to stimulated emission, it is necessary for the electron beam to respond in a collective manner to the radiation field and to form coherent bunches. This can occur when a light wave traverses an undulatory magnetic field such as a wiggler because the spatial variations of the wiggler and the electromagnetic wave combine to produce a beat wave, which is essentially an interference pattern. It is the interaction between the electrons and this beat wave which gives rise to the stimulated emission in free-electron lasers. In the case of a magnetostatic wiggler, this beat wave has the same frequency as the light wave, but its wavenumber is the sum of the wavenumbers of the electromagnetic and wiggler fields. As a result, the phase velocity of the beat wave is less than that of the electromagnetic wave, and it is called a *ponderomotive* wave. Since the ponderomotive wave propagates at less than the speed of light in vacuo, it can be in synchronism with electrons that are limited by that velocity.

Our purpose in this chapter is to give a detailed discussion of the free-electron laser as a linear gain medium as well as to provide a comprehensive derivation of the relevant formulae for the gain in various configurations in both the idealized one-dimensional and the realistic three-dimensional limits. To this end, we derive the expressions for the gain in both the low- and high-gain regimes. The low-gain regime is relevant to short-wavelength free-electron laser oscillators driven by high-energy but low current electron beams. In contrast, the results in the high (exponential)-gain regime are usually described in terms of a dispersion equation and are appropriate to free-electron laser amplifiers and SASE driven by intense relativistic electron beams. The linear theory of interaction has been extensively discussed in the literature. Early work concentrated upon the idealized one-dimensional regime [1–19] and dealt predominantly with helical wiggler configurations. Additional work included planar wiggler configurations as well and dealt with the effects of a beam thermal spread [20–22] and an axial solenoidal field [23–31], as well as realistic multidimensional models of the interaction [32–46].

H. P. Freund, T. M. Antonsen, Jr., *Principles of Free Electron Lasers*,
https://doi.org/10.1007/978-3-031-40945-5_4

In this chapter, we present a derivation and discussion of the linear gain for both helical and planar wiggler configurations. To that end, we first discuss the phase space dynamics and how they are described by the nonlinear pendulum equation in Sect. 4.1. The analysis of the linear gain in the idealized one-dimensional limit is presented in Sect. 4.2 based upon a solution of the Vlasov-Maxwell equations for both helical and planar wigglers. This is not the only technique used in the literature to obtain the linear gain; however, it has the advantage of providing a unified formalism which can treat both configurations of interest in the low- and high-gain regimes, as well as the inclusion of the effects of an electron beam thermal spread. An analysis of the linear stability in three dimensions is given in Sect. 4.3.

4.1 Phase Space Dynamics and the Pendulum Equation

The interaction between the electron beam and the ponderomotive wave is governed by a nonlinear pendulum equation. In order to derive the dynamical equations, we consider the Lorentz force equations of an electron subject to idealized helical wiggler field (i.e., Eq. (2.5)), a uniform axial solenoidal field ($=B_0\hat{\mathbf{e}}_z$), and to a circularly polarized plane wave of frequency ω and wavenumber k given by the vector potential

$$\delta\mathbf{A}(z,t) = \delta A\left[\hat{\mathbf{e}}_x\cos(kz-\omega t) - \hat{\mathbf{e}}_y\sin(kz-\omega t)\right], \tag{4.1}$$

where δA denotes the amplitude. As a result, the orbit equations in the wiggler frame are of the form

$$\frac{d}{dt}\upsilon_1 = -(\Omega_0 - k_w\upsilon_3)\upsilon_2 + \frac{e\delta A}{\gamma m_e c}\left[\left(\omega - \omega\frac{\upsilon_1^2}{c^2} - k\upsilon_3\right)\sin\psi - \omega\frac{\upsilon_1\upsilon_2}{c^2}\cos\psi\right], \tag{4.2}$$

$$\frac{d}{dt}\upsilon_2 = (\Omega_0 - k_w\upsilon_3)\upsilon_1 - \Omega_w\upsilon_3 + \frac{e\delta A}{\gamma m_e c}\left[\left(\omega - \omega\frac{\upsilon_2^2}{c^2} - k\upsilon_3\right)\cos\psi - \omega\frac{\upsilon_1\upsilon_2}{c^2}\sin\psi\right], \tag{4.3}$$

$$\frac{d}{dt}\upsilon_3 = \Omega_w\upsilon_2 + \frac{e\delta A}{\gamma m_e c}\left(k - \omega\frac{\upsilon_3}{c^2}\right)(\upsilon_1\sin\psi + \upsilon_2\cos\psi), \tag{4.4}$$

and

$$\frac{d\gamma}{dt} = \frac{\omega}{c}\frac{e\delta A}{\gamma m_e c^2}(\upsilon_1\sin\psi + \upsilon_2\cos\psi), \tag{4.5}$$

where **p** and **v** are the momentum and velocity, respectively, $\Omega_{0,w} = eB_{0,w}/\gamma m_e c$, γ is the relativistic factor, and $\psi = (k + k_w)z - \omega t$ is the ponderomotive phase.

Equations (4.2, 4.3, 4.4, and 4.5) are solved by perturbation about the steady-state trajectories to first order in the electromagnetic field δA. To this end, we expand $v_1 = v_w + \delta v_1$, $v_2 = \delta v_2$, $v_3 = v_\parallel + \delta v_3$, $\gamma = \gamma_0 + \delta\gamma$, where $(\gamma_0, v_w, v_\parallel)$ are the constants which characterize the steady-state trajectory. The first-order equations, therefore, are

$$\frac{d}{dt}\delta v_1 = -\left(\Omega_0 - k_w v_\parallel\right)\delta v_2 + \frac{e\delta A}{\gamma_0 m_e c}\left(\omega - \omega\frac{v_w^2}{c^2} - kv_\parallel\right)\sin\psi, \tag{4.6}$$

$$\frac{d}{dt}\delta v_2 = \left(\Omega_0 - k_w v_\parallel\right)\delta v_1 - \Omega_0\frac{v_w}{v_\parallel}\delta v_3 - k_w v_\parallel v_w\frac{\delta\gamma}{\gamma_0} + \frac{e\delta A}{\gamma_0 m_e c}\left(\omega - kv_\parallel\right)\cos\psi, \tag{4.7}$$

$$\frac{d}{dt}\delta v_3 - \Omega_w\delta v_2 + \frac{e\delta A}{\gamma_0 m_e c}\left(k - \omega\frac{v_\parallel}{c^2}\right)v_w\sin\psi, \tag{4.8}$$

and

$$\frac{d}{dt}\delta\gamma = \frac{\omega}{c}\frac{e\delta A}{\gamma_0 m_e c^2}v_w\sin\psi. \tag{4.9}$$

Differentiation of Eq. (4.7) allows us to eliminate δv_1, δv_3, and $\delta\gamma$; hence, we obtain the following second-order differential equation for δv_2

$$\left(\frac{d^2}{dt^2} + \Omega_r^2\right)\delta v_2 - \frac{e\delta A}{\gamma_0 m_e c}\left\{\Omega_0\left[\omega - \left(1 + \beta_w^2\right)kv_\parallel\right] + \left(\omega - kv_\parallel\right)\left[\omega - \left(k + 2k_w\right)v_\parallel\right]\right\}\sin\psi, \tag{4.10}$$

where Ω_r^2 is defined in Eq. (2.21). Observe that this procedure is formally identical to the orbital stability analysis described in Chap. 2 and that the homogeneous form of Eq. (4.10) is identical to Eq. (2.20). Substitution of the particular solution to Eq. (4.10) into Eq. (4.8) under the assumption that the electrons are in near resonance with the wave [i.e., $\omega \approx (k + k_w)v_\parallel$] yields

$$\frac{d}{dt}\delta v_3 = \frac{e\delta A}{\gamma_0\gamma_\parallel^2 m_e c}(k + k_w)v_w\Phi\sin\psi, \tag{4.11}$$

where $\gamma_\parallel^2 = (1 - v_\parallel^2/c^2)^{-1}$ and Φ is defined in Eq. (2.24).

The pendulum equation that governs the evolution of the ponderomotive phase follows readily from Eq. (4.11) since

$$\frac{d}{dt}\delta v_3 \cong \frac{v_\parallel^2}{k+k_w}\frac{d^2}{dz^2}\psi. \tag{4.12}$$

As a result, we find that

$$\frac{d^2}{dz^2}\psi = -\kappa^2 \sin\psi, \tag{4.13}$$

where

$$\kappa^2 \equiv -\frac{e\delta A}{\gamma_0\gamma_\parallel^2 m_e c v_\parallel}(k+k_w)^2\frac{v_w}{v_\parallel}\Phi \tag{4.14}$$

Observe that the pendulum constant is proportional to the product of the wiggler and electromagnetic field amplitudes. In the absence of the axial solenoidal field, the pendulum constant reduces to

$$\kappa^2 \equiv \frac{e\delta A}{\gamma_0^4 m_e v_\parallel^2}(k+k_w)^2 K\left(1+K^2\right) \cong 4\frac{e\delta A}{m_e c^2}\frac{K}{1+K^2}, \tag{4.15}$$

where K ($\equiv eB_w/m_e c^2 k_w$) is the wiggler strength parameter (Eq. (1.3)). The presence of the solenoidal field can enhance the magnitude of the pendulum constant. For the Group I trajectories (which includes the limit in which the solenoidal field vanishes), v_w is less than zero and increases in magnitude with B_0 as the magnetic resonance at $\Omega_0 \approx k_w v_\parallel$ is approached. In addition, Φ is greater than unity and exhibits a singularity at the transition to orbital instability (i.e., when $\Omega_r = 0$). As a consequence, the pendulum constant increases rapidly with the solenoidal field which gives rise to significant enhancements in the axial bunching of the electrons. The situation is similar for Group II orbits in the vicinity of the magnetic resonance. In this case, the phase of the bunching process is preserved since v_w is greater than zero, while Φ is less than zero. However, as the axial solenoidal field continues to increase, Φ vanishes when

$$\gamma_\parallel^2\Omega_0 = \gamma_0^2 k_w v_\parallel, \tag{4.16}$$

and the bunching process creases. At this point, the coherent interaction disappears. At still higher values of the axial solenoidal field, both v_w and Φ are positive, and the ponderomotive phase that controls the bunching process is shifted 180° in phase.

The pendulum equation can be reduced to

$$\frac{1}{2}\left(\frac{d\psi}{dz}\right)^2 + U(\psi) = H, \tag{4.17}$$

where H has the form of the Hamiltonian or total energy of the system and

$$U(\psi) = -\kappa^2 \cos\psi \tag{4.18}$$

is the ponderomotive potential. The electron trajectories through the wiggler, therefore, may be expressed as

$$\frac{d\psi}{dz} = \pm\sqrt{2H + 2\kappa^2\cos\psi}. \tag{4.19}$$

Observe that the first derivative of the phase [i.e., $d\psi/dz = (k + k_w) - \omega/v_z$] is a measure of the electron streaming velocity; hence, this equation effectively describes the electron velocity as a function of the phase of the ponderomotive wave.

There are two classes of trajectory: trapped and untrapped. The *separatrix* describes the transition between trapped and untrapped orbits and occurs when $H = \kappa^2$. Hence, the separatrix is defined by a pair of curves in the phase space for $(d\psi/dz, \psi)$

$$\frac{d\psi}{dz} = \pm\begin{cases} 2\kappa\cos(\psi/2) & ;\kappa^2 > 0 \\ 2|\kappa|\sin(\psi/2) & ;\kappa^2 < 0 \end{cases}, \tag{4.20}$$

and free-streaming, untrapped orbits are characterized by $H > |\kappa^2|$ and occupies that region of the phase space outside of the separatrix

$$\left|\frac{d\psi}{dz}\right| > \begin{cases} 2\kappa\cos(\psi/2) & ;\kappa^2 > 0 \\ 2|\kappa|\sin(\psi/2) & ;\kappa^2 < 0 \end{cases}. \tag{4.21}$$

The trapped orbits are those for which $H < |\kappa^2|$ within the bounds of the separatrix. The free-streaming orbits correspond to the case in which the pendulum swings through the full 360° cycle. The electrons pass over the crests of many waves traveling fastest at the bottom of the troughs and slowest at the crests of the ponderomotive wave. In contrast, the electrons are confined within the trough of a single wave in the trapped orbits. This corresponds to the motion of a pendulum that does not rotate a full circle but is confined to oscillate about the lower equilibrium point.

The dynamical evolution of the electron phase space during the coherent emission process is illustrated in Fig. 4.1 and is one in which the pendulum constant evolves during the course of the interaction. We assume that the electrons are initially characterized by the same axial velocity and describe a horizontal line in phase

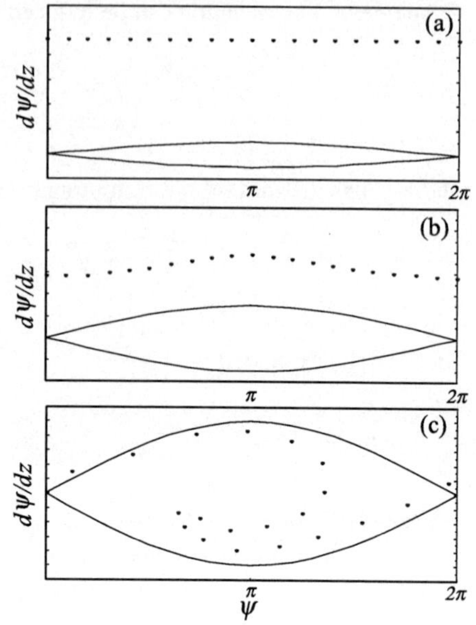

Fig. 4.1 Schematic of the electron phase space evolution in free-electron lasers. The electron beam is initially (**a**) monoenergetic. During the linear phase of the interaction (**b**), the wave grows in amplitude and the separatrix expands. The bulk of the beam executes trapped orbits at saturation (**c**). The phase space distribution shown represents an electron beam where the electrons that are still losing energy to the wave are balanced by those gaining energy from the wave

space. Since the amplification process has only just begun, the wave is of small amplitude, and the separatrix encloses a small area of phase space. The electrons lose energy as the wave is amplified; hence, the electrons decelerate, and both the pendulum constant and separatrix grow. During the linear phase of the interaction, as illustrated in Fig. 4.1b, the electrons have only begun to form bunches and remain on untrapped trajectories outside of the separatrix. Ultimately, the electrons cross the growing separatrix from untrapped to trapped orbits. The interaction saturates after the electrons have executed approximately half of an oscillation in the ponderomotive well. At this point, the electrons that are still losing energy to the wave are balanced by those electrons that are gaining energy at the expense of the wave. In this chapter, we deal with the amplification mechanism during the linear phase of the interaction.

4.2 Linear Stability in the Idealized Limit

In this section, we shall derive the linearized dispersion equation for the free-electron laser in the idealized one-dimensional representation within the context of a linearized Vlasov-Maxwell formalism. The first step in this process is the development of a general formalism that is applicable to both the helical and planar configurations. The Vlasov equation in the combined wiggler and axial magnetostatic fields and the electromagnetic fields is

$$\left[\frac{\partial}{\partial t}+\mathbf{v}\cdot\nabla-e\Big(\delta\mathbf{E}(z,t)+\frac{1}{c}\mathbf{v}\times[B_0\hat{\mathbf{e}}_z+\mathbf{B}_w(z)+\delta\mathbf{B}(z,t)]\Big)\cdot\nabla_{\mathbf{p}}\right]f_b(z,\mathbf{p},t)=0, \tag{4.22}$$

where $f_b(z,\mathbf{p},t)$ is the distribution function of the electron beam, $\delta\mathbf{E}(z,t)$ and $\delta\mathbf{B}(z,t)$ denote the fluctuating electric and magnetic fields of the wave, and

$$\nabla_{\mathbf{p}}=\hat{\mathbf{e}}_x\frac{\partial}{\partial p_x}+\hat{\mathbf{e}}_y\frac{\partial}{\partial p_x}+\hat{\mathbf{e}}_z\frac{\partial}{\partial p_z}. \tag{4.23}$$

The Vlasov equation is linearized by expanding the distribution in powers of the fluctuating fields. To this end, we write $f_b(z,\mathbf{p},t) = F_b(z,\mathbf{p}) + \delta f_b(z,\mathbf{p},t)$ where F_b and δf_b are the equilibrium and perturbed components of the distribution, and it is assumed that the perturbed distribution is of the order of the fluctuating fields and $|\delta f_b| << |F_b|$. The equilibrium distribution must satisfy the lowest-order Vlasov equation

$$\left[\frac{\partial}{\partial t}+\mathbf{v}\cdot\nabla-\frac{e}{c}\mathbf{v}\times[B_0\hat{\mathbf{e}}_z+\mathbf{B}_w(z)]\cdot\nabla_{\mathbf{p}}\right]F_b(z,\mathbf{p})=0. \tag{4.24}$$

This is satisfied for any equilibrium distribution that is a function of the constants of the motion. As discussed in Chap. 2, these constants are the total energy (or momentum) as well as the canonical momenta for both helical and planar wigglers in the one-dimensional representation; hence, we may express the equilibrium distribution in the form $F_b(z,\mathbf{p}) = F_b(P_x,P_y,p)$. It should be remarked, however, that the canonical momenta are only approximate constants of the motion in the presence of an axial solenoidal magnetic field. We shall discuss this in more detail later. Correct to first order in the fluctuation fields, the perturbed distribution satisfies

$$\left[\frac{\partial}{\partial t}+\mathbf{v}\cdot\nabla-\frac{e}{c}\mathbf{v}\times[B_0\hat{\mathbf{e}}_z+\mathbf{B}_w(z)]\cdot\nabla_{\mathbf{p}}\right]\delta f_b(z,\mathbf{p},t)=$$
$$=e\Big(\delta\mathbf{E}(z,t)+\frac{1}{c}\mathbf{v}\times\delta\mathbf{B}(z,t)\Big)\cdot\nabla_{\mathbf{p}}F_b(z,\mathbf{p}), \tag{4.25}$$

The perturbed Vlasov equation may be solved by the method of characteristics in which we integrate

$$\delta f_b(z,\mathbf{p},\tau(z))=e\int_0^z\frac{dz'}{v_z(z')}\left[\delta\mathbf{E}(z',\tau(z'))+\frac{1}{c}\mathbf{v}(z')\times\delta\mathbf{B}(z',\tau(z'))\right]\cdot\nabla_{\mathbf{p}}F_b(z,\mathbf{p}), \tag{4.26}$$

over the unperturbed trajectories under the assumption that the perturbations are negligibly small at time $t = 0$. Observe that we treat the case of spatial growth and have adopted Lagrangian coordinates in which $\mathbf{v}(z)$ denotes the unperturbed velocity of an electron as a function of the axial position, and

$$\tau(z) = t_0 + \int_0^z \frac{dz'}{v_z(z', t_0)} \tag{4.27}$$

represents the time it takes an electron to reach a particular axial position after crossing the $z = 0$ plane at time t_0.

The solution to the perturbed Vlasov equation is solved in conjunction with Maxwell's equations. We choose to deal with the scalar $\delta\varphi(z,t)$ and vector potentials $\delta\mathbf{A}_\perp(z,t)$ in the Coulomb gauge. Note that since we treat a one-dimensional model, the scalar and vector potentials describe plane waves. Hence, the vector potential represents a purely transverse electromagnetic wave. In terms of this representation, Maxwell's equations are

$$\left(\nabla^2 - \frac{1}{c^2}\frac{\partial^2}{\partial t^2}\right)\delta\mathbf{A}_\perp = -\frac{4\pi}{c}\delta\mathbf{J}_\perp, \tag{4.28}$$

and

$$\frac{\partial^2}{\partial t \partial z}\delta\varphi = 4\pi\delta J_z. \tag{4.29}$$

Observe that the scalar potential is described in terms of the z-component of Ampere's law rather than with Poisson's equation. The perturbed source current is given in terms of the perturbed distribution function as follows:

$$\delta J_\perp(z,t) = -\frac{e}{m_e}\int d^3p \frac{1}{\gamma}\mathbf{p}\delta f_b(z, \mathbf{p}, t). \tag{4.30}$$

The dispersion equation governing the growth and/or damping of the electromagnetic field is obtained by the simultaneous solution of Eqs. (4.26), (4.28), and (4.29).

4.2.1 *Helical Wiggler Configurations*

In treating a helical wiggler configuration, we shall include the effect of the axial solenoidal magnetic field in the interest of generality. However, the limit in which the axial field vanishes is evident and will be explicitly discussed. In addition, the discussion of the effect of beam thermal effects on the gain is restricted to the limit in which the axial field vanishes. The formulation itself follows that described in Refs. [9, 18]. As discussed in Chap. 2, the effect of the axial field is to enhance both the transverse wiggler-induced velocity and the gain of the free-electron laser when the solenoidal field is strong enough that $\Omega_0 \approx k_w v_{\|}$. However, it should be noted that the canonical momenta are only approximate constants of the motion in the presence of the axial field. Hence, we used Eqs. (2.42, 2.43, and 2.44) to describe the unperturbed trajectories and make the implicit assumption that $V_\perp << v_w$ $[V_\perp^2 = (P_x^2 + P_y^2)/\gamma^2 m_e^2]$. Within the context of this orbit representation, the canonical momenta describe the Larmor rotation of the electrons in the solenoidal field.

In treating the helical wiggler geometry, we begin with the application of the Vlasov-Maxwell formalism to this configuration under the assumption of a plane wave representation for the vector and scalar potentials. This general analysis provides the framework necessary to treat both the low- and exponential gain regimes. The gain in the low-gain regime is presented in Eq. (4.59), in which we have neglected the space-charge fields. This is usually appropriate since the low-gain regime is generally applicable to free-electron laser oscillators that are driven by high-energy but low current accelerators such as rf linacs, microtrons, and storage rings.

We subsequently treat the high-gain regime that applies to free-electron lasers driven by intense beam accelerators such as pulse-line accelerators and induction linacs. A derivation of the general dispersion equation including the effect of an axial magnetic field is given in the ideal beam limit in which thermal effects can be neglected (see Eqs. (4.76) and (4.83)). This includes both the high-gain Compton and collective Raman regimes. The effect of the axial magnetic field on the stability of the beam space-charge wave is discussed, as is the general effect of the axial magnetic field on the structure of the interacting electromagnetic modes (see Figs. 4.4 and 4.5). The transition between the high-gain Compton and collective Raman regimes as a function of the beam density is illustrated in Fig. 4.6, and the expressions for the gain in these two regimes are given in Eqs. (4.94) and (4.97), respectively, for the case in which the orbit parameter $\Phi > 0$ (see Eq. (2.24)). The case in which $\Phi < 0$ represents a negative-mass regime in which the axial velocity accelerates as the electrons lose energy. This regime is also discussed, and it is shown that the bandwidth of the interaction is substantially broadened in this regime.

The effect of an electron beam thermal spread is treated for a configuration in which the axial field is absent in the interest of simplicity. The analysis is presented under the assumption that the electron beam is monoenergetic but exhibits a spread in pitch angle. This is adequate to treat the effect of an axial energy [or momentum] spread. A distribution function that describes such a beam is given in Eqs. (4.113) and (4.114), and a general dispersion equation is derived. The general dispersion Eq. (4.128) describes the effect of the axial energy spread in terms of a generalized thermal function T (see Eq. (4.125)). It is shown that this function approaches unity in the limit in which the axial energy spread vanishes and the ideal beam limit is recovered.

4.2.1.1 The Source Currents

Under the assumption of plane wave solutions, the vector and scalar potentials for a wave with angular frequency ω are of the form

$$\delta \mathbf{A}_{\perp}(z,t) = \frac{1}{2}\delta \hat{\mathbf{A}}_{\perp}(z)\exp(-i\omega t) + c.c., \tag{4.31}$$

and

$$\delta\varphi(z,t) = \frac{1}{2}\delta\hat{\varphi}(z)\exp(i\omega t) + c.c.. \tag{4.32}$$

After transformation to the basis

$$\hat{\boldsymbol{e}}_{\pm} \equiv \frac{1}{2}(\hat{\boldsymbol{e}}_x \pm i\hat{\boldsymbol{e}}_y), \tag{4.33}$$

that is, convenient for the description of left- and right-hand circularly polarized electromagnetic waves, the perturbed distribution function can be written as

$$\delta f_b(z,\mathbf{p},\tau(z)) \equiv \delta\hat{f}_b(z,\mathbf{p})\exp(-i\omega\tau(z)) + c.c., \tag{4.34}$$

where

$$\delta\hat{f}_b(z,\mathbf{p}) = \frac{e}{2c}\left[D_+\left(\frac{\partial}{\partial P_x} + i\frac{\partial}{\partial P_x}\right) + D_-\left(\frac{\partial}{\partial P_x} - i\frac{\partial}{\partial P_x}\right) + D_z\frac{\partial}{\partial p}\right]F_b(P_x,P_y,p). \tag{4.35}$$

The orbit integrals in Eq. (4.35) are defined as

$$D_{\pm} \equiv \int_0^z dz' \frac{\exp(i\omega\tau(z,z'))}{v_z(z')}\exp(\pm i\Omega_0\tau(z'))\left(i\omega - v_z(z')\frac{\partial}{\partial z'}\right)\delta\hat{A}_{\pm}(z'), \tag{4.36}$$

and

$$D_z \equiv \frac{1}{p}\int_0^z dz' \frac{\exp(i\omega\tau(z,z'))}{v_z(z')}\left(-cp_z(z')\delta\hat{\varphi}(z') + i\omega\left[p_-(z')\delta\hat{A}_+(z') + p_+(z')\delta\hat{A}_-(z')\right]\right), \tag{4.37}$$

where $p_{\pm} = p_x \mp ip_y$, $\tau(z,z') = \tau(z) - \tau(z')$,

$$\delta\hat{A}_{\pm} \equiv \frac{1}{2}\left(\delta\hat{A}_x \mp i\delta\hat{A}_y\right) \tag{4.38}$$

denotes the amplitudes of the circularly polarized electromagnetic waves.

The source current

$$\delta\mathbf{J}(z,t) = \left[\delta\hat{J}_+(z)\hat{\boldsymbol{e}}_+ + \delta\hat{J}_-(z)\hat{\boldsymbol{e}}_- + \delta\hat{J}_z(z)\hat{\boldsymbol{e}}_z\right]\exp(-i\omega t) + c.c. \tag{4.39}$$

is determined by means of the perturbed distribution as follows:

$$\begin{pmatrix} \delta\hat{J}_{\pm}(z) \\ \delta\hat{J}_z(z) \end{pmatrix} = -\frac{e}{m_e}\int dP_x dP_y dp \frac{p}{\gamma p_z}\delta\hat{f}_b(z,\mathbf{p})\begin{pmatrix} p_{\pm} \\ p_z \end{pmatrix}. \tag{4.40}$$

Substitution of the solution for the perturbed distribution from Eq. (4.35), we obtain

$$\delta\hat{J}_{\pm}(z) = \frac{e^2}{2m_e c}\int dPdP_y dp\frac{p}{\gamma}\left[\left(\frac{\partial}{\partial P_x} + i\frac{\partial}{\partial P_y}\right)\left(\frac{p_{\pm}}{p_z}D_+\right)\right. \tag{4.41}$$
$$\left. + \left(\frac{\partial}{\partial P_x} - i\frac{\partial}{\partial P_y}\right)\left(\frac{p_{\pm}}{p_z}D_-\right) - \frac{p_{\pm}}{p_z}D_z\frac{\partial}{\partial p}\right]F_b(P_x,P_y,p),$$

and

$$\delta\hat{J}_z(z) = \frac{e^2}{2m_e c}\int dPdP_y dp\frac{p}{\gamma}\left[\left(\frac{\partial}{\partial P_x} + i\frac{\partial}{\partial P_y}\right)D_+ + \left(\frac{\partial}{\partial P_x} - i\frac{\partial}{\partial P_y}\right)D_-\right.$$
$$\left. - D_z\frac{\partial}{\partial p}\right]F_b(P_x,P,p). \tag{4.42}$$

The dispersion equation is obtained by substitution of the source current into the wave equations

$$\left(\frac{\partial^2}{\partial z^2} + \frac{\omega^2}{c^2}\right)\delta\hat{A}_{\pm}(z) = -\frac{4\pi}{c}\delta\hat{J}_{\pm}(z), \tag{4.43}$$

and

$$\frac{\partial}{\partial z}\delta\hat{\varphi}(z) = \frac{4\pi i}{\omega}\delta\hat{J}_z(z). \tag{4.44}$$

An ideal electron beam is both monoenergetic and has a vanishing pitch angle spread (i.e., $P_x = P_y = 0$). We first impose the condition that the pitch angle spread of the beam vanishes and assume an equilibrium distribution of the form

$$F_b(P_x, P_y p) n_b \delta(P_x)\delta(P_y) G_b(p), \tag{4.45}$$

where n_b denotes the ambient density of the electron beam, and the distribution in the total momentum satisfies the normalization condition

$$\int_0^\infty dp \frac{p}{p_z} G_b(p) = 1. \tag{4.46}$$

As a consequence, the source currents take the form

$$\delta\hat{J}_\pm(z) = \frac{\omega_b^2}{8\pi c} \int_0^\infty dp \frac{p}{\gamma p_z} \left[\exp[\mp i\Omega_0 \tau(z)] \left(2 + \frac{p_+ p_-}{p_z^2}\right) D_\pm + \exp[\pm i\Omega_0 \tau(z)] \frac{p_\pm^2}{p_z^2} \right.$$
$$\left. + p_\pm \left(\frac{\partial}{\partial P_x} + i \frac{\partial}{\partial P_x}\right) D_+ + p_\pm \left(\frac{\partial}{\partial P_x} - i \frac{\partial}{\partial P_x}\right) D_- - p_\pm D_z \frac{\partial}{\partial p} \right]_{P_x = P_y = 0} G_b(p), \tag{4.47}$$

and

$$\delta\hat{J}_z(z) = \frac{\omega_b^2}{8\pi c} \int_0^\infty dp \frac{p}{\gamma} \left[\left(\frac{\partial}{\partial P_x} + i \frac{\partial}{\partial P_x}\right) D_+ + \left(\frac{\partial}{\partial P_x} - i \frac{\partial}{\partial P_x}\right) D_- - D_z \frac{\partial}{\partial p} \right]_{P_x = P_y = 0} G_b(p), \tag{4.48}$$

where $\omega_b{}^2 = 4\pi e^2 n_b/m_e$ is the square of the electron beam-plasma frequency. If we require that $P_x = P_y = 0$, then $p_\pm = p_w \exp(\mp i k_w z)$, $\tau(z,z') = (z - z')/\upsilon_\parallel$,

$$D_\pm = -\exp[\pm i\Omega_0 \tau(z)] \Big\{ \delta\hat{A}_\pm(z) - \delta\hat{A}_\pm(0) \exp\left[i(\omega \mp \Omega_0) z/\upsilon_\parallel\right]$$
$$\mp i \frac{\Omega_0}{\upsilon_\parallel} \int_0^z dz' \delta\hat{A}_\pm(z') \exp[i(\omega \pm \Omega_0)\tau(z,z')] \Big\}, \tag{4.49}$$

$$D_z = \frac{\gamma m_e c}{p} \int_0^z dz' \exp[i\omega\tau(z,z')] \left[-\frac{\partial}{\partial z'} \delta\hat{\varphi}(z') + \frac{i\omega}{c} \left(\frac{p_-}{p_\parallel} \delta\hat{A}_+(z') + \frac{p_+}{p_\parallel} \delta\hat{A}_-(z') \right) \right], \tag{4.50}$$

and $p_{w,\|} = \gamma m_e v_{w,\|}$ is determined by the steady-state trajectories. In addition, we observe that

$$\begin{aligned}\left(\frac{\partial}{\partial P_x} \pm i\frac{\partial}{\partial P_x}\right)D_\pm\Big|_{P_= = P_= = 0} &= -\frac{p_w}{p_\|^2}\frac{\exp\left[\pm i\left(k_w - \Omega_0/v_\|\right)z\right] - 1}{k_w v_\| - \Omega_0} \\ &\times \left[\Omega_0 \delta\hat{A}_\pm(z)\exp\left(\pm i\Omega_0 z/v_\|\right) \mp \omega\delta\hat{A}_\pm(0)\exp\left(\pm i\omega z/v_\|\right)\right] \\ &\pm \frac{i\Omega_0}{v_\|}\frac{p_w}{p_\|^2}\exp\left(i\omega z/v_\|\right)\int_0^z dz'\delta\hat{A}_\pm(z')\exp\left[-i\left(\omega/v_\| - k_w\right)z'\right] \\ &\times \left\{1 \pm \frac{\omega}{k_w v_\| - \Omega_0}\left[\exp\left(\pm i\left(k_w - \Omega_0/v_\|\right)(z - z')\right) - 1\right]\right. \\ &\left. - \frac{\Omega_0}{k_w v_\| - \Omega_0}\left[\exp\left(\mp i\left(k_w - \Omega_0/v_\|\right)z'\right) - 1\right]\right\},\end{aligned} \tag{4.51}$$

The calculation of the linear dispersion requires the evaluation of the source currents in Eqs. (4.47) and (4.48). In the remainder of this section, we develop the solutions in both the low- and high-gain regimes.

4.2.1.2 The Pierce Parameter

The Pierce parameter, denoted by ρ for free-electron lasers, was originally formulated to characterize the interaction in traveling wave tubes [47] and has been adapted to deal with the traveling wave interaction in free-electron lasers. In the case of a helical wiggler, the Pierce parameter is defined as

$$\rho^3 = \frac{v_w^2}{16 v_\|^2}\frac{\omega_b^2}{\gamma_0 k_w^2 v_\|^2}\Phi, \tag{4.52}$$

where Φ is given in Eq. (2.24). This can be expressed in terms of the beam current, I_b, and the beam radius, R_b, as

$$\rho^3 \equiv \frac{v_w^2}{4 v_\|^2}\frac{1}{\gamma_0 k_w^2 R_b^2}\frac{I_b}{I_A}\Phi, \tag{4.53}$$

where I_A is the Alfvén current. Finally, noting that $v_w{}^2 = K^2 c^2/\gamma_0{}^2$ in the absence of an axial magnetic field, the Pierce parameter can also be written as

$$\rho^3 \equiv \frac{K^2}{4\gamma_0^3 \beta_\|^2}\frac{1}{k_w^2 R_b^2}\frac{I_b}{I_A}. \tag{4.54}$$

4.2.1.3 The Low-Gain Regime

The low-gain, tenuous-beam limit is relevant to free-electron laser configurations in which the electron beam current is low and the gain of the signal in a single pass through the wiggler is less than unity. In this regime, the beam-plasma frequency $\omega_b << \omega$ and collective effects due to beam space-charge waves are negligible. In addition, the coupling between the left- and right-hand circularly polarized electromagnetic mode scales with $(v_w/v_\parallel)^2$, which is assumed to be small in the low-gain limit. Hence, we can focus attention on either of these modes. We choose to consider the right-hand circularly polarized mode that can be represented in the low-gain regime as [9]

$$\delta\hat{A}_+(z) = \delta\hat{A}_+(z)\exp\left[i\int_0^z dz' k_+(z')\right], \tag{4.55}$$

where we assume that $|\mathrm{Im}\, k_+(z)| << |\mathrm{Re}\, k_+(z)|$. Under the additional assumption that $k_+(z) = k + \delta k_+(z)$, where k is independent of axial position and $|\delta k_+(z)| << k$, the dispersion equation for the real part of the wavenumber is

$$\frac{\omega^2}{c^2} - k^2 \cong 0, \tag{4.56}$$

and for the growth rate, we obtain

$$\begin{aligned}\mathrm{Im}k_+(z) \cong -\frac{\omega_b^2}{4c}\int_0^\infty dp\beta_w^2\frac{\sin\Delta kz}{\Delta k}&\left[m_e\frac{\partial}{\partial p}G_b(p)\right.\\&\left.+\frac{p}{\gamma p_\parallel}\frac{\Delta k}{k_w v_\parallel - \Omega_0}\left(1+\frac{\Omega_0\left[\left(k_w v_\parallel - \Omega_0\right)+c\Delta k\right]}{c\Delta k\left(\omega - \Omega_0 - k_w v_\parallel\right)}\right)G_b(p)\right],\end{aligned} \tag{4.57}$$

where

$$\Delta k = \frac{\omega}{v_\parallel} - k - k_w \tag{4.58}$$

is the detuning parameter and $\Omega_{0,w} = eB_{0,w}/\gamma m_e c$. Observe that $\Delta k = 0$ frequencies in the vicinity of the free-electron laser resonance $\omega = (k + k_w)v_\parallel$.

The gain in power over an interaction length L is defined as

$$G(L) \equiv -2\int_0^L dz\mathrm{Im}\delta k_+(z). \tag{4.59}$$

It should be remarked that it has been implicitly assumed in the derivation that the gain is weak and that $G(L) < 1$. Integration of the growth rate gives

$$G(L) \cong -\frac{\omega_b^2 L}{2c}\int_0^\infty dpG_b(p)\left[\frac{m_e L}{2}\frac{\partial}{\partial p}\left(\beta_w^2\frac{\sin^2\Theta}{\Theta^2}\right)\right.$$
$$\left. -\beta_w^2\frac{p}{\gamma p_\parallel}\frac{\sin^2\Theta}{\Theta^2}\frac{1}{k_w v_\parallel-\Omega_0}\left(1+\frac{\Omega_0\left[\left(k_w v_\parallel-\Omega_0\right)+c\Delta k\right]}{c\Delta k\left(\omega-\Omega_0-k_w v_\parallel\right)}\right)\right], \tag{4.60}$$

where $\Theta = \Delta kL/2$.

Under the assumption that the electron beam is cold (i.e., monoenergetic), the distribution function can be expressed in the form

$$G_b(p) = \frac{p_\parallel}{p}\delta(p - p_0), \tag{4.61}$$

where p_0 denotes the bulk momentum of the beam. The gain in this limit can be expressed as [9]

$$G(L) = \frac{\omega_b^2 L^3 k}{8\gamma_0\gamma_\parallel^2 v_\parallel^2}\beta_w^2\Phi F(\Theta) \approx 4(2\pi\rho)^3\left(\frac{L}{\lambda_w}\right)^3 F(\Theta), \tag{4.62}$$

where $\gamma_0 = (1 + p_0^2/m_e^2c^2)^{1/2}$, $\beta_w = v_w/v_\parallel$, $(v_w, v_\parallel)$ are the transverse and axial velocities of the steady-state orbits, $\gamma_\parallel^2 = (1 - v_\parallel^2/c^2)^{-1}$, Φ is defined in Eq. (2.24), and

$$F(\Theta) \equiv \frac{d}{d\Theta}\left(\frac{\sin\Theta}{\Theta}\right)^2 \tag{4.63}$$

defines the spectral function. The spectral function is shown in Fig. 4.2 and exhibits the extrema $F(\pm 1.3) \approx \mp 0.54$. Observe that $\Phi = 1$ and $\beta_w = -\Omega_w/k_w v_\parallel$ in the limit in which the axial solenoidal magnetic field vanishes. Observe that the spectral variation of the emission in the low-gain regime as shown in Fig. 4.2 is proportional to the derivative of the spectral dependence of the spontaneous emission (see Chap. 3). This property of the gain is referred to as Madey's theorem [4].

The maximum permissible thermal spread under which this expression for the gain (4.59) is valid can be estimated by comparison of the bandwidth characterized by the spectral function with the bandwidth induced by a thermal spread. The frequency spread characterized by the spectral function is similar to that found for the spontaneous emission from a cold beam (Eq. (3.25)), that is, $\Delta\omega/\omega \approx \lambda_w/L$. The effect of a thermal spread in the axial velocity of the beam is to cause a shift in the resonant frequency of the order of $\Delta\omega/\omega \approx 2\gamma_\parallel^2\Delta v_\parallel/v_\parallel$. As a consequence, the cold-beam approximation is valid whenever

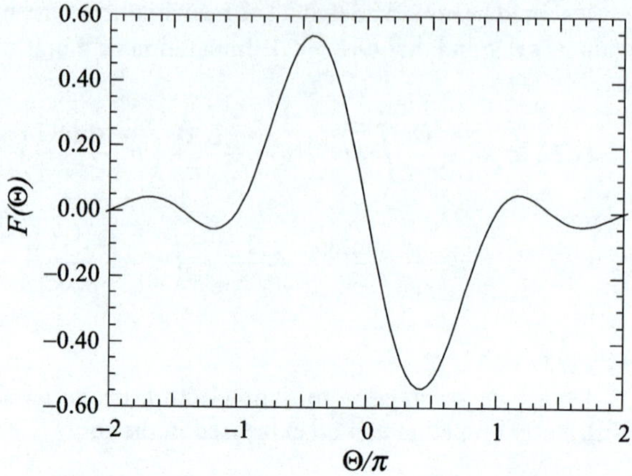

Fig. 4.2 Graph of the free-electron laser spectral function

$$\frac{\Delta v_{\parallel}}{v_{\parallel}} << \frac{\lambda}{L}, \tag{4.64}$$

which is identical to the condition found to obtain for the spontaneous emission for a cold beam (3.23).

In the case of either (1) weak axial magnetic fields corresponding to the Group I trajectories [which applies in the limit in which the axial field vanishes] or (2) strong magnetic fields corresponding to Group II orbits with $\Phi > 0$ the peak gain occurs for $\Theta \approx -1.3$ which corresponds to a resonant frequency

$$\omega \cong \frac{k_w v_{\parallel}}{1-\beta_{\parallel}}\left(1-\frac{2.6}{k_w L}\right), \tag{4.65}$$

where $\beta_{\parallel} = v_{\parallel}/c$. However, for solenoidal fields corresponding to Group II steady-state trajectories for which $\Phi < 0$, there is a phase shift in the interaction due to the negative-mass effect. In this case, peak gain occurs for $\Theta \approx 1.3$ which defines a resonant frequency of

$$\omega \cong \frac{k_w v_{\parallel}}{1-\beta_{\parallel}}\left(1+\frac{2.6}{k_w L}\right). \tag{4.66}$$

The fundamental physics contained in the result for the low-gain regime (4.59) is explained in the introductory material for this chapter and is summarized in Eq. (1.8) in units of dB. The result is identical to that found by a single-particle analysis either from quantum mechanical principles [4] or from classical considerations [8, 15] and describes the effect of electron bunching in the ponderomotive potential formed by the beating of the wiggler and radiation fields. The gain in the low-gain regime will be rederived later in this chapter in the analysis of three-dimensional effects. The derivation in that case will rely upon a phase average over the untrapped electron trajectories, which makes plain the essential physics of the interaction.

4.2.1.4 The High-Gain Regime

The high-gain regime is applicable to intense beam free-electron lasers which are usually operated in an amplifier or SASE mode. The fields in this regime exhibit exponential growth of the fluctuation fields, and we assume that

$$\delta\hat{A}_\pm(z) = \delta\hat{A}_\pm(0)\exp(ik_\pm z), \tag{4.67}$$

and

$$\delta\hat{\varphi}(z) = \delta\hat{\varphi}(0)\exp(ikz). \tag{4.68}$$

As a result, the source currents can be expressed as

$$\begin{aligned}\delta\hat{J}_\pm(z) = &-\frac{\omega k}{8\pi c}\delta\hat{A}_\pm(z)\left[2\left\langle\frac{p}{\gamma p_\parallel}\frac{\omega - k_\pm v_\parallel}{\omega - \Omega_0 - k_\pm v_\parallel}\right\rangle + a_w^2 S_\pm(k_\pm,\omega)\right] \\ &- a_w^2\frac{ck^2}{8\pi}\delta\hat{\varphi}(z)\exp(\mp ik_w z)\chi_a(k,\omega),\end{aligned} \tag{4.69}$$

and

$$\begin{aligned}\delta\hat{J}_z(z) = &-\frac{\omega k}{8\pi c}\delta\hat{\varphi}(z)\chi(k,\omega) \\ &-\frac{\omega k_w}{8\pi}a_w\left[\delta\hat{A}_+(z)\exp(ik_w z)\sigma_+(k_+,\omega) + \delta\hat{A}_-(z)\exp(-ik_w z)\sigma_-(k_-,\omega)\right],\end{aligned} \tag{4.70}$$

where $a_w = eA_w/m_ec^2 = K$ (the wiggler strength parameter), the average is defined as

$$\langle(\cdots)\rangle \equiv \int_0^\infty dp G_b(p)(\cdots), \tag{4.71}$$

the dielectric function of the uncoupled beam-plasma waves is

$$\chi(k,\omega) \equiv \frac{\omega_b^2}{k}\int_0^\infty dp\frac{m_e}{\omega - kv_\parallel}\frac{\partial}{\partial p}G_b(p), \tag{4.72}$$

the second-order dielectric function of the electromagnetic waves is

$$S_{\pm}(k_{\pm},\omega) \equiv \int_0^\infty dp \frac{p}{\gamma^3 p_\|} G_b(p) \frac{k_w^2 c^2}{(k_w v_\| - \Omega_0)^2} \left[1 + \frac{\Omega_0}{k_w v_\| - \Omega_0}\left(\frac{\omega \mp \Omega_0 \pm k_w v_\|}{\omega \mp \Omega_0 - k_\pm v_\|} - \frac{\omega \pm k_w v_\|}{\omega \mp k_w v_\| - k_\pm v_\|}\right)\right] - \omega c^2 \int_0^\infty dp \frac{k_w^2 v_\|}{\gamma^2 (k_w v_\| - \Omega_0)^2} \frac{m_e}{\omega \mp k_w v_\| - k_\pm v_\|} \frac{\partial}{\partial p} G_b(p), \tag{4.73}$$

and the dielectric functions which describe the mode-mode couplings are

$$\chi_a(k,\omega) \equiv \frac{\omega_b^2}{k} \int_0^\infty dp \frac{k_w v_\|}{k_w v_\| - \Omega_0} \frac{m_e}{\gamma(\omega - k v_\|)} \frac{\partial}{\partial p} G_b(p) \tag{4.74}$$

and

$$\sigma_{\pm}(k_{\pm},\omega) \equiv \frac{\omega_b^2}{\omega} \int_0^\infty dp \frac{p}{\gamma^2 p_\|} G_b(p) \frac{\Omega_0}{(k_w v_\| - \Omega_0)^2} \left(\frac{\omega \mp k_w v_\|}{\omega \mp k_w v_\| - k_\pm v_\|} - \frac{\omega}{\omega \mp \Omega_0 - k_\pm v_\|}\right) + \frac{\omega_b^2}{\omega} \int_0^\infty dp \frac{\omega}{\gamma (k_w v_\| - \Omega_0)} \frac{m_e v_\|}{\omega \mp k_w v_\| - k_\pm v_\|} \frac{\partial}{\partial p} G_b(p). \tag{4.75}$$

Substitution of these expressions for the source current into Maxwell's equations (Eqs. (4.43) and (4.44)) yields three coupled equations for the amplitudes of the vector and scalar potentials

$$\Lambda_{\pm}(k \mp k_w,\omega)\delta\hat{A}_{\pm}(0) + \frac{a_w}{2} \frac{c^2 k^2}{\omega^2} \chi_a(k,\omega)\delta\hat{\varphi}(0) = 0, \tag{4.76}$$

$$[1 + \chi(k,\omega)]\delta\hat{\varphi}(0) + a_w \frac{k_w}{k} \left[\sigma_+(k - k_w,\omega)\delta\hat{A}_+(0) + \sigma_-(k + k_w,\omega)\delta\hat{A}_-(0)\right] = 0, \tag{4.77}$$

where

$$\Lambda_{\pm}(k \mp k_w,\omega) \equiv 1 - \frac{c^2 k_\pm^2}{\omega^2} - \frac{\omega_b^2}{\omega^2}\left\langle \frac{p}{\gamma p_\|} \frac{\omega - k_\pm v_\|}{\omega \mp \Omega_0 - k_\pm v_\|}\right\rangle - \frac{\omega_b^2}{2\omega^2} a_w^2 S_{\pm}(k_\pm,\omega) \tag{4.78}$$

describes the dispersion of the pure electromagnetic modes in the presence of the wiggler. Observe that the wavelength matching conditions between the scalar potential and the right- and left-hand circularly polarized vector potentials imply that $k_{\pm} = k \mp k_w$.

4.2.1.4.1 The General Dispersion Equation

If we now impose the cold-beam distribution function (Eq. (4.61)) in the evaluation of the coupled dispersion equations and take the determinant of the matrix of coefficients, then we obtain a dispersion equation of the form

$$\left(\omega - kv_{\parallel}\right)^2 - \kappa_b^2 v_{\parallel}^2 = -\frac{\beta_w^2}{2}\frac{\omega_b^2}{\gamma_0}\left[\frac{\alpha_+(k-k_w,\omega)}{\varepsilon_+(k-k_w,\omega)} + \frac{\alpha_-(k+k_w,\omega)}{\varepsilon_-(k+k_w,\omega)}\right], \tag{4.79}$$

to second order in the wiggler amplitude, where all orbit quantities denote the steady-state trajectories computed using a total energy given by $\gamma_0 = (1 + p_0^2/m_e^2c^2)^{1/2}$,

$$\kappa_b^2 v_{\parallel}^2 \equiv \frac{\omega_b^2}{\gamma_0 \gamma_{\parallel}^2}\Phi, \tag{4.80}$$

$$\varepsilon_{\pm}(k_{\pm},\omega) \equiv 1 - \frac{\omega_b^2\left(\omega - k_{\pm}v_{\parallel}\right)}{\gamma_0\left(\omega^2 - k_{\pm}^2c^2\right)\left(\omega \mp \Omega_0 - k_{\pm}v_{\parallel}\right)}, \tag{4.81}$$

$$\begin{aligned}
\alpha_{\pm}(k_{\pm},\omega) \equiv{}& \frac{1}{\omega^2 - k_{\pm}^2c^2}\left\{\beta_{\parallel}^2\left(\omega^2 - k^2c^2 - \frac{\omega_b^2}{\gamma_0}\right) + \beta_{\parallel}^2\frac{\omega_b^2}{\gamma_0\left(\omega - kv_{\parallel}\right)}\right.\\
&\times\left[(1-\Psi)\left(kc - \omega\frac{v_{\parallel}}{c}\right)^2 + \frac{1}{\gamma_{\parallel}^2}(1-\Phi)\left(\omega^2 - c^2k^2\right)\right]\\
&+\frac{\omega_b^2}{\gamma_0\omega}\beta_{\parallel}\Psi\frac{\Omega_0}{kv_{\parallel}-\Omega_0}\left(\frac{\omega \mp kv_{\parallel}}{\omega - kv_{\parallel}} - \frac{\omega}{\omega \mp \Omega_0 - k_{\pm}v_{\parallel}}\right)\\
&+\frac{\Omega_0}{k_wv_{\parallel}-\Omega_0}\left[\left(\omega - kv_{\parallel}\right)^2 - \kappa_b^2v_{\parallel}^2\right]\left(\frac{3\omega \mp k_wv_{\parallel}}{\omega - k_wv_{\parallel}} - \frac{\omega \mp \Omega_0 \pm k_wv_{\parallel}}{\omega \mp \Omega_0 - k_{\pm}v_{\parallel}}\right)\\
&\left.-\frac{1}{\gamma_{\parallel}^2}(1-\Phi)\frac{\omega\left[\left(\omega - kv_{\parallel}\right)^2 - \kappa_b^2v_{\parallel}^2\right]}{\omega - k_wv_{\parallel}}\left(\frac{\omega \mp 2kv_{\parallel}}{\omega - kv_{\parallel}} - \frac{2\Omega_0}{k_wv_{\parallel}-\Omega_0}\right)\right\},
\end{aligned} \tag{4.82}$$

and we have defined

$$\Psi \equiv 1 - \frac{\Omega_0}{\beta_{\parallel}\left(kc - \omega v_{\parallel}/c\right)}\frac{\left(1+\beta_w^2\right)k_wv_{\parallel} - \omega}{\left(1+\beta_w^2\right)\Omega_0 - k_wv_{\parallel}}. \tag{4.83}$$

It should be observed that the left-hand side of Eq. (4.79) describes the uncoupled dispersion equation for the beam-plasma waves in the presence of the wiggler, while $\varepsilon_{\pm}(k_{\pm},\omega)$ describes the dispersion functions for the circularly polarized electromagnetic modes in the absence of the wiggler. Hence, this equation describes the coupling of the electromagnetic modes to the electrostatic beam-plasma waves due to the presence of the wiggler field.

4.2.1.4.2 The Stability of Beam-Plasma Modes

Before proceeding further, we observe that the electrostatic beam-plasma modes are themselves modified in the presence of the wiggler field. If we neglect the coupling to the electromagnetic waves, then the beam-plasma modes satisfy the dispersion equation [28]

$$\left(\omega - kv_{\parallel}\right)^2 = \frac{\omega_b^2}{\gamma_0 \gamma_{\parallel}^2}\Phi. \tag{4.84}$$

This implies that the effect of the wiggler is to modify the beam-plasma frequency based upon the response of the beam to the axial electric field perturbations (see Eq. (2.23)). As a consequence, the effective plasma frequency is increased by the combined presence of the wiggler and solenoidal fields for the Group I trajectories since $\Phi > 1$ for these orbits. In contrast, $|\Phi| < 1$ for the Group II trajectories, and the beam-plasma frequency decreases relative to that found in the absence of either magnetostatic field. Indeed, the effective plasma frequency is imaginary for those Group II orbits for which $\Phi < 0$, which corresponds to an unstable beam-plasma wave in which the growth rate is

$$\mathrm{Im}k = \frac{\omega_b}{\gamma_0^{1/2}\gamma_{\parallel}}|\Phi|^{1/2}. \tag{4.85}$$

This is a particularly interesting regime for the free-electron laser because the coupling of the electromagnetic wave is to an unstable beam-plasma mode. This has the effects of increasing the free energy available to the interaction and broadening the bandwidth of the instability. This regime will be discussed in more detail in regard to both the linear and nonlinear analyses of the interaction.

4.2.1.4.3 A Reduced Form of the Dispersion Equation

The wiggler field provides for the coupling between the beam space-charge wave and either polarization state of the electromagnetic field, as well as a growth mechanism for the electromagnetic waves in the absence of the beam-plasma waves. In the former case, coherent amplification occurs by three-wave coherent Raman scattering in which the magnetostatic wiggler represents the pump wave, the beam-plasma mode represents the idler, and the output signal is the daughter wave. The latter case is coherent Compton scattering in which the wiggler (that appears as a backward-propagating electromagnetic wave in the beam frame) scatters off the electron beam to produce the output signal. We choose, without loss of generality, to focus on the right-hand mode. As a consequence, we assume that $|\varepsilon_+(k_+,\omega)| << |\varepsilon_-(k_-,\omega)|$ and neglect the term in $[\varepsilon_-(k_-,\omega)]^{-1}$ in Eq. (4.79). If we assume in addition that $\omega_b{}^2/\gamma_0\omega^2 << 1$, then the dispersion equation can be cast into the substantially simpler form [18]

$$\left[\left(\omega-kv_\parallel\right)^2-\kappa_b^2v_\parallel^2\right]\left[\omega^2-k_+^2c^2-\frac{\omega_b^2\left(\omega-k_+v_\parallel\right)}{\gamma_0\left(\omega-\Omega_0-k_+v_\parallel\right)}\right]$$
$$\cong-\frac{\beta_w^2}{2}\frac{\omega_b^2}{\gamma_0}\left[\beta_\parallel^2\left(\omega^2-k^2c^2-\frac{\omega_b^2\left(\omega-k_+v_\parallel\right)}{\gamma_0\left(\omega-\Omega_0-k_+v_\parallel\right)}\right)+\frac{\omega^2}{\gamma_\parallel^2}(1-\Phi)\right], \tag{4.86}$$

where we have imposed the wavenumber matching condition $k_+ = k - k_w$. This equation describes the coupling of the electromagnetic modes with the positive and negative energy beam-plasma waves [denoted by $\omega = (k_+ + k_w)v_\parallel \pm \kappa_b v_\parallel$, respectively, when $\kappa_b > 0$]. The peak gain occurs near the intersections of the uncoupled dispersion curves of the beam-plasma and electromagnetic modes; hence, we may make the replacements

$$\omega^2-k^2c^2-\frac{\omega_b^2\left(\omega-k_+v_\parallel\right)}{\gamma_0\left(\omega-\Omega_0-k_+v_\parallel\right)}\cong-k_w(2k_++k_w)c^2 \tag{4.87}$$

and $\omega \approx (k_+ + k_w)v_\parallel$ on the right-hand side of Eq. (4.86). As a result, the dispersion equation becomes

$$\left[\left(\omega-kv_\parallel\right)^2-\kappa_b^2v_\parallel^2\right]\left[\omega^2-k_+^2c^2-\frac{\omega_b^2\left(\omega-k_+v_\parallel\right)}{\gamma_0\left(\omega-\Omega_0-k_+v_\parallel\right)}\right]\cong\frac{\beta_w^2}{2}\frac{\omega_b^2}{\gamma_0\gamma_\parallel^2}\omega^2\Phi. \tag{4.88}$$

It is important to observe that growth is found near the intersections of the electrostatic beam-plasma mode and the circularly polarized electromagnetic mode, in particular, in the neighborhood of the intersection between the electromagnetic wave and the negative energy beam-plasma mode.

4.2.1.4.4 Dispersive Effects on the Interaction

In the absence of a solenoidal magnetic field, both polarization states of the electromagnetic modes have the same dispersion relation [i.e., $\omega^2 = k^2c^2 + \omega_b{}^2/\gamma_0$] that is shown schematically in Fig. 4.3 in the first quadrant of the (ω,k) plane. The situation

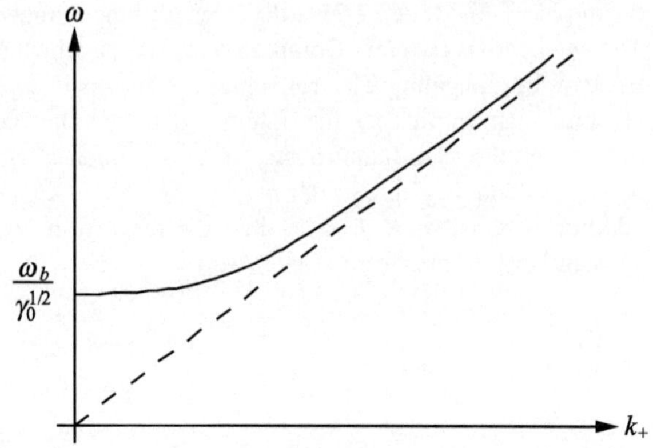

Fig. 4.3 Schematic drawing of the circularly polarized electromagnetic modes in the absence of a solenoidal magnetic field

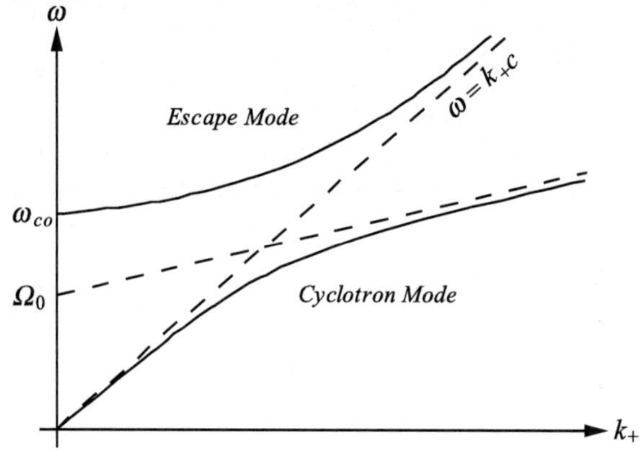

Fig. 4.4 Schematic of the dispersion relation of the right-hand circularly polarized electromagnetic mode in the presence of a solenoidal magnetic field showing both the escape and cyclotron branches

is more complex in the presence of the solenoidal magnetic field in which both escape and cyclotron branches exist (see Fig. 4.4) and the escape branch is characterized by a cutoff at

$$\omega_{co} = \frac{1}{2}\Omega_0\left[1 + \sqrt{1 + \frac{4\omega_b^2}{\gamma_0\Omega_0^2}}\right]. \tag{4.89}$$

Since the slope of the negative energy beam-plasma mode is identical to that of the cyclotron resonance line (i.e., $\omega \approx \Omega_0 + k_+ v_\parallel$), the particular branch of the electromagnetic dispersion curve that is resonant depends upon the magnitude of the axial velocity. When $\Omega_0 < (k_w - \kappa_b)v_\parallel$, the intersection is possible only with the escape branch of the electromagnetic waves for which $\omega > \Omega_0 + k_+ v_\parallel$. This case includes the limit in which the axial solenoidal field vanishes. Observe that this condition is satisfied for Group I trajectories because for most cases of practical interest $\kappa_b << k_w$. In contrast, when $\Omega_0 > (k_w - \kappa_b)v_\parallel$ corresponding to the Group II trajectories, the intersection is possible only with the electromagnetic cyclotron mode.

4.2.1.4.5 The Compton and Raman Regimes

The terminology of a weak axial solenoidal magnetic field refers to the regime in which $\Omega_0 \leq k_w v_\parallel$. This corresponds to electrons executing Group I trajectories in which $\Phi \geq 1$ and includes the limit in which the axial solenoidal field vanishes. As a consequence, the beam space-charge mode is stable, and the only effect of the solenoidal magnetic field is to enhance both the transverse wiggler-induced velocity and the effective plasma frequency.

The simplest case is that in which the solenoidal field vanishes and the dispersion equation reduces to

$$\left[\left(\omega - kv_\parallel\right)^2 - \kappa_b^2 v_\parallel^2\right]\left(\omega^2 - k_+^2 c^2 - \frac{\omega_b^2}{\gamma_0}\right) \cong \frac{\beta_w^2}{2}\frac{\omega_b^2}{\gamma_0 \gamma_\parallel^2}\omega^2, \tag{4.90}$$

which is a quartic equation for k. This describes a dispersion equation in which the electromagnetic mode is coupled to positive and negative energy beam-plasma modes. The form of the solutions for the coupled modes is shown schematically in Fig. 4.5. Growth is found in the vicinity of the intersection of the negative energy beam-plasma wave and the forward-propagating electromagnetic mode; hence, the resonant frequency of the interaction occurs in the vicinity of two intersection points

$$\omega \cong \gamma_\parallel^2 (k_w - \kappa_b) v_\parallel \left[1 \pm \beta_\parallel \sqrt{1 - \frac{\kappa_b^2}{(k_w - \kappa_b)^2}}\right], \tag{4.91}$$

which reduces to the well-known free-electron laser resonance condition (i.e., $\omega = 2\gamma_\parallel^2 k_w v_\parallel$) in the limit in which $\kappa_b << k_w$. In addition, since we are interested in the propagation of waves in the first quadrant (i.e., $\omega > 0$, $k_+ > 0$), we may make the approximation

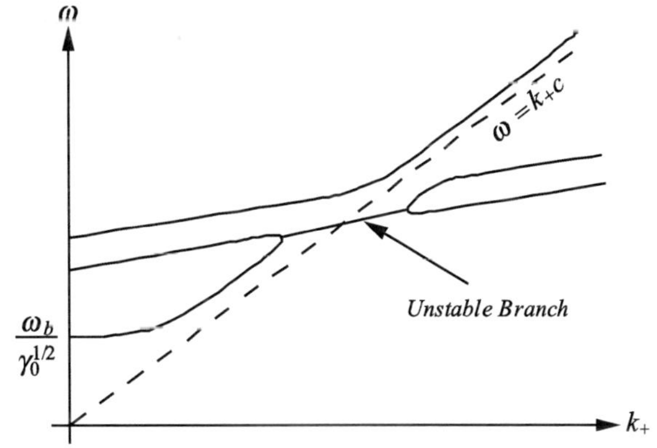

Fig. 4.5 Schematic illustration of the coupled dispersion relation for the electromagnetic mode and the positive and negative energy beam-plasma modes. As shown, it is the coupling with the negative energy beam-plasma mode that leads to instability

$$\omega^2 - k_+^2 c^2 - \frac{\omega_b^2}{\gamma_0} \cong -2c^2 k_+ (k_+ - K_b) \tag{4.92}$$

on the right-hand side of Eq. (4.87), where $K_b \equiv (\omega^2 - \omega_b{}^2/\gamma_0{}^2)^{1/2}/c$. Therefore, under the transformation

$$k = \frac{\omega}{\upsilon_\parallel} + \kappa_b + \delta k, \tag{4.93}$$

the dispersion equation reduces to a cubic in δk

$$\delta k(\delta k + 2\kappa_b)(\delta k - \Delta k) \cong -\frac{\beta_w^2}{2} \frac{\omega_b^2}{\gamma_0 c^2} \frac{k_w}{\beta_\parallel} = -(2\rho k_w)^3, \tag{4.94}$$

where

$$\Delta k \equiv k_w + K_b - \frac{\omega}{\upsilon_\parallel} - \kappa_b \tag{4.95}$$

defines the wavenumber mismatch parameter. Equation (4.94) is analogous to Pierce's dispersion equation for traveling wave tubes. There are two principal regimes of interest in the solution of the dispersion equation (Eq. (4.94)) corresponding to the low- and high-density regimes.

The *high-gain Compton regime* (sometimes called the *strong-pump* regime) is found when the effect of the ponderomotive potential is greater than that of the space-charge potential of the beam-plasma waves. In this limit, $|\delta k| >> 2\kappa_b$, and the dispersion equation reduces to a cubic equation

$$\delta k^2(\delta k - \Delta k) \cong -(2\rho k_w)^3. \tag{4.96}$$

This is the dominant regime for most short-wavelength free-electron lasers driven by high-energy electron beams.

Complex roots of the dispersion equation are found (i.e., growth) when $\Delta k > -3\sqrt[3]{2}\rho$, which implies that $\omega < \left(1 + 3\sqrt[3]{2}\rho\right)\omega_{res}$ when $\omega_b/\gamma_0 << \omega$ and $\omega_{res} = 2\gamma_\parallel{}^2 k_w \upsilon_\parallel$. Maximum growth is found to correspond to zero detuning (i.e., $\Delta k = 0$) for which

$$\frac{\delta k_{\max}}{k_w} \cong \left(1 \pm i\sqrt{3}\right)\rho. \tag{4.97}$$

As a result, the power gain (exponentiation) length, $L_G = (2\text{Im}\ \delta k_{\max})^{-1}$, is given by

$$L_G = \frac{\lambda_w}{4\sqrt{3}\pi\rho}. \tag{4.98}$$

In order to neglect the effect of the space-charge fields, therefore, we must require that $|\delta k_{max}| >> 2\kappa_b$; hence, $\rho >> \kappa_b/2$ or

$$\frac{\omega_b}{\gamma_0^{1/2} c k_w} << \frac{\gamma_\parallel^3}{2} \frac{v_w^2}{c^2}. \tag{4.99}$$

The *Raman* (or collective) regime is found when the space-charge fields dominate over the ponderomotive potential. This regime is generally found in the long-wavelength, microwave regime where the electron beam energy is relatively low. In this limit, $|\delta k| << 2\kappa_b$ and the dispersion equation reduces to a quadratic equation for δk

$$\delta k(\delta k - \Delta k) \cong -\frac{\beta_w^2}{4} \frac{\omega_b}{\gamma_0^{1/2} c} k_w, \tag{4.100}$$

where peak growth is again found for zero detuning. This peak growth rate is given by (see Eq. (1.11))

$$\frac{\delta k_{max}}{k_w} \cong i \frac{\beta_w}{2} \sqrt{\frac{\omega_b}{\gamma_0^{1/2} c k_w}}, \tag{4.101}$$

which is valid for sufficiently high beam densities that

$$\frac{\omega_b}{\gamma_0^{1/2} c k_w} >> \frac{\gamma_\parallel^3}{2} \frac{v_w^2}{c^2}. \tag{4.102}$$

4.2.1.4.6 The Transition Between the Compton and Raman Regimes

The detailed effect of increasing beam density upon the growth rate of the instability is shown in Fig. 4.6 in which we plot the magnitude of the growth rate versus the beam density for $\Omega_w/ck_w = 0.05$ and $\gamma_0 = 2.957$ (corresponding to a beam energy of 1 MeV). The dashed line represents the growth rate as calculated for the high-gain Compton regime (4.94) which scales as $\omega_b^{2/3}$. For this choice of parameters, $\beta_w^2 \approx 0.00282$, $\gamma_\parallel^2 \approx 8.72$, and the transition between the high-gain Compton and the collective Raman regimes occurs for $\omega_b/ck_w \approx 0.004$. It is evident from the figure that the growth rate tracks the solution for the high-gain Compton up to this point but diverges from it as the beam density continues to increase. For beam densities such that $\omega_b/\gamma_0^{1/2} ck_w > 0.01$, the solution follows the Raman solution (Eq. (4.97)) in which the growth rate scales as $\omega_b^{1/2}$. The discrepancy between the actual solution for the growth rate and the prediction for the high-gain Compton regime in the dashed line indicates that collective effects involved in the stimulated Raman scattering process tend to reduce the linear growth rate.

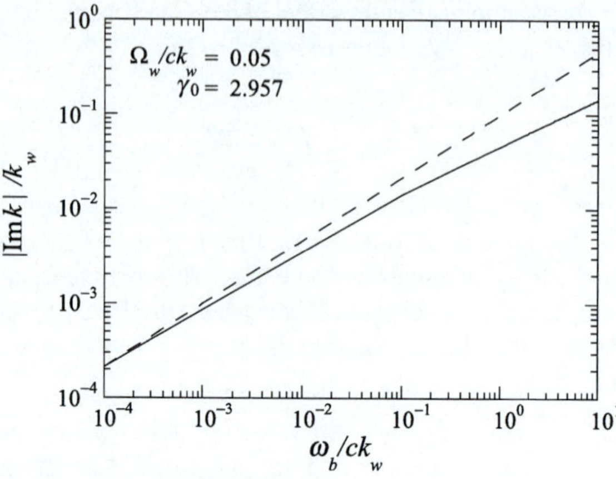

Fig. 4.6 Graph of the peak growth rate for zero detuning (i.e., $\Delta k = 0$) is shown in the solid line as a function of the beam-plasma frequency. The dashed line shows an increase as the 2/3 power of the beam-plasma frequency as found in the high-gain Compton regime. The onset of the Raman regime is evident from the decrease in the slope of the curve

4.2.1.5 The Effect of an Axial Magnetic Field

The dispersion equation (Eq. (4.88)) is, in general, a quintic polynomial in k when an axial solenoidal field is present. If we make the restriction that $k_+ > 0$, then the dispersion equation can be reduced to a quartic polynomial

$$\begin{aligned}\left[(\omega - kv_\parallel)^2 - \kappa_b^2 v_\parallel^2\right](k_+ - K_+)(k_- - K_-)\\ \cong -\beta_w^2 \frac{\omega_b^2}{\gamma_0 c^2}\frac{\omega^2}{4\gamma_\parallel^2 K_b}\Phi\left(k_+ - \frac{\omega - \Omega_0}{v_\parallel}\right),\end{aligned} \tag{4.103}$$

where

$$K_\pm \equiv \frac{1}{2}\left(K_b + \frac{\omega - \Omega_0}{v_\parallel}\right) \pm \frac{1}{2}\sqrt{\Delta K^2 + 2\frac{\omega_b^2}{\gamma_0 c^2}\frac{\Omega_0}{K_b v_\parallel}} \tag{4.104}$$

and $\Delta K = K_b - (\omega - \Omega_0)/v_\parallel$. It is clear that the nature of the instability depends upon the sign of Φ, which determines the effective plasma frequency and mediates the ponderomotive potential.

4.2.1.5.1 The Case of a Weak Magnetic Field

In the weak magnetic field regime, we deal with the Group I orbits for which $\Phi \geq 1$ and the intersection between the beam-plasma mode and the electromagnetic wave occurs on the escape branch. Substantial simplification of the dispersion equation (Eq. (4.100)) is found in the limit in which the cyclotron resonance effects on the dispersion of the electromagnetic wave can be neglected. This implies that the wave frequency $\omega >> \omega_{co}$. In this regime, the dispersion equation can be approximated by

$$\delta k(\delta k + 2\kappa_b)(\delta k - \Delta k) \cong -\frac{\beta_w^2}{2}\frac{\omega_b^2}{\gamma_0 c^2}\frac{k_w}{\beta_\parallel}\Phi = -(2\rho k_w)^3, \tag{4.105}$$

as in Eq. (4.96) and in which the effective plasma frequency [i.e., $\omega_b{}^2\Phi$] is used in κ_b. In addition, the resonant frequency is given approximately by Eq. (4.102) subject to the substitution of the effective plasma frequency. As a consequence, the dispersion equation in the *high-gain Compton regime* is

$$\delta k^2(\delta k - \Delta k) \cong -(2\rho k_w)^3. \tag{4.106}$$

The criterion for the high-gain Compton regime is now

$$\frac{\omega_b}{\gamma_0^{1/2} c k_w}\Phi^{1/2} << \frac{\gamma_\parallel^3}{16}\frac{v_w^2}{c}, \tag{4.107}$$

and at maximum growth

$$\frac{\delta k_{max}}{k_w} \cong \left(1 \pm i\sqrt{3}\right)\rho. \tag{4.108}$$

The Raman regime is found in the opposite limit to satisfy

$$\delta k(\delta k - \Delta k) \cong -\frac{\beta_w^2}{4}\frac{\omega_b}{\gamma_0^{1/2} c} k_w \Phi^{1/2}, \tag{4.109}$$

which has the solution at maximum growth of

$$\frac{\delta k_{max}}{k_w} \cong i\frac{\beta_w}{2}\sqrt{\frac{\omega_b}{\gamma_0^{1/2} c k_w}\Phi^{1/2}}. \tag{4.110}$$

The effect of the axial solenoidal field in this regime is to enhance the growth rates due to enhancements in both the transverse wiggler-induced velocity v_w and the effective plasma frequency through Φ. However, since this enhancement occurs at the expense of a reduction in $v_\parallel$, the interaction frequency at fixed energy must decrease. This is illustrated in Fig. 4.7 in which we plot the magnitude of the maximum growth rate and the corresponding frequency as a function of the solenoidal magnetic field for $\Omega_w/ck_w = 0.05$, $\gamma_0 = 2.957$ (for a beam energy of 1 MeV), and $\omega_b/ck_w = 0.1$. For these parameters, which correspond to the steady-state orbit obtained in Chap. 2, the Group I orbits exhibit the transition to orbital instability at $\Omega_0/ck_w \approx 0.750$, and this defines the maximum value of the solenoidal field for which an interaction is possible. Observe that the growth rate increases to a singularity at this transition (which is due to the singularity in Φ and the effective plasma frequency), but that this singularity is an artifact of the approximations implicit in the

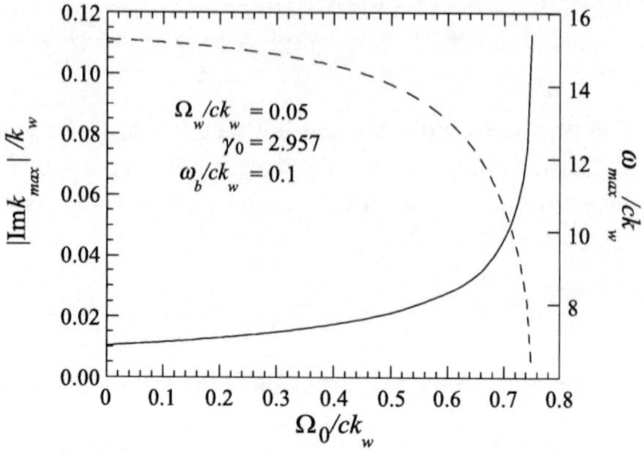

Fig. 4.7 Graph of the magnitude of the maximum growth rate (solid line) and the frequency corresponding to maximum growth (dashed line) as a function of the axial solenoidal field for Group I trajectories

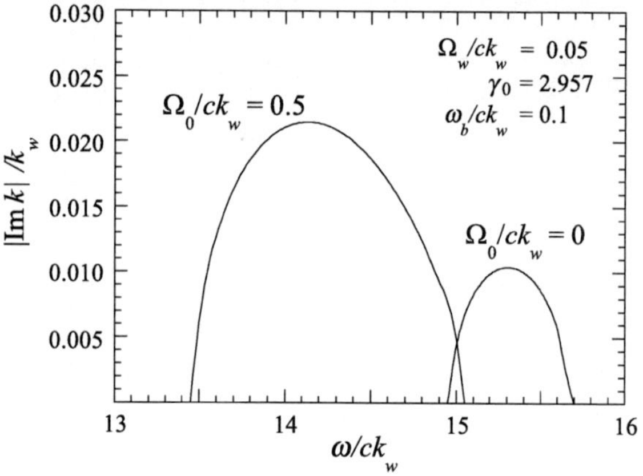

Fig. 4.8 Graph of the magnitude of the growth rate versus frequency for $\Omega_w/ck_w = 0.05$, $\gamma_0 = 2.957$, and $\Omega_0/ck_w = 0$ and 0.05

linearization. It is also evident that the frequency corresponding to peak growth decreases with increasing Ω_0 over this range. The increase in the growth rate and the decrease in the frequency of the interaction are accompanied by an increase in the bandwidth over which growth is found. This is illustrated in Fig. 4.8 in which we plot the magnitude of the growth rate versus frequency for $\Omega_0/ck_w = 0$ and 0.5.

4.2.1.5.2 Case of a Strong Magnetic Field

The terminology of strong axial magnetic field refers to the case in which $\Omega_0 \geq k_w\upsilon_\parallel$ and the electrons are on Group II orbits. There are two distinct regimes for these orbits that relate to (1) the negative-mass limit in the vicinity of the magnetic resonance at $\Omega_0 \approx k_w\upsilon_\parallel$ in which the axial velocity of the beam increases with decreases in the electron energy and (2) a more usual positive-mass case found at still higher axial magnetic fields. The dispersion equation in the positive-mass limit is

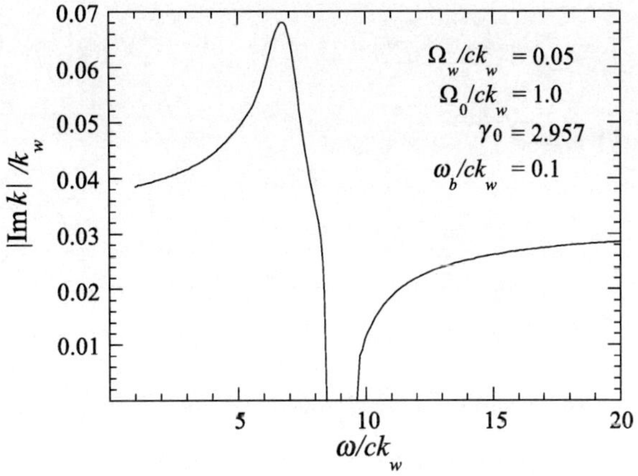

Fig. 4.9 Graph of the magnitude of the growth rate as a function of frequency for a strong axial magnetic field corresponding to the negative-mass regime of the Group II electron orbits

identical to that found for the weak axial magnetic field limit. However, since $\lim_{B0\to\infty} v_w = 0$ and $\lim_{B0\to\infty}\Phi = 1$, the growth rate in this limit is small. As a consequence, we shall focus our attention on the negative-mass regime.

The character of the beam-plasma mode in the negative-mass regime has already been shown to be unstable. As a consequence, the circularly polarized electromagnetic mode couples with an unstable beam-plasma wave, which greatly enhances the bandwidth of the instability. In general, the growth rate must be found by solution of the full quartic dispersion equation (Eq. (4.100)) by numerical methods. This solution of the growth rate as a function of frequency is shown in Fig. 4.9 for $\Omega_w/ck_w = 0.05$, $\gamma_0 = 2.957$, $\omega_b/ck_w = 0.1$, and $\Omega_0/ck_w = 1.0$. These parameters correspond to both the idealized steady-state trajectories discussed in Chap. 2, and we find that $\beta_\parallel = 0.8744$ and $\Phi = -1.3685$. Note that at the point at which $\Phi = 0$ (which occurs for $\Omega_0/ck_w \approx 1.2$), the coupling coefficient on the right-hand side of the dispersion equation (Eq. (4.88)) vanishes along with the interaction. It is evident form the figure that there are two unstable bands. For the negative-mass regime, the frequency of the beam-plasma mode varies as $\omega = (k + k_w)v_\parallel$, and there is no contribution due to the effective plasma frequency. Hence, the intersection point between the electromagnetic mode and the beam-plasma mode occurs at

$$\omega \cong \gamma_\parallel^2 k_w v_\parallel \left[1 + \beta_\parallel \sqrt{1 - \frac{\omega_b^2}{\gamma_0 \gamma_\parallel^2 k_w^2 v_\parallel^2}}\right]. \tag{4.111}$$

For the present choice of parameters, this frequency occurs at $\omega/ck_w \approx 6.94$ that corresponds to the point of maximum gain shown in the figure.

The additional unstable band that is found for $\omega/ck_w > 9.6$ arises from the unstable nature of the beam-plasma wave. For this frequency range in which $|\Delta k| >> |\delta k|$, the dispersion equation may be approximated by

$$k^2 - 2\frac{\omega}{v_\parallel}k + \frac{\omega^2}{v_\parallel^2} + |\kappa_b|^2\left[1 - \frac{\beta_w^2}{4}\beta_\parallel\gamma_\parallel^2\left(1 + \beta_\parallel\right)\right] \cong 0, \tag{4.112}$$

for which

$$k \cong \frac{\omega}{v_\parallel} \pm \frac{\omega^2}{v_\parallel^2} + i|\kappa_b|\sqrt{1 - \frac{\beta_w^2}{4}\beta_\parallel\gamma_\parallel^2\left(1 + \beta_\parallel\right)}. \tag{4.113}$$

This mode is clearly a modified beam-plasma wave, where it should be observed that the wiggler acts as a stabilizing influence that reduces the effective growth rate of the instability. Indeed, this instability vanishes whenever

$$\frac{\beta_w^2}{4}\beta_\parallel\gamma_\parallel^2\left(1 + \beta_\parallel\right) > 1. \tag{4.114}$$

4.2.1.6 Thermal Effects on the Instability

In the treatment of thermal effects upon the growth rate of the free-electron laser, we consider the regime in which the axial solenoidal field is absent. In general, however, the resonance condition $\omega = (k + k_w)v_\parallel$ implies that thermal effects become important when the ratio of the axial velocity spread $\Delta v_\parallel$ to the bulk axial velocity of the beam is $\Delta v_\parallel / v_\parallel \approx \text{Im}\, k/(\text{Re}\, k + k_w)$. In the high-gain Compton regime, this condition implies that thermal effects cannot be neglected when

$$\frac{\Delta v_\parallel}{v_\parallel} \approx \frac{\sqrt{3}}{4\gamma_\parallel^2}\left(\frac{\beta_w^2}{2\beta_\parallel^2}\frac{\omega_b^2}{\gamma_0 k_w^2 v_\parallel^2}\right)^{1/3} \approx \frac{\sqrt{3}\rho}{2\gamma_\parallel^2}. \tag{4.115}$$

Similarly, thermal effects are important in the collective Raman regime when

$$\frac{\Delta v_\parallel}{v_\parallel} \approx \frac{\beta_w}{4\gamma_\parallel^2\beta_\parallel}\sqrt{\frac{\omega_b}{\gamma_0^{1/2} c k_w}}. \tag{4.116}$$

The aforementioned formulations of the growth rate in the Raman and high-gain Compton regimes break down when these conditions are satisfied, and a more general formulation must be developed.

Since the free-electron laser operates by means of an axial bunching mechanism, it is the axial velocity spread that is most important. As a consequence, in the treatment of thermal effects on the linear stability properties, we shall impose the simplification that the electron beam is monoenergetic but exhibits a pitch angle spread [23]. The effect of the pitch angle spread is to include velocity spreads in both the axial and transverse directions and may be described by a distribution function of the form

$$F_b(P_x, P_y, p) = n_b G_\perp(P_x, P_y) \frac{p_z}{p} \delta(p - p_0), \tag{4.117}$$

where $G_\perp(P_x, P_y)$ represents the transverse distribution. For convenience, we shall assume that this transverse distribution takes the form of a Gaussian

$$G_\perp(P_x, P_y) = \frac{1}{\pi \Delta P^2} \exp(-P_\perp^2/\Delta P^2), \tag{4.118}$$

where $P_\perp^2 = P_x^2 + P_y^2$ and ΔP represents the thermal spread.

Under the assumption that $\Delta P << p_0$, the electron orbits can be represented in terms of the perturbed steady-state trajectories given in Eqs. (2.42, 2.43, and 2.44). In the absence of an axial solenoidal field, these orbits are of the form

$$p_\pm = (P_x \mp iP_y) + p_w \exp(\mp ik_w z), \tag{4.119}$$

and

$$p_z = \sqrt{P_\parallel^2 - 2p_w(P_x \cos k_w z + P_y \sin k_w z)}, \tag{4.120}$$

where $\gamma_0 = (1 + p_0^2/m_e^2 c^2)^{1/2}$, $p_w = \gamma_0 m_e v_w$, $P_\parallel^2 = p_0^2 - p_w^2 - P_x^2 - P_y^2$, and $v_w = -cK/\gamma$.

For simplicity, we neglect the left-hand circular polarization state and treat the high-gain regime in which the vector and scalar potentials may be expressed as in Eqs. (4.67) and (4.68). As a consequence, the source currents can be written as

$$\delta \hat{J}_+(z) = \frac{e^2}{2m_e c} \int dP_x dP_y dp \frac{p}{\gamma p_z} \left[\left(2 + \frac{p_+ p_-}{p_z^2} \right) D_+ + p_+ \left(\frac{\partial}{\partial P_x} + i \frac{\partial}{\partial P_y} \right) D_+ \right.$$
$$\left. - p_+ D_z \frac{\partial}{\partial p} \right] F_b(P_x, P_y, p), \tag{4.121}$$

and

$$\delta \hat{J}_z(z) = \frac{e}{2m_e c} \int dP_x dP_y dp \frac{p}{\gamma} \left[\left(\frac{\partial}{\partial P_x} + i \frac{\partial}{\partial P_y} \right) D_+ - D_z \frac{\partial}{\partial p} \right] F_b(P_x, P_y, p), \tag{4.122}$$

where the orbit integrals are

$$D_{\pm} = \delta\hat{A}_{\pm}(0)\exp[i\omega\tau(z,0)] - \delta\hat{A}_{\pm}(z) \tag{4.123}$$

and

$$D_z = i\frac{\gamma m_e c}{p}\int_0^z dz' \exp[i\omega\tau(z,z')]\left[-k\delta\hat{\varphi}(z') + \frac{\omega}{c}\frac{p_-}{p_z}\delta\hat{A}_{\pm}(z')\right]. \tag{4.124}$$

The orbit integrals may be evaluated using the perturbed steady-state trajectories (Eqs. (4.119) and (4.120)). We retain only the lowest-order contributions due to the perturbed steady-state trajectories in P_x and P_y that appear in the resonance condition. In this limit, the source currents become

$$\begin{aligned}\delta\hat{J}_+(z) \cong &-\frac{e^2}{m_e c}\delta\hat{A}_+(z)\int dP_x dP_y dp\left[\frac{p}{\gamma p_z}\left(1+\frac{\beta_w^2}{2}\right) - \frac{\beta_w^2}{2}\frac{\omega m_e v_\parallel}{\omega - kV_\parallel}\frac{\partial}{\partial p}\right]F_b(P_x,P_y,p)\\ &-\frac{e^2}{2m_e}p_w\delta\hat{\varphi}(z)\exp(-ik_w z)\int dP_x dP_y dp\frac{k}{\gamma(\omega - kV_\parallel)}\frac{\partial}{\partial p}F_b(P_x,P_y,p),\end{aligned} \tag{4.125}$$

and

$$\begin{aligned}\delta\hat{J}_z(z) \cong &\frac{e^2}{2m_e c}p_w\delta\hat{A}_+(z)\exp(ik_w z)\int dP_x dP_y dp\frac{\omega}{\gamma(\omega - kV_\parallel)}\frac{\partial}{\partial p}F_b(P_x,P_y,p)\\ &-\frac{e^2}{2m_e}\delta\hat{\varphi}(z)\exp(ik_w z)\int dP_x dP_y dp\frac{m_e k v_\parallel}{\omega - kV}\frac{\partial}{\partial p}F_b(P_x,P_y,p),\end{aligned} \tag{4.126}$$

where $V_\parallel = P_\parallel/\gamma_0 m_e$ is the axial velocity corresponding to the generalized steady-state trajectory, $v_\parallel = ({p_0}^2 - {p_w}^2)/\gamma_0 m_e$, and ${\beta_w}^2 = {v_w}^2/{v_\parallel}^2$. Note that $v_\parallel$ and v_w denote the axial and transverse velocities for the steady-state trajectory corresponding to γ_0.

The derivatives of the distribution (4.118) with respect to p which appear in the above expressions for the source currents may be integrated by parts and the results substituted into Maxwell's equations to give

$$\left[\omega^2 - k_+^2c^2 - \frac{\omega_b^2}{\gamma_0}\left(1 - \frac{a_w^2}{\gamma_0^2}\frac{\omega^2 - k^2c^2}{(\omega - kv_\parallel)^2}T(\zeta)\right)\right]\delta\hat{A}_+(0) = \frac{\omega_b^2}{2\gamma_0}\frac{a_w}{\gamma_0}\frac{ck\left(ck - \omega\beta_\parallel\right)}{(\omega - kv_\parallel)^2}T(\zeta)\delta\hat{\varphi}(0), \tag{4.127}$$

and

$$\left[(\omega - kv_{\parallel})^2 - \frac{\omega_b^2}{\gamma_0 \gamma_{\parallel}^2} T(\zeta)\right] \delta\hat{\varphi}(0) = \frac{\omega_b^2}{\gamma_0} \frac{a_w}{\gamma_0} \left(1 - \frac{\omega}{ck}\beta_{\parallel}\right) T(\zeta) \delta\hat{A}_{+}(0), \tag{4.128}$$

where we identify $k = k_+ + k_w$ from the wavenumber matching condition, $\beta_{\parallel} = v_{\parallel}/c$, ${\gamma_{\parallel}}^2 = (1 - {v_{\parallel}}^2/c^2)^{-1}$, and $T(\zeta)$ defines the thermal function. The thermal function that describes the effect of the pitch angle spread on the instability is defined as [23]

$$T(\zeta) \equiv \zeta[1 - \zeta \exp(\zeta) E_1(\zeta)], \tag{4.129}$$

for argument

$$\zeta \equiv \frac{\gamma_0^2 m_e^2}{\Delta P^2}\left(\frac{\omega^2}{k^2} - v_{\parallel}^2\right), \tag{4.130}$$

where

$$E_1(\zeta) \equiv \int_{\zeta}^{\infty} dt \frac{\exp(-t)}{t} \tag{4.131}$$

denotes the exponential integral function defined over the domain $|\arg \zeta| < \pi$.

The dispersion equation that results from this formulation is

$$\left[(\omega - kv_{\parallel})^2 - \frac{\omega_b^2}{\gamma_0 \gamma_{\parallel}^2} T(\zeta)\right]\left(\omega^2 - k_+^2 c^2 - \frac{\omega_b^2}{\gamma_0}\right) = -\frac{\omega_b^2}{2\gamma_0} \frac{a_w^2}{\gamma_0^2} T(\zeta)\left(\omega^2 - k^2c^2 - \frac{\omega_b^2}{\gamma_0}T(\zeta)\right), \tag{4.132}$$

correct to lowest nontrivial order in a_w. In order to verify that this dispersion equation reproduces that found in the idealized beam limit (Eq. (4.90)), we observe that $\lim_{\Delta P \to 0} |\zeta| = \infty$. Expanding the exponential integral function in the asymptotic limit, therefore, we find that $\lim_{\Delta P \to 0} T(\zeta) = 1$ and that the ideal beam dispersion equation (Eq. (4.90)) is recovered. Thermal effects become dominant whenever Im $k/(\mathrm{Re}\, k + k_w) \approx \Delta v_{\parallel}/v_{\parallel}$, where the wavenumber is to be evaluated at the peak growth rate in the ideal beam limit. On the basis of the perturbed trajectories, it is clear that $\Delta v_{\parallel}/v_{\parallel} \approx \Delta P^2/2{p_0}^2$; hence, thermal effects are important when

$$\frac{\Delta P^2}{p_{\parallel}^2} \approx \frac{\sqrt{3}}{2\gamma_{\parallel}^2}\left(\frac{\beta_w^2}{2\beta_{\parallel}^2} \frac{\omega_b^2}{\gamma_0 k_w^2 v_{\parallel}^2}\right)^{1/3} = \frac{\sqrt{3}}{\gamma_{\parallel}^2}\rho, \tag{4.133}$$

in the high-gain Compton regime, and

$$\frac{\Delta P^2}{p_\parallel^2} \approx \frac{\beta_w}{2\gamma_\parallel^2 \beta_\parallel} \sqrt{\frac{\omega_b}{\gamma_0^{1/2} ck_w}}, \tag{4.134}$$

in the collective Raman regime.

The effect of the thermal spread on the linear growth rate is threefold. In the first place, the wider range of axial velocities introduced thereby results in a broader resonance condition in which the unstable frequency band increases. In the second place, the fact that the bulk axial velocity decreases means that the center frequency of the gain band also decreases. In the third place, the peak growth rate decreases with increasing ΔP. Each of these properties is illustrated in Fig. 4.10 in which we solve Eq. (4.132) numerically for the growth rate and plot the magnitude of the growth rate as a function of the frequency for $\Omega_w/ck_w = 0.05$, $\gamma_0 = 2.957$, and $\omega_b/ck_w = 0.1$. Observe that the growth rate peaks for $|\mathrm{Im}\ k|/k_w \approx 0.011$ in the absence of the thermal spread and decreases by over 100% as the thermal spread increases to $\Delta P/p_0 = 5\%$. Comparison with Fig. 4.6 for this value of the plasma frequency indicates that this example is in the intermediate range between the high-gain Compton and the collective Raman regimes.

The detailed variation in the peak growth rate and the frequency corresponding to peak growth as a function of ΔP is illustrated in Fig. 4.11. As shown in the figure, the peak growth rate remains relatively constant for $\Delta P/p_0 < 2\%$ and decreases rapidly thereafter. As a consequence, thermal effects become dominant for $\Delta P/p_0 \geq 3\%$. The reason for the decline in the growth rate is that progressively more electrons are driven out of resonance with the wave as the thermal spread in the axial energy (i.e., velocity) increases. This effect can be quantified by consideration of the resonance condition $\omega \approx (k + k_w)v_z$, from which it is apparent that thermal effects dominate whenever $(\mathrm{Im}\ k)v_\parallel \approx \mathrm{Re}\ (k + k_w)\Delta v_\parallel$. Since this example is neither in the Raman nor the high-gain Compton regimes, we expect that for thermal effects to be important

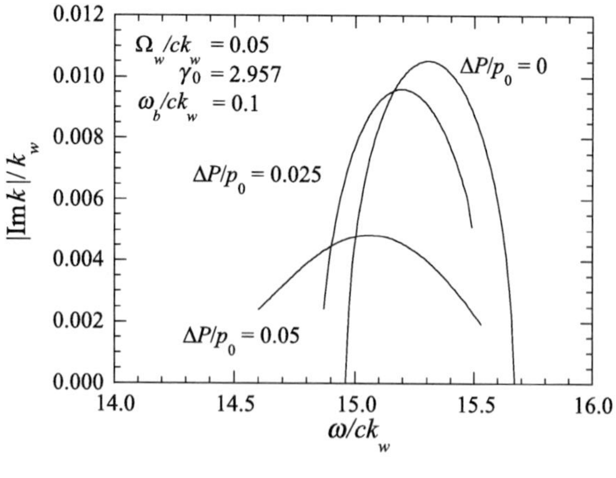

Fig. 4.10 Graph of the magnitude of the growth rate versus frequency for $\Delta P/p_0 = 0$, 0.025, and 0.05 and $\Omega_w/ck_w = 0.05$, $\gamma_0 = 2.957$, and $\omega_b/ck_w = 0.1$

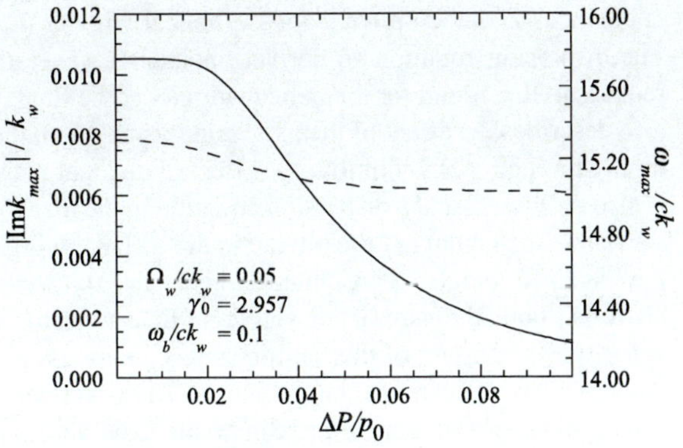

Fig. 4.11 Graph of the magnitude of the maximum growth rate (solid line) and the corresponding frequency (dashed line) as a function of $\Delta P/p_0$ for $\Omega_w/ck_w = 0.05$, $\gamma_0 = 2.957$, and $\omega_b/ck_w = 0.1$

$\Delta \upsilon_{\|}/\upsilon_{\|} \approx \text{Im}\, k/(1 + \text{Re}\, k) \approx 0.067\%$ [where Im $k/k_w \approx 0.011$ and Re $k/k_w \approx 15.4$ at peak growth]. This yields an estimate of $\Delta P/p_0 \approx 3.7\%$, which is in reasonable agreement with the numerical solution for the growth rate shown in the figure. The frequency at which peak growth is found drops approximately 2% over this range.

4.2.2 Planar Wiggler Configurations

The physics of the free-electron laser interaction in planar wiggler configurations is similar but not identical to that for helical wiggler configurations. The principal difference arises due to the oscillations in the axial velocity that were discussed in Chap. 2 (see Eq. (2.104)), which have two major effects on the interaction. The first effect is that the bulk resonance condition and the bulk gain are determined by the average value of the axial and transverse velocities. Hence, the effective wiggler field is reduced from the peak on-axis field by a factor of $\sqrt{2}$ (i.e., the root-mean-square value of the wiggler amplitude). The second effect is that the oscillation in the axial velocity results in an interaction at the even harmonics [i.e., $\omega \approx (k + lk_w)\upsilon_{\|}$, where l is an even integer]. The oscillation in the axial velocity also results in an oscillation in the gain about the bulk value. This effect is beyond the scope of this chapter, however, and will be discussed in some detail in Chap. 5 in which a nonlinear analysis and simulation of the interaction in the presence of a planar wiggler is presented.

The analysis of planar wiggler configurations is presented in a similar manner to that for the helical wiggler. The low-gain regime is discussed after a basic discussion of the application of the Vlasov-Maxwell formalism to this wiggler geometry. The expression for the gain in this regime (4.159) includes the effects of the oscillation in the axial velocity both to the bulk gain and to the excitation of harmonics.

A derivation of the general dispersion equation in the high-gain regime is also presented for the case of an ideal beam (see Eq. (4.174)), and the similarities to that found for the helical wiggler are discussed. A derivation of the dispersion equation subject to the inclusion of beam thermal effects follows. The beam distribution used is identical to that employed for the helical wiggler (see Eqs. (4.117) and (4.118)). The dispersion equation so derived includes a generalized thermal function which reduces to that found for the helical wiggler at the fundamental resonance, but which also describes the effect of the axial energy spread on the harmonic interaction in the high-gain regime. In addition, the effect of an axial magnetic field on the interaction is also discussed, and a dispersion equation in the ideal beam limit is derived (4.209).

Within the context of the interaction in a planar wiggler configuration, we employ the idealized wiggler representation given in Eq. (2.99) and perform a perturbation analysis about the associated single-particle trajectories. The y-component of the velocity is a constant of the motion in this geometry, $\upsilon_y = P_y/\gamma m_e$ where P_y is the y-component of the canonical momentum. The x-component of the velocity is given in Eq. (2.101), and we may approximate the axial velocity as

$$\upsilon_z = V_\parallel \sqrt{1 - \frac{\upsilon_w^2}{2V_\parallel^2}\cos 2k_w z - 2\frac{P_x}{P_\parallel}\frac{\upsilon_w}{V_\parallel}\cos k_w z}, \tag{4.135}$$

where P_x denotes the canonical momentum in the x-direction, $\upsilon_w = -cK/\gamma$ is the wiggler-induced velocity,

$$V_\parallel \equiv c\sqrt{1 - \frac{1}{\gamma^2} - \frac{\upsilon_w^2}{2c^2} - \frac{P_\perp^2}{\gamma^2 m_e^2 c^2}}, \tag{4.136}$$

${P_\perp}^2 = {P_x}^2 + {P_y}^2$, and we define $P_\parallel = \gamma m_e V_\parallel$. Note that this expression reduces to the bulk axial velocity (2.105) found in the limit in which the canonical momentum vanishes.

4.2.2.1 The Source Currents

The interaction occurs principally for plane waves polarized in the direction of the wiggler-induced oscillation, and the vector and scalar potentials for a wave with angular frequency ω are of the form

$$\delta\mathbf{A}(z,t) = \frac{1}{2}\delta\hat{A}(z)\hat{\mathbf{e}}_x \exp(-i\omega t) + c.c., \tag{4.137}$$

and

$$\delta\varphi(z,t) = \frac{1}{2}\delta\hat{\varphi}(z)\exp(-i\omega t) + c.c.. \tag{4.138}$$

As in the case of the helical wiggler analysis, the perturbed distribution function takes the form

$$\delta f_b(z,\mathbf{p}) = \frac{e}{2c}\left[D_x\frac{\partial}{\partial P_x} + D_z\frac{\partial}{\partial p}\right]F_b(P_xP_y,p), \tag{4.139}$$

where the orbit integrals are defined as

$$D_x \equiv \int_0^z dz'\frac{\exp[i\omega\tau(z,z')]}{v_z(z')}\left(i\omega - v_z(z')\frac{\partial}{\partial z'}\right)\delta\hat{A}(z'), \tag{4.140}$$

and

$$D_z \equiv \frac{1}{p}\int_0^z dz'\frac{\exp[i\omega\tau(z,z')]}{v_z(z')}\left[-cp_z(z')\frac{\partial}{\partial z'}\delta\hat{\varphi}(z') + i\omega p_x(z')\delta\hat{A}(z')\right], \tag{4.141}$$

$\tau(z,z') = \tau(z) - \tau(z')$, and $\tau(z)$ is the Lagrangian time variable defined in Eq. (4.27) and is defined as in Eq. (4.36). Observe that D_x may be integrated directly to give

$$D_x = -\delta\hat{A}(z) + \delta\hat{A}(0)\exp[i\omega\tau(z,0)]. \tag{4.142}$$

The source current

$$\delta\mathbf{J}(z,t) = \left[\delta\hat{J}_x(z)\hat{\mathbf{e}}_x + \delta\hat{J}_z(z)\hat{\mathbf{e}}_z\right]\exp(-i\omega t) + c.c., \tag{4.143}$$

is determined by integration over the perturbed distribution

$$\delta\hat{\mathbf{J}}(z) = -\frac{e}{m_e}\int dP_x dP_y dp\frac{p}{\gamma}\delta f_b\left(\frac{p_x}{p_z}\hat{\mathbf{e}}_x + \hat{\mathbf{e}}_z\right). \tag{4.144}$$

Substitution of the perturbed distribution (4.139) yields, after integration by parts over P_x,

$$\delta\hat{J}_x(z) = \frac{e}{m_e}\int dP_x dP_y dp\frac{p}{\gamma}\left[\frac{\partial}{\partial P_x}\left(\frac{p_x}{p_z}D_x\right) - \frac{p_x}{p_z}D_z\frac{\partial}{\partial p}\right]F_b(P_x,P_y,p), \tag{4.145}$$

and

$$\delta\hat{J}_z(z) = \frac{e}{m_e}\int dP_x dP_y dp\frac{p}{\gamma}\left[\frac{\partial}{\partial P_x}D_x - D_z\frac{\partial}{\partial p}\right]F_b(P_x,P_y,p). \tag{4.146}$$

The determination of the linear dispersion equation requires the evaluation of the source currents for representative distribution functions. The orbit integrals D_x and D_z that appear in the source currents represent integration over the unperturbed electron trajectories in the planar wiggler. The characteristic trajectories in a planar wiggler differ from those in a helical wiggler in that the magnitudes of the axial and transverse velocity components are not constant but, rather, oscillate at harmonics of the wiggler period. This, in turn, introduces harmonic components into the dynamics of the interaction. In particular, we observe that in a planar wiggler, the Lagrangian time variable characteristic of the electron trajectories is of the approximate form

$$\tau(z) \cong t_0 + \frac{z}{V_\parallel} + \frac{P_x}{P_\parallel}\frac{v_w}{V_\parallel}\frac{1}{k_w V_\parallel}\sin k_w z + \frac{v_w^2}{8V_\parallel^2}\frac{1}{k_w V_\parallel}\sin 2k_w z, \tag{4.147}$$

where it is assumed that both $v_w < V_\parallel$ and $P_x < P_\parallel$. Observe that the existence of a nonvanishing canonical momentum introduces an oscillation at the wiggler period into the trajectory.

4.2.2.2 The Pierce Parameter and the *JJ*-Factor

The interaction in a planar wiggler differs from that in a helical wiggler in two respects. First, the axial and transverse velocities are oscillatory so that the average axial and transverse velocities are given by the rms magnitude of the wiggler field. Second, the interaction derives from the beating of the wiggler and radiation fields. In general, there are upper and lower beat waves. The lower beat wave is suppressed due to the symmetry of a helical wiggler, but it is present in the case of a planar wiggler and acts to reduce the strength of the interaction.

The *JJ*-factor describes the impact of the lower beat wave on the interaction. In this work, we define

$$JJ_n(b) \equiv J_n(b) - J_{n+1}(b). \tag{4.148}$$

It should be remarked, however, that the *JJ*-factor is sometimes defined as the product $K_{rms}[J_n(b) - J_{n+1}(b)]$ in the literature.

In order to illustrate the source of the *JJ*-factor, we observe that the interaction is determined by the average $<\mathbf{v}\cdot\delta\mathbf{E}>$ over the electrons in the beam, where $\mathbf{v}$ is the electron velocity and $\delta\mathbf{E}$ is the radiation electric field. If we assume a planar undulator oriented in the *y*-direction as shown in Eq. (2.99), then the wiggle-velocity in the *x*-direction is given by $v_x = v_w\cos k_w z$ (Eq. (2.103)). The interaction will be with an plane-polarized electric field in the *x*-direction, and we assume that $\delta E_x = \delta E_0\sin(kz - \omega t)$. As a result, the inner product contains the sum of the upper and lower beat waves

$$v_x\delta E_x = \frac{1}{2}v_w\delta E_0\{\sin[(k+k_w)z - \omega t] + \sin[(k-k_w)z - \omega t]\}, \tag{4.149}$$

However, under the assumption that $v_w << c$, Eq. (2.104) implies that

$$v_z \cong v_\parallel \left(1 - \frac{v_w^2}{4c^2}\cos 2k_w v_\parallel t\right), \tag{4.150}$$

and

$$z \cong z_0 + v_\parallel t - \frac{v_w^2}{8k_w c^2}\sin 2k_w v_\parallel t. \tag{4.151}$$

Note that $v_w^2/c^2 = 2K_{rms}^2/\gamma^2$. For simplicity, we also assume that $z_0 = 0$, so that

$$v_x \delta E_x = \frac{v_w \delta E_0}{4i}\left\{\exp\left[i\left(\psi - b\sin 2k_w v_\parallel t\right)\right] + \exp\left[i\left(\psi - 2k_w v_\parallel t - b\sin 2k_w v_\parallel t\right)\right]\right\} + c.c., \tag{4.152}$$

where $\psi = [(k + k_w)v_\parallel - \omega]t$ is the usual ponderomotive phase that arises due to the upper beat wave and $b = (k + k_w)v_w^2/8k_w c^2$. Observe that the first (second) term on the right-hand side corresponds to the upper (lower) beat wave. Using the Bessel function identity

$$\exp(ib\sin\theta) = \sum_{n=-\infty}^{\infty} J_n(b)\exp(in\theta), \tag{4.153}$$

we have that

$$v_x \delta E_x = \frac{v_w \delta E_0}{4i}\sum_{n=-\infty}^{\infty}(-1)^n[J_n(b) - J_{n+1}(b)] \times \exp\left\{i\left[(k + (1+2n)k_w)v_\parallel - \omega\right]t\right\} + c.c., \tag{4.154}$$

where the term in $J_{n+1}(b)$ comes from the lower beat wave. Observe that the JJ-factor describes the effect of the lower beat wave on the $(n+1)^{\text{th}}$ harmonic resonance.

The Pierce parameter for the fundamental resonance in a planar wiggler is given in terms of the rms value of the wiggler field strength and the JJ-factor

$$\rho^3 \equiv \frac{K_{rms}^2 JJ_0^2(b)}{4\gamma_0^3 \beta_\parallel^2}\frac{1}{k_w^2 R_b^2}\frac{I_b}{I_A}. \tag{4.155}$$

where $b = K_{rms}^2/2(1 + K_{rms}^2)$.

4.2.2.3 The Low-Gain Regime

The analysis of the low-gain regime for planar wiggler configurations is analogous to that for the helical wiggler system. We assume that the beam is sufficiently diffuse that space-charge forces are negligible and focus attention on the linearly polarized electromagnetic wave that can be represented in the form

$$\delta\hat{A}(z) = \sum_{n=-\infty}^{\infty} \delta\hat{A}(0)\exp\left[i\int_0^z dz' k_n(z')\right], \tag{4.156}$$

where $k_n(z) = k(z) + nk_w$, and we assume that $|\mathrm{Im}\, k(z)| << |\mathrm{Re}\, k(z)|$. We impose the additional assumption that $k(z) = k_0 + \delta k(z)$, where k_0 is real and independent of z and $|\delta k(z)| << k_0$. In the tenuous beam limit, the dispersion of the wave is given approximately by the free-space value

$$\frac{\omega^2}{c^2} - k_{0n}^2 \cong 0, \tag{4.157}$$

where $k_{0n} = k_0 + nk_w$, and the gain is determined by the imaginary part of $\delta k(z)$. Gain is driven by the x-component of the source current, and from Maxwell's equations, we obtain to lowest order in $\delta k(z)$ that

$$\sum_{n=-\infty}^{\infty} k_{0n}\delta k(z)\delta\hat{A}_n(z) \cong \frac{4\pi}{c}\delta\hat{J}_x(z). \tag{4.158}$$

The imaginary part of the x-component of the source current is determined primarily by the term in D_z, which can be written approximately as

$$D_z \cong \frac{ie}{2cp}\gamma m_e\omega \sum_{n=-\infty}^{\infty} \delta\hat{A}_n(z)\exp[-ik_{0n}z + i\omega\tau(z)] \times \int_0^z dz' \frac{v_x(z')}{v_z(z')}\exp[ik_{0n}z - i\omega\tau(z')]. \tag{4.159}$$

Under the assumption that P_x and $P_y < |\gamma m_e v_w|$, the orbit integral in Eq. (4.159) can be reduced to

$$\int_0^z dz' \frac{v_x(z')}{v_z(z')}\exp[ik_{0n}z - i\omega\tau(z')] \cong i\frac{v_w}{V_\parallel}\exp(-i\omega t_0)$$
$$\times \sum_{l,m=-\theta}^{\infty} (-1)^{l+m}\frac{l}{b_1}J_l(b_1)J_m(b_2)\frac{\exp(-ik_{n+l+2m}z) - 1}{\Delta k_{n+l+2m}}, \tag{4.160}$$

where

$$\Delta k_n \equiv \frac{\omega}{V_\parallel} - k_{0n}, \tag{4.161}$$

$$b_1 \equiv \frac{\omega}{k_w V_\parallel} \frac{v_w}{V_\parallel} \frac{P_x}{P_\parallel}, \tag{4.162}$$

$$b_2 \equiv \frac{\omega}{k_w V_\parallel} \frac{v_w^2}{8V_\parallel^2}, \tag{4.163}$$

and we have made use of the Bessel function identity (Eq. (4.153)). As a consequence,

$$\frac{p_x}{p_z} D_z \cong -\frac{e\omega}{2cp}\gamma m_e \frac{v_w^2}{4V_\parallel^2} \sum_{n,l,m=-\infty}^{\infty} \delta\hat{A}_n J_l^2(b_1) K_m^{(\pm)}(b_2) \frac{1-\exp(i\Delta k_{n+l+2m\pm 1} z)}{\Delta k_{n+l+2m\pm 1}}, \tag{4.164}$$

where

$$K_m^{(\pm)}(b) \equiv [J_m(b) - J_{m\pm 1}(b)]^2, \tag{4.165}$$

and we have made use of the Bessel function recursion relation

$$\frac{2n}{x} J_n(x) = J_{n-1}(x) + J_{n+1}(x). \tag{4.166}$$

Observe that $K_m^{(+)}(b) = JJ_m^2(b)$ and that this expression for D_z contains harmonic contributions to the interaction. The presence of a nonvanishing canonical momentum is related to a beam pitch angle spread and results in resonant interactions at all harmonics [i.e., $\omega \approx (k_0 + lk_w)V_\parallel$, for integer l]. Harmonics arise in the absence of a pitch angle spread due to the oscillation of the axial velocity due to the planar geometry. In this case, odd harmonics are excited in which $\omega \approx [k_0 + (1 + 2m)k_w]V_\parallel$.

The gain in power over an interaction length L is given in Eq. (4.59) which can be expressed as

$$G(L) = \frac{\omega_b^2 L^3}{16c} \sum_{l,m=-\infty}^{\infty} \iint dP_x dP_y G_\perp(P_x, P_y) \int_0^\infty dp \frac{p}{\gamma p_z} G_b(p) \times \frac{v_w^2}{\gamma_\parallel^2 V_\parallel^4} J_l^2(b_1) K_m^{(\pm)}(b_2) F(\Theta_{l+2m\pm 1}), \tag{4.167}$$

where we have assumed a distribution function of the form

$$F_b(P_x, P_y, p) = n_b G_\perp(P_x, P_y) G_b(p), \tag{4.168}$$

for an ambient density n_b, $\omega_b{}^2 = 4\pi e^2 n_b/m_e$, $\gamma_\| = (1 - V_\|{}^2/c^2)^{-1/2}$, and $F(\Theta_{l+2m\pm1})$ is the spectral function defined in Eq. (4.63) for

$$\Theta_n \equiv \frac{L}{2}\left(\frac{\omega}{V_\|} - k_0 - nk_w\right). \tag{4.169}$$

Equation (4.167) describes the small-signal gain corresponding to two Doppler-shifted frequencies, that is, $\omega \approx [k_0 + (l + 2m + 1)k_w]V_\|$ and $\omega \approx [k_0 + (l + 2m - 1)k_w] V_\|$. The former case corresponds to the frequency upshift associated with the slowly varying ponderomotive potential, while the latter case describes a frequency downshift. In order to understand the underlying physical mechanism for these two resonance conditions, recall that the free-electron laser operates by means of a beating between the wiggler and radiation fields. The upper beat wave is associated with the slowly varying ponderomotive potential and a frequency upshift. In contrast, the lower beat wave is associated with a frequency downshift. In particular, for a wave satisfying the upshifted resonance condition, the lower beat wave describes a rapid oscillation with a period half that of the wiggler. Since this rapid oscillation is nonresonant, it will produce no net gain of the signal and will be neglected, henceforth, in the linear formulation. However, we shall return to this issue within the context of the nonlinear formulation.

Restricting the discussion to the resonant frequency upshift, we consider the small-signal gain in the cold-beam limit for which $G_\perp(P_x,P_y) = \delta(P_x)\delta(P_y)$ and $G_b(p)$ is given in Eq. (4.61). In this regime, $b_1 = 0$ and only the $l = 0$ term survives. Therefore, the gain for the fundamental and each even harmonic contribution in the cold regime is

$$G(L) \cong \frac{\omega_b^2 L^3 k_0}{16\gamma_0\gamma_\|^2 v_\|^2}\beta_w^2 K_m^{(+)}(b_2)F(\Theta_{2m+1}), \tag{4.170}$$

or

$$G(L) \cong 4(2\pi\rho)^3\left(\frac{L}{\lambda_w}\right)^3 K_m^{(+)}(b_2)F(\Theta_{2m+1}), \tag{4.171}$$

where $v_\|$ is the bulk axial velocity defined in Eq. (2.105), $\gamma_0 = (1 + p_0{}^2/m_e{}^2c^2)^{1/2}$, $\beta w = |v_w/v_\||$, and

$$b_2 \cong \frac{a_w^2(2m+1)}{2(2+a_w^2)}, \tag{4.172}$$

for $a_w = eA_w/m_ec^2 = K$. The small-signal gain for the planar wiggler configuration, therefore, is similar to that found for a helical wiggler (4.62). There are, however, three distinctions that merit discussion. In the first place, the magnitudes of the transverse and axial components of the velocity are oscillatory for planar wigglers; hence, the average value of the transverse velocity is reduced by a factor of $\sqrt{2}$ with respect to that found for a helical wiggler of comparable magnitude. This implies, in turn, that for a fixed magnitude of the wiggler field the gain is reduced by half with respect to that of a helical wiggler. In the second place, since the axial velocity is also oscillatory, the electrons will maintain no fixed resonance with the wave. Hence, the gain will be modulated, and the average gain will be reduced. This effect is described by $K_m^{(+)}$ or alternately by the *JJ*-factor. In the third and last place, the oscillatory nature of the axial velocity permits resonant interactions to occur at the odd harmonics. In contrast, harmonic interactions do not appear within the context of the idealized one-dimensional formulation of the interaction for a helical wiggler since the axial velocity is constant in the limit in which the canonical momenta vanish. This last distinction, however, is somewhat arbitrary since harmonic radiation does appear for a helical wiggler in the three-dimensional formulation and will be discussed for both linear and nonlinear formulations.

4.2.2.4 The High-Gain Regime

The analysis for the high-gain regime follows that discussed in Ref. [23]. We express the vector and scalar potentials by application of Floquet's theorem for periodic systems in the form

$$\delta\hat{A}(z) = \sum_{n=-\infty}^{\infty} \delta\hat{A}_n \exp(ik_nz), \tag{4.173}$$

and

$$\delta\hat{\varphi}(z) = \sum_{n=-\infty}^{\infty} \delta\hat{\varphi}_n \exp(ik_nz), \tag{4.174}$$

where $k_n = k + nk_w$. We first impose the condition that the pitch angle spread of the beam vanishes, and assume a distribution of the form given in Eq. (4.45). Since the gain is exponential in this regime, we may neglect the initial value contributions to the orbit integrals. As a consequence, the source current can be expressed as

$$\delta\hat{J}_x(z) = -\frac{\omega_b^2}{8\pi c}\sum_{n=-\infty}^{\infty}\delta\hat{A}_n\exp(ik_nz)\int_0^\infty dp\frac{p}{\gamma p_z}G_b(p)\left(1+\frac{p_x^2}{p_z^2}\right)$$
$$+\frac{\omega_b^2}{32\pi c}\sum_{m,n=-\infty}^{\infty}\delta\hat{A}_n\exp(ik_nz)\int_0^\infty dpK_m^{(\pm)}(b_2)\frac{m_e\omega\beta_w^2 v_\|}{\omega-k_{n+2m\pm1}v_\|}\frac{\partial}{\partial p}G_b(p)$$
$$-\frac{\omega_b^2}{16\pi}\sum_{m,n=-\infty}^{\infty}\delta\hat{\varphi}_n\exp(ik_{n\mp1}z)\int_0^\infty dpJ_m(b_2)[J_m(b_2)-J_{m\pm1}(b_2)]$$
$$\times\frac{m_ek_nv_w}{\omega-k_{n+2m}v_\|}\frac{\partial}{\partial p}G_b(p), \tag{4.175}$$

and

$$\delta\hat{J}_z(z) = -\frac{\omega_b^2}{8\pi}\sum_{m,n=-\infty}^{\infty}\delta\hat{\varphi}_n\exp(ik_nz)J_m^2(b_2)\frac{m_ek_nv_w}{\omega-k_{n+2m}v_\|}\frac{\partial}{\partial p}G_b(p)$$
$$+\frac{\omega_b^2}{16\pi c}\sum_{m,n=-\infty}^{\infty}\delta\hat{A}_{n\mp1}\exp(ik_nz)\int_0^\infty dpJ_m(b_2)[J_m(b_2)-J_{m\pm1}(b_2)]$$
$$\times\frac{m_e\omega v_\|}{\omega-k_{n+2m}v_\|}\frac{\partial}{\partial p}G_b(p). \tag{4.176}$$

We now assume a cold-beam distribution as in Eq. (4.61) and integrate by parts in p. Retaining only the terms that correspond to the Doppler frequency upshift, we find that the source currents may be approximated by

$$\delta\hat{J}_x(z) \cong -\frac{\omega_b^2}{8\pi\gamma_0 c}\sum_{n=-\infty}^{\infty}\delta\hat{A}_n\exp(ik_nz)\left(1-\frac{v_w^2}{4c^2}\sum_{m=-\infty}^{\infty}K_m^{(+)}(b_2)\frac{\omega^2-c^2k_{n+2m+1}^2}{\left(\omega-k_{n+2m+1}v_\|\right)^2}\right)$$
$$+\frac{\omega_b^2}{16\pi\gamma_0}\sum_{m,n=-\infty}^{\infty}k_n\delta\hat{\varphi}_n\exp(ik_{n-1}z)J_m(b_2)[J_m(b_2)-J_{m+1}(b_2)]$$
$$\times\frac{v_w\left(k_{n+2m}-\omega v_\|/c^2\right)}{\left(\omega-k_{n+2m}v_\|\right)^2}, \tag{4.177}$$

and

$$\delta \hat{J}_z(z) \cong \frac{\omega_b^2}{8\pi\gamma_0\gamma_\parallel^2} \sum_{m,n=-\infty}^{\infty} k_n \delta\hat{\varphi}_n \exp(ik_n z) J_m^2(b_2) \frac{\omega}{\left(\omega - k_{n+2m}v_\parallel\right)^2}$$
$$- \frac{\omega_b^2}{16\pi\gamma_0} \frac{\omega}{c} \sum_{m,n=-\infty}^{\infty} \delta\hat{A}_{n-1} \exp(ik_n z) J_m(b_2)[J_m(b_2) - J_{m+1}(b_2)] \quad (4.178)$$
$$\times \frac{v_w\left(k_{n+2m} - \omega v_\parallel/c^2\right)}{\left(\omega - k_{n+2m}v_\parallel\right)^2}.$$

Note that in the derivation of Eq. (4.176), we have made use of the identity

$$\sum_{m=-\infty}^{\infty} [J_m(x) - J_{m\pm 1}(x)]^2 = 2. \quad (4.179)$$

Substitution of these forms for the source currents into Maxwell's equations yields the following set of coupled equations for the vector and scalar potentials

$$\left[\left(\omega - k_{n+2m+1}v_\parallel\right)^2 - \frac{\omega_b^2}{\gamma_0\gamma_\parallel^2}\right] \delta\hat{\varphi}_n J_m(b_2)$$
$$\cong -\frac{\omega_b^2}{\gamma_0 k_n} \frac{v_w}{2c} \delta\hat{A}_{n-1}[J_m(b_2) - J_{m+1}(b_2)]\left(k_{n+2m} - \omega\frac{v_\parallel}{c^2}\right), \quad (4.180)$$

and

$$\left[\omega^2 - c^2k_{n-1}^2 - \frac{\omega_b^2}{\gamma_0}\left(1 - \frac{v_w^2}{4c^2} K_m^{(+)}(b_2) \frac{\omega^2 - c^2k_{n+2m}^2}{\left(\omega - k_{n+2m}v_\parallel\right)^2}\right)\right] \delta\hat{A}_{n-1}$$
$$\cong -\frac{\omega_b^2}{2\gamma_0} \frac{k_n v_w\left(k_{n+2m} - \omega v_\parallel/c^2\right)}{\left(\omega - k_{n+2m}v_\parallel\right)^2} J_m(b_2)[J_m(b_2) - J_{m+1}(b_2)]\delta\hat{\varphi}_n, \quad (4.181)$$

where we have focused on a single harmonic contribution. The dispersion equation is obtained by the requirement that the determinant of the coefficients vanishes, which gives

$$\left[\left(\omega - k_{n+2m+1}v_\parallel\right)^2 - \frac{\omega_b^2}{\gamma_0\gamma_\parallel^2}\right]\left(\omega^2 - c^2k_{n-1}^2 - \frac{\omega_b^2}{\gamma_0}\right)$$
$$\cong -\frac{v_w^2}{4c^2} \frac{\omega_b^2}{\gamma_0} K_m^{(+)}(b_2)\left(\omega^2 - c^2k_{n+2m}^2 - \frac{\omega_b^2}{\gamma_0}\right). \quad (4.182)$$

This dispersion equation is similar to that found for the helical wiggler configuration in Eq. (4.86). The differences associated with the planar wiggler are the same as that found for low-gain regime; specifically, (1) the wiggler field is replaced by the root-mean-square amplitude (i.e., B_w is replaced by $B_w/\sqrt{2}$), (2) the inclusion of the $K_m^{(+)}$ factor describes the effect of the lower beat wave, and (3) the amplification of signals at harmonic frequencies is present in the idealized one-dimensional formulation. As such, the dispersion equation for the fundamental resonance can be reduced to a cubic equation

$$\delta k(\delta k + 2\kappa_b)(\delta k - \Delta k) \cong -(2\rho k_w)^3, \tag{4.183}$$

which is directly analogous to Eq. (4.93) for the helical wiggler, except where ρ is now given by Eq. (4.155), κ_b is given by Eq. (4.80) for $\Phi = 1$, δk is given in Eq. (4.94), and Δk is given in Eq. (4.95). As such, the solutions for the growth rates and gain lengths carry over directly from the solutions to Eq. (4.93).

4.2.2.4.1 The Compton and Raman Regimes

The *high-gain Compton regime* (sometimes called the *strong-pump* regime) is found when the effect of the ponderomotive potential is greater than that of the space-charge potential of the beam-plasma waves. In this limit, $|\delta k| >> 2\kappa_b$, and the dispersion equation reduces to a cubic equation

$$\delta k^2(\delta k - \Delta k) \cong -(2\rho k_w)^3, \tag{4.184}$$

This is the dominant regime for most short-wavelength free-electron lasers driven by high-energy electron beams.

As in the case of a helical wiggler, complex roots of the dispersion equation are found (i.e., growth) when $\Delta k > -3\sqrt[3]{2}\rho$, which implies that $\omega < \left(1 + 3\sqrt[3]{2}\rho\right)\omega_{res}$ when $\omega_b/\gamma_0 << \omega$ and $\omega_{res} = 2\gamma_\parallel^2 k_w v_\parallel$. Maximum growth is found to correspond to zero detuning (i.e., $\Delta k = 0$) for which

$$\frac{\delta k_{\max}}{k_w} \cong \left(1 \pm i\sqrt{3}\right)\rho. \tag{4.185}$$

As a result, the power gain (exponentiation) length, $L_G = (2\text{Im}\ \delta k_{\max})^{-1}$, is given by

$$L_G = \frac{\lambda_w}{4\sqrt{3}\pi\rho}. \tag{4.186}$$

In order to neglect the effect of the space-charge fields, therefore, we must require that

$$\frac{\omega_b}{\gamma_0^{1/2} c k_w} << \frac{\gamma_\parallel^3}{4} \frac{v_w^2}{c^2} JJ_0{}^2(b_2). \tag{4.187}$$

The *Raman* (or collective) regime is found when the space-charge fields dominate over the ponderomotive potential. This regime is generally found in the long-wavelength, microwave regime where the electron beam energy is relatively low. In this limit, $|\delta k| << 2\kappa_b$, and the dispersion equation reduces to a quadratic equation for δk

$$\delta k(\delta k - \Delta k) \cong -\frac{\beta_w^2}{8} JJ_0{}^2(b_2) \frac{\omega_b^2}{\gamma_0 c^2} k_w, \tag{4.188}$$

where peak growth is again found for zero detuning. This peak growth rate is given by (see Eq. (1.11))

$$\frac{\delta k_{\max}}{k_w} \cong i \frac{\beta_w}{2\sqrt{2}} JJ_0(b_2) \sqrt{\frac{\omega_b}{\gamma_0^{1/2} c k_w}}, \tag{4.189}$$

which is valid for sufficiently high beam densities that

$$\frac{\omega_b}{\gamma_0^{1/2} c k_w} >> \frac{\gamma_\parallel^3}{4} \frac{v_w^2}{c^2} JJ_0{}^2(b_2). \tag{4.190}$$

4.2.2.5 Thermal Effects on the Instability

In the analysis of the thermal regime in planar wiggler configurations, we employ the same distribution (4.117) used in the treatment of thermal effects in helical wiggler geometries, which describes a monoenergetic beam with a pitch angle spread. The dominant contribution of the axial thermal spread occurs within the resonance condition; hence, if we restrict the analysis to the resonance associated with the Doppler upshift in frequency, then the source currents can be expressed in the form

$$\begin{aligned}\delta\hat{J}_x(z) \cong & -\frac{\omega_b^2}{8\pi\gamma_0 c}\sum_{n=-\infty}^{\infty}\delta\hat{A}_n\exp(ik_n z)\\ & \times\left[1-\frac{v_w^2}{4c^2}\sum_{l,m=-\infty}^{\infty}K_m^{(+)}(b_2)\int\int dP_x dP_y G_\perp(P_x,P_y)J_l^2(b_1)\frac{\omega^2-c^2k_{n+l+2m+1}^2}{\left(\omega-k_{n+l+2m+1}V_\parallel\right)^2}\right]\\ & +\frac{\omega_b^2}{16\pi\gamma_0}\sum_{l,m,n=-\infty}^{\infty}k_n\delta\hat{\varphi}_n\exp(ik_{n-1}z)J_m(b_2)[J_m(b_2)-J_{m+1}(b_2)]\\ & \times\int\int dP_x dP_y G_\perp(P_x,P_y)J_l^2(b_1)\frac{v_w\left(k_{n+l+2m}-\omega v_\parallel/c^2\right)}{\left(\omega-k_{n+l+2m}V_\parallel\right)^2}\end{aligned} \tag{4.191}$$

and

$$\begin{aligned}\delta\hat{J}_z(z) \cong & \frac{\omega_b^2}{8\pi\gamma_0\gamma_\parallel^2}\sum_{m,n=-\infty}^{\infty}k_n\delta\hat{\varphi}_n\exp(ik_n z)J_m^2(b_2)\\ & \times\int\int dP_x dP_y G_\perp(P_x,P_y)J_l^2(b_1)\frac{\omega}{\left(\omega-k_{n+l+2m}V_\parallel\right)^2}\\ & -\frac{\omega_b^2}{16\pi\gamma_0}\frac{\omega}{c}\sum_{l,m,n=-\infty}^{\infty}\delta\hat{A}_{n-1}\exp(ik_n z)J_m(b_2)[J_m(b_2)-J_{m+1}(b_2)]\\ & \times\int\int dP_x dP_y G_\perp(P_x,P_y)J_l^2(b_1)\frac{v_w\left(k_{n+l+2m}-\omega v_\parallel/c^2\right)}{\left(\omega-k_{n+l+2m}V_\parallel\right)^2}\end{aligned} \tag{4.192}$$

where $V_\parallel$ and $v_\parallel$ are the bulk axial velocities with Eq. (4.136) and without Eq. (2.105) the pitch angle spread, respectively. The integrals over the canonical momenta in Eqs. (4.184) and (4.192) may be evaluated using the transverse distribution function (4.114), and we find that in the limit in which $P_\perp << p_\parallel$

$$\iint dP_x dP_y G_\perp(P_x,P_y)\frac{J_l^2(b_1)}{\left(\omega-kV_\parallel\right)^2}\cong\frac{T_l(\zeta)}{\left(\omega-kv_\parallel\right)^2}, \tag{4.193}$$

where

$$T_l(\zeta)\equiv\frac{\zeta^2}{2\pi}\int_0^{2\pi}d\phi\int_0^{\infty}dz\exp(-z)\frac{J_l^2(b_1)}{(z-\zeta)^2}, \tag{4.194}$$

$$\zeta\equiv\frac{\gamma_0^2 m_e^2}{\Delta P^2}\left(\frac{\omega^2}{k^2}-v_\parallel^2\right), \tag{4.195}$$

and we may write that

$$b_1 = \frac{\omega}{k_w v_\parallel} \frac{v_w}{v_\parallel} \frac{\Delta P}{p_\parallel} z^{1/2} \cos\phi. \tag{4.196}$$

If we now select a specific harmonic (i.e., for fixed l and m), then the coupled mode equations may be written as

$$\begin{aligned} &\left[\left(\omega - k_{n+l+2m} v_\parallel\right)^2 - \frac{\omega_b^2}{\gamma_0 \gamma_\parallel^2} T_l(\zeta_{n+l+2m}) \right] \delta\hat{\varphi}_n J_m(b_2) \\ &\quad \cong - \frac{\omega_b^2}{\gamma_0 k_n} \frac{v_w}{2c} \delta\hat{A}_{n-1} T_l(\zeta_{n+l+2m}) [J_m(b_2) - J_{m\pm 1}(b_2)] \left(k_{n+l+2m} - \omega \frac{v_\parallel}{c^2} \right) \end{aligned} \tag{4.197}$$

and

$$\begin{aligned} &\left[\omega^2 - c^2 k_{n-1}^2 - \frac{\omega_b^2}{\gamma_0} \left(1 - \frac{v_w^2}{4c^2} T_l(\zeta_{n+l+2m}) K_m^{(+)}(b_2) \frac{\omega^2 - k_{n+l+2m}^2 c^2}{\left(\omega - k_{n+l+2m} v_\parallel\right)^2} \right) \right] \delta\hat{A}_{n-1} \\ &\quad \cong - \frac{\omega_b^2}{2\gamma_0} \delta\hat{\varphi}_n J_m(b_2) T_l(\zeta_{n+l+2m}) [J_m(b_2) - J_{m\pm 1}(b_2)] \\ &\quad \times \frac{k_n v_w \left(k_{n+l+2m} - \omega v_\parallel / c^2\right)}{\left(\omega - k_{n+l+2m} v_\parallel\right)^2} \end{aligned} \tag{4.198}$$

where

$$\zeta_{n+l+2m} \equiv \frac{\gamma_0^2 m_e^2}{\Delta P^2} \left(\frac{\omega^2}{k_{n+l+2m}^2} - v_\parallel^2 \right). \tag{4.199}$$

The dispersion equation is found by requiring that the determinant of the coefficients vanishes. Therefore, for a specific choice of harmonic interaction, the dispersion equation which results is expressed as a straightforward generalization of that found in the cold-beam limit (Eq. (4.189)); specifically

$$\begin{aligned} &\left[\left(\omega - k_{n+l+2m} v_\parallel\right)^2 - \frac{\omega_b^2}{\gamma_0 \gamma_\parallel^2} T_l(\zeta_{n+l+2m}) \right] \left(\omega^2 - c^2 k_{n-1}^2 - \frac{\omega_b^2}{\gamma_0} \right) \\ &\quad \cong - \frac{v_w^2}{4c^2} \frac{\omega_b^2}{\gamma_0} T_l(\zeta_{n+l+2m}) K_m^{(+)}(b_2) \left(\omega^2 - c^2 k_{n+l+2m}^2 - \frac{\omega_b^2}{\gamma_0} T_l(\zeta_{n+l+2m}) \right) \end{aligned} \tag{4.200}$$

This dispersion equation, which includes the effect of an axial energy spread for a planar configuration, is similar to that found for the corresponding case for a helical wiggler geometry (Eq. (4.132)). The differences are as stated previously in that (1) the wiggler amplitude is replaced by the root-mean-square value, (2) the oscillation in the axial velocity introduces modifications in $K_m^{(+)}$ and $J_l^2(b_1)$, and (3) harmonic amplification is found in the one-dimensional formulation. The effect of the pitch angle spread on the axial velocity is the source of the lth harmonic contribution, which has the effect of modifying the thermal function T_l. In order to describe this effect in more detail, we assume that $b_1 << 1$ which is valid as long as $P_\perp << p_\parallel$. As a result, we expand

$$J_l(b_1) \cong \frac{1}{(l!)^2}\left(\frac{\upsilon_w}{2\upsilon_\parallel}\frac{\omega}{k_w\upsilon_\parallel}\frac{\Delta P}{p_\parallel}\right)^{2l} z^l \cos^{2l}\phi. \tag{4.201}$$

As a consequence

$$T_l(\zeta) \cong \frac{(2l)!}{(l!)^4}\left(\frac{\upsilon_w}{2\upsilon_\parallel}\frac{\omega}{k_w\upsilon_\parallel}\frac{\Delta P}{p_\parallel}\right)^{2l} \zeta^2 \exp(\zeta) \int_\zeta^\infty dt \frac{\exp(-t)}{t^2}(t-\zeta)^l. \tag{4.202}$$

As in the case of the helical wiggler, the thermal function may be expressed in terms of the exponential integral function (Eq. (4.128)). To this end, we observe that

$$\zeta^2 \exp(\zeta) \int_\zeta^\infty dt \frac{\exp(-t)}{t^2}(t-\zeta)^l = (-1)^l \zeta^{l+1}[1-(\zeta+l)\exp(\zeta)E_1(\zeta)] + U_l(\zeta), \tag{4.203}$$

where

$$U_l(\zeta) \equiv \begin{cases} 0 & ; l<2 \\ \sum_{k=2}^{l}\sum_{n=0}^{k-2} \frac{(-1)^{k-n} l! n! \zeta^{l-n}}{k!(l-k)!(k-2-n)!} & ; l \geq 2 \end{cases}. \tag{4.204}$$

As a consequence

$$T_l(\zeta) \cong \frac{(2l)!}{(l!)^4}\left(\frac{\upsilon_w}{2\upsilon_\parallel}\frac{\omega}{k_w\upsilon_\parallel}\frac{\Delta P}{p_\parallel}\right)^{2l} \left\{(-1)^l \zeta^{l+1}[1-(\zeta+l)\exp(\zeta)E_1(\zeta)] + U_l(\zeta)\right\}. \tag{4.205}$$

We observe that for $l = 0$ the planar wiggler thermal function reduces to that found for the helical wiggler (Eq. (4.125)), that is,

$$T_0(\zeta) \cong \zeta[1 - \zeta \exp(\zeta) E_1(\zeta)]. \tag{4.206}$$

This will reproduce a thermal response for the interaction at the fundamental which is similar to that found for the helical geometry. The interaction at harmonics will be discussed in a later chapter, in which it will be shown that the harmonic response is more sensitive to thermal effects.

4.2.2.6 The Effect of an Axial Magnetic Field

It has been shown in Chap. 2 that the single-particle orbits in a combined planar wiggler and axial guide field configuration exhibit a drift motion which can result in the loss of the electron beam to the walls of the drift tube. As a consequence, great care must be exercised in the design of such as system to ensure that substantial beam loss does not occur over the interaction length of the device. Under the assumption, therefore, that substantial beam loss does not occur, the presence of the axial magnetic field can substantially modify many of the characteristics of the interaction [31]. In the first place, the electrons execute elliptic trajectorics in the transverse plane [as opposed to the circular trajectories in helical wiggler geometries] that results in the excitation of elliptically polarized electromagnetic waves. In the second place, the enhancements in the gain that are present in the combined helical wiggler and axial guide field configurations when the Larmor period and wiggler periods are comparable are also present for the planar wiggler systems.

In order to treat the essential features of this configuration, it is instructive to simplify the analysis to the cold-beam regime in which $\upsilon_w << \upsilon_{\|}$ and to neglect the oscillatory components of the axial velocity. The source currents and charge density can be obtained from a fluid analysis of the interaction. The fundamental equations describing the beam are those of continuity

$$\frac{\partial}{\partial t} n + \nabla \cdot (n\mathbf{v}) = 0, \tag{4.207}$$

momentum transfer

$$\frac{d}{dt}\mathbf{v} = -\frac{e}{\gamma m_e}\left[\left(\mathrm{I} - \frac{1}{c^2}\mathbf{v}\mathbf{v}\right) \cdot \delta\mathbf{E} + \frac{1}{c}\mathbf{v} \times (\mathbf{B}_0 + \mathbf{B}_w + \delta\mathbf{B})\right], \tag{4.208}$$

and energy balance

$$\frac{d\gamma}{dt} = -\frac{e}{m_e c^2}\mathbf{v} \cdot \delta\mathbf{E}, \tag{4.209}$$

where n and $\mathbf{v}$ are the macroscopic density and velocity of the beam, γ is the relativistic factor corresponding to the bulk energy of the beam, $\mathbf{I}$ is the unit dyadic, $\mathbf{B}_0$ is the axial solenoidal field, $\mathbf{B}_w$ is the idealized planar wiggler representation (Eq. (2.99)), and $\delta\mathbf{E}$ and $\delta\mathbf{B}$ are the fluctuating electric and magnetic fields.

These equations are solved by means of a perturbation expansion to first order in the electromagnetic fields. To this end, we write $n = n_b + \delta n$, $\gamma = \gamma_0 + \delta\gamma$, and

$$\mathbf{v} = \alpha_x \hat{\mathbf{e}}_x \cos k_w z + \alpha_y \hat{\mathbf{e}}_y \sin k_w z + v_\parallel \hat{\mathbf{e}}_z + \delta\mathbf{v}, \tag{4.210}$$

where n_b describes the ambient beam density, $\alpha_{x,y}$ are defined in Eqs. (2.112) and (2.113), and $v_\parallel$ denotes the unperturbed bulk axial velocity of the beam. Observe that the unperturbed velocity is described by the single-particle electron trajectory in the combined magnetostatic field. We choose to express the electromagnetic field in terms of the vector and scalar potentials [in the Coulomb gauge] and employ Floquet's theorem for periodic systems to write

$$\delta\mathbf{A}(z) = \sum_{n=-\infty}^{\infty} \delta\hat{\mathbf{A}}_n \exp(ik_n z), \tag{4.211}$$

for the vector potential, where $k_n = k + nk_w$. The scalar potential is written as in Eq. (4.173). Observe that in the Coulomb gauge the z-component of the vector potential must vanish.

Solution of the perturbed fluid equations shows that the density perturbation is given by

$$\delta\hat{n}_m = n_b \frac{k_m}{\Delta\omega_m} \delta\hat{v}_{z,m}, \tag{4.212}$$

where $\Delta\omega_m = \omega - k_m v_\parallel$. The perturbed velocity, in turn, is given approximately by

$$\begin{aligned}
\left(\Delta\omega^2 - \Omega_0^2\right)\delta\hat{v}_{x,m\pm1} \cong{}& \frac{e}{2\gamma_0\gamma_\parallel^2 m_e v_\parallel}\frac{k_m}{\Delta\omega_m}\delta\hat{\varphi}_m \\
&\times\left[\alpha_x\beta_\parallel^2\gamma_\parallel^2\Delta\omega_{m\pm1}^2 \mp \alpha_y\Omega_0\left(\Phi_p\Delta\omega_{m\pm2} + \beta_\parallel^2\gamma_\parallel^2\Delta\omega_{m\pm1}\right)\right] \\
&+ \frac{e}{\gamma_0 m_e c}\delta\hat{A}_{x,m\pm1}\left\{\Delta\omega_{m\pm1}^2 - \frac{\alpha_x^2}{2c^2}\frac{\omega\Delta\omega_{m\pm1}\left(\Delta\omega_{m\pm1}^2 + \Omega_0^2\right)}{\Delta\omega_m\Delta\omega_{m\pm2}}\right. \\
&\left.\pm \frac{\alpha_y}{4c}\frac{\Omega_0\Delta\omega_{m\pm2}}{\Delta\omega_m}\left[\alpha_x\left(k_{m\pm1} - \omega\frac{v_\parallel}{c^2}\right) - \Omega_w\Delta_m U_m^{(\pm)}\right]\right\} \\
&+ \frac{ie}{\gamma_0 m_e c}\Omega_0\delta\hat{A}_{y,m\pm1}\left\{\Delta\omega_{m\pm1}\left(1 - \frac{\alpha_x^2 + \alpha_y^2}{2c^2}\frac{\omega\Delta\omega_{m\pm1}}{\Delta\omega_m\Delta\omega_{m\pm2}}\right)\right. \\
&\left.- \frac{\alpha_y}{4c}\frac{\Delta\omega_{m\pm2}}{\Delta\omega_m}\left[\alpha_y\left(k_{m\pm1} - \omega\frac{v_\parallel}{c^2}\right) \mp \Omega_w\Delta_m V_m^{(\pm)}\right]\right\}
\end{aligned} \tag{4.213}$$

$$
\begin{aligned}
\left(\Delta\omega^2 - \Omega_0^2\right)\delta\hat{v}_{y,m\pm1} \cong &\mp \frac{e}{2\gamma_0\gamma_\parallel^2 m_e}\frac{k_m}{v_\parallel \Delta\omega_m}\delta\hat{\varphi}_m \\
&\times \left[\alpha_y \beta_\parallel^2 \gamma_\parallel^2 \Delta\omega_{m\pm1}\Delta\omega_{m\pm2} \mp \Phi_p \Omega_0 \left(\alpha_x \Delta\omega_{m\pm1} \mp \alpha_y \Omega_0\right)\right] \\
&+ \frac{ie}{\gamma_0 m_e c}\Omega_0 \delta\hat{A}_{x,m\pm1}\left\{\Delta\omega_{m\pm1}\left(1 - \frac{\alpha_x^2}{2c^2}\frac{\omega\Delta\omega_{m\pm1}}{\Delta\omega_m \Delta\omega_{m\pm2}}\right)\right. \\
&\left. - \frac{\alpha_x \Delta\omega_{m\pm1} \mp \alpha_y \Omega_0}{4v_\parallel \Delta\omega_m}\left[\alpha_x\left(k_{m\pm1} - \omega\frac{v_\parallel}{c^2}\right) - \Omega_w \Delta_m U_m^{(\pm)}\right]\right\} \\
&+ \frac{e}{\gamma_0 m_e c}\delta\hat{A}_{y,m\pm1}\left\{\Delta\omega_{m\pm1}^2 - \frac{\alpha_y^2}{2c^2}\frac{\omega\Delta\omega_{m\pm1}\left(\Delta\omega_{m\pm1}^2 + k_w^2 v_\parallel^2\right)}{\Delta\omega_m \Delta\omega_{m\pm2}}\right. \\
&\left. \pm \frac{\Omega_0\left(\alpha_x \Delta\omega_{m\pm1} \mp \alpha_y \Omega_0\right)}{4v_\parallel \Delta\omega_m}\left[\alpha_y\left(k_{m\pm1} - \omega\frac{v_\parallel}{c^2}\right) \mp \Omega_w \Delta_m V_m^{(\pm)}\right]\right\}
\end{aligned}
\tag{4.214}
$$

and

$$
\begin{aligned}
\Delta\omega_m \delta\hat{v}_{z,m} \cong &- \frac{e}{\gamma_0 \gamma_\parallel^2 m_e} k_m \Phi_p \delta\hat{\varphi}_m \\
&+ \frac{e}{2\gamma_0 m_e c}\delta\hat{A}_{x,m+1}\left[\alpha_x\left(ck_{m+1} - \omega\beta_\parallel\right) - \Omega_w \Delta_m U_m^{(+)}\right] \\
&+ \frac{e}{2\gamma_0 m_e c}\delta\hat{A}_{x,m-1}\left[\alpha_x\left(ck_{m-1} - \omega\beta_\parallel\right) - \Omega_w \Delta_m U_m^{(-)}\right] \\
&+ \frac{ie}{2\gamma_0 m_e c}\delta\hat{A}_{y,m+1}\left[\alpha_y\left(ck_{m+1} - \omega\beta_\parallel\right) - \Omega_w \Delta_m V_m^{(+)}\right] \\
&- \frac{ie}{2\gamma_0 m_e c}\delta\hat{A}_{y,m-1}\left[\alpha_y\left(ck_{m-1} - \omega\beta_\parallel\right) - \Omega_w \Delta_m V_m^{(-)}\right]
\end{aligned}
\tag{4.215}
$$

where Φ_p is given in Eq. (2.117) and $\Omega_{0,w} = eB_{0,w}/\gamma_0 m_e c$. The coefficients that appear in these expressions for the perturbed velocity are defined as

$$
\begin{aligned}
U_m^{(\pm)} \equiv &\mp \frac{\Delta\omega_{m\pm1}}{\Delta\omega_{m\pm1}^2 - \Omega_0^2}\left[\Delta\omega_{m\pm1} - \frac{\alpha_x^2}{2c^2}\frac{\omega\left(\Delta\omega_{m\pm1}^2 + \Omega_0^2\right)}{\Delta\omega_m \Delta\omega_{m\pm2}}\right] \\
&\mp \frac{\alpha_x^2}{4c^2}\left(\Delta\omega_{m\mp1} \pm \frac{\alpha_x}{\alpha_y}\Omega_0\right)\frac{\omega\Delta\omega_{m\mp1}}{\Delta\omega_m\left(\Delta\omega_{m\mp1}^2 - \Omega_0^2\right)} \\
&+ \frac{\alpha_x^2}{2c^2}\left(k_{m\pm1} - \omega\frac{v_\parallel}{c^2}\right)\frac{\beta_\parallel \Omega_0^2\left(\Omega_0^2 + 3k_w^2 v_\parallel^2\right)}{\left(\Delta\omega_{m+1}^2 - \Omega_0^2\right)\left(\Delta\omega_{m-1}^2 - \Omega_0^2\right)},
\end{aligned}
\tag{4.216}
$$

$$
\begin{aligned}
V_m^{(\pm)} \equiv & \pm \frac{\Omega_0 \Delta\omega_{m\pm1}}{\Delta\omega_{m\pm1}^2 - \Omega_0^2}\left[1 - \frac{\alpha_x^2 + \alpha_y^2}{2c^2}\frac{\omega\Delta\omega_{m\pm1}}{\Delta\omega_m \Delta\omega_{m\pm2}}\right] \\
& - \frac{\alpha_x \alpha_y}{4c^2}\left(\Delta\omega_{m\mp1} \pm \frac{\alpha_x}{\alpha_y}\Omega_0\right)\frac{\omega\Delta\omega_{m\mp1}}{\Delta\omega_m\left(\Delta\omega_{m\mp1}^2 - \Omega_0^2\right)} \\
& \pm \frac{\alpha_y^2}{2c^2}\left(k_{m\pm1} - \omega\frac{v_\parallel}{c^2}\right)\frac{\beta_\parallel \Omega_0^2\left(\Omega_0^2 + 3k_w^2 v_\parallel^2\right)}{\left(\Delta\omega_{m+1}^2 - \Omega_0^2\right)\left(\Delta\omega_{m-1}^2 - \Omega_0^2\right)},
\end{aligned}
\tag{4.217}
$$

and

$$
\Delta_m \equiv \frac{2v_\parallel\left(\Delta\omega_{m+1}^2 - \Omega_0^2\right)\left(\Delta\omega_{m-1}^2 - \Omega_0^2\right)}{2v_\parallel\left(\Delta\omega_{m+1}^2 - \Omega_0^2\right)\left(\Delta\omega_{m-1}^2 - \Omega_0^2\right) + \alpha_y \Omega_0 \Omega_w\left(\Omega_0^2 + 3k_w^2 v_\parallel^2\right)}. \tag{4.218}
$$

Observe that the y-component of the perturbed velocity vanishes in the limit in which the axial magnetic field vanishes.

The source current may be expressed in terms of the perturbed density and velocity in the following manner:

$$
\delta\hat{\mathbf{J}}_{\perp,m\pm1} = -en_b\left[\delta\hat{\mathbf{v}}_{\perp,m\pm1} + \frac{1}{2}\left(\alpha_x\hat{\mathbf{e}}_x + \alpha_y\hat{\mathbf{e}}_y\right)\frac{k_m}{\Delta\omega_m}\delta\hat{v}_{z,m}\right]. \tag{4.219}
$$

The perturbed source current and charge density are substituted into the Maxwell-Poisson equations

$$
\left(\frac{\omega^2}{c^2} - k_{m\pm1}^2\right)\delta\hat{\mathbf{A}}_{\perp,m\pm1} = -\frac{4\pi}{c}\delta\hat{\mathbf{J}}_{\perp,m\pm1}, \tag{4.220}
$$

and

$$
k_m^2\delta\hat{\varphi}_m = 4\pi e\delta\hat{n}_m, \tag{4.221}
$$

to obtain the dispersion equation. Since the interaction excites elliptically polarized waves, it is convenient to transform to the basis that describes right- and left-hand circularly polarized waves, and we define

$$
\delta\hat{\mathbf{A}}_m^{(\pm)} \equiv \delta\hat{A}_{x,m} \mp i\delta\hat{A}_{y,m}. \tag{4.222}
$$

As a consequence, the dispersion equations describing these modes corresponding to the Doppler upshifted resonance are given by

$$\left(\Delta\omega_m^2-\frac{\omega_b^2}{\gamma_0\gamma_\parallel^2}\Phi_p\right)\left(\omega^2-c^2k_{m-1}^2-\frac{\omega_b^2\Delta\omega_{m-1}}{\gamma_0(\Delta\omega_{m-1}\mp\Omega_0)}\right)\cong\frac{\omega_b^2\Omega_w\Psi}{\gamma_0\left(\Omega_0\mp k_w v_\parallel\right)^2}k_wk_{m-1}v_\parallel^2, \tag{4.223}$$

where the "$\mp$" refers to the right- and left-hand modes, respectively, and

$$\Psi\equiv 1-\frac{\alpha_y^2\left(\Omega_0^2+3k_w^2v_\parallel^2\right)}{2v_\parallel^2\left(\Omega_0^2-k_w^2v_\parallel^2\right)+\alpha_y^2\left(\Omega_0^2+3k_w^2v_\parallel^2\right)}, \tag{4.224}$$

which reduces to unity in the limit in which the axial solenoidal field vanishes. Observe that the interaction differs from that found for a helical wiggler, in which the transverse projection of the single-particle orbits is circular. In that case, the amplified wave has a right-hand-circular polarization only. In contrast, the amplified wave has an elliptic polarization for a planar wiggler, and the dispersion equations describe the amplification of both right- and left-hand modes.

In the limit in which the axial solenoidal field vanishes, we expect to recover the previously derived results (Eq. (4.182)) for the dispersion equation at the fundamental resonance for a planar wiggler geometry. Indeed, it is clear that the dispersion equations (Eq. (4.209)) are identical for the right- and left-hand modes. This implies that these modes have identical growth rates and that, as expected, the aggregate wave is linearly polarized in the x-direction. The dispersion equation for this mode reduces to

$$\left(\Delta\omega_m^2-\frac{\omega_b^2}{\gamma_0\gamma_\parallel^2}\right)\left(\omega^2-c^2k_{m-1}^2-\frac{\omega_b^2}{\gamma_0}\right)\cong\frac{v_w^2}{2c^2}\frac{\omega_b^2}{\gamma_0}k_wk_{m-1}, \tag{4.225}$$

which corresponds to Eq. (4.182) for the fundamental resonance in the limit in which the canonical momentum vanishes and $v_w << c$.

4.3 Linear Stability in Three Dimensions

In treating the linear stability in three dimensions, we present the theory for helical wigglers. Typically, helical wigglers were more commonly used in long-wavelength, high-current applications where a waveguide mode analysis is appropriate, and the analyses we present are relevant to the interaction in this regime with a helical wiggler in a cylindrical waveguide. While planar wigglers are more commonly used for short-wavelength free-electron lasers at the present time, variable polarization wigglers are coming into more common use, and we describe a linear stability theory for optical modes in the presence of a helical wiggler. We note, however, that the corresponding results for planar wigglers are obtained by replacing the wiggler strength parameter by the rms value and the inclusion of the JJ-factor.

4.3.1 Waveguide Mode Analysis

Free-electron lasers that operate in the long-wavelength regime (submillimeter wavelengths and longer) typically employ electron beams with relatively low energies, high currents, and large transverse scale sizes. As a result, both the resonant wavelengths and electron beam radii are substantial fractions of the drift tube radius. Because of this, the boundary conditions of the electromagnetic field on the drift tube walls become important, and the drift tube acts as a waveguide. Furthermore, in view of the relatively low-energy/high-current electron beams used in many of these experiments, the interaction can reach saturation in relatively short wigglers with lengths on the order of meters. Often, therefore, helical undulators with weak, two-plane focusing are used.

In the formulation of the linear stability theory of free-electron lasers in realizable (i.e., three-dimensional) helical wiggler configurations, we consider the treatment of helical wiggler/axial solenoidal magnetic field configurations as illustrated in Eq. (2.1), in which the electron beam propagates through a cylindrical drift tube. The analysis [34, 35] involves a perturbation expansion of the Lorentz force equations about the single-particle trajectories to first order in the electromagnetic field. To this end, we employ the nonhelical, axis-encircling trajectories described in Eqs. (2.71, 2.72, 2.73, 2.74, 2.75, 2.76, 2.77, and 2.78). These orbits represent perturbations about the helical steady-state trajectories that include oscillatory components at both the betatron and Larmor periods. For the sake of simplicity, if we neglect the effect of the betatron oscillations on the trajectory, then these orbits can be expressed in rectangular coordinates in the form

$$\begin{aligned} p_x = p_w \cos k_w z &+ \alpha_+ \left[P_x \cos(k_w z - \Omega_- t) - P_y \sin(k_w z - \Omega_- t)\right] \\ &+ \alpha_- \left[P_x \cos(k_w z + \Omega_- t) + P_y \sin(k_w z + \Omega_- t)\right] \end{aligned} \tag{4.226}$$

$$\begin{aligned} p_y = p_w \sin k_w z &+ \alpha_+ \left[P_x \sin(k_w z - \Omega_- t) + P_y \cos(k_w z - \Omega_- t)\right] \\ &+ \alpha_- \left[P_x \sin(k_w z + \Omega_- t) - P_y \cos(k_w z + \Omega_- t)\right], \end{aligned} \tag{4.227}$$

and

$$p = p_\parallel - \frac{p_w}{p_\parallel \sqrt{1+\lambda_0^2}} \left(P_x \cos \Omega_- t + P_y \sin \Omega_- t\right), \tag{4.228}$$

where $(p_w, p_\parallel)$ denote the transverse and axial momenta corresponding to the steady-state trajectory, Ω_- is given in Eq. (2.62), $\lambda_0 = |p_w/p_\parallel|$, P_x and P_y represent the canonical momenta, and

$$\alpha_\pm = \frac{1 \pm \sqrt{1+\lambda_0^2}}{2\sqrt{1+\lambda_0^2}}. \tag{4.229}$$

Observe that $\lim_{\lambda 0 \to 0} \Omega_- = k_w v_\parallel - \Omega_0$, $\lim_{\lambda 0 \to 0} \alpha_+ = 1$, and $\lim_{\lambda 0 \to 0} \alpha_- = 0$. As a consequence, these orbits reduce to Eqs. (2.42) and (2.43) in the idealized one-dimensional limit and describe the effect of the combined wiggler and Larmor oscillations.

4.3.1.1 The Low-Gain Regime

The analysis of the low-gain regime for a three-dimensional realizable configuration differs from that used for the idealized one-dimensional treatments of both the helical and planar wiggler geometries. In those cases, we employed a unified Vlasov-Maxwell treatment for both the low- and high-gain regimes. However, we employ the more usual phase-average technique to treat the more general three-dimensional case. This technique results in an identical expression for the gain, at considerable reduction in mathematical complexity. The general expression obtained in this case (4.235) reduces to the previously discussed one-dimensional result (4.59) in the limit in which the three-dimensional effects are negligible. The principal differences between the one- and three-dimensional results are twofold. In the first place, the orbital parameters contained in these expressions are those given by the one- and three-dimensional steady-state trajectories discussed in Chap. 2. In the second place, the three-dimensional result contains a *filling factor* that describes the effect of the overlap between the electron beam and the radiation field. In the one-dimensional limit, this filling factor is unity.

It should be observed that we employ a waveguide mode decomposition for the electromagnetic field rather than a Gaussian mode decomposition which is often more appropriate to short-wavelength free-electron laser oscillators. However, the generalization to the Gaussian mode case can be effected merely by the modification of the filling factor. We have chosen to employ the waveguide mode decomposition in the interest of providing a unified formalism between the low- and high-gain regimes. The high-gain regime is more appropriate to the treatment of longer wavelength free-electron laser amplifiers [submillimeter wavelengths and longer] driven by intense electron beam sources. As a consequence, the waveguide is the dominant dispersive effect in these configurations.

Since it is our intention to treat the effects of transverse inhomogeneities in a self-consistent manner [34], these effects must also be included in the electromagnetic field. In this regard, we observe that the drift tube acts as a waveguide and that the electromagnetic field must satisfy the boundary conditions on the walls. In order to ensure that these boundary conditions are satisfied, we represent the electromagnetic field in terms of an ensemble of the TE and TM modes in the vacuum waveguide. The vector potential of the electromagnetic field can, therefore, be expressed in cylindrical coordinates as

$$\delta \mathbf{A}(\mathbf{x},t)=\sum_{\substack{l=0\\n=1}}^{\infty}\delta A_{ln}(z)\left[\frac{l}{\kappa_{ln}r}J_l(\kappa_{ln}r)\hat{\mathbf{e}}_r\sin\alpha_l+J_l{}'(\kappa_{ln}r)\hat{\mathbf{e}}_\theta\cos\alpha_l\right], \tag{4.230}$$

for the TE modes, and

$$\delta \mathbf{A}(\mathbf{x},t)=\sum_{\substack{l=0\\n=1}}^{\infty}\delta A_{ln}(z)\left[J_l{}'(\kappa_{ln}r)\hat{\mathbf{e}}_r\cos\alpha_l-\frac{l}{\kappa_{ln}r}J_l(\kappa_{ln}r)\hat{\mathbf{e}}_\theta\sin\alpha_l+\frac{\kappa_{ln}}{k}J_l(\kappa_{ln}r)\hat{\mathbf{e}}_z\sin\alpha_l\right], \tag{4.231}$$

for the TM modes, where for frequency ω and wavenumber k the phase is defined as

$$\alpha_l\equiv\int_0^z dz'k(z')+l\theta-\omega t. \tag{4.232}$$

J_l and J_l' denote the regular Bessel function of the first kind and its derivative, and κ_{ln} defines the cutoff wavenumber. For convenience, we have suppressed the mode indices on the wavenumber k that, for fixed frequency, will differ for each mode. In this representation, the cutoff wavenumbers for the TE$_{ln}$ mode are given by $\kappa_{ln}=x_{ln}'/R_g$ where $J_l'(x_{ln}')=0$ and R_g denotes the radius of the waveguide. The cutoff wavenumber for the TM$_{ln}$ mode is given by $\kappa_{ln}=x_{ln}/R_g$ where $J_l(x_{ln})=0$. The coherent amplification process is included under the assumption that the amplitude and wavenumber vary slowly with respect to the wavelength of the radiation.

In order to determine the gain, we make use of Maxwell's equations under the assumption that the beam density is sufficiently diffuse that the space-charge effects can be ignored. Observe that the divergence of the vector potential vanishes in the vacuum waveguide, as well as for the TE modes under the assumption of a slowly varying amplitude and wavenumber. However, the divergence of the vector potential does not vanish within the context of this formulation for the TM modes due to the presence of an axial component of the electric field. In this case,

$$\nabla\cdot\delta\mathbf{A}=\sum_{\substack{l=0\\n=1}}^{\infty}\kappa_{ln}\frac{d}{dz}\left(\frac{1}{k}\delta A_{ln}\right)J_l(\kappa_{ln}r)\sin\alpha_l. \tag{4.233}$$

The effect of the divergence of the vector potential on the evolution of the TM modes will be discussed in detail in the context of the nonlinear analysis of the interaction in Chap. 5. For the present discussion, we shall assume that the wave is far above cutoff (i.e., $\kappa_{ln} << k$) and neglect this contribution. The vector potential, therefore, satisfies Maxwell's equations

$$\left(\nabla^2 - \frac{1}{c^2}\frac{\partial^2}{\partial t^2}\right)\delta\mathbf{A}(\mathbf{x},t) = -\frac{4\pi}{c}\delta\mathbf{J}(\mathbf{x},t), \tag{4.234}$$

where $\delta\mathbf{J}$ represents the source current.

Observe that these assumptions are equivalent to the condition that $kR_g << 1$, and the principal characteristic of the waveguide structure included in the analysis is the radial localization of the modes. Since we are concerned with the low-gain regime, we make the further assumption that the beam is sufficiently tenuous that that it only provides for a gain medium and that the dispersion relation for both the TE and TM modes is given approximately by the free-space limit

$$\omega \cong ck. \tag{4.235}$$

This is valid as long as the frequency is much greater than both the beam-plasma frequency and the waveguide cutoff frequencies. The growth of each TE or TM mode is determined by the diagonalization of Maxwell's equations in view of the azimuthal and radial mode structure of the waveguide modes. Using the orthogonality properties of the Bessel functions, it can be shown that the evolution of the slowly varying amplitude is governed by

$$\frac{d}{dz}\delta A_{ln} = -\frac{1}{ck}\frac{x_{ln}'^{2}}{\left(x_{ln}'^{2} - l^2\right)J_l^2\left(x_{ln}'\right)}\frac{\omega}{\pi R_g^2}\int\limits_{-\pi/\omega}^{\pi/\omega} dt \int\limits_0^{2\pi} d\theta \int\limits_0^{\infty} drr[C_l\delta J_1 - S_l\delta J_2], \tag{4.236}$$

for the TE$_{ln}$ mode, and

$$\frac{d}{dz}\delta A_{ln} = -\frac{1}{ck}\frac{1}{J_l'(x_{ln})}\frac{\omega}{\pi R_g^2}\int\limits_{-\pi/\omega}^{\pi/\omega} dt \int\limits_0^{2\pi} d\theta \int\limits_0^{\infty} drr[S_l\delta J_1 + C_l\delta J_2], \tag{4.237}$$

for the TM$_{ln}$ mode, where we have neglected second-order derivatives of the amplitude and phase (i.e., terms of order dk/dz), $(\delta J_1, \delta J_2)$ denotes the transverse components of the source current in the wiggler frame, $\chi = \theta - k_w z$

$$S_l \equiv J_{l-1}(\kappa_{ln}r)\sin[\psi_l + (l-1)\chi] - J_{l+1}(\kappa_{ln}r)\sin[\psi_l + (l+1)\chi], \tag{4.238}$$

$$C_l \equiv J_{l-1}(\kappa_{ln}r)\cos[\psi_l + (l-1)\chi] + J_{l+1}(\kappa_{ln}r)\cos[\psi_l + (l+1)\chi], \tag{4.239}$$

and

$$\psi_l = -\omega t_0 + \int_0^z dz' \left(k(z') + lk_w - \frac{\omega}{v_z(z')} \right) \tag{4.240}$$

defines the ponderomotive phase. Note that t_0 denotes the entry time at which the electron crosses the $z = 0$ plane.

We now assume the existence of an idealized electron beam [i.e., monoenergetic with a vanishing pitch angle spread] that is executing axi-centered, steady-state trajectories. Hence, $v_1 = v_w$, $v_2 = 0$, $v_3 = v_{\|}$, $r = r_0 = \mp v_w/k_w v_{\|}$, and $\chi = \pm\,\pi/2$ for the Group I and Group II orbits, respectively. The beam configuration is that of a thin axi-centered helix in which the beam density is uniform within the range $r_0 - \Delta R \leq r \leq r_0$ for $\Delta R << r_0$, and the azimuthal position rotates with $\theta = k_w z \pm \pi/2$. The source current which corresponds to this case is characterized by $\delta J_2 = 0$ and

$$\delta J_1 = -2\pi e n_b v_w \Delta R \delta(r - r_0)\delta(\chi \mp \pi/2) \int_0^\infty dt_0 \delta(t - t_0 - z/v_{\|}), \tag{4.241}$$

where n_b is the ambient density of the beam, and we have assumed that the beam is continuous (i.e., uniform in entry time). Substitution of this source current into Eqs. (4.236) and (4.230) indicates that the amplitude evolves as

$$\frac{d}{dz}\delta a_{ln} = \pm \frac{\omega_b^2}{c^2 k} \frac{v_w}{c} \frac{r_0 \Delta R}{R_g^2} \frac{x'^2_{ln} J_l'(\kappa_{ln} r_0)}{(x'^2_{ln} - l^2) J_l^2(x'_{ln})} \left\langle \sin\left(\psi_l \pm \frac{l\pi}{2}\right) \right\rangle, \tag{4.242}$$

for the TE$_{ln}$ mode, and

$$\frac{d}{dz}\delta a_{ln} = \pm \frac{\omega_b^2}{c^2 k} \frac{v_w}{c} \frac{r_0 \Delta R}{R_g^2} \frac{l}{\kappa_{ln} r_0} \frac{J_l(\kappa_{ln} r_0)}{J_l'^2(x_{ln})} \left\langle \cos\left(\psi_l \pm \frac{l\pi}{2}\right) \right\rangle, \tag{4.243}$$

for the TM$_{ln}$ mode. Here $\delta a_{ln} = e\delta A_{ln}/m_e c^2$, $\omega_b{}^2 = 4\pi e^2 n_b/m_e$, and the averaging operator is defined over the initial phase $\psi_{l0} = -\,\omega t_0$ as

$$\langle(\cdots)\rangle \equiv \frac{1}{2\pi} \int_0^{2\pi} d\psi_0 \sigma_{\|}(\psi_0)(\cdots), \tag{4.244}$$

where $\sigma_{\|}$ defines the electron distribution in initial phase.

In order to determine the gain for each waveguide mode, we must first obtain solutions for the ponderomotive phase to first order in the mode amplitudes. Generalization of the derivation of the pendulum equation to three dimensions by perturbation about the axi-centered, steady-state trajectories to first order in each waveguide mode yields

$$\frac{d^2}{dz^2}\psi_l = \mp\kappa_l^2 J_l'(\kappa_{ln}r_0)\sin\left(\psi_l \pm \frac{l\pi}{2}\right), \tag{4.245}$$

for the TE modes,

$$\frac{d^2}{dz^2}\psi_l = \mp\kappa_l^2 \frac{l}{\kappa_{ln}r_0} J_l(\kappa_{ln}r_0)\cos\left(\psi_l \pm \frac{l\pi}{2}\right), \tag{4.246}$$

for the TM modes, where

$$\kappa_l^2 \equiv \frac{cv_w(k+lk_w)^2}{\gamma_0\gamma_\parallel^2 v_\parallel^2}\Phi(k_w r)\delta a_{ln}, \tag{4.247}$$

and Φ is defined in Eq. (2.70).

In the linear gain regime, we seek to solve the pendulum equations for untrapped trajectories. These solutions are of the form $\psi_l = \psi_{l0} + \Delta k_l z + \delta\psi_l$, where $\Delta k_l = (k + lk_w) - \omega/v_\parallel$, and we assume that $|\delta\psi_l/\Delta k_l z| << 1$. Expansion of the pendulum equations to first order in the electromagnetic field and $\delta\psi_l$ yields

$$\frac{d^2}{dz^2}\delta\psi_l \cong \mp\kappa_l^2 J_l'(\kappa_{ln}r_0)\sin\left(\psi_{l0} + \Delta k_l z \pm \frac{l\pi}{2}\right), \tag{4.248}$$

for the TE_{ln} mode,

$$\frac{d^2}{dz^2}\delta\psi_l \cong \mp\kappa_l^2 \frac{l}{\kappa_{ln}r_0} J_l(\kappa_{ln}r_0)\cos\left(\psi_{l0} + \Delta k_l z \pm \frac{l\pi}{2}\right), \tag{4.249}$$

for the TM_{ln} mode. Hence, the solutions are

$$\delta\psi_l \cong \mp\frac{\kappa_l^2}{\Delta k_l^2} J_l'(\kappa_{ln}r_0)\left[\sin\left(\psi_{l0} + \Delta k_l z \pm \frac{l\pi}{2}\right) - \sin\left(\psi_{l0} \pm \frac{l\pi}{2}\right) - \Delta k_l z\cos\left(\psi_{l0} \pm \frac{l\pi}{2}\right)\right], \tag{4.250}$$

for the TE_{ln} mode, and

$$\delta\psi_l \cong \pm\frac{\kappa_l^2}{\Delta k_l^2}\frac{l}{\kappa_{ln}r_0} J_l(\kappa_{ln}r_0) \times\left[\cos\left(\psi_{l0} + \Delta k_l z \pm \frac{l\pi}{2}\right) - \cos\left(\psi_{l0} \pm \frac{l\pi}{2}\right) + \Delta k_l z\sin\left(\psi_{l0} \pm \frac{l\pi}{2}\right)\right], \tag{4.251}$$

for the TM_{ln} mode, subject to the initial conditions $\delta\psi_l(0) = 0$ and $d\delta\psi_l(0)/dz = 0$. As a consequence of these untrapped solutions, the phase averages appearing in Eqs. (4.242) and (4.243) are

$$\left\langle \sin\left(\psi_{l0} \pm \frac{l\pi}{2}\right)\right\rangle \cong \pm \frac{\kappa_l^2}{2\Delta k_l^2} J_l'(\kappa_{ln} r_0)[\sin \Delta k_l z - \Delta k_l z \cos \Delta k_l z], \tag{4.252}$$

for the TE_{ln} mode, and

$$\left\langle \cos\left(\psi_{l0} \pm \frac{l\pi}{2}\right)\right\rangle \cong \pm \frac{\kappa_l^2}{2\Delta k_l^2} \frac{l}{\kappa_{ln} r_0} J_l(\kappa_{ln} r_0)[\sin \Delta k_l z - \Delta k_l z \cos \Delta k_l z], \tag{4.253}$$

for the TM_{ln} mode in the random phase limit [i.e., $\sigma_{\|}(\psi_{l0}) = 1$].

The linear power gain over an axial distance L is obtained by integration of the derivatives of the field amplitudes

$$G(L) \equiv \frac{2}{\delta a_{ln}(0)} \int_0^L dz \frac{d}{dz} \delta a_{ln}(z), \tag{4.254}$$

under the assumption that $G(L)$ is much less than unity. Substitution of the expressions for the phase averages for the untrapped trajectories into Eqs. (4.242) and (4.243), therefore, yields

$$G(L) \cong \frac{\omega_b^2 L^3 k}{8\gamma_0 \gamma_{\|}^2 \upsilon_{\|}^2} \beta_w^2 \Phi(k_w r_0) T_{ln} F(\Theta_l), \tag{4.255}$$

where F is the spectral function defined in Eq. (4.60), $\Theta_l = \Delta k_l L/2$, $\beta_w{}^2 = \upsilon_w{}^2/\upsilon_{\|}{}^2$, and

$$T_{ln} \equiv \begin{cases} \dfrac{2r_0\Delta R}{R_g^2} \dfrac{x_{ln}'^2 J_l'(\kappa_{ln} r_0)}{(x_{ln}'^2 - l^2) J_l^2(x_{ln}')} & ;\text{TE}_{ln}\text{ mode} \\[2ex] \dfrac{2r_0\Delta R}{R_g^2} \dfrac{l^2}{\kappa_{ln}^2 r_0^2} \dfrac{J_l^2(\kappa_{ln} r_0)}{J_l'^2(x_{ln})} & ;\text{TM}_{ln}\text{ mode} \end{cases} \tag{4.256}$$

is a mode- and beam-dependent geometric factor which describes the overlap of the electron beam and the radiation field and is sometimes referred to as the *filling factor*. Observe that this expression for the gain (4.234) is similar to that found in the idealized one-dimensional limit (4.59) where the electron trajectories may be approximated by the idealized steady-state orbits and the filling factor is unity. The spectral features of the linear gain [i.e., the bandwidth and the frequency corresponding to the peak gain] are similar to those found in the idealized description.

One feature of the three-dimensional analysis that does not appear in one-dimension, however, is the appearance of gain at harmonics of the fundamental resonance frequency. Typically, peak gain is found for $\Theta_l \approx \mp 1.3$ corresponding to Group I and Group II orbits, respectively. This implies that growth is found at the fundamental and all harmonics and that the peak gain occurs for

$$\omega \cong \frac{lk_w v_\parallel}{1-\beta_\parallel}\left(1 \mp \frac{2.6}{lk_w L}\right). \tag{4.257}$$

Indeed, this implies the existence of a *selection rule* whereby a TE or TM mode with the azimuthal mode number l is resonant at the lth harmonic and will support gain. The physical basis for the appearance of harmonics in the three-dimensional analysis of free-electron lasers based upon helical wiggler fields is the azimuthal variation of the electromagnetic field. Since the electrons execute a helical trajectory, they excite circularly polarized waves. Resonant modes are those in which the polarization vector rotates in synchronism with the electrons. If the polarization vector rotates once per wiggler period, then the interaction is resonant at the fundamental. The lth harmonic interaction corresponds to waves in which the polarization vector rotates l times per wiggler period. The phase of the TE and TM modes of a cylindrical waveguide varies as $\exp(ikz + il\theta - i\omega t)$. Since the azimuthal trajectory of the electrons in a helical wiggler varies as $\theta \approx k_w z$, this implies that a TE_{ln} or TM_{ln} mode will be resonant [i.e., the electron will experience a near-constant phase] when $\omega \approx (k + lk_w)v_\parallel$. The growth of harmonics for a helical wiggler configuration, therefore, arises from a different mechanism than is found for a planar wiggler and does not require harmonic oscillations in the electron trajectories. The question of harmonic radiation will be discussed in some depth in Chap. 6 for both planar and helical wigglers.

4.3.1.2 The High-Gain Regime

The high-gain regime is distinguished from the previously described low-gain regime in that the dielectric effects of the electron beam are sufficiently strong that the mode structure of the electromagnetic wave can differ significantly from that of the vacuum waveguide. In this regime, a self consistent model of the interaction must be used to describe both the resonant amplification of the signal and the beam- and wiggler-driven modifications to the vacuum dispersion [35]. The configuration we consider for this purpose is that of a relativistic electron beam propagating through a cylindrical waveguide in the presence of a helical wiggler (2.1) and an axial guide magnetic field. Within this context, we shall consider that the electrons are executing the helical, steady-state trajectories (described in Eqs. (2.53), (2.54), and (4.226, 4.227, and 4.228)) and consider the first-order perturbations to these trajectories due to the electromagnetic field.

We analyze the spatial amplification of a signal, for which the source current and charge density are obtained from the moments of the perturbed distribution function

$$\delta f_b[\mathbf{r}(z), z, \mathbf{p}, \tau(z)] = e \int_0^z \frac{dz'}{v_z[\mathbf{r}(z'), z']} \{\delta \mathbf{E}[\mathbf{r}(z'), z', \tau(z')]$$
$$+ \frac{1}{c} \mathbf{v}[\mathbf{r}(z'), z', \tau(z')] \times \delta \mathbf{B}[\mathbf{r}(z'), z', \tau(z')]\} \cdot \nabla_p F_b(P_x, P_y, p), \tag{4.258}$$

where the equilibrium distribution F_b is a function of the constants of the motion (P_x, P_y,p), $\delta\mathbf{E}$ and $\delta\mathbf{B}$ are the fluctuating electric and magnetic fields, $\mathbf{r}(z)$ is the position of the electron in the xy-plane at the axial position z, $\mathbf{v}[\mathbf{r}(z),z]$ is the electron velocity, and

$$\tau[\mathbf{r}(z'), z'] \equiv t_0 + \int_0^z \frac{dz'}{v_z[\mathbf{r}(z'), z']} \tag{4.259}$$

is the sum of the entry time t_0 and the time required for an electron to travel from ($\mathbf{r}(z = 0)$, $z = 0$) at the start of the interaction region to ($\mathbf{r}(z)$,z). In addition, we work with the vector and scalar potentials that are written as

$$\delta \mathbf{A}(\mathbf{x}, t) = \frac{1}{2} \delta \hat{\mathbf{A}}(\mathbf{x}) \exp(-i\omega t) + c.c., \tag{4.260}$$

and

$$\delta \varphi(\mathbf{x}, t) = \frac{1}{2} \delta \hat{\varphi}(\mathbf{x}) \exp(-i\omega t) + c.c.. \tag{4.261}$$

Integration of Eq. (4.258) with respect to the vector basis that describes circularly polarized electromagnetic waves (Eq. (4.33)) yields a perturbed distribution of the form

$$\delta \hat{f}_b[\mathbf{r}(z), z, \mathbf{p}] = \frac{e}{2c} \left[D_+ \left(\frac{\partial}{\partial P_x} + i \frac{\partial}{\partial P_y} \right) + D_- \left(\frac{\partial}{\partial P_x} - i \frac{\partial}{\partial P_y} \right) + D_z \frac{\partial}{\partial p} \right] F_b(P_x, P_y, p), \tag{4.262}$$

which is formally identical to Eq. (4.35) derived for the idealized one-dimensional limit. In the three-dimensional regime, however,

$$D_z \equiv \frac{1}{2p} \int_0^z dz' \frac{\exp[i\omega\tau(z, z')]}{v_z[\mathbf{r}(z')z']} [-c(2p_z \nabla_{z'} + p_- \nabla_+ + p_+ \nabla_-)\delta\hat{\varphi}$$
$$+2i\omega(p_z \delta\hat{A}_z + p_- \delta\hat{A}_+ + p_+ \delta\hat{A}_-)] \tag{4.263}$$

and

$$D_{\pm} \equiv \int_0^z dz' \frac{\exp[i\omega\tau(z,z')]}{v_z[1r(z')z']} \left\{ \alpha_+ \exp(\pm i\theta_-) \left[-\frac{c}{2}\nabla_{\pm}\delta\hat{\varphi} + \left(i\omega - v_z\frac{\partial}{\partial z'}\right)\delta\hat{A}_{\pm} \right. \right.$$
$$\left. + \frac{v_z}{2}\nabla_{\pm}\delta\hat{A}_z \mp \frac{v_{\pm}}{2}(\nabla_-\delta\hat{A}_+ - \nabla_+\delta\hat{A}_-)\right]$$
$$- \alpha_- \exp(\mp i\theta_+)\left[-\frac{c}{2}\nabla_{\mp}\delta\hat{\varphi} + \left(i\omega - v_z\frac{\partial}{\partial z'}\right)\delta\hat{A}_{\mp}\right.$$
$$\left.\left. + \frac{v_z}{2}\nabla_{\mp}\delta\hat{A}_z \pm \frac{v_{\mp}}{2}(\nabla_-\delta\hat{A}_+ - \nabla_+\delta\hat{A}_-)\right]\right\}, \tag{4.264}$$

where $\tau(z,z') = \tau(z) - \tau(z')$, $\theta_{\pm} = k_w z \pm \Omega_-\tau(z)$, and

$$\nabla_{\pm} \equiv \frac{\partial}{\partial x} \mp i\frac{\partial}{\partial y}. \tag{4.265}$$

4.3.1.2.1 Source Currents

The source current and charge density are found by computation of the appropriate moments of the perturbed distribution as

$$\delta\hat{J}_{\pm} = -\frac{e}{m_e}\iiint dP_x dP_y dp \frac{pp_{\pm}}{\gamma p_z\sqrt{1+\lambda_0^2}}\delta\hat{f}_b, \tag{4.266}$$

$$\delta\hat{J}_z = -\frac{e}{m_e}\iiint dP_x dP_y dp \frac{p}{\gamma\sqrt{1+\lambda_0^2}}\delta\hat{f}_b, \tag{4.267}$$

for the components of the current, and

$$\delta\hat{\rho} = -e\iiint dP_x dP_y dp \frac{p}{p_z\sqrt{1+\lambda_0^2}}\delta\hat{f}_b, \tag{4.268}$$

for the charge density. By application of Floquet's theorem for periodic systems, the axial and azimuthal structure of the fields and sources are expressed in the form

$$f(r,\theta,z) = \sum_{l,n=-\infty}^{\infty} f_{l,n}(r)\exp[i(k+nk_w)z + il\theta], \tag{4.269}$$

for some arbitrary function $f(r,\theta,z)$. Substitution of the perturbed distribution function into the expressions for the source currents and charge densities results in harmonic contributions to the source current $\delta\hat{J}_{l,n}$ and charge density $\delta\hat{\rho}_{l,n}$ each of which depends upon many harmonics of $\delta_{\hat{A}l,n}$ and $\delta\hat{\varphi}_{l,n}$. However, in the limit in which $\omega >> \Omega_-$ and the vector and scalar potentials vary slowly in r with respect to the wavelength of the radiation, then the sources assume the comparatively simple forms

$$\delta\hat{J}_{l,n}^{(\pm)} \cong -\frac{i\omega_b^2}{8\pi c}\int_0^\infty dp\frac{1}{\gamma}\left\{\frac{p_w}{\omega - k_{l+n}v_\parallel}\left[p_w\left(\hat{H}_{l,n}^{(\pm)} + \hat{H}_{l\mp 2,n\pm 2}^{(\pm)}\right)\right.\right.$$
$$\left. - ip_\parallel\left(ck_{n\pm 1}\delta\hat{\varphi}_{l\mp 1,n\pm 1} - \omega\delta\hat{A}_{l\mp 1,n\pm 1,z}\right)\right]\frac{1}{p}\frac{\partial}{\partial p} - \frac{2\alpha_+^2}{\omega \pm \Omega_- - k_{l+n}v_\parallel}\hat{L}_{l,n}^{(\pm)}$$
$$\left. \mp \frac{\lambda_0^2}{\sqrt{1+\lambda_0^2}}\frac{\omega}{\Omega_-}\left(R_{l,n}^{(\pm)}\hat{L}_{l,n}^{(\pm)} - R_{l,n}^{(\mp)}\hat{L}_{l\mp 2,n\pm 2}^{(\mp)}\right)\right\}G_b(p),$$
(4.270)

and

$$\delta\hat{\rho}_{l,n} \cong -\frac{i\omega_b^2}{8\pi c}\int_0^\infty dp\frac{1}{\gamma v_w}\left\{\frac{p_w}{\omega - k_{l+n}v_\parallel}\left[p_w\left(\hat{H}_{l-1,n+1}^{(-)} + \hat{H}_{l\mp 2,n\pm 2}^{(-)}\right) - ip_\parallel\left(ck_n\delta\hat{\varphi}_{l,n} - \omega\delta\hat{A}_{l,n,z}\right)\right]\right.$$
$$\left.\times\frac{1}{p}\frac{\partial}{\partial p} + \frac{\lambda_0^2}{\sqrt{1+\lambda_0^2}}\frac{\omega}{\Omega_-}\left(R_{l,n}^{(-)}\hat{L}_{l-1,n+1}^{(-)} - R_{l,n}^{(+)}\hat{L}_{l+1,n-1}^{(+)} + \frac{v_w}{2}\hat{K}_{l,n}S_{l,n}\right)\right\}G_b(p),$$
(4.271)

where we have assumed a distribution function of the form

$$F_b(P_x,P_y,p) = n_b\delta(P_x)\delta(P_y)G_b(p), \tag{4.272}$$

for an ambient beam density n_b, $\omega_b{}^2 = 4\pi e^2 n_b/m_e$, $\delta J_{l,n}{}^{(\pm)} = (\delta J_{l,n})_r \mp i(\delta J_{l,n})_\theta$, and

$$\delta\hat{A}_{l,n}^{(\pm)} \equiv \frac{1}{2}\left[\left(\delta\hat{A}_{l,n}\right)_r \mp i\left(\delta\hat{A}_{l,n}\right)_\theta\right]. \tag{4.273}$$

In addition, we have defined

$$\hat{L}_{l,n}^{(\pm)} \equiv i\left(\omega - k_n v_\parallel\right)\delta\hat{A}_{l,n}^{(\pm)} - \frac{1}{2}\nabla_l^{(\pm)}\left(c\delta\hat{\varphi}_{l,n} - v_\parallel\delta\hat{A}_{l,n,z}\right), \tag{4.274}$$

$$\hat{H}^{(\pm)}_{l,n} \equiv i\omega\delta\hat{A}^{(\pm)}_{l,n} - \frac{c}{2}\nabla^{(\pm)}_{l}\delta\hat{\varphi}_{l,n}, \tag{4.275}$$

$$\hat{K}_{l,n} \equiv \nabla^{(-)}_{l-1}\delta\hat{A}^{(+)}_{l,n} - \nabla^{(+)}_{l+1}\delta\hat{A}^{(-)}_{l,n}, \tag{4.276}$$

$$R^{(\pm)}_{l,n} \equiv \frac{\alpha_+}{\omega \pm \Omega_- - k_{l+n}v_\parallel} - \frac{\alpha_-}{\omega \mp \Omega_- - k_{l+n}v_\parallel} - \frac{1}{\omega - k_{l+n}v_\parallel}, \tag{4.277}$$

and

$$S_{l,n} \equiv \frac{1}{\omega \pm \Omega_- - k_{l+n}v_\parallel} - \frac{1}{\omega \mp \Omega_- - k_{l+n}v_\parallel} - \frac{2}{\omega - k_{l+n}v_\parallel}, \tag{4.278}$$

where $\nabla_l^{(\pm)} = \partial/\partial r \pm l/r$. Observe that $\delta_{\hat{J}l,n,z}$ has been omitted because the specification of a gauge condition allows us to eliminate one of the components of the four-vector formed by the vector and scalar potentials, and we choose to deal with $\delta_{\hat{A}}^{(\pm)}$ and $\delta\hat{\varphi}$.

4.3.1.2.2 The Dispersion Equation

The dispersion equation is obtained by substitution of the sources into the Maxwell-Poisson equations

$$\left[\frac{1}{r}\frac{d}{dr}r\frac{d}{dr} + p_n^2 - \frac{(l \mp 1)^2}{r^2}\right]\delta\hat{A}^{(\pm)}_{l,n} = -\frac{4\pi}{c}\delta\hat{J}^{(\pm)}_{l,n}, \tag{4.279}$$

$$\left[\frac{1}{r}\frac{d}{dr}r\frac{d}{dr} + p_n^2 - \frac{l^2}{r^2}\right]\delta\hat{\varphi}_{l,n} = -8\pi\delta\hat{\rho}_{l,n}, \tag{4.280}$$

as well as the Lorentz gauge condition

$$k_n\delta\hat{A}_{l,n,z} - \frac{\omega}{c}\delta\hat{\varphi}_{l,n} = i\left(\nabla^{(-)}_{l-1}\delta\hat{A}^{(+)}_{l,n} + \nabla^{(+)}_{l+1}\delta\hat{A}^{(-)}_{l,n}\right), \tag{4.281}$$

where $p_n^2 = \omega^2/c^2 - k_n^2$. In order to proceed, we must specify the distribution function $G_b(p)$. In this regard, we observe that there is a one-to-one correspondence between the total energy and the radius of the trajectory for the steady-state orbits. Hence, a small energy spread in the distribution of steady-state orbits is equivalent to a narrow radial profile. For simplicity, therefore, we shall assume a monoenergetic distribution with a sharp radial profile and choose $G_b(p) = N(p)\delta[p - p(r)]$, where

$$p(r) = m_e c\sqrt{1+\lambda_0^2}\left[\frac{\Omega_0}{ck_w} \pm 2\frac{\Omega_w}{ck_w}\frac{1+\lambda_0^2}{\lambda_0^2}I_1(\lambda_0)\right] \tag{4.282}$$

defines the relationship between the momentum and the orbit radius, and $N(p)$ models the density profile. In general, the Maxwell-Poisson equations define a system of equations in which each harmonic contribution is coupled to many others. However, a substantial simplification is found by retaining only the dominant coupling terms, which is valid as long as $\omega_b/\gamma^{1/2}ck_w << 1$. In this case, we find that

$$\delta\hat{J}_{l,n}^{(\pm)} \cong \frac{c}{4\pi}\left[\Lambda_{l,n}^{(\pm)}\delta\hat{A}_{l,n}^{(\pm)} + T_{l\mp1,n\pm1}\delta\hat{\varphi}_{l\mp1,n\pm1} + V_{l\mp2,n\pm2}\delta\hat{A}_{l\mp2,n\pm}^{(\mp)}\right], \tag{4.283}$$

and

$$\delta\hat{\rho}_{l,n} \cong \frac{1}{4\pi}\left[\chi_{l,n}\delta\hat{\varphi}_{l,n} + W_{l+1,n-1}\delta\hat{A}_{l+1,n-1}^{(+)} + W_{l-1,n+1}\delta\hat{A}_{l-1,n+1}^{(-)}\right], \tag{4.284}$$

where we note that $N(p)$ is an implicit function of r

$$\begin{aligned}\Lambda_{l,n}^{(\pm)} \equiv & -\frac{\omega_b^2}{\gamma c^2}N(r)\left[\omega - k_n v_\parallel\left(1+\frac{l^2}{2k_n^2r^2}\right)\right] \\ & \times\left[\frac{\alpha_+^2}{\omega\pm\Omega_- - k_{l+n}v_\parallel} - \frac{\alpha_-^2}{\omega\mp\Omega_- - k_{l+n}v_\parallel}\right] + V_{l\pm,n\mp1},\end{aligned} \tag{4.285}$$

$$\chi_{l,n} \equiv -\frac{\omega_b^2}{\gamma^3c^2}\Phi(\lambda_0)N(r)\frac{k_{l,n}}{k_n}\frac{\omega^2 - c^2k_n^2}{\left(1+\lambda_0^2\right)\left(\omega - k_{l+n}v_\parallel\right)^2}, \tag{4.286}$$

$$T_{l,n}^{(\pm)} \equiv -\frac{\omega_b^2}{2\gamma^3c^2}\frac{p_w}{p_\parallel}\Phi(\lambda_0)N(r)\frac{\omega}{ck_n}\frac{\omega^2 - c^2k_n^2}{\left(1+\lambda_0^2\right)\left(\omega - k_{l+n}v_\parallel\right)^2}, \tag{4.287}$$

$$W_{l\pm1,n\mp1} \equiv -\frac{\omega_b^2}{\gamma^3c^2}\frac{p_w}{p_\parallel}\Phi(\lambda_0)N(r)\frac{\omega k_{l+n}c}{\left(1+\lambda_0^2\right)\left(\omega - k_{l+n}v_\parallel\right)^2}, \tag{4.288}$$

and

$$V_{l\pm1,n\mp1} \equiv -\frac{\omega_b^2}{2\gamma^3c^2}\Phi(\lambda_0)N(r)\frac{\lambda_0^2\omega^2}{\left(1+\lambda_0^2\right)\left(\omega - k_{l+n}v_\parallel\right)^2}, \tag{4.289}$$

where $\Phi(\lambda_0)$ is given in by Eq. (2.70). As a consequence, we obtain the following set of coupled differential equations

$$\left[\frac{1}{r}\frac{d}{dr}r\frac{d}{dr}+p_{n\mp1}^2-\frac{l^2}{r^2}\right]\delta\hat{A}^{(\pm)}_{l\pm1,n\mp1}=-\Lambda^{(\pm)}_{l\pm1,n\mp1}\delta\hat{A}^{(\pm)}_{l\pm1,n\mp1}-T^{(\pm)}_{l,n}\delta\hat{\varphi}_{l,n}$$
$$-V_{l\mp1,n\pm1}\delta\hat{A}^{(\mp)}_{l\mp1,n\pm1}, \tag{4.290}$$

and

$$\left[\frac{1}{r}\frac{d}{dr}r\frac{d}{dr}+p_n^2-\frac{l^2}{r^2}\right]\delta\hat{\varphi}_{l,n}=-\chi_{l,n}\delta\hat{\varphi}_{l,n}-W_{l+1,n-1}\delta\hat{A}^{(+)}_{l+1,n-1}-W_{l-1,n+1}\delta\hat{A}^{(-)}_{l-1,n+1}. \tag{4.291}$$

In order to solve these coupled second-order differential equations, we must specify the boundary conditions appropriate for a cylindrical waveguide of radius R_g. We assume the walls to be grounded at zero potential, and we may express the boundary conditions in the form

$$\delta A^{(+)}_{l,n}(R_g)=\delta A^{(-)}_{l,n}(R_g), \tag{4.292}$$

and

$$\delta\hat{\varphi}_{l,n}(R_g)=\frac{d}{dr}\left[r\left(\delta A^{(+)}_{l,n}(R_g)+\delta A^{(-)}_{l,n}(R_g)\right)\right]\bigg|_{r=R_g}. \tag{4.293}$$

In the thin beam limit, we assume, as in the previous section, that the electron beam density is within the range $r_0-\Delta R\leq r\leq r_0$ [where $\Delta R << r_0$], and we write $N(r)=\Delta R\,\delta(r-r_0)$ where r_0 denotes the radius of the steady-state orbit. Hence, we shall also assume that the vector and scalar potentials are continuous at $r=r_0$ and use the Maxwell-Poisson equations to obtain jump conditions for the derivatives of the potentials across the beam. This thin beam condition is equivalent to a small energy spread on the beam, which is of the order of

$$\frac{\Delta\gamma}{\gamma}=\frac{\gamma^2-1}{\Phi(\lambda_0)-1}\frac{\lambda_0^2}{1+\lambda_0^2}\frac{\Delta R}{r_0} \tag{4.294}$$

and is assumed to be small.

Equations (4.289) and (4.291) describe the coupling between five harmonic components: $\delta\hat{\varphi}_{l,n}$, $\delta\hat{A}^{(\pm)}{}_{l+1,n-1}$, and $\delta\hat{A}^{(\pm)}{}_{l-1,n+1}$. The solutions are given in terms of the Bessel and Neumann functions and are of the form

$$\delta\hat{\varphi}_{l,n}=A_{l,n}J_l(p_nr), \tag{4.295}$$

$$\delta\hat{A}^{(\pm)}_{l,n}=A^{(\pm)}_{l,n}J_{l\mp1}(p_nr), \tag{4.296}$$

for $0\leq r<r_0$, and

$$\delta\hat{\varphi}_{l,n} = B_{l,n}J_l(p_n r) + C_{l,n}Y_l(p_n r), \tag{4.297}$$

$$\delta\hat{A}^{(\pm)}_{l,n} = B^{(\pm)}_{l,n}J_{l\mp 1}(p_n r) + C^{(\pm)}_{l,n}Y_{l\mp 1}(p_n r), \tag{4.298}$$

for $r > r_0$, where J_l and Y_l denote the Bessel and Neumann functions. Thus, each harmonic component requires three coefficients to characterize the solution throughout the waveguide, for a total of 15 coefficients in all. Two of these coefficients for each component are determined from the boundary and continuity conditions at the waveguide wall and the electron beam. The third coefficient is found from the jump condition at the beam, which is obtained by multiplying the Maxwell-Poisson equations by r and integrating over $r_0 - \varepsilon \le r \le r_0 + \varepsilon$ in the limit as $\varepsilon \to 0^+$. The jump condition results in a 5 × 5 matrix equation for the coefficients: $A_{l,n}$, $A^{(+)}{}_{l+1,n\text{-}1}$, $A^{(-)}{}_{l\text{-}1,n+1}$, $A^{(+)}{}_{l\text{-}1,n+1}$, and $A^{(-)}{}_{l+1,n\text{-}1}$.

Observe that the coupling of the field components through their coefficients $A^{(\pm)}{}_{lm1,n\pm 1}$ occurs through the boundary conditions and not through the source terms. Eliminating these coefficients, therefore, we may reduce the problem to a 3 × 3 matrix equation

$$\begin{bmatrix} \varepsilon_{l,n} & -\frac{\pi}{2}r_0\Delta R\overline{W}_{l+1,n-1} & -\frac{\pi}{2}r_0\Delta R\overline{W}_{l-1,n+1} \\ \frac{\pi}{2}r_0\Delta R\overline{T}_{l,n} & \varepsilon^{(+)}_{l+1,n-1} & \frac{\pi}{2}r_0\Delta R\overline{V}_{l-1,n+1} \\ \frac{\pi}{2}r_0\Delta R\overline{T}_{l,n} & \frac{\pi}{2}r_0\Delta R\overline{V}_{l+1,n-1} & \varepsilon^{(-)}_{l-1,n+1} \end{bmatrix} \begin{bmatrix} A_{l,n} \\ A^{(+)}_{l+1,n-1} \\ A^{(-)}_{l-1,n+1} \end{bmatrix} = 0, \tag{4.299}$$

where

$$\varepsilon_{l,n} \equiv D_{l,n} - \frac{\pi}{2}r_0\Delta R\overline{\chi}_{l,n}, \tag{4.300}$$

$$\varepsilon^{(\pm)}_{l\pm 1,n\mp 1} \equiv D^{(\pm)}_{l\pm 1,n\mp 1} - \frac{\pi}{2}r_0\Delta R\left(\overline{\Lambda}^{(\pm)}_{l\pm 1,n\mp 1} + \overline{\Lambda}^{(\mp)}_{l\pm 1,n\mp 1}M^{(l\pm 2)}_{l\mp 1,n\pm 1}\right), \tag{4.301}$$

and $\overline{W}_{l\pm 1,n\mp 1}$, $\overline{V}_{l\pm 1,n\mp 1}$, $\overline{\Lambda}^{(\pm)}_{l\pm 1,n\mp 1}$, $\overline{T}_{l,n}$, and $\overline{\chi}_{l,n}$ denote those quantities specified in Eqs. (4.284, 4.285, 4.286, 4.287, and 4.288) where the factor $N(r)$ has been removed. In addition

$$D_{l,n} \equiv \frac{J_l(\zeta_n)}{J_l(\xi_n)[J_l(\xi_n)Y_l(\zeta_n) - J_l(\zeta_n)Y_l(\xi_n)]}, \tag{4.302}$$

$$D^{(\pm)}_{l\pm 1,n\mp 1} \equiv \frac{2J_{l\pm 1}(\zeta_{n\mp 1})J'_{l\pm 1}(\zeta_{n\mp 1})}{J_l(\xi_{n\mp 1})S^{(l\pm 2)}_{l\pm 1,n\mp 1}}, \tag{4.303}$$

$$M^{(\pm)}_{l\mp 1,n\pm 1} \equiv \frac{J_{l\pm 2}(\zeta_{n\mp 1})\Gamma^{(l\pm 2)}_{l\pm}(\zeta_{n\mp 1})}{J_l(\xi_{n\mp 1})S^{(l\pm 2)}_{l\pm 1,n\mp 1}}, \tag{4.304}$$

where $\zeta_n = p_n R_g$ and $\xi_n = p_n r_0$. In addition

$$\Gamma^{(l)}_{k,m} \equiv Y_l(\xi_m)\frac{d}{d\zeta_m}J^2_k(\zeta_m) - J_l(\xi_m)\frac{d}{d\zeta_m}[J_k(\zeta_m)Y_k(\zeta_m)], \tag{4.305}$$

$$S^{(l\pm 2)}_{k,m} \equiv \Gamma^{(l)}_{k,m} + \frac{\pi}{2}r_0\Delta R\overline{\Lambda}^{(\mp)}_{k,m}J_{l\pm 2}(\xi_m)\Psi^{(l\pm 2)}_{k,m}, \tag{4.306}$$

$$\Psi^{(l\pm 2)}_{k,m} \equiv Y_{l\pm 2}(\xi_m)\Gamma^{(l)}_{k,m} - J_{l\pm 2}(\xi_m)\left[Y_l(\xi_m)\frac{d}{d\zeta_m}[J_k(\zeta_m)Y_k(\zeta_m)] - J_l(\xi_m)\frac{d}{d\zeta_m}Y^2_k(\zeta_m)\right], . \tag{4.307}$$

The dispersion equation is found by setting the determinant of this interaction matrix to zero.

Substantial simplification occurs in the resonant approximation for which $|\omega - k_{l+n}v_{\|}| << \omega$, $|k_{l+n}v_{\|}|$ and

$$\overline{\chi}_{l,m} \simeq \frac{\omega_b^2}{\gamma^3 c^2}\Phi(\lambda_0)\frac{\omega}{k_n v_{\|}}\frac{\omega^2 - c^2k_n^2}{\left(1+\lambda_0^2\right)\left(\omega - k_{l+n}v_{\|}\right)^2}, \tag{4.308}$$

$$\overline{W}_{l\pm 1,n\mp 1} \cong -\frac{\omega_b^2}{\gamma^3 c^2}\frac{p_w}{p_{\|}}\Phi(\lambda_0)\frac{\omega^2}{\beta_{\|}\left(1+\lambda_0^2\right)\left(\omega - k_{l+n}v_{\|}\right)^2}, \tag{4.309}$$

$$\begin{aligned}\overline{\Lambda}^{(\pm)}_{l\pm,n\mp 1} \cong &-\frac{\omega_b^2}{\gamma c^2}\frac{p_w}{p_{\|}}\left[\omega - k_n v_{\|}\left(1+\frac{l^2}{2k_n^2 r^2}\right)\right]\times\left[\frac{\alpha_+^2}{\omega\pm\Omega_- - k_{l+n}v_{\|}} - \frac{\alpha_-^2}{\omega\mp\Omega_- - k_{l+n}v_{\|}}\right]\\ &-\frac{\omega_b^2}{2\gamma^3c^2}\Phi(\lambda_0)\frac{\lambda_0^2\omega^2}{\beta_{\|}\left(1+\lambda_0^2\right)\left(\omega - k_{l+n}v_{\|}\right)^2},\end{aligned} \tag{4.310}$$

$$\overline{T}_{l,n} \cong \frac{v_w}{2c}\overline{\chi}_{l,n}, \tag{4.311}$$

$$\overline{V}_{l,n} \cong \frac{v_w}{2c}\overline{W}_{l\pm 1,n\mp 1}. \tag{4.312}$$

For all cases of practical interest, $v_w << c$ and $\left|\overline{V}_{l\pm 1,n\mp 1}\right| << \left|\overline{W}_{l\pm 1,n\mp 1}\right|$. As a consequence, the terms in $\overline{V}_{l\pm 1,n\mp 1}$ can be ignored. This is equivalent to the neglect of any direct coupling between the electromagnetic modes given by $_{\delta\hat{A}}{}^{(\pm)}{}_{l\pm 1,n\mp 1}$. In addition, we can neglect the coupling to the $_{\delta\hat{A}}{}^{(\pm)}{}_{l\mp 1,n\pm 1}$, so that

$$\varepsilon^{(\pm)}_{l\pm,n\mp 1} \cong \frac{2J_{l\pm 1}(\zeta_{n\mp 1})J'_{l\pm 1}(\zeta_{n\mp 1})}{J_l(\xi_{n\mp 1})S^{(l\pm 2)}_{l\pm 1,n\mp 1}} + \frac{\pi}{2}r_0\Delta R\overline{\Lambda}^{(\pm)}_{l\pm,n\mp 1}. \tag{4.313}$$

Within the context of this approximation, the dispersion equation is of the form

$$\varepsilon_{l,n} \cong \left(\frac{\pi}{2}r_0\Delta R\right)^2 \frac{\omega_b^2}{2\gamma^3c^2}\frac{\lambda_0^2}{1+\lambda_0^2}\frac{\omega^2\Phi(\lambda_0)}{\beta_\parallel\left(\omega - k_{l+n}v_\parallel\right)^2}\overline{\chi}_{l,n}\left[\frac{1}{\varepsilon^{(+)}_{l+1,n-1}} + \frac{1}{\varepsilon^{(-)}_{l-1,n+1}}\right]. \tag{4.314}$$

Finally, if the solution is restricted to the first quadrant in the (ω,k_n) plane, the $\left|\varepsilon^{(-)}_{l-1,n+1}\right| >> \left|\varepsilon^{(+)}_{l+1,n-1}\right|$, and the dispersion equation can be approximated by

$$\varepsilon_{l,n}\varepsilon^{(+)}_{l+1,n-1} \cong \left(\frac{\pi}{2}r_0\Delta R\right)^2 \frac{\omega_b^2}{2\gamma^3c^2}\frac{\lambda_0^2}{1+\lambda_0^2}\frac{\omega^2\Phi(\lambda_0)}{\beta_\parallel\left(\omega - k_{l+n}v_\parallel\right)^2}\overline{\chi}_{l,n}. \tag{4.315}$$

This equation describes the coupling between the beam space-charge modes and a circularly polarized electromagnetic mode (either TE or TM modes of the waveguide).

4.3.1.2.3 Numerical Solution of the Dispersion Equation

Equation (4.308) describes the coupling between the beam space-charge mode and a right-hand circularly polarized waveguide mode. In order to illustrate the nature of the gain in the free-electron laser, the complete dispersion Eq. (4.307) is solved for a representative set of parameters: $\gamma = 3.5$, $\omega_b/\gamma^{1/2}ck_w = 0.1$, $\Omega_w/ck_w = 0.05$, $\Delta R/r_0 = 0.1$, $k_wR_g = 1.5$, and a wide range of axial guide fields corresponding to both Group I and Group II orbits. For this choice of parameters, the transitions to orbital instability (and the corresponding singularities in Φ) occur for $\Omega_0/ck_w \approx 0.75$ for the Group I and $\Omega_0/ck_w \approx 0.62$ for the Group II trajectories. Hence, there is a substantial overlap in the range of axial fields for which these trajectories occur. While the growth rates found at these orbital stability transitions are also singular, it should be recognized that the linear stability analysis breaks down in this regime. In addition, both Φ and the growth rate vanish for Group II orbits when $\Omega_0/ck_w \approx 1.25$.

The normalized magnitude of the growth rate $|\mathrm{Im}\ k_n|/k_w$ of the TE_{11} mode is plotted in Fig. 4.12 for these parameters with $\Omega_0/ck_w = 0.0$ and 0.5. This corresponds to Group I orbits. The waveguide cutoff occurs at a frequency $\omega_{co}/ck_w \approx 1.23$ for this mode, and the two peaks correspond to the upper and lower intersections between the space-charge and waveguide modes. An approximate expression for the location of these intersection points can be obtained on the basis of the dispersion equation

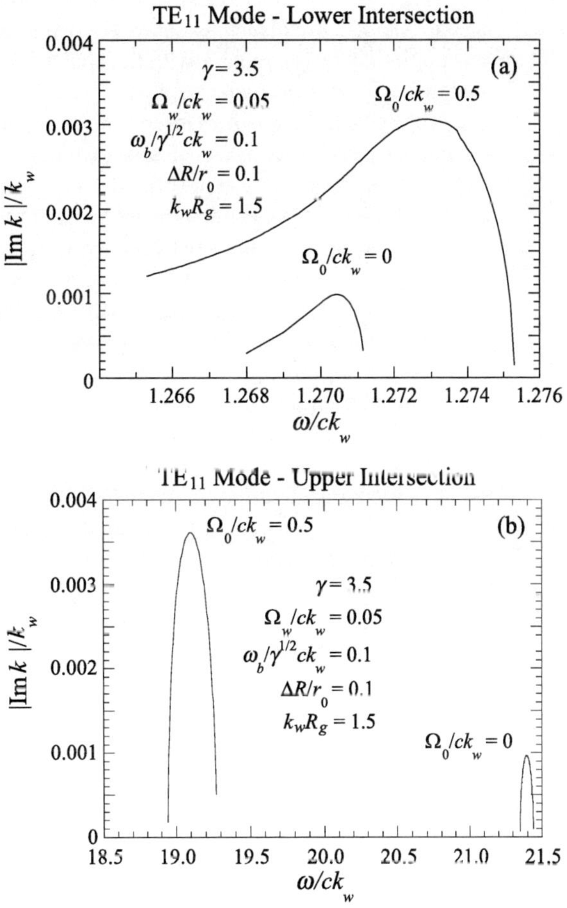

Fig. 4.12 Plot of the normalized growth rate at the lower (**a**) and upper (**b**) intersections as a function of frequency for the TE_{11} mode and Group I orbits

for the vacuum waveguide modes [i.e., in the absence of the electron beam] and the negative-energy space-charge wave. In this limit, it can be shown that the intersections occur at

$$\frac{\omega}{ck_w} \cong \gamma_\parallel^2 \beta_\parallel \left[\left(1 - \frac{\omega_b}{\gamma^{1/2}\gamma_\parallel k_w v_\parallel} \right) \pm \beta_\parallel \sqrt{\left(1 - \frac{\omega_b}{\gamma^{1/2}\gamma_\parallel k_w v_\parallel} \right)^2 - \frac{\omega_{co}^2}{\gamma_\parallel^2 k_w^2 v_\parallel^2}} \right], \quad (4.316)$$

where $\beta_\parallel = v_\parallel/c$. It should be emphasized that this is only an approximate expression for the intersection points because (1) the dielectric effect of the beam can significantly modify the dispersion relation of the waveguide mode and (2) the waveguide geometry reduces the effective plasma frequency. As shown in the figure, the peak growth rate increases rapidly with the axial magnetic field corresponding to the increase (decrease) in the transverse (axial) beam velocity and the increase in Φ. In addition, the resonant frequency is sensitive to the axial velocity, and the upper intersection point decreases rapidly with the change in the axial velocity associated with the increasing axial magnetic field.

The peak growth rates for the Group I trajectories, as well as the frequencies corresponding to peak growth, are plotted in Fig. 4.13 as a function of the axial solenoidal field. The singularity at the orbital instability transition ($\Omega_0/ck_w \approx 0.75$) is evident from the figure. It is clear that the frequency of the upper (lower) intersection point decreases (increases) as the axial velocity decreases with increasing axial field strength. As a result, the intersections tend to coalesce with decreasing $v_\parallel$; however, the TE_{11} mode cutoff for the chosen parameters is sufficiently low that coalescence does not occur for the Group I orbits, and the two lines remain well separated. While the growth rate corresponding to the upper intersection exceeds that of the lower intersection over the entire range of the Group I trajectories for the parameters illustrated, it should not be construed that this holds in general.

The maximum growth rates and corresponding frequencies are plotted in Fig. 4.14 as a function of the axial magnetic field for Group II trajectories. The growth rate for the Group II trajectories exhibits characteristics similar to that found for Group I trajectories, specifically, (1) that the growth rate tends to increase in the vicinity of the singularity in Φ at $\Omega_0/ck_w \approx 0.62$ and (2) that the upper and lower intersections tend to coalesce as the axial velocity decreases in the neighborhood of the magnetic resonance. Differences between the case of Group I and Group II trajectories, however, arise from several effects. In the first place, the resonance at $\Omega_0 \approx k_w v_\parallel$ can be more closely approached for the Group II trajectories, and the growth rate is correspondingly larger than that found for the Group I case. In the second place, Φ vanishes at the transition between the negative- and positive-mass regimes ($\Omega_0/ck_w \approx 1.25$ for the chosen parameters), and the growth rate vanishes at this point. In the third place, the space-charge wave is itself unstable in the negative-

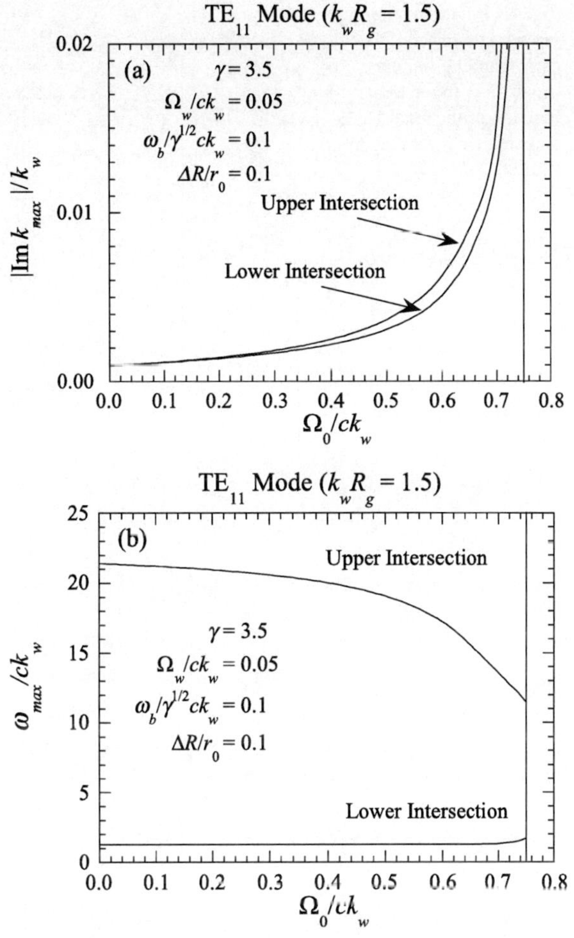

Fig. 4.13 The maximum growth rate (**a**) and corresponding frequency (**b**) of the TE_{11} mode as a function of the axial magnetic field for the Group I orbits

mass regime (i.e., $\Phi < 0$). In this case, the interaction is found at the approximate intersections between the vacuum waveguide modes and the beam resonance line $[\omega = (k + k_w)v_{\parallel}]$, which yields

$$\frac{\omega}{ck_w} \cong \gamma_{\parallel}^2\beta_{\parallel}\left[1 \pm \beta_{\parallel}\sqrt{1 - \frac{\omega_{co}^2}{\gamma_{\parallel}^2 k_w^2 v_{\parallel}^2}}\right]. \tag{4.317}$$

Finally, an interaction is possible for the TM_{11} mode as well for the chosen parameters. Indeed, the growth rates are comparable to those found for the TE_{11} mode, and the character of the interaction differs only insofar as the cutoff frequency is higher.

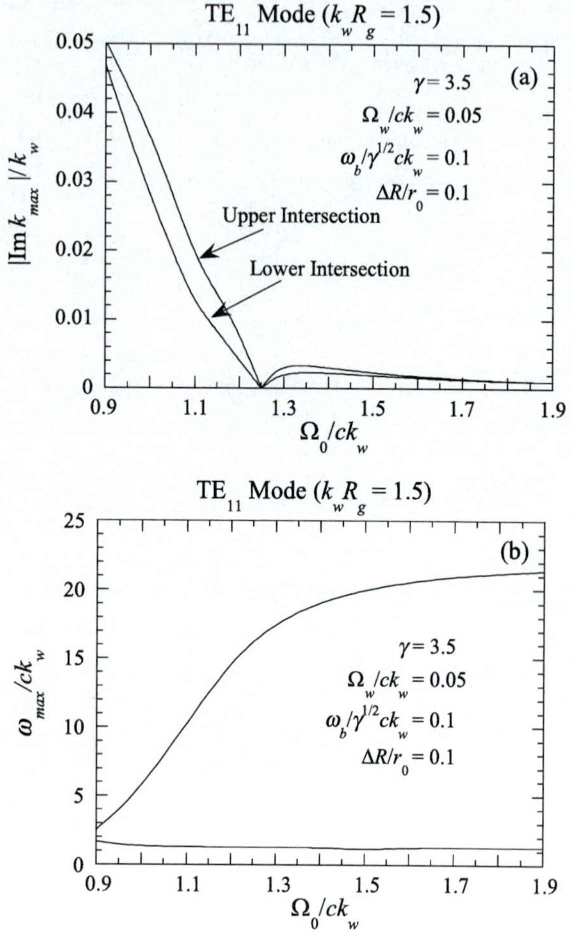

Fig. 4.14 The maximum growth rate (**a**) and corresponding frequency (**b**) of the TE_{11} mode as a function of the axial magnetic field for the Group II orbits

4.3.1.2.4 Comparison with Experiment

The aforementioned dispersion equation has been applied to the analysis of collective Raman free-electron laser experiments conducted at Columbia University [48] and the Massachusetts Institute of Technology [49]. Both of these experiments employed helical wigglers and axial solenoidal magnetic fields generated by current-carrying coils and, hence, were variable over a wide range of parameters. In addition, the principal resonance in each case was with the TE_{11} mode. The principal distinctions between these experiments involved the differences in the parameter ranges that could be accessed. As a result, these experiments together covered an extremely wide range of operating parameters.

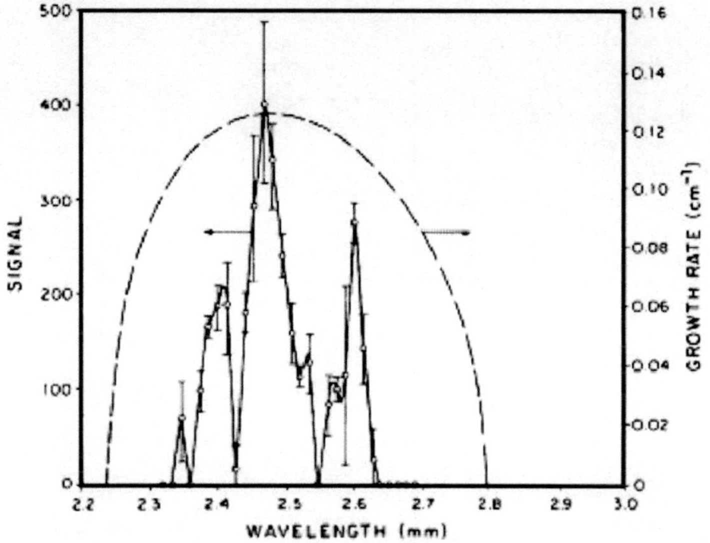

Fig. 4.15 Comparison of the observed spectrum of Columbia University Raman free-electron laser experiment [48] and the theoretical prediction of the growth rate as a function of wavelength

The free-electron laser at Columbia University employed a beam with an energy of 700 keV and a current of 200 A. The wiggler was characterized by a period of 1.45 cm. A specific comparison with the three-dimensional analysis dealt with the case of a wiggler field of 760 G and a solenoidal field of 9.5 kG. As a consequence, the experiment operated with Group I trajectories for which $v_{\|}/c \approx 0.900$ and $\lambda_0 \approx 0.124$. Since the waveguide radius was 0.3 cm, the high frequency intersection occurred at a frequency of approximately 124 GHz at a wavelength of approximately 2.5 mm. Agreement between the observed spectrum (both as regards the frequency and growth rate) and the three-dimensional linear theory was good. A comparison of the observed spectrum and the calculated growth rates is given in Fig. 4.15 as a function of wavelength. As shown in the figure, the peak in the spectrum is found at a wavelength of approximately 2.47 mm in both the experiment and the theory. The observed exponential growth rate was approximately 0.1 cm^{-1}, which is also in good agreement with the theoretical result of 0.12 cm^{-1}.

The experiment at the Massachusetts Institute of Technology was conducted at much lower voltages and currents, although the current density was sufficiently high for this experiment to be in the collective Raman regime as well. The specific parameters of this experiment involved a beam voltage and current of 160 keV and 4.8 A, respectively, with a beam radius of approximately 2.6 mm. The wiggler field amplitude was variable up to approximately 1.5 kG with a period of 3.3 cm, and the axial solenoidal field could achieve a maximum field strength of approximately 7 kG. As in the case of the experiment at Columbia University, the interaction was

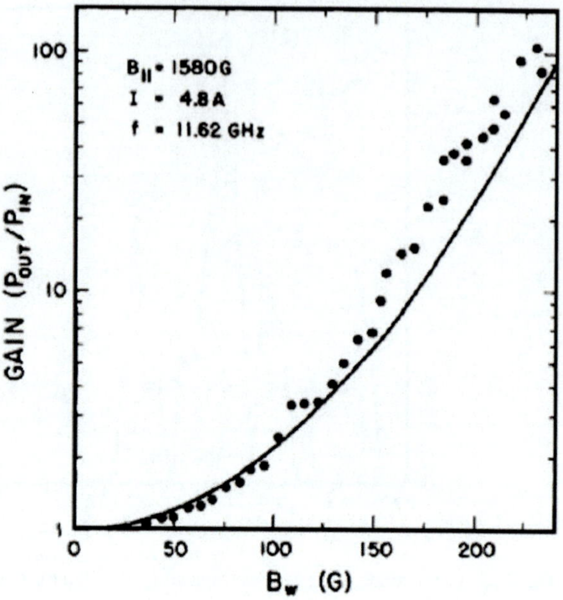

Fig. 4.16 Comparison of the observed spectrum of the Raman free-electron laser experiment at M.I.T. and the theoretical prediction of the growth rate as a function of wiggler amplitude [49]

resonant with the TE_{11} mode, which had a cutoff frequency of approximately 6.92 GHz for the choice of a drift tube with a 2.54 cm inner diameter. An example of the comparison between the theoretically predicted growth rates and those measured in the experiment is shown in Fig. 4.16. The specific parameters used in this figure involved a beam current of 4.8 A and an axial guide field of 1.58 kG. The free-electron laser was run in an amplifier mode with a drive signal at a frequency of 11.62 GHz. The power gain was then measured as a function of the wiggler field amplitude over a range of up to 300 G. As is evident from the figure, the comparison between the experiment and the theoretical prediction is quite good over the entire range of wiggler field amplitudes studied.

4.3.2 Optical Mode Analysis

The linear stability theory for optical modes that we present here is based upon the analyses by Chin, Kim, and Xie [45, 46]. This theory is developed for the high-gain regime using a Vlasov-Maxwell analysis for a Gaussian electron beam including realistic models for the energy spread, emittance, and betatron oscillations for weak focusing wigglers. The low-gain regime is appropriate for short-wavelength oscillators and will be dealt with within the context of numerical simulations in a later chapter.

4.3.2.1 Helical Wiggler Configurations

In this section, we describe the derivation of the dispersion equation in the high-gain limit by the solution of the Vlasov-Maxwell equations for optical modes. The solution to the Vlasov equation is typically found by the method of characteristics where the distribution is a function of the constants of the motion.

4.3.2.1.1 The Electron Trajectories

As discussed in Chap. 2, any given electron trajectory through the wiggler is characterized by a slow betatron oscillation in addition to the rapid wiggle motion. We consider a quasi-one-dimensional orbit analysis where the rapid wiggler motion is treated in the one-dimensional limit, with the addition of the betatron motion. This betatron motion described by a harmonic oscillator where the transverse motion is described by the canonical coordinates ($\mathbf{x}_\beta$,$\mathbf{p}_\beta$) which satisfy the equations

$$\frac{d}{dz}\mathbf{x}_\beta = \mathbf{p}_\beta, \tag{4.318}$$

$$\frac{d}{dz}\mathbf{p}_\beta = -k_\beta^2 \mathbf{x}_\beta, \tag{4.319}$$

where $k_\beta = \Omega_w/\sqrt{2}c$ and $\Omega_w = eB_w/\gamma m_e c$. As a result, the betatron motion is given by

$$\mathbf{x}_\beta = \mathbf{x}_{\beta 0}\cos k_\beta z + \frac{\mathbf{p}_{\beta 0}}{k_\beta}\sin k_\beta z, \tag{4.320}$$

$$\mathbf{p}_\beta = -k_\beta \mathbf{x}_{\beta 0}\sin k_\beta z + \mathbf{p}_{\beta 0}\cos k_\beta z, \tag{4.321}$$

where $\mathbf{x}_{\beta 0}$ and $\mathbf{p}_{\beta 0}$ are the initial values at $z = 0$ that characterize each betatron trajectory. We now assume that the transverse components of the trajectory of each electron can be described by a superposition of the helical and betatron motion, $\mathbf{x} = \mathbf{x}_w + \mathbf{x}_\beta$, where $\mathbf{x}_w$ denotes the helical wiggler motion

$$\mathbf{x}_w = r_w\left(-\hat{\mathbf{e}}_x \sin k_w z + \hat{\mathbf{e}}_y \cos k_w z\right), \tag{4.322}$$

where $r_w = |v_w|/k_w v_\parallel$ is the radius of the helical wiggler motion about the orbit center. As a consequence, the Vlasov-Maxwell analysis is based upon modified one-dimensional helical trajectories that neglect the transverse gradient of the helical wiggler. The longitudinal motion of the electrons can be described in terms of a Lagrangian time coordinate defined relative to the resonant energy ($\tau = t - z/v_\parallel$) as

$$\frac{d\tau}{dz} \cong \frac{1}{c}\left[-2\frac{k_{res}}{k_w}\frac{\gamma-\gamma_{res}}{\gamma_{res}}+\frac{1}{2}\left(\mathbf{p}_\beta^2+k_\beta^2\mathbf{x}_\beta^2\right)\right], \tag{4.323}$$

where γ_{res} is the resonant energy and $k_{res} = 2k_w(1 + K^2)/\gamma_{res}^2$ is the resonant wavenumber corresponding to the wiggler strength parameter, K, and the bulk axial velocity is $v_\parallel/c = [1 - (1 + K^2)/\gamma_{res}^2]^{1/2}$. These equations describe the unperturbed, steady-state trajectories.

4.3.2.1.2 The Dispersion Equation

The electron distribution is assumed to be of the functional form $f(\mathbf{x}_\beta,\mathbf{p}_\beta,\tau,\gamma;z)$ which satisfies the Vlasov equation

$$\left[\frac{\partial}{\partial z}+\mathbf{p}_\beta\cdot\nabla_{\mathbf{x}_\beta}-k_\beta^2\mathbf{x}_\beta\cdot\nabla_{\mathbf{p}_\beta}+\frac{d\tau}{dz}\frac{\partial}{\partial\tau}+\frac{d\gamma}{dz}\frac{\partial}{\partial\gamma}\right]f=0, \tag{4.324}$$

where the distribution is normalized so that

$$\int_1^\infty d\gamma\int_{-T/2}^{T/2}d\tau\int_{-\infty}^{\infty}d^2\mathbf{x}_\beta\int_{-\infty}^{\infty}d^2\mathbf{p}_\beta f(\mathbf{x}_\beta,\mathbf{p}_\beta,\tau,\gamma;z)=N_e, \tag{4.325}$$

for a pulse time T which is very long and the total number of electrons N_e. It is further assumed that the radiation field is purely transverse (i.e., a TEM mode). If we denote the vector potential as $\delta\mathbf{A}(\mathbf{x},z,t)$, then

$$\frac{d\gamma}{dz}=-\frac{e}{m_ec^2}\frac{d}{dz}\mathbf{x}_w\cdot\frac{\partial}{\partial t}\delta\mathbf{A}. \tag{4.326}$$

The Vlasov equation is solved by a perturbation analysis where we expand the distribution in powers of the radiation field. To first order, we write $f = f_0 + \delta f$.

If the electron beam is matched into the wiggler yielding a near-constant beam radius, then the equilibrium distribution is a function solely of $\mathbf{x}_\beta^2 + \mathbf{p}_\beta^2/k_\beta^2$ and γ which we write as the product

$$f_0(\mathbf{x}_\beta,\mathbf{p}_\beta,\gamma)=\frac{N_e}{T}G_\perp\left(\mathbf{x}_\beta^2+\mathbf{p}_\beta^2/k_\beta^2\right)G_\parallel(\gamma). \tag{4.327}$$

This describes a flattop temporal beam profile. Since the betatron motion describes simple harmonic oscillations, we can express the trajectories in cylindrical coordinates as

$$x_\beta = r_x \cos \phi_x, y_\beta = r_y \sin \phi_y, \tag{4.328}$$

and

$$p_{\beta x} = k_\beta r_x \sin \phi_x, p_{\beta y} = k_\beta r_y \sin \phi_y. \tag{4.329}$$

As a result, $\mathbf{x}_\beta{}^2 + \mathbf{p}_\beta{}^2/k_\beta{}^2 = r_x{}^2 + r_y{}^2 = r^2$, and we can write $G_\perp(\mathbf{x}_b{}^2 + \mathbf{p}_b{}^2/k_b{}^2) = G_\perp(r^2)$. These distributions are subject to the normalization conditions

$$\int_1^\infty d\gamma G_\parallel(\gamma) = 1, \tag{4.330}$$

and

$$\iint d^2\mathbf{x}_\beta \iint d^2\mathbf{p}_\beta G_\perp\left(\mathbf{x}_\beta^2 + \mathbf{p}_\beta^2/k_\beta^2\right) = 2\pi^2 k_\beta^2 \int_0^\infty dr r^3 G_\perp\left(r^2\right) = 1. \tag{4.331}$$

We consider Gaussian distributions in energy and transverse extent. The energy distribution is given by

$$G_\parallel(\gamma) = \frac{1}{\sqrt{2\pi}\Delta\gamma} \frac{1}{2 - \text{erf}\left[(\gamma_{res} - 1)/\sqrt{2}\Delta\gamma\right]} \exp\left[-(\gamma - \gamma_{res})^2/2\Delta\gamma^2\right], \tag{4.332}$$

where $\Delta\gamma$ denotes the rms energy spread. In the limit in which $\gamma_{res} >> 1$, this is given approximately by

$$G_\parallel(\gamma) \cong \frac{1}{\sqrt{2\pi}\Delta\gamma} \exp\left[-(\gamma - \gamma_{res})^2/2\Delta\gamma^2\right]. \tag{4.333}$$

The transverse Gaussian distribution is

$$G_\perp\left(r^2\right) = \frac{1}{\left(2\pi k_\beta \Delta R^2\right)^2} \exp\left(-r^2/2\Delta R^2\right), \tag{4.334}$$

where ΔR denotes the rms beam size.

The first-order perturbed distribution satisfies

$$\left[\frac{\partial}{\partial z} + \mathbf{p}_\beta \cdot \nabla_{\mathbf{x}_\beta} - k_\beta^2 \mathbf{x}_\beta \cdot \nabla_{\mathbf{p}_\beta} + \frac{d\tau}{dz}\frac{\partial}{\partial \tau}\right]\delta f = -\frac{d\gamma}{dz}\frac{\partial}{\partial \gamma} f_0, \tag{4.335}$$

where $d\gamma/dz$ is given in Eq. (4.326).

The vector potential satisfies the wave equation

$$\left(\nabla^2 - \frac{1}{c^2}\frac{\partial^2}{\partial t^2}\right)\delta\mathbf{A} = -\frac{4\pi}{c}\delta\mathbf{J}_\perp, \tag{4.336}$$

where the source current is

$$\delta\mathbf{J}_\perp(\mathbf{r},t) \cong -e\frac{d\mathbf{x}_w}{dz}\int_1^\infty d\gamma \int_{-\infty}^\infty d^2\mathbf{p}_\beta \delta f\left(\mathbf{x}_\beta, \mathbf{p}_\beta, \gamma, \tau; z\right), \tag{4.337}$$

and we observe that

$$\frac{d\mathbf{x}_w}{dz} = -\frac{v_w}{v_\parallel}\left(\hat{\boldsymbol{e}}_x \cos k_w z + \hat{\boldsymbol{e}}_y \sin k_w z\right). \tag{4.338}$$

The solution to the wave equation can be found in terms of a Green's function

$$\delta\mathbf{A}(\mathbf{r},t) = \frac{c}{4\pi}\iiint d^3r' \int_{-\infty}^\infty dt' \, \mathbf{G}(\mathbf{r},t\,|\,\mathbf{r}',t')\delta\mathbf{J}_\perp(\mathbf{r}',t'), \tag{4.39}$$

where the Green's function is the free-space solution to the equation

$$\left(\nabla^2 - \frac{1}{c^2}\frac{\partial^2}{\partial t^2}\right)\mathbf{G}(\mathbf{r},t\,|\,\mathbf{r}',t') = -\mathrm{I}\delta(\mathbf{r}-\mathbf{r}')\delta(t-t'), \tag{4.340}$$

and **I** is the unit dyadic. Hence, the Green's function is

$$\mathbf{G}(\mathbf{r},t\,|\,\mathbf{r}',t') = \frac{1}{(2\pi)^4}\mathrm{I}\iiint d^3\mathbf{k}\exp[i\mathbf{k}\cdot(\mathbf{r}-\mathbf{r}')]\int_{-\infty}^\infty d\omega\frac{\exp[-i\omega(t-t')]}{\mathbf{k}^2-(\omega/c)^2}, \tag{4.341}$$

and $\mathbf{k} = k_x\hat{\boldsymbol{e}}_x + k_y\hat{\boldsymbol{e}}_y + k_z\hat{\boldsymbol{e}}_z$. Converting the integration $d^3\mathbf{r}'dt' \to d^2\mathbf{x}'_\beta dz' d\tau'$, we write

$$\delta\mathbf{A}(\mathbf{r},t) = -\frac{ec}{4\pi}\int_0^z dz'\frac{d\mathbf{x}'_w}{dz'}\int_{-\infty}^\infty d\tau' \iint d^2\mathbf{x}'_\beta \mathbf{G}(\mathbf{r},t\,|\,\mathbf{r}',t')\delta\rho\left(\mathbf{x}'_\beta,\tau';z'\right), \tag{4.342}$$

where

$$\delta\rho(\mathbf{x}_\beta, \tau; z) \equiv \int_1^\infty d\gamma \int_{-\infty}^\infty d^2\mathbf{p}_\beta \delta f(\mathbf{x}_\beta, \mathbf{p}_\beta, \gamma, \tau; z). \tag{4.343}$$

We now introduce the Fourier transform with respect to the transverse coordinates

$$\delta\hat{\rho}(\omega, \mathbf{k}_\perp; z) = \frac{1}{(2\pi)^3} \int_{-\infty}^\infty d\tau \iint d^2\mathbf{x}_\beta \delta\rho(\mathbf{x}_\beta, \tau; z) \exp(i\omega\tau - i\mathbf{k}_\perp \cdot \mathbf{x}_\beta), \tag{4.344}$$

and

$$\begin{aligned}\hat{\mathbf{G}}(\omega, \mathbf{k}_\perp, z|z') &= \int_{-\infty}^\infty d\hat{t} \int_{-\infty}^\infty d\hat{x} \int_{-\infty}^\infty d\hat{y}\, \mathbf{G}(\mathbf{r}, t|\mathbf{r}', t') \exp(i\omega\hat{t} - i\mathbf{k}_\perp \cdot \hat{\mathbf{x}}) \\ &= \frac{1}{(2\pi)^4} \mathrm{I} \int_{-\infty}^\infty dk_z \frac{\exp[ik_z(z - z')]}{\mathbf{k}^2 - \omega^2/c^2},\end{aligned} \tag{4.345}$$

where $\hat{t} = t - t'$, $\hat{x} = x - x'$, $\hat{y} = y - y'$, and $\mathbf{k}_\perp = k_x\hat{\boldsymbol{e}}_x + k_y\hat{\boldsymbol{e}}_y$. We introduce further the Laplace transform of the perturbed density

$$\delta\hat{\rho}(\omega, \mathbf{k}_\perp; q) = \int_0^\infty dz \exp(-qz)\delta\hat{\rho}(\omega, \mathbf{k}_\perp; z), \tag{4.346}$$

with the associated inverse transform

$$\delta\hat{\rho}(\omega, \mathbf{k}_\perp; z) = \frac{1}{2\pi i} \int_{q_0 - i\infty}^{q_0 + i\infty} dq \exp(qz)\delta\hat{\rho}(\omega, \mathbf{k}_\perp; q). \tag{4.347}$$

As a result, we have that

$$\begin{aligned}\delta\mathbf{A}(\mathbf{r}, t) = -\frac{ec}{4\pi} \int_{-\infty}^\infty d\omega \int_{q_0 - i\infty}^{q_0 + i\infty} dq \iint d^2\mathbf{k}_\perp \delta\hat{\rho}(\omega, \mathbf{k}_\perp; q) \\ \times \hat{H}(\omega, \mathbf{k}_\perp; q, z) \exp(i\mathbf{k}_\perp \cdot \mathbf{x} - i\omega\tau + qz),\end{aligned} \tag{4.348}$$

where

$$\hat{H}(\omega,\mathbf{k}_\perp;q,z)=\int_0^z dz'\,\hat{\mathbf{G}}(\omega,\mathbf{k}_\perp,z|z')\frac{d\mathbf{x}'_w}{dz'}\exp\left[-i\mathbf{k}_\perp\cdot\mathbf{x}'_w+\left(i\omega/v_\parallel+q\right)(z'-z)\right].\tag{4.349}$$

It was shown by Chin, Kim, and Xie [45] that

$$\hat{H}(\omega,\mathbf{k}_\perp;q,z)=-\frac{i\pi}{(2\pi)^4}\frac{v_w v_\parallel}{ck_z}\sum_{n=-\infty}^{\infty}\frac{\mathbf{V}_n(\mathbf{k}_\perp)\exp(-ink_w z)}{q+i\left[\omega-(k_z+nk_w)v_\parallel\right]},\tag{4.350}$$

where

$$\begin{aligned}\mathbf{V}_n(\mathbf{k}_\perp) &= (-1)^{n-1}\exp(-in\phi_k)\left\{\frac{1}{2}\hat{\mathbf{e}}_x\left[\exp(-i\phi_k)J_{n+1}(k_\perp r_w)\right.\right.\\ &\quad\left.+\exp(i\phi_k)J_{n-1}(k_\perp r_w)\right]+\frac{1}{2i}\hat{\mathbf{e}}_y\left[\exp(-i\phi_k)J_{n+1}(k_\perp r_w)\right.\\ &\quad\left.\left.-\exp(i\phi_k)J_{n-1}(k_\perp r_w)\right]\right\}\end{aligned}\tag{4.351}$$

and $\phi_k = \tan^{-1}(k_y/k_x)$. Having obtained $\delta\mathbf{A}$, it may be shown after substantial manipulation that [45]

$$\begin{aligned}\frac{d\gamma}{dz}&=\frac{1}{2\pi}\int_{-\infty}^{\infty}d\omega\frac{1}{2\pi i}\int_{q_0-i\infty}^{q_0+i\infty}dq\iint d^2\mathbf{k}_\perp\exp\left[i\mathbf{k}_\perp\cdot\mathbf{x}_\beta-i\omega\tau+qz\right]\\ &\quad\times\hat{\Gamma}(\omega,\mathbf{k}_\perp;q)\delta\hat{\rho}(\omega,\mathbf{k}_\perp;q),\end{aligned}\tag{4.352}$$

where

$$\hat{\Gamma}(\omega,\mathbf{k}_\perp;q)\equiv\frac{r_e}{2\pi}\frac{v_w^2}{c^3}\sum_{n=-\infty}^{\infty}(-1)^n\frac{k}{k_z}\frac{\left[(n/k_\perp r_w)^2J_n^2(k_\perp r_w)+J_n'^2(k_\perp r_w)\right]}{q+i\left[\omega/v_\parallel-(k_z+nk_w)\right]},\tag{4.353}$$

$k_z = (k^2 - k_\perp{}^2)^{1/2}$, and $r_e = e^2/m_ec^2$ is the classical electron radius. The expression for the rate of energy transfer, $d\gamma/dz$, depends upon the perturbed density, $\delta\hat{\rho}$, which is given in terms of the perturbed distribution. We remark that substantial simplification is found for near-paraxial propagation, which is the dominant mode for emission in a free-electron laser. In this case, $k_\perp = k\sin\theta$, where $\theta << 1$, and the term corresponding to the fundamental resonance (i.e., $n = 1$) is dominant, so that

$$\hat{\Gamma}(\omega, \mathbf{k}_\perp; q) \cong -\frac{r_e}{2\pi}\frac{v_w^2}{c^3}\frac{\left[(k_\perp r_w)^{-2}J_n^2(k_\perp r_w) + J_n'^2(k_\perp r_w)\right]}{q + i\left[\omega/v_\parallel - (k_{res} + k_w)\right] + i\frac{k}{2}\theta^2}. \tag{4.354}$$

As a consequence, substitution of Eq. (4.352) into Eq. (4.335) permits the solution of the Vlasov equation for the perturbed distribution. Taking the Fourier-Laplace transform of Eq. (4.352), we obtain

$$\left[\left(q - i\omega\frac{d\tau}{dz}\right) + \mathbf{p}_\beta\cdot\nabla_{\mathbf{x}_\beta} - k_\beta^2\mathbf{x}_\beta\cdot\nabla_{\mathbf{p}_\beta}\right]\delta\hat{f}\left(\mathbf{x}_\beta, \mathbf{p}_\beta, \gamma; \omega, q\right)$$
$$= -G_\perp\left(r^2\right)\frac{\partial}{\partial\gamma}G_\parallel(\gamma)\iint d^2\mathbf{k}_\perp \exp\left(i\mathbf{k}_\perp\cdot\mathbf{x}_\beta\right)\hat{\Gamma}(\omega, \mathbf{k}_\perp; q)\delta\hat{\rho}(\omega, \mathbf{k}_\perp; q), \tag{4.355}$$

where

$$\delta\hat{f}\left(\mathbf{x}_\beta, \mathbf{p}_\beta, \gamma; \omega, q\right) = \int_{-\infty}^{\infty} d\tau \int_0^{\infty} dz \exp(i\omega\tau - qz)\delta f\left(\mathbf{x}_\beta, \mathbf{p}_\beta, \tau, \gamma; z\right) \tag{4.356}$$

is the Fourier-Laplace transform of the perturbed distribution function. Note that $d\tau/dz$ is a function of γ and r. In view of the betatron trajectories (Eqs. (4.328) and (4.329))

$$\left(\mathbf{p}_\beta\cdot\nabla_{\mathbf{x}_\beta} - k_\beta^2\mathbf{x}_\beta\right)\delta\hat{f} = -k_\beta\left(\frac{\partial}{\partial\phi_x} + \frac{\partial}{\partial\phi_x}\right)\delta\hat{f}, \tag{4.357}$$

and we can expand the perturbed distribution as a superposition of azimuthal modes

$$\delta\hat{f}\left(\mathbf{x}_\beta, \mathbf{p}_\beta, \gamma; \omega, q\right) = \sum_{m,n=-\infty}^{\infty} \delta\hat{F}_{m,n}\left(r_x, r_y, \gamma; \omega, q\right)\exp\left(im\phi_x + in\phi_y\right). \tag{4.358}$$

As a result

$$\left[q - i\omega\frac{d\tau}{dz} - ik_\beta(m+n)\right]\delta\hat{F}_{m,n}\left(r_x, r_y, \gamma; \omega, q\right) = -G_\perp\left(r^2\right)\frac{\partial}{\partial\gamma}G_\parallel(\gamma)$$
$$\times\int_1^{\infty} d\gamma' \sum_{m',n'=-\infty}^{\infty}\int_0^{\infty} dr_x' r_x' \int_0^{\infty} dr_y' r_y' \delta\hat{F}_{m',n'}\left(r_x', r_y', \gamma; \omega, q\right)K_{m,n}^{m',n'}\left(r_x, r_y \middle| r_x', r_y'\right), \tag{4.359}$$

where

$$K_{m,n}^{m',n'}\left(r_x, r_y \middle| r'_x, r'_y\right) \equiv \left(2\pi k_\beta\right)^2 i^{|m|+|n|-|m'|-|n'|} \iint d^2\mathbf{k}_\perp \hat{\Gamma}(\omega, \mathbf{k}_\perp; q) \times J_{|m|}(k_x r_x) J_{|n|}\left(k_y r_y\right) J_{|m|}\left(k_x r'_x\right) J_{|n|}\left(k_y r'_y\right) \quad (4.360)$$

If we divide Eq. (4.359) by $[q - i\omega(d\tau/dz) - ik_\beta(m + n)]$ and integrate over γ, then we obtain

$$\hat{R}_{m,n}\left(r_x, r_y; \omega, q\right) = -G_\perp\left(r^2\right) \int_1^\infty d\gamma \frac{dG_\parallel(\gamma)/d\gamma}{q - i\omega(d\tau/dz) - ik_\beta(m+n)} \times \sum_{m',n'=-\infty}^{\infty} \int_0^\infty dr'_x r'_x \int_0^\infty dr'_y r'_y K_{m,n}^{m',n'}\left(r_x, r_y \middle| r'_x, r'_y\right) \hat{R}_{m,n}\left(r'_x, r'_y; \omega, q\right), \quad (4.361)$$

where

$$\hat{R}_{m,n}\left(r_x, r_y; \omega, q\right) \equiv \int_1^\infty d\gamma \delta\hat{F}_{m,n}\left(r_x, r_y, \gamma; \omega, q\right) \quad (4.362)$$

describes the radial dependence of the perturbed distribution.

The radial function in Eq. (4.362) can be expressed as a superposition of Gaussian-Laguerre modes. If we restrict the analysis to the lowest-order radial mode, then we can write

$$\hat{R}_{m,n}\left(r_x, r_y; \omega, q\right) = \left(2\pi\Delta R k_\beta\right)^2 G_\perp\left(r^2\right) \delta a_{m,n}(\omega, q) S_{m,n} r_x^{|m|} r_y^{|n|}, \quad (4.363)$$

where

$$S_{m,n} = \frac{1}{\Delta R^{|m|+|n|}} \frac{1}{2^{(|m|+|n|)/2}\sqrt{|m|!|n|!}} \quad (4.364)$$

and $\delta a_{m,n}$ are the modal coefficients. Since $S_{m,n}$ is a constant, we can write the Bessel functions in Eq. (4.360) as

$$J_{|m|}(k_x r_x) J_{|n|}\left(k_y r_y\right) = C_{m,n}\left(k_x, k_y\right) S_{m,n} r_x^{|m|} r_y^{|n|}, \quad (4.365)$$

where

$$C_{m,n}\left(k_x, k_y\right) = \frac{1}{\sqrt{|m|!|n|!}} \left(\frac{k_x \Delta R}{\sqrt{2}}\right)^{|m|} \left(\frac{k_y \Delta R}{\sqrt{2}}\right)^{|n|} \exp\left(-\frac{1}{2} k_\perp^2 \Delta R^2\right). \quad (4.366)$$

The dispersion equation is obtained after substitution of Eqs. (4.363) and (4.365) into Eq. (4.361). Multiplying the resultant equation by $C_{m,n}r_x^{|m|+1}r_y^{|n|+1}$ and integrating over r_x and r_y, yields

$$\delta a_{m,n} + B_{m,n} \sum_{m',n'=-\infty}^{\infty} M_{m,n,m',n'} \delta a_{m',n'} = 0, \tag{4.367}$$

where

$$B_{m,n} = \frac{S_{m,n}^2}{\Delta R^4} \times \int_1^\infty d\gamma \int_{-\infty}^{\infty} dr_x r_x^{2|m|+1} \int_{-\infty}^{\infty} dr_y r_y^{2|n|+1} \frac{\exp(-r^2/2\Delta R^2) dG_\parallel(\gamma)/d\gamma}{q - i\omega(d\tau/dz) - ik_\beta(m+n)} \tag{4.368}$$

and

$$M_{m,n,m',n'} = i^{|m|+|n|-|m'|-|n'|} \int_{-\infty}^{\infty} dk_x \int_{-\infty}^{\infty} dk_y C_{m,n}(k_x,k_y) C_{m',n'}(k_x,k_y) \hat{\Gamma}(\omega, \mathbf{k}_\perp; q). \tag{4.369}$$

The dispersion equation is found by setting the determinant of the matrix

$$D_{m,n} = \delta_{m,n} + B_{m,n} \sum_{m',n'=-\infty}^{\infty} M_{m,n,m',n'}, \tag{4.370}$$

equal to zero.

The lowest-order dispersion equation is found for the case of a single Gaussian TEM_{00} mode for which we assume that $m = n = m' = n' = 0$. In this case, the dispersion equation becomes

$$1 = \frac{i}{\sqrt{2\pi}} (2\rho k_w)^3 \frac{k}{2k_{res}} \int_{-\infty}^{\infty} dt \int_0^\infty dx x^3 \frac{\exp(-x^2/2)\exp(-t^2/2)}{\left[q + 2i\Delta\gamma k_w kt/\gamma_{res} k_{res} - ikk_\beta^2 \Delta R^2 x^2/2\right]^2} \times \int_0^{\pi/2} d\theta\theta \frac{(k\Delta R)^2 \exp(-k^2\Delta R^2\theta^2)}{q + ik_w(k-k_{res})/k_{res} + ik\theta^2/2}. \tag{4.371}$$

Equation (4.371) depends upon four independent parameters. To see this, we note first that the one-dimensional gain length is given by (Eq. (4.98)) $L_G^{(1D)} = 1/2\sqrt{3}\rho k_w$. The Rayleigh range, z_R, which will be discussed further in a later chapter, describes the distance over which the area of a Gaussian mode expands by a factor of two is given by $z_R = kw_0^2/2$, where w_0 is the radius of the mode waist. Since the spot size in a free-electron laser is expected to be comparable to the electron beam radius, we approximate $z_R \approx 2k\Delta R^2$. Finally, the emittance of a beam matched into the natural focusing of a helical undulator is $\varepsilon = \pi\gamma_{res}k_\beta\Delta R^2/2$. In view of this, we can rewrite this equation as

$$1 = \frac{i}{24\sqrt{6\pi}}\int_0^\infty dx\int_{-\infty}^\infty dt\frac{x\exp(-x/2)\exp(-t^2/2)}{\left[\overline{q} + i\frac{\Delta\gamma/\gamma_{res}}{\sqrt{3}\rho}t - i\frac{4\pi\varepsilon}{\gamma_{res}\hat{\beta}\lambda_{res}}L_G^{(1D)}x\right]^2} \times \int_0^{(\pi k_{res}\Delta R)^2/4} dy\frac{\exp(-y)}{\overline{q} + \frac{i\delta k}{2\sqrt{3}\rho k_{res}} + iy\frac{L_G^{(1D)}}{z_R}}, \tag{4.372}$$

where $\overline{q} = q/L_G^{(1D)}$, $\hat{\beta} = 2\pi/k_\beta$ is the Twiss β-parameter, and we have approximated $k \approx k_{\text{res}}$ except where $\delta k = k - k_{res}$ describes the detuning. We can define four parameters, therefore, which determine the dispersion relation that describe the effects of (1) the axial energy spread $[\eta_\gamma \equiv (\Delta\gamma/\gamma_{res})/\sqrt{3}\rho]$, (2) the emittance $[\eta_\varepsilon \equiv 4\pi\varepsilon_n L_G^{(1D)}/\gamma_{res}\hat{\beta}\lambda_{res}]$, (3) the diffraction of the optical field $[\eta_d \equiv L_G^{(1D)}/z_R]$, and (4) the detuning of the resonance condition $[\eta_\delta \equiv \delta k/2\sqrt{3}\rho k_{res}]$.

4.3.2.1.3 The Idealized, One-Dimensional Limit

In the idealized, one-dimensional limit, thermal effects and diffraction are neglected. In this case, the Rayleigh range is much longer than the gain length, which implies that the extent of the beam in the transverse direction is much greater than the resonant wavelength and the direction of propagation is along the axis of symmetry; hence, $k_{res}\Delta R >> 1$. As a result, the integral over θ in Eq. (4.371) can be approximated by

$$\int_0^{\pi/2} d\theta\theta\frac{(k\Delta R)^2\exp(-k^2\Delta R^2\theta^2)}{q + ik_w(k - k_{res})/k_{res} + ik\theta^2/2} \cong \frac{1}{q + i\Delta k}\int_0^\infty du u\exp(-u^2) = \frac{1}{2(q + i\Delta k)}, \tag{4.373}$$

where $\Delta k = k_w(k - k_{res})/k_{res}$. We also neglect thermal effects by neglecting the axial energy spread ($\Delta\gamma = 0$) and the emittance ($\varepsilon = 0$). Observe that neglecting the emittance is equivalent to neglecting the betatron motion, and

$$\int_{-\infty}^{\infty} dt \int_{0}^{\infty} dx x^3 \frac{\exp(-x^2/2)\exp(-t^2/2)}{\left[q + 2i\Delta\gamma k_w kt/\gamma_{res}k_{res} - ikk_\beta^2\Delta R^2x^2/2\right]^2} \cong \frac{\sqrt{2\pi}}{q^2}. \tag{4.374}$$

As a result, replacing q by ik, we find that the dispersion equation reduces to

$$k^2(k - \Delta k) \cong -(2\rho k_w)^3, \tag{4.375}$$

which is the cubic equation found in the high-gain Compton regime (Eq. (4.106)).

4.3.2.1.4 Ming Xie's Parameterization

The solution to the dispersion Eq. (4.372) can be a complicated numerical process. In order to simplify the determination of the gain length, Ming Xie [50] has developed a parametric scaling function that is widely used in the free-electron laser community, based upon a parametric fit to the numerical solution of the complete dispersion equation. Since the peak growth rate is found in the exponential regime at zero detuning, it is assumed that $\eta_\delta = 0$. As a result, this scaling function for the power gain Length, L_G, is written as

$$L_G = \frac{L_G^{(1D)}}{1 + \eta_{total}}, \tag{4.376}$$

where

$$\begin{aligned}\eta_{total} = {} & 0.45\eta_d^{0.57} + 0.55\eta_\varepsilon^{1.6} + 3\eta_\gamma^2 + 0.35\eta_\varepsilon^{2.9}\eta_\gamma^{2.4} \\ & + 51\eta_d^{0.95}\eta_\gamma^3 + 5.4\eta_d^{0.7}\eta_\varepsilon^{1.9} + 1140\eta_d^{2.2}\eta_\varepsilon^{2.9}\eta_\gamma^{3.2}.\end{aligned} \tag{4.377}$$

This parameterization has shown good agreement with both simulations and experiments over a wide range of parameters, as will be discussed further in subsequent chapters. It represents a near-universal scaling law that is applicable for a wide range of free-electron laser configurations. It has been shown to apply to planar wiggler geometries subject to the use of the appropriate values for ρ and the Twiss $\hat{\beta}$-function. It has also been successfully applied to segmented wiggler configurations where external strong focusing is employed using magnetic quadrupoles by substitution of the appropriate $\hat{\beta}$-function.

We now consider an example of an infrared free-electron laser operating at a wavelength of 1 μm with a helical wiggler with a period of 3.5 cm and an on-axis amplitude of 3.0 kG corresponding to a wiggler strength parameter $K = 0.98$. In order to achieve a resonance at this wavelength, the electron beam energy must be 81.7 MeV. This yields a Twiss $\hat{\beta} = 8.1$ m. If we further assume a current of 300 A and an emittance of 4.0 mm-mrad, then this yields an rms beam radius of 450 μm, a Pierce parameter of $\rho = 4.27 \times 10^{-3}$, and a one-dimensional gain length of $L_G^{(1D)} = 0.37$ m. If we now assume reasonable values for the emittance of 4.0 mm-mrad and an energy spread of 0.1%, then the three-dimensional gain length we obtain is $L_G = 0.46$ m.

The interaction sensitivity to energy spread is illustrated in Fig. 4.17 where we plot the gain length (left axis) and the growth rate (right axis) versus the rms energy spread (i.e., $\Delta\gamma/\gamma_{res}$). It is evident in the figure that initially, for relatively small energy spread, the growth rate remains relatively constant, after which the rate of decrease of the growth rate with increases in the energy spread increases. This was also found in the one-dimensional theory (Fig. 4.11). It was demonstrated on the basis of that one-dimensional theory that thermal effects become important when (Eq. (4.133)) $\Delta\gamma/\gamma_{res} \approx \left[\sqrt{3}\rho\left(1 + K^2\right)/\gamma_{res}^2\right]^{1/2} \approx 0.075\%$, which is in reasonable agreement with the three-dimensional theory as shown in the figure.

The sensitivity of the interaction in a free-electron laser to increasing emittance is illustrated in Fig. 4.18, where the gain length is shown versus the normalized emittance for constant energy spread and current. The emittance provides for an effective increase in the axial energy spread, and increases in the emittance are expected to result in a decrease in the interaction strength and increases in the gain length. In addition, variations in the emittance give rise to changes in the rms beam size for a matched beam; in particular, the larger the emittance, the larger the rms beam size. This means that the Pierce parameter and the current density decrease with increases in the emittance, and this is expected to result in increases in the gain length. In contrast, the Rayleigh range will increase with increases in the rms beam size, and this can be expected to have the effect of making diffraction less important,

Fig. 4.17 Plots if the gain length and normalized growth rate versus rms energy spread

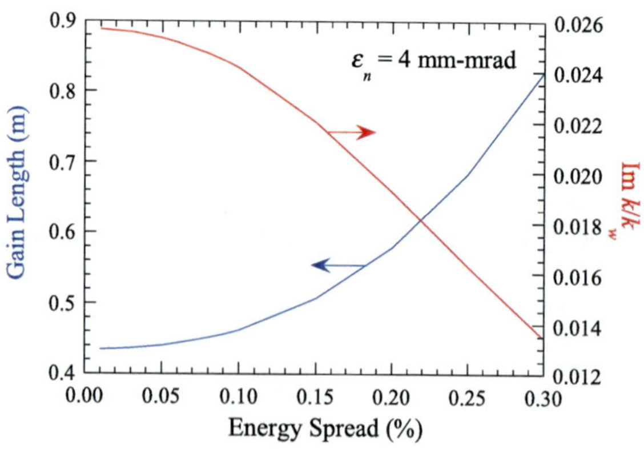

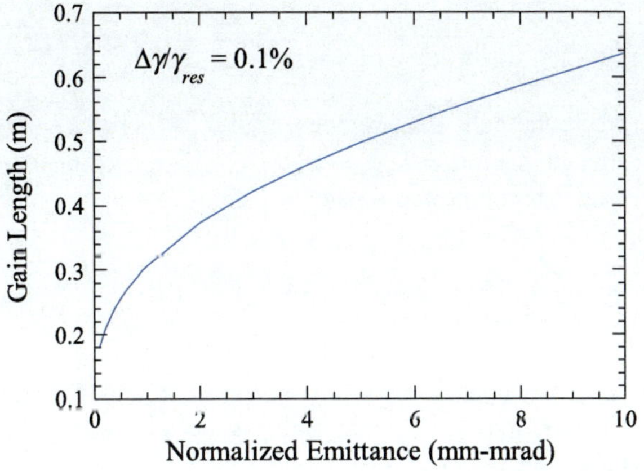

Fig. 4.18 Plot of the gain length versus the normalized emittance

thereby contributing to a decrease in the gain length. These competing effects are included in the parameterization. As shown in the figure, there is no plateau in the gain length for small emittance such as found in Fig. 4.17 for the variation with respect to the energy spread. As a result, the gain length is found to increase sharply with increases in the emittance.

4.3.2.2 Planar Wiggler Configurations

Following Chin, Kim, and Xie [45], the analysis for the case of a planar wiggler follows the formulation presented above for the helical wiggler. As in the above analysis, we consider a quasi-one-dimensional orbit analysis, followed by a Vlasov-Maxwell treatment of the dispersion equation.

4.3.2.2.1 The Electron Trajectories

The trajectories are treated for a weak, equal two-plane focusing wiggler (Eq. (2.3)), which is referred to as the parabolic-pole-face wiggler in Chap. 2. Following the analysis for the helical wiggler, the trajectories can be expressed as the sum of the wiggler and betatron motions, $\mathbf{x} = \mathbf{x}_w + \mathbf{x}_\beta$.

The one-dimensional limit of the parabolic-pole-face wiggler model is given by $\mathbf{B}_w = B_w \hat{\mathbf{e}}_y \cos k_w z$ so that the transverse wiggler motion is

$$\mathbf{x}_w = r_w \hat{\mathbf{e}}_x \cos k_w z, \tag{4.378}$$

where $r_w = |\upsilon_w|/k_w \upsilon_\parallel$ describes the amplitude of the wiggler motion. The axial velocity corresponding to this transverse motion is

$$\frac{v_z}{c} \cong \frac{v_\parallel}{c} - \frac{v_w^2}{4c^2} \cos 2k_w z, \tag{4.379}$$

where the bulk axial velocity $v_\parallel/c = \left[1 - \left(1 + K_{rms}^2\right)/\gamma_{res}^2\right]^{1/2}$ and the rms wiggler strength parameter $K_{rms} = |v_w|/\sqrt{2}c$. The axial velocity can be integrated to yield the Lagrangian time coordinate

$$\frac{}{d\tau} \cong \frac{1}{c}\left[\frac{K_{rms}^2}{2\gamma_{res}^2} \cos 2k_w z - 2\frac{k_{res}}{k_w}\frac{\gamma - \gamma_{res}}{\gamma_{res}}\left(1 + \frac{K_{rms}^2}{1 + K_{rms}^2} \cos 2k_w z\right) + \frac{1}{2}\left(\mathbf{p}_\beta^2 + k_\beta^2 \mathbf{x}_\beta^2\right)\right]. \tag{4.380}$$

where $k_{res} = 2k_w\left(1 + K_{rms}^2\right)/\gamma_{res}^2$ is the resonant wavenumber corresponding to the bulk axial velocity.

The betatron motion in the parabolic-pole-face wiggler has been discussed in Chap. 2 (see Eq. (2.145)) and can be written in the form given by Eqs. (4.320) and (4.321) for a helical wiggler

$$\mathbf{x}_\beta = \mathbf{x}_{\beta 0} \cos k_\beta z + \frac{\mathbf{p}_{\beta 0}}{k_\beta} \sin k_\beta z, \tag{4.381}$$

and

$$\mathbf{p}_\beta = -k_\beta \mathbf{x}_{\beta 0} \sin k_\beta z + \mathbf{p}_{\beta 0} \cos k_\beta z, \tag{4.382}$$

Except that for the parabolic-pole-face wiggler $k_\beta = \Omega_w/2c$ for $\Omega_w = eB_w/\gamma m_e c$. In this case, we treat $x_\beta^2 + p_{\beta x}^2/k_\beta^2$ and $y_\beta^2 + p_{\beta y}^2/k_\beta^2$ separately as constants of the motion, and we can express them in polar coordinates as in the case of a helical wiggler as

$$x_\beta = r_x \cos \phi_x, y_\beta = r_y \sin \phi_y \tag{4.383}$$

and

$$p_{\beta x} = k_\beta r_x \sin \phi_x, p_{\beta y} = k_\beta r_y \sin \phi_y \tag{4.384}$$

Hence, $x_\beta^2 + p_{\beta x}^2/k_\beta^2 = r_x^2$ and $y_\beta^2 + p_{\beta y}^2/k_\beta^2 = r_y^2$.

4.3.2.2.2 The Dispersion Equation

As in the case of the helical wiggler, we solve the Vlasov-Maxwell equations by a perturbation analysis to first order in the optical field; hence, we write $f = f_0 + \delta f$.

The equilibrium distribution is a function of the constants of the motion which may be expressed as

$$f_0(\mathbf{x}_\beta, \mathbf{p}_\beta, \gamma) = \frac{N_e}{T} G_\perp(r_x, r_y) G_\parallel(\gamma). \tag{4.385}$$

This describes a flattop temporal beam profile. The distribution in energy, $G_\parallel$, is given by Eq. (4.333). The transverse distribution is

$$G_\perp(r_x, r_y) = \frac{1}{(2\pi k_\beta)^2 \Delta R_x^2 \Delta R_y^2} \exp\left(-r_x^2/2\Delta R_x^2 - r_y^2/2\Delta R_y^2\right), \tag{4.386}$$

which satisfies the normalization condition

$$\iint d^2\mathbf{x}_\beta \iint d^2\mathbf{p}_\beta G_\perp\left(\mathbf{x}_\beta^2 + \mathbf{p}_\beta^2/k_\beta^2\right) = (2\pi k_\beta)^2 \int_0^\infty dr_x r_x \int_0^\infty dr_y r_y G_\perp(r_x, r_y) = 1. \tag{4.387}$$

The perturbed distribution is a solution to the linearized Vlasov equation, which may be written following Eq. (4.355) for the helical wiggler as

$$\left[\left(q - i\omega\frac{dt}{dz}\right) + k_\beta\left(\frac{\partial}{\partial\phi_x} + \frac{\partial}{\partial\phi_x}\right)\right]\delta\hat{f}(\mathbf{x}_\beta, \mathbf{p}_\beta, \gamma; \omega, q)$$
$$= -G_\perp(r^2)\frac{\partial}{\partial\gamma}G_\parallel(\gamma)\iint d^2\mathbf{k}_\perp \exp(i\mathbf{k}_\perp \cdot \mathbf{x}_\beta)\hat{\Gamma}(\omega, \mathbf{k}_\perp; q)\delta\hat{\rho}(\omega, \mathbf{k}_\perp; q), \tag{4.388}$$

where $\delta\hat{f}(\mathbf{x}_\beta, \mathbf{p}_\beta, \gamma; \omega, q)$ is the Fourier-Laplace transform of the perturbed distribution (Eq. (4.356)) and $\delta\hat{\rho}(\omega, \mathbf{k}_\perp; q)$ is the Fourier-Laplace transform of the perturbed charge density (see Eqs. (4.346) and (4.347)). The work done by the fields on the particles describes the electron energy change as given in Eq. (4.352) and which depends on $\hat{\Gamma}(\omega, \mathbf{k}_\perp; q)$. In the case of the parabolic-pole-face wiggler and the limit in which $k_w r_w << 1$, which is appropriate for the quasi-one-dimensional approximation, this function was shown to be [45]

$$\hat{\Gamma}(\omega, \mathbf{k}_\perp; q) \cong -\frac{r_e}{4\pi c}\frac{K_{rms}^2}{\gamma_{res}^2}\sum_{n=-\infty}^{\infty}\frac{k}{k_z}\frac{JJ_n^2(b)}{q + i\left[\omega/v_\parallel - (k_z + (2n+1)k_w)\right]}, \tag{4.389}$$

where $b \cong K_{rms}^2/\left[2\left(1 + K_{rms}^2\right)\right]$ and $r_e = e^2/m_e c^2$ is the classical electron radius. If we restrict the analysis to the fundamental harmonic ($n = 0$), then this expression simplifies further to

$$\hat{\Gamma}(\omega, \mathbf{k}_\perp; q) \cong -\frac{r_e}{4\pi c}\frac{K_{rms}^2}{\gamma_{res}^2}\frac{k}{k_z}\frac{JJ_0^2(b)}{q + i\left[\omega/v_\parallel - (k_z + k_w)\right]}. \tag{4.390}$$

Following the development for the helical wiggler, the dispersion equation becomes

$$1 = i\frac{(2\rho k_w)^3}{2\sqrt{2\pi}}\int_{-\infty}^{\infty} dt \int_0^{\infty} dxx \int_0^{\infty} dyy \frac{\exp[-(x^2+y^2)/2]\exp(-t^2/2)}{\left[q + 2i\Delta\gamma k_w kt/\gamma_{res}k_{res} - ikk_\beta^2\left(\Delta R_x^2 x^2 + \Delta R_y^2 y^2\right)/2\right]^2}$$
$$\times \frac{\Delta R_x \Delta R_y}{\pi}\int_{-\infty}^{\infty} dk_x \int_{-\infty}^{\infty} dk_y \frac{\exp\left[-k_x^2\Delta R_x^2 - k_y^2 \Delta R_y^2\right]}{q + ik_w(k - k_{res})/k_{res} + i\left(k_x^2 + k_y^2\right)/2k}, \tag{4.391}$$

where we have approximated $k \approx k_{res}$ except in $\delta k = k - k_{res}$, and we have also assumed that $k_x r_x << 1$ and $k_y r_y << 1$

$$1 = \frac{i}{3\sqrt{6\pi}}\int_{-\infty}^{\infty} dt \int_0^{\infty} d\overline{x} \int_0^{\infty} d\overline{y} \frac{\exp[-(\overline{x}+\overline{y})/2]\exp(-t^2/2)}{\left[\overline{q} + i\frac{\Delta\gamma/\gamma_{res}}{\sqrt{3}\rho}t - i\frac{4\pi\left(\varepsilon_x\overline{x}+\varepsilon_y\overline{y}\right)}{\gamma_{res}\hat{\beta}\lambda_{res}}L_G^{(1D)}\right]^2}$$
$$\times \frac{\Delta R_x \Delta R_y}{\pi}\int_{-\infty}^{\infty} dk_x \int_{-\infty}^{\infty} dk_y \frac{\exp\left[-k_x^2\Delta R_x^2 - k_y^2\Delta R_y^2\right]}{\overline{q} + \frac{i\delta k}{2\sqrt{3}\rho k_w} + i\left(k_x^2 + k_y^2\right)\Delta R_x \Delta R_y \frac{L_G^{(1D)}}{z_R}}, \tag{4.392}$$

where $\overline{q} = q/L_G^{(1D)}$, $\hat{\beta} = 2\pi/k_\beta$ is the Twiss β-parameter, and we have approximated $k \approx k_{\text{res}}$ except where $\delta k = k - k_{res}$ describes the detuning. It should also be remarked that the area rms cross-sectional area of the beam is $\sigma_\perp = 4\Delta R_x \Delta R_y$ so that the Rayleigh range is $z_R = k_{res}\sigma_\perp/2$.

4.3.2.2.3 The Idealized, One-Dimensional Approximation

As in the case of the helical wiggler, thermal effects and diffraction are neglected in the idealized, one-dimensional limit. In this case, the Rayleigh range is much longer than the gain length, which implies that the extent of the beam in the transverse direction is much greater than the resonant wavelength and the direction of propagation is along the axis of symmetry; hence, $k_{res}\Delta R_{x,y} >> 1$. As a result

$$\frac{\Delta R_x \Delta R_y}{\pi} \int_{-\infty}^{\infty} dk_x \int_{-\infty}^{\infty} dk_y \frac{\exp\left[-k_x^2 \Delta R_x^2 - k_y^2 \Delta R_y^2\right]}{q + ik_w(k - k_{res})/k_{res} + i\left(k_x^2 + k_y^2\right)/2k} \cong \frac{1}{q + ik_w(k - k_{res})/k_{res}}. \quad (4.393)$$

We also neglect thermal effects by setting the axial energy spread ($\Delta\gamma = 0$) and the emittance ($\varepsilon_{x,y} = 0$) to zero. Observe that neglecting the emittance is equivalent to neglecting the betatron motion, so that as in the case of a helical wiggler

$$\int_{-\infty}^{\infty} dt \int_{0}^{\infty} dxx \int_{0}^{\infty} dyy \frac{\exp[-(x^2+y^2)/2]\exp(-t^2/2)}{\left[q + 2i\Delta\gamma k_w kt/\gamma_{res}k_{res} - ikk_\beta^2\left(\Delta R_x^2 x^2 + \Delta R_y^2 y^2\right)/2\right]^2} \cong \frac{2\sqrt{2\pi}}{q^2}. \quad (4.394)$$

As a result, replacing q by ik, we find that the dispersion equation reduces to

$$k^2(k - \Delta k) \cong -(2\rho k_w)^3, \quad (4.395)$$

where $\Delta k = k_w(k - k_{res})/k_{res}$. Just as in the case of the helical wiggler, (Eq. (4.375)), this is the cubic equation found in the high-gain Compton regime (Eq. (4.106)), except that the Pierce parameter must correspond to that for the planar wiggler.

4.3.2.2.4 Ming Xie's Parameterization

A full solution to the dispersion Eq. (4.392) requires a numerical solution to the dispersion equation. However, a relatively simple fit to the numerical solution has been obtained by Ming Xie [50] that represents an extension of the one-dimensional theory using a polynomial expansion based upon the scaling parameters shown in Eq. (4.392) for the energy spread, emittance, and diffraction. For the case of a symmetric electron beam (i.e., $\Delta R_x = \Delta R_y$), this parameterization is

$$L_G = \frac{L_G^{(1D)}}{1 + \eta_{total}}, \quad (4.396)$$

where

$$\begin{aligned} \eta_{total} = {} & 0.45\eta_d^{0.57} + 0.55\eta_\varepsilon^{1.6} + 3\eta_\gamma^2 + 0.35\eta_\varepsilon^{2.9}\eta_\gamma^{2.4} \\ & + 51\eta_d^{0.95}\eta_\gamma^3 + 5.4\eta_d^{0.7}\eta_\varepsilon^{1.9} + 1140\eta_d^{2.2}\eta_\varepsilon^{2.9}\eta_\gamma^{3.2}. \end{aligned} \quad (4.397)$$

This parameterization has shown good agreement with both simulations and experiments over a wide range of parameters, as will be discussed further in subsequent chapters. It represents a near-universal scaling law that is applicable for a wide range of free-electron laser configurations. It has been shown to apply to planar wiggler geometries subject to the use of the appropriate values for ρ and the Twiss β-function. It has also been successfully applied to segmented wiggler configurations where external strong focusing is employed using magnetic quadrupoles by substitution of the appropriate β-function.

References

1. V.P. Sukhatme, P.A. Wolff, Stimulated Compton scattering as a radiation source - theoretical limitations. J. Appl. Phys. **44**, 2331 (1973)
2. T.J.T. Kwan, J.M. Dawson, A.T. Lin, Free-electron laser. Phys. Fluids **20**, 581 (1977)
3. N.M. Kroll, W.A. McMullin, Stimulated emission from relativistic electrons passing through a spatially periodic transverse magnetic field. Phys. Rev. A **17**, 300 (1978)
4. J.M.J. Madey, Relationship between mean radiated energy, mean squared radiated energy and spontaneous power spectrum in a power series expansion of the equations of motion in a free-electron laser. Nuovo Cimento **50B**, 64 (1979)
5. T.J.T. Kwan, J.M. Dawson, Investigation of the free-electron laser with a guide magnetic field. Phys. Fluids **22**, 1089 (1979)
6. I.B. Bernstein, J.L. Hirshfield, Amplification on a relativistic electron beam in a spatially periodic transverse magnetic field. Phys. Rev. A **20**, 1661 (1979)
7. P. Sprangle, R.A. Smith, V.L. Granatstein, Free-electron lasers and stimulated scattering from relativistic electron beams, in *Infrared and Millimeter Waves*, ed. by K.J. Button, vol. 1, (Academic Press, New York, 1979), p. 279
8. L. Friedland, J.L. Hirshfield, Free-electron laser with a strong axial magnetic field. Phys. Rev. Lett. **44**, 1456 (1980)
9. H.P. Freund, P. Sprangle, D. Dillenburg, E.H. da Jornada, B. Liberman, R.S. Schneider, Coherent and Incoherent radiation from free-electron lasers with an axial guide field. Phys. Rev. A **24**, 1965 (1981)
10. B. Bernstein, L. Friedland, Theory of free-electron laser in combined helical pump and axial guide magnetic fields. Phys. Rev. A **23**, 816 (1981)
11. A. Gover, P. Sprangle, A generalized formulation of free-electron lasers in the low-gain regime including transverse velocity spread and wiggler incoherence. J. Appl. Phys. **52**, 599 (1981)
12. S.T. Stenholm, A. Bambini, Single-particle theory of the free-electron laser in a moving frame. IEEE J. Quantum Electron. **QE-17**, 1363 (1981)
13. C.C. Shih, A. Yariv, Inclusion of space-charge effects with Maxwell's equations in the single-particle analysis of free-electron lasers. IEEE J. Quantum Electron. **QE-17**, 1387 (1981)
14. R. Coisson, Energy loss calculation of gain in a plane sinusoidal free-electron laser. IEEE J. Quantum Electron. **QE-17**, 1409 (1981)
15. W.B. Colson, The nonlinear wave equation for higher harmonics in free-electron lasers. IEEE J. Quantum Electron. **QE-17**, 1417 (1981)
16. N.M. Kroll, P.L. Morton, M.N. Rosenbluth, Free-electron lasers with variable parameter wigglers. IEEE J. Quantum Electron. **QE-17**, 1436 (1981)
17. H.S. Uhm, R.C. Davidson, Free-electron laser instability for a relativistic annular electron beam in a helical wiggler field. Phys. Fluids **24**, 2348 (1981)

18. H.P. Freund, P. Sprangle, D. Dillenburg, E.H. da Jornada, R.S. Schneider, B. Liberman, Collective effects on the operation of free-electron lasers with an axial guide field. Phys. Rev. A **26**, 2004 (1982)
19. J.A. Davies, R.C. Davidson, G.L. Johnston, Compton and Raman free-electron laser stability properties for a cold electron beam propagating through a helical magnetic field. Aust. J. Plant Physiol. **33**, 387 (1985)
20. L.K. Grover, R.H. Pantell, Simplified analysis of free-electron lasers using Madey's theorem. IEEE J. Quantum Electron. **QE-21**, 944 (1985)
21. L.F. Ibanez, S. Johnston, Finite-temperature effects in free-electron lasers. IEEE J. Quantum Electron. **QE-19**, 339 (1983)
22. E. Jerby, A. Gover, Investigation of the gain regimes and gain parameters of the free-electron laser dispersion equation. IEEE J. Quantum Electron. **QE-21**, 1041 (1985)
23. H.P. Freund, R.C. Davidson, D.A. Kirkpatrick, Thermal effects on the linear gain in free-electron lasers. IEEE J. Quantum Electron. **27**, 2550 (1991)
24. L. Friedland, A. Fruchtman, Amplification on relativistic electron beams in combined helical and axial magnetic fields. Phys. Rev. A **25**, 2693 (1982)
25. H.S. Uhm, R.C. Davidson, Helically distorted relativistic beam equilibria for free-electron laser applications. J. Appl. Phys. **53**, 2910 (1982)
26. W.A. McMullin, R.C. Davidson, Low-gain free-electron laser near cyclotron resonance. Phys. Rev. A **25**, 3130 (1982)
27. H.S. Uhm, R.C. Davidson, Free electron laser instability for a relativistic solid electron beam in a helical wiggler field. Phys. Fluids **26**, 288 (1983)
28. H.P. Freund, P. Sprangle, Unstable electrostatic beam modes in free-electron laser systems. Phys. Rev. A **28**, 1835 (1983)
29. C. Grebogi, H.S. Uhm, Vlasov susceptibility of relativistic magnetized plasma and application to free-electron lasers. Phys. Fluids **29**, 1748 (1986)
30. N.S. Ginzburg, Diamagnetic and paramagnetic effects in free-electron lasers. IEEE Trans. Plasma. Sci **PS-15**, 411 (1987)
31. H.P. Freund, R.C. Davidson, G.L. Johnston, Linear theory of the collective Raman interaction in a free-electron laser with a planar wiggler and an axial guide field. Phys. Fluids B **2**, 427 (1990)
32. J.R. Cary, T.J.T. Kwan, Theory of off-axis mode production by free-electron lasers. Phys. Fluids **24**, 729 (1981)
33. T.J.T. Kwan, J.R. Cary, Absolute and convective instabilities in two-dimensional free-electron lasers. Phys. Fluids **24**, 899 (1981)
34. H.P. Freund, S. Johnston, P. Sprangle, Three-dimensional theory of free-electron lasers with an axial guide field. IEEE J. Quantum Electron. **QE-19**, 322 (1983)
35. H.P. Freund, A.K. Ganguly, Three-dimensional theory of the free-electron laser in the collective regime. Phys. Rev. A **28**, 3438 (1983)
36. P. Luchini, S. Solimeno, Gain and mode-coupling in a three-dimensional free-electron laser: A generalization of Madey's theorem. IEEE J. Quantum Electron. **QE-21**, 952 (1985)
37. M.N. Rosenbluth, Two-dimensional effects in free-electron lasers. IEEE J. Quantum Electron. **QE-21**, 966 (1985)
38. B.Z. Steinberg, A. Gover, S. Ruschin, Three-dimensional theory of free-electron lasers in the collective regime. Phys. Rev. A **36**, 147 (1987)
39. C.J. Elliot, M.J. Schmitt, Small-signal gain for a planar free-electron laser with a period magnetic field. IEEE Trans. Plasma. Sci **PS-15**, 319 (1987)
40. A. Fruchtman, High-density thick beam free-electron laser. Phys. Rev. A **37**, 4259 (1988)
41. T.M. Antonsen, P.E. Latham, Linear theory of a sheet beam free-electron laser. Phys. Fluids **31**, 3379 (1988)
42. V.K. Tripathi, C.S. Liu, A slow wave free-electron laser. IEEE Trans. Plasma. Sci **PS-17**, 583 (1989)

43. A. Fruchtman, H. Weitzner, Raman free-electron laser with transverse density gradients. Phys. Rev. A **39**, 658 (1989)
44. L.H. Yu, S. Krinsky, R.L. Gluckstern, Calculation of universal scaling function for free-electron laser gain. Phys. Rev. Lett. **64**, 3011 (1990)
45. Y.H. Chin, K.-J. Kim, M. Xie, Three-dimensional theory of the small-signal, high-gain free-electron laser including betatron oscillations. Phys. Rev. A **46**, 6662 (1992)
46. Y.H. Chin, K.-J. Kim, M. Xie, Three-dimensional free-electron laser dispersion relation including betatron oscillations. Nucl. Instr. Meth. **A318**, 481 (1992)
47. J.R. Pierce, *Traveling Wave Tubes* (Van Nostrand, New York, 1950)
48. J. Masud, T.C. Marshall, S.P. Schlesinger, F.G. Yee, W.M. Fawley, E.T. Scharlemann, S.S. Yu, A.M. Sessler, E.J. Sternbach, Sideband control in a millimeter-wave free-electron laser. Phys. Rev. Lett. **58**, 763 (1987)
49. J. Fajans, G. Bekefi, Measurements of amplification and phase shift in a free-electron laser. Phys. Fluids **29**, 3461 (1986)
50. M. Xie, Design optimization for an x-ray free electron laser driven by the SLAC linac. Proc. IEEE 1995 particle accelerator conference, Vol. 183, IEEE Cat. No. 95CH35843, 1995

Chapter 5
Nonlinear Theory: Guided Mode Analysis

The self-consistent nonlinear theory of the free-electron laser describes the interaction through the linear regime and includes the saturation of the growth mechanism. Saturation can occur through a variety of mechanisms. For an ideal beam that is both monoenergetic and with a vanishing pitch angle spread, saturation occurs by means of electron trapping in the ponderomotive potential. In the thermal regime, saturation occurs through a different process. In this case, the axial energy spread of the beam [which can arise due to either a distribution in the total energy of the beam electrons or pitch angle spread] gives rise to a broadband emission spectrum. As a result, a *quasilinear* saturation mechanism is operative in which the beam undergoes turbulent diffusion in momentum space. Since the growth rate in this regime is proportional to the slope of the distribution function, the turbulent diffusion acts to form a plateau in momentum space that *flattens out* the distribution of the beam. As a result, the axial energy spread of the beam increases, and the instability is quenched when the slope of the distribution falls to zero. As might be expected, however, the saturation efficiency in the thermal regime is greatly reduced relative to that found for a sufficiently cold beam in which saturation occurs through the particle-trapping mechanism. As a consequence, we shall focus attention on the latter case in this chapter.

This chapter will describe the development of *slowly varying envelope approximation* (SVEA) formulations in the steady-state regime, as well as the application of the analyses to the description of the fundamental physics of the nonlinear saturation mechanism. The SVEA was originally developed for the treatment of continuous wave (CW) traveling wave tubes and has been adapted to the treatment of long-pulse, long-wavelength free-electron lasers or free-electron masers at millimeter wavelengths or longer that are driven by relatively low-energy accelerators such as pulse-line accelerators, modulators, and induction linear accelerators. These accelerators produce relatively low-energy but long-pulse electron beams that are used primarily to generate relatively long wavelengths which are comparable to the transverse dimensions of the drift tube. As a consequence, the description of these free-electron lasers requires a guided mode analysis.

H. P. Freund, T. M. Antonsen, Jr., *Principles of Free Electron Lasers*,
https://doi.org/10.1007/978-3-031-40945-5_5

We are primarily concerned in this chapter with the interaction during a single pass of the electron beam and the electromagnetic wave through the wiggler, where the electromagnetic wave starts from some initial seed. Hence, it is more concerned with free-electron laser amplifiers, and we will discuss the application, and validation, of the analyses to specific experiments.

The origins of the *steady-state* formalism date to the early development of traveling wave tubes for radar applications, where it was often referred to as a "moving window" model because it treats the co-propagation of a "beamlet" of electrons that is one wavelength long with the corresponding electromagnetic field. These tubes employed thermionic cathodes that operated in continuous mode with the corresponding injection of a low-power continuous seed. As a result, a single-frequency approximation gave a good approximation to the wave-particle dynamics, and the electromagnetic field was represented using amplitude that varied slowly with position, and the fast timescale was removed by averaging Maxwell's equations, hence, the SVEA. The structures used in traveling wave tubes produce a subluminous wave that can interact directly with the electron beam and induce axial bunching. It was realized early [1] in the development of free-electron lasers that the ponderomotive wave in a free-electron laser resulted in a similar axial bunching mechanism and a free-electron laser might be thought of as a variant on a traveling wave tube. The steady-state formulations used to model free-electron lasers bear many similarities to these "moving-window" models and were originally developed to treat long-wavelength free-electron lasers that employed long-pulse accelerators. It is important to remark that this formulation is based on the propagation of a single wave frequency; hence, it is applicable for narrow bandwidth interactions.

5.1 The Phase-Trapping Efficiency

The interaction in a free-electron laser is mediated by the beating of the wiggler and radiation fields that gives rise to a ponderomotive wave. For electron beams with a narrow axial energy spread, the interaction saturates due to the particle trapping in the ponderomotive wave. This interaction is described by a nonlinear pendulum equation for the ponderomotive phase, ψ $(= k + k_w - \omega/v_{\|})$, and the interaction through saturation is illustrated in Fig. 4.1. As shown in the figure, the separatrix grows during the linear phase of the interaction up to the point at which the beam electrons cross the separatrix. This point occurs approximately when the separatrix encloses the initial phase-space position [i.e., in $d\psi/dz$] of the beam. The trapped electrons subsequently undergo a rotation in the ponderomotive well during which they give up energy to the wave during alternate half cycles. As a result, the growth of the wave continues until the beam electrons reach the bottom of the trough. At this point, the leading electrons begin to absorb energy from the wave, and saturation occurs when this energy absorption is balanced by the energy still being lost by the trailing electrons.

An ad hoc estimate of the saturation efficiency obtained by this process is found under the assumption that the individual electrons in the beam lose an amount of energy comparable to twice the difference in the bulk axial velocity of the beam, v_b, and the phase velocity of the trapping potential, v_{ph}, which is formed by the aggregate of the ponderomotive and space-charge potentials. The loss in beam kinetic energy due to this deceleration $\Delta v_b \approx -2(v_b - v_{ph})$ is given by

$$\Delta T \approx 2\gamma\gamma_{\|}^2 m_e v_b \left(v_b - v_{ph}\right), \tag{5.1}$$

where $\gamma_{\|} = (1 - {v_b}^2/c^2)^{-1/2}$. The interaction efficiency, η, is defined by the ratio of the energy lost by the beam to the initial beam kinetic energy; specifically

$$\eta \approx \frac{2\gamma\gamma_{\|}^2}{\gamma - 1}\frac{v_b}{c}\frac{v_b - v_{ph}}{c}. \tag{5.2}$$

The phase velocity is given by $v_{ph} \approx \omega/(\mathrm{Re}\ k + k_w)$, where the real part of the wavenumber differs in the collective Raman and high-gain Compton regimes. Typically

$$\mathrm{Re}\, k \approx \frac{\omega}{v_b} + \frac{\omega_b}{\gamma^{1/2}\gamma_{\|} v_b} + \mathrm{Re}\, \delta k, \tag{5.3}$$

where δk is given by solutions to Eqs. (4.93) and (4.96) for the high-gain Compton and collective Raman regimes, respectively, in the idealized one-dimensional regime. In the *high-gain Compton regime*, the ponderomotive potential is dominant, and

$$\mathrm{Re}\, k \approx \frac{\omega}{v_b} + \rho k_w, \tag{5.4}$$

at the frequency corresponding to maximum growth. Equation (1.20) for the efficiency in the high-gain Compton regime is recovered using this expression to calculate the phase velocity of the trapping ponderomotive potential. In contrast, the space-charge potential dominates in the *collective Raman regime*, and from Eq. (4.97)

$$\mathrm{Re}\, k \approx \frac{\omega}{v_b} + \frac{\omega_b}{\gamma^{1/2}\gamma_{\|} v_b}, \tag{5.5}$$

at the frequency of maximum growth. In this case, the efficiency is determined by Eq. (1.15). While these estimates of the saturation efficiency were made on the basis of the idealized one-dimensional model of the linear stability analysis, the ad hoc phase trapping argument may be generalized to the three-dimensional as well by the simple expedient of using the results of the theory for the real part of the wavenumber.

The estimates of the saturation efficiency obtained from the phase-trapping arguments are applicable to configurations employing uniform wigglers [i.e., wigglers with constant amplitudes and periods]. However, as first demonstrated experimentally by Phillips [1] and subsequently by others [2–5], it is possible to enhance the saturation efficiency by using a nonuniform wiggler in which either the amplitude or period is tapered in the direction of the symmetry axis [6–13]. In addition, for configurations which include an axial solenoidal magnetic field, the efficiency may be enhanced by tapering this field as well [14]. The effect of tapering either the wiggler or solenoidal magnetic fields is to alter both the transverse and axial velocities of the beam electrons. The fundamental concept underlying this mechanism is based upon the fact that as the beam gives up energy to the wave, it decelerates in the axial direction until it ultimately drops out of resonance. The fields must be tapered, therefore, in such a way as to accelerate the beam and maintain the phase relationship necessary for the resonant interaction. It should be remarked that in the negative-mass regime appropriate to strong axial solenoidal fields and Group II orbits, the effect of the loss of energy by the beam to the wave is to accelerate the beam in the direction of the symmetry axis. In this regime, therefore, the fields must be tapered in the opposite sense so as to decelerate the beam in order to achieve an enhancement in the saturation efficiency. In either case, however, the process is effective only if the beam has crossed the separatrix and has become trapped in the ponderomotive potential. For this reason, the field taper must begin at the point at which the beam crosses the separatrix, and the total efficiency is the sum of the efficiency in the uniform-wiggler region and that determined by the tapered fields.

In order to treat the effect of the field taper analytically, we perturb the orbit equations about the steady-state trajectories to first order in both the radiation field and the gradient in the wiggler and solenoidal magnetic fields. For the sake of simplicity, we consider the case of a helical wiggler field in the idealized one-dimensional limit and the amplification of a circularly polarized electromagnetic wave as given by Eq. (4.1). The combined effect of the electromagnetic wave and the gradients in the magnetostatic fields is obtained by the combination of Eqs. (2.154) and (4.11), and the nonlinear pendulum equation may be expressed as

$$\frac{d^2}{dz^2}\psi = -\kappa^2(\sin\psi - \sin\psi_{res}), \tag{5.6}$$

where κ^2 is given by Eq. (4.15), and the resonant phase ψ_{res} is

$$\sin\psi_{res} = -\frac{k+k_w}{\kappa^2}\,\frac{\beta_w^2}{(1+\beta_w^2)\Omega_0 - k_w v_\parallel} \times \left(\frac{\Omega_0 - k_w v_\parallel}{B_w}\frac{d}{dz}B_w - \frac{k_w v_\parallel}{\lambda_w}\frac{d}{dz}\lambda_w - \frac{\Omega_0}{B_0}\frac{d}{dz}B_0\right), \tag{5.7}$$

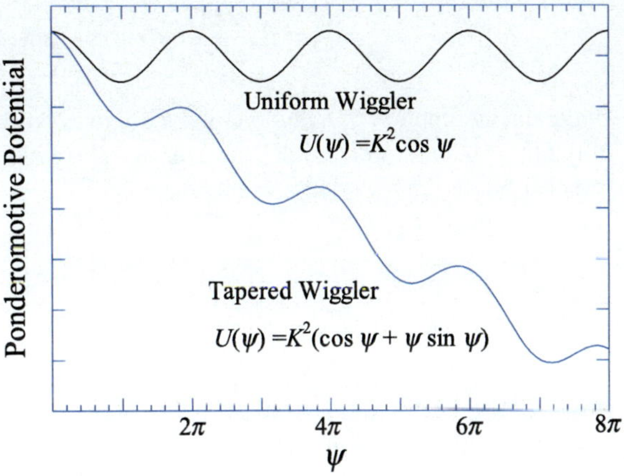

Fig. 5.1 A comparison of the ponderomotive potential for uniform and tapered magnetostatic fields

where $\beta_w{}^2$ is determined by the rms magnitude of the wiggler field for a planar wiggler.

Integration of this equation results in an equation similar to that found for uniform magnetostatic field (4.17), with a ponderomotive potential

$$U(\psi) = -\kappa^2(\cos\psi + \psi\cos\psi_{res}). \tag{5.8}$$

The tapered field, therefore, results in either a linear increase or decrease in the ponderomotive potential. The difference between the ponderomotive potential for a uniform and tapered wiggler is illustrated in Fig. 5.1 for a case in which $\sin\psi_{res} < 0$. As a result, the motion is similar to that of a ball rolling down a bumpy hill and accelerating as it falls. For Group I orbits in which $\Omega_0 < k_w v_\|$, this acceleration implies that the wiggler field amplitude and period and the axial solenoidal field must be tapered downward for efficiency enhancement to occur. The situation is more complicated for Group II orbits in which $\Omega_0 > k_w v_\|$. In this case, the wiggler period and axial solenoidal field must be tapered downward when $\Phi < 0$, while the wiggler field amplitude must be tapered upward in order to achieve acceleration. As a consequence, efficiency enhancement in the negative-mass regime requires either an upward taper in the wiggler amplitude or a downward taper in the wiggler period and the axial solenoidal field. In contrast, when $\Phi > 0$, acceleration is found when either the wiggler period is tapered downward or the axial field and wiggler period are tapered upward.

The magnitude of the efficiency enhancement can be calculated by noting that

$$\frac{d\gamma}{dz} = \frac{\omega}{v_\|^2}\left(\frac{dv_\|}{d\gamma}\right)^{-1}\frac{d^2}{dz^2}\psi. \tag{5.9}$$

Substitution of Eqs. (2.23) and (5.6), therefore, gives

$$\frac{d\gamma}{dz} = \beta_w \frac{e\delta A}{m_e c^2} \frac{\omega}{c} (\sin\psi - \sin\psi_{es}). \tag{5.10}$$

Under the assumption of a random phase approximation [which is appropriate for a fully filled ponderomotive well], the average rate of energy loss is determined by the resonant phase [i.e., $<\sin\psi> = 0$]; hence

$$\left\langle \frac{d\gamma}{dz} \right\rangle = -\frac{\gamma\gamma_\parallel^2}{\Phi} \frac{v_w^2/c^2}{(1+\beta_w^2)\Omega_0 - k_w v_\parallel} \left(\frac{\Omega_0 - k_w v_\parallel}{B_w} \frac{d}{dz} B_w - \frac{k_w v_\parallel}{\lambda_w} \frac{d}{dz} \lambda_w - \frac{\Omega_0}{B_0} \frac{d}{dz} B_0 \right). \tag{5.11}$$

Observe that in the absence of an axial magnetic field, $\Phi = 1$ and that the rms wiggler field amplitude must be imposed for a planar wiggler. As a consequence, the efficiency enhancement, $\Delta\eta$, corresponding to changes in the field amplitudes ΔB_w and ΔB_0 and the wiggler period $\Delta\lambda_w$ over some axial length Δz is

$$\Delta\eta = -\frac{\gamma_\parallel^2}{\Phi} \frac{v_w^2/c^2}{(1+\beta_w^2)\Omega_0 - k_w v_\parallel} \left(\frac{\Omega_0 - k_w v_\parallel}{B_w} \Delta B_w - \frac{k_w v_\parallel}{\lambda_w} \Delta\lambda_w - \frac{\Omega_0}{B_0} \Delta B_0 \right). \tag{5.12}$$

In the limit in which the axial solenoidal field vanishes and the wiggler period is fixed, this result reduces to that given in Eq. (1.25).

5.2 One-Dimensional Analysis: Helical Wigglers

There is an extensive literature dealing with idealized one-dimensional models of the interaction in free-electron lasers [6–27] including both helical and planar wiggler configurations, both amplifiers [6, 14–17, 22–24, 26, 27] and oscillators [13, 18–21, 25], and with slow-time-scale formulations [6, 14, 17, 24] as well as a full-scale one-dimensional particle-in-cell simulations [15, 26, 27]. Slow-time-scale formulations are suitable for the description of free-electron laser amplifiers in which it is reasonable to assume that the wave phase and amplitude vary slowly with respect to the wavelength and frequency; hence, Maxwell's equations may be averaged over a wave period. This results in an enormous increase in the computational efficiency since the integration of the Maxwell-Lorentz equations need not resolve the fast timescale corresponding to the wave frequency. In contrast, the Maxwell-Lorentz equations must be integrated on a fast timescale corresponding to the wave period in particle-in-cell techniques. In this section, we shall concentrate on a one-dimensional SVEA formulation of free-electron laser amplifiers with helical wigglers in the steady-state regime where slippage is not important.

5.2.1 *The Dynamical Equations*

The physical configuration used to develop a self-consistent, one-dimensional nonlinear formulation of the free-electron laser interaction consists of a magnetostatic field which is composed of both a solenoidal guide magnetic field and a helical wiggler field; specifically

$$\mathrm{B}(z) = B_0\hat{e}_z + B_w(z)\left[\hat{e}_x \cos k_w z + \hat{e}_y \sin k_w z\right], \tag{5.13}$$

where B_0 and $B_w(z)$ denote the amplitudes of the solenoidal and wiggler fields and k_w is the wiggler wavenumber. The wiggler amplitude has been assumed to be a slowly varying function of axial position in order to treat either the injection of the beam into the wiggler or efficiency enhancement by means of a tapered wiggler. The problem of the injection of the beam into the wiggler is of considerable practical importance, because it is necessary to inject the fluctuations of the beam about the steady-state trajectories in order to maintain a uniformly resonant interaction. For this purpose, therefore, the wiggler field is assumed to increase adiabatically from zero to a constant value over N_w wiggler periods as follows [17]:

$$B_w(z) = \begin{cases} B_w \sin^2\left(\dfrac{k_w z}{4N_w}\right) & ; 0 \le z \le N_w\lambda_w \\ B_w & ; N_w\lambda_w \le z \end{cases}. \tag{5.14}$$

Observe that in order to neglect the fringing fields associated with the wiggler gradient, we implicitly assume that the wiggler amplitude varies slowly over a wiggler [i.e., $N_w >> 1$]. Efficiency enhancement by means of a tapered magnetostatic field will be treated within the context of the three-dimensional analysis.

5.2.1.1 The Field Equations

The fluctuating electromagnetic and electrostatic fields are treated using the vector and scalar potentials in the Coulomb gauge, and we assume that these fields are of the form

$$\delta\mathrm{A}(z,t) = \delta A(z)\left[\hat{e}_x \cos\alpha_+(z,t) - \hat{e}_y \sin\alpha_+(z,t)\right], \tag{5.15}$$

and

$$\delta\Phi(z,t) = \delta\Phi(z)\cos\alpha(z,t), \tag{5.16}$$

where the phase of an electromagnetic wave of frequency ω and wavenumber k_+ is given by

$$\alpha_+(z,t) = \int_0^z dz' k_+(z') - \omega t, \tag{5.17}$$

and

$$\alpha(z,t) = \int_0^z dz' k(z') - \omega t, \tag{5.18}$$

is the phase of the space-charge wave with the same frequency but wavenumber k. This is equivalent to a WKB formulation in which it is implicitly assumed that the amplitudes and wavenumbers vary slowly over a wavelength. This requires that

$$\left|\frac{1}{\delta A}\frac{d}{dz}\delta A\right| << k_+ \text{ and } \left|\frac{1}{\delta \Phi}\frac{d}{dz}\delta \Phi\right| << k, \tag{5.19}$$

which describes slowly varying envelopes.

In the Coulomb gauge, the Maxwell-Poisson equations are of the form

$$\left(\frac{\partial^2}{\partial z^2} - \frac{1}{c^2}\frac{\partial^2}{\partial t^2}\right)\delta \mathrm{A}(z,t) = -\frac{4\pi}{c}\delta \mathrm{J}_\perp(z,t), \tag{5.20}$$

$$\frac{\partial^2}{\partial z \partial t}\delta\Phi(z,t) = 4\pi \delta J_z(z,t), \tag{5.21}$$

and

$$\frac{\partial^2}{\partial z^2}\delta\Phi(z,t) = -4\pi\delta\rho(z,t), \tag{5.22}$$

where $\delta\mathbf{J}(z,t)$ and $\delta\rho(z,t)$ are the nonlinear current and charge densities. The charge and current densities can be written [17] as averages over the entry time t_0 (defined as the time at which an electron crosses the $z = 0$ plane)

$$\delta \mathrm{J}(z,t) = -en_b v_{z0} \int_{-\infty}^{\infty} dt_0 \sigma(t_0) \mathrm{v}(t,t_0) \frac{\delta[t-\tau(z,t_0)]}{|v_z(t,t_0)|}, \tag{5.23}$$

and

$$\delta\rho(z,t) = -en_b v_{z0} \int_{-\infty}^{\infty} dt_0 \sigma(t_0) \frac{\delta[t - \tau(z,t_0)]}{|v_z(t,t_0)|}, \tag{5.24}$$

where n_b is the ambient electron density, v_{z0} is the initial axial velocity, $\mathbf{v}(t,t_0)$ is the velocity of an electron at time t which crossed the entry plane at time t_0, $\sigma(t_0)$ is the distribution in entry times, and

$$\tau(z,t_0) = t_0 + \int_0^z \frac{dz'}{v_z(z',t_0)}, \tag{5.25}$$

is a Lagrangian time coordinate which measures the time at which an electron reaches the axial position z and $v_z(z,t_0)$ denotes the velocity of an electron at axial position z which crossed the entry plane at time t_0. It should be remarked that it is implicitly assumed in the charge and current densities that the electron beam is monoenergetic and with a vanishing pitch angle spread. This is not essential to the formulation, and a more general momentum space distribution will be employed in the context of the three-dimensional formulation. In addition, it should be noted that the average over entry times is equivalent to a discrete numbering system for the electrons.

A set of coupled nonlinear differential equations for the slowly varying amplitudes and phases is obtained by the straightforward substitution of the representations for the fields in Eqs. (5.15) and (5.16) into the Maxwell-Poisson equations. Substitution of the representation of the vector potential into Maxwell's equations yields

$$\left(\frac{d^2}{dz^2} + \frac{\omega^2}{c^2} - k_+^2\right)\delta A \cos\alpha_+ - 2k_+^{1/2}\frac{d}{dz}\left(k_+^{1/2}\delta A\right)\sin\alpha_+ = -\frac{4\pi}{c}\delta J_x, \tag{5.26}$$

and

$$\left(\frac{d^2}{dz^2} + \frac{\omega^2}{c^2} - k_+^2\right)\delta A \sin\alpha_+ + 2k_+^{1/2}\frac{d}{dz}\left(k_+^{1/2}\delta A\right)\cos\alpha_+ = \frac{4\pi}{c}\delta J_y. \tag{5.27}$$

We now multiply Eq. (5.26) by $\cos\alpha_+$, Eq. (5.27) by $\sin\alpha_+$, average over a wave period, and add the resulting equations to obtain

$$\left(\frac{d^2}{dz^2}+\frac{\omega^2}{c^2}-k_+^2\right)\delta a=\frac{\omega_b^2}{c^2}\beta_{z0}\frac{\omega}{2\pi}\int_0^{2\pi/\omega}dt\int_{-\infty}^{\infty}dt_0\sigma(t_0)\frac{\delta[t-\tau(z,t_0)]}{|v_z(t,t_0)|}$$
$$\times\left(v_x\cos\alpha_+-v_y\sin\alpha_+\right), \tag{5.28}$$

after substitution of the source currents. By an analogous procedure, we obtain the complementary equation

$$2k_+^{1/2}\frac{d}{dz}\left(k_+^{1/2}\delta a\right)=-\frac{\omega_b^2}{c^2}\beta_{z0}\frac{\omega}{2\pi}\int_0^{2\pi/\omega}dt\int_{-\infty}^{\infty}dt_0\sigma(t_0)\frac{\delta[t-\tau(z,t_0)]}{|v_z(t,t_0)|}$$
$$\times\left(v_x\sin\alpha_++v_y\cos\alpha_+\right), \tag{5.29}$$

where $\beta_{z0}=v_{z0}/c$, ω_b is the plasma frequency denoted by the ambient beam density and $\delta a=e\delta A(z)/m_ec^2$ is the normalized amplitude of the vector potential. Similarly, Poisson's equation yields the following equations for the evolution of the amplitude and wavenumber of the space-charge potential

$$\left(\frac{d^2}{dz^2}-k^2\right)\delta\varphi=2\frac{\omega_b^2}{c^2}v_{z0}\frac{\omega}{2\pi}\int_0^{2\pi/\omega}dt\int_{-\infty}^{\infty}dt_0\sigma(t_0)\frac{\delta[t-\tau(z,t_0)]}{|v_z(t,t_0)|}\cos\alpha, \tag{5.30}$$

and

$$2k^{1/2}\frac{d}{dz}\left(k^{1/2}\delta\varphi\right)=-2\frac{\omega_b^2}{c^2}v_{z0}\frac{\omega}{2\pi}\int_0^{2\pi/\omega}dt\int_{-\infty}^{\infty}dt_0\sigma(t_0)\frac{\delta[t-\tau(z,t_0)]}{|v_z(t,t_0)|}\sin\alpha, \tag{5.31}$$

where $\delta\varphi=e\delta\Phi(z)/m_ec^2$ is the normalized amplitude of the space-charge potential.

5.2.1.2 The Quasi-Static Assumption

The double integrals that appear in Eqs. (5.29)–(5.32) are reducible to a single integral over the entry time of a *beamlet* of electrons which cross the entry plane within one wave period under a *quasi-static* assumption. This is the essential approximation in the development of the SVEA approximation. In order to understand this concept, observe that all the beam electrons execute identical trajectories in the absence of the fluctuating fields. For a steady-state amplifier model in which the fluctuating fields vary in axial position, but not in time, two electrons that enter the interaction region separated in time by integral multiples of the wave period will execute identical trajectories. As a consequence, we assume that

$$\mathrm{v}(t+2\pi N/\omega, t_0+2\pi N/\omega)=\mathrm{v}(t,t_0), \tag{5.32}$$

and

$$\tau(z,t_0+2\pi N/\omega)=\tau(z,t_0), \tag{5.33}$$

for integer N. The integrands in Eqs. (5.28)–(5.31) are of the generic form $G(z,t,t_0)\delta[t-\tau(z,t_0)]$, where

$$G(z,t+2\pi N/\omega,t_0+2\pi N/\omega)=G(z,t,t_0), \tag{5.34}$$

On the basis of this symmetry condition, the abovementioned double integrals reduce to

$$\begin{aligned}\int_0^{2\pi/\omega} dt \int_{-\infty}^{\infty} dt_0 G(z,t,t_0)\delta[t-\tau(z,t_0)] &= \int_0^{2\pi/\omega} dt_0 \int_{-\infty}^{\infty} dt G(z,t,t_0)\delta[t-\tau(z,t_0)] \\ &= \int_0^{2\pi/\omega} dt_0 G(z,\tau(z,t_0),t_0).\end{aligned} \tag{5.35}$$

Hence, the problem is reduced to the consideration of only one *beamlet* of electrons that enter the interaction region during a wave period. This has the effect of greatly reducing the computational requirements since there is no necessity of filling the entire interaction region with electrons.

The equations governing the evolution of the slowly varying amplitudes and phases are considerably simplified using this symmetry property of the integrands (5.36). Transforming to the wiggler frame in which $v_x = v_1 \cos k_w z - v_2 \sin k_w z$, $v_y = v_1 \sin k_w z + v_2 \cos k_w z$, and $v_3 = v_z$, we find that [17]

$$\left(\frac{d^2}{dz^2}+\frac{\omega^2}{c^2}-k_+^2\right)\delta a=\frac{\omega_b^2}{c^2}\beta_{z0}\left\langle\frac{v_1\cos\psi - v_2\sin\psi}{|v_3|}\right\rangle, \tag{5.36}$$

$$2k_+^{1/2}\frac{d}{dz}\left(k_+^{1/2}\delta a\right)=-\frac{\omega_b^2}{c^2}\beta_{z0}\left\langle\frac{v_1\sin\psi - v_2\cos\psi}{|v_3|}\right\rangle, \tag{5.37}$$

$$\left(\frac{d^2}{dz^2}-k^2\right)\delta\varphi=2\frac{\omega_b^2}{c^2}v_{z0}\left\langle\frac{\cos\psi_{sc}}{|v_3|}\right\rangle, \tag{5.38}$$

and

$$2k^{1/2}\frac{d}{dz}\left(k^{1/2}\delta\varphi\right) = -2\frac{\omega_b^2}{c^2}\upsilon_{z0}\left\langle\frac{\sin\psi_{sc}}{|\upsilon_3|}\right\rangle, \tag{5.39}$$

where the ponderomotive phase is defined in Lagrangian coordinates as

$$\psi = \psi_0 + \int_0^z dz'\left(k_+ + k_w - \frac{\omega}{\upsilon_3}\right), \tag{5.40}$$

the phase of the space-charge wave is

$$\psi_{sc} = \psi_0 + \int_0^z dz'\left(k - \frac{\omega}{\upsilon_3}\right). \tag{5.41}$$

Observe that in the Raman regime, the wavenumber of the space-charge mode and the electromagnetic mode differs by the wiggler wavenumber (i.e., $k \approx k_+ + k_w$). As a consequence, $\psi \approx \psi_{sc}$ during the course of the interaction. The averaging operator is defined over the initial phase $\psi_0 = -\omega t_0$ as

$$\langle(\cdots)\rangle \equiv \frac{1}{2\pi}\int_0^{2\pi} d\psi_0\sigma(\psi_0)(\cdots). \tag{5.42}$$

5.2.1.3 The Physical Meaning of the Lagrangian Model

Since the Lagrangian model is not intuitively obvious, some discussion of the physical interpretation of this formulation is useful. A schematic illustration of the difference between the Lagrangian and Eulerian formulations is shown in Fig. 5.2, where the Eulerian model is shown on the left and the Lagrangian model is shown on the right. The figure includes the macro-electrons in the simulation as shown by the circles as well as one wavelength of the optical field. The Eulerian formulation is commonly used in time-dependent simulations of plasmas, particle beams, and fluids, where the motion of the particles and/or fluid elements is tracked in both space and time. The Lagrangian formulation as used in free-electron lasers was originally developed to treat steady-state traveling wave tube amplifiers so that the particle motion and field evolution are tracked only in distance.

The figure shows the treatment of a single *slice* of the electron beam that is one wavelength long in both the Eulerian and Lagrangian formulations. In the Eulerian formulation, both the particles and fields that enter the interaction region at $z = 0$ are propagated both in space and time. As a result, the slippage of the optical field relative to the electrons due to the difference between the group velocity of the optical field and the bulk electron beam velocity is included self-consistently.

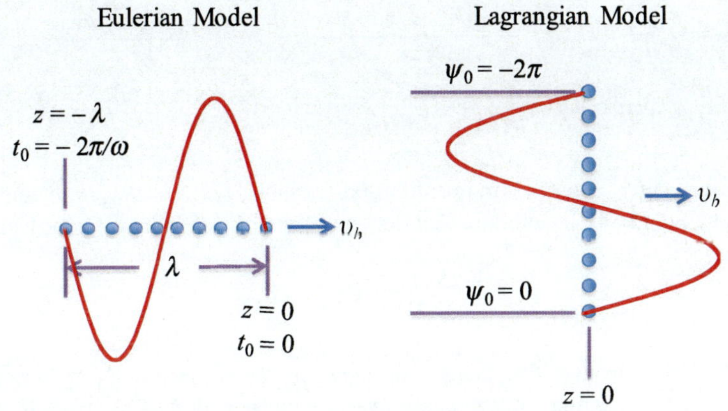

Fig. 5.2 Schematic illustration of the Eulerian and Lagrangian formulations

In contrast, the Lagrangian formulation treats the "fiction" that the electrons in one *slice* of the beam all propagate from $z \rightarrow z + \Delta z$ simultaneously. This "fiction" is compensated for since the electrons experience different phases of the optical field as specified by the assignment of the initial phases ψ_0. This formulation is valid, strictly speaking, when the interaction in *each* slice is identical so that the slippage of the optical field from one *slice* into the next *slice* is compensated for by the slippage of the field from the preceding *slice*. This implies that the electron beam parameters are constants.

As we will discuss in the next chapter, the Lagrangian formulation has been adapted to treat time dependence in free-electron lasers; however, the underlying assumption in this process is that the electron beam parameters vary slowly along the electron beam.

5.2.1.4 The Electron Orbit Equations

In order to complete the formulation, the electron orbit equations in the presence of the static and fluctuating fields must be specified. Since we treat the case of a steady-state amplifier configuration, we integrate in z and write the Lorentz force equations in the form [17]

$$\frac{d}{dz}u_1 = -\left(\widehat{\Omega}_0 - k_w u_3\right)\frac{u_2}{u_3} + c\left[\delta a\left(\frac{\omega}{\upsilon_3} - k_+\right)\sin\psi + \cos\psi\frac{d}{dz}\delta a\right], \quad (5.43)$$

$$\frac{d}{dz}u_2 = \left(\widehat{\Omega}_0 - k_w u_3\right)\frac{u_1}{u_3} - \widehat{\Omega}_w + c\left[\delta a\left(\frac{\omega}{\upsilon_3} - k_+\right)\cos\psi - \sin\psi\frac{d}{dz}\delta a\right], \quad (5.44)$$

and

$$\frac{d}{dz}u_3 = \widehat{\Omega}_w \frac{u_2}{u_3} + ck_+\delta a \frac{u_1 \sin\psi + u_2 \cos\psi}{u_3} - c\frac{d}{dz}\delta a \frac{u_1 \cos\psi - u_2 \sin\psi}{u_3}$$
$$- \frac{c^2}{\upsilon_3}\left(k\delta\varphi \sin\psi_{sc} - \frac{d}{dz}\delta\varphi \cos\psi_{sc}\right), \tag{5.45}$$

where $\mathbf{u} = \mathbf{p}/m_e$ denotes a momentum-like variable, $\widehat{\Omega}_{0,w} = eB_{0,w}/m_e c$, and B_w is given by Eq. (5.14). In addition, we integrate a time-like coordinate for each electron

$$\frac{d\psi}{dz} = k_+ + k_w - \frac{\omega}{\upsilon_3}. \tag{5.46}$$

It is important to note that there is no necessity to obtain a reduced form of the Lorentz force equations by averaging over a wiggler period. It has been implicitly assumed that the amplitude and phase are slowly varying functions of position, and this occurs only in the vicinity of the wave-particle resonance at $\omega \approx (k_+ + k_w)\upsilon_3$. When this is satisfied, the ponderomotive phase is also a slowly varying quantity [i.e., $d\psi/dz \approx 0$], and the Lorentz force equations describe trajectories which vary slowly with respect to the wave period as well. Hence, the orbit equations implicitly describe a slowly varying interaction near resonance, and no wiggler average is required.

5.2.1.5 The Numerical Procedure

Together, Eqs. (5.36)–(5.39) for the fields and (5.43)–(5.45) for an ensemble of electrons represent a complete and self-consistent formulation of the interaction in the free-electron laser. The field equations constitute a set of four second-order differential equations for the amplitudes and phases of the fluctuating fields. This may be reduced to a set of eight first-order differential equations for δa, Γ_+, k_+, $\Delta\phi_+$, $\delta\varphi$, Γ, k, and $\Delta\phi$, where $\Delta\phi_+$ and $\Delta\phi$ denote the relative phases defined by

$$\Delta\phi_+ \equiv \int_0^z dz'[k_+(z') - k_+(0)], \tag{5.47}$$

$$\Delta\phi \equiv \int_0^z dz'[k(z') - k(0)], \tag{5.48}$$

and Γ_+ and Γ define the growth rates [i.e., the logarithmic derivatives] of the field vector and scalar potentials. These equations may be expressed as

$$\frac{d}{dz}\delta a = \Gamma_+ \delta a, \tag{5.49}$$

$$\frac{d}{dz}\Gamma_+ = -\left(\frac{\omega^2}{c^2} + \Gamma_+^2 - k_+^2\right) + \frac{\omega_b^2}{c^2}\frac{\beta_0}{\delta a}\left\langle \frac{\upsilon_1 \cos\psi - \upsilon_2 \sin\psi}{|\upsilon_3|}\right\rangle, \tag{5.50}$$

$$\frac{d}{dz}\Delta\phi_+ = k_+, \tag{5.51}$$

$$\frac{d}{dz}k_+ = -2k_+\Gamma_+ - \frac{\omega_b^2}{c^2}\frac{\beta_0}{\delta a}\left\langle \frac{\upsilon_1 \sin\psi + \upsilon_2 \cos\psi}{|\upsilon_3|}\right\rangle, \tag{5.52}$$

$$\frac{d}{dz}\delta\varphi = \Gamma\delta\varphi, \tag{5.53}$$

$$\frac{d}{dz}\Gamma = k^2 - \Gamma^2 + 2\frac{\omega_b^2}{c^2}\frac{\upsilon_{z0}}{\delta\varphi}\left\langle \frac{\cos\psi_{sc}}{|\upsilon_3|}\right\rangle, \tag{5.54}$$

$$\frac{d}{dz}\Delta\phi = k, \tag{5.55}$$

$$\frac{d}{dz}k = -2k\Gamma - 2\frac{\omega_b^2}{c^2}\frac{\upsilon_{z0}}{\delta\varphi}\left\langle \frac{\sin\psi_{sc}}{|\upsilon_3|}\right\rangle. \tag{5.56}$$

The problem, therefore, consists in the simultaneous solution of $4N + 8$ first-order differential equations, where N denotes the total number of electrons, subject to the initial conditions. The numerical procedure involves the evaluation of the particle averages for the source current and charge density at each step in the integration and the subsequent integration of the field quantities and the electron trajectories. This may be accomplished by a variety of integration schemes including Adams-Moulton predictor/corrector as well as Runge-Kutta algorithms. The quantities that are averaged in the dynamical equations are functions of the instantaneous values of the momenta and phases, which are *implicit* functions of the *initial* phases of the electrons. It is the *initial* phase that appears in the averaging operator and represents a convenient method of discretizing (i.e., counting) the electrons based upon their initial phase relative to the wave. The most efficient method of discretization for a free-electron laser is to use an Nth order Gaussian quadrature technique to distribute the electrons in their initial phases. Within this context, "efficient" refers to the convergence of a discrete sum over particles to the continuous integral with the fewest number of particles in the simulation. In comparison, Simpson's rule requires at least twice as many electrons to achieve the same degree of accuracy.

5.2.1.6 The Initial Conditions

The specific initial conditions appropriate to the analysis are that of a continuous and uniform electron beam, and we choose $\sigma(\psi_0) = 1$ for $-\pi < \psi_0 < \pi$. This describes an electron beam with an initial random phase distribution. The beam is also assumed to be monoenergetic and propagating paraxially (i.e., with a zero emittance). As a consequence, $\upsilon_{z0} = (1 - \gamma_0^{-2})^{1/2}c$, where γ_0 is the relativistic factor corresponding to the total beam energy. It is important to recognize, however, that the subsequent evolution of the beam is integrated self-consistently, and the beam may bunch in axial phase as well as develop both energy and pitch-angle spreads due to the nature of the interaction. The initial conditions of the wave fields are chosen as follows. Since the wiggler field increases adiabatically from zero at the entry plane, the growth rate of the vector potential is initially zero as well. The wavenumbers of the vector and scalar potentials are chosen to satisfy the uncoupled dispersion equations for a circularly polarized electromagnetic wave

$$\omega^2 - c^2k_+^2 - \frac{\omega_b^2}{\gamma_0(\omega - \Omega_0 - k_+\upsilon_{z0})} = 0, \tag{5.57}$$

and the beam-plasma mode

$$(\omega - k_+\upsilon_{z0})^2 - \kappa_b^2\upsilon_{z0}^2 = 0, \tag{5.58}$$

of angular frequency ω. Since we treat the case of a steady-state amplifier model, the initial amplitude of the vector potential can be arbitrarily selected to represent the amplitude of the injected signal. However, the scalar potential must be chosen to represent the initial level of fluctuations (i.e., noise) in the beam. This is conveniently determined by means of the z-component of Maxwell's equations. Following a procedure analogous to that used for Poisson's equation, we find that the scalar potential and its derivative must also satisfy

$$\delta\varphi = -2\frac{\omega_b^2}{\omega k c^2}\upsilon_{z0}\langle \cos\psi_{sc}\rangle, \tag{5.59}$$

and

$$\frac{d}{dz}\delta\varphi = -2\frac{\omega_b^2}{\omega k c^2}\upsilon_{z0}\langle \sin\psi_{sc}\rangle. \tag{5.60}$$

For a random initial phase distribution modeled by a discrete and finite number of electrons, the phase averages in these equations are not identically zero, and they may be used to define a convenient initial fluctuation level for the space-charge modes. Typically, the initial average over the $<\sin\psi_{sc}> \approx 0$, and we may choose $\Gamma(z = 0) = 0$. However, the average over $\cos\psi_{sc}$ can be relatively large due to the discretization, and we choose

$$\delta\varphi(z=0) = -2\frac{\omega_b^2}{\omega k c^2}\upsilon_{z0}\langle \cos\psi_{sc}\rangle. \tag{5.61}$$

5.2.2 Electron Beam Injection

Before proceeding to the complete solution of the dynamical equations, we turn to the question of the injection of the electron beam into the wiggler. This is an issue of practical importance since the beam is not created in the wiggler field, and the electron optics required to transport the beam from the accelerator to the wiggler can often be a complicated system in itself. The specific question we examine here is the effect of an adiabatic increase in the wiggler amplitude such as that described in Eq. (5.16) on the trajectories of electrons that are propagating paraxially upon entry to the wiggler.

The appropriate equations of motion are a subset of the complete orbit equations given in Eqs. (5.43)–(5.45) and are given as

$$\frac{d}{dz}\upsilon_1 = -(\Omega_0 - k_w\upsilon_3)\frac{\upsilon_2}{\upsilon_3}, \tag{5.62}$$

$$\frac{d}{dz}\upsilon_2 = (\Omega_0 - k_w\upsilon_3)\frac{\upsilon_2}{\upsilon_3} - \Omega_w, \tag{5.63}$$

and

$$\frac{d}{dz}\upsilon_3 = \Omega_w\frac{\upsilon_2}{\upsilon_3}, \tag{5.64}$$

where $\Omega_{0,w} = |eB_{0,w}/\gamma m_e c|$ and γ is a constant of the motion. These equations are formally identical to Eqs. (2.10)–(2.12) except that we integrate over axial position rather than time and that the wiggler amplitude is not constant but is given in Eq. (5.14).

The desired result is to inject the beam onto the steady-state trajectories in which the axial velocity is a constant. This ensures that the wave-particle resonance will be maintained over an extended interaction length and maximizes the gain and efficiency of the interaction. In order to determine the efficacy of the adiabatic entry taper, the first example we consider is one in which the axial solenoidal field is absent, and the steady-state trajectories in the wiggler correspond to the Group I orbits. The specific parameters of this example are a beam energy corresponding to $\gamma = 3.5$ and a wiggler field in which $\Omega_w/ck_w = 0.05$ and the entry taper is effected over ten wiggler periods (i.e., $N_w = 10$). The results of the integration of Eqs. (5.62)–(5.64) for these parameters are shown in Fig. 5.3 in which we plot υ_1 and υ_2 versus axial distance. In this case, the injection process works well, and the electron trajectory exhibits small oscillations about the steady-state orbit for which $\upsilon_1/c \approx -0.05$, $v_2 \approx 0$, and $\upsilon_3/c \approx 0.957$. As is evident from the figure, the bulk value for the magnitude of υ_1 increases with the adiabatic rise in B_w, after which small oscillations about the mean value corresponding to the steady-state trajectory are found in the constant-B_w region. The value of υ_2 remains small over the entire course

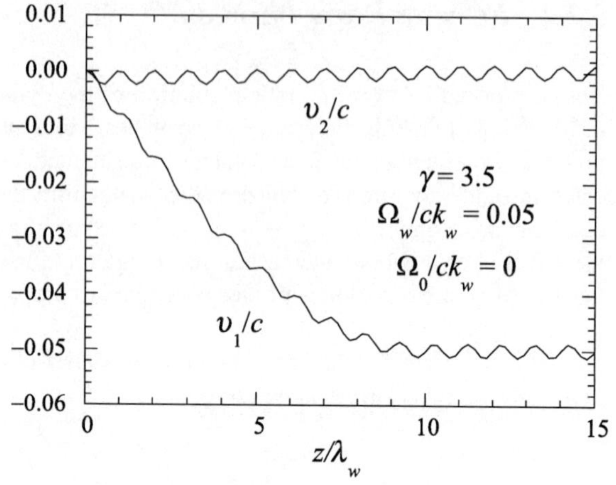

Fig. 5.3 Plot of the single-particle trajectories versus axial distance

of the trajectory. The behavior of v_3, while not shown explicitly, also exhibits small oscillations (of the order of 1%) about the value for the steady-state orbit.

Thus, we conclude that the adiabatic wiggler entry taper is capable of injecting electrons on near-steady-state trajectories. However, it should be observed that the magnitude of the fluctuation about the steady-state orbit increases as the axial solenoidal field approaches values near the resonance for $\Omega_0 \approx k_w v_{\|}$. Indeed, it becomes increasingly difficult to inject electrons onto these near-steady-state trajectories as the orbital instability point is approached ($\Omega_0/ck_w \approx 0.76$ for the present choice of parameters). At this point, the orbits differ widely from the steady-state trajectories and exhibit large fluctuations in the velocity. As a result, it becomes increasingly difficult to either inject or propagate a coherent beam through the system near the gyroresonance. An example of injection for parameters corresponding to Group II orbits is shown in Fig. 5.4 in which we plot the individual components of the velocity as a function of axial position for $\Omega_0/ck_w = 1.0$, $\Omega_w/ck_w = 0.05$, $\gamma = 3.5$, and $N_w = 10$. These parameters correspond to Group II orbits in the negative-mass regime close to the gyroresonance. Although the orbital instability is not found for Group II orbits in the idealized one-dimensional limit, the axial velocity of a steady-state orbit near the gyroresonance is significantly less than the initial velocity of paraxial flow, and the orbits which result exhibit large fluctuations about mean values corresponding to the steady-state orbit.

In view of these results regarding the injection of an electron beam into a combined helical wiggler and axial solenoidal field, we conclude that large-scale fluctuations in the electron velocity accompany injection whenever the axial field is close to gyroresonance. Indeed, these fluctuations act to counter the enhancements found in the linear gain of the interaction demonstrated in Chap. 4 for free-electron laser operation near the gyroresonance.

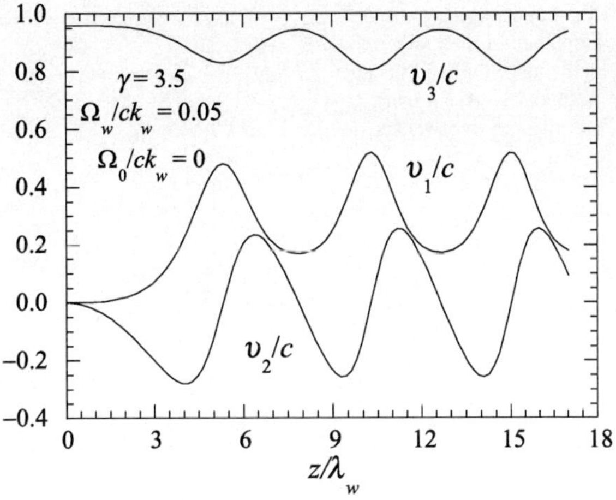

Fig. 5.4 Plot of the single-particle trajectories versus axial distance

5.2.3 Numerical Solution of the Dynamical Equations

The dynamical equations given in Eqs. (5.49)–(5.56) for the fields and Eqs. (5.43)–(5.45) for each electron have been solved numerically. In this case of the idealized one-dimensional model, the ensemble of electrons is discretized using Simpson's rule with a total of 61 electrons per wave period. The use of larger numbers of electrons was found to result in discrepancies of considerably less than 1%. As in the case of the single-particle trajectories, the wiggler field amplitude increases adiabatically to a constant level, and the effect of the injection mechanism on the gain and saturation of the interaction can be studied directly. Particular interest is centered on the degradation in the gain and saturation efficiency due to large-amplitude fluctuations in the electron trajectories in the vicinity of the gyroresonance.

The first example we consider corresponds to field parameters shown in Fig. 5.3, and we choose $\Omega_0/ck_w = 0$, $\Omega_w/ck_w = 0.05$, $N_w = 10$, $\gamma = 3.5$, and $\omega_b/\gamma^{1/2}ck_w = 0.1$. These parameters correspond to the boundary between the high-gain Compton and the collective Raman regimes. We arbitrarily initialize the vector potential in this case at $\delta a(z = 0) = 1.0 \times 10^{-7}$ and find from Eq. (5.61) that the initial value of the scalar potential is $\delta\varphi(z = 0) = 1.09 \times 10^{-6}$. Since we are primarily interested in the electromagnetic radiation, we confine our attention henceforth to the vector potential. Since the trajectories within the uniform wiggler obtained for these parameters are close to the steady-state trajectories, we expect to find that the growth rate in the linear phase of the interaction is close to that found from the linear dispersion analysis in Chap. 4. The results of the simulation for these parameters are shown in Fig. 5.5 in which we plot both the amplitude δa and the growth rate Γ_+ versus axial position for a frequency $\omega/ck_w = 21.6$ which corresponds to maximum growth. Observe that $k/k_w \approx 21.6$ for this case as well. It is evident from the figure that an extended region of exponential growth occurs after an initial transient phase (for $z/\lambda_w \leq 29$). The growth rate during this phase of the interaction as computed from the

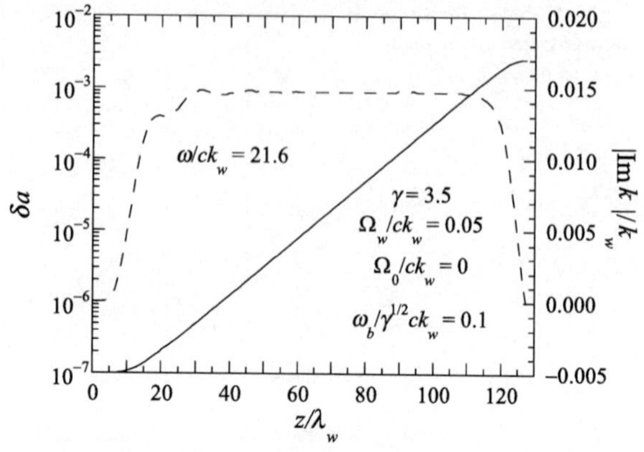

Fig. 5.5 Graphs of (**a**) the amplitude of the radiation field (solid line) and (**b**) the magnitude of the growth rate (dashed line) versus axial position

simulation is $\Gamma_+/k_w \approx 0.0146$, which is in good agreement with the growth rate of $\Gamma_{lin}/k_w \approx 0.0145$ computed on the basis of the linear dispersion Eq. (4.83).

Fluctuations in the growth rate are of the order of $\Delta\Gamma_+ \approx \pm 0.0002$ which is to be expected based upon the orbit calculation. Saturation begins to occur at $z/\lambda_w \approx 114.1$, after which the growth rate rapidly decreases to zero at $z/\lambda_w \approx 127.3$. At saturation, the radiation field amplitude is $\delta a_{sat} \approx 2.56 \times 10^{-3}$. The saturation efficiency is given in terms of these results by

$$\eta \cong \frac{1}{(\gamma-1)\beta_\parallel}\frac{ck}{\omega}\left(\frac{\omega}{\omega_b}\delta a_{sat}\right)^2, \tag{5.65}$$

which implies that $\eta \approx 3.65\%$ for this case. Observe that for these parameters, $\upsilon_\parallel/c \approx 0.957$, $\upsilon_{ph}/c \approx 0.956$, and the estimate of the saturation efficiency made on the basis of the ad hoc phase trapping arguments in Eq. (5.2) is $\eta_{est} \approx 3.2\%$, which is in good agreement with the simulation results.

Increases in the axial solenoidal field result in increasing levels of fluctuations in the electron velocities about the steady-state orbits, and this effect will compete with the increases in the growth rate and efficiency that correspond to increases in the transverse wiggler-induced velocity due to operation near the gyroresonance. In order to illustrate this effect, we consider the case for which $\Omega_0/ck_w = 0.5$, $\Omega_w/ck_w = 0.05$, $N_w = 10$, $\gamma = 3.5$, and $\omega_b/\gamma^{1/2}ck_w = 0.1$. These parameters correspond to Group I trajectories with $\upsilon_\parallel/c \approx 0.953$ and $\upsilon_w/c \approx -0.105$. In this case, both the growth rate and the level of oscillations in the growth rate have increased relative to that shown in Fig. 5.5. The results of the simulation yield an average growth rate of $\Gamma_+/k_w \approx 0.030$ with a fluctuation of $\Delta\Gamma_+/k_w \approx \pm 0.003$ at the frequency of peak growth $\omega/ck_w = 19.4$. This is also in substantial agreement with the linear dispersion analysis which yields a growth rate of $\Gamma_{lin}/k_w \approx 0.029$. The increased growth rate results in a shorter distance to saturation of $z/\lambda_w \approx 67.5$ at a field level of $\delta a_{sat} \approx 3.30 \times 10^{-3}$. This corresponds to an efficiency of 4.92%.

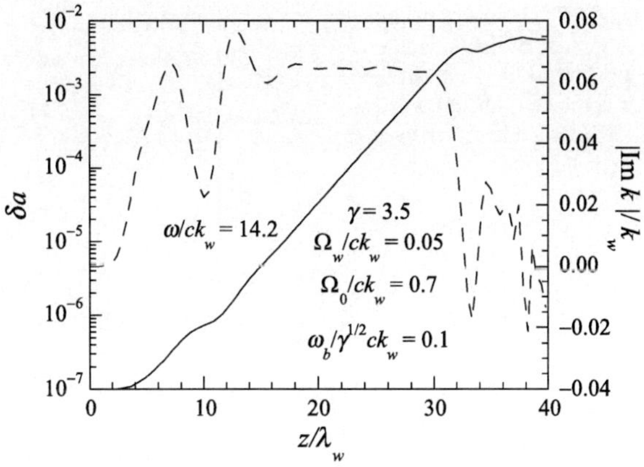

Fig. 5.6 Graphs of (**a**) the amplitude of the radiation field (solid line) and (**b**) the magnitude of the growth rate (dashed line) versus axial position

Further increases in the axial magnetic field above this level (but still corresponding to Group I trajectories) lead to still larger fluctuations in both the orbits and the linear growth rate, until a point of diminishing returns is reached at which the orbital fluctuations result in a degradation in both the growth rate and the efficiency. A transitional case is illustrated in Fig. 5.6, in which we plot the amplitude and the growth rate of the vector potential for $\Omega_0/ck_w = 0.7$, $\Omega_w/ck_w = 0.05$, $N_w = 10$, $\gamma = 3.5$, and $\omega_b/\gamma^{1/2}ck_w = 0.1$ at the frequency of maximum growth of $\omega/ck_w = 14.2$. Observe that this is close to, but still below, the orbital instability transition at $\Omega_0/ck_w \approx 0.76$. The increased level of fluctuations in the growth rate is apparent from the figure although there is an extended range for $20 < z/\lambda_w < 30$ over which the growth rate is relatively constant. The growth rate found in simulation for this example is $\Gamma_+/k_w \approx 0.063$ (which is close to the result of the linear theory of $\Gamma_{lin}/k_w \approx 0.060$) and is still increased relative to that found for $\Omega_0/ck_w = 0.5$, but the saturation efficiency has decreased to $\eta \approx 4.02\%$ corresponding to a field level of $\delta a_{sat} \approx 4.09 \times 10^{-3}$.

For higher levels of the axial solenoidal field such that $\Omega_0/ck_w > 0.76$, the orbits correspond to perturbations about the Group II trajectories. However, as seen in Fig. 5.4, large oscillations about the steady-state orbits are characteristic of the low bulk axial velocities found near gyroresonance in this regime. The implication of such orbital behavior is that (1) the growth rate must also exhibit large scale oscillations and (2) the resonant frequency is relatively low. An example of this regime is given in Fig. 5.7 in which we plot the amplitude and growth rate of the vector potential for $\Omega_0/ck_w = 1.1$, $\Omega_w/ck_w = 0.05$, $N_w = 10$, $\gamma = 3.5$, and $\omega_b/\gamma^{1/2}ck_w = 0.1$ at the frequency of maximum growth of $\omega/ck_w = 11.4$. The system exhibits the expected large-scale fluctuation in the growth rate of $\Delta\Gamma_+/k_w \approx \pm 0.036$ about a mean value of $\Gamma_+/k_w \approx 0.072$ after the transients corresponding to launching losses have decayed away. Note that the linear theory predicts a growth rate of $\Gamma_{lin}/k_w \approx 0.056$ and that this is within the range of fluctuation found in simulation. Saturation occurs at $z/\lambda_w \approx 41.1$ for $\delta a_{sat} \approx 4.91 \times 10^{-3}$ corresponding to an efficiency of 3.88%.

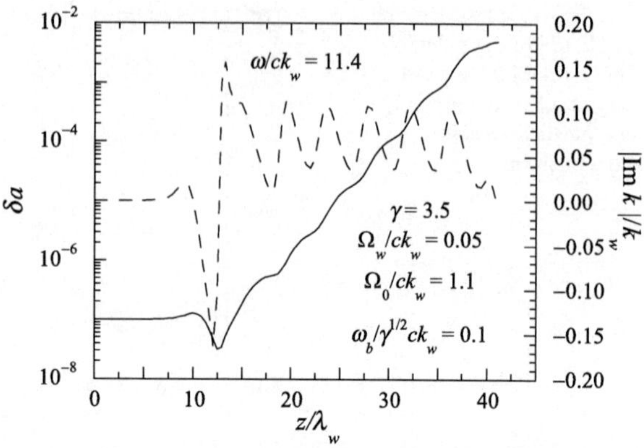

Fig. 5.7 Graphs of (**a**) the amplitude of the radiation field (solid line) and (**b**) the magnitude of the growth rate (dashed line) versus axial position

Further increases in the axial solenoidal field correspond to increases in the resonant frequency and decreases in the oscillations of the orbits about the steady-state trajectories. As a consequence, the evolution of the radiation field becomes more regular, and saturation efficiency rises with the improved wave-particle resonance. For example, when $\Omega_0/ck_w = 1.5$, the frequency of peak growth occurs for $\omega/ck_w = 20.3$, and a growth rate of $\Gamma_+/k_w \approx 0.021$ is found in simulation with a fluctuation of only approximately 1% about this value. It should be remarked that a growth rate of $\Gamma_{lin}/k_w \approx 0.021$ is also recovered from the linear theory. The saturation efficiency found for this case has risen to a level of 5.02%, which is comfortably higher than the peak efficiency observed in simulation for parameters corresponding to Group I orbits.

An overview of the effect of the axial magnetic field on the interaction frequency and the linear growth rate is shown in Fig. 5.8 in which ω/ck_w and Γ_+/k_w are plotted as functions of the axial field strength. The curves for the frequency represent the variation in the resonant frequency at maximum growth found from the intersection of the uncoupled dispersion curves for the appropriate value of the axial velocity of the steady-state trajectory. These values represent values used in the simulations. The solid lines in the plot of the growth rate represent the calculated value from the linear theory for the appropriate steady-state orbit, while the dots are used to denote the results found in the simulation. As seen in the figure, the agreement between the linear and nonlinear analyses is excellent. For all cases shown, only the values of the growth rate for $\Omega_0/ck_w = 1.1$ (note that the bars indicated for this point denote an oscillation range) for the linear and nonlinear analyses differ by more than 2%. This is due to the relatively large oscillation in the trajectory about the steady-state orbit corresponding to this case. Note also that the zero for $\Omega_0/ck_w \approx 1.25$ corresponds to the upper bound of the Group II orbit negative-mass regime at which Φ (2.24) and the ponderomotive potential vanish.

The variation in the saturation efficiency with the axial magnetic field is shown in Fig. 5.9 in which we plot the distance to saturation and the saturation efficiency versus Ω_0/ck_w. It is clear from the figure that substantial enhancements in the efficiency are possible over that found in the absence of an axial magnetic field.

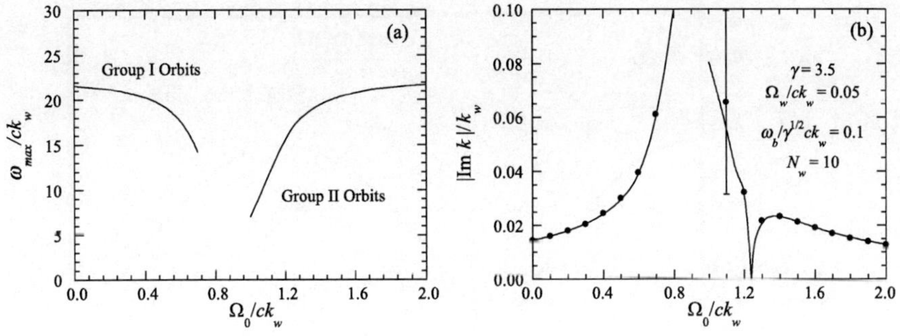

Fig. 5.8 Graphs of (**a**) the resonant frequency corresponding to peak growth and (**b**) the magnitude of the peak growth rate versus Ω_0/ck_w [17]

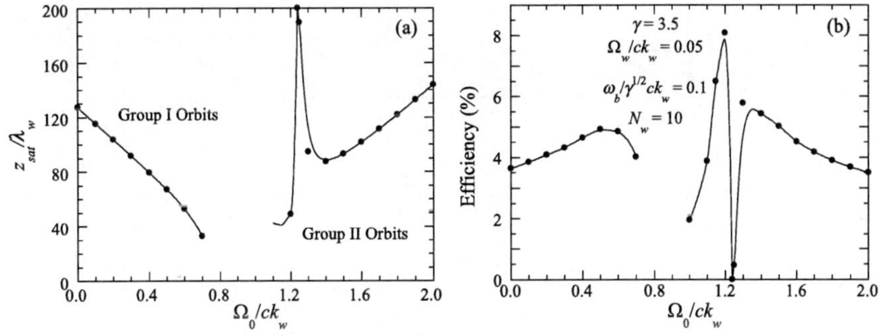

Fig. 5.9 Graphs of (**a**) the distance to saturation and (**b**) the saturation efficiency versus Ω_0/ck_w

However, a point of diminishing returns is found as the axial magnetic field approaches the gyroresonance, and the interaction efficiency declines due to the oscillations in the electron trajectories for axial fields closer to the resonance. For parameters corresponding to Group I orbits, the maximum efficiency is approximately 5% for $\Omega_0/ck_w = 0.5$ for the chosen parameters which constitutes a 37% enhancement over the efficiency found when no axial field is present. However, the greatest enhancements in the efficiency correspond to axial magnetic fields associated with Group II trajectories in the negative-mass regime for which the maximum efficiency approaches 8.1%. It is important to bear in mind, however, that these enhancements in the efficiency occur at the expense of a reduced interaction frequency.

5.2.4 The Phase-Space Evolution of the Electron Beam

The saturation mechanism is particle trapping in the combined ponderomotive and space-charge fields. An example of this is shown in Figs. 5.10 and 5.11 in which the

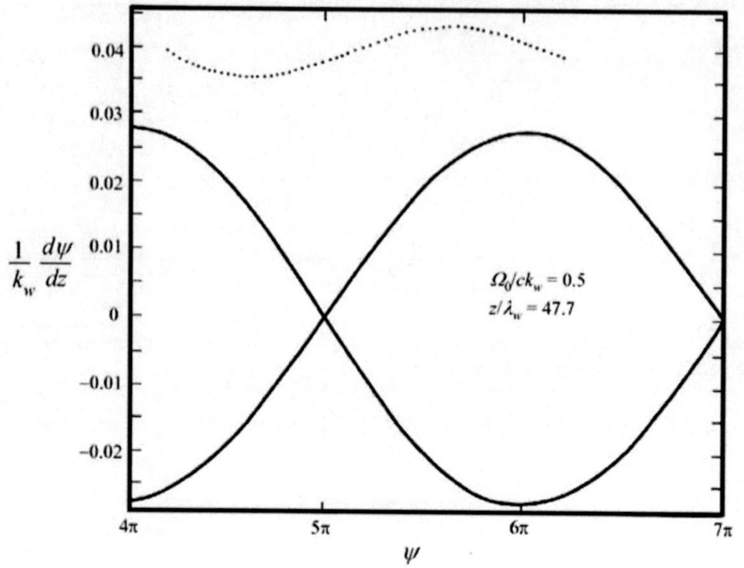

Fig. 5.10 Plot of the electron phase space during the linear exponential growth phase of the interaction for $\Omega_0/ck_w = 0.5$

positions of the electrons in the phase space ($\psi, d\psi/dz$) are plotted for the case in which $\Omega_0/ck_w = 0.5$, which corresponds to Group I trajectories. The initial phase-space distribution reflects a uniform phase-space distribution in which the electrons are propagating paraxially with identical axial velocities; hence, the initial phase-space distribution describes a horizontal line for which $d\psi/dz = k_+ + k_w - \omega/v_{z0}$. For the parameters appropriate to this example, the initial value of $d\psi/dz \approx 0.149$. Figure 5.10 represents the phase space during the linear phase of the interaction after the electrons have lost energy but still prior to saturation. The solid lines in the figure represent the approximate separatrix as determined from Eq. (4.20), which is included in the figure for illustrative purposes.

The dynamics of the nonlinear pendulum motion is included implicitly in the simulation by means of the integration of the Lorentz force equations. It is evident from the figure that the electrons lose energy as they decelerate and fall in phase space (i.e., $d\psi/dz$ decreases). The early phases of the process of axial bunching in the ponderomotive potential are shown in the figure, but the electrons have not yet crossed the separatrix onto trapped orbits. The phase-space distribution at saturation is shown in Fig. 5.12, which shows a snapshot of the electrons as they execute a clockwise rotation within the ponderomotive well. At this point, the energy transfer from the electrons to the wave in the first half cycle of the oscillation is exactly balanced by the energy transfer from the wave to the electrons in the second half cycle. Note that the entire beam has not been trapped.

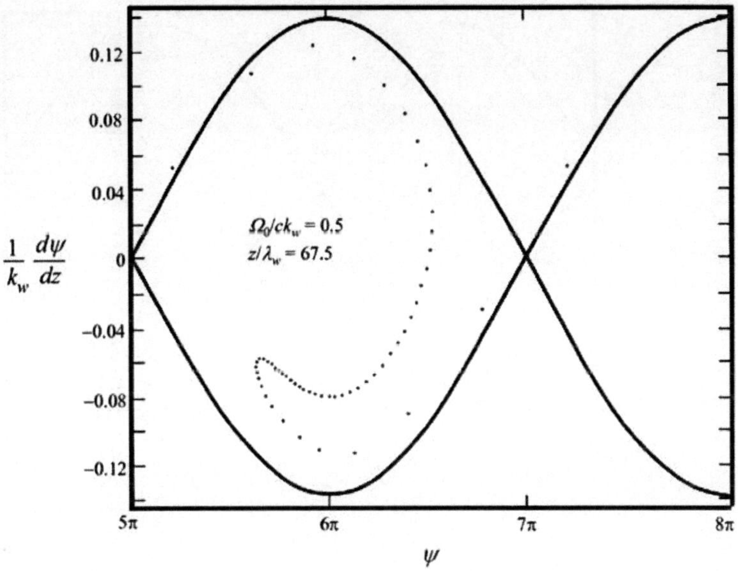

Fig. 5.11 Plot of the electron phase space at saturation of the interaction for $\Omega_0/ck_w = 0.5$

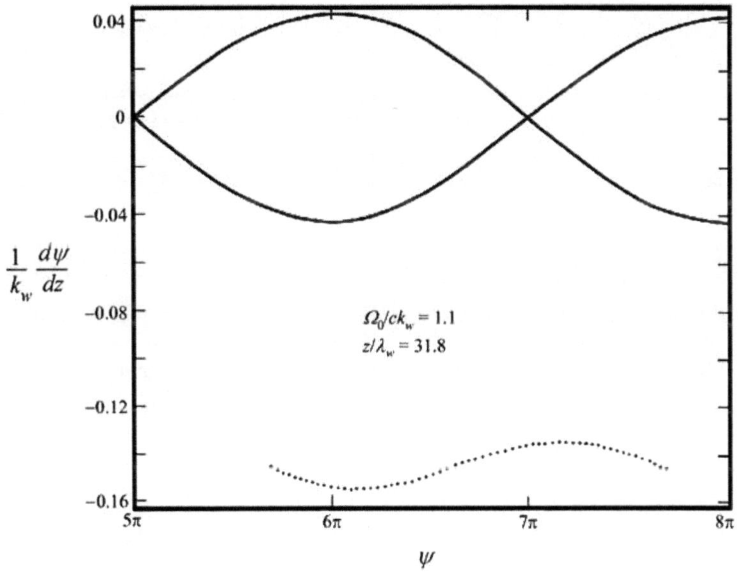

Fig. 5.12 Plot of the electron phase space during the linear exponential growth phase of the interaction for $\Omega_0/ck_w = 1.1$

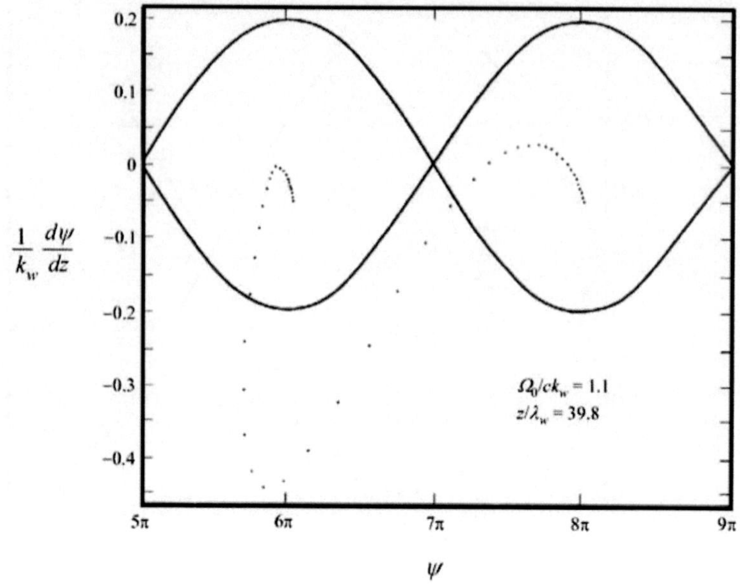

Fig. 5.13 Plot of the electron phase space at saturation of the interaction for $\Omega_0/ck_w = 1.1$

An example of the phase-space evolution for an axial magnetic field corresponding to Group II trajectories is shown in Figs. 5.12 and 5.13, where the solid line again indicates the analytic approximation for the separatrix and not a result from the simulation. This case corresponds to the choice of $\Omega_0/ck_w = 1.1$ discussed in regard to Fig. 5.7. Note that while the initial phase-space distribution corresponds to $d\psi/dz \approx 0.149$ as in the case of $\Omega_0/ck_w = 0.5$, this example describes the negative-mass regime in which the axial velocity in the uniform-wiggler region is much smaller than the paraxial velocity of the beam on entry to the wiggler. For this reason, the bulk axial velocity in the uniform-wiggler region is $v_{\|}/c \approx 0.922$, corresponding to $d\psi/dz \approx -0.063$. Hence, the electron phase-space distribution describes a line in the lower half-plane at the start of the uniform-wiggler region, and, since this is in the negative-mass regime, the electrons accelerate in the axial direction (i.e., $d\psi/dz$ increases) as energy is lost to the wave. This process is clearly indicated in Fig. 5.12, which illustrates the electron phase-space distribution at $z/\lambda_w = 31.8$ which is a late stage of the linear phase of the interaction. The electron phase space at saturation is shown in Fig. 5.13.

5.2.5 *Comparison with Experiment*

Experimental observations of the variation of the saturation efficiency with the axial magnetic field have been obtained at the Naval Research Laboratory [28]. This experiment employed an intense relativistic electron beam with a voltage of 1.35 MeV and a current of 1.5 kA and a helical wiggler field with a 3.0 cm period and an amplitude variable up to approximately 4.0 kG. The axial solenoidal field was variable up to approximately 20 kG. This experiment was operated as a superradiant amplifier in which the signal grew from noise on the beam and reached saturated efficiencies of the order of 2–3%. It should be noted that the term "superradiant amplifier" has been superseded by the term self-amplified spontaneous emission (SASE). This experiment employed a pulse line accelerator with a 30–40 ns flat top. In this experiment, the slippage time was approximately 7–8 ps, which is much less than the bunch duration; hence, slippage is a negligible effect, and the steady-state formulation is appropriate. An example of the output power as a function of the axial field strength is shown in Fig. 5.14 for a wiggler field strength of 630 G, and it is evident that the peak emission was found for strong axial fields corresponding to Group II orbit parameters. As seen in the figure, there is no emission in the range of 10–13 kG, which corresponds to the vicinity of the gyroresonance in which the orbits executed extremely large-amplitude oscillations. The lack of emission for low axial magnetic fields was due to the fact that it was impossible to propagate such an intense beam without a sufficiently high axial magnetic field to confine the beam against the effects of self-electric and magnetic fields. In addition, the cutoff in the emission for strong axial magnetic fields corresponds to the transition between the negative-mass and positive-mass regimes of the Group II trajectories.

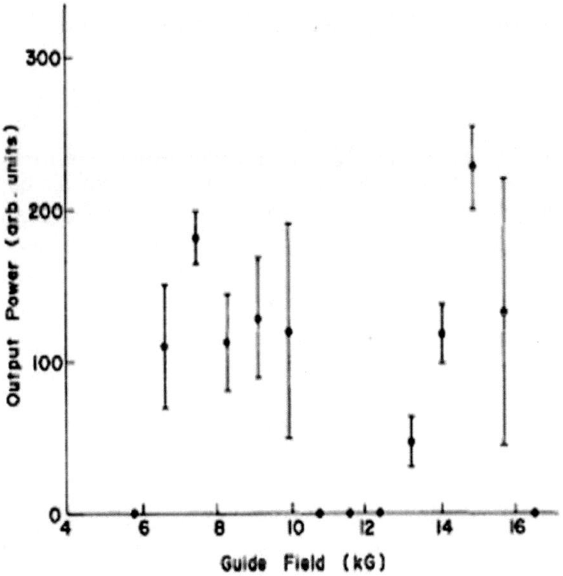

Fig. 5.14 Measured variations in the output power with the axial magnetic field for experiments performed at the Naval Research Laboratory [28] with a 1.35 MeV/1.5 kA electron beam and a wiggler field with a period of 3.0 cm and an amplitude of 630 G

5.3 One-Dimensional Analysis: Planar Wigglers

We now turn to a slow-time-scale nonlinear analysis of planar wiggler configurations. In many ways, the analysis is similar to that described for the helical wiggler; however, there are interesting differences in the interaction dynamics for planar wigglers. In the first place, we do not include an axial guide field in the analysis due to the existence, as described in Chap. 2, of a $\mathbf{B}_0 \times \nabla\mathbf{B}_w$ drift which causes the beam to diverge. In the second place, in contrast to the case of a helical wiggler, there are no orbits in a planar wiggler for which the axial electron velocity is constant. This leads (1) to the presence of harmonic interactions within the context of a one-dimensional analysis and (2) results in the instantaneous oscillations in the power and phase of the wave due to the lower beat wave. Harmonic interactions are possible with both helical and planar wigglers but do not appear in the one-dimensional treatment of the interaction in a helical wiggler. The effect of the lower beat wave is suppressed by the symmetry properties of a helical wiggler but results in an oscillation with a period of $\lambda_w/2$ in a planar wiggler in both the power and phase of the wave.

5.3.1 *The Dynamical Equations*

The physical configuration employed in the treatment of a self-consistent model of the interaction in a planar wiggler-based free-electron laser is based upon an idealized representation of the planar magnetostatic field given in Eq. (2.2). Under the assumption that the electron displacement form the plane of symmetry is small so that $|k_w y| << 1$, this field can be approximated as

$$\mathrm{B}_w(z) = B_w(z)\hat{\mathbf{e}}_y \sin k_w z, \tag{5.66}$$

where the wiggler amplitude is assumed to increase adiabatically as in Eq. (5.14). As in the case of the helical wiggler, we shall model the injection of the beam into the wiggler. Observe that the wiggler-induced transverse velocity is in the x-direction; hence, the beam will couple to an electromagnetic wave polarized in the x-direction as well.

The formulation governing a planar wiggler configuration is closely similar to those governing the helical wiggler geometry.

5.3.1.1 The Field Equations

The vector and scalar potentials of the fluctuating electromagnetic and electrostatic waves are given in terms of a plane wave approximation in the form

$$\delta \mathbb{A}(z,t) = \delta A(z)\widehat{e}_x \cos \alpha_+(z,t), \tag{5.67}$$

and

$$\delta \Phi(z,t) = \delta \Phi(z) \cos \alpha(z,t), \tag{5.68}$$

where the amplitudes and phases are assumed to vary slowly in z and the phases are defined as in Eqs. (5.17) and (5.18). Following the procedure described in Sect. 5.1, substitution of these fields into the Maxwell-Poisson equations yields

$$\left(\frac{d^2}{dz^2} + \frac{\omega^2}{c^2} - k_+^2\right)\delta A \cos \alpha_+ - 2k_+^{1/2}\frac{d}{dz}\left(k_+^{1/2}\delta A\right) \sin \alpha_+ = -\frac{4\pi}{c}\delta J_x, \tag{5.69}$$

and

$$\left(\frac{d^2}{dz^2} - k^2\right)\delta \Phi \cos \alpha - 2k^{1/2}\frac{d}{dz}\left(k^{1/2}\delta \Phi\right) \sin \alpha = -4\pi\delta\rho, \tag{5.70}$$

where the source current and charge density are defined in Eqs. (5.23) and (5.24).

Multiplying Eq. (5.69) by cos α_+ and sin α_+ and averaging over a wave period yields

$$\left(\frac{d^2}{dz^2} + \frac{\omega^2}{c^2} - k_+^2\right)\delta A = -\frac{4\omega}{c}\int_0^{2\pi/\omega} dt\delta J_x \cos \alpha_+, \tag{5.71}$$

and

$$2k_+^{1/2}\frac{d}{dz}\left(k_+^{1/2}\delta A\right) = \frac{4\omega}{c}\int_0^{2\pi/\omega} dt\delta J_x \sin \alpha_+, \tag{5.72}$$

Substitution of the x-component of the source current into these equations yields the dynamical equations for the vector potential

$$\left(\frac{d^2}{dz^2} + \frac{\omega^2}{c^2} - k_+^2\right)\delta a = 2\frac{\omega_b^2}{c^2}\beta_{z0}\left\langle \frac{\upsilon_x}{|\upsilon_z|} \cos \alpha_+ \right\rangle, \tag{5.73}$$

and

$$2k_+^{1/2}\frac{d}{dz}\left(k_+^{1/2}\delta a\right)=-2\frac{\omega_b^2}{c^2}\beta_{z0}\left\langle\frac{v_x}{|v_z|}\sin\alpha_+\right\rangle, \tag{5.74}$$

where $\delta a = e\delta A/m_e c^2$, ω_b denotes the beam plasma frequency, $\beta_{z0} = v_{z0}/c$ (v_{z0} denotes the initial electron velocity), and the averaging operator is defined in Eq. (5.42). A similar procedure using the cos α and sin α on Poisson's equation yields the dynamical equations for the scalar potential

$$\left(\frac{d^2}{dz^2}-k^2\right)\delta\varphi=2\frac{\omega_b^2}{c^2}v_{z0}\left\langle\frac{\cos\alpha}{|v_z|}\right\rangle, \tag{5.75}$$

and

$$2k^{1/2}\frac{d}{dz}\left(k^{1/2}\delta\varphi\right)=-2\frac{\omega_b^2}{c^2}v_{z0}\left\langle\frac{\sin\alpha}{|v_z|}\right\rangle, \tag{5.76}$$

where $\delta\varphi = e\delta\Phi/m_e c^2$.

5.3.1.2 The Electron Orbit Equations

Equations (5.73)–(5.76) describe the evolution of the vector and scalar potentials and are the analogs of Eqs. (5.36)–(5.39) derived for the helical wiggler configuration. In order to complete the formulation, we must also specify the Lorentz force equations governing the trajectories of an ensemble of electrons. The Lorentz force equations can be written as

$$\frac{1}{c}\frac{d}{dz}u_x=k_wK\sin k_wz+\left(\frac{\omega}{v_z}-k_+\right)\delta a\sin\alpha_+ + \cos\alpha_+\frac{d}{dz}\delta a, \tag{5.77}$$

$$\frac{d}{dz}u_y=0, \tag{5.78}$$

$$\frac{1}{c}\frac{d}{dz}u_z=-k_wK\frac{v_x}{v_z}\sin k_wz+\frac{v_x}{v_z}\left(k_+\delta a\sin\alpha_+ - \cos\alpha_+\frac{d}{dz}\delta a\right)-\frac{c}{v_z}\times\left(k\delta\varphi\sin\alpha-\cos\alpha\frac{d}{dz}\delta\varphi\right), \tag{5.79}$$

where $K = eB_w/m_e c^2 k_w$ is the wiggler strength parameter, $\mathbf{u} = \mathbf{p}/m_e$ denotes a normalized momentum, and

$$\frac{d\psi}{dz} = k_+ + k_w - \frac{\omega}{v_z}, \tag{5.80}$$

describes the evolution of the ponderomotive phase.

5.3.1.3 The Numerical Procedure and Initial Conditions

The numerical procedure and the initial conditions associated with the integration of these dynamical equations follow analogously to that described for the helical configuration. That is, we rewrite the two sets of second-order differential equations for the amplitude and phase of the scalar and vector potentials in terms of eight first-order differential equations that are integrated simultaneously with the orbit equations for an ensemble of electrons. The initial conditions of the particles and fields are chosen as in the case of the helical wiggler as well; thus, we also use the noise level associated with the discrete initial electron distribution to determine the initial phase averages of the beam and, hence, the initial value of the scalar potential.

One difference with the numerical procedure used for the helical wiggler geometry is that Gaussian quadrature rather than Simpson's rule is used for the particle averages. The reason for this is that the Gaussian quadrature is more computationally efficient and the same accuracy is obtained using 20 electrons with Gaussian quadrature as Simpson's rule produces with 61 electrons. For this reason, Gaussian quadrature is employed in all subsequent numerical discussions.

5.3.2 *Numerical Solutions of the Dynamical Equations*

We consider an analogous case to that studied for the helical wiggler and shown in Fig. 5.4. Specifically, we choose $\Omega_0/ck_w = 0$, $N_w = 10$, $\gamma = 3.5$, and $\omega_b/\gamma^{1/2}ck_w = 0.1$. However, the wiggler field value must be increased relative to that used for the helical wiggler due to the oscillation in the wiggler-induced transverse velocity associated with the planar symmetry. Noting that it is the rms wiggler field amplitude which is significant, we choose $\Omega_w/ck_w = 0.071$, which is equivalent to the value of $\Omega_w/ck_w = 0.05$ used for the helical wiggler in the preceding section. The initial level of the vector potential is, as for the helical wiggler, chosen to be $\delta a(z = 0) = 10^{-7}$. The bulk resonance condition for these parameters is $\omega/ck_w \approx 21.6$.

The results of the simulation are shown in Figs. 5.14 and 5.15. The initial value of the scalar potential, however, is $\delta\varphi(z = 0) \approx 2.04 \times 10^{-12}$ which is much smaller than the value obtained in the helical wiggler example [for which $\delta\varphi(z = 0) \approx 1.04 \times 10^{-6}$]. The reason for this is that the initial value for the scalar potential is dependent upon the accuracy of the initial phase average and the Gaussian quadrature technique being used for the planar wiggler is more accurate than Simpson's rule that was used for the helical wiggler. Be that as it may, the results for the planar wiggler are in substantial agreement with those found for the

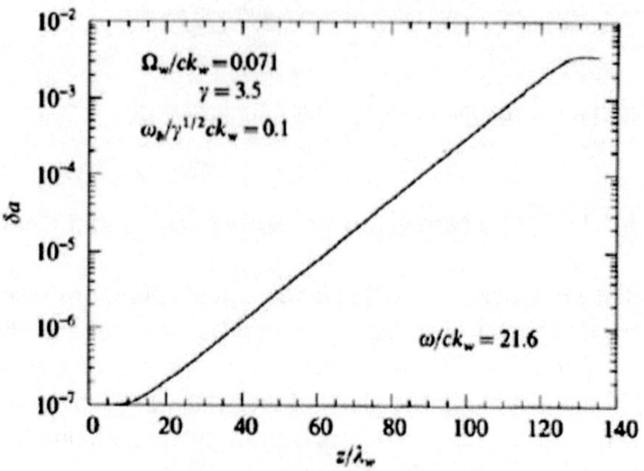

Fig. 5.15 Evolution of the amplitude of the vector potential versus axial position

helical wiggler. The evolution of the vector potential versus axial position is shown in Fig. 5.15 in which it is evident that the bulk growth of the vector potential is exponential and the amplitude grows from its' initial value to a saturation value of $\delta a_{sat} \approx 3.6 \times 10^{-3}$ over a length of 131 wiggler periods. The efficiency for planar polarization is given in terms of the normalized amplitude by one half the value shown in Eq. (5.65). This yields a maximum extraction efficiency of 3.68%. These results are comparable to that found for the helical wiggler where saturation occurred at $z_{sat}/\lambda_w \approx 127$ for $\delta a_{sat} \approx 2.6 \times 10^{-3}$ corresponding to an efficiency of 3.65%. Indeed, the bulk growth rate obtained from the figure in the exponential growth region of $30 \leq z/\lambda_w \leq 100$ is $\Gamma_{avg}/k_w \approx 0.0147$ which is very close to the value of 0.0146 found for the helical wiggler. However, a close examination of the data indicates that a fast oscillation is present that occurs with a period of $\lambda_w/2$ which corresponds to the lower beat wave. This effect will be described in more detail in regard to three-dimensional simulations of planar wiggler geometries.

5.4 Three-Dimensional Analysis: Helical Wigglers

The three-dimensional analyses of free-electron laser amplifiers have also received considerable attention [5, 29–50] from a variety of formulations. One technique for dealing with the transverse gradients associated with the wiggler and radiation fields is to use a numerical field solver for the transverse variation of the electromagnetic radiation [5, 29, 31, 32, 39]. An alternate technique is to expand the field in terms of the normal modes determined by the specific geometry of the free-electron laser. This can involve the normal modes supported by a waveguide [34–38, 40–50] or the Gaussian resonator modes supported by confocal mirrors [30, 33]. The electron trajectories have been treated both by means of a *wiggler-averaged* approximation in which the Lorentz force equations are reduced to a simplified form by means of an average over a wiggler period [5, 29–33, 39, 50] as well as by integration over the complete Lorentz force equations [34–38, 40–49]. Due to the expanded computational requirements, the number of particle-in-cell analyses of free-electron laser amplifiers in three dimensions is limited [35]. In the remainder of this section, we shall rely on a slow-timescale, steady-state formulation in which the electromagnetic wave is treated by a normal mode analysis based upon a waveguide mode expansion in conjunction with a non-wiggler-averaged orbit analysis. The formalism we adopt is described in Refs. [34, 36–38, 40] and deals with the steady-state model similar to those described in Sects. 5.1 and 5.2.

5.4.1 The General Formulation

The single-frequency analysis described in the preceding section is readily generalizable to three dimensions. The wiggler field model we employ is the general form for a helical wiggler given in Eq. (2.1)

$$\mathrm{B}_w(\mathrm{x}) = 2B_w\left[I_1'(\lambda)\widehat{\boldsymbol{e}}_r\cos\chi - \frac{1}{\lambda}I_1(\lambda)\widehat{\boldsymbol{e}}_\theta\sin\chi + I_1(\lambda)\widehat{\boldsymbol{e}}_z\sin\chi\right], \tag{5.81}$$

where the wiggler amplitude is assumed to vary adiabatically to model both the injection of the beam and the enhancement of the interaction efficiency by means of a tapered wiggler amplitude. Hence, as in Eq. (5.16), we chose

$$B_w(z) = \begin{cases} B_w\sin^2\left(\dfrac{k_w z}{4N_w}\right) & ;0 \le z \le N_w\lambda_w \\ B_w & ;N_w\lambda_w \le z \le z_0 \\ B_w[1 + k_w\varepsilon_w(z - z_0)] & ;z_0 < z \end{cases}. \tag{5.82}$$

In addition, we include the presence of an axial solenoidal magnetic field that can also be tapered to achieve an enhancement of the efficiency. A self-consistent model of a uniformly tapered solenoidal field that is both curl- and divergence-free is

$$\mathrm{B}_0(z) = \begin{cases} B_0\widehat{\boldsymbol{e}}_z & ;z \le z_0 \\ B_0\widehat{\boldsymbol{e}}_z[1 + k_w\varepsilon_0(z - z_0)] - \dfrac{1}{2}B_0k_w r\varepsilon_0\widehat{\boldsymbol{e}}_r & ;z < z_0 \end{cases}, \tag{5.83}$$

In Eqs. (5.83) and (5.84), the scale lengths for variation of the field amplitudes are assumed to be constant [i.e., linear variation and defined as

$$\varepsilon_{0,w} \equiv \frac{1}{k_w}\frac{d}{dz}B_{0,w}. \tag{5.84}$$

As a consequence, it is possible to vary the degree of taper as well as the point at which the taper begins. It should be remarked that the wiggler field model is valid [i.e., curl- and divergence-free] only as long as $\varepsilon_w << 1$; however, the tapered axial solenoidal field model is valid for arbitrary ε_0.

The radiation field can be expressed as a superposition of any convenient orthogonal basis such as waveguide or Gaussian resonator modes. We observe that the drift tube through which the electron beam passes in a free-electron laser constitutes a waveguide that specifies the boundary conditions that the electromagnetic must satisfy. For the purposes of the present analysis, therefore, we employ waveguide modes.

5.4.1.1 The Field Equations

In the treatment of a helical wiggler configuration, we assume the drift tube is cylindrical and expand the electromagnetic field in terms of a superposition of the TE and TM modes of a loss-free cylindrical waveguide in the absence of the electron beam. Thus, we write the vector potential of the radiation field in cylindrical coordinates in the form

$$\delta \mathbb{A}(\mathrm{x},t)=\sum_{\substack{l=0\\n=1}}^{\infty}\delta A_{ln}(z)\left[\frac{l}{\kappa_{ln}r}J_l(\kappa_{ln}r)\widehat{e}_r\sin\alpha_{ln}+J_l{}'(\kappa_{ln}r)\widehat{e}_\theta\cos\alpha_{ln}\right], \tag{5.85}$$

for the TE modes, and

$$\delta \mathbb{A}(\mathrm{x},t)=\sum_{\substack{l=0\\n=1}}^{\infty}\delta A_{ln}(z)\left[J_l{}'(\kappa_{ln}r)\widehat{e}_r\cos\alpha_{ln}-\frac{l}{\kappa_{ln}r}J_l(\kappa_{ln}r)\widehat{e}_\theta\sin\alpha_{ln}\right.$$
$$\left.\times+\frac{\kappa_{ln}}{k}J_l(\kappa_{ln}r)\widehat{e}_z\sin\alpha_{ln}\right], \tag{5.86}$$

for the TM modes, where for frequency ω and wavenumber $k_{ln}(z)$

$$\alpha_{ln}\equiv\int_0^z dz'k_{ln}(z')+l\theta-\omega t. \tag{5.87}$$

In Eqs. (5.85) and (5.86), J_l and $J_l^{'}$ denote the regular Bessel function of the first kind and its derivative and κ_{ln} describes the cutoff wavenumber of each mode. In the case of the TE modes, $\kappa_{ln}=x_{ln}^{'}/R_g$, where $J_l^{'}(x_{ln}^{'})=0$ and R_g is the waveguide radius. For the TM modes, $\kappa_{ln}=x_{ln}/R_g$, where $J_l(x_{ln})=0$. It is implicitly assumed that the mode amplitudes $\delta A_{ln}(z)$ and wavenumbers $k_{ln}(z)$ vary slowly over a wavelength so that

$$\left|\frac{1}{\delta A_{ln}}\frac{d}{dz}\delta A_{ln}\right| << k_{ln} \text{ and } \left|\frac{dk_{ln}}{dz}\right| << k_{ln}^2. \tag{5.88}$$

Space-charge fields are neglected at the present stage of the analysis. It should be remarked that while the inclusion of space-charge fields in the one-dimensional formulations was reasonably straightforward, that is not the case in three dimensions. The reason for the difficulty in treating space-charge waves in three dimensions is that the boundary conditions vary along the axis of symmetry with variations in the transverse cross section of the electron beam. Nevertheless, a model for treating the space-charge fields in three dimensions is discussed later in this chapter.

The microscopic source current can be written in terms of the initial conditions of the beam as in the idealized one-dimensional analysis. A three-dimensional representation of the source current is of the form

$$\begin{aligned}\delta \mathbf{J}(\mathbf{x},t) = & -en_b \iiint d^3p v_{z0} F_b(\mathbf{p}_0) \iint_{A_g} dx_0 dy_0 \sigma_\perp(x_0,y_0)\\ & \times \int_{-T/2}^{T/2} dt_0 \sigma_\parallel(t_0) \mathbf{v}(z;x_0,y_0,t_0,\mathbf{p}_0)\delta[x - x(z;x_0,y_0,t_0,\mathbf{p}_0)]\\ & \times \delta[y - y(z;x_0,y_0,t_0,\mathbf{p}_0)] \frac{\delta[t - \tau(z;x_0,y_0,t_0,\mathbf{p}_0)]}{|v_z(z;x_0,y_0,t_0,\mathbf{p}_0)|},\end{aligned} \tag{5.89}$$

where n_b is the ambient beam density, A_g $(= \pi R_g{}^2)$ denotes the cross-sectional area of the waveguide, T denotes the duration of the electron beam pulse, v_{z0} is the initial axial velocity of the electron, (x_0,y_0) describes the cross-sectional position of the electron as it crosses the $z = 0$ plane at time t_0, $\sigma_\perp(x_0,y_0)$ is the cross-sectional distribution of the beam, $\sigma_\parallel(t_0)$ is the initial distribution in entry times, $\mathbf{p}_0$ is the initial momentum of the electron, $F_b(\mathbf{p}_0)$ is the momentum space distribution of the beam, and the Lagrangian time coordinate is defined as

$$\tau(z;x_0,y_0,t_0,\mathbf{p}_0) \equiv t_0 + \int_0^z \frac{dz'}{v_z(z;x_0,y_0,t_0,\mathbf{p}_0)}. \tag{5.90}$$

The initial distributions are assumed to be subject to the normalization conditions

$$\iint_{A_g} dx_0 dy_0 \sigma_\perp(x_0,y_0) = A_g, \tag{5.91}$$

$$\int_{-T/2}^{T/2} dt_0 \sigma_{\|}(t_0) = T, \tag{5.92}$$

and

$$\iiint d^3 p v_{z0} F_b(\mathrm{p}_0) = 1. \tag{5.93}$$

Finally, we note that the generalized form of the *quasi-static assumption* is also imposed, specifically, that $\mathbf{v}(z;x_0,y_0,\mathbf{p}_0,t_0) = \mathbf{v}(z;x_0,y_0,\mathbf{p}_0,t_0 + 2\pi N/\omega)$ and $\mathbf{x}_\perp(z;x_0,y_0,\mathbf{p}_0,t_0) = \mathbf{x}_\perp(z;x_0,y_0,\mathbf{p}_0,t_0 + 2\pi N/\omega)$ for integer N.

The dynamical equations for the evolution of the slowly varying amplitude and wavenumber of each mode are obtained by direct substitution of the representations of the vector potentials into Maxwell's equations

$$\left(\nabla^2 - \frac{1}{c^2}\frac{\partial^2}{\partial t^2}\right)\delta \mathrm{A} - \nabla(\nabla \cdot \delta \mathrm{A}) = -\frac{4\pi}{c}\delta \mathrm{J}. \tag{5.94}$$

Observe that the divergence of the vector potential vanishes for the vacuum state [i.e., in the absence of the electron beam]. In the case of the TE modes, the divergence of the vector potential vanishes even in the presence of the electron beam, and substitution of the waveguide mode representation (5.84) into Maxwell's equations gives

$$\sum_{\substack{l=0\\n=1}}^{\infty} \frac{l}{\kappa_{ln} r} J_l(\kappa_{ln} r)\left[\left(\frac{d^2}{dz^2} + \frac{\omega^2}{c^2} - k_{ln}^2 - \kappa_{ln}^2\right)\delta A_{ln} \sin \alpha_{ln} \right.$$
$$\left. \times + 2k_{ln}^{1/2}\frac{d}{dz}\left(k_{ln}^{1/2}\delta A_{ln}\right)\cos \alpha_{ln}\right] = -\frac{4\pi}{c}\delta J_r, \tag{5.95}$$

and

$$\sum_{\substack{l=0\\n=1}}^{\infty} J_l'(\kappa_{ln} r)\left[\left(\frac{d^2}{dz^2} + \frac{\omega^2}{c^2} - k_{ln}^2 - \kappa_{ln}^2\right)\delta A_{ln} \cos \alpha_{ln} \right.$$
$$\left. \times - 2k_{ln}^{1/2}\frac{d}{dz}\left(k_{ln}^{1/2}\delta A_{ln}\right)\cos \alpha_{ln}\right] = -\frac{4\pi}{c}\delta J_\theta. \tag{5.96}$$

Equations (5.95) and (5.96) are reduced further by averaging them over a wave period, as in the case of the idealized one-dimensional formulation, and orthogonalization over the both the azimuthal angle and the radial position. For the TE modes, the relevant Bessel function orthogonality relation is

$$\int_0^{R_g} drr\left[J_l'(\kappa_{ln}r)J_l'(\kappa_{lm}r)+\frac{l}{\kappa_{ln}r}J_l(\kappa_{ln}r)\frac{l}{\kappa_{lm}r}J_l(\kappa_{lm}r)\right]=\frac{x_{ln}'^2-l^2}{2\kappa_{ln}^2}\delta_{n,m}J_l^2(x_{ln}'). \quad (5.97)$$

As a consequence, the dynamical equations take the form in cylindrical coordinates

$$\left(\frac{d^2}{dz^2}+\frac{\omega^2}{c^2}-k_{ln}^2-\kappa_{ln}^2\right)\delta A_{ln}=-\frac{4\kappa_{ln}^2}{c\left(x_{ln}'^2-l^2\right)J_l^2\left(x_{ln}'\right)}\frac{\omega}{\pi}\int_{-\pi/\omega}^{\pi/\omega}dt\int_0^{2\pi}d\theta\int_0^{R_g}drr$$
$$\times\left[\frac{l}{\kappa_{ln}r}J_l(\kappa_{ln}r)\delta J_r\sin\alpha_{ln}+J_l'(\kappa_{ln}r)\delta J_\theta\cos\alpha_{ln}\right], \quad (5.98)$$

and

$$2k_{ln}^{1/2}\frac{d}{dz}\left(k_{ln}^{1/2}\delta A_{ln}\right)=-\frac{4\kappa_{ln}^2}{c\left(x_{ln}'^2-l^2\right)J_l^2\left(x_{ln}'\right)}\frac{\omega}{\pi}\int_{-\pi/\omega}^{\pi/\omega}dt\int_0^{2\pi}d\theta\int_0^{R_g}drr \quad (5.99)$$
$$\times\left[\frac{l}{\kappa_{ln}r}J_l(\kappa_{ln}r)\delta J_r\cos\alpha_{ln}-J_l'(\kappa_{ln}r)\delta J_\theta\sin\alpha_{ln}\right],$$

for each TE_{ln} mode. Observe also that in the absence of the electron beam, the source currents vanish and these equations reduce to the vacuum solutions; specifically, $\omega^2/c^2 = k_{ln}^2 + \kappa_{ln}^2$ and $d(\delta A_{ln})/dz = 0$.

The analysis is more complicated for the TM modes because the assumption of a slowly varying amplitude implies that the divergence of the vector potential no longer vanishes. This arises due to the z-component of the vector potential, and we find that

$$\nabla\cdot\delta\mathbf{A}=\sum_{\substack{l=0\\n=1}}^{\infty}\kappa_{ln}J_l(\kappa_{ln}r)\frac{d}{dz}\left(\frac{1}{k_{ln}}\delta A_{ln}\right)\sin\alpha_{ln}. \quad (5.100)$$

As a consequence, substitution of the TM mode representation into Maxwell's equations yields equations of the form

$$\sum_{\substack{l=0 \\ n=1}}^{\infty} J_l'(\kappa_{ln}r)\left[\left(\frac{d^2}{dz^2}+\frac{\omega^2}{c^2}-k_{ln}^2-\kappa_{ln}^2\right)\delta A_{ln}\cos\alpha_{ln}\right.$$
$$\left.-\left(2k_{ln}^{1/2}\frac{d}{dz}\left(k_{ln}^{1/2}\delta A_{ln}\right)+\kappa_{ln}^2\frac{d}{dz}\left(\frac{1}{k_{ln}}\delta A_{ln}\right)\right)\right]\sin\alpha_{ln}=-\frac{4\pi}{c}\delta J_r, \tag{5.101}$$

$$\sum_{\substack{l=0 \\ n=1}}^{\infty} \frac{l}{\kappa_{ln}r}J_l(\kappa_{ln}r)\left[\left(\frac{d^2}{dz^2}+\frac{\omega^2}{c^2}-k_{ln}^2-\kappa_{ln}^2\right)\delta A_{ln}\sin\alpha_{ln}\right.$$
$$\left.+\left(2k_{ln}^{1/2}\frac{d}{dz}\left(k_{ln}^{1/2}\delta A_{ln}\right)+\kappa_{ln}^2\frac{d}{dz}\left(\frac{1}{k_{ln}}\delta A_{ln}\right)\right)\cos\alpha_{ln}\right]=\frac{4\pi}{c}\delta J_\theta, \tag{5.102}$$

and

$$\sum_{\substack{l=0 \\ n=1}}^{\infty} \frac{\kappa_{ln}}{k_{ln}}J_l(\kappa_{ln}r)\left[\left(\frac{\omega^2}{c^2}-k_{ln}^2-\kappa_{ln}^2\right)\delta A_{ln}\sin\alpha_{ln}+k_{ln}\frac{d}{dz}\delta A_{ln}\cos\alpha_{ln}\right]=-\frac{4\pi}{c}\delta J_z. \tag{5.103}$$

Note that the z-component of Maxwell's equations cannot be ignored in the derivation of the dynamical equations. The transverse components represented in Eqs. (5.101) and (5.102) are averaged over a wave period and orthogonalized in r and θ in a manner analogous to that used for the TE modes. In the case of the TM modes, however, the Bessel function orthogonality relation is of the form

$$\int_0^{R_g} drr\left[J_l'(\kappa_{ln}r)J_l'(\kappa_{lm}r)+\frac{l}{\kappa_{ln}r}J_l(\kappa_{ln}r)\frac{l}{\kappa_{lm}r}J_l(\kappa_{lm}r)\right]=\frac{x_{ln}^2}{2\kappa_{ln}^2}\delta_{n,m}J_l'^2(x_{ln}). \tag{5.104}$$

As a consequence, the transverse components of Maxwell's equations reduce to equations of the form

$$\left[\frac{d^2}{dz^2}+\left(1+\frac{\kappa_{ln}^2}{k_{ln}^2}\right)\left(\frac{\omega^2}{c^2}-k_{ln}^2-\kappa_{ln}^2\right)\right]\delta A_{ln}=-\frac{4}{cR_g^2J_l'^2(x_{ln})}\frac{\omega}{\pi}\int_{-\pi/\omega}^{\pi/\omega}dt\int_0^{2\pi}d\theta\int_0^{R_g}drr$$
$$\times\left[J_l'(\kappa_{ln}r)\delta J_r\cos\alpha_{ln}-\frac{l}{\kappa_{ln}r}J_l(\kappa_{ln}r)\delta J_\theta\sin\alpha_{ln}\right], \tag{5.105}$$

and

$$2\left(k_{ln}+\frac{\kappa_{ln}^2}{k_{ln}}\right)^{1/2}\frac{d}{dz}\left[\left(k_{ln}+\frac{\kappa_{ln}^2}{k_{ln}}\right)^{1/2}\delta A_{ln}\right]-\frac{\kappa_{ln}^2}{k_{ln}}\frac{d}{dz}\delta A_{ln}=\frac{4}{cR_g^2 J_l'^2(x_{ln})}\frac{\omega}{\pi}\int\limits_{-\pi/\omega}^{\pi/\omega}dt\int\limits_0^{2\pi}d\theta\int\limits_0^{R_g}drr$$
$$\times\left[J_l'(\kappa_{ln}r)\delta J_r\sin\alpha_{ln}+\frac{l}{\kappa_{ln}r}J_l(\kappa_{ln}r)\delta J_\theta\cos\alpha_{ln}\right]. \tag{5.106}$$

The z-component of Maxwell's equations as represented by Eq. (5.103) must also be averaged over a wave period and orthogonalized in r and θ. In this case, the Bessel function orthogonality relation is of the form

$$\int\limits_0^{R_g}drrJ_l(\kappa_{ln}r)J_l(\kappa_{lm}r)=\frac{R_g^2}{2}\delta_{n,m}J_l'^2(x_{ln}), \tag{5.107}$$

and we obtain the equations

$$\left(\frac{\omega^2}{c^2}-k_{ln}^2-\kappa_{ln}^2\right)\delta A_{ln}=-\frac{4k_{ln}}{c\kappa_{ln}R_g^2 J_l'^2(x_{ln})}\frac{\omega}{\pi}\int\limits_{-\pi/\omega}^{\pi/\omega}dt\int\limits_0^{2\pi}d\theta\int\limits_0^{R_g}drr$$
$$\times J_l(\kappa_{ln}r)\delta J_z\sin\alpha_{ln}, \tag{5.108}$$

and

$$k_{ln}\frac{d}{dz}\delta A_{ln}=-\frac{4}{cR_g^2 J_l'^2(x_{ln})}\frac{\omega}{\pi}\int\limits_{\pi/\omega}^{\pi/\omega}dt\int\limits_0^{2\pi}d\theta\int\limits_0^{R_g}drrJ_l(\kappa_{ln}r)\delta J_z\cos\alpha_{ln}. \tag{5.109}$$

Multiplying Eqs. (5.108) and (5.109) by ${\kappa_{ln}}^2/{k_{ln}}^2$ and adding the results to Eqs. (5.105) and (5.106), we obtain

$$\left[\frac{d^2}{dz^2}+\left(1+\frac{\kappa_{ln}^2}{k_{ln}^2}\right)\left(\frac{\omega^2}{c^2}-k_{ln}^2-\kappa_{ln}^2\right)\right]\delta A_{ln}=-\frac{4}{cR_g^2 J_l'^2(x_{ln})}\frac{\omega}{\pi}\int\limits_{-\pi/\omega}^{\pi/\omega}dt\int\limits_0^{2\pi}d\theta\int\limits_0^{R_g}drr$$
$$\times\left[J_l'(\kappa_{ln}r)\delta J_r\cos\alpha_{ln}-\left(\frac{l}{\kappa_{ln}r}\delta J_\theta-2\frac{\kappa_{ln}}{k_{ln}}\delta J_z\right)J_l(\kappa_{ln}r)\delta J_\theta\sin\alpha_{ln}\right], \tag{5.110}$$

and

$$2\left(k_{ln}+\frac{\kappa_{ln}^2}{k_{ln}}\right)^{1/2}\frac{d}{dz}\left[\left(k_{ln}+\frac{\kappa_{ln}^2}{k_{ln}}\right)^{1/2}\delta A_{ln}\right]=\frac{4}{cR_g^2J_l'^2(x_{ln})}\frac{\omega}{\pi}\int_{-\pi/\omega}^{\pi/\omega}dt\int_0^{2\pi}d\theta\int_0^{R_g}drr$$
$$\times\left[J_l'(\kappa_{ln}r)\delta J_r\sin\alpha_{ln}+\left(\frac{l}{\kappa_{ln}r}\delta J_\theta-2\frac{\kappa_{ln}}{k_{ln}}\delta J_z\right)J_l(\kappa_{ln}r)\delta J_\theta\cos\alpha_{ln}\right]. \tag{5.111}$$

Equations (5.110) and (5.111) are the dynamical equations for each TM$_{ln}$ mode and are analogous to Eqs. (5.98) and (5.99) derived for the TE$_{ln}$ modes.

The final form for the dynamical equations is obtained by substitution of the source current. It should be remarked at this point that the δ-functions in the source current are used to eliminate the integrals over r, θ, and t, and that the symmetry property illustrated in Eqs. (5.33)–(5.35) also applies in three dimensions. As a result, the dynamical equations for the TE$_{ln}$ mode take the form

$$\left(\frac{d^2}{dz^2}+\frac{\omega^2}{c^2}-k_{ln}^2-\kappa_{ln}^2\right)\delta a_{ln}=\frac{\omega_b^2}{c^2}H_{ln}\left\langle\frac{v_1T_{ln}^{(+)}+v_2W_{ln}^{(+)}}{|v_3|}\right\rangle, \tag{5.112}$$

and

$$2k_{ln}^{1/2}\frac{d}{dz}\left(k_{ln}^{1/2}\delta a_{ln}\right)=\frac{\omega_b^2}{c^2}H_{ln}\left\langle\frac{v_1W_{ln}^{(-)}-v_2T_{ln}^{(-)}}{|v_3|}\right\rangle, \tag{5.113}$$

where $\delta a_{ln}=e\delta A_{ln}/m_ec^2$, $\omega_b{}^2=4\pi e^2n_b/m_e$, and (v_1,v_2,v_3) are the components of the electron velocity in the frame rotating with the wiggler. The analogous equations for the TM$_{\text{ln}}$ mode take the form

$$\left[\frac{d^2}{dz^2}+\left(1+\frac{\kappa_{ln}^2}{k_{ln}^2}\right)\left(\frac{\omega^2}{c^2}-k_{ln}^2-\kappa_{ln}^2\right)\right]\delta a_{ln}$$
$$=\frac{\omega_b^2}{c^2}H_{ln}\left\langle\frac{v_1T_{ln}^{(+)}+v_2W_{ln}^{(+)}}{|v_3|}+2\frac{\kappa_{ln}}{k_{ln}}J_l(\kappa_{ln}r)\sin\alpha_{ln}\right\rangle, \tag{5.114}$$

and

$$2\left(k_{ln}+\frac{\kappa_{ln}^2}{k_{ln}}\right)^{1/2}\frac{d}{dz}\left[\left(k_{ln}+\frac{\kappa_{ln}^2}{k_{ln}}\right)^{1/2}\delta a_{ln}\right]$$
$$=\frac{\omega_b^2}{c^2}H_{ln}\left\langle\frac{v_1W_{ln}^{(-)}-v_2T_{ln}^{(-)}}{|v_3|}+2\frac{\kappa_{ln}}{k_{ln}}J_l(\kappa_{ln}r)\cos\alpha_{ln}\right\rangle. \tag{5.115}$$

In the preceding equations, H_{ln}, $T_{ln}{}^{(\pm)}$, and $W_{ln}{}^{(\pm)}$ are mode-dependent quantities defined as

$$H_{ln}\equiv\begin{cases}\dfrac{x_{ln}'^{\,2}}{(x_{ln}'^{\,2}-l^2)J_l^2(x_{ln}')} & ;TE_{ln}\ mode\\[2ex] \dfrac{1}{J_l'^2(x_{ln})} & ;TM_{ln}\ mode\end{cases}, \tag{5.116}$$

and

$$T_{ln}^{(\pm)}\equiv\begin{cases}F_{ln}^{(\pm)}\sin\psi_{ln}+G_{ln}^{(\pm)}\cos\psi_{ln} & ;TE_{ln}\ mode\\ F_{ln}^{(\mp)}\cos\psi_{ln}-G_{ln}^{(\mp)}\sin\psi_{ln} & ;TM_{ln}\ mode\end{cases}, \tag{5.117}$$

$$W_{ln}^{(\pm)}\equiv\begin{cases}F_{ln}^{(\mp)}\cos\psi_{ln}-G_{ln}^{(\mp)}\sin\psi_{ln} & ;TE_{ln}\ mode\\ -F_{ln}^{(\pm)}\sin\psi_{ln}-G_{ln}^{(\pm)}\cos\psi_{ln} & ;TM_{ln}\ mode\end{cases}, \tag{5.118}$$

where

$$\psi_{ln}\equiv\psi_0+\int_0^z dz'\left(k_{ln}+lk_w-\frac{\omega}{v_z}\right), \tag{5.119}$$

denotes the ponderomotive phase, $\psi_0=-\omega t_0$ is the initial phase,

$$F_{ln}^{(\pm)}\equiv J_{l-1}(\kappa_{ln}r)\cos[(l-1)\chi]\pm J_{l+1}(\kappa_{ln}r)\cos[(l+1)\chi], \tag{5.120}$$

and

$$G_{ln}^{(\pm)}\equiv J_{l-1}(\kappa_{ln}r)\sin[(l-1)\chi]\pm J_{l+1}(\kappa_{ln}r)\sin[(l+1)\chi]. \tag{5.121}$$

The particle average defined in Eqs. (5.112)–(5.115) is defined over the *initial* conditions of the beam. Observe that this formulation includes the effect of an arbitrary number of TE and TM modes as long as they all have identical frequencies.

In order to specify the particle average, we first choose an initial momentum space distribution that describes a monoenergetic beam with a pitch angle spread. This distribution corresponds to the physical model of beam thermal effects employed in the linearized theory described in Chap. 4. The difference between the present model and the linear theory is that we specify only the *initial* momentum space distribution here, which describes the beam *prior* to injection into the wiggler. The complete nonlinear formulation includes the description of the evolution of the beam in its injection, propagation through the wiggler, and interaction with the electromagnetic field. Hence, the subsequent evolution of the beam, including the possible growth of the thermal spread, is included self-consistently in the formulation. The specific distribution we choose is of the form

$$F_b(p_{z0}) = A\exp\left[-(p_{z0}-p_0)^2/\Delta p_{\|}^2\right]\delta\left(p_0^2-p_\perp^2-p_{z0}^2\right)H(p_{z0}), \tag{5.122}$$

where $H(p_{z0})$ is the Heaviside function, the normalization constant is given by

$$A = \left[\pi\int_0^{p_0} dp_{z0}\exp\left[-(p_{z0}-p_0)^2/\Delta p_{\|}^2\right]\right]^{-1}, \tag{5.123}$$

p_0 describes the total momentum of the beam, and $\Delta p_{\|}$ describes the axial momentum spread. This axial momentum spread can be related to the axial energy spread by the relation

$$\frac{\Delta\gamma_z}{\gamma_0} = 1 - \frac{1}{\sqrt{1+2\left(\gamma_0^2-1\right)\frac{\Delta p_{\|}}{p_0}}}, \tag{5.124}$$

where $\gamma_0 = (1 + p_0^2/m_e^2c^2)^{1/2}$ is the relativistic factor corresponding to the total energy. The averaging operator is defined in terms of this momentum space distribution as

$$\langle\langle(\cdots)\rangle\rangle = \frac{A}{4\pi A_g}\int_0^{2\pi} d\phi_0\int_0^{p_0} dp_{z0}\beta_{z0}\exp\left[-(p_{z0}-p_0)^2/\Delta p_{\|}^2\right]$$
$$\times\iint_{A_g} dx_0dy_0\sigma_\perp(x_0,y_0)\int_0^{2\pi} d\psi_0\sigma_{\|}(\psi_0)(\cdots), \tag{5.125}$$

where $\beta_{z0} = v_{z0}/c$, $\phi_0 = \tan^{-1}(p_{y0}/p_{x0})$, and p_{z0} defines the initial axial momentum of the electron. It is important to recognize that this average implicitly includes the effect of the overlap of the electron beam with the transverse mode structure of the radiation field [often included in one-dimensional analyses in an ad hoc manner by the inclusion of a filling-factor] in a self-consistent manner.

It is important to emphasize here that this distribution describes the effect of an *initial* momentum spread on the beam. The effective momentum spread which develops during the course of the interaction due to (1) the injection of the beam into the wiggler, (2) wiggler inhomogeneities [i.e., Larmor and betatron oscillations], and (3) the interaction of the electrons with the electromagnetic field is included implicitly in the formulation. The implicit inclusion of these effects is one of the advantages of this formulation in which the complete set of three-dimensional Lorentz force equations is integrated for the ensemble of electrons and is carried over in the discussions of planar wiggler configurations in Sect. 5.4 and the inclusion of collective effects in Sect. 5.5

5.4.1.2 The Electron Orbit Equations

The formulation is inherently multimode in the sense that an arbitrary number of TE and TM modes may be treated using the dynamical equations. Coupling between the various modes is included through the effect of the mode ensemble on the electron trajectories. Hence, in order to complete the formulation, the electron orbit equations in the presence of the static and fluctuation fields must be specified. Since we deal with an amplifier model, we choose to integrate in z and write the Lorentz force equations in the form

$$\upsilon_z \frac{d}{dz}\mathrm{p} = -e\delta\mathrm{E} - \frac{e}{c}\mathrm{v}\times(\mathrm{B}_0 + \mathrm{B}_w + \delta\mathrm{B}), \tag{5.126}$$

where the electric and magnetic fields

$$\delta\mathrm{E} = -\frac{1}{c}\frac{\partial}{\partial t}\delta\mathrm{A} \text{ and } \delta\mathrm{B} = \nabla\times\delta\mathrm{A}, \tag{5.127}$$

are derivable from the vector potentials and consist in the sum of all the TE and/or TM modes included in any specific simulation. Substitution of the appropriate forms for the vector potentials shows that in the wiggler frame

$$\begin{aligned}\upsilon_3\frac{d}{dz}p_1 = &-\{\Omega_0[1 + k_w\varepsilon_0(z - z_0)] - k_w\upsilon_3 + 2\Omega_w I_1(\lambda)\sin\chi\}p_2 \\ &+\Omega_w p_2 I_2(\lambda)\sin 2\chi - \frac{\varepsilon_0}{2}\Omega_0 p_3\lambda\sin\chi \\ &-\frac{m_e c}{2}\sum_{\text{TE modes}}\delta a_{ln}\left[(\omega - k_{ln}\upsilon_3)W_{ln}^{(-)} - 2\kappa_{ln}\upsilon_2 J_l(\kappa_{ln}r)\cos\alpha_{ln} - \Gamma_{ln}\upsilon_3 T_{ln}^{(+)}\right] \\ &-\frac{m_e c}{2}\sum_{\text{TM modes}}\delta a_{ln}\left[\left(\omega - \frac{k_{ln}^2 + \kappa_{ln}^2}{k_{ln}}\upsilon_3\right)W_{ln}^{(-)} + \Gamma_{ln}\upsilon_3 T_{ln}^{(+)}\right],\end{aligned} \tag{5.128}$$

$$
\begin{aligned}
v_3 \frac{d}{dz} p_2 = &\{\Omega_0[1 + k_w \varepsilon_0 (z - z_0)] - k_w v_3 + 2\Omega_w I_1(\lambda) \sin \chi\} p_1 \\
&- \Omega_w p_3 [I_0(\lambda) + I_2(\lambda) \cos 2\chi] + \frac{\varepsilon_0}{2} \Omega_0 p_3 \lambda \cos \chi \\
&+ \frac{m_e c}{2} \sum_{TE\ modes} \delta a_{ln} \left[(\omega - k_{ln} v_3) T_{ln}^{(-)} - 2\kappa_{ln} v_2 J_l(\kappa_{ln} r) \cos \alpha_{ln} - \Gamma_{ln} v_3 W_{ln}^{(+)} \right] \\
&+ \frac{m_e c}{2} \sum_{TM\ modes} \delta a_{ln} \left[\left(\omega - \frac{k_{ln}^2 + \kappa_{ln}^2}{k_{ln}} v_3 \right) T_{ln}^{(-)} + \Gamma_{ln} v_3 W_{ln}^{(+)} \right],
\end{aligned}
\tag{5.129}
$$

and

$$
\begin{aligned}
v_3 \frac{d}{dz} p_3 = &\Omega_w p_2 [I_0(\lambda) + I_2(\lambda) \cos 2\chi] - \Omega_w p_1 I_1(\lambda) \sin 2\chi \\
&+ \frac{\varepsilon_0}{2} \Omega_0 \lambda (p_1 \sin \chi - p_2 \cos \chi) \\
&- \frac{m_e c}{2} \sum_{TE\ modes} \delta a_{ln} \left[k_{ln} \left(v_1 W_{ln}^{(-)} - v_2 T_{ln}^{(-)} \right) + \Gamma_{ln} \left(v_1 T_{ln}^{(+)} + v_2 W_{ln}^{(+)} \right) \right] \\
&- \frac{m_e c}{2} \sum \delta a_{ln} \left[\frac{k_{ln}^2 + \kappa_{ln}^2}{k_{ln}} \left(v_1 W_{ln}^{(-)} - v_2 T_{ln}^{(-)} \right) + \Gamma_{ln} \left(v_1 T_{ln}^{(+)} + v_2 W_{ln}^{(+)} \right) \right] \\
&- \frac{m_e c}{2} \sum_{TM\ modes} \delta a_{ln} \left[\frac{k_{ln}^2 + \kappa_{ln}^2}{k_{ln}} \left(v_1 W_{ln}^{(-)} - v_2 T_{ln}^{(-)} \right) + \Gamma_{ln} \left(v_1 T_{ln}^{(+)} + v_2 W_{ln}^{(+)} \right) \right. \\
&\left. + 2\omega \frac{\kappa_{ln}}{k_{ln}} J_l(\kappa_{ln} r) \cos \alpha_{ln} \right],
\end{aligned}
\tag{5.130}
$$

where $\Omega_0 = eB_0/\gamma m_e c$, $\Omega_w = eB_w(z)/\gamma m_e c$ is determined by the variation in the wiggler amplitude as given in Eq. (5.67), and

$$
\Gamma_{ln} \equiv \frac{1}{\delta a_{ln}} \frac{d}{dz} \delta a_{ln},
\tag{5.131}
$$

defines the growth rate of the wave mode. Observe that this formulation of the Lorentz force equations includes the effects of tapered wiggler and axial solenoidal fields, and we neglect the terms in ε_0 for a uniform solenoidal field. In addition, we have that for either the TE or TM modes

$$
v_3 \frac{dx}{dz} = v_1 \cos k_w z - v_2 \sin k_w z,
\tag{5.132}
$$

$$
v_3 \frac{dy}{dz} = v_1 \sin k_w z + v_2 \cos k_w z,
\tag{5.133}
$$

and

$$\frac{d}{dz}\psi_{ln} = k_{ln} + lk_w - \frac{\omega}{v_3}. \tag{5.134}$$

Together, Eqs. (5.112)–(5.115) for an ensemble of modes and Eqs. (5.128)–(5.130) and (5.132)–(5.134) for an ensemble of particles completely specify the nonlinear problem which governs the self-consistent evolution of the wave modes and particle trajectories. As described for the case of the idealized one-dimensional formulation, the dynamical equations for the fields constitute a set of two second-order differential equations for the amplitude and phase of each mode, which can be converted into a set of four first-order differential equations. As a consequence, this formulation describes an initial value problem that consists of a set of $4N_m + 6N_p$ coupled first-order differential equations, where N_m and N_p denote the number of modes and electrons, respectively.

5.4.1.3 The First-Order Field Equations

However, a simplification of the field equations can be made under the assumption that the second-order derivatives of the amplitude and phase [which is given by the derivative of the wavenumber] can be neglected. This is generally valid within the context of the implicit assumption of a slowly varying amplitude and wavenumber. The dynamical equations for the electromagnetic wave modes can be expressed in this manner as

$$k_{ln}^2 = \frac{\omega^2}{c^2} - \kappa_{ln}^2 - \frac{\omega_b^2}{c^2}\frac{H_{ln}}{\delta a_{ln}}\left\langle \frac{v_1 T_{ln}^{(+)} + v_2 W_{ln}^{(+)}}{|v_3|} \right\rangle, \tag{5.135}$$

$$2k_{ln}\frac{d}{dz}\delta a_{ln} = \frac{\omega_b^2}{c^2}H_{ln}\left\langle \frac{v_1 W_{ln}^{(-)} - v_2 T_{ln}^{(-)}}{|v_3|} \right\rangle, \tag{5.136}$$

for the TE modes, and

$$\left(1 + \frac{\kappa_{ln}^2}{k_{ln}^2}\right)\left(\frac{\omega^2}{c^2} - k_{ln}^2 - \kappa_{ln}^2\right) = \frac{\omega_b^2}{c^2}\frac{H_{ln}}{\delta a_{ln}}$$
$$\times \left\langle \frac{v_1 T_{ln}^{(+)} + v_2 W_{ln}^{(+)}}{|v_3|} + 2\frac{\kappa_{ln}}{k_{ln}} J_l(\kappa_{ln} r)\sin\alpha_{ln} \right\rangle, \tag{5.137}$$

$$2\left(k_{ln} + \frac{\kappa_{ln}^2}{k_{ln}}\right)\frac{d}{dz}\delta a_{ln} = \frac{\omega_b^2}{c^2}H_{ln}\left\langle \frac{v_1 W_{ln}^{(-)} - v_2 T_{ln}^{(-)}}{|v_3|} + 2\frac{\kappa_{ln}}{k_{ln}} J_l(\kappa_{ln} r)\cos\alpha_{ln} \right\rangle, \tag{5.138}$$

for the TM modes. These two pairs of equations for the TE and TM modes represent an algebraic equation for the wavenumbers [Eqs. (5.135) and (5.137)] and a first-order differential equation for the amplitudes [Eqs. (5.136) and (5.138)] of each mode. In order to complete the formulation in order to specify the evolution of the phase, we define the relative phase of each mode in the form

$$\Delta\phi_{ln}(z) \equiv \int_0^z dz' \left(k_{ln}(z') - \sqrt{\frac{\omega^2}{c^2} - \kappa_{ln}^2} \right), \tag{5.139}$$

which describes the dielectric effect of the electron beam in the shift of the wavenumber from the vacuum value. The evolution of the relative phase is described by an additional first-order differential equation

$$\frac{d}{dz}\Delta\phi_{ln}(z) = k_{ln}(z) - \sqrt{\frac{\omega^2}{c^2} - \kappa_{ln}^2}. \tag{5.140}$$

This alternate formulation, therefore, consists of the simultaneous solution of $2N_m + 6N_p$ first-order differential equations as well as the evaluation of an algebraic equation for the wavenumber of each mode. As a practical consequence, this results in a reduction in computational requirements with only a minimal loss in accuracy. Indeed, for all subsequent cases considered herein, the discrepancy between these two formulations of the problem is substantially less than 10%.

It should be remarked before we proceed to the discussion of numerical examples that the time-averaged Poynting flux for each mode, P_{ln}, is related to the field amplitudes and wavenumbers through the relations

$$P_{ln} = \frac{m_e^2 c^4}{8c^2} \omega k_{ln} \frac{R_g^2}{H_{ln}} \delta a_{ln}^2, \tag{5.141}$$

for the TE modes, and

$$P_{ln} = \frac{m_e^2 c^4}{8c^2} \omega \left(\frac{k_{ln}^2 + \kappa_{ln}^2}{k_{ln}} \right) \frac{R_g^2}{H_{ln}} \delta a_{ln}^2, \tag{5.142}$$

for the TM modes. Comparison of these expressions with Eqs. (5.136) and (5.138) reveals that the dynamical equations describe the evolution of the time-averaged Poynting flux, which is a constant in the absence of the electron beam. Further, the source terms in the dynamical equations constitute a calculation of the average of $\delta\mathbf{J} \cdot \delta\mathbf{E}_{ln}$ for each mode.

5.4.1.4 The Initial Conditions

The set of coupled nonlinear differential equations can be solved numerically for an amplifier configuration in which a single wave of frequency ω is injected into the system at $z = 0$ in concert with the electron beam. The solution to this initial value problem can be accomplished by a variety of different algorithms including Adams-Moulton predictor/corrector and Runge-Kutta techniques. The advantage of the Adams-Moulton technique is that it is more stable than the Runge-Kutta algorithm; however, this occurs at the practical cost of a greatly increased memory requirement. In practice, it is found that the fourth-order Runge-Kutta-Gill technique leads to no serious numerical instabilities. Hence, either algorithm may be employed depending upon the available computational facilities. The averages in the dynamical equations for the wave modes are performed by means of an Nth order Gaussian quadrature technique in each of the variables (r_0, θ_0, ψ_0, p_{z0}, ϕ_0). In the absence of an energy spread, the number of degrees of freedom in the initial conditions reduces to (r_0, θ_0, ψ_0), and the choice of $N_r = N_\theta = N_\psi = 10$ (corresponding to 1000 particles) is typically made. The inclusion of an energy spread requires additional electrons in the numerical simulation. In this case, we find that a reduction in the number of electrons corresponding to the radial positions to $N_r = 4$ does not result in a severe loss of accuracy, and we choose a sixth- and fourth-order Gaussian quadrature to resolve the p_{z0} and ϕ_0 distributions. This gives a total number of 9600 electrons in the simulation.

The initial state of the electron beam is chosen to model the injection of a monoenergetic, uniform, axisymmetric electron beam with a flattop density profile (i.e., $\sigma_\perp = \sigma_\parallel = 1$). The electron positions are chosen by means of the Gaussian algorithm within the ranges $-\pi \le \psi_0 \le \pi$, $0 \le \theta_0 \le 2\pi$, and $R_{min} \le r_0 \le R_{max}$. In the absence of an energy spread, the momenta are chosen so that $\mathbf{p}_{\perp 0} = 0$ and $p_{z0} = m_e c(\gamma_0^2 - 1)^{1/2}$, where γ_0 is the relativistic factor corresponding to the total beam energy. This model can treat a beam with either an annular or a solid cross section. Within the context of this geometry, the plasma frequency is related to the total beam current, I_b, by means of the relation

$$\omega_b^2 = \frac{4e}{m_e v_{z0}} \frac{I_b}{R_{max}^2 - R_{min}^2}. \tag{5.143}$$

It is also required in order to self-consistently satisfy the condition necessary to the neglect of the beam space charge waves in the high-gain Compton regime that

$$\frac{\omega_b^2}{\gamma_0^{1/2} c^2 k_w^2} << \gamma_z^3 \frac{v_\perp^2}{c^2}. \tag{5.144}$$

within the uniform-wiggler region. The initial conditions on the radiation fields are chosen such that the initial amplitude of each TE or TM mode included in the simulation is chosen to correspond to an arbitrary input power via Eqs. (5.141) or (5.142) and the initial value of the wavenumber is chosen to correspond to the vacuum value [i.e., $k_{ln}(z = 0) = (\omega^2/c^2 - \kappa_{ln}^2)^{1/2}$]. Note also that the initial value of the growth rate of each mode is zero [i.e., $\Gamma_{ln}(z = 0) = 0$], which is consistent with the choice of a wiggler field amplitude that increases from zero adiabatically.

It is important to bear in mind that although the analysis is restricted to the high-gain Compton regime, collective effects are included through the dielectric response of the beam to the waveguide mode and the analysis is not purely in the single-particle regime. Thus, while the wavenumbers are initially set to the vacuum dispersion relations, the system evolves into a fully self-consistent, dielectrically loaded waveguide mode.

5.4.2 *Simulation for Group I Orbit Parameters*

The first case under consideration is that of a wideband 35 GHz amplifier operating in the TE_{11} mode. To this end, the wiggler field is assumed to be characterized by an amplitude and period of $B_w = 2$ kG and $\lambda_w = 1.175$ cm, respectively, and an entry taper of $N_w = 10$ wiggler periods is chosen for the smooth injection of the beam. The axial solenoidal magnetic field is chosen to be $B_0 = 1.3$ kG. This corresponds to beam propagation on Group I trajectories comfortably far from the transition to orbital instability that occurs at $B_0 \approx 3.9$ kG. The electron beam configuration is that of a solid (i.e., pencil) beam with an energy of 250 keV, a current of 35 A, and an initial radius $R_{max} = 0.155$ cm. These parameters ensure a relatively broad resonance band about 35 GHz in the TE_{11} mode for a waveguide radius of $R_g = 0.36626$ cm. Note that only one mode is included in this numerical example since all other modes are below cutoff in the vicinity of the TE_{11} resonance.

We first consider the case of an ideal beam with a vanishing axial energy spread (i.e., $\Delta p_{\|} = 0$). The initial distribution in the axial phase space (ψ_{ln}, $d\psi_{ln}/dz$) is shown in Fig. 5.16. Each dot in the figure represents a *phase sheet* composed of 100 electrons distributed throughout the cross section of the beam. Each phase sheet, therefore, represents a cross-sectional slice of the beam, which is chosen initially as shown in Fig. 5.17. The circle shown in Fig. 5.17 represents the waveguide wall. Each phase sheet is initially chosen to be identical; however, the subsequent evolution of the electron trajectories in the presence of the electromagnetic fields is followed self-consistently. The positions in (r_0,θ_0,ψ_0) were chosen by means of a ten-point Gaussian weighting scheme, and the nonuniformity in the positions is compensated for by a nonuniform weighting of the electrons.

The evolution of the TE_{11} waveguide mode is illustrated in Fig. 5.18 in which we plot both the power (dashed line) and growth rate (solid line) of the mode as a function of axial position for an input signal of frequency $\omega/ck_w = 1.3$ (corresponding to a frequency of 33.2 GHz) and a power of 10 W. As shown in the figure, the growth of the wave mode was approximately exponential after an initial transient period of $k_w z \leq 80$ that includes the entry taper region. During the linear phase of the interaction, the growth rate was $\Gamma_{11}/k_w \approx 0.029$, and a small increase is observed prior to saturation at $k_w z \approx 267$. The radiation power at saturation was approximately 1.87 MW for an overall efficiency of $\eta \approx 21.4\%$. A complete spectrum of the efficiency at saturation is shown in Fig. 5.19 as a function

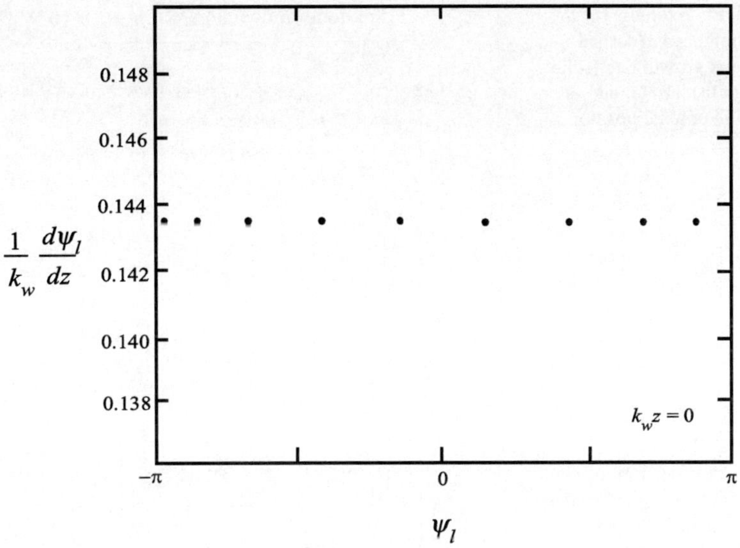

Fig. 5.16 Initialization of the axial phase space. Each point represents the superposition of 100 particles distributed throughout the cross section of the beam

Fig. 5.17 Initialization of the cross section of the beam

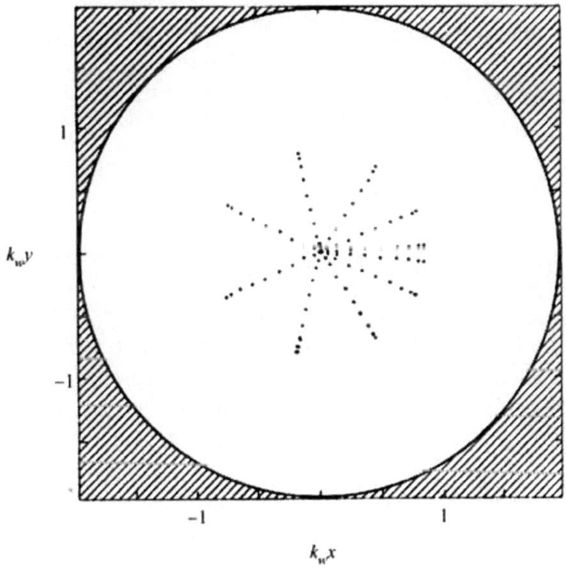

of frequency, in which the dots represent the results of the simulation. Observe that the peak efficiency occurs at $\omega_{max}/ck_w \approx 1.3$, with a bandwidth of $\Delta\omega/\omega_{max} \approx 54\%$.

The consistency of the simulation procedure has been verified in several ways. The most fundamental is the requirement of energy conservation between the

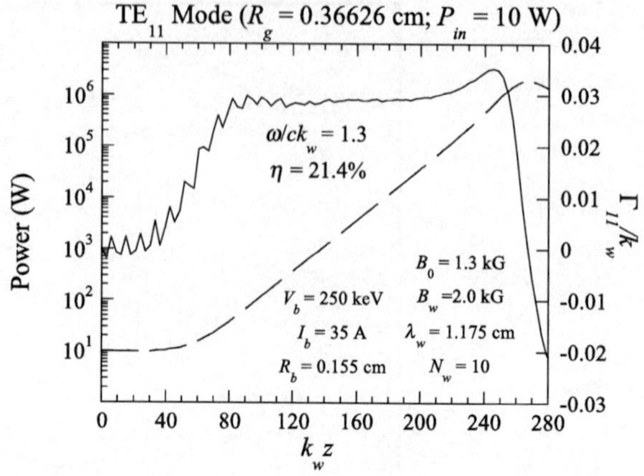

Fig. 5.18 Evolution of the radiation power (dashed line) and growth rate (solid line) of the TE_{11} mode as a function of axial position

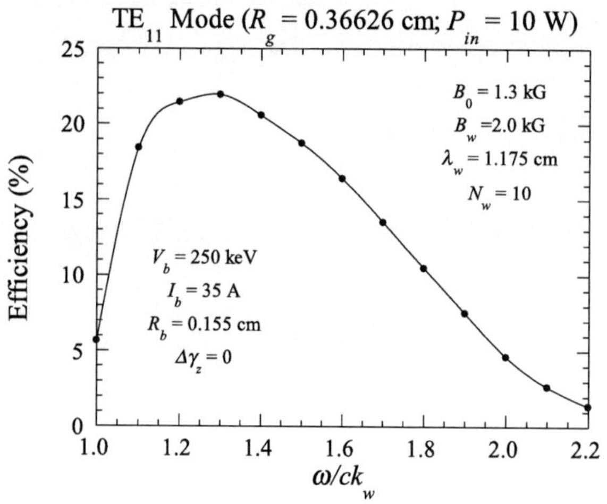

Fig. 5.19 Spectrum of the interaction efficiency of the TE_{11} mode as a function of frequency. The dots represent the results of the nonlinear simulation

electrons and the wave, and agreement between the energy lost by the beam and that gained by the wave was found to be significantly better than 0.1% in all cases considered. In addition, the growth rate in the linear regime has been checked against the results of the three-dimensional linear theory described in Chap. 4. Although the linear theory represents an idealized model in which the entry taper is not included and all electrons execute the ideal helical trajectories, agreement between the nonlinear simulation and the linear theory is good. At the frequency corresponding to peak growth (i.e., $\omega/ck_w = 1.3$), the linear theory predicts a growth rate of $\Gamma_{lin}/k_w \approx 0.031$, which compares well with the simulation result of $\Gamma_{11}/k_w \approx 0.029$. Finally, an estimate of the efficiency based upon the ad hoc phase trapping model yields an estimate of $\eta \approx 19.1\%$, which is close to the 21.4% efficiency found in the simulation.

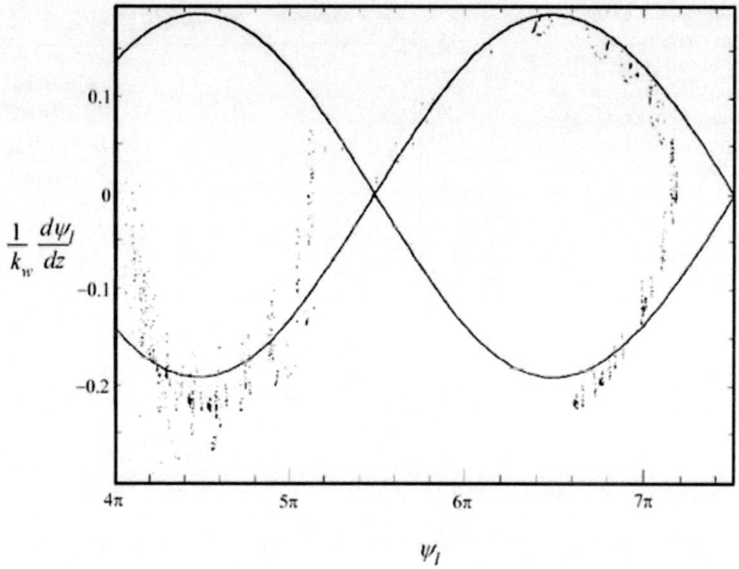

Fig. 5.20 Illustration of the axial phase space at saturation. The solid line represents an analytic approximation of the separatrix

Saturation is by means of particle trapping in the ponderomotive wave. This is clearly shown in Fig. 5.20 in which we plot the axial phase space at saturation for the case of $\omega/ck_w = 1.3$. The solid lines in the figure represent *approximate* separatrices calculated under the assumption that all electrons are executing the ideal helical trajectories. Thus, the actual number of electrons trapped may differ slightly from that shown in the figure, but the essential conclusion remains valid. Note that the entire beam has not been trapped even at saturation.

The effect of the *initial* axial energy spread upon the interaction at the frequency which corresponds to maximum efficiency is shown in Fig. 5.21, in which we plot the efficiency versus $\Delta\gamma_z/\gamma_0$. The decrease in the efficiency with the increase in the axial energy spread is quite rapid, and the efficiency is found to decrease by an order of magnitude as for an energy spread of 2.6%. In addition, the slope of this decrease is seen to change abruptly for $\Delta\gamma_z/\gamma_0 \approx 1.1$–$1.5\%$ corresponding to the point at which thermal effects become important. In order to understand this, we observe that since $|\mathrm{Im}\, k|/k_w \approx 0.029$ and $\mathrm{Re}\, k/k_w \approx 0.88$ for this case, the critical axial velocity spread for which thermal effects become important is estimated to be $\Delta\upsilon_{\|}/\upsilon_{\|} \approx |\mathrm{Im}\, k|/(k_w + \mathrm{Re}\, k) \approx 1.5\%$. Using Eq. (5.108) to relate this axial velocity spread to the axial energy spread, we find that this corresponds to a $\Delta\gamma_z/\gamma_0 \approx 1.8\%$ which is in reasonable agreement with the results of the nonlinear simulation.

The detailed variation of the relative phase is shown in Fig. 5.22 as a function of axial position for a range of frequencies spanning the interaction spectrum, and the arrows indicate the positions at which the power saturates. Two features should be noted from the figure. The first is that it is possible to find a frequency that

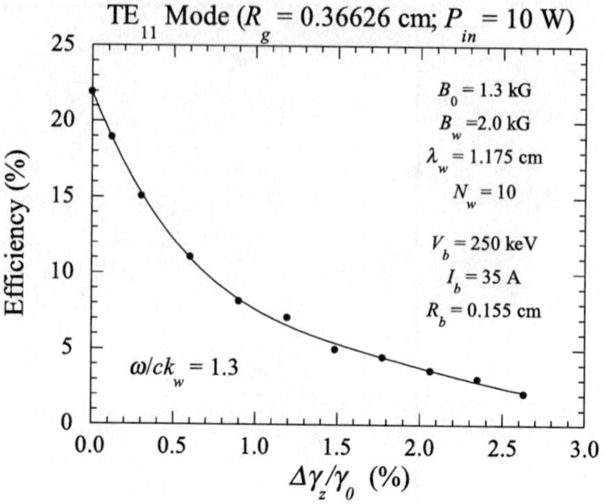

Fig. 5.21 Graph of the efficiency versus axial energy spread at the frequency of maximum efficiency. The dots represent the nonlinear simulation

corresponds to a relatively small change in the relative phase over the course of the interaction. As shown in the figure, this corresponds to a frequency of $\omega/ck_w = 1.6$ at which the wavenumber is shifted relatively little by the dielectric effect of the beam in the wiggler. While this frequency is higher than the frequency of maximum saturation efficiency (i.e., $\omega/ck_w = 1.3$), it is slightly below the frequency of maximum growth rate, which occurs in the range of $\omega/ck_w \approx 1.7$–1.8 for the present choice of parameters. The second feature is that the frequency at which the phase remains relatively constant represents a transition point. Above this frequency, the relative phase increases monotonically with axial position reflecting an increase in the wavenumber with respect to the vacuum mode. In contrast, the relative phase decreases with axial position up to the point at which saturation occurs at lower frequencies. This decrease in the relative phase occurs because the wavenumber is downshifted with respect to the vacuum wavenumber in this regime.

On the basis of the rapid variation in the phase with axial position shown in Fig. 5.22 for the bulk of the interaction spectrum, it is to be expected that the phase observed at saturation can vary widely with beam voltage. Indeed, the typical variation in the phase at saturation with beam voltage across the unstable frequency spectrum ranges from a low of 44° per 1% variation in the beam voltage at $\omega/ck_w = 1.7$ to a high of 63° per 1% variation in the beam voltage at $\omega/ck_w = 1.3$.

The beam dynamics within the interaction region are illustrated by the evolution of the beam cross section over the course of the interaction. The beam is spun-up during the entry taper region ($k_w z \leq 62.8$), and the beam cross section at a point just after the start of the uniform-wiggler region is shown in Fig. 5.23. It is evident from this figure that, in comparison with the initial state shown in Fig. 5.16, the center of the beam is displaced from the waveguide symmetry axis. The injection process yields orbits close to the ideal helical trajectories and $k_w r_{center} \approx 0.25$ for the beam that is in good agreement with the calculated value for the Group I orbit for these

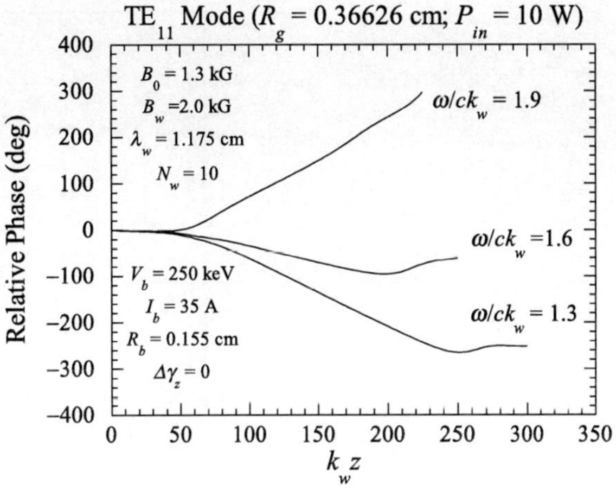

Fig. 5.22 Graph of the relative phase versus axial position at the three frequencies. The arrows indicate the saturation points for the interaction

parameters. In addition, the beam has been substantially compressed due to the focusing effect of the helical wiggler. Finally, the beam executes rotational motion about the centroid that occurs due to the combined effects of a variation in the wiggler-induced transverse velocity across the beam and the betatron oscillation due to the transverse wiggler gradient.

The overall bulk motion of the beam is expected to display a periodicity at the wiggler period and to twist the beam into a helix about the axis of symmetry. Such a motion is observed as shown in Figs. 5.24, 5.25, 5.26, and 5.27, which displays the beam cross section as it evolves through one wiggler period from $k_w z = 150$ to 156. Note that this is within the linear [i.e., exponential growth] regime of the interaction and substantial amplification of the signal has occurred.

The case of the TM_{11} mode can also be investigated. In order to facilitate comparison with the TE_{11} mode, all the parameters used in Fig. 5.18 have been retained. The only alteration is in the choice of the waveguide radius in order to shift the waveguide dispersion curve relative to the beam resonance line [i.e., $\omega \approx (k + k_w)v_\parallel$] so as to maintain the interaction in the same frequency range. We choose $R_g = 0.76223$ cm so that the waveguide cutoff is identical to that of the TE_{11} mode for $R_g = 0.36626$ cm. Numerical results indicate that the peak efficiency $\eta \approx 5.92\%$ occurs at a frequency $\omega/ck_w \approx 1.78$. This is significantly lower than the peak efficiency found for the TE_{11} mode and is shifted relative to the TE_{11} mode peak (at $\omega/ck_w \approx 1.3$). The growth rate was also significantly lower, and we observe a growth rate of $\Gamma_{11}/k_w \approx 0.012$ at the frequency corresponding to peak efficiency.

5.4.3 Numerical Simulation for Group II Orbit Parameters

The second case under consideration also corresponds to a 35 GHz amplifier but for a set of parameters consistent with Group II orbits in the negative-mass

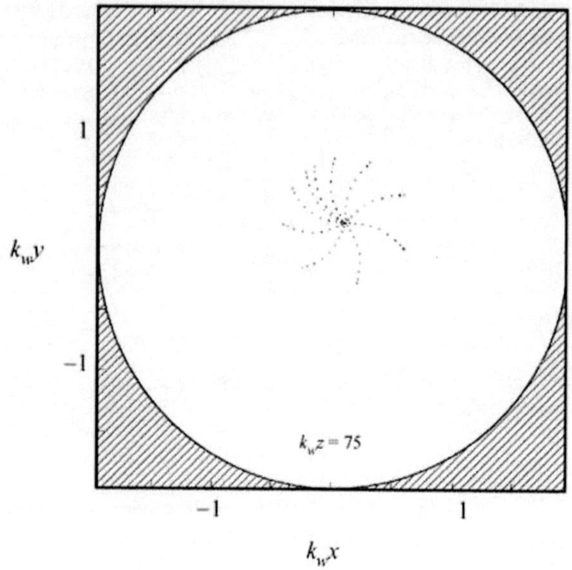

Fig. 5.23 The beam cross section at $k_w z = 75$ shortly after the start of the uniform-wiggler region

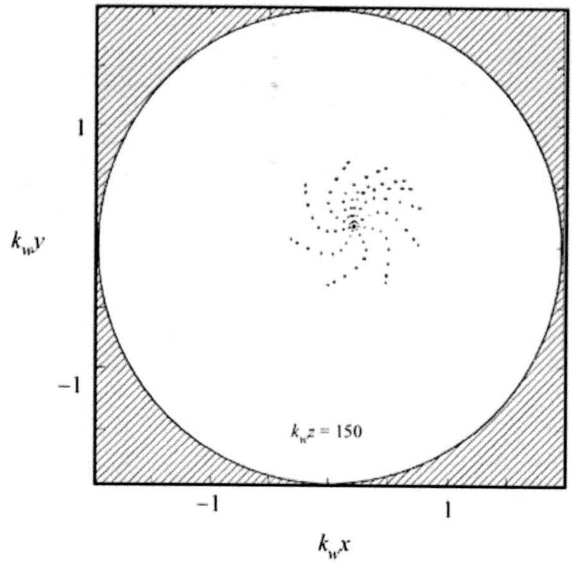

Fig. 5.24 The beam cross section at $k_w z = 150$ in the linear phase of the interaction

regime [34]. In this case, we choose a wiggler field with an amplitude of $B_w = 1.0$ kG, a period of $\lambda_w = 3.0$ cm, and an entry taper of $N_w = 10$ wiggler periods. The beam is characterized by an energy of 1.0 MeV, a current of 50 A, and a beam radius of 0.2 cm. The sole interaction is with the TE_{11} mode of a waveguide with a radius $R_g = 0.5$ cm. For these parameters, the Group II negative-mass regime is found for

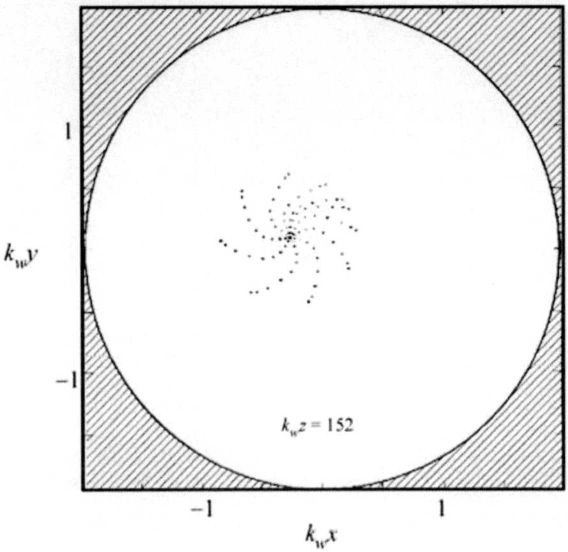

Fig. 5.25 Representation of the beam cross section at $k_w z = 152$

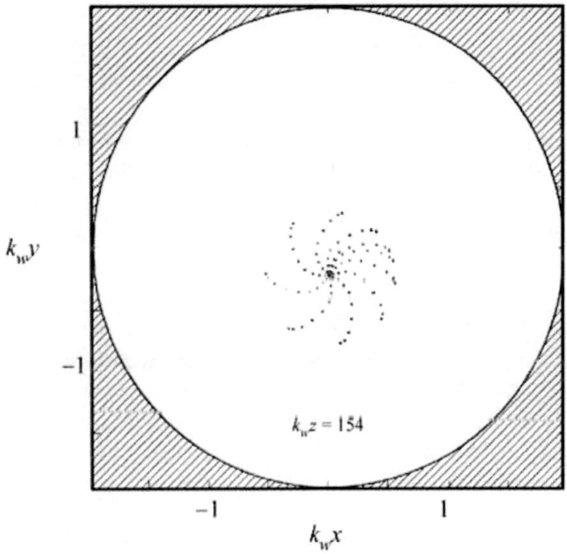

Fig. 5.26 Representation of the beam cross section at $k_w z = 154$

axial solenoidal fields in the range $B_0 \approx 8.2$–12.0 kG. It is difficult to inject a beam for axial magnetic field levels that are extremely close to the magneto-resonance at $\Omega_0 \approx k_w v_\parallel$; hence, we choose to operate at $B_0 = 11.75$ kG which is close to the extreme high-field region of the negative-mass regime.

The results of a series of numerical simulations for these parameters are shown in Fig. 5.28 in which we plot the saturation efficiency as a function of frequency for an

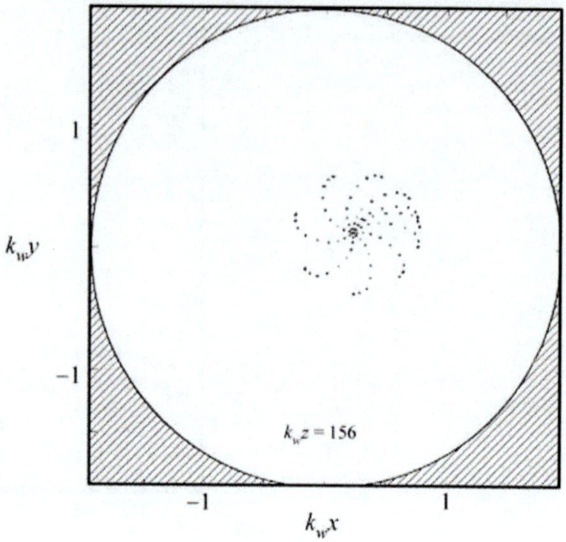

Fig. 5.27 Representation of the beam cross section at $k_w z = 156$

ideal beam in which $\Delta\gamma_z = 0$. It should be remarked that the transverse wiggler-induced velocity in this case $|v_w/v_{\|}| \approx 0.45$ is very large and the beam is shifted very close to the waveguide wall. As a result, a small fraction of the beam is lost to the wall. The procedure followed when this occurs is to eject such particles from the simulation, and the efficiencies shown reflect this effect. The peak efficiency obtained for this example was $\eta \approx 47\%$ at a frequency of $\omega/ck_w = 3.0–3.1$ (corresponding to approximately 31 GHz), with a bandwidth of approximately $\Delta\omega/\omega \approx 57\%$ extending over the range of $\omega/ck_w \approx 28–45$ GHz. It should be remarked for future reference that this efficiency has been obtained without recourse to a tapered wiggler field.

The small degree of variation in the efficiency about the line of best fit shown in the figure is difficult to account for precisely due to the complexity of the system of coupled equations. Some of the causes of the additional structure include (1) variations in the number of particles which are lost to the wall and the axial positions at which they are lost, (2) the variation in the ponderomotive potential with the frequency at which the interaction occurs, and (3) variations in the evolution of the radial profile of the beam near saturation which have an effect upon both the effective filling factor and trapping fraction.

The effect of an axial energy spread upon the interaction is illustrated in Fig. 5.29 in which the efficiency is plotted as a function of $\Delta\gamma_z$ at the frequency of maximum efficiency (i.e., $\omega/ck_w = 3.1$). The efficiency decreases much more slowly with the axial energy spread than was found in Figs. 5.20 and 5.21 corresponding to Group I orbit parameters. In this case, an axial energy spread of greater than 20% is required for the efficiency to decrease by an order of magnitude. The reason for this is that the transverse wiggler-induced velocity and, hence, the growth rate are extremely large;

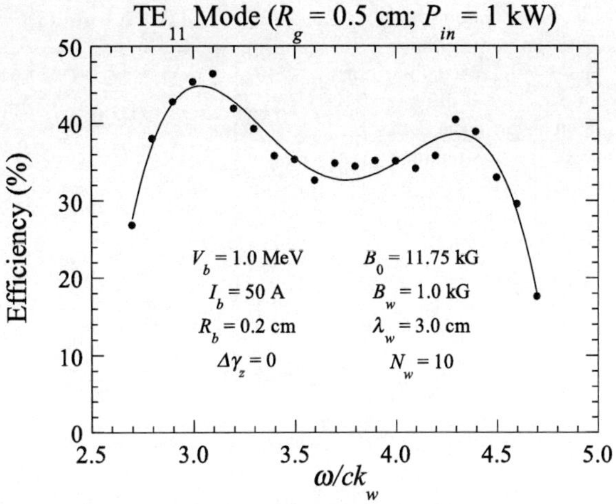

Fig. 5.28 Spectrum of the interaction efficiency as a function of the frequency for the TE_{11} mode corresponding to Group II negative-mass parameters

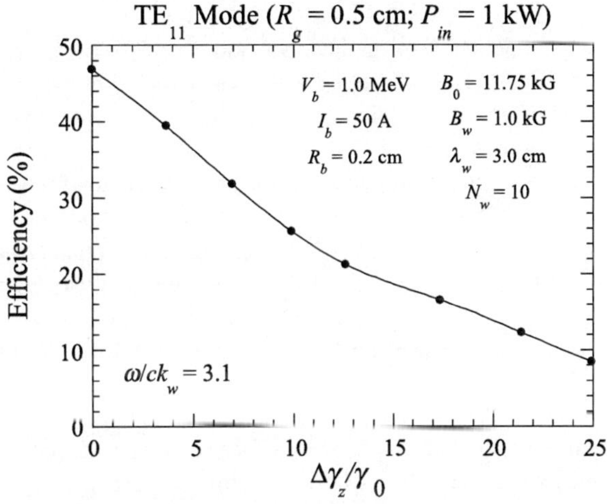

Fig. 5.29 Graph of the variation in the saturation efficiency with the axial energy spread for the Group II negative-mass regime

this is close to the magneto-resonance. Indeed, the growth rate and wavenumber corresponding to the case of $\omega/ck_w = 3.1$ were $|\mathrm{Im}\, k|/k_w \approx 0.045$ and $|\mathrm{Re}\, k|/k_w \approx 2.56$ which implies that thermal effects become important when $\Delta v_\parallel/v_\parallel \approx 1.26\%$. This is equivalent to an axial energy spread of $\Delta\gamma_z/\gamma_0 \approx 8.5\%$. As in the preceding case shown for the Group I orbit parameters, Fig. 5.29 shows that the decrease in the efficiency with the axial energy spread is monotonic but exhibits a discontinuity in the slope that corresponds to the onset of the thermally dominated regime.

The variation in the relative phase with axial position for a variety of frequencies ($\omega/ck_w = 3.1, 3.5, 4.0$, and 4.3) across the unstable spectrum is shown in Fig. 5.30 for the ideal beam limit. As in Fig. 5.22, the arrows in the figure denote the points at which the power saturates. The sudden variations that are evident in the relative phase over the entry taper region ($k_w z \le 62.8$) are associated with the rapid and

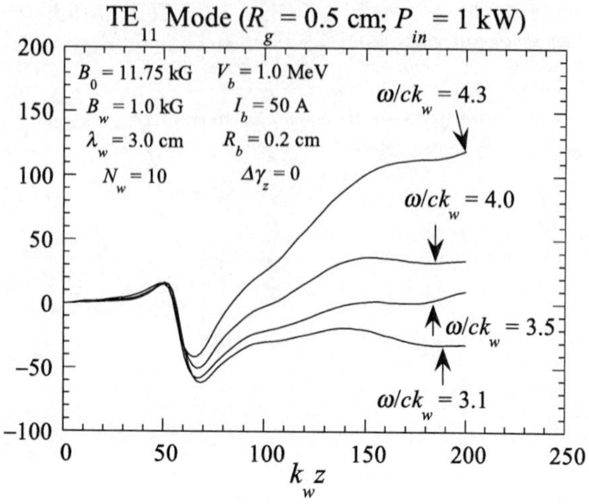

Fig. 5.30 Graph of the variation in the relative phase with the axial position for a variety of different frequencies in the Group II negative-mass regime

substantial decrease in the axial velocity of the beam with the increase in the wiggler amplitude. The subsequent variation shows that the relative phase increases up to a point just prior to that at which the power saturates, after which the phase displays an oscillatory character.

It is important to note that the specific values for the relative phase at any given stage of the interaction are highly sensitive to both the wave frequency. An alternate way of viewing this is to hold the frequency fixed but to vary the beam energy. In this way, the phase stability of the interaction against fluctuations in the beam voltage can be determined. For this example, the dependence of the relative phase on beam voltage (at the position corresponding to the saturation of the signal) varies with frequency. The best case (found for $\omega/ck_w = 3.1$) exhibits a 33° variation in the relative phase per 1% shift in the beam voltage, while the worst case shows a 60° variation in the relative phase. These estimates are comparable to those found in the preceding example for Group I orbit parameters.

The extremely high efficiencies and relative insensitivities to thermal effects that are evident in the Group II negative-mass regime raise the question of the essential limitations of this operating regime. On the one hand, the Group II negative-mass regime requires sufficiently strong axial magnetic fields that $\Omega_0 \geq ck_w$. As a consequence, the axial magnetic field increases in direct proportion with beam voltage and in inverse proportion with the wiggler period. In practical terms, this prohibits the use of this operational regime for wavelengths in the submillimeter regime with the presently available magnet technology because the required beam voltages are too high. On the other hand, the negative-mass regime contracts to an extremely narrow range of axial magnetic fields (in the vicinity of the magneto-resonance) as the voltage decreases. This implies that it will be difficult to inject the beam onto stable helical trajectories if the beam voltage is too low. Since many applications also require a low beam voltage, the essential constraints in this limit are of great interest.

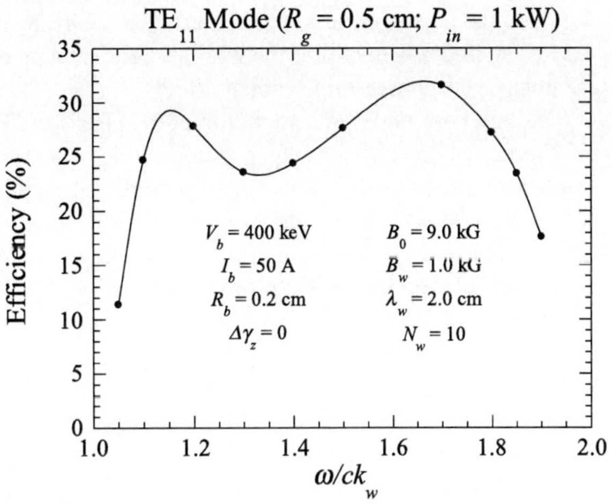

Fig. 5.31 Graph of the efficiency versus frequency for the Group II negative-mass regime corresponding to a beam energy of 400 keV

At low beam voltages, the narrow range of axial magnetic fields under which the negative-mass regime can be accessed implies that the interaction will be extremely sensitive to fluctuations in all the operating parameters. For operation in the vicinity of 35 GHz as described above, the low-energy limit on stable operation in this regime appears to be in the neighborhood of 400 keV. However, for operation above this energy, this resonant interaction appears to be a realizable approach to high-efficiency free-electron laser amplifiers without the necessity of tapered wiggler fields. An example of the spectrum of the efficiency versus frequency is shown in Fig. 5.31 for the case of a 400 keV electron beam. Observe that for the parameters indicated in the figure, the efficiency approaches 30% for frequencies in the range of $\omega/ck_w \approx 1.0$–3.0 (corresponding to between 5 and 15 GHz).

5.4.4 Numerical Simulation for the Case of a Tapered Wiggler

The nonlinear formulation can also treat the case of efficiency enhancement by either a tapered wiggler or axial solenoidal field. As shown in Eq. (5.12) for an idealized one-dimensional model, the sense of the taper of the wiggler and/or axial magnetic fields depends upon whether the electrons trajectories are of Group I or Group II. For small axial magnetic fields and Group I trajectories, both the wiggler field and axial magnetic fields must be tapered downward in order to achieve an efficiency enhancement. In contrast, for strong axial magnetic fields and Group II trajectories, the situation is more complicated, and two distinct regimes are found. When $\Phi > 0$ [see Eq. (2.70)], the axial field must be tapered upward in order to achieve an efficiency enhancement, while the wiggler field must be tapered downward. In contrast, when

$\Phi < 0$ in the negative-mass regime, efficiency enhancement is found when the axial field is tapered downward or the wiggler field is tapered upward. These conclusions are unaltered by three-dimensional effects.

The first case under consideration corresponds to Group I orbit parameters, and we choose a wiggler field with an amplitude $B_w = 1.0$ kG, a period $\lambda_w = 4.0$ cm, and an entry taper region $N_w = 6$ wiggler periods in length. The axial solenoidal field is assumed to be absent. An electron beam with an energy of 750 keV, a current of 200 A, and an initial beam radius of 0.5 cm is assumed to propagate through a waveguide of radius $R_g = 1.5$ cm. The resonant interaction is with the TE_{11} mode in the vicinity of $7.7 < \omega/ck_w < 9.3$. We consider the case in which $\omega/ck_w = 8.3$ (corresponding to 62.3 GHz). All other modes are either evanescent or nonresonant in this frequency range, and we may restrict the numerical analysis solely to the TE_{11} mode. The evolution of the power with axial position in the TE_{11} mode for a uniform and tapered wiggler [with $\varepsilon_w = -0.006$] is shown in Fig. 5.32 for the case of an ideal beam with $\Delta\gamma_z = 0$. In the case of a uniform wiggler with an input power of 10 W, the signal grows exponentially to saturation at $k_w z \approx 229$ with a power level of 10.7 MW. This corresponds to an efficiency of $\eta \approx 7.1\%$. In the tapered wiggler cases, the signal is allowed to grow from an initial level of 10 W to a point just short of saturation before the taper is begun. It is found that the maximum efficiency enhancement is sensitive to the *start point* of the taper. In practice, the taper must begin at the point at which the bulk of the electrons cross the separatrix onto trapped orbits. The ultimate efficiency enhancement is degraded if the start point is chosen either (1) prior to the point at which the electrons become trapped or (2) if the uniform-wiggler interaction has become saturated (i.e., if the electrons have undergone a rotation through half of the cycle in the ponderomotive wave). For the present case, the ideal start point is located at $k_w z \approx 220$. In addition, the maximum efficiency is also dependent upon the slope of the taper, and too high a slope of the taper is as bad as too low. The evolution of the power for the tapered wiggler case yielding the optimum efficiency enhancement is shown in the figure. In this case, the power can be enhanced almost double to approximately 20.2 MW for an efficiency of $\eta \approx 13.4\%$.

We now consider the case of Group II negative-mass orbit parameters for an axial magnetic field of $B_0 = 14.0$ kG, and a wiggler field of amplitude $B_w = 1.0$ kG, a period $\lambda_w = 3.0$ cm, and an entry taper region $N_w = 10$ wiggler periods. The electron beam parameters for this case are an energy of 1.25 MeV, a current of 50 A, and an initial radius of 0.25 cm. As before, we consider the TE_{11} mode but with a waveguide radius of $R_g = 1.0$ cm, which results in amplification over the frequency range of $7.5 < \omega/ck_w < 8.5$. We choose $\omega/ck_w = 8.0$ (corresponding to 80 GHz) for purposes of illustration. The saturation efficiency for a uniform wiggler is found to be approximately $\eta \approx 5.47\%$ for a total power of 3.42 MW. For an input power of 10 W, the saturation point was found to be $k_w z = 253$. A near-optimum example of efficiency enhancements by means of a tapered wiggler is shown in Fig. 5.33 in which we plot the evolution of the power as a function of the axial position for both a uniform and tapered wiggler ($\varepsilon_w = 0.001$ and $k_w z_0 = 247$). Observe that in the negative-mass regime the wiggler must, indeed, be tapered upward in order to

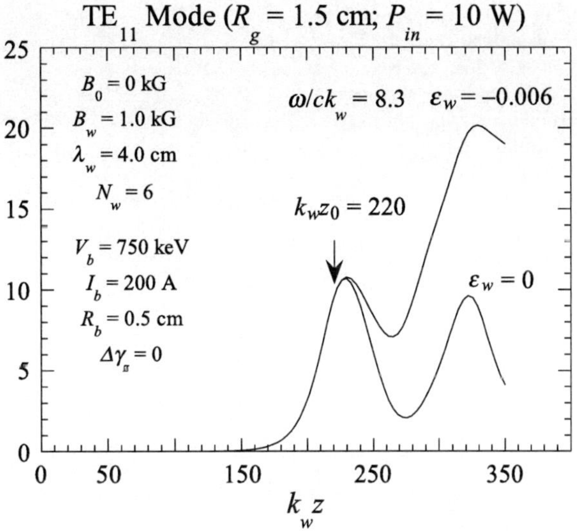

Fig. 5.32 Evolution of the radiation power in the TE_{11} mode versus axial position for Group I orbit parameters and uniform and tapered wigglers

enhance the efficiency. As shown in the figure, the efficiency may be enhanced to a level of approximately $\eta \approx 37.0\%$ for an output power of 23.1 MW, an enhancement of almost 700%.

The case of a tapered axial magnetic field is shown in Fig. 5.34 for the identical parameters illustrated in Fig. 5.33. In this case, however, we plot the evolution of the power with axial position for uniform fields and a tapered axial magnetic field with a normalized slope of $\varepsilon_0 = -0.0005$ and a start point of $k_w z_0 = 247$. As is evident in the figure, the enhanced efficiency reaches a level of $\eta \approx 16.9\%$ for a maximum power of 10.6 MW. While this is substantially lower than that found for a tapered wiggler in Fig. 5.33, it should be observed that in this regime the wiggler must be tapered upward, while the axial field must be tapered downward. As a result, there is a maximum efficiency enhancement possible for the tapered axial field associated with the point at which the axial magnetic field vanishes.

5.4.5 *Comparison with Experiment: A Submillimeter Free-Electron Laser*

The nonlinear formulation has been compared with an intense beam free-electron laser experiment conducted at the Massachusetts Institute of Technology with good results [51, 52]. This experiment operated in what was called at the time a *superradiant* amplifier in which the signal grows from noise. This experiment employed a helical wiggler field with a period of 3.14 cm, an amplitude that could be varied up to almost 2 kG, and an entry taper region six wiggler periods in length. No axial magnetic field was imposed. It should be remarked that the term

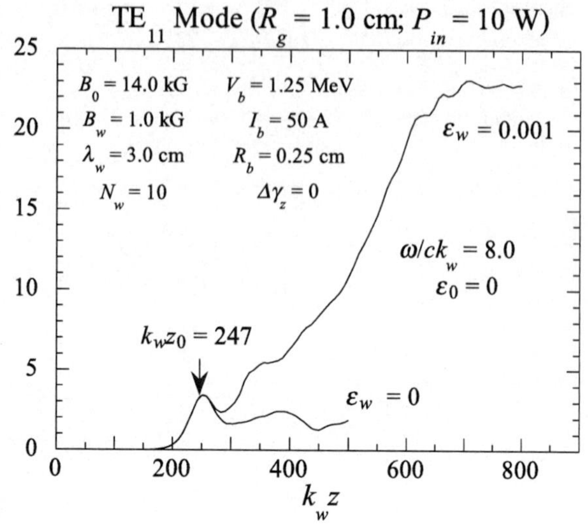

Fig. 5.33 Evolution of the radiation power in the TE_{11} mode versus axial position for Group II negative-mass orbit parameters and uniform and tapered wigglers

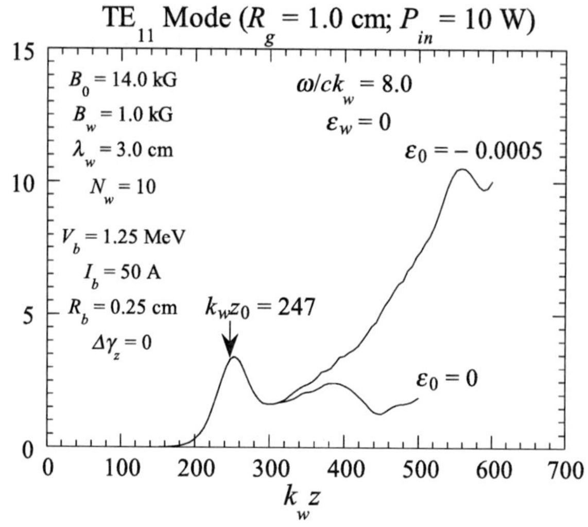

Fig. 5.34 Evolution of the radiation power in the TE_{11} mode versus axial position for Group II negative-mass orbit parameters and uniform and tapered axial fields

superradiant amplifier was used at the time of the experiment to denote SASE; however, the term has subsequently gone out of use. The electron beam was produced by a Physics International Pulserad 110 A accelerator in concert with a five-stage multielectrode field-emission electron gun to produce a beam with a pulse length of 30 ns, a maximum energy of up to 2.5 MeV, and a maximum current of up to 7 kA. The slippage time for this experiment was about 89 ps, which is much less than the electron bunch time; hence, the steady-state approximation is suitable for this experiment. The beam was then focused and apertured to produce a high-quality beam with a current of up to approximately 1 kA, which was then propagated through a drift tube with an inner diameter of 0.8 cm. The output of the free-electron laser was found to range in frequency form approximately 200–500 GHz at power levels of the order of 18 MW in the neighborhood of 470 GHz. Estimates of the beam quality were made on the basis of the electron gun geometry and produced estimates of the axial energy spread of approximately $\Delta\gamma_z/\gamma_0 \approx 0.22$–$0.40\%$.

Measurements of the growth rate were performed by means of a kicker magnet that dumped the electron beam to the wall at various points along the drift tube. Two examples are shown in Fig. 5.35 for wiggler field amplitudes of 1510 G and 1275 G, respectively, for a beam energy of 2.0 MeV, current 780 A, and equilibrium radius of approximately 0.4 cm. As shown in the figures, the gains for these two cases were in the neighborhood of 70 ± 6 dB/m and 56 ± 3 dB/m, respectively. The polarization of the output radiation has been determined to be primarily in the fundamental TE_{11} mode of the waveguide.

The variation in the gain with axial energy spread for these two cases has been studied using the numerical simulation, and the results are shown in Fig. 5.36. These calculations were made using only the TE_{11} mode and with an initial beam radius of 0.4 cm. No phenomenological fit parameters are necessary. The maximum gain is found with a vanishing axial energy spread. For the case of $B_w = 1275$ G, the calculated gain is found to be approximately 62 dB/m for $\Delta\gamma_z/\gamma_0 = 0$ and decreases to 56 dB/m for $\Delta\gamma_z/\gamma_0 = 0.25\%$. Similarly, for $B_w = 1510$ G, the gains calculated for $\Delta\gamma_z/\gamma_0 = 0$ and $\Delta\gamma_z/\gamma_0 = 0.25\%$ are 70 dB/m and 64 dB/m, respectively. When compared with the measured gains, these results are consistent with the interpretation of an axial beam energy spread of less than or of the order of 0.25%. This is also in excellent agreement with the estimates of beam quality to be obtained from the electron gun. Hence, the value of $\Delta\gamma_z/\gamma_0 = 0.25\%$ is used henceforth.

The measured output spectrum for the case of a 2.3 MeV, 930 A electron beam and a wiggler field amplitude of 1275 G is shown in Fig. 5.37, together with the results of the nonlinear simulation. Total measured output power is of the order of 18 MW for this example. In this case, the simulation included the TE_{11} and TM_{11} modes. Since the nonlinear formulation is restricted to single-frequency propagation, the numerical procedure used to generate the theoretical curve was to consider each frequency separately and to plot the combined output spectrum normalized to the peak power. As can be seen in the figure, the spectral agreement is excellent, and the measured and computed spectral peak differed by less than 2%.

Finally, it should be observed that space-charge effects were not required to treat this intense beam experiment, even though it should fall in the borderline region between the high-gain Compton and collective Raman regimes based upon the well-known criterion derived from the idealized one-dimensional formulation [Eqs. (4.98) and (4.102)]. We conclude, therefore, that three-dimensional effects imposed by the wiggler and bounded geometry of the waveguide act to reduce the effective plasma frequency and shift the experiment closer to the high-gain Compton regime.

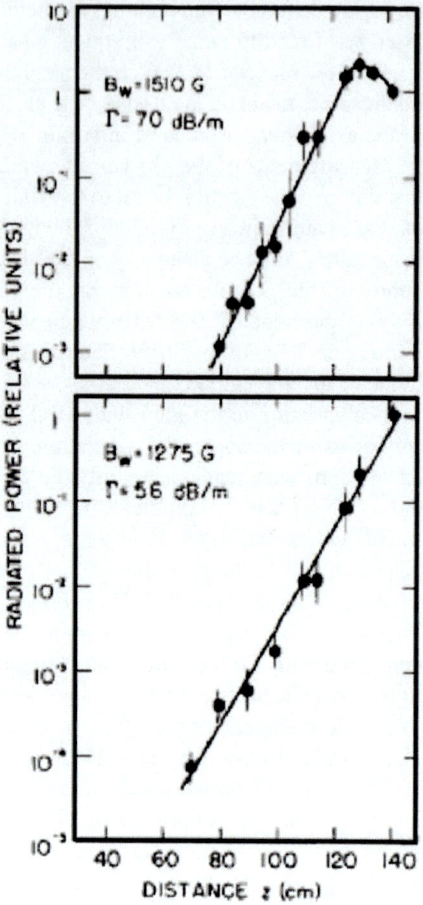

Fig. 5.35 Observations of the output power as a function of axial position for two wiggler fields (**a**) 1510 G and (**b**) 1275 G with a beam energy of 2.0 MeV and current of 780 A [51, 52]

5.5 Three-Dimensional Analysis: Planar Wigglers

Planar wiggler configurations have been treated from a variety of standpoints. Due to their ease of construction and adjustment in comparison with helical wigglers [which can only be adjusted as to their bulk field magnitude], planar wigglers are employed in a wide variety of experimental configurations including both amplifiers and oscillators and low- and high-gain systems. Here, we concentrate on high-gain amplifier configurations. Nonlinear amplifier analyses in both two and three dimensions based upon planar wiggler designs have been largely based upon an analysis in which the orbit equations are averaged over a wiggler period [29–31, 33, 39, 49, 53].

The formulation employed herein follows that described in Refs. [42–46, 48] and is similar to that described for helical wigglers in that the electron orbit equations are not averaged over a wiggler period. In this section, we treat the interaction by means

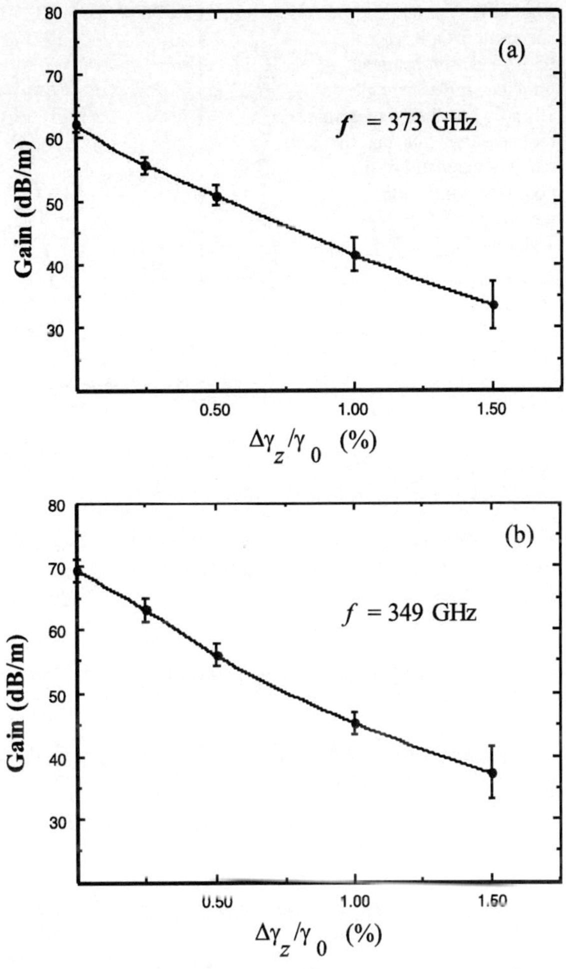

Fig. 5.36 Calculated gains versus axial energy spread for the two wiggler amplitudes shown in Fig. 5.33 [51, 52]

of a steady-state, single-frequency formulation. Recall that the free-electron laser operates by means of the beating of the wiggler and radiation fields. In normal operation, it is the upper-frequency beat wave that results in the Doppler-upshifted interaction. In the case of helical wiggler geometries, the lower-frequency beat wave vanishes due to the symmetry of the configuration. This is not the case for planar geometries. In the planar geometry, the lower beat wave is present and results in an oscillation in both the power and phase with a period approximately equal to one half that of the wiggler period. Such effects are not included in descriptions that are based upon a wiggler-averaged orbit analysis. In most cases, the effects of the lower beat wave do not contribute substantially to the bulk evolution of the wave; however, the instantaneous values of the power and phase can be markedly altered by this effect. In the cases considered in this chapter, it will be shown that the overall magnitude of

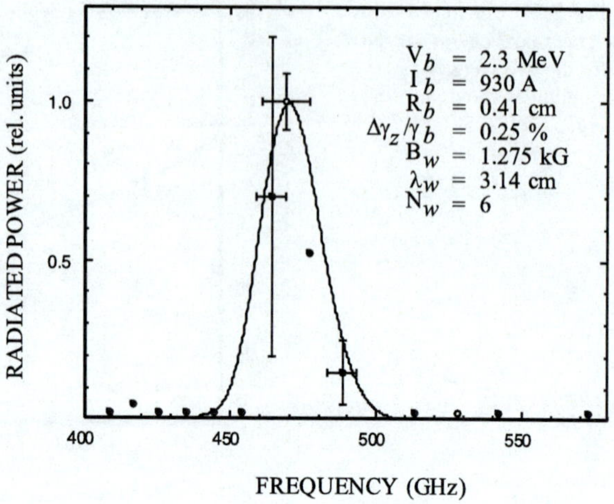

Fig. 5.37 The output spectrum for a 2.3 MeV, 930 A electron beam an on-axis wiggler strength of 1275 G. The dots represent the measurements, and the curve is the result of the nonlinear simulation including the TE_{11} and TM_{11} modes [51, 52]

the oscillations in the power and phase of the signal can approach 10–20% of the overall bulk values for these quantities.

5.5.1 *The General Configuration*

The wiggler model we employ in the treatment of the three-dimensional nonlinear formulation is that given in Eq. (2.3)

$$\begin{aligned}\mathbf{B}_w(\mathbf{x}) &= B_w\left\{\cos k_w z\left[\widehat{\boldsymbol{e}}_x \sinh\left(\frac{k_w x}{\sqrt{2}}\right)\sinh\left(\frac{k_w y}{\sqrt{2}}\right) + \widehat{\boldsymbol{e}}_y \cosh\left(\frac{k_w x}{\sqrt{2}}\right)\cosh\left(\frac{k_w y}{\sqrt{2}}\right)\right]\right.\\ &\times\left. -\sqrt{2}\widehat{\boldsymbol{e}}_z \cosh\left(\frac{k_w x}{\sqrt{2}}\right)\sinh\left(\frac{k_w y}{\sqrt{2}}\right)\sin k_w z\right\},\end{aligned} \tag{5.145}$$

which includes the effect of tapered pole faces for enhanced focusing. No axial guide magnetic field is included in this formulation due to the deleterious drift associated with the combined planar wiggler and axial field. As in the case of the helical wiggler, we permit the amplitude to vary adiabatically in z as given in Eq. (5.83) in order to model both beam injection into the wiggler and tapered wiggler efficiency enhancement.

5.5.1.1 The Field Equations

In treating the planar wiggler, we assume that the waveguide geometry corresponds to a loss-free rectangular waveguide, and the boundary conditions of the radiation field at the waveguide wall are satisfied by expansion of the field in terms of the vacuum waveguide modes. The vector potential in a rectangular waveguide bounded by $-a/2 \leq x \leq a/2$ and $-b/2 \leq y \leq b/2$ may be written in the form

$$\delta \mathbb{A}(\mathbf{x},t) = \sum_{l,n=0}^{\infty} \delta A_{ln}(z)\, \boldsymbol{e}_{ln}^{(1)}(x,y) \cos \alpha_{ln}(z,t), \tag{5.146}$$

for the TE modes where l and n are not both zero, and

$$\delta \mathbb{A}(\mathbf{x},t) = \sum_{l,n=1}^{\infty} \delta A_{ln}(z) \Big[\boldsymbol{e}_{ln}^{(2)}(x,y) \cos \alpha_{ln}(z,t)$$
$$\times + \frac{\kappa_{ln}}{k_{ln}} \boldsymbol{e}_z \sin\left(\frac{l\pi x}{a}\right) \cos\left(\frac{n\pi y}{b}\right) \sin \alpha_{ln}(z,t) \Big], \tag{5.147}$$

for the TM modes, where the phase for frequency ω and wavenumber $k_{ln}(z)$ is given by

$$\alpha_{ln}(z,t) = \int_0^z dz' k_{ln}(z') - \omega t, \tag{5.148}$$

$$\boldsymbol{e}_{ln}^{(1)}(x,y) = \frac{n\pi}{\kappa_{ln} b} \widehat{\boldsymbol{e}}_x \cos\left(\frac{l\pi X}{a}\right) \sin\left(\frac{n\pi Y}{b}\right) - \frac{l\pi}{\kappa_{ln} a} \widehat{\boldsymbol{e}}_y \sin\left(\frac{l\pi X}{a}\right) \cos\left(\frac{n\pi Y}{b}\right), \tag{5.149}$$

$$\boldsymbol{e}_{ln}^{(2)}(x,y) = \frac{l\pi}{\kappa_{ln} a} \widehat{\boldsymbol{e}}_x \cos\left(\frac{l\pi X}{a}\right) \sin\left(\frac{n\pi Y}{b}\right) + \frac{n\pi}{\kappa_{ln} b} \widehat{\boldsymbol{e}}_y \sin\left(\frac{l\pi X}{a}\right) \cos\left(\frac{n\pi Y}{b}\right), \tag{5.150}$$

where $X = x + a/2$ and $Y = y + b/2$ and the cutoff wavenumber of the mode is given by

$$\kappa_{ln} = \pi \sqrt{\frac{l^2}{a^2} + \frac{n^2}{b^2}}. \tag{5.151}$$

Observe that, unlike the case of a helical wiggler, the TE_{ln} and TM_{ln} modes have identical cutoff wavenumbers. As in the preceding cases, we implicitly assume that the amplitudes and wavenumbers of each mode vary slowly with respect to the wavelength.

The equations governing the evolution of the slowly varying amplitude and phase of each of the TE modes in a rectangular waveguide follow in a similar manner as in the case of a cylindrical waveguide, and we assume the identical model for the source current and the electron beam distribution function. The principal difference is that the orthogonalization is now over the x and y coordinates (rather than r and θ). The orthogonality integrals over the x-coordinate are

$$\int_0^a dx \cos\left(\frac{l\pi x}{a}\right) \cos\left(\frac{n\pi x}{a}\right) = \frac{a}{2}\delta_{l,n}(1+\delta_{l,0}), \tag{5.152}$$

$$\int_0^a dx \sin\left(\frac{l\pi x}{a}\right) \sin\left(\frac{n\pi x}{a}\right) = \frac{a}{2}\delta_{l,n}(1+\delta_{l,0}), \tag{5.153}$$

and similar relations hold for the y-orthogonalization. As a consequence, we find that the dynamical equations for each TE_{ln} mode are given by

$$\left(\frac{d^2}{dz^2} + \frac{\omega^2}{c^2} - k_{ln}^2 - \kappa_{ln}^2\right)\delta a_{ln} = \frac{8\omega_b^2}{c^2} F_{ln}\left\langle \frac{\mathbf{v}\cdot\boldsymbol{e}_{ln}^{(1)}}{|v_z|} \cos\alpha_{ln}\right\rangle, \tag{5.154}$$

and

$$2k_{ln}^{1/2}\frac{d}{dz}\left(k_{ln}^{1/2}\delta a_{ln}\right) = -\frac{8\omega_b^2}{c^2} F_{ln}\left\langle \frac{\mathbf{v}\cdot\boldsymbol{e}_{ln}^{(1)}}{|v_z|} \sin\alpha_{ln}\right\rangle, \tag{5.155}$$

where the averaging operator is the same as that defined in Eq. (5.125) for a cross-sectional area $A_g = ab$ and $F_{ln} = 1/2$ whenever either $l = 0$ or $n = 0$ and unity otherwise. The dynamical equations for each TM_{ln} mode are governed by a similar set of equations

$$\left[\frac{d^2}{dz^2} + \left(1+\frac{\kappa_{ln}^2}{k_{ln}^2}\right)\left(\frac{\omega^2}{c^2} - k_{ln}^2 - \kappa_{ln}^2\right)\right]\delta a_{ln} = \frac{8\omega_b^2}{c^2} F_{ln}\left\langle \frac{\mathbf{v}\cdot\boldsymbol{e}_{ln}^{(2)}}{|v_z|} \cos\alpha_{ln}\right.$$
$$\left. \times + \frac{v_z}{|v_z|}\frac{\kappa_{ln}}{k_{ln}} \sin\left(\frac{l\pi X}{a}\right) \sin\left(\frac{n\pi Y}{b}\right) \sin\alpha_{ln}\right\rangle, \tag{5.156}$$

and

$$2\left(k_{ln}+\frac{\kappa_{ln}^2}{k_{ln}^2}\right)^{1/2}\frac{d}{dz}\left[\left(k_{ln}+\frac{\kappa_{ln}^2}{k_{ln}^2}\right)^{1/2}\delta a_{ln}\right]=-\frac{8\omega_b^2}{c^2}F_{ln}\left\langle\frac{\mathbf{v}\cdot\boldsymbol{e}_{ln}^{(2)}}{|v_z|}\sin\alpha_{ln}\right.$$
$$\left.\times-\frac{v_z}{|v_z|}\frac{\kappa_{ln}}{k_{ln}}\sin\left(\frac{l\pi X}{a}\right)\sin\left(\frac{n\pi Y}{b}\right)\cos\alpha_{ln}\right\rangle. \tag{5.157}$$

These equations describe the evolution of the wavenumber and amplitude for each TE and TM mode and must be integrated for each mode included in the simulation. A similar analysis has also been performed for the slow-time-scale equations for the modes in a dielectric-lined rectangular waveguide [54].

5.5.1.2 The Electron Orbit Equations

In order to complete the formulation, the trajectories of an ensemble of electrons must be determined using the Lorentz force equations in the aggregate fields. For the planar wiggler model and the TE and TM modes of the rectangular waveguide, we obtain

$$v_z\frac{d}{dz}p_x=\Omega_w\cosh\left(\frac{k_wx}{\sqrt{2}}\right)\left[\sqrt{2}p_y\sinh\left(\frac{k_wx}{\sqrt{2}}\right)\sin k_wz+p_z\cosh\left(\frac{k_wx}{\sqrt{2}}\right)\cos k_wz\right]$$
$$-m_ec\sum_{TE\,modes}\delta a_{ln}\cos\left(\frac{l\pi x}{a}\right)\left\{\frac{n\pi}{\kappa_{ln}b}(\omega-k_{ln}v_z)\sin\left(\frac{n\pi y}{b}\right)\sin\alpha_{ln}\right.$$
$$\left.+\left[\kappa_{ln}v_y\cos\left(\frac{n\pi y}{b}\right)+\frac{n\pi}{\kappa_{ln}b}\Gamma_{ln}v_z\sin\left(\frac{n\pi y}{b}\right)\right]\cos\alpha_{ln}\right\}$$
$$+m_ec\sum_{TM\,modes}\delta a_{ln}\frac{l\pi}{\kappa_{ln}a}\cos\left(\frac{l\pi x}{a}\right)\sin\left(\frac{n\pi y}{b}\right)$$
$$\times\left\{\sin\alpha_{ln}\left[\omega-\left(k_{ln}+\frac{\kappa_{ln}^2}{k_{ln}}\right)v_z\right]+\Gamma_{ln}v_z\cos\alpha_{ln}\right\}, \tag{5.158}$$

$$
\begin{aligned}
\upsilon_z \frac{d}{dz} p_y = & -\Omega_w \sinh\left(\frac{k_w x}{\sqrt{2}}\right)\left[\sqrt{2} p_x \cosh\left(\frac{k_w x}{\sqrt{2}}\right) \sin k_w z + p_z \sinh\left(\frac{k_w x}{\sqrt{2}}\right) \cos k_w z\right] \\
& - m_e c \sum_{TE\ modes} \delta a_{ln} \cos\left(\frac{n\pi y}{b}\right)\left\{\frac{l\pi}{\kappa_{ln} a}(\omega - k_{ln}\upsilon_z) \sin\left(\frac{l\pi x}{a}\right) \sin\alpha_{ln}\right. \\
& \left. + \left[\kappa_{ln}\upsilon_x \cos\left(\frac{l\pi x}{a}\right) + \frac{l\pi}{\kappa_{ln} a}\Gamma_{ln}\upsilon_z \sin\left(\frac{l\pi x}{a}\right)\right] \cos\alpha_{ln}\right\} \\
& + m_e c \sum_{TM\ modes} \delta a_{ln} \frac{n\pi}{\kappa_{ln} b} \sin\left(\frac{l\pi x}{a}\right) \cos\left(\frac{n\pi y}{b}\right) \\
& \times \left\{\sin\alpha_{ln}\left[\omega - \left(k_{ln} + \frac{\kappa_{ln}^2}{k_{ln}}\right)\upsilon_z\right] + \Gamma_{ln}\upsilon_z \cos\alpha_{ln}\right\},
\end{aligned}
\tag{5.159}
$$

and

$$
\begin{aligned}
\upsilon_z \frac{d}{dz} p_z = & -\Omega_w \cos k_w z\left[p_x \cosh\left(\frac{k_w x}{\sqrt{2}}\right) \cosh\left(\frac{k_w x}{\sqrt{2}}\right) - p_y \sinh\left(\frac{k_w x}{\sqrt{2}}\right) \sinh\left(\frac{k_w x}{\sqrt{2}}\right)\right] \\
& - m_e c \sum_{TE\ modes} \delta a_{ln}(k_{ln} \sin\alpha_{ln} - \Gamma_{ln} \cos\alpha_{ln}) \\
& \times \left[\frac{n\pi}{\kappa_{ln} b}\upsilon_x \cos\left(\frac{l\pi x}{a}\right) \sin\left(\frac{n\pi y}{b}\right) - \frac{l\pi}{\kappa_{ln} a}\upsilon_y \sin\left(\frac{l\pi x}{a}\right) \cos\left(\frac{n\pi y}{b}\right)\right] \\
& + m_e c \sum_{TM\ modes} \delta a_{ln}\left[\left(k_{ln} + \frac{\kappa_{ln}^2}{k_{ln}}\right) \sin\alpha_{ln} - \Gamma_{ln} \cos\alpha_{ln}\right] \\
& \times \left[\frac{l\pi}{\kappa_{ln} a}\upsilon_x \cos\left(\frac{l\pi x}{a}\right) \sin\left(\frac{n\pi y}{b}\right) + \frac{n\pi}{\kappa_{ln} b}\upsilon_y \sin\left(\frac{l\pi x}{a}\right) \cos\left(\frac{n\pi y}{b}\right)\right] \\
& - \omega m_e c \sum_{TM\ modes} \delta a_{ln} \frac{\kappa_{ln}}{k_{ln}} \sin\left(\frac{l\pi x}{a}\right) \sin\left(\frac{n\pi y}{b}\right) \cos\alpha_{ln},
\end{aligned}
\tag{5.160}
$$

where $\Omega_w = eB_w/\gamma m_e c$ and Γ_{ln} denotes the growth rate of the mode. In addition, we also integrate

$$\frac{dx}{dz} = \frac{\upsilon_x}{\upsilon_z}, \tag{5.161}$$

$$\frac{dy}{dz} = \frac{\upsilon_y}{\upsilon_z}, \tag{5.162}$$

and

$$\frac{d}{dz}\psi_{ln} = k_{ln} + k_w - \frac{\omega}{\upsilon_z}. \tag{5.163}$$

It should be remarked that these equations are implicitly slowly varying functions of axial position as long as the waves are near resonance [i.e., $\omega \approx (k + mk_w)\upsilon_z$, where m is an odd integer]. Hence, no average over the wiggler period need be performed.

5.5.1.3 The First-Order Field Equations

The second-order differential equations for the amplitude and phase of each mode can be simplified under the neglect of the second-order derivatives of the amplitude and phase. This yields algebraic equations for the wavenumber and amplitude

$$\frac{\omega^2}{c^2} - k_{ln}^2 - \kappa_{ln}^2 = \frac{8\omega_b^2}{c^2}\frac{F_{ln}}{\delta a_{ln}}\left\langle \frac{\mathrm{v}\cdot e_{ln}^{(1)}}{|\upsilon_z|}\cos\alpha_{ln}\right\rangle, \tag{5.164}$$

$$2k_{ln}\frac{d}{dz}\delta a_{ln} = -\frac{8\omega_b^2}{c^2}F_{ln}\left\langle \frac{\mathrm{v}\cdot e_{ln}^{(1)}}{|\upsilon_z|}\sin\alpha_{ln}\right\rangle, \tag{5.165}$$

of each TE mode. The analogous equations for the TM modes are

$$\left(1+\frac{\kappa_{ln}^2}{k_{ln}^2}\right)\left(\frac{\omega^2}{c^2} - k_{ln}^2 - \kappa_{ln}^2\right) = \frac{8\omega_b^2}{c^2}\frac{F_{ln}}{\delta a_{ln}}\left\langle \frac{\mathrm{v}\cdot e_{ln}^{(2)}}{|\upsilon_z|}\cos\alpha_{ln}\right.$$
$$\times\left. + \frac{\upsilon_z}{|\upsilon_z|}\frac{\kappa_{ln}}{k_{ln}}\sin\left(\frac{l\pi X}{a}\right)\sin\left(\frac{n\pi Y}{b}\right)\sin\alpha_{ln}\right\rangle, \tag{5.166}$$

$$2\left(k_{ln}+\frac{\kappa_{ln}^2}{k_{ln}^2}\right)\frac{d}{dz}\delta a_{ln} = -\frac{8\omega_b^2}{c^2}F_{ln}\left\langle \frac{\mathrm{v}\cdot e_{ln}^{(2)}}{|\upsilon_z|}\sin\alpha_{ln}\right.$$
$$\times\left. - \frac{\upsilon_z}{|\upsilon_z|}\frac{\kappa_{ln}}{k_{ln}}\sin\left(\frac{l\pi X}{a}\right)\sin\left(\frac{n\pi Y}{b}\right)\cos\alpha_{ln}\right\rangle. \tag{5.167}$$

In addition, Eq. (5.140) is integrated for the relative phase of each mode. As in the case of the helical wiggler geometry, the discrepancy between this reduced set of dynamical equations and the complete set of second-order differential equations is under 10% for typical parameters of most free-electron lasers.

The power in the TE_{ln} mode is given by

$$P_{ln} = \frac{m_e^2c^4}{e^2}\frac{\omega k_{ln}}{16\pi}F_{ln}A_g\delta a_{ln}^2, \tag{5.168}$$

and the power in the TM_{ln} mode is

$$P_{ln} = \frac{m_e^2 c^4}{e^2} \frac{\omega k_{ln}}{32\pi} \left(1 + \frac{\kappa_{ln}^2}{k_{ln}^2}\right) A_g \delta a_{ln}^2. \tag{5.169}$$

5.5.2 The Initial Conditions

The algorithms and initial conditions employed in the numerical simulation for the rectangular waveguide and planar wiggler configuration are identical to those described for the helical wiggler geometry.

5.5.3 Numerical Simulation: Single-Mode Limit

The particular example under consideration is that of a 35 GHz amplifier employing an electron beam with an energy of 3.5 MeV, a current of 800 A, and an initial radius of 1.0 cm which propagates through a waveguide characterized by $a = 9.8$ cm and $b = 2.9$ cm. In order to obtain peak growth rates in the vicinity of 35 GHz, we choose a wiggler field with an amplitude of $B_w = 3.72$ kG, a period of $\lambda_w = 9.8$ cm, and an entry taper region $N_w = 5$ wiggler periods in length. For purposes of illustration, we first consider the case of an ideal beam (i.e., $\Delta\gamma_z = 0$). For these parameters, there are three resonant modes: the TE_{01}, TE_{21}, and TM_{21} modes. Before proceeding to the analysis of the complete multimode problem, we shall first consider the properties of the interaction solely in the presence of each of these modes. Bear in mind that these parameters are relevant to an experiment conducted at the Lawrence Livermore National Laboratory [5, 32, 39] and that a detailed discussion of the comparison of the results of the numerical simulation with the experiment will follow (Fig. 5.38).

The initial electron beam distributions in axial phase space and in cross section are similar to those used for the helical wiggler geometry. The axial phase space is shown in Fig. 5.38. Each dot in the figure represents an entire phase sheet that represents a cross-sectional slice of the beam upon entry to the wiggler. The cross-sectional distribution of the beam is illustrated in Fig. 5.39 that represents the injection of a solid (i.e., pencil) beam with a flattop density profile. As in the case of a helical wiggler, the nonuniform spacings between the electrons are artifacts of the Gaussian algorithm and are compensated for by nonuniform weights assigned to each electron.

The detailed evolution of the TE_{01} mode as a function of axial position is shown in Fig. 5.40 for the injection of a signal at a normalized frequency of $\omega/ck_w = 11.5$ (34.6 GHz) at a power level of 50 kW. As shown in the figure, saturation occurs at $k_w z \approx 115$ (1.79 m) at a power level of approximately 214 MW corresponding to an efficiency $\eta \approx 7.75\%$. Wave amplification occurs principally within the uniform-wiggler region, which is 1.30 m in length for this example. Hence, there is an

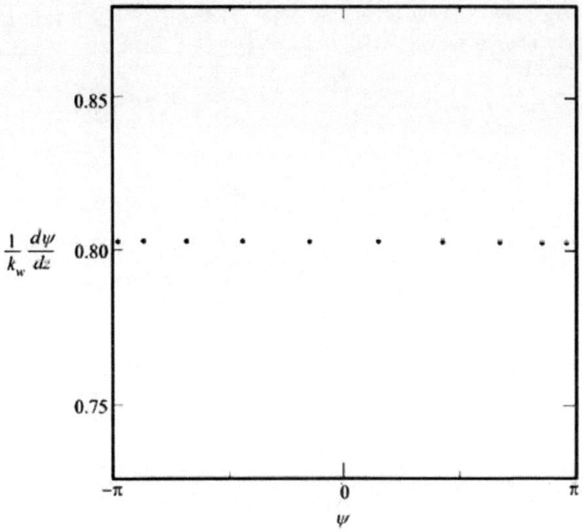

Fig. 5.38 Initial distribution of the axial phase space of the beam. Each point represents a phase sheet of electrons distributed throughout the cross section of the beam

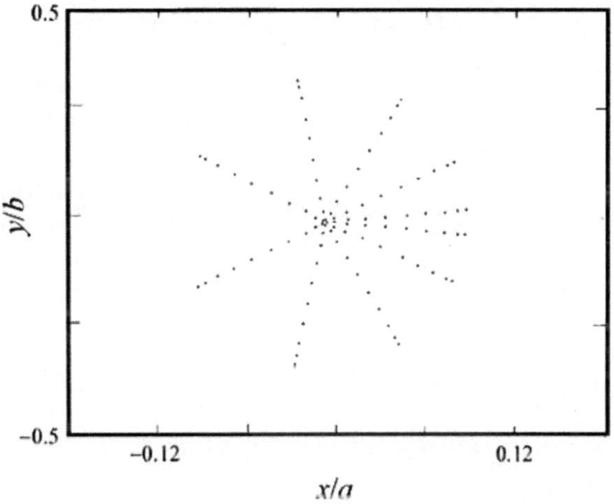

Fig. 5.39 Initial cross-sectional distribution of the electron beam

average gain of 28 dB/m throughout the uniform-wiggler region, which corresponds to an average normalized growth rate of $\Gamma_{01}/k_w \approx 0.05$.

One feature of interest shown in the figure that merits discussion is the oscillation in the instantaneous power. This oscillation occurs at a period of $\lambda_w/2$ and is not generally found for helical wiggler configurations. In order to explain this oscillation, recall that the interaction in a free-electron laser arises due to the beating of the wiggler and radiation fields. The slowly varying ponderomotive wave that results in the bulk amplification of the signal corresponds to the upper beat wave, while the rapid oscillation shown in the figure is due to the lower beat wave. In order to

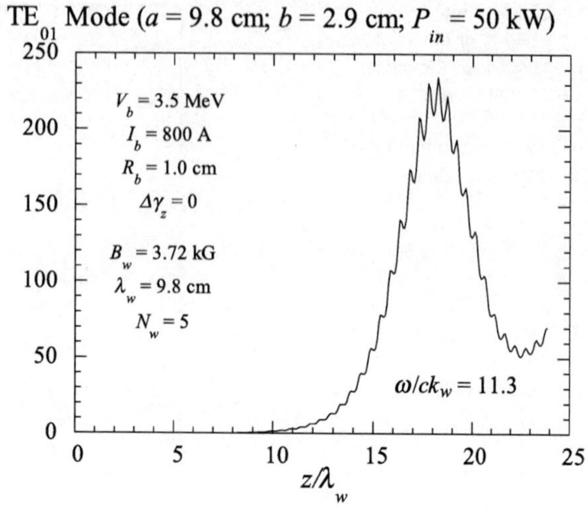

Fig. 5.40 The evolution of the TE_{01} mode with axial position

illustrate this, observe that for the present planar configuration the bulk transverse velocity is aligned along the *x*-axis and varies approximately as

$$v_w \approx \frac{\Omega_w}{k_w}\widehat{e}_z \sin k_w z. \tag{5.170}$$

The source terms contained in the dynamical equations are derived essentially from a calculation of $<\mathbf{J} \cdot \delta\mathbf{E}>$; hence, the principal wave-particle coupling is with the *x*-component of the radiation field. If we assume that $\delta E_x \approx \delta E_0 \sin(kz - \omega t)$, then it is evident that

$$\langle \delta \mathrm{J} \cdot \delta \mathrm{E} \rangle \approx \frac{\Omega_w}{2k_w} \delta E_0 [\langle \cos \psi \rangle - \langle \cos(\psi - 2k_w z) \rangle]. \tag{5.171}$$

The first average in the square brackets represents the upper beat wave, while the second term represents the lower beat wave. Since amplification is found when the ponderomotive phase ψ is a slowly varying quantity, the lower beat wave describes an oscillation with a period half that of the wiggler period. Although the contribution from the lower beat wave provides no contribution to the bulk growth of the wave, the instantaneous magnitude of this contribution is comparable to that of the upper beat wave, and it does affect the instantaneous values of both the power and phase of the signal. This effect is not found in helical wiggler geometries because the wiggler-induced transverse velocity describes a helix with near-constant magnitudes that drives an interaction with a circularly polarized wave. As a result, the symmetry of the helical interaction suppresses the lower beat wave.

A full spectrum of the TE_{01} mode is shown in Fig. 5.41 in which we plot the saturation efficiency and the distance to saturation as a function of frequency within the unstable band. As shown in the figure, wave amplification is found for

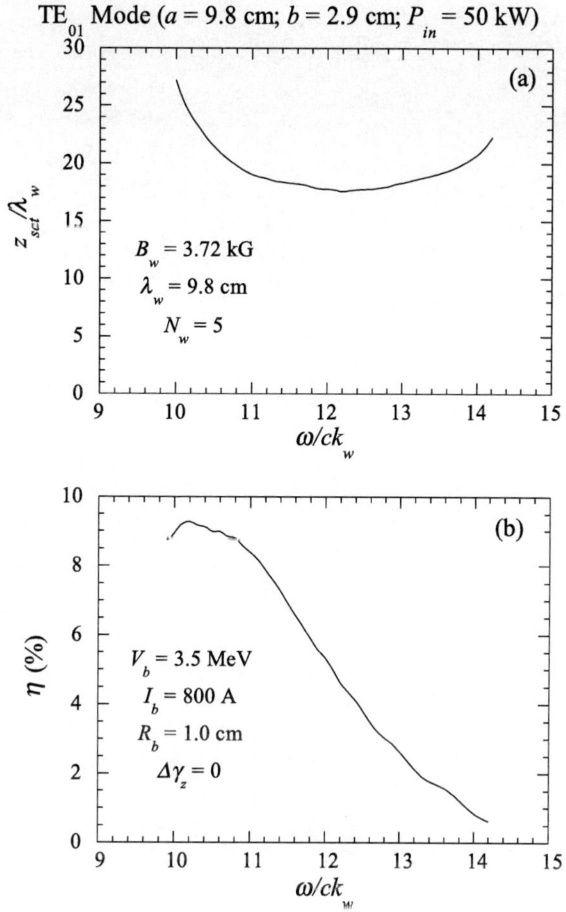

Fig. 5.41 Graphs showing the distance to saturation and the efficiency of the TE_{01} mode as a function of frequency

frequencies extending from $\omega/ck_w \approx 10$ (30.6 GHz) to $\omega/ck_w \approx 14.2$ (43.5 GHz) with a peak efficiency of the order of $\eta \approx 9.8\%$. The peak growth rate by contrast (as measured by the shortest distance to saturation) occurs for $\omega/ck_w \approx 12.3$ (37.7 GHz).

The variation in the relative phase as a function of axial position is illustrated in Fig. 5.42 for $\omega/ck_w = 10.4$, 10.7 11.0, 11.3, and 11.9. As is evident in the figure, the oscillation at one half the wiggler period due to the lower beat wave is also manifested in the evolution of the relative phase. The bulk variation [i.e., averaged over a wiggler period] shows the same qualitative behavior as that found for a helical wiggler. Specifically, for frequencies at the low end of the gain band, the relative phase decreases up to a point just prior to saturation (indicated in the figure by an arrow), after which the relative phase remains relatively constant. As the frequency increases, the variation in the relative phase decreases until a critical frequency is reached ($\omega/ck_w \approx 11$, e.g., under consideration) at which the phase varies little over

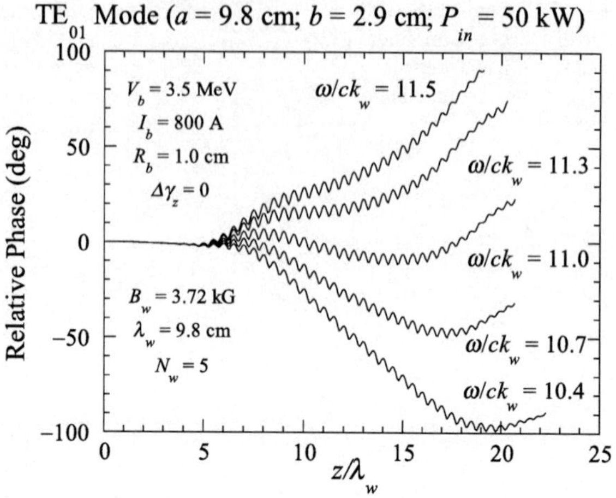

Fig. 5.42 Plots of the relative phase of the TE_{01} mode at a variety of frequencies within the gain band

the course of the interaction. This frequency is typically found to be approximately 10% below the frequency of peak growth rate. For still higher frequencies, the bulk phase increases monotonically.

The axial phase space of the electron beam at saturation is illustrated in Fig. 5.43. The dashed line represents an approximate separatrix calculated for electrons located at the center of the beam; hence, many of the electrons that appear to be executing untrapped trajectories may instead be on trapped orbits at the edge of the beam. It should be remarked that this phase-space distribution is located in the lower half of the phase plane, which contrasts with the typical behavior of the nonlinear pendulum equation in which the separatrix is centered at $d\psi/dz = 0$. The reason for this behavior is that the single-particle oscillation in the axial velocity at half the wiggler period results in a large-amplitude oscillation in the axial phase of the electrons as well. Indeed, the pendulum equation for a planar wiggler takes the form

$$\frac{d^2}{dz^2}\psi_{ln} + \kappa_{ln}^2 \cos\psi_{ln} = -\kappa^2 \sin 2k_w z, \tag{5.172}$$

where for the TE mode polarization

$$\kappa_{ln}^2 \equiv \frac{\omega^2}{2\gamma_0\gamma_\parallel^2 v_\parallel^2}\left(\frac{e\langle B_w\rangle}{m_e k_w c^2}\right)\left(\frac{e\langle \delta A_{ln}\rangle}{m_e c^2}\right), \tag{5.173}$$

and

$$\kappa^2 \equiv \frac{k_w^2}{2}\left(\frac{\omega}{k_w v_\parallel}\right)\left(\frac{e\langle B_w\rangle}{m_e k_w c^2}\right)^2, \tag{5.174}$$

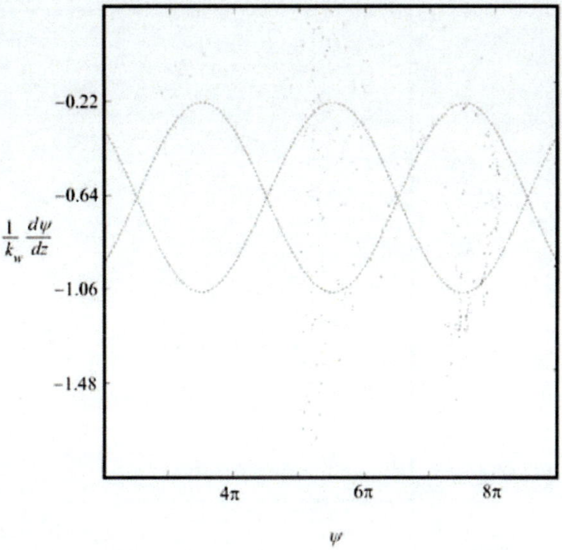

Fig. 5.43 The axial phase space of the TE_{01} mode at saturation

where $<B_w>$ and $<\delta A_{ln}>$ represent the average fields found at the center of the beam and $v_{\|}$ is the bulk axial velocity. The term in $\kappa^2 \sin 2k_w z$ arises from the oscillation in the axial velocity and generally dominates over the effect of the ponderomotive potential (i.e., $\kappa^2 >> \kappa_{ln}{}^2$). In principle, therefore, the effect of this term is to give rise to a general oscillation in the entire axial phase space of the electron beam.

The evolution of the cross-sectional distribution of the beam is illustrated in Figs. 5.44, 5.45, 5.46, 5.47, 5.48, and 5.49. The cross-sectional distribution at the start of the uniform-wiggler region is shown in Fig. 5.44. The bulk motion exhibits four essential features. The first is the primary wiggler-induced oscillation that is aligned along the x-axis, and this is clearly shown in the figure. The second feature is that the transverse wiggler gradient has a focusing effect on the beam that results in a reduction in the maximum beam radius relative to the initial value. The third feature is that the transverse wiggler gradient introduces a betatron oscillation that causes a macroscopic scalloping of the beam envelope. The betatron oscillation for the present example occurs over a distance of $k_w \Delta z_b \approx 19.3$ and is illustrated in the sequence of Figs. 5.44, 5.45, 5.46, 5.47, and 5.48. In addition, on the microscopic level, the individual electrons come into a focus and out again on the opposite side of the beam. Lastly, the geometry of the wiggler tends to distort the overall beam into an elliptical cross section. This is more clearly shown in Fig. 5.48, which depicts the cross section of the beam at saturation.

The question of the effect of the injection process of the electron beam into the wiggler can be addressed by varying the length of the entry taper region. The results of this analysis are shown in Fig. 5.50, in which the saturation efficiency and the distance to saturation are plotted as functions of the length of the entry taper region

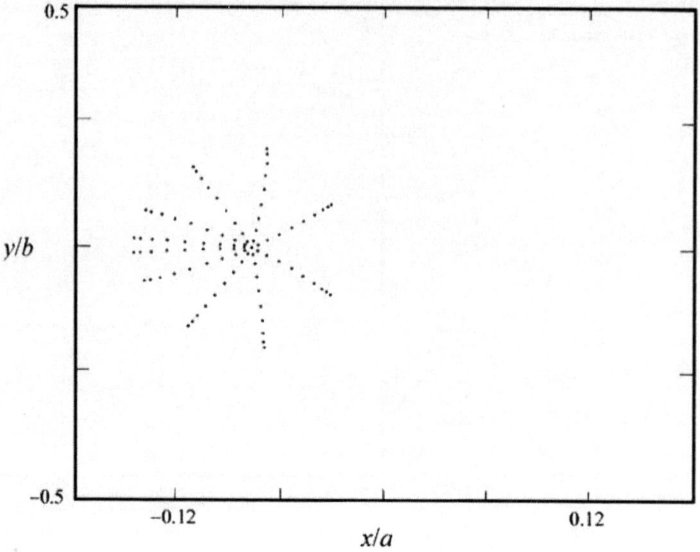

Fig. 5.44 The beam cross-sectional distribution at the start of the uniform-wiggler region (i.e., $k_w z \approx 31$)

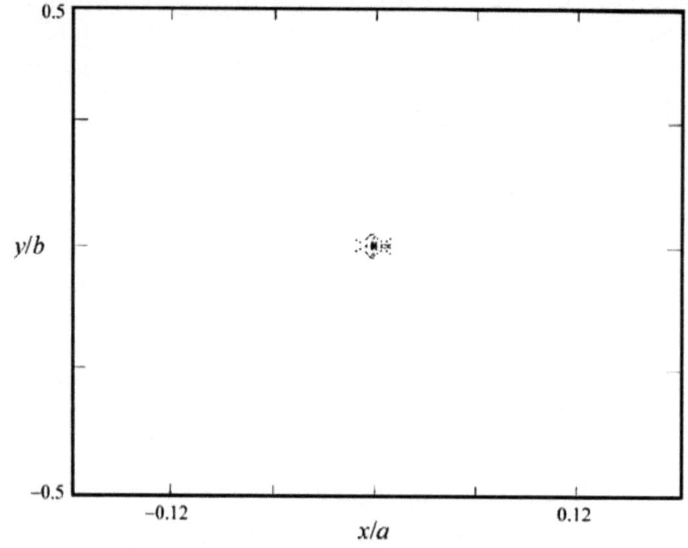

Fig. 5.45 The beam cross-sectional distribution at $k_w z = 36$

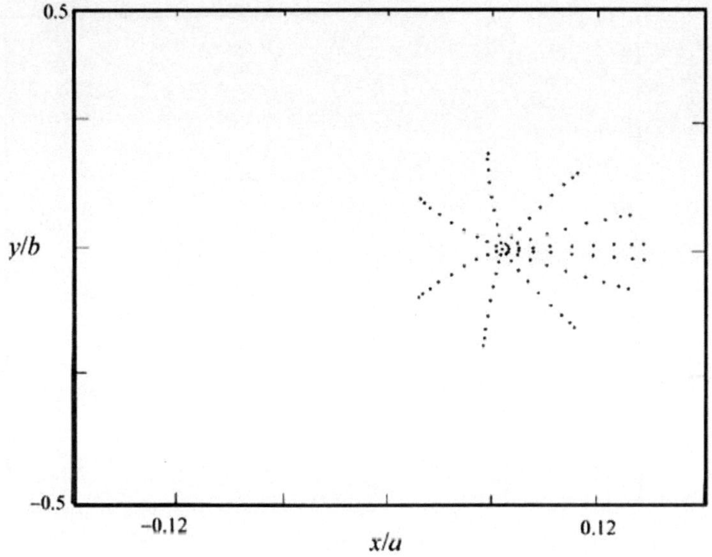

Fig. 5.46 The beam cross-sectional distribution at $k_w z = 41$

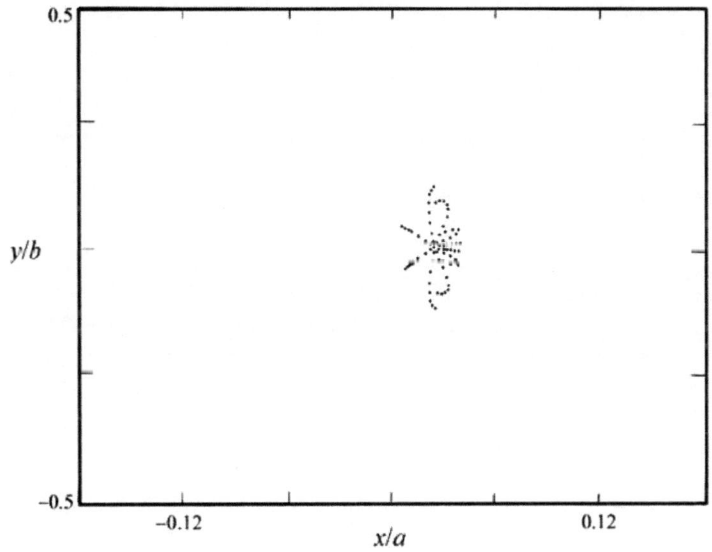

Fig. 5.47 The beam cross-sectional distribution at $k_w z = 46$

for $N_w \geq 3$. The minimum entry length of the entry taper region has been chosen to ensure that the fringing fields associated with the tapered wiggler amplitude may be neglected. The results indicate that the saturation efficiency increases markedly with

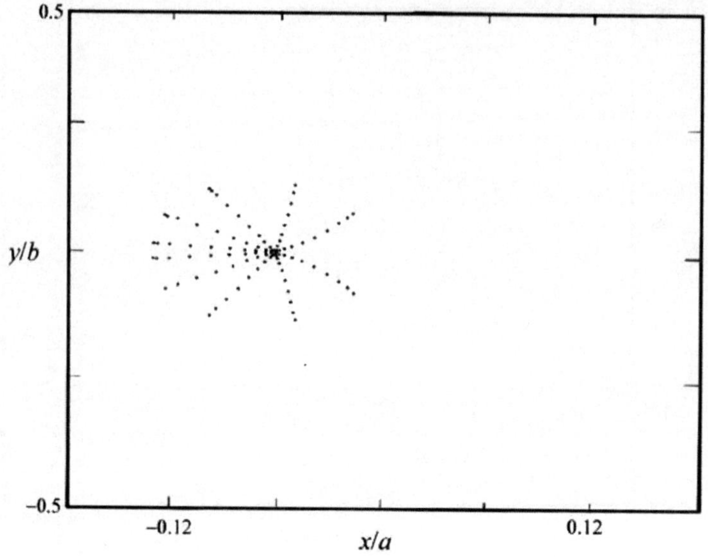

Fig. 5.48 The beam cross-sectional distribution at $k_w z = 51$

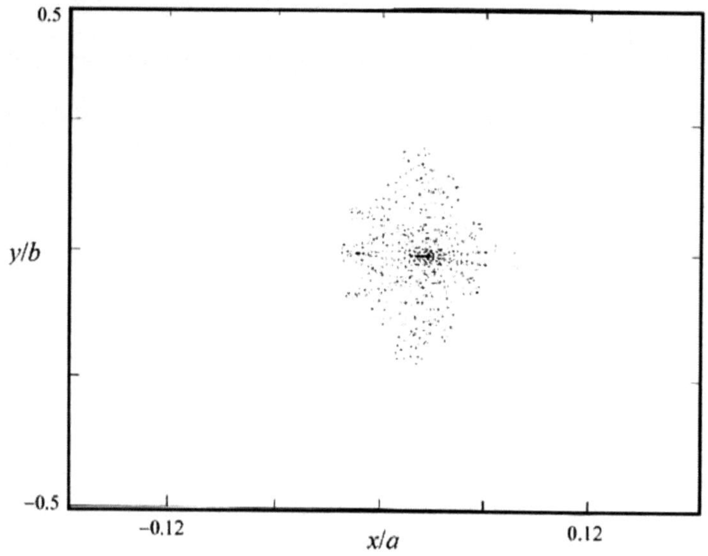

Fig. 5.49 The beam cross-sectional distribution at saturation for $k_w z \approx 115$

Fig. 5.50 Plot of the distance to saturation and the saturation efficiency of the TE_{01} mode as functions of the length of the entry taper region

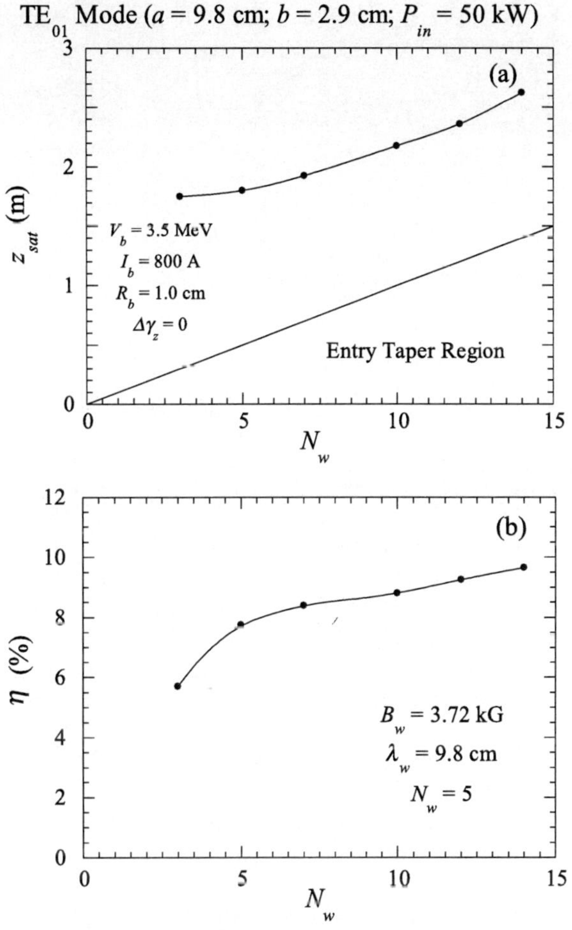

the length of the entry taper region up to approximately $N_w = 6$, after which the increase is more gradual. This increase in the efficiency can be attributed to a decrease in the effective axial energy spread induced by the injection process as the wiggler gradient becomes more adiabatic. It should also be noted that the increase in the distance to saturation is approximately linear for $N_w \geq 7$ and closely corresponds to the increase in the length of the entry taper. This implies that the length of the uniform-wiggler region over which the bulk of the amplification occurs remains constant.

The evolution of the TE_{21} mode is shown in Fig. 5.51 in which the power is plotted versus axial position for the same parameters as shown for the TE_{01} mode. It is evident from the figure that the power saturates at $k_w z \approx 104$ at a power level of 194 MW for an efficiency of $\eta \approx 6.85\%$. In comparison with the TE_{01} mode, therefore, we conclude that the average growth rate of the TE_{21} mode is somewhat higher and the efficiency lower than for the lower-order mode.

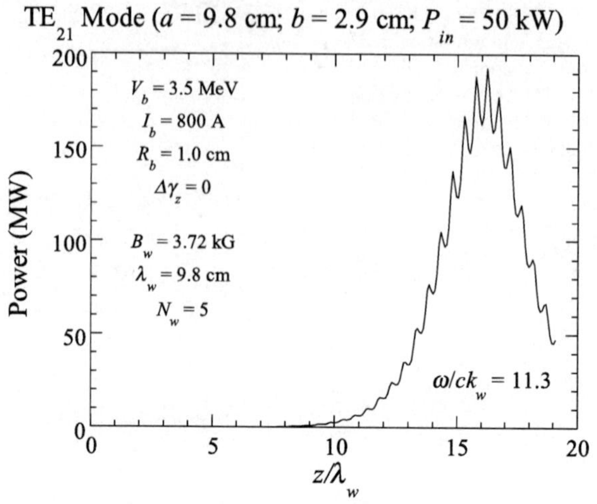

Fig. 5.51 The evolution of the TE_{21} mode with axial position

A complete spectrum of the TE_{21} mode is shown in Fig. 5.52 in which the distance to saturation and the efficiency are plotted versus frequency. As shown in the figure, gain is found for frequencies ranging from $\omega/ck_w \approx 8.9$ to $\omega/ck_w \approx 14$, with a peak efficiency of approximately 12%. As a result, both the bandwidth and efficiency of the TE_{21} are greater than that found for the TE_{01} mode (Fig. 5.50).

The evolution of the wave power with axial distance for the TM_{21} mode is shown in Fig. 5.53 for parameters identical to those used for the TE_{01} and TE_{21} modes. As shown in the figure, the power saturates at $k_w z \approx 237$ with a power level of approximately 68.5 MW and an efficiency of $\eta \approx 2.45\%$. This represents a much lower growth rate and efficiency than found for either of the TE modes, despite the fact that the cutoff frequency and dispersion curves are degenerate for the TE_{21} and TM_{21} modes. The difference between the two modes lies in the transverse mode structure. As mentioned previously, the principal component of the wiggler-induced motion is aligned along the x-axis; hence, the wave-particle interaction is governed largely by the x-component of the electric field. Comparison of the mode structures for the TE and TM modes shows that for a given mode amplitude the ratio of the x-component of the electric field of the TM_{21} mode to that of the TE_{21} mode is approximately $2b/a \approx 0.59$. As a result, the wave-particle coupling is much weaker for the TM mode in the present configuration. A consequence of the weaker coupling of the TM mode is that the efficiency and gain are lower and the bandwidth narrower than for either TE mode.

The effect of an initial axial energy spread on each mode is shown in Fig. 5.54 in which the saturation efficiency is plotted versus the $\Delta\gamma_z/\gamma_0$ for the TE_{01}, TE_{21}, and TM_{21} modes. Using the average growth rates found for each mode, the point at which thermal effects become important can be estimated to be $\Delta\gamma_z/\gamma_0 \approx 18\%$ for the TE_{01} and TE_{21} modes and $\Delta\gamma_z/\gamma_0 \approx 8\%$ for the TM_{21} mode. As a consequence, the cases shown in the figure are well short of the thermal-beam regime. The efficiency is found to decrease in an approximately linear fashion with increasing axial energy

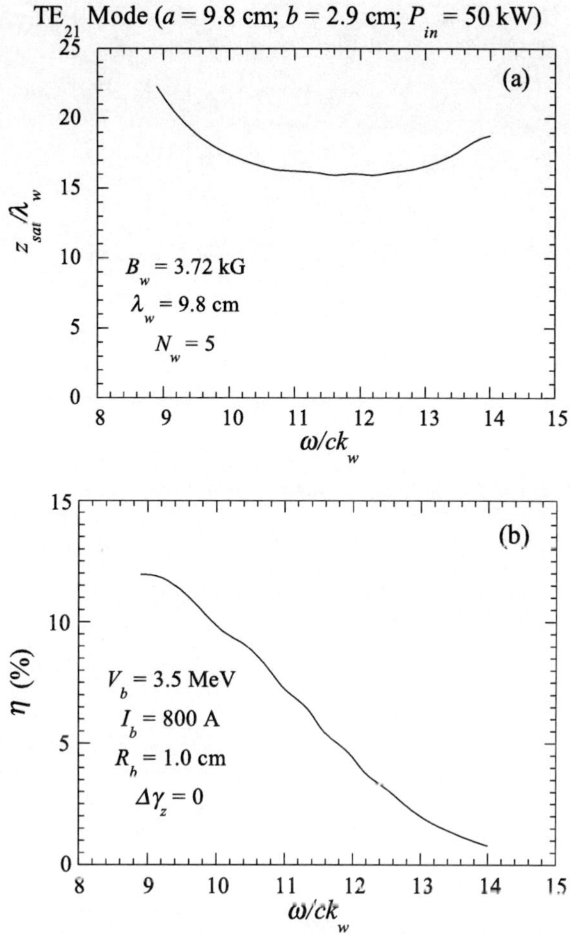

Fig. 5.52 Plot of the distance to saturation and the saturation efficiency of the TE_{21} mode as functions of the length of the entry taper region

spread for each of these modes in correspondence with the results found for helical wiggler geometries. Note, however, that these results are obtained for the same model distribution and the detailed variation of the efficiency with axial energy spread may vary with the choice of distribution. Be that as it may, it is found that the efficiency drops from $\eta \approx 7.75\%$ to $\eta \approx 4.98\%$ for the TE_{01} mode as the $\Delta\gamma_z/\gamma_0$ increases to approximately 2.3%. The efficiency of the TE_{21} mode drops from $\eta \approx 6.85\%$ to $\eta \approx 3.76\%$ for the same increase in the axial energy spread. The efficiency of the TM_{21} mode decreases from $\eta \approx 2.45\%$ to $\eta \approx 1.27\%$ as $\Delta\gamma_z/\gamma_0$ increases to only 1% in keeping with the weakened wave-particle coupling.

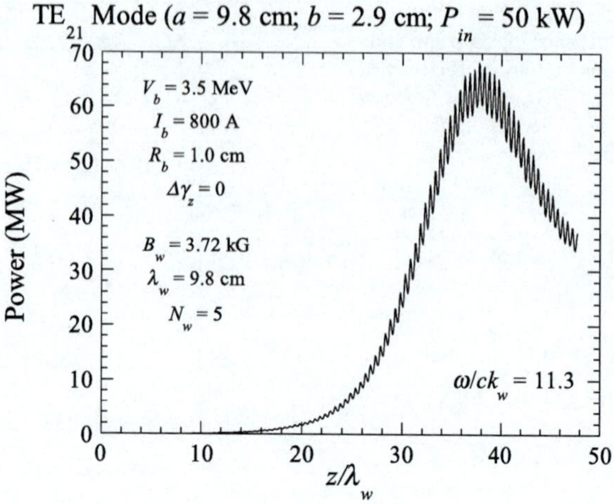

Fig. 5.53 The evolution of the TM_{21} mode with axial position

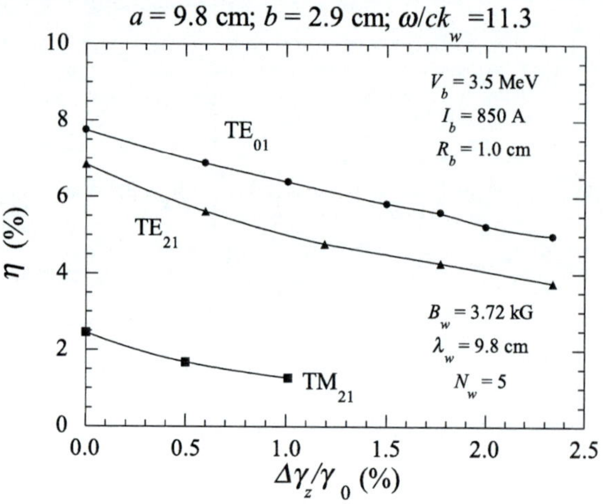

Fig. 5.54 Plot of the variation in the saturation efficiencies of the TE_{01}, TE_{21}, and TM_{21} modes with the initial axial energy spread of the beam

5.5.4 *Numerical Simulation: Multiple Modes*

The multiple-mode analysis proceeds in the same fashion as that for single modes, except that all resonant modes are simultaneously included in the simulation. For the example under consideration, this is restricted to the TE_{01}, TE_{21}, and TM_{21} modes. The initialization of these modes is chosen to correspond to the injection of a signal at a frequency of $\omega/ck_w = 11.3$ which consists of 50 kW in the TE_{01} mode, 500 W in the TE_{21} mode, and 100 W in the TM_{21} mode. The detailed evolution of the total wave power with axial position for this example is shown in Fig. 5.54 for the case of an ideal beam. As seen in the figure, the total power saturates at a level of approximately 260 MW after a distance of $k_w z \approx 98$ (1.53 m). It is also evident

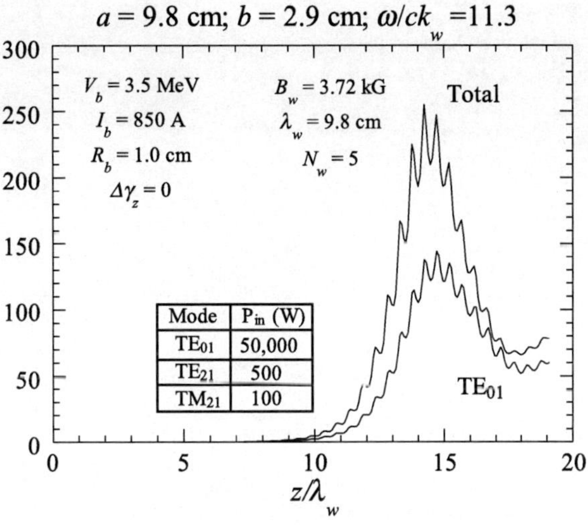

Fig. 5.55 Evolution of the wave power with axial position

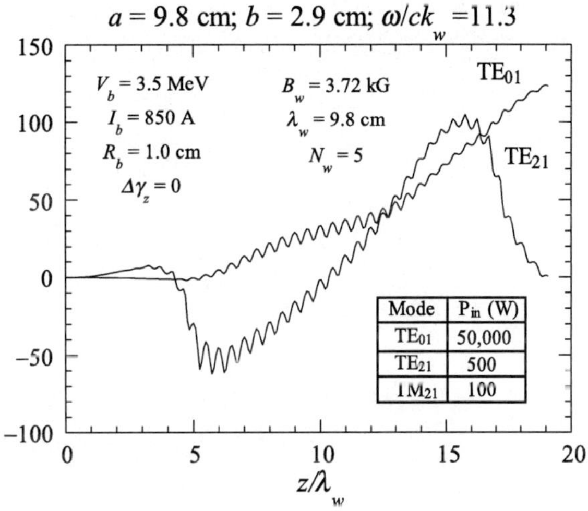

Fig. 5.56 The evolution of the relative phases of the TE_{01} and TE_{21} modes

that although the TE_{01} mode was overwhelmingly dominant upon the injection of the signal, it comprises only approximately 60% of the power at saturation. The remaining power is composed primarily of the TE_{21} mode (approximately 37%). This is due to the fact that the growth rate of the TE_{21} mode is higher than that of the TE_{01} mode at this frequency (Fig. 5.55).

The phase variation of each of these modes is illustrated in Figs. 5.56 and 5.57, where the arrow indicates the point at which the power saturates. It is evident from these figures that the bulk evolution (i.e., averaged over a wiggler period) of the relative phase of the TE_{01} mode increases monotonically with axial position in much the same manner as in the single-mode example (see Fig. 5.41). In contrast, the

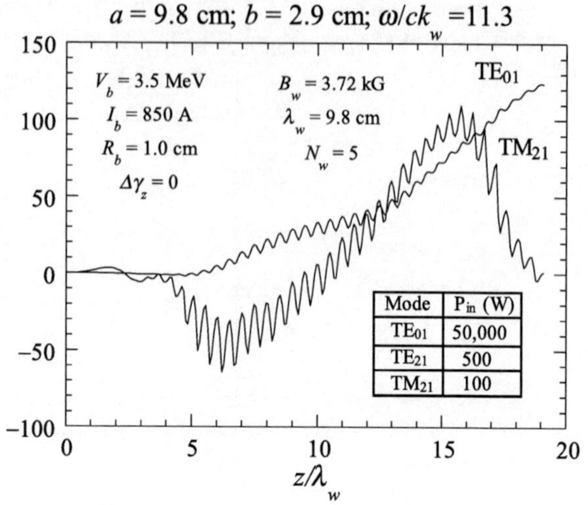

Fig. 5.57 The evolution of the relative phases of the TE_{01} and TM_{21} modes

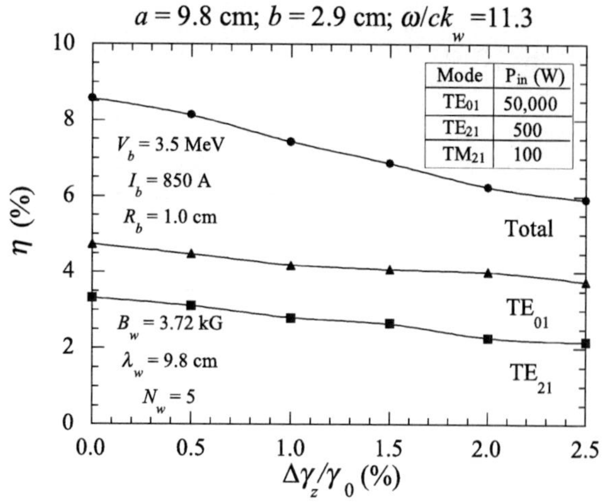

Fig. 5.58 Variation in the saturation efficiency of the total signal and the TE modes with the initial axial energy spread

relative phases of both the TE_{21} and TM_{21} modes are decidedly not monotonic and exhibit a decrease with axial position after a point somewhat beyond saturation. This is a multiple-mode effect since the relative phases of each of these modes also exhibit a monotonic increase with axial position in the single-mode analysis. Finally, we observe that the relative phase varies in an almost identical manner for the TE_{21} and TM_{21} modes. This is attributed to the fact that the vacuum dispersion curves for these modes are degenerate.

The effect of an initial axial energy spread on the saturation efficiency of the total signal as well as on the TE_{01} and TE_{21} modes is shown in Fig. 5.58. The TM_{21} mode is excluded from the figure because it composes a small fraction of the signal. As shown in the figure, the saturation efficiency is relatively insensitive to the axial

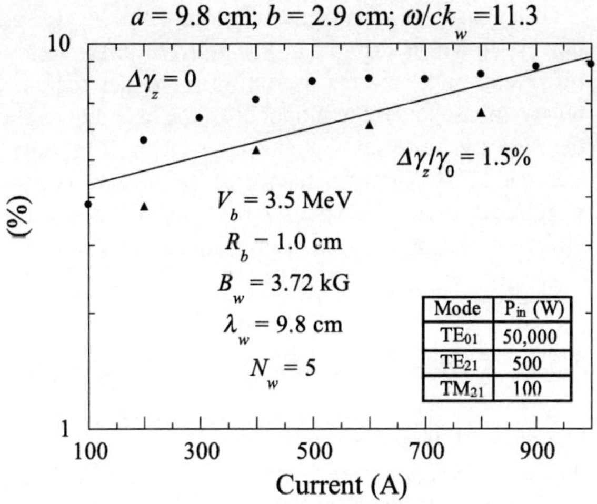

Fig. 5.59 The scaling of the saturation efficiency with electron beam current

energy spread over the range of $\Delta\gamma_z/\gamma_0 \leq 2.5\%$ and decreases from $\eta \approx 8.6\%$ at $\Delta\gamma_z = 0$ to $\eta \approx 5.9\%$ at $\Delta\gamma_z/\gamma_0 \approx 2.5\%$. The reason for this is, as demonstrated in the single-mode analysis, that the growth rate is large and the thermal regime occurs for $\Delta\gamma_z/\gamma_0 \approx 18\%$.

The saturation efficiency in the high-gain Compton regime scales as the cube root of the beam current at the frequency of maximum growth in the idealized one-dimensional model [see Eq. (1.14)]. This type of scaling law is also found to be largely valid in three dimensions, as illustrated in Fig. 5.59. In this figure, the saturation efficiency is plotted as a function of beam current for an ideal beam ($\Delta\gamma_z = 0$) as well as for the case of $\Delta\gamma_z/\gamma_0 = 1.5\%$. Observe that in both cases the efficiency scales approximately as $I_b^{1/3}$.

Turning to the question of the enhancement of the efficiency by means of a tapered wiggler field, we plot the evolution of the power versus axial position in Fig. 5.59 for parameters consistent with the uniform-wiggler examples. In this case, however, we assume that the axial energy spread is $\Delta\gamma_z/\gamma_0 = 1.5\%$ and taper the wiggler downward with a slope $\varepsilon_w = -0.0078$. The saturation point for this case occurs at $k_w z \approx 96$ for a uniform wiggler, and the optimum start-taper point is found to be $k_w z \approx 86$. As is evident from the figure, the total power of the signal increases to a level of approximately 1.2 GW for an efficiency of $\eta \approx 40.6\%$ if the wiggler field is tapered to zero. This represents an enhancement of more than 600% relative to the total efficiency for the uniform wiggler. In addition, it should be remarked that virtually the entire signal is in the TE_{01} mode; hence, we conclude that it is possible to selectively enhance a single mode through the tapered wiggler interaction. Observe that the uniform-wiggler interaction for these parameters yields a total efficiency of approximately 76.8% of which the TE_{01} mode comprises some 60% of the total. In contrast, the TE_{01} mode comprises some 99% of the signal at the end of the tapered wiggler interaction. Both the TE_{21} and TM_{21} [not shown in the figure] ultimately decay to extremely low intensities.

The phase variation of the TE_{01} mode in the tapered wiggler multimode interaction is shown in Fig. 5.61. The relative phase follows the variation indicated in Figs. 5.56 and 5.57 for the uniform-wiggler region. Subsequent variation of the relative phase under the influence of the tapered wiggler shows a relative slowing of the increase up to an asymptotic level that is approximately 120° for the present example. This type of behavior of the relative is consistently found for a tapered wiggler whether (1) the configuration is that of a helical or planar wiggler or (2) the interaction is single mode or multimode.

5.5.5 Comparison with the ELF Experiment at LLNL

The aforementioned example nominally corresponds to an experiment conducted at the Lawrence Livermore National Laboratory [5, 32], which employed an induction linac with a 20–40 ns flat top beam. As such, the slippage time was of the order of 10–20 ps, which is much greater than the electron bunch time; hence, the steady-state formulation is appropriate for this case. The principal differences between the analysis and the experiment are that in the experiment (1) the beam was injected into the wiggler through an entry taper region only one wiggler period in length and (2) a quadrupole field was used to provide for enhanced focusing instead of parabolically tapered pole faces. Since the fringing fields associated with the wiggler field in the entry taper region were not included in the analytical model for this example, it is invalid to apply the numerical simulation for $N_w = 1$. However, a choice of $N_w = 5$ is made as a compromise and gives good agreement with the experiment, subject to the additional assumption of an initial axial energy spread of 1.5%. This is within an upper bound of 2% on the axial energy spread established by means of an electron spectrometer measurement.

The experimental measurement for a uniform-wiggler interaction resulted in a saturated power level of 180 MW over a length of 1.3 m. The simulation for this case gives a peak power of 204 MW and a saturated average power (i.e., averaged over the $\lambda_w/2$ oscillation due to the lower beat wave) of 185 MW. Given the experimental uncertainties in high-power measurements, the latter figure is more relevant for comparison and is in substantial agreement with the observations. In addition, the saturation length found in simulation (i.e., the length of the uniform-wiggler region plus one wiggler period for the entry taper) is 1.1 m, which is also in good agreement and shows that the average growth rate from the nonlinear simulation is close to that found in the experiment. Note that the rapid oscillation due to the lower beat wave introduces a 10–20% uncertainty in the measurement of the power and phase. A comparison can also be made with the experiment for a tapered wiggler. In this case, the wiggler field was tapered downward by 55% (i.e., $\Delta B_w/B_w \approx 0.55$) over a length of 1.1 m in the experiment, and the efficiency was observed to increase to 34% for a total power of approximately 1 GW. This corresponds to the slope employed in the calculation shown in Figs. 5.60 and 5.61. The calculation shown in these figures represents a tapering of the wiggler field to zero. However, the calculation is in

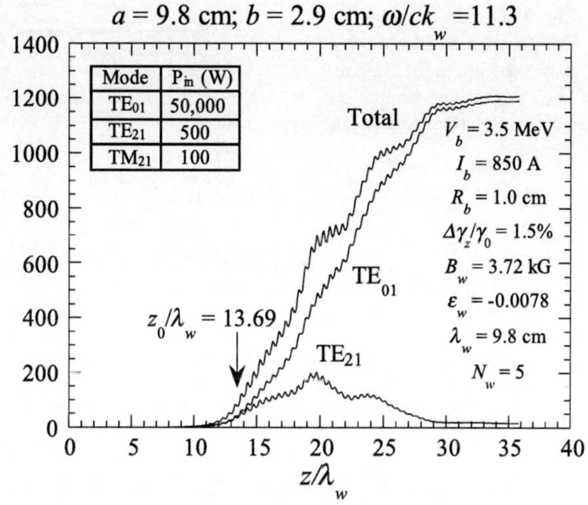

Fig. 5.60 The evolution of the total signal and the TE modes for a tapered wiggler

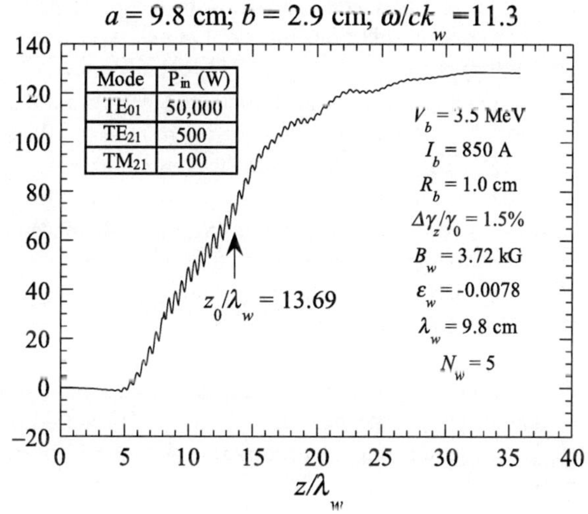

Fig. 5.61 The evolution of the axial phase of the TE_{01} mode for the multimode tapered wiggler interaction

substantial agreement with the experiment over a length of 1.1 m, for which a total power level of 1 GW is also recovered. The evolution of the relative phase is also in good agreement with the reported measurements of this quantity [39]. Thus, within the uncertainties imposed by the choices of the length of the entry taper region and the initial axial energy spread, the essential physics of the interaction are included within the formulation.

This experiment is transitional between the Compton and Raman regimes on the basis of the idealized criterion (4.102), where $\omega_b/\gamma_0^{1/2}ck_w \approx 0.25$ and $\gamma_z^3 v_w^2/16c^2 \approx 0.17$. As a result, it might be expected that space-charge effects play some role. However, the linewidth determined from the nonlinear simulation as shown in

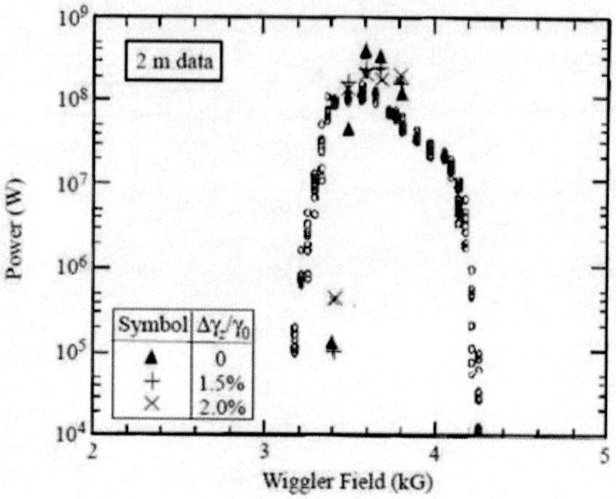

Fig. 5.62 Comparison of the tuning curves for the ELF experiment [3] shown as circles and the simulation over a 2 m interaction length

Fig. 5.41 is approximately 15 GHz, which is much greater than the plasma frequency ($\approx$ 2.2 GHz). In addition, the axial energy spread ($\Delta\gamma_z/\gamma_0 \leq 2\%$) yields a Debye length of approximately 0.8 cm. This is comparable to the space-charge wavelength of about 0.8 cm; hence, the space-charge waves are strongly damped. The issue, therefore, is not whether this experiment is in the Raman regime (it is not) but the extent to which space-charge effects were important. This question can be addressed by comparison of the nonlinear theory with a detailed experimental spectrum.

Such a comparison is shown in Fig. 5.62 where we plot output power after 2 m as calculated by the nonlinear simulation (for three choices of the axial energy spread) with an experimentally measured tuning curve. The data here are taken from Ref. [55]. Note that (1) since the experiment was a 34.6 GHz amplifier driven by a magnetron, the tuning is accomplished by varying the magnetic field and (2) saturation was found to occur over a length of 1.4 m. In view of the latter point, a detailed comparison of the spectral width is not valid because sidebands are expected to result in spectral broadening after saturation, and the nonlinear analysis as described herein does not include sidebands in the formulation. Be that as it may, the agreement between the spectral peak predicted by the simulation and found in the experiment is excellent and does not vary greatly with the choices of axial energy spread. The agreement between the peaks from the simulation and the experiment is to within approximately 30 G, which is well within the experimental uncertainty. As a result, we conclude that space-charge effects did not play a role in this experiment.

5.6 The Inclusion of Space-Charge Waves in Three Dimensions

The treatment of collective effects will be discussed for a helical wiggler/axial solenoidal magnetic field configuration in which the electron the electron beam propagates through a cylindrical waveguide. Within the context of this formulation [discussed in Ref. [41]], the space-charge waves are treated in an approximate fashion by means of an expansion of the electrostatic potential in terms of the Gould-Trivelpiece modes of a fully filled cylindrical waveguide. The axial component of these electrostatic modes exhibits the same transverse mode variation as the TM modes of the waveguide, and the eigenvalue (i.e., cutoff wavenumber) is also identical. Hence, the Gould-Trivelpiece modes [56] are orthogonal to the TE modes in the sense that there are no direct mode-mode coupling terms present in the dynamical equations for the fields and the coupling occurs through the particle dynamics. In contrast, there are direct mode-mode coupling terms between the TM and space-charge modes. Because of this, the admixture of TE and TM modes is neglected from the analysis, and the formulation is restricted to the collective interaction in the presence of TE modes only. In addition, collective effects on the interaction are typically important only in the Raman regime. Raman free-electron lasers generally operate in the mm-wave spectrum and employ high-current, long-pulse accelerators; hence, slippage is not important, and a steady-state formulation is appropriate.

5.6.1 The Raman Criterion

For convenience, we restate the condition required for the dominance of the Raman regime as

$$\frac{\omega_b}{\gamma_0^{1/2} c k_w} >> \frac{\gamma_z^3}{16} \frac{v_w^2}{c^2}, \tag{5.175}$$

where ${\omega_b}^2 = 4\pi e^2 n_b/m_e$, n_b is the ambient beam density, γ_0 is the relativistic factor corresponding to the bulk beam energy, and ${\gamma_z}^2 = (1 - {v_z}^2/c^2)^{-1}$ for a bulk axial velocity v_z. In addition, $v_w = -\Omega_w/k_w$ is the transverse wiggle velocity, where $\Omega_w = eB_w/\gamma_0 m_e c$ for a wiggler amplitude B_w and k_w is the wiggler wavenumber for a period λ_w. For a planar wiggler, the rms wiggler amplitude must be used in v_w.

This criterion is derived by examining the relative importance of the ponderomotive and space-charge forces on the linear growth rate in the idealized one-dimensional regime and is a measure of the dominance of the space-charge forces. However, this criterion must be used with some caution in characterizing free-electron laser experiments, and its application to real systems is clouded by several factors. Firstly, the boundary conditions imposed by the drift tube walls reduce the effective plasma frequency. Secondly, the bulk characteristics of the electron orbits are modified by wiggler inhomogeneities, beam thermal effects, and the use of an axial guide magnetic field. Planar wiggler configurations introduce further difficulties since, in contrast to a helical wiggler, the axial and transverse electron velocities are oscillatory. Due to these difficulties, a full three-dimensional nonlinear analysis is often required to characterize space-charge effects in any given experiment.

In addition to the abovementioned criterion for the importance of space-charge effects, a second criterion required for space-charge effects to be important is that the Raman frequency shift be comparable to or greater than the linewidth of the free-electron laser. The physical interpretation of this criterion is that the wiggler must be long enough for several plasma oscillations during the course of the interaction. Of course, realistic three-dimensional effects can be expected to modify this condition as well.

Finally, a third criterion required for space-charge effects to play an important role is that Landau damping of the space-charge waves due to the thermal spread of the beam must be small. In general, Landau damping of space-charge waves is important for wavelengths less than the Debye length.

5.6.2 The Field Equations

When expressed in terms of the Gould-Trivelpiece modes of a fully filled cylindrical waveguide, the space-charge potential can be written in the form

$$\delta\Phi(\mathrm{x},t)=\sum_{\substack{l=0\\n=1}}^{\infty}\delta\Phi_{ln}(z)J_l(\kappa_{ln}r)\cos\alpha_{ln}^{(sc)}(z,t), \tag{5.176}$$

where $\kappa_{ln}=x_{ln}/R_g$ for $J_l(x_{ln})=0$ is the same cutoff wavenumber as obtained for a TM_{ln} mode and the phase is given by

$$\alpha_{ln}^{(sc)}\equiv\int_0^z dz' k_{ln}^{(sc)}(z')+l\theta-\omega t, \tag{5.177}$$

for a wave of frequency ω and wavenumber $k_{ln}{}^{(sc)}(z)$. Observe that the amplitude and wavenumber are implicitly assumed to vary slowly with respect to the wavelength. At the outset (i.e., at $z=0$), the wavenumbers of each mode satisfy a dispersion equation

$$\left(\omega - k_{ln}^{(sc)} v_{z0}\right)^2 - \frac{\omega_b^2}{\gamma_0^3} = -\frac{\kappa_{ln}^2}{k_{ln}^{(sc)2}} \frac{\left(\omega - k_{ln}^{(sc)} v_{z0}\right)^2}{\left(\omega - k_{ln}^{(sc)} v_{z0}\right)^2 - \Omega_0^2}$$

$$\times \left[\left(\omega - k_{ln}^{(sc)} v_{z0}\right)^2 - \Omega_0^2 - \frac{\omega_b^2}{\gamma_0} \frac{\omega - k_{ln}^{(sc)} v_{z0}}{\omega}\right], \tag{5.178}$$

where γ_0 denotes the bulk initial relativistic factor for the beam and v_{z0} is the initial axial velocity. Observe that this dispersion equation reduces to that of the space-charge modes in the idealized one-dimensional regime in the limit in which the wavenumber is much larger than the cutoff (i.e., $k_{ln}{}^{(sc)} >> \kappa_{ln}$).

The Maxwell-Poisson equations for this configuration are of the form

$$\left(\nabla^2 - \frac{1}{c^2}\frac{\partial^2}{\partial t^2}\right)\delta\mathbf{A}(\mathbf{x},t) - \frac{1}{c}\frac{\partial}{\partial t}\nabla\delta\Phi(\mathbf{x},t) = -\frac{4\pi}{c}\delta\mathbf{J}(\mathbf{x},t), \tag{5.179}$$

and

$$\nabla^2\delta\Phi(\mathbf{x},t) = -4\pi\delta\rho(\mathbf{x},t), \tag{5.180}$$

where the vector potential is assumed to be given in terms of a superposition of TE modes of the vacuum waveguide [Eq. (5.85)], the source current is defined in Eq. (5.89), and the charge density is defined as

$$\delta\rho(\mathbf{x},t) = -en_b \iiint d^3p v_{z0} F_b(\mathbf{p}_0) \iint_{A_g} dx_0 dy_0 \sigma_\perp(x_0 y_0) \int_{-T/2}^{T/2} dt_0 \sigma_\parallel(t_0)$$

$$\times \delta[x - x(z; x_0, y_0, t_0, \mathbf{p}_0)]\delta[y - y(z; x_0, y_0, t_0, \mathbf{p}_0)] \frac{\delta[t - \tau(z; x_0, y_0, t_0, \mathbf{p}_0)]}{|v_z(z; x_0, y_0, t_0, \mathbf{p}_0)|}. \tag{5.181}$$

Due to the complexity of the numerical solution for Poisson's equation, it is more convenient to describe the evolution of the space-charge modes in terms of the z-component of Maxwell's equations (5.178). Since the z-component of the vector potential for the TE modes vanishes, this equation is

$$\sum_{\substack{l=0 \\ n=1}}^{\infty} J_l(\kappa_{ln} r)\left(\frac{d}{dz}\delta\Phi_{ln}\sin\alpha_{ln}^{(sc)} + k_{ln}^{(sc)}\delta\Phi_{ln}\cos\alpha_{ln}^{(sc)}\right) = \frac{4\pi}{\omega}\delta J_z. \tag{5.182}$$

Orthogonalization of this equation in the azimuthal angle yields two equations

$$\sum_{\substack{l=0\\n=1}}^{\infty} J_l(\kappa_{ln}r)\left(\frac{d}{dz}\delta\Phi_{ln}\sin\Theta_{ln}^{(sc)} + k_{ln}^{(sc)}\delta\Phi_{ln}\cos\Theta_{ln}^{(sc)}\right) = \frac{4\pi}{\omega}\int_0^{2\pi} d\theta\delta J_z\cos l\theta, \tag{5.183}$$

and

$$\sum_{\substack{l=0\\n=1}}^{\infty} J_l(\kappa_{ln}r)\left(\frac{d}{dz}\delta\Phi_{ln}\cos\Theta_{ln}^{(sc)} - k_{ln}^{(sc)}\delta\Phi_{ln}\sin\Theta_{ln}^{(sc)}\right) = \frac{4\pi}{\omega}\int_0^{2\pi} d\theta\delta J_z\sin l\theta, \tag{5.184}$$

where

$$\Theta_{ln}^{(sc)} \equiv \int_0^z dz' k_{ln}^{(sc)}(z') - \omega t. \tag{5.185}$$

Orthogonalization in r using Eq. (5.107) and the subsequent time average over a wave period yields the following dynamical equations governing the evolution of the amplitude and wavenumber of the space-charge modes

$$k_{ln}^{(sc)}\delta\varphi_{ln} = -2\frac{\omega_b^2}{\omega c}\frac{1}{J_{l+1}^2(x_{ln})}\left\langle J_l(\kappa_{ln}r)\cos\Theta_{ln}^{(sc)}\right\rangle, \tag{5.186}$$

and

$$\frac{d}{dz}\delta\varphi_{ln} = -2\frac{\omega_b^2}{\omega c}\frac{1}{J_{l+1}^2(x_{ln})}\left\langle J_l(\kappa_{ln}r)\sin\Theta_{ln}^{(sc)}\right\rangle, \tag{5.187}$$

where ω_b is the beam plasma frequency corresponding to the ambient beam density, $\delta\varphi_{ln} = e\delta\Phi_{ln}/m_e c^2$, and the average is defined as in Eq. (5.125) and makes use of the momentum space distribution in (5.122).

The transverse components of Maxwell's equations may be orthogonalized in the same manner as that employed previously for the TE modes in the high-gain Compton regime. In this case, the contributions due to the space-charge modes vanish, and Eqs. (5.98) and (5.99) [or, alternately, Eqs. (5.135) and (5.136)] which govern the evolution of the TE modes are recovered. Hence, the dynamical equations which determine the evolution of the amplitude and wavenumber of each electromagnetic TE mode [Eqs. (5.135) and (5.136)] and space-charge mode [Eqs. (5.185) and (5.186)] are determined. The detailed coupling between each of these modes is governed by the orbit equations for the ensemble of electrons.

5.6.3 The Electron Orbit Equations

The electron trajectories are governed by the complete Lorentz force equations in the combined magnetostatic, electromagnetic, and electrostatic fields and may be written as a generalization of Eqs. (5.128)–(5.130)

$$\begin{aligned} \upsilon_3 \frac{d}{dz} p_1 = &- \{\Omega_0[1 + k_w \varepsilon_0(z - z_0)] - k_w \upsilon_3 + 2\Omega_w I_1(\lambda) \sin\chi\} p_2 \\ &+ \Omega_w p_2 I_2(\lambda) \sin 2\chi - \frac{\varepsilon_0}{2} \Omega_0 p_3 \lambda \sin\chi \\ &- \frac{m_e c}{2} \sum_{TE\ modes} \delta a_{ln} \left[(\omega - k_{ln}\upsilon_3) W_{ln}^{(-)} - 2\kappa_{ln}\upsilon_2 J_l(\kappa_{ln} r) \cos\alpha_{ln} - \Gamma_{ln}\upsilon_3 T_{ln}^{(+)}\right] \\ &+ \frac{m_e c^2}{2} \sum_{sc\ modes} \kappa_{ln} \delta\varphi_{ln} \left(\overline{F}_{ln}^{(-)} \cos\overline{\psi}_{ln} - \overline{G}_{ln}^{(-)} \sin\overline{\psi}_{ln}\right), \end{aligned} \tag{5.188}$$

$$\begin{aligned} \upsilon_3 \frac{d}{dz} p_2 = &\{\Omega_0[1 + k_w \varepsilon_0(z - z_0)] - k_w \upsilon_3 + 2\Omega_w I_1(\lambda) \sin\chi\} p_1 \\ &- \Omega_w p_3 [I_0(\lambda) + I_2(\lambda) \cos 2\chi] + \frac{\varepsilon_0}{2} \Omega_0 p_3 \lambda \cos\chi \\ &+ \frac{m_e c}{2} \sum_{TE\ modes} \delta a_{ln} \left[(\omega - k_{ln}\upsilon_3) T_{ln}^{(-)} - 2\kappa_{ln}\upsilon_2 J_l(\kappa_{ln} r) \cos\alpha_{ln} - \Gamma_{ln}\upsilon_3 W_{ln}^{(+)}\right] \\ &- \frac{m_e c^2}{2} \sum_{sc\ modes} \kappa_{ln} \delta\varphi_{ln} \left(\overline{F}_{ln}^{(+)} \sin\overline{\psi}_{ln} + \overline{G}_{ln}^{(+)} \cos\overline{\psi}_{ln}\right), \end{aligned} \tag{5.189}$$

and

$$\begin{aligned} \upsilon_3 \frac{d}{dz} p_3 = &\Omega_w p_2 [I_0(\lambda) + I_2(\lambda) \cos 2\chi] - \Omega_w p_1 I_1(\lambda) \sin 2\chi \\ &+ \frac{\varepsilon_0}{2} \Omega_0 \lambda (p_1 \sin\chi - p_2 \cos\chi) \\ &- \frac{m_e c}{2} \sum_{TE\ modes} \delta a_{ln} \left[k_{ln}\left(\upsilon_1 W_{ln}^{(-)} - \upsilon_2 T_{ln}^{(-)}\right) + \Gamma_{ln}\left(\upsilon_1 T_{ln}^{(+)} + \upsilon_2 W_{ln}^{(+)}\right)\right] \\ &+ m_e c^2 \sum_{sc\ modes} J_l(\kappa_{ln} r) \left(\frac{d}{dz} \delta\varphi_{ln} \cos\alpha_{ln}^{(sc)} - k_{ln}^{(sc)} \delta\varphi_{ln} \sin\alpha_{ln}^{(sc)}\right). \end{aligned} \tag{5.190}$$

Observe that these equations include the effects of both tapered axial and wiggler magnetic fields and that

$$\overline{F}_{ln}^{(\pm)} \equiv J_{l-1}(\kappa_{ln} r) \cos[(l - 1)\chi] \pm J_{l+1}(\kappa_{ln} r) \cos[(l + 1)\chi], \tag{5.191}$$

$$\overline{G}_{ln}^{(\pm)} \equiv J_{l-1}(\kappa_{ln}r)\sin[(l-1)\chi] \pm J_{l+1}(\kappa_{ln}r)\sin[(l+1)\chi], \tag{5.192}$$

and

$$\overline{\Psi}_{ln} \equiv \int_0^z dz' \left[k_{ln}^{(sc)}(z') + lk_w\right] - \omega t. \tag{5.193}$$

Observe that the appropriate cutoffs must be employed in the summations over the TE and space-charge modes. In addition, Eqs. (5.132)–(5.134) for the coordinates must also be integrated, as well as an equation for the phase of each space-charge mode

$$\frac{d}{dz}\overline{\Psi}_{ln} = k_{ln}^{(sc)} + lk_w - \frac{\omega}{\upsilon_z}. \tag{5.194}$$

As a result, the problem in its reduced form consists in the integration of equations for the amplitude and phase of each TE and space-charge mode included in the simulation and six equations per electron, as well as the evaluation of an algebraic equation for the wavenumber of each mode. It is important to bear in mind that the collective wave-particle resonance occurs for ${k_{ln}}^{(sc)} \approx k_{ln} + k_w$ so that the interaction is found when the space-charge mode has an azimuthal mode number which is one unit less than the TE mode. Specifically, the collective interaction of a TE_{11} mode and a space-charge mode will occur for an azimuthally symmetric space-charge mode (i.e., $l = 0$); however, an arbitrary number of radial eigenmodes may participate in the interaction.

5.6.4 Numerical Examples

The initial conditions for the TE modes and the electrons considered in the collective regime are identical to those employed in the high-gain Compton limit and are chosen to model the injection of an axisymmetric beam into the waveguide in conjunction with an ensemble of TE modes of arbitrary initial powers. The space-charge modes describe the effect of shot noise on the beam and must be initialized in a self-consistent fashion. For this purpose, we assume that the initial wavenumber is given approximately by

$$k_{ln}^{(sc)}(z=0) = \frac{\omega}{\upsilon_{z0}} + \frac{\omega_b}{\gamma_0^{3/2}\upsilon_{z0}}\sqrt{1 - \frac{\kappa_{ln}^2\upsilon_{z0}^2}{\omega^2}}, \tag{5.195}$$

The initial amplitude of the space-charge modes is determined by means of the initial phase average

$$k_{ln}^{(sc)}(z=0)\delta\varphi_{ln}(z=0) = -2\frac{\omega_b^2}{\omega c}\frac{1}{J_{l+1}^2(x_{ln})}\left\langle J_l(\kappa_{ln}r)\cos\Theta_{ln}^{(sc)}\right\rangle, \tag{5.196}$$

which is a measure of the noise introduced by the discretization of the electron beam.

The first case we consider is that of the 35 GHz amplifier operating for Group I orbit parameters which was studied in the high-gain Compton regime and shown in Fig. 5.17. The specific parameters of this example involve the propagation of a 250 keV/35 A electron beam with an initial beam radius of 0.155 cm through a waveguide with a radius 0.36626 cm in the presence of a wiggler with an amplitude of 2.0 kG and a period of 1.175 cm and an axial magnetic field of 1.3 kG. The interaction for this case was with the TE_{11} mode. In view of the relatively low current, the high-gain Compton limit is expected to be an adequate approximation for this example. As shown in Fig. 5.17, a maximum efficiency of approximately 21.4% is found at a frequency of 33.2 GHz in the high-gain Compton regime. The collective effects upon this interaction are illustrated in Fig. 5.63, in which we plot the efficiency as a function of frequency for this case for both the high-gain Compton and collective Raman limits. The essential physics of the collective interaction are treated by means of a single space-charge mode with $l = 0$ and $n = 1$. It is evident from the figure that collective effects reduce both the bandwidth and efficiency of the interaction. However, as expected, the reductions are not large. In addition, a comparison of the linear growth rates found from the simulation and the linear theory described in Chap. 4 is shown in Fig. 5.64. The dashed line in the figure represents the linearized analysis, and good agreement is also found near the center of the gain band. The discrepancies that appear near the edges of the gain band result from the approximation imposed by the use of Gould-Trivelpiece modes for a fully filled waveguide which have the effect of underestimating the beam-plasma frequency by a small amount.

Collective effects for a 35 GHz amplifier operating with Group II negative-mass orbit parameters have also been investigated. The magnetostatic fields for this example involve a wiggler field amplitude of 1.0 kG and period of 3.0 cm and an axial magnetic field of 11.75 kG. The electron beam is characterized by an energy of 1.0 MeV, a current of 50 A, and an initial beam radius of 0.2 cm. The interaction is with a single TE_{11} mode in a waveguide of radius 0.5 cm. The results for the high-gain Compton regime simulation are shown in Fig. 5.65 in which the efficiency is plotted versus frequency and a peak efficiency of the order of 47% is found at a frequency of 31 GHz. As in the previous case, the current for this example was chosen to be small enough that collective effect would not be important, and this is borne out by the simulation. The collective effects upon this case are adequately handled by inclusion of a single space-charge mode [$l = 0$ and $n = 1$], and the results of the simulation are shown in Fig. 5.65. As shown in the figure, the effect of the inclusion of the space-charge modes is to reduce both the efficiency and bandwidth

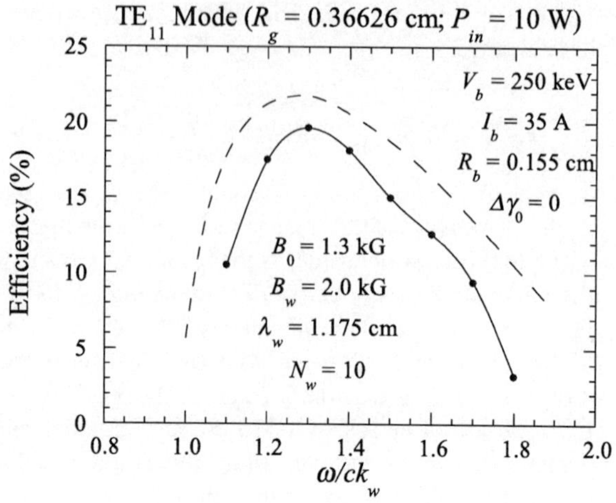

Fig. 5.63 Graph of the interaction efficiency versus frequency for both the high-gain Compton (dashed line) and collective Raman simulations of a 35 GHz amplifier operating on Group I orbit parameters

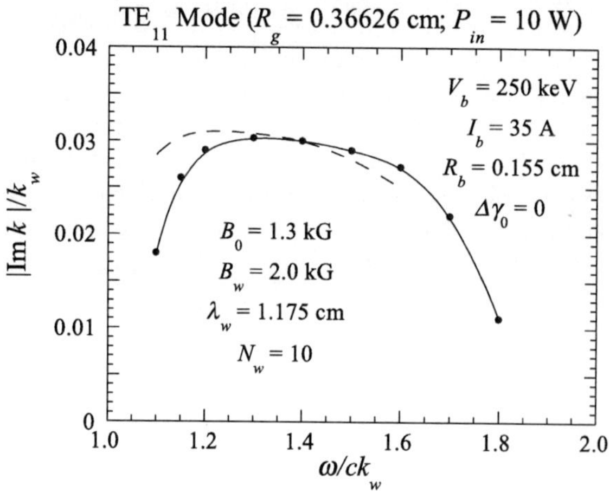

Fig. 5.64 Comparison of the growth rates found from simulation and the linearized theory (dashed line) for 35 GHz amplifier operating on Group I orbit parameters

of the interaction; however, these reductions are not large and the simulation in the high-gain Compton limit represents an adequate approximation for all practical purposes.

5.6.5 *Comparison with Experiments*

Two experiments dealing with 35 GHz amplifiers in the Raman regime have been conducted at the Massachusetts Institute of Technology. These experiments were conducted in the same laboratory and used helical wigglers and axial guide magnetic fields. However, one experiment operated at X-band frequencies between 9 and 10 GHz in which the wiggler and axial magnetic fields were oriented in the same

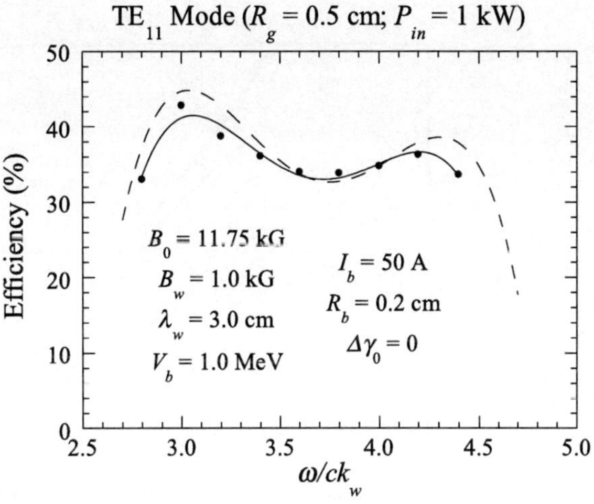

Fig. 5.65 Graph of the interaction efficiency versus frequency for both the high-gain Compton (dashed line) and collective Raman simulations of a 35 GHz amplifier operating on Group II negative-mass orbit parameters

direction. The second experiment was a 35 GHz amplifier in which the wiggler and axial magnetic fields were oriented in opposite directions; hence, we shall refer to this as the reversed-field amplifier.

5.6.5.1 An X-Band Amplifier

An example of an experiment conducted at the Massachusetts Institute of Technology [54] in which collective effects were important is shown in Figs. 5.64 and 5.65 in which we plot a comparison of the simulation with experimental measurements of both the output power and phase of the signal as functions of beam energy. The electron beam in this experiment ranged in energy from about 143 to 159 keV with a current of approximately 4.1 A and an initial beam radius of 0.25 cm. The best estimate of the axial velocity spread of the beam in the experiment yielded a value of $\Delta v_z/v_0 \approx 0.5\%$, and this value is employed in the simulation. The electron beam propagated through a drift tube with a radius of 1.27 cm, and the interaction was with a single TE_{11} mode. This experiment employed a wiggler field with an amplitude of 1.45 kG, a period of 3.3 cm, and an entry taper region of six wiggler periods in length. The axial magnetic field for the specific case shown was 1.45 kG which corresponds to Group I orbit parameters. The experiment was operated as an amplifier with an input signal of 27 kW [linearly polarized] at a frequency of 9.318 GHz. As in the previous experiments described, this experiment employed a pulse-line accelerator with a 20–40 ns flat top so that the slippage time was much less than the bunch duration.

The output power is plotted as a function of beam energy in Fig. 5.67 in which the solid line represents the simulation and the dashed line represents the experimental measurements. The agreement between the simulation and the experiment is excellent for beam energies in the vicinity of the peak output power that corresponds to

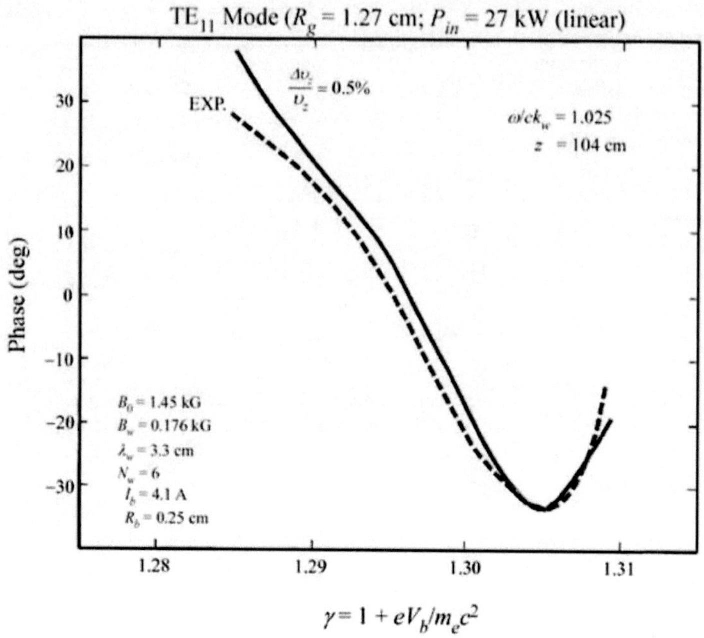

Fig. 5.66 Comparison of the output power versus beam energy from experiment and simulation

the center of the gain band. However, it is evident that the simulation shows a secondary peak at lower beam energies. The reason for this peak stems from the approximate representation of the electrostatic potential that, as noted previously, underestimates the value of beam plasma frequency by a small amount. This has the effect of separating the two intersection points between the TE mode and space-charge mode dispersion curves which, in the present case, is enough to introduce two peaks corresponding to each intersection point. However, the phase of the output signal as calculated in the simulation, as plotted in Fig. 5.66 versus beam energy, is seen to be in good agreement with the measured value over the entire range of energies studied (Fig. 5.67).

5.6.5.2 The Reversed-Field Configuration

Thus far, the primary consideration in the treatment of free-electron lasers with the combination of helical wigglers and axial guide fields has been devoted to configurations in which the guide field is directed parallel to the wiggler since the bulk of free-electron laser experiments conducted to date have employed this geometry. However, the physical mechanism of the free-electron laser will also operate when the axial guide field is directed antiparallel to the wiggler field. We shall refer to this as a *reversed-field* configuration.

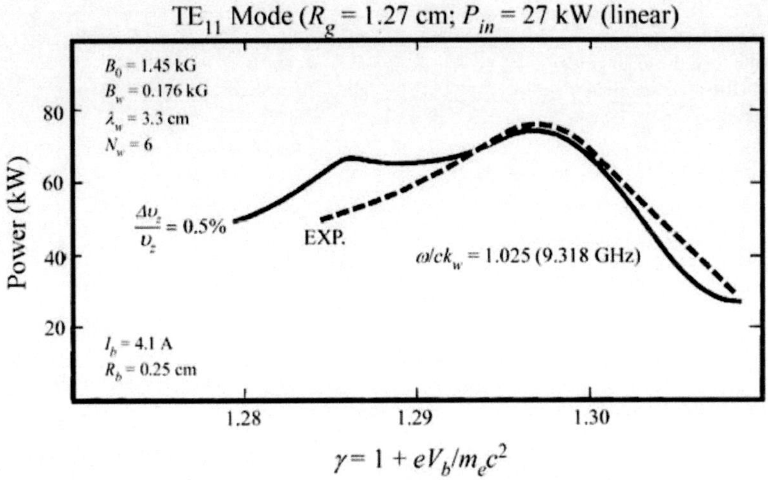

Fig. 5.67 Comparison of the phase of the output signal versus beam energy from experiment and simulation

One difference that arises from the reversed-field geometry is that the wiggler-induced transverse velocity is reduced relative to that found in the absence of a guide field. Hence, the linearized gain for the interaction is reduced relative to typical values expected for the Group I and II regimes. However, this does not necessarily imply that the saturation efficiency in the reversed-field configuration is also reduced. Indeed, one experiment that studied both orientations of the guide field found higher efficiencies for the reversed-field orientation [57, 58].

The basic orbital dynamics of electrons in reversed-field configurations follow routinely from the discussion in Chap. 2. The steady-state orbits for this configuration are governed by Eqs. (2.53) and (2.54) as in the case in which the axial magnetic field is directed parallel to the wiggler field. The principal distinction that obtains in the reversed-field configuration is that there is no resonant enhancement in the wiggler-induced transverse velocity v_w when the Larmor period associated with the axial field is comparable to the wiggler period. As a result, there is only one class of trajectory for this configuration, and it is a stable one. An example of the variation in the axial velocity as a function of the magnitude of the reversed axial field is shown in Fig. 5.68 in which we plot $v_{\|}/c$ versus the magnitude of the axial field for a beam energy of 750 keV and a wiggler field with a magnitude of 1.47 kG and a period of 3.18 cm. As shown in the figure, the axial velocity is relatively constant over the entire range of guide fields shown. Note that the parameters used in the figure correspond to a reversed-field free-electron laser amplifier experiment [57, 58] that will be analyzed in detail.

However, the electron dynamics in the reversed-field configuration is far more complex than is indicated by the study of the steady-state trajectories. The reason for this is that the inhomogeneity in the wiggler field introduces a sinusoidal driving

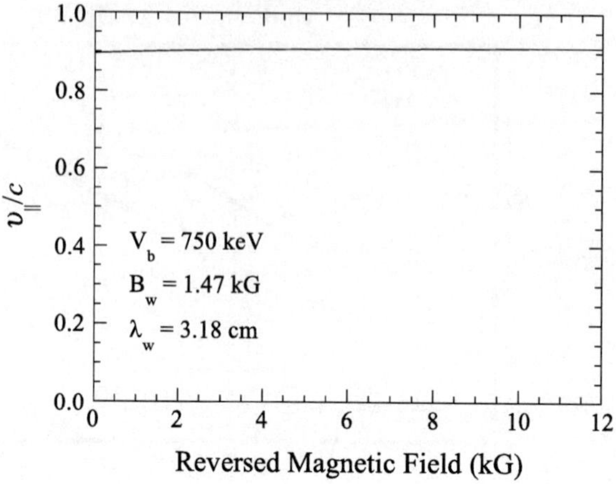

Fig. 5.68 Plot of the axial velocity versus the magnitude of the axial guide field in the reversed-field orientation

term to the electron orbit equations [59–62]. This term arises from the fact that an electron on a helical orbit centered off the axis of symmetry experiences a sinusoidally varying wiggler field which acts to drive the electrons at a period close to the wiggler period. Hence, this effect becomes important for electron beams that are large enough that a substantial fraction of the electrons is located relatively far from the symmetry axis. In addition, there is a resonant enhancement in this effect when the Larmor period associated with the axial field is comparable to the wiggler period. We shall discuss this *antiresonance* in more detail in the comparison of the simulation with the aforementioned experiment, since the orbital irregularities introduced by this effect have a significant impact on the output of the free-electron laser.

The fundamental physics of this effect of the wiggler inhomogeneity and the antiresonant enhancement process can be understood by means of a relatively simple treatment of the single-particle orbit dynamics [61]. In order to accomplish this, we turn to a guiding-center formalism. In this treatment, the electron position and velocity are written as $\mathbf{x} = \mathbf{x}_c + \mathbf{x}_{osc}$, where the subscript "*c*" denotes the guiding-center position and "*osc*" denotes the various oscillatory motions. Under the assumption that the guiding-center position is fixed [i.e., $\mathbf{v} = \mathbf{v}_{osc}$], expansion of the orbit equations about the guiding-center position results in the following equations for the electron velocity [where we drop the subscript *osc* for convenience]:

$$\frac{d}{dt}v_1 = -(\Omega_0 - k_w v_3)v_2 + \Omega_w v_3 I_2(\lambda_c)\sin 2\chi_c, \tag{5.197}$$

$$\frac{d}{dt}v_2 = (\Omega_0 - k_w v_3)v_1 - \Omega_w v_3[I_0(\lambda_c) + I_2(\lambda_c)\cos 2\chi_c], \tag{5.198}$$

$$\frac{d}{dt}v_3 = \Omega_w v_2[I_0(\lambda_c) + I_2(\lambda_c)\cos 2\chi_c] - \Omega_w v_1 I_2(\lambda_c)\sin 2\chi_c, \tag{5.199}$$

where (r_c,θ_c) denote the guiding-center position in cylindrical coordinates and $\lambda_c = k_w r_c$ and $\chi_c = \theta_c - k_w z$. If we now expand about the steady-state orbits via $\upsilon_1 = \upsilon_w + \delta\upsilon_1$, $\upsilon_2 = \delta\upsilon_2$, and $\upsilon_3 = \upsilon_\parallel + \delta\upsilon_3$, where

$$\upsilon_w \equiv \frac{\Omega_w \upsilon_\parallel}{\Omega_0 - k_w \upsilon_\parallel} I_0(\lambda_c). \tag{5.200}$$

then the equations for the perturbations are

$$\frac{d}{dt}\delta\upsilon_1 = -(\Omega_0 - k_w\upsilon_3)\delta\upsilon_2 + \Omega_w\upsilon_\parallel I_2(\lambda_c)\sin 2\chi_c, \tag{5.201}$$

$$\frac{d}{dt}\delta\upsilon_2 = (\Omega_0 - k_w\upsilon_3)\delta\upsilon_1 - \frac{\upsilon_w}{\upsilon_\parallel}\Omega_w\delta\upsilon_3 - \Omega_w\upsilon_\parallel I_2(\lambda_c)\cos 2\chi_c, \tag{5.202}$$

$$\frac{d}{dt}\delta\upsilon_3 = \Omega_w\delta\upsilon_2 I_0(\lambda_c) - \Omega_w\upsilon_w I_2(\lambda_c)\sin 2\chi_c, \tag{5.203}$$

Note that we have also neglected terms which vary as $\delta\mathbf{v}I_2(\lambda_c)$ under the assumption that $\lambda_c < 1$ as well. In this representation, the electrons execute a helical trajectory centered about the guiding center. In addition, this representation is quasi-idealized in the sense that the transverse velocity includes three-dimensional effects only in the inclusion of the $I_0(\lambda_c)$ function that describes the effect of the off-axis increase in the magnitude of the field at the guiding center. These equations may be reduced to a set of second-order differential equations

$$\left(\frac{d^2}{dt^2} + \Omega_r^2\right)\begin{bmatrix}\delta\upsilon_1\\ \delta\upsilon_2\\ \delta\upsilon_3\end{bmatrix} = \Omega_w\upsilon_\parallel I_2(\lambda_c)\begin{bmatrix}(\Omega_0 - 3k_w\upsilon_\parallel)\cos 2\chi_c\\ \left[(1+\beta_w^2)\Omega_0 - 3k_w\upsilon_\parallel\right]\sin 2\chi_c\\ -\beta_w(\Omega_0 - 3k_w\upsilon_\parallel)\cos 2\chi_c\end{bmatrix}, \tag{5.204}$$

where $\beta_w = \upsilon_w/\upsilon_\parallel$ and Ω_r is given by Eq. (2.21). The particular solutions of these equations are

$$\begin{bmatrix}\delta\upsilon_1\\ \delta\upsilon_2\\ \delta\upsilon_3\end{bmatrix} = \frac{\Omega_w\upsilon_\parallel I_2(\lambda_c)}{(\Omega_0 + k_w\upsilon_\parallel)(\Omega_0 - 3k_w\upsilon_\parallel) + \beta_w^2\Omega_0(\Omega_0 - 3k_w\upsilon_\parallel)}$$
$$\times\begin{bmatrix}(\Omega_0 - 3k_w\upsilon_\parallel)\cos 2\chi_c\\ \left[(1+\beta_w^2)\Omega_0 - 3k_w\upsilon_\parallel\right]\sin 2\chi_c\\ -\beta_w(\Omega_0 - 3k_w\upsilon_\parallel)\cos 2\chi_c\end{bmatrix}. \tag{5.205}$$

In order to demonstrate the antiresonant nature of this effect, consider the limit in which $\beta_w^2 << 1$ for which these particular solutions reduce to the simpler form

$$\begin{bmatrix} \delta v_1 \\ \delta v_2 \\ \delta v_3 \end{bmatrix} = \frac{\Omega_w v_\parallel I_2(\lambda_c)}{(\Omega_0 + k_w v_\parallel)} [\cos 2\chi_c \sin 2\chi_c - \beta_w \cos 2\chi_c], \tag{5.206}$$

which clearly shows the antiresonant enhancement in the perturbation when $\Omega_0 \approx -k_w v_\parallel$. However, since the particular solution also depends upon $I_2(\lambda_c)$, this effect will not become appreciable unless the electron guiding center is located relatively far from the symmetry axis.

Note, however, that the effect of the inhomogeneity in the wiggler field, even near the antiresonance, is not as serious a problem for beam transport than is the Group I and II orbital instabilities which occur when the axial guide field is oriented parallel to the wiggler field.

The basic configuration used in the experiment is that of an amplifier in which a weakly relativistic electron beam is injected into a cylindrical waveguide in the presence of both a helical wiggler field and an axial guide solenoidal field. The wave-particle interaction is with the fundamental TE_{11} mode of the waveguide at a frequency of 33.39 GHz, which corresponds to the frequency of the magnetron used to drive the amplifier.

The electron beam is generated using a Physics International Pulserad 110A by means of field emission from a graphite cathode, and the beam energy used in the experiment is 750 keV (±50 keV). The beam quality [i.e., the emittance and energy spread] delivered to the interaction region is controlled by scraping the beam with a shaped graphite anode. This technique was originally pioneered at the Naval Research Laboratory for use in a free-electron laser experiment driven by the VEBA accelerator [6]. In the experiment of present interest, the shaped anode-cathode geometry results in a beam with a radius of 0.25 cm [corresponding to the radius of the anode aperture] and an axial energy spread estimated to be approximately $\Delta\gamma_z/\gamma_0 \approx 1.5\%$. This energy spread corresponds to a normalized RMS beam emittance of $\varepsilon_n \leq 4.4 \times 10^{-2}$ cm-rad.

The current available using this configuration was of the order of 300 A (±30 A) as delivered to the interaction region at the entrance to the wiggler. However, the amount of current that could be propagated through the wiggler/guide field configuration varied based upon the stability of the electron trajectories. Current propagation data indicated quite different results depending upon the orientation of the axial guide field. The results found for the current propagating through the wiggler as a function of the magnitude of the axial guide field for orientations in which the guide field is directed parallel and antiparallel (referred to as the reversed-field configuration) to the wiggler are shown in Figs. 5.69 and 5.70, respectively (data courtesy of Conde and Bekefi). In each case, the amount of current that could be propagated generally increased with increases in the magnitude of the guide field. An exception to this, however, was found when the guide field was oriented parallel to the wiggler in the vicinity of the magnetic resonance at which the Larmor period associated with the guide field is comparable to the wiggler period. In this case, the orbital instability in the electron trajectories described in Chap. 2 prevents propagation of the beam. In

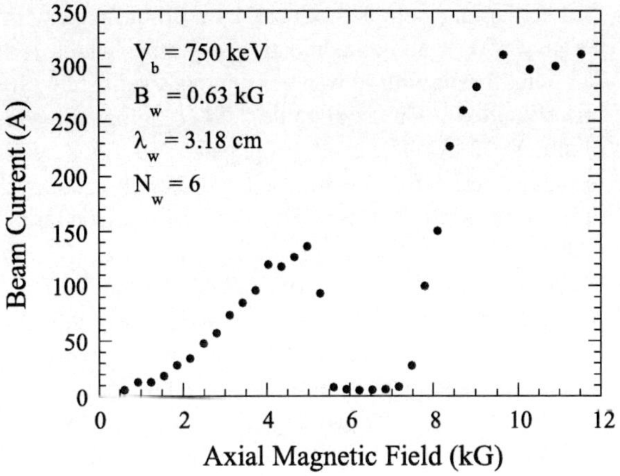

Fig. 5.69 Current propagation as a function of the axial guide field for the parallel orientation of the field

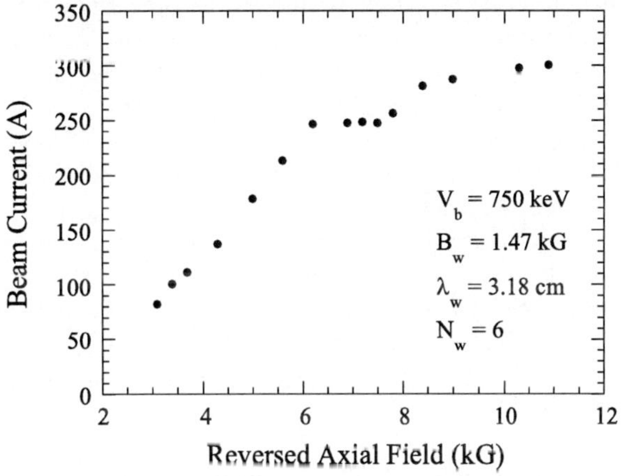

Fig. 5.70 Current propagation as a function of the axial guide field for the reversed-field configuration

addition, the increase in the propagated current leveled off in the reversed-field configuration when the Larmor and wiggler periods were comparable. For convenience, we shall refer to this as the *antiresonance*.

The explanation for this antiresonant effect is that the inhomogeneity in the wiggler field introduces a sinusoidal driving term to the electron orbit equations discussed in the previous section which arises from the fact that an electron on a helical orbit centered off the axis of symmetry experiences a sinusoidally varying wiggler field which acts to drive the electrons at a period close to the wiggler period. Hence, this effect becomes important for electron beams that are big enough that a substantial fraction of the electrons is located relatively far from the symmetry axis. In addition, it is resonant for axial fields close to the antiresonance. We shall discuss this in more detail during the comparison of the simulation with the power measurements, since the orbital irregularities introduced by this effect have a significant impact on the growth of the signal.

The wiggler field is produced by a bifilar helix with a period of $\lambda_w = 3.18$ cm, a length of $50\lambda_w$, and an adiabatic entry taper which is six wiggler periods in length. The wiggler amplitude was continuously adjustable up to an amplitude of approximately 1.8 kG. The axial guide field could be adjusted up to a maximum amplitude of almost 12 kG. The beam propagated through a cylindrical waveguide of 0.51 cm in radius, which provided for a wave-particle resonance with the fundamental TE_{11} mode in the vicinity of 35 GHz. The free-electron laser was operated as an amplifier, and a magnetron that produced approximated 17 kW ($\pm 10\%$) at a frequency of 33.39 GHz was used as a driver. Since the output from the magnetron was linearly polarized, this corresponded to approximately 8.5 kW in the right-hand-circularly polarized state that was capable of interacting with the helical wiggler geometry.

The output from the amplifier showed the greatest efficiency for the reversed-field configuration. In this case, a peak power of 61 MW for a conversion efficiency of 27% was found for a wiggler field magnitude of approximately 1.47 kG and an axial magnetic field of 10.92 kG. The current which could be propagated in these fields was near the maximum of 300 A. The output power for the reversed-field configuration also showed a severe decrease in the vicinity of the antiresonance, dropping by more than three orders of magnitude. The power observed when the axial magnetic field was oriented parallel to the wiggler was much less than for the reversed-field configuration and showed a maximum measured power of approximately 4 MW. Details of the output power spectra will be presented in the comparison with the theoretical results.

The experiment has been operated with the axial magnetic field oriented both parallel and antiparallel with the wiggler field, and we shall discuss the comparison with each of these regimes separately. Features common to both regimes, however, involve the choice of various system parameters as well as the initialization of the modes included in the simulation. The basic analysis for the nonlinear Raman regime is described in Sect. 5.4.

Features common to all cases studied derive from the geometry of the system. Specifically, we take the waveguide radius to be $R_g = 0.51$ cm, the wiggler period to be $\lambda_w = 3.18$ cm, and the wiggler entry taper as $N_w = 6$ wiggler periods in length. In addition, while the beam current varies with the magnitude of the axial guide field, the beam energy is 750 keV, and the radius is fixed at the aperture of the anode to $R_b = 0.25$ cm.

Since the frequency of the amplifier experiment is fixed by the 33.39 GHz magnetron, the beam energy of 750 keV and the waveguide radius of 0.51 cm ensure that a resonant interaction is possible only with the fundamental TE_{11} mode of the guide. Since the magnetron produces approximately 17 kW with a linear polarization, we assume that only half of this power is available with the correct circular polarization to interact with the beam. Hence, the initial power of the TE_{11} mode is chosen to be 8.5 kW. The collective Raman interaction in a free-electron laser couples the TE_{11} mode, in principle, with each of the Gould-Trivelpiece modes having an azimuthal mode number of $l = 0$. In practice, however, we find that inclusion of only the lowest-order radial mode is required to give reasonable agreement with the experiment. Note that the axial electric field of this mode has

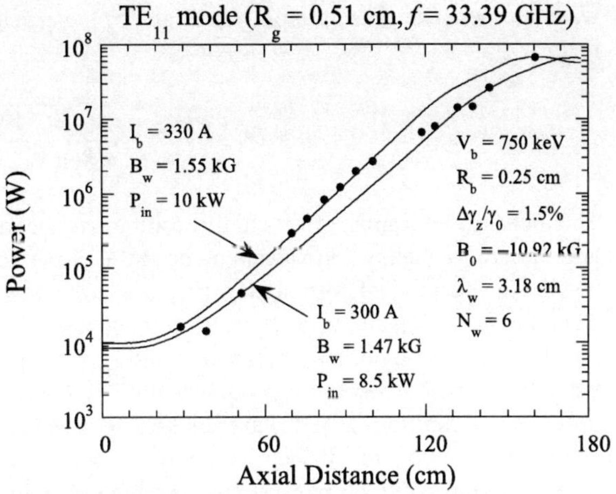

Fig. 5.71 The evolution of the power with axial position as determined from the theory and from the experiment (dots) for a reversed-field configuration

the same transverse variation as the TM_{01} waveguide mode. Hence, the following simulations have been performed using only one waveguide mode and one Gould-Trivelpiece mode.

The first case we consider is that of a reversed-field configuration in which the nominal experimental magnetic field parameters were an axial field magnitude of 10.92 kG and a wiggler field of 1.47 kG. The transmitted current for these field parameters was 300 A [±10%], and the axial energy spread of the beam is assumed to be 1.5% as indicated in the experiment. These parameters represent the case of the peak power observed in the experiment of 61 MW.

The comparison of the theoretical formulation and simulation and the experiment is shown in Fig. 5.71 in which we plot the power as a function of axial position and in which the dots represent the power as measured in the experiment. As shown in the figure, the theoretical analysis was performed for two sets of parameters. The first corresponds to the nominal experimental values given above, and the second corresponds to the upper limits on the (1) current, (2) wiggler field, and (3) input power (due to the experimental uncertainties) of 330 A, 1.55 kG, and 10 kW, respectively. As is evident in the figure, the agreement between the experimental measurements and the theory is good, and the virtually, all the data points fall between these two curves. The saturated power for these two choices of the current, wiggler field, and input power differ only marginally and are close to the 61 MW measured in the experiment. The principal difference is in the saturation length that is due to a small discrepancy in the growth rates for these two cases.

The interaction efficiency in this case is approximately 27% and is comparable to that which is expected due to the phase trapping of the electron beam in the ponderomotive wave formed by the beating of the wiggler and radiation fields. The efficiency estimated by this technique represents the energy lost by the electron as the axial velocity decreases by an amount $\Delta v_z = 2(v_z - v_{ph})$, where $v_{ph} = \omega/(k + k_w)$ is the phase velocity of the ponderomotive wave. Here, (ω,k) are the angular frequency and wavenumber of the electromagnetic wave, and k_w is the wiggler wavenumber. The phase trapping estimate for the efficiency can be expressed in the form

$$\eta \approx 2\gamma_{\parallel}^{2}\frac{v_{\parallel}}{c}\left(\frac{v_{\parallel}}{c} - \frac{\omega}{c(k+k_w)}\right). \tag{5.207}$$

The inclusion of Raman effects in this estimate is accomplished by the choice of the appropriate frequency and wavenumber in the phase velocity of the ponderomotive wave. The theoretical formulation includes collective Raman effects and for the example shown results in a normalized wavenumber of $k/k_w \approx 2.98$ for the TE_{11} mode at a frequency of 33.39 GHz [i.e., $\omega/ck_w \approx 3.5$]. As a consequence, the phase velocity of the ponderomotive wave is $v_{ph}/c \approx 0.879$. Note that this wavenumber differs from the normalized wavenumber for a vacuum TE_{11} mode for which $k/k_w \approx 2.985$ and that the dielectric loading of the waveguide due to the interaction in either the Raman or Compton regimes is included in the theory. In addition, the axial electron velocity for the steady-state orbit in this field configuration is $v_z/c \approx 0.911$ which gives $\gamma_z^2 \approx 5.89$. As a result, the estimate of the phase trapping efficiency is approximately 34%. It is important to recognize that estimates such as this must be employed with caution and should be taken in the present case to indicate the possibility of high-efficiency operation. However, it should also be noted that while the estimate is higher than that found in either the simulation or the experiment, it does not include the effects of an axial energy spread.

The variation in the output power over an interaction length of 150 cm as a function of the magnitude of the reversed magnetic field is shown in Fig. 5.72. Again, the dots represent the experimentally measured power, and the curve is the result from the theory. The current used in the simulation for each value of the axial field corresponds to the transmitted current shown in Fig. 5.67. Agreement between the experiment and theory is good across the entire range studied. Of particular note, however, is the sharp decrease in the output power in the vicinity of the antiresonance at axial field magnitudes between approximately 7 and 8.5 kG.

The source of this antiresonant decrease in the interaction efficiency is the irregularities introduced into the electron trajectories by the transverse inhomogeneity in the wiggler. For this particular example, the radius of the wiggler-induced motion [i.e., the radius of the helical steady-state trajectory] is approximately 0.04 cm. However, the beam radius is 0.25 cm in this experiment. As a consequence, the electrons at the outer regions of the beam are quite sensitive to the wiggler inhomogeneity and experience a sinusoidally varying wiggler field during the course of their trajectories. These effects are implicitly included in the theoretical formulation, and we can illustrate their effect on the electron beam by examining the orbits of selected electrons in the simulation.

The first case we shall consider is that of an electron which is located near the center of the beam upon entry to the wiggler for a reversed axial field of 7.2 kG, which is in the center of the antiresonance region. The evolution of the trajectory in the transverse plane is shown in Fig. 5.73, in which the jaggedness is an artifact

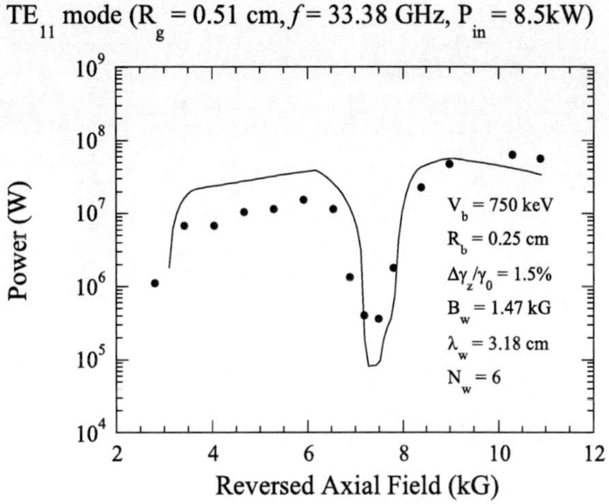

Fig. 5.72 The variation in the output power versus the magnitude of the reversed axial field as measured in the experiment (dots) over an interaction length of 150 cm and in simulation

introduced into the figure by plotting only every tenth point in the integration. The orbit shown in the figure exhibits the expected *spin-up* of the electron trajectory due to the adiabatic injection into the wiggler field, and the electron executes a near-helical steady-state orbit upon transition to the uniform-wiggler region (i.e., after the six wiggler periods of the entry taper region). The principal characteristics of such an orbit are the regular wiggler-induced transverse velocity that mediates the interaction and a near-uniform axial velocity which permits the resonant wave-particle interaction to occur over an extended interaction length. This behavior is also found for electrons at the center of the beam for axial fields away from the antiresonance. The differences occur principally for the edge electrons.

In order to show the nature of these differences, we focus on a characteristic electron that is initially located at the edge of the beam ($x \approx 0.25$ cm and $y \approx 0$) upon entry to the wiggler. The cross-sectional evolution of the trajectory of such an electron for a reversed axial field of 10.92 kG is shown in Fig. 5.74. This case is not in the vicinity of the antiresonance, and the orbit illustrates several features. The predominant feature is the aforementioned spin-up of the electron due to the bulk wiggler motion. However, the electron also executes slower motion that corresponds to betatron and Larmor motion due to the wiggler inhomogeneity that manifests as a guiding-center drift in the counterclockwise direction. This orbit is fairly regular and does not result in any significant degradation in the interaction efficiency.

However, the situation is quite different for an edge electron in the vicinity of the antiresonance. The cross-sectional evolution of such a trajectory is shown in Fig. 5.75 for an axial field magnitude of 7.2 kG. The orbit in this case exhibits the initial spin-up due to the bulk wiggler action but subsequently undergoes what appears to quite irregular motion. The effect of this motion on the axial momentum is shown in Fig. 5.76 in which we plot the axial momentum versus axial position for the central and the edge electrons. It is clear from the figure that the axial momentum exhibits regular oscillations about the bulk value for the central electron but not for

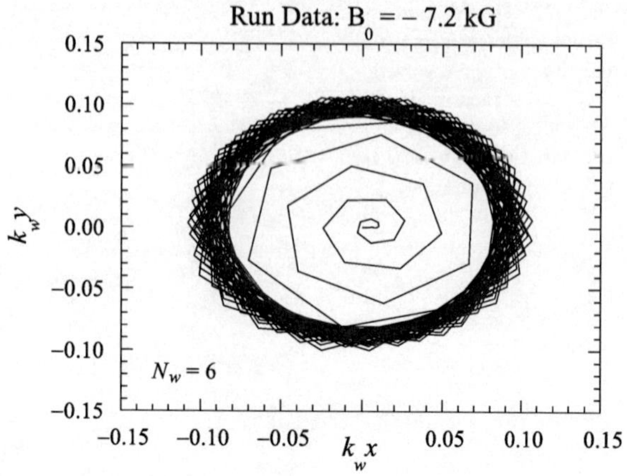

Fig. 5.73 The cross-sectional evolution of the trajectory of an electron injected near the center of the beam for an axial field in the vicinity of the antıresonance

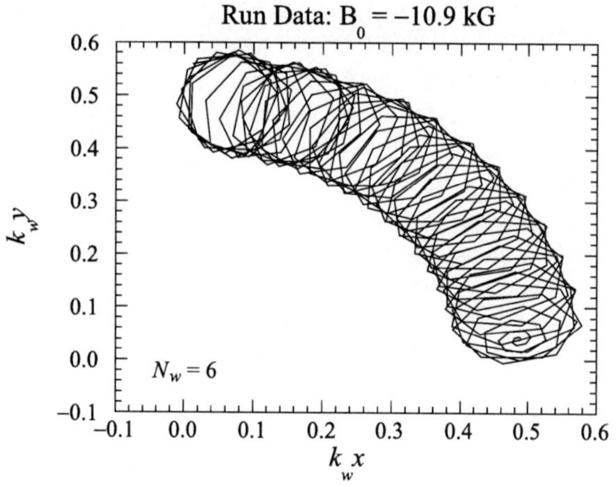

Fig. 5.74 The cross-sectional evolution of the trajectory of an electron injected at the edge of the beam for an axial field far from the antiresonance

the edge electron. In the latter case, the motion exhibits far more structure reflecting the betatron and Larmor motions as well as sudden transitions. Similar behavior is found for the transverse components of the momentum as well. These rapid and large variations in the axial momentum are the major cause of the degradation in the interaction efficiency, since they act to disrupt the resonant wave-particle interaction.

Finally, in regard to the issue of beam transport in the reversed-field configuration, we remark that the entire beam was able to propagate through the wiggler for $B_0 = -10.92$ kG which is far from the antiresonance. However, substantial beam loss was found in the vicinity of the antiresonance. For example, approximately 60% of the beam was found in the simulation to be lost to the waveguide wall for $B_0 = -7.2$ kG. Such massive beam loss is a contributing factor to the low saturation efficiencies found near the antiresonance.

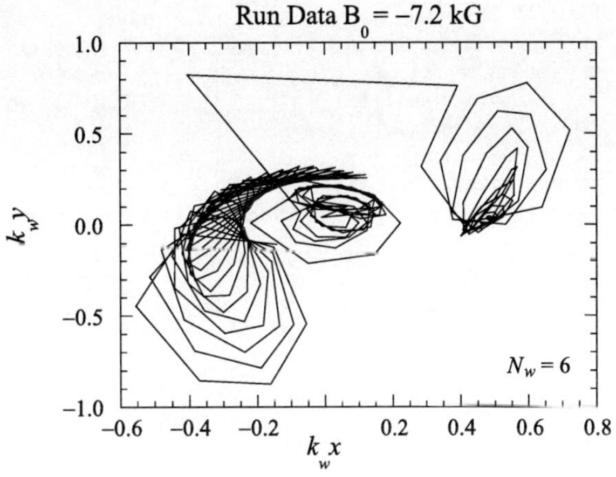

Fig. 5.75 The cross-sectional evolution of the trajectory of an electron injected at the edge of the beam for an axial field in the vicinity of the antiresonance

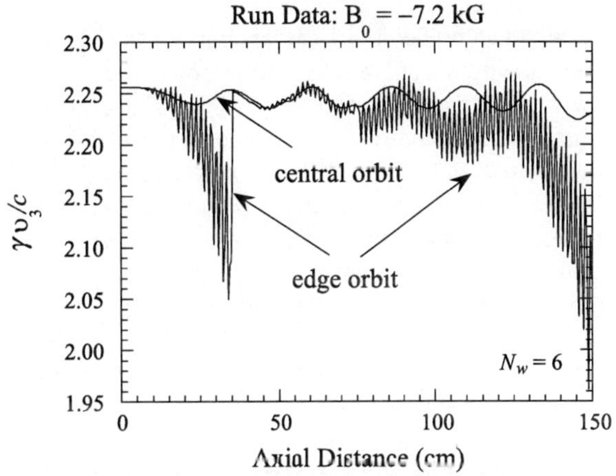

Fig. 5.76 The evolution of the axial momentum versus axial position for electrons injected at the edge of the beam and the center of the beam in the vicinity of the antiresonance

The agreement between the simulation and the experimental measurements in the Group I and II regimes is not as good as that found for the reversed-field configuration. In the cases in which the axial guide field is oriented parallel to the wiggler field, we find that a much larger energy spread than that estimated for the experiment is required in order to account for the measured power levels. Indeed, we find that the assumption of an energy spread of 1.5% results in efficiencies comparable to that found for the reversed-field configuration, and we note that the efficiency predicted in simulation does not vary appreciably for either orientation of the magnetostatic fields for comparable beam energy spreads. In order to account for the Group I experimental data, we must assume an energy spread of approximately 6.25% to account for the measured power. The Group II data are more difficult to explain. The assumption of a comparable energy spread of 6.4% results in reasonable agreement for the power at the end of the interaction region but not for the detailed evolution of the signal [i.e., the launching loss and the instantaneous growth rate] during the course of the interaction.

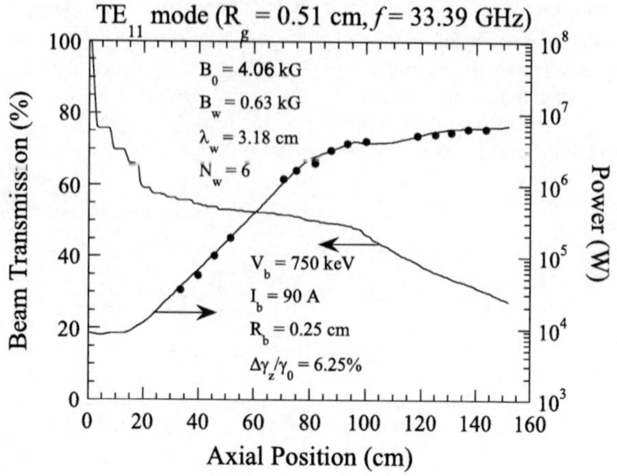

Fig. 5.77 Plot of the evolution of the transmitted beam and power as seen both in simulation and in the experiment (as indicated by the dots) versus axial position for Group I parameters

We have no definitive explanation for this discrepancy but merely suggest that there might be some misalignment or other beam transport problems from the gun to the wiggler that are exacerbated by the orientation of the axial guide field. A possible source for such a discrepancy could be the existence of irregularities in the wiggler field upstream from the entrance due to the sudden termination of the coils. Such irregularities might give rise to orbital instabilities for a parallel alignment of the wiggler and axial guide fields that result in enhanced emittance growth. A detailed evaluation of these suggestions, however, can only be accomplished by means of a thorough analysis of the experimental configuration.

With all of this in mind, we plot the evolution of the beam transmission and power versus axial position as measured in the experiment and as determined with the simulation in Fig. 5.77 for wiggler and axial guide fields of 0.63 kG and 4.06 kG, respectively, and for an axial energy spread of 6.25%. These fields correspond to Group I operation, and the initial current at the entrance to the wiggler is assumed to be 90 A. As shown in the figure, the simulation is in substantial agreement with the experiment for the presumed energy spread of 6.25% both as regards the linear growth rate and the saturation efficiency.

It is evident from the figure that a substantial fraction of the beam is lost in the entry taper region but that the little beam is lost thereafter until the electromagnetic power reaches a level of approximately 4–5 MW after which beam loss occurs at a faster rate. We conclude from this that the saturation mechanism in this case is not phase trapping in the ponderomotive wave as in the reversed-field example at $B_0 = -10.92$ kG. Rather, saturation occurs in this case through loss of the beam to the waveguide wall.

The Group II data are more difficult to explain than the Group I case. The assumption of a comparable energy spread of 6.4% results in reasonable agreement for the power at the end of the interaction region but not for the detailed evolution of the signal (i.e., the launching loss and the instantaneous growth rate) during the course of the interaction.

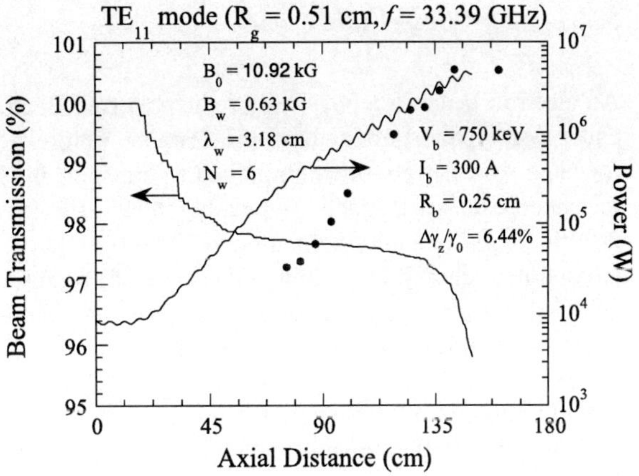

Fig. 5.78 Plot of the evolution of the transmitted beam and power as seen both in simulation and in the experiment (as indicated by the dots) versus axial position for Group II parameters

Comparison of the simulation with experiment in the case of wiggler and guide fields of 0.63 kG and 10.92 kG, respectively, and a transmitted current of 300 A is shown in Fig. 5.78 in which we plot the evolution of the transmitted current and the power both from the simulation and from the experiment (shown by the dots). This case shows rough agreement in the case of a 6.4% energy spread for the power [≈ 4 MW] and saturation efficiency but not the growth rate. In addition, the launching loss observed in the experiment is much higher than that seen in simulation, as evidenced by the fact that there is negligible growth in the observed power until after an axial position of 70 cm after the wiggler entrance. Beam loss is not found in simulation to be a major factor for this case until a power level of approximately 4 MW is reached, after which beam loss occurs at a very rapid rate. Hence, beam loss appears to be the saturation mechanism for this case as well as for the Group I cases.

In general, therefore, we find that particle loss provides a significant, if not dominant, contribution to the saturation mechanism for the Group I and Group II regimes for this experiment.

5.7 DC Self-Field Effects in Free-Electron Lasers

DC self-field effects have been treated in one-dimensional analyses of free-electron lasers in both linear theory [63, 64] and by means of a nonlinear particle-in-cell simulation [64]. Here, we describe a nonlinear model of the DC self-fields in three dimensions [65]. Of course, the most general treatment of the self-fields arises in the context of a fully three-dimensional particle-in-cell simulation of the free-electron laser. In many cases, however, this poses an unacceptably high computational burden. It is important to remark that the effects described here are applicable to long-pulse or DC electron beams. The self-field effects are significantly reduced for short-pulse, high-energy electron beams.

5.7.1 The Self-Fields

An electron beam in a physically realizable free-electron laser is born, accelerated, and possibly transported for some distance before it enters the wiggler. Various focusing schemes are often employed to transport the beam to the wiggler that can rely upon external magnetic (typically either solenoids or magnetic quadrupoles) or electric fields. In addition, many experiments employ some form of beam scraping (sometimes referred to as emittance selection) to ensure a beam with a small axial velocity spread [3, 14]. Hence, the electron beam, and therefore the self-fields, can exhibit a complex structure at the entrance to the wiggler, and a complete treatment of the initial conditions and self-fields in the electron beam in a free-electron laser would require a full-scale particle-in-cell simulation for each specific configuration including the accelerator and beam transport system as well as the free-electron laser. Instead, we adopt a simpler approach here and develop a model of the self-fields based upon the treatment of an electron beam derived from an idealized model of a beam with uniform, azimuthally symmetric profiles in both the density and velocity. This describes the case of the injection of a uniform parallel-propagating beam.

In such a case, the beam density is given by $n_b(r) = n_b$ for $r \leq R_b$ and zero otherwise (where R_b denotes the beam radius), and the self-electric field, $\mathbf{E}^{(s)}$, is determined by Poisson's equation

$$\frac{1}{r}\frac{\partial}{\partial r}\left(rE_r^{(s)}\right) = -4\pi e n_b(r), \tag{5.208}$$

which has the solution

$$\mathrm{E}^{(s)} = -\frac{m_e}{2e}\omega_b^2 r\hat{e}_r, \tag{5.209}$$

where e and m_e are the electronic charge and mass and ${\omega_b}^2 = 4\pi e^2 n_b/m_e$ is the square of the beam-plasma frequency. Energy conservation for this configuration is given by the sum of the kinetic and potential (due to the self-electric field) energies. Within the beam, the Lorentz force equations yield

$$\frac{d}{dt}\left(\gamma - \frac{\omega_b^2}{4c^2}r^2\right) = 0, \tag{5.210}$$

where γ is the relativistic factor. This results in a *space-charge depression* in the kinetic energy across the beam that may be expressed as

$$\gamma(r) = \gamma_0 + \frac{\omega_b^2}{4c^2}\left(r^2 - R_b^2\right). \tag{5.211}$$

In Eq. (5.211), γ_0 denotes the kinetic energy at the edge of the beam or, alternatively, the total energy.

The lowest-order representation for the self-magnetic field is obtained under the assumption that the beam propagates paraxially with $\mathbf{v} = v_z\hat{\mathbf{e}}_z$ for $r \leq R_b$ and zero otherwise. This assumption requires that the space-charge depression across the beam be small. Observe that the space-charge depression in the kinetic energy at the center of the beam depends upon $\omega_b{}^2 R_b{}^2/4c^2$ and is proportional to the total beam current (I_b) through

$$\frac{\Delta\gamma_{self}}{\gamma_0} \approx 5.88 \times 10^{-5} \frac{I_b[\text{Amperes}]}{\sqrt{\gamma_0^2 - 1}}. \tag{5.212}$$

For the specific case under consideration here, $\Delta\gamma_{self}/\gamma_0 < 1\%$, and the assumption of a uniform axial velocity provides a good approximation for the self-magnetic field. In this case, the self-magnetic field is determined by Ampere's law

$$\frac{1}{r}\frac{\partial}{\partial r}\left(rB_\theta^{(s)}\right) = -\frac{4\pi e}{c} n_b(r) v_z, \tag{5.213}$$

hence,

$$\mathrm{B}^{(s)} = -\frac{m_e}{2e}\omega_b^2 \beta_z r \hat{\mathbf{e}}_\theta, \tag{5.214}$$

where $\beta_z = v_z/c$. Note that the effect of the self-electric and magnetic fields will cancel to order $1/\gamma_z{}^2$ in the Lorentz force equations and so will have a small effect at high energies.

It is important to observe that several implicit assumptions underlie this approximation for the self-fields in the wiggler. The first is that even while the beam density is assumed to be a uniform flattop profile with a circular cross section upon entry to the wiggler, it does not necessarily remain either uniform or circular during the course of the interaction. The second is that the self-magnetic field has been derived under the additional assumptions of uniform paraxial motion of the beam. However, the effect of the wiggler is to induce a bulk transverse wiggle motion and a velocity shear due to the wiggler field gradients, and these distortions to the beam current can affect the self-magnetic field in the wiggler. In order to estimate the magnitude of this effect, consider the inclusion of the lowest-order wiggler-induced motion in an idealized one-dimensional helical wiggler in the source current for the self-magnetic field.

Using the idealized one-dimensional representation of a helical wiggler given in Eq. (2.5) and the helical steady-state trajectories, Ampere's law can be written in the form

$$\frac{1}{r}\frac{\partial}{\partial\theta}B_z^{(s)} - \frac{\partial}{\partial z}B_\theta^{(s)} = -\frac{4\pi e}{c}n_b(r)v_w\cos(k_wz-\theta), \tag{5.215}$$

$$\frac{\partial}{\partial z}B_r^{(s)} - \frac{\partial}{\partial r}B_z^{(s)} = -\frac{4\pi e}{c}n_b(r)v_w\sin(k_wz-\theta). \tag{5.216}$$

$$\frac{1}{r}\frac{\partial}{\partial r}\left(rB_\theta^{(s)}\right) - \frac{1}{r}\frac{\partial}{\partial\theta}B_r^{(s)} = -\frac{4\pi e}{c}n_b(r)v_\parallel. \tag{5.217}$$

These equations have solutions of the form

$$B_r^{(s)} = \widehat{B}_r(r)\cos(k_wz-\theta), \tag{5.218}$$

$$B_\theta^{(s)} = \widehat{B}_\theta(r)\sin(k_wz-\theta) + \overline{B}_\theta(r), \tag{5.219}$$

$$B_z^{(s)} = \widehat{B}_z(r)\sin(k_wz-\theta). \tag{5.220}$$

As a result, we have that

$$k_wr\widehat{B}_\theta + \widehat{B}_z = \frac{4\pi en_b}{c}rv_w, \tag{5.221}$$

$$k_w\widehat{B}_r + \frac{d}{dr}\widehat{B}_z = \frac{4\pi en_b}{c}v_w, \tag{5.222}$$

$$\frac{d}{dr}\left(r\overline{B}_\theta\right) + \left[\frac{d}{dr}\left(r\widehat{B}_\theta\right) - \widehat{B}_r\right]\sin(k_wz-\theta) = -\frac{4\pi en_b}{c}rv_\parallel. \tag{5.223}$$

The last of these equations implies that

$$\frac{d}{dr}\left(r\overline{B}_\theta\right) = -\frac{4\pi en_b}{c}rv_\parallel, \tag{5.224}$$

and

$$\frac{d}{dr}\left(r\widehat{B}_\theta\right) = \widehat{B}_r, \tag{5.225}$$

which can also be obtained by elimination of $\widehat{B}_z$ from Eqs. (5.221) and (5.222). Equation (5.224) can be integrated immediately to give

$$\overline{B}_\theta = \frac{m_e}{2e}\omega_b^2 r\frac{v_\parallel}{c}. \tag{5.226}$$

The requirement that the divergence of this field vanishes yields

$$\frac{1}{r}\frac{d}{dr}\left(r\widehat{B}_r\right) - \frac{1}{r}\widehat{B}_\theta + k_w\widehat{B}_z = 0. \tag{5.227}$$

Elimination of $\widehat{B}_z$ and $\widehat{B}_r$ from this equation gives

$$\frac{d^2}{dr^2}\widehat{B}_\theta + \frac{3}{r}\frac{d}{dr}\widehat{B}_\theta - k_w^2\widehat{B}_\theta = -\frac{m_e}{e}\frac{\omega_b^2}{c}k_w v_w. \tag{5.228}$$

The simplest solution to Eq. (5.228) is

$$\widehat{B}_\theta = \frac{\omega_b^2}{k_w^2c^2}\frac{k_w v_\parallel}{\Omega_0 - k_w v_\parallel}B_w, \tag{5.229}$$

which implies that

$$\widehat{B}_r = \frac{\omega_b^2}{k_w^2c^2}\frac{k_w v_\parallel}{\Omega_0 - k_w v_\parallel}B_w, \tag{5.230}$$

and

$$\widehat{B}_z = 0. \tag{5.231}$$

Finally, this results in a correction to the self-magnetic field of the form

$$\mathrm{B}^{(s)} = -\frac{m_e}{2e}\omega_b^2\beta_\parallel r\widehat{e}_\theta + \frac{\omega_b^2}{\gamma_0 k_w^2c^2}\frac{k_w v_\parallel}{\Omega_0 - k_w v_\parallel}\mathrm{B}_w, \tag{5.232}$$

where $\Omega_0 = eB_0/\gamma_0 m_e c$. The second term above can describe either diamagnetic ($\Omega_0 < k_w v_\parallel$) or paramagnetic ($\Omega_0 > k_w v_\parallel$) corrections to the wiggler field but is negligible as long as $k_w R_b > 2\ |v_w/v_\parallel|$, where v_w is the wiggler-induced transverse velocity. Hence, Eqs. (5.209) and (5.214) represent a reasonable approximation for the self-fields in the wiggler as long as the transverse electron displacement due to the wiggler is less than the beam radius. Once again, it should be emphasized that this diamagnetic or paramagnetic effect is applicable only to long-wavelength electron beams and the effect is greatly reduced in short-wavelength, high-energy electron beams as employed in X-ray free-electron lasers [66].

5.7.2 *The Nonlinear Formulation*

The analysis employed is that presented in this chapter which describes intense beams propagating through cylindrical waveguides in the presence of helical wigglers. Within the context of this formulation, each mode interacts resonantly with the electrons and is coupled via the Lorentz force equations in the combined static and fluctuating fields that include the self-electric and magnetic fields of the beam. The description of the self-fields given in Eqs. (5.209) and (5.214) is modified to allow for the motion of the beam centroid in the wiggler and uses an average axial velocity in the self-magnetic field. As a result, the self-electric and magnetic fields are represented as

$$\mathrm{E}^{(s)} = -\frac{m_e}{2c}\omega_b^2\left[(x-\langle x\rangle)\widehat{\boldsymbol{e}}_x + (y-\langle y\rangle)\widehat{\boldsymbol{e}}_y\right], \tag{5.233}$$

and

$$\mathrm{B}^{(s)} = -\frac{m_e}{2c}\omega_b^2\langle\beta_z\rangle\left[(y-\langle y\rangle)\widehat{\boldsymbol{e}}_x - (x-\langle x\rangle)\widehat{\boldsymbol{e}}_y\right]. \tag{5.234}$$

Given these self-fields, as well as the external fields for the axial guide and wiggler fields, the Lorentz force equations take the form

$$\upsilon_z\frac{d}{dz}\mathrm{p} = -e\left(\mathrm{E}^{(s)} + \delta\mathrm{E}\right) - \frac{e}{c}\mathrm{v}\times\left(B_0\widehat{\boldsymbol{e}}_z + \mathrm{B}_w + \mathrm{B}^{(s)} + \delta\mathrm{B}\right), \tag{5.235}$$

where $\delta\mathbf{E}$ and $\delta\mathbf{B}$ represent the aggregate electric and magnetic fields from each TE, TM, and Gould-Trivelpiece mode. The magnetostatic fields employed are given by Eq. (2.1) for the helical wiggler.

5.7.3 *The Numerical Analysis*

The initial state of the electron beam is chosen to model the injection of a monoenergetic, uniform, axisymmetric electron beam with a flattop density profile for $r_0 \leq R_b$. The effect of the self-electric field on the initial kinetic energy mirrors the space-charge depression, where γ_0 describes the total energy that is the initial kinetic energy at the edge of the beam. As a result, we scale the initial momentum

$$\mathrm{p}_0(r_0) \cong \mathrm{p}_0\sqrt{1 + \frac{\Delta\gamma(r_0)[2\gamma_0 + \Delta\gamma(r_0)]}{\gamma_0^2 - 1}}, \tag{5.236}$$

where

$$\Delta\gamma(r_0) \equiv \frac{\omega_b^2}{4c^2}\left(r_0^2 - R_b^2\right), \tag{5.237}$$

and $|\mathbf{p}_0|^2/m_e^2c^2 = \gamma_0^2 - 1$. Thus, the initial kinetic energy of the particles increases with radius from the beam center.

It is important to observe here that no attempt is made to match the beam into the wiggler in order to achieve a beam envelope with a relatively constant radius throughout the wiggler. We treat a simpler model in which a paraxially propagating beam is injected into the wiggler, and the subsequent motion is calculated for the assumed electrostatic, magnetostatic (including the self-magnetic), and electromagnetic fields.

5.7.4 Comparison with Experiment

We compare the analysis with an experiment conducted at the Naval Research Laboratory [67, 68]. The experiment at the Naval Research Laboratory makes use of a 250 keV modulator capable of producing currents in excess of 100 A, and the electron gun is designed to produce a beam with a radius of 0.4 cm and an axial energy spread of 0.3%. The wiggler field is generated by a bifilar helical coil with a period of 2.54 cm and a total length of 33 wiggler periods including both entrance and exit tapers of 5 and 3 wiggler periods, respectively, in length. The wiggler amplitude can be varied up to approximately 500 G. In addition, the axial guide field can be varied up to 3.2 kG. This ensures operation in the Group I regime. The experiment is configured as an amplifier driven by a coupled cavity TWT that injects up to 5 kW of power over the K_u-band from 12 to 18 GHz. Hence, since the circular waveguide has a radius of 0.815 cm, the interaction is predominantly with the TE_{11} mode. Observe that the space-charge depression for this experiment, $\Delta\gamma_{self}/\gamma_0 \approx 0.53\%$, is comparable to the other two experiments despite the much lower beam current.

In order to simulate this experiment, we consider operation at 16 GHz and use the following model for the wiggler field amplitude for the helical wiggler

$$B_w(z) = \begin{cases} B_w \sin^2\left(\dfrac{k_w z}{20}\right) & ; 0 \leq z \leq 5\lambda_w \\ B_w & ; 5\lambda_w < z \leq 30\lambda_w \\ B_w \cos^2\left\{\dfrac{k_w(z - 30\lambda_w)}{12}\right\} & ; 30\lambda_w < z \leq 33\lambda_w \end{cases} . \tag{5.238}$$

Preliminary simulations without the inclusion of the self fields indicated that extremely high efficiencies would be possible. Both the efficiency and beam transmission are plotted as functions of the axial guide field in Fig. 5.79 for the case of a wiggler field amplitude of 300 G. It is clear that the efficiency varies over a wide range from 3% to 33% as the axial guide field increases from 2.2 to 2.6 kG. However, the beam transmission falls precipitously with the increase in the efficiency from a value of about 99% at an axial field of 2.2 kG to approximately 5% at an axial field of 2.6 kG. This decline in the beam transmission is due to two factors. The first factor is that the loss of up to 30% of the beam energy to the TE_{11} mode

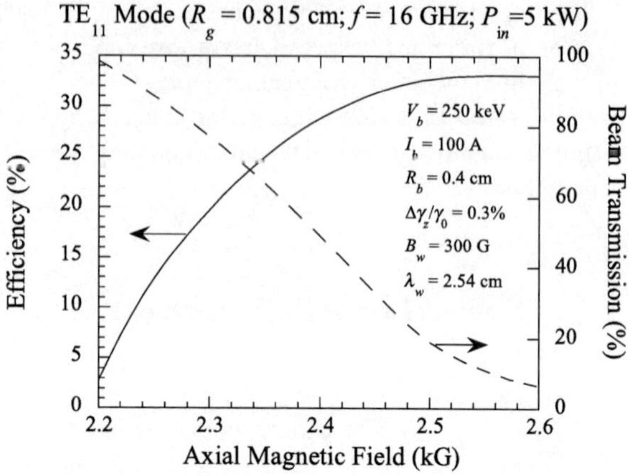

Fig. 5.79 Plots of the variation in the efficiency and beam transmission versus the axial magnetic field in the absence of the self-fields

implies that the beam undergoes massive deceleration that is accompanied by an increase in the radius of the wiggler-induced trajectory. The second factor is that the high-power electromagnetic wave acts to kick the beam away from the axis. It should be noted that, as in the case of the Group I and II regimes in the experiment at the Massachusetts Institute of Technology, saturation in this experiment occurs due to beam loss rather than the more familiar phase trapping of the beam in the ponderomotive potential formed by the beating of the wiggler and radiation fields. Operation with acceptable levels of beam loss, therefore, was expected to restrict the experiment to efficiencies of approximately 20%.

As might be expected, the effect of the self-fields can act to enhance the beam losses. The effects of the self-fields are more pronounced in this experiment than in the previously analyzed experiment at the Massachusetts Institute of Technology (note that the space-charge depressions are comparable for the two experiments) since the beam voltage and axial guide field are lower in the experiment at the Naval Research Laboratory. The efficiency and beam transmission are plotted in Fig. 5.80 as functions of the axial guide field for a wiggler amplitude of 300 G subject to the inclusion of the space-charge fields. It is evident from the figure that for strong axial guide fields in excess of approximately 2.5 kG the efficiency and beam transmission do not differ greatly from those found in the absence of the self-fields. This is because the axial field acts to confine the beam against the spreading induced by the self-fields. In contrast, both the efficiency and beam transmission are substantially less than that found in the absence of the self-fields for weak axial guide fields below about 2.3 kG. In the intermediate regime for axial guide fields in the range of 2.3–2.5 kG, however, the beam transmission is enhanced relative to both the weak and strong guide field cases. This occurs for two reasons. First, the axial guide field is strong enough to provide appreciable confinement of the electron beam. Second, the self-fields are strong enough to cause a reduction in the interaction efficiency; hence, the beam has not lost enough energy, and the radiation has not gained enough energy to kick the beam appreciably toward the wall. As a result, we expect that operation

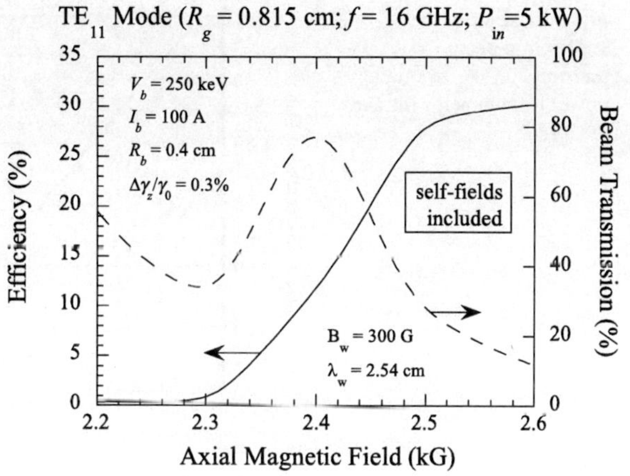

Fig. 5.80 Plots of the variation in the efficiency and beam transmission versus the axial magnetic field in the presence of the self-fields

with axial guide fields in the neighborhood of 2.4 kG for a wiggler field of 300 G is preferred and will result in efficiencies of approximately 10–15%.

With some modification to the original design, this experiment demonstrated high power and broadband operation with a maximum output power of 4.2 MW for an efficiency of 18%, a gain of 29 dB, and a large signal bandwidth (not saturated) greater than 22%. Thus, the experiment met the fundamental design performance goals, specifically, an output power of 1–5 MW, an efficiency greater than 15%, a large-signal gain of 25–30 dB, and a large-signal bandwidth greater than 20%. Experimental results are in good agreement with the theoretical predictions, and it is important to note that, as predicted, the DC self-fields of the beam played an important role in the interaction.

This is illustrated in Fig. 5.81 in which the variation in beam transmission is plotted versus output power [68]. The solid triangles, circles, and diamonds in the figure represent experimental measurements of beam transmission obtained from a wide range of runs in which the dependence of the output power was studied as functions of wiggler field, guide field, and beam voltage. The hollow triangles are from the theory, and the solid line is merely a smooth curve showing the overall variation. It is clear from the figure that good agreement exists between theory and experiment. Note that comparisons of the theoretical predictions without the inclusion of the self-fields exhibit wide discrepancies with the experiment and predict much higher levels of beam transmission.

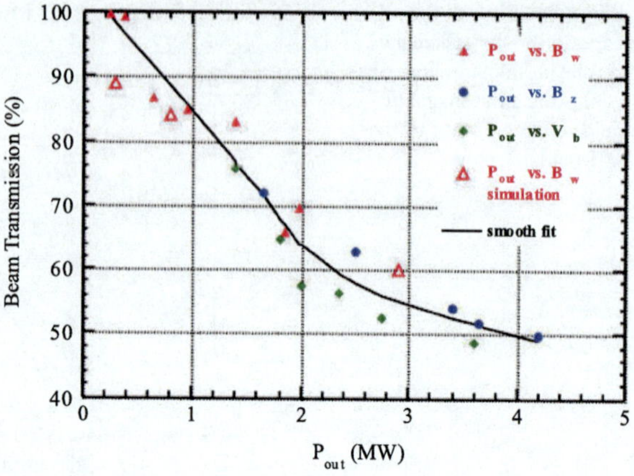

Fig. 5.81 Plot of the variation in beam transmission versus output power from experiment and theory [68]

References

1. R.M. Phillips, History of the ubitron. Nucl. Instrum. Methods **A272**, 1 (1988)
2. H. Boehmer, M.Z. Caponi, J. Edighoffer, S. Fornaca, J. Munch, G.R. Neil, B. Saur, C. Shih, Variable-wiggler free-electron laser experiment. Phys. Rev. Lett. **48**, 141 (1982)
3. R.W. Warren, B.E. Newnam, J.G. Winston, W.E. Stein, L.M. Young, C.A. Brau, Results of the Los Alamos free-electron laser experiment. IEEE J. Quantum Electron. **QE-19**, 391 (1983)
4. J. Edighoffer, G.R. Neil, C.E. Hess, T.I. Smith, S.W. Fornaca, H.A. Schwettman, Variable-wiggler free-electron laser oscillation. Phys. Rev. Lett. **52**, 344 (1984)
5. T.J. Orzechowski, B.R. Anderson, J.C. Clark, W.M. Fawley, A.C. Paul, D. Prosnitz, E.T. Scharlemann, S.M. Yarema, D.B. Hopkins, A.M. Sessler, J.S. Wurtele, High-efficiency extraction of microwave radiation from a tapered-wiggler free-electron laser. Phys. Rev. Lett. **57**, 2172 (1986)
6. P. Sprangle, C.M. Tang, W.M. Manheimer, Nonlinear formulation and efficiency enhancement of free-electron lasers. Phys. Rev. Lett. **43**, 1932 (1979)
7. N.M. Kroll, P.L. Morton, M.N. Rosenbluth, A variable parameter free-electron laser, in *The Physics of Quantum Electronics: Free-Electron Generators of Coherent Radiation*, ed. by S.F. Jacobs, H.S. Pilloff, M. Sargent, M.O. Scully, R. Spitzer, vol. 7, (Addison-Wesley, Reading, 1980), p. 89
8. D. Prosnitz, A. Szöke, V.K. Neil, One-dimensional computer simulation of the variable wiggler free-electron laser, A variable parameter free-electron laser, in *The Physics of Quantum Electronics: Free-Electron Generators of Coherent Radiation*, ed. by S.F. Jacobs, H.S. Pilloff, M. Sargent, M.O. Scully, R. Spitzer, vol. 7, (Addison-Wesley, Reading, 1980), p. 175
9. S.A. Mani, Free-electron laser interaction in a variable pitch wiggler, A variable parameter free-electron laser, in *The Physics of Quantum Electronics: Free-Electron Generators of Coherent Radiation*, ed. by S.F. Jacobs, H.S. Pilloff, M. Sargent, M.O. Scully, R. Spitzer, vol. 7, (Addison-Wesley, Reading, 1980), p. 589
10. W.H. Louisell, C.D. Cantrell, W.A. Wegener, Single-particle approach to free-electron lasers with tapered wigglers, A variable parameter free-electron laser, in *The Physics of Quantum Electronics: Free-Electron Generators of Coherent Radiation*, ed. by S.F. Jacobs, H.S. Pilloff, M. Sargent, M.O. Scully, R. Spitzer, vol. 7, (Addison-Wesley, Reading, 1980), p. 623
11. C.A. Brau, R.K. Cooper, Variable wiggler optimization, A variable parameter free-electron laser, in *The Physics of Quantum Electronics: Free-Electron Generators of Coherent*

Radiation, ed. by S.F. Jacobs, H.S. Pilloff, M. Sargent, M.O. Scully, R. Spitzer, vol. 7, (Addison-Wesley, Reading, 1980), p. 647
12. J.M. Slater, J. Adamski, D.C. Quimby, W.M. Grossman, T.L. Churchill, R.E. Center, Tapered wiggler free-electron laser demonstration, in *Proceedings of the International Conference on Lasers '82*, ed. by R.C. Powell, (STS Press, McLean, 1982), p. 212
13. J.C. Goldstein, Evolution of long pulses in a tapered wiggler free-electron laser, in *Free-Electron Generators of Coherent Radiation*, ed. by C.A. Brau, S.F. Jacobs, M.O. Scully, vol. 453, (Proc. SPIE, 1984), p. 2
14. H.P. Freund, S.H. Gold, Efficiency enhancement in free-electron lasers using a tapered axial guide field. Phys. Rev. Lett. **52**, 926 (1984)
15. T.J.T. Kwan, J.M. Dawson, A.T. Lin, Free-electron laser. Phys. Fluids **20**, 581 (1977)
16. T. Taguchi, K. Mima, T. Mochizuki, Saturation mechanism and improvement of conversion efficiency of the free-electron laser. Phys. Rev. Lett. **46**, 824 (1981)
17. H.P. Freund, Nonlinear analysis of free-electron laser amplifiers with axial guide fields. Phys. Rev. A **27**, 1977 (1983)
18. W.B. Colson, J.L. Richardson, Multimode theory of free-electron lasers. Phys. Rev. Lett. **50**, 1050 (1983)
19. T.M. Antonsen Jr., B. Levush, Mode competition and control in free-electron laser oscillators. Phys. Rev. Lett. **62**, 1488 (1989)
20. T.M. Antonsen Jr., B. Levush, Mode competition and suppression in free-electron laser oscillators. Phys. Fluids B **1**, 1097 (1989)
21. T.M. Antonsen Jr., B. Levush, Spectral characteristics of a free-electron laser with time-dependent beam energy. Phys. Fluids B **2**, 2791 (1990)
22. Y. Kishimoto, H. Oda, M. Shiho, K. Odajima, H. Maeda, Effect of electrostatic field on energy conversion efficiency in high current Raman regime free-electron laser. J. Phys. Soc. Jpn. **59**, 118 (1990)
23. Y. Kishimoto, H. Oda, M. Shiho, Parasitic wave excitation by multimode coupling in a Raman-regime free-electron laser. Phys. Rev. Lett. **65**, 851 (1990)
24. V.A. Bazylev, A.V. Tilupov, Enhancement of the efficiency of a free-electron laser in a longitudinal magnetic field. Sov. J. Quantum Electron. **20**, 115 (1990)
25. D. Iracane, J.L. Ferrer, Stability of a free-electron laser spectrum in the continuous beam limit. Phys. Rev. Lett. **66**, 33 (1991)
26. T.J.T. Kwan, C.M. Snell, Efficiency of free-electron lasers with a scattered electron beam. Phys. Fluids **26**, 835 (1983)
27. A.T. Lin, C.C. Lin, T. Taguchi, W.W. Cheng, Nonlinear saturation of free-electron lasers around gyroresonance. Phys. Fluids **26**, 3 (1983)
28. R.K. Parker, R.H. Jackson, S.H. Gold, H.P. Freund, V.L. Granatstein, P.C. Efthimion, M. Herndon, A.K. Kinkead, Axial magnetic-field effects in a collective-interaction free-electron laser at millimeter wavelengths. Phys. Rev. Lett. **48**, 238 (1982)
29. D. Prosnitz, R.A. Haas, S. Doss, R.J. Galinas, A two-dimensional numerical model of the tapered wiggler free-electron laser, in *Physics of Quantum Electronics: Free-Electron Generators of Coherent Radiation*, ed. by S.F. Jacobs, H.S. Pilloff, M. Sargent, M.O. Scully, R. Spitzer, vol. 9, (Addison-Wesley, Reading, 1981), p. 1047
30. P. Sprangle, C.M. Tang, Three-dimensional nonlinear theory of the free-electron laser. Appl. Phys. Lett. **39**, 677 (1981)
31. E.T. Scharlemann, Wiggler plane focussing in linear wigglers. J. Appl. Phys. **58**, 2154 (1985)
32. T.J. Orzechowski, B.R. Anderson, W.M. Fawley, D. Prosnitz, E.T. Scharlemann, S.M. Yarema, D.B. Hopkins, A.C. Paul, A.M. Sessler, J.S. Wurtele, Microwave radiation from a high-gain free-electron laser amplifier. Phys. Rev. Lett. **54**, 889 (1985)
33. C.M. Tang, P. Sprangle, Three-dimensional numerical simulation of free-electron lasers by the transverse mode spectral method. IEEE J. Quantum Electron. **QE-21**, 970 (1985)
34. A.K. Ganguly, H.P. Freund, Nonlinear analysis of free-electron lasers in three dimensions. Phys. Rev. A **32**, 2275 (1985)

35. A.T. Lin, C.C. Lin, Mode competition in Raman free-electron lasers. Nucl. Instrum. Methods **A250**, 1373 (1986)
36. H.P. Freund, A.K. Ganguly, Nonlinear analysis of efficiency enhancement in free-electron laser amplifiers. Phys. Rev. A **33**, 1060 (1986)
37. H.P. Freund, A.K. Ganguly, Effect of beam quality on the free-electron laser. Phys. Rev. A **34**, 1242 (1986)
38. H.P. Freund, A.K. Ganguly, Phase variation in free-electron laser amplifiers. IEEE J. Quantum Electron. **QE-23**, 1657 (1987)
39. T.J. Orzechowski, E.T. Scharlemann, D.B. Hopkins, Measurement of the phase of the electromagnetic wave in a free-electron laser amplifier. Phys. Rev. A **35**, 2184 (1987)
40. A.K. Ganguly, H.P. Freund, High-efficiency operation of free-electron laser amplifiers. IEEE Trans. Plasma Sci. **PS-16**, 167 (1988)
41. A.K. Ganguly, H.P. Freund, Three-dimensional simulation of the Raman free-electron laser. Phys. Fluids **31**, 387 (1988)
42. H.P. Freund, H. Bluem, C.L. Chang, Three-dimensional analysis of free-electron laser amplifiers with planar wigglers. Phys. Rev. A **36**, 2182 (1987)
43. H.P. Freund, C.L. Chang, H. Bluem, Harmonic generation in free-electron lasers. Phys. Rev. A **36**, 3218 (1987)
44. H.P. Freund, H. Bluem, C.L. Chang, Three-dimensional simulation of free-electron lasers with planar wigglers. Nucl. Instrum. Methods **A272**, 556 (1988)
45. H.P. Freund, Multimode nonlinear analysis of free-electron laser amplifiers in three dimensions. Phys. Rev. A **37**, 3371 (1988)
46. H. Bluem, H.P. Freund, C.L. Chang, Harmonic content in a planar wiggler based free-electron laser amplifier. Nucl. Instrum. Methods **A272**, 579 (1988)
47. S.F. Chang, O.C. Eldridge, J.E. Sharer, Analysis and nonlinear simulation of a quadrupole wiggler free-electron laser at millimeter wavelengths. IEEE J. Quantum Electron. **QE-24**, 2309 (1988)
48. H.P. Freund, H. Bluem, R.H. Jackson, Nonlinear theory and design of a harmonic ubitron/free-electron laser. Nucl. Instrum. Methods **A285**, 169 (1989)
49. S.F. Chang, J. Joe, J.E. Sharer, Nonlinear analysis of wiggler taper, mode competition, and space-charge effects for a 280 GHz free-electron laser. IEEE Trans. Plasma Sci. **PS-18**, 451 (1990)
50. J.S. Wurtele, R. Chu, J. Fajans, Nonlinear theory and experiment of collective free-electron lasers. Phys. Fluids B **2**, 1626 (1990)
51. D.A. Kirkpatrick, G. Bekefi, A.C. DiRienzo, H.P. Freund, A.K. Ganguly, A millimeter and submillimeter wavelength free-electron laser. Phys. Fluids B **1**, 1511 (1989)
52. D.A. Kirkpatrick, G. Bekefi, A.C. DiRienzo, H.P. Freund, A.K. Ganguly, A high-power 600μm wavelength free-electron laser. Nucl. Instrum. Methods **A285**, 43 (1989)
53. J. Byers, R.H. Cohen, A microwave free-electron laser code using waveguide modes. Nucl. Instrum. Methods **A272**, 595 (1988)
54. J. Fajans, J.S. Wurtele, G. Bekefi, D.S. Knowles, K. Xu, Nonlinear power saturation and phase in a collective free-electron laser amplifier. Phys. Rev. Lett. **57**, 579 (1986)
55. T.J. Orzechowski, B.R. Anderson, W.M. Fawley, D. Prosnitz, E.T. Scharlemann, S. Yarema, High gain and high extraction efficiency from a free-electron laser operating in the millimeter wave regime. Nucl. Instrum. Methods **A250**, 144 (1986)
56. N.A. Krall, A.W. Trivelpiece, *Principle of Plasma Physics* (San Francisco Press, San Francisco, 1986), p. 202
57. M.E. Conde, G. Bekefi, Experimental study of a 33.3 GHz free-electron laser amplifier with a reversed axial guide magnetic field. Phys. Rev. Lett. **67**, 3082 (1991)
58. M.E. Conde, G. Bekefi, Amplification and superradiant emission from a 33.3 GHz free-electron laser with a reversed axial guide magnetic field. IEEE Trans. Plasma Sci. **20**, 240 (1992)
59. K.R. Chu, A.T. Lin, Harmonic gyroresonance of electrons in combined helical wiggler and axial guide magnetic fields. Phys. Rev. Lett. **67**, 3235 (1991)

60. H.P. Freund, A.K. Ganguly, Nonlinear simulation of a high-power collective free-electron laser. IEEE Trans. Plasma Sci. **20**, 245 (1992)
61. V.A. Bazylev, A. Bourdier, P. Gouard, J.M. Buzzi, Electron trajectories in a free-electron laser with a reversed axial guide field. Phys. Rev. E **48**, 3959 (1993)
62. H.P. Freund, Beam transmission in a high-power collective free-electron laser. Phys. Fluids B **5**, 1869 (1993)
63. N.S. Ginzburg, Diamagnetic and paramagnetic effects in free-electron lasers. IEEE Trans. Plasma Sci **PS-15**, 411 (1987)
64. T.J.T. Kwan, J.M. Dawson, Investigation of the free-electron laser with a guide magnetic field. Phys. Fluids **22**, 1089 (1979)
65. H.P. Freund, R.H. Jackson, D.E. Pershing, The nonlinear analysis of self-field effects in free-electron lasers. Phys. Fluids B **5**, 2318 (1993)
66. H.P. Freund, R. Tatchyn, Collective effects in short wavelength free-electron lasers. Nucl. Instrum. Methods **A429**, 76 (1999)
67. D.E. Pershing, R.H. Jackson, H. Bluem, H.P. Freund, Improved amplifier performance of the NRL ubitron. Nucl. Instrum. Methods **A304**, 127 (1991)
68. D.E. Pershing, R.D. Seeley, R.H. Jackson, H.P. Freund, Amplifier performance of the NRL ubitron. Nucl. Instrum. Methods **A358**, 104 (1995)

Chapter 6
Nonlinear Theory: Optical Mode Analysis

The previous chapter dealt with the nonlinear theory in the steady-state regime based on the slowly-varying envelope approximation (SVEA). Most of the time-dependent free-electron laser simulation codes that are in use at the present time deal either with an extension of the SVEA in order to solve the wave equation [1–10] or a particle in cell simulation [11] where Maxwell's equations are solved using a finite-difference time-domain (FDTD) algorithm [12]. The time-dependent formulation presented in this chapter is an extension of the SVEA, in which the SVEA is extended by allowing the slowly-varying amplitude to vary in both axial position and time. However, polychromatic formulations using an SVEA where the slowly-varying amplitudes vary only in position can also describe time dependence. This can be accomplished by employing a superposition of modes that are harmonics of an underlying frequency (ω_0). As a result, a time-average over the period $2\pi/\omega_0$ will orthogonalize the dynamical equations of the mode amplitudes and allow each harmonic component to be treated separately; hence, this polychromatic approach is equivalent to an explicit Fourier decomposition of the optical field. These two techniques have been formally shown to be equivalent [13]; however, in practice the polychromatic approach is least computationally efficient of the two and is not commonly used.

In contrast to the steady-state formulations, a time-dependent formulation is necessary in the simulation of short wavelength free-electron lasers that employ radio frequency linear accelerators (RF linacs) or storage rings. Radio frequency linacs produce high-energy beams with pulse times of the order of 1–10 psec and bunch charges of at most several nano-Coulombs. In X-ray free-electron lasers, the actual bunch charge used is about 250 pC or less. Since the growth rate depends upon the peak current, it is desirable to produce bunches with peak currents of several hundred to several thousand Amperes, and this requires compression of the bunch to sub-picosecond pulse times. For example, the optical field in a free-electron laser operating in the ultraviolet at a wavelength of 300 nm would slip ahead of the

H. P. Freund, T. M. Antonsen, Jr., *Principles of Free Electron Lasers*,
https://doi.org/10.1007/978-3-031-40945-5_6

electron bunch at a rate of 1 fsec per wiggler period. As a result, the total slippage in a wiggler with a period of 3 cm and a length of 2 m would be about 67 fsec, which is significant for a sub-picosecond bunch.

In addition to describing the slippage of the optical pulse, time dependence is also needed to study the spectral properties of the optical field such as the temporal coherence, linewidth, sideband production, etc. Furthermore, in contrast to the guided mode analysis used for the steady-state formulation presented in the preceding chapter, the three-dimensional formulations presented in this chapter make use of superpositions of Gaussian optical modes to represent the radiation fields.

The optical field representation employed in the previous editions was based on a phasor representation in which the polarization of the optical field is fixed throughout the simulation [8]. However, interest has developed in the ability to generate different polarization states, particularly in X-ray FELs that employ an aggregate line of relatively short wigglers in sequence. This can be accomplished by a variety of techniques including, but not limited to, (1) alternating the orientations of successive planar wigglers in a long propagation line, (2) using variable polarization wigglers, and (3) using an afterburner wiggler with a different polarization state than those in the preceding sequence. In order to better accommodate a variety of polarization states where the polarizations of the wigglers may vary along the transport line, the three-dimensional formulation presented here treats the x- and y-components of the optical field independently [10]. Note that the one-dimensional formulation is still treated using a phasor representation.

6.1 Optical Guiding

The previous chapter dealt with long wavelength interactions in waveguides where the electromagnetic field is confined by a waveguide. However, free-space modes can also be guided during the interaction in a free-electron laser. Optical guiding in free-electron lasers refers to the self-focusing of the electromagnetic wave by the electron beam [14–43]. Optical guiding of the signal occurs by two related mechanisms referred as gain and refractive guiding. Gain guiding describes the preferential amplification of radiation in the region occupied by the electron beam. Therefore, an optical ray will undergo amplification as long as it is coincident with the beam. If it propagates out of the beam, then the interaction will cease. Refractive guiding describes the focusing (or defocusing) of the radiation by means of the shift in the refractive index due to the dielectric response of the electron beam. In particular, if the wavenumber is shifted upward due to the interaction with respect to the vacuum state, then the phase velocity of the wave decreases, and the beam acts as an optical guide. It should be remarked, however, that gain and refractive guiding are intimately linked and are not independent processes.

The process of refractive guiding is related to variation in the relative phase [43], since this quantity measures the shift in the wavenumber due to the dielectric effect of the beam. As shown in the nonlinear simulations of both the helical and planar wiggler configurations, the relative phase decreases with axial position at the low-frequency portion of the gain spectrum. This decrease occurs because the dielectric shift induced by the beam reduces the wavenumber below that of the vacuum state and corresponds to a defocusing of the signal. As the frequency increases, however, the downshift in the wavenumber decreases until a critical frequency is reached at which the relative phase remains approximately constant. This corresponds to a wavenumber that is comparable to the vacuum state, and for which there is no refraction of the signal. The frequency at which this is found is, typically, below the frequency of peak growth rate. The relative phase increases with axial position for frequencies higher than the critical point, corresponding to the guiding of the signal.

The mechanism of optical guiding has been studied both by analytical means [15, 19, 22, 30, 31, 36, 40, 41] and numerical simulations [14, 16, 17, 21, 23, 24, 26–29, 33, 39, 41, 43]. These analyses indicate that the optical guiding of the beam can result in both a focusing of the electromagnetic signal to smaller or larger areas depending upon the cross-section of the electron beam and the steering of the radiation in which the centroid of the electromagnetic signal follows that of the electron beam. It is also important to note in this regard that the relative phase has been found to reach a saturation level in tapered wiggler configurations in both experiment and simulation [44, 45]. Beyond this saturation point, the relative phase varies little with axial position. As a consequence, the optical guiding mechanism is weakly operative in the presence of a tapered wiggler [41].

6.1.1 Optical Guiding and the Relative Phase

The physical basis of the optical guiding mechanism can be best understood in terms of the behavior of the relative phase, which is defined as the integrated difference between the wavenumber in the interaction region and the free-space wavenumber. This can be understood most clearly on the basis of the idealized one-dimensional analysis. We first consider the high-gain regime. Under the assumption that $|v_w/v_\|| << 1$, the dispersion equations for both the helical and planar wiggler geometries can be expressed as

$$\left(\left[\omega-(k+k_w)v_\|\right]^2-\frac{\omega_b^2}{\gamma_0\gamma_\|^2}\right)\left(\omega^2-c^2k^2-\frac{\omega_b^2}{\gamma_0}\right)\cong\frac{v_w^2}{c^2}\frac{\omega_b^2}{\gamma_0}\omega ck_w, \tag{6.1}$$

where ω and k are the angular frequency and wavenumber of the electromagnetic wave, γ_0 denotes the relativistic factor corresponding to the bulk energy of the beam, $v_\|$ is the bulk axial velocity of the beam $\gamma_\| = (1 - v_\|^2/c^2)^{-1/2}$, ω_b is the beam plasma frequency, and

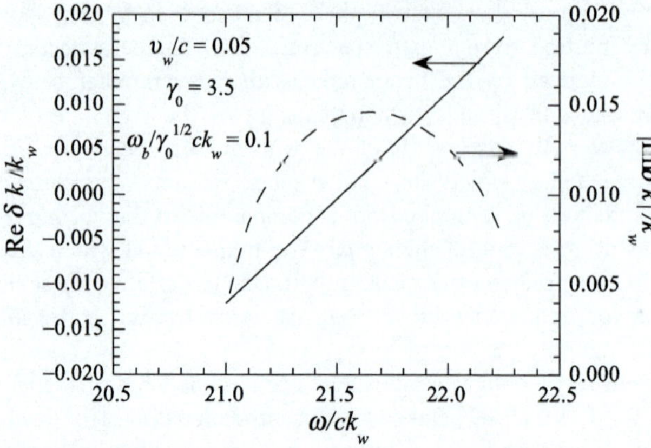

Fig. 6.1 Graph showing the growth rate (dashed line) and the shift in the wavenumber (solid line) as functions of frequency

$$\left|\frac{\upsilon_w}{c}\right| \equiv \begin{cases} \dfrac{\Omega_w}{ck_w} & ;\text{Helical Wiggler} \\ \dfrac{\Omega_w}{\sqrt{2}ck_w} & ;\text{Planar Wiggler} \end{cases}, \tag{6.2}$$

denotes the bulk wiggler-induced transverse velocity corresponding to $\Omega_w = eB_w/\gamma_0 m_e c$ for a wiggler amplitude B_w and wavenumber k_w. In order to illustrate the refractive shift in the wavenumber, we transform the wavenumber in Eq. (6.1) to $\Delta k = k - \omega/c$, which measures the shift from the vacuum wavenumber. Under the assumptions that $|\Delta k| << k_0 + (\omega^2 - \omega_b{}^2/\gamma_0)^{1/2}/c$, and $\omega_b/\gamma_0{}^{1/2} << \omega$, the dispersion equation can be written in the form of a cubic equation:

$$(\delta k - k_0)\left[\delta k^2 \upsilon_\parallel^2 - 2\Delta\omega\delta k \upsilon_\parallel - \left(\Delta\omega^2 - \frac{\omega_b^2}{\gamma_0 \gamma_\parallel^2}\right)\right] \cong -\frac{\upsilon_w^2}{2c^2}\frac{\omega_b^2}{\gamma_0}k_w, \tag{6.3}$$

where $\Delta\omega = (1 - \upsilon_\parallel/c)\omega - k_w \upsilon_\parallel$.

The solution to Eq. (6.3) for the real (solid line) and imaginary (dashed line) parts of Δk as functions of frequency is shown in Fig. 6.1 for $\gamma_0 = 3.5$, $\upsilon_w/c = 0.05$, and $\omega_b/\gamma_0{}^{1/2}ck_w = 0.1$. The frequency corresponding to the peak growth rate for this choice of parameters is $\omega/ck_w \approx 21.6$. It is clear that the wavenumber is shifted downward from ω/c in the absence of the wiggler (i.e., $\Delta k < 0$) for frequencies below $\omega/ck_w \leq 21.5$, which is below the frequency of peak growth. In contrast, for frequencies above this critical value, including at peak growth, the wavenumber is shifted upward from the vacuum wavenumber. Hence, the qualitative behavior for the wavenumber in the idealized one-dimensional analysis is the same as that from the three-dimensional nonlinear formulation. The conclusion to be drawn from these results is that the refractive effect of the wave-particle interaction in the wiggler can either guide or defocus the electromagnetic wave depending upon the interaction frequency.

In order to understand the process of optical guiding in the low-gain regime, we consider an analysis of the evolution of the relative phase in an idealized one-dimensional model in which the vector potential of the optical signal is represented in the form

$$\delta \mathbf{A}(z,t) = \delta A(z)\hat{\boldsymbol{e}}_x \cos[\omega(z/c - t) + \Delta\phi(z)] \tag{6.4}$$

where $\Delta\phi$ denotes the relative phase and both $\Delta\phi$ and the amplitude ΔA are assumed to be real. The wiggler field is assumed to be the idealized one-dimensional representation in Eq. (2.99). Substitution of this form for the vector potential into Maxwell's Eq. (5.19) yields an equation of the form

$$\delta k(z)\delta A(z)\cos[\omega(z/c - t)] + \frac{d}{dz}\delta A(z)\sin[\omega(z/c - t)] \cong \frac{2\pi}{\omega}\delta J_x, \tag{6.5}$$

where the source current is given in Eq. (5.23), $\Delta k = (d/dz)\Delta\phi$ denotes the perturbed wavenumber, and it is assumed that $|\Delta k| << \omega/c$. Note that second-order derivatives of the amplitude and phase have been neglected in Eq. (8.5). Multiplication by $\cos[\omega(z/c - t)]$ and subsequent averaging of the resulting equation over a wave period yield

$$\delta k \cong \frac{2}{\delta A}\int_0^{2\pi/\omega} dt\delta J_x \cos[\omega(z/c - t)], \tag{6.6}$$

where the source current is given by Eq. (5.22). Under the assumption that $v_x \approx v_w \cos k_w z$, therefore, the perturbed wavenumber becomes

$$\delta k \simeq \frac{\omega_b^2}{2\omega c}\frac{v_w}{c}\frac{1}{\delta a}\langle \cos\psi\rangle, \tag{6.7}$$

where ψ denotes the ponderomotive phase, ω_b is the beam plasma frequency, $\Delta a = e\Delta A/m_e c^2$ is the normalized amplitude, and the average is taken over the initial phase as defined in Eq. (5.41). Observe that the equation for the small-signal gain is obtained by multiplication by $\sin[\omega(z/c - t)]$.

The phase average may be determined by the solution of the pendulum equation in the untrapped limit. The evolution of the ponderomotive phase in the combined electromagnetic and magnetostatic fields is governed by the nonlinear pendulum equation

$$\frac{d^2}{dz^2}\psi = \kappa^2 \sin\psi, \tag{6.8}$$

where

$$\kappa^2 \equiv \delta a \frac{\upsilon_w}{\upsilon_\parallel} \frac{\omega^2/c^2}{2\gamma_0 \gamma_\parallel^2 \beta_\parallel^3}. \tag{6.9}$$

In the linear regime, the solutions describe untrapped trajectories which may be determined by a perturbative solution to Eq. (8.8). In this case, we expand $\psi = \psi_0 + \Delta kz + \Delta\psi$, where ψ_0 is the initial phase,

$$\Delta k \equiv -\frac{\omega}{\upsilon_\parallel}\left(1 - \frac{\upsilon_\parallel}{c}\right) + k_w \tag{6.10}$$

describes the mismatch parameter, and it is assumed that $|\Delta\psi/\psi| < < 1$. To lowest order in the perturbation, therefore, the pendulum equation can be expressed in the form

$$\frac{d^2}{dz^2}\delta\psi \cong \kappa^2 \sin(\psi_0 + \Delta kz). \tag{6.11}$$

Subject to the initial conditions that $\Delta\psi(z = 0) = 0$ and $(d/dz)\Delta\psi(z = 0) = 0$, this equation has the solution

$$\delta\psi = -\frac{\kappa^2}{\Delta k^2}[\sin(\psi_0 + \Delta kz) - \sin\psi_0 - \Delta kz\cos\psi_0]. \tag{6.12}$$

As a consequence,

$$\langle \cos\psi \rangle \cong \frac{\kappa^2}{2\Delta k^2}[1 - \cos\Delta kz - \Delta kz\sin\Delta kz], \tag{6.13}$$

and

$$\langle \sin\psi \rangle \cong -\frac{\kappa^2}{2\Delta k^2}[\sin\Delta kz - \Delta kz\cos\Delta kz]. \tag{6.14}$$

Hence, the perturbed wavenumber is given by

$$\delta k(z) \cong \frac{\omega_b^2}{8\gamma_0\gamma_\parallel^2 \upsilon_\parallel^2} \frac{\upsilon_w^2}{\upsilon_\parallel^2} \frac{\omega}{c\Delta k^2}[1 - \cos\Delta kz - \Delta kz\sin\Delta kz]. \tag{6.15}$$

The relative phase is found by integration of (Eq. 8.15) over axial position; hence, the relative phase at the end of the wiggler (i.e., $z = L_w$) is given by

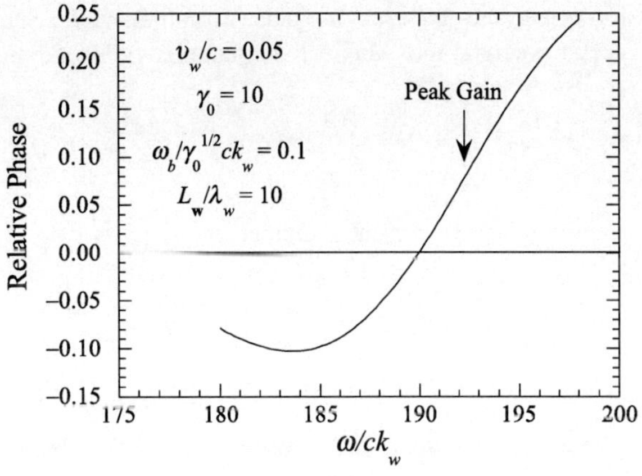

Fig. 6.2 The relative phase as a function of frequency within the gain band as calculated from the idealized one-dimensional model of the low-gain regime

$$\delta\phi(L_w) \cong \frac{\omega_b^2 k L_w^3}{16\gamma_0\gamma_\parallel^2 v_\parallel^2} \frac{v_w^2}{2v_\parallel^2} \frac{4}{\Delta k^2 L_w^2}\left[1 + \cos\Delta k L_w - \frac{2}{\Delta k L_w}\sin\Delta k L_w\right]. \tag{6.16}$$

The relative phase has been evaluated for a choice of parameters in which $\gamma_0 = 10$, $v_w/c = 0.05$, $\omega_b/\gamma_0^{1/2}ck_w = 0.1$, and a wiggler which is $L_w = 10\lambda_w$ in length, and the results are illustrated in Fig. 6.2. The gain band for this choice of parameters ranges over $\omega/ck_w \approx 180$–200, and a peak gain of approximately 20% is found at $\omega/ck_w \approx 192$. As shown in the figure, the relative phase is positive at the high-frequency portion of the gain band that includes the frequency of peak gain.

6.1.2 The Separable Beam Limit

The fundamentals of the optical guiding process can best be understood in terms of a separable beam approximation [22, 30, 39, 40, 43] in which the wiggler-driven source current is decomposed into a product of functions depending upon radius and axial position. Hence, the electron beam is assumed to be cylindrically symmetric. The electron trajectories are treated in the context of an idealized one-dimensional approximation of the planar wiggler geometry. This model has the considerable advantage of allowing an analytic solution for the radiation spot size and is generally valid as long as the beam radius is much less than either the spot size of the radiation and the wiggler period.

The idealized planar wiggler geometry implies that the interaction will be with a cylindrically symmetric plane polarized wave in which the vector potential of the electromagnetic field is expressed in the form

$$\delta\mathbf{A}(r,z,t) = \frac{1}{2}\delta A(r,z)\hat{e}_x \exp[i\omega(z/c - t)] + c.c., \tag{6.17}$$

for an angular frequency ω and wavenumber ω/c. Substitution of this form of the vector potential into Maxwell's equations yields a dynamical equation of the form

$$\left(\frac{1}{r}\frac{\partial}{\partial r}r\frac{\partial}{\partial r}+2i\frac{\omega}{c}\frac{\partial}{\partial z}\right)\delta A(r,z)=-\frac{8\pi}{c}\delta J_x(r,z)\exp[-i\omega(z/c-t)], \tag{6.18}$$

where the source current is defined as

$$\delta J_x(r,z)=-en_b(r)v_{z0}\int_{-\infty}^{\infty}dt_0\sigma(t_0)\frac{v_x}{|v_z|}\delta[t-\tau(z,t_0)], \tag{6.19}$$

where $n_b(r)$ notes the radial variation of the ambient beam density, $\tau(z,t_0)$ is the Lagrangian time coordinate defined in Eq. (5.24) of an electron which enters the interaction region at time t_0, and initial axial velocity v_{z0}. As a consequence, averaging Eq. (8.18) over a wave period subject to the quasi-static approximation for the electron trajectories gives a dynamical equation of the form

$$\left(\frac{1}{r}\frac{\partial}{\partial r}r\frac{\partial}{\partial r}+2i\frac{\omega}{c}\frac{\partial}{\partial z}\right)\delta a(r,z)=\frac{\omega_b^2(r)}{c^2}\beta_{z0}\frac{v_w}{v_\parallel}\langle\exp(-i\psi)\rangle, \tag{6.20}$$

where $\beta_{z0} = v_{z0}/c$, $\omega_b{}^2(r) = 4\pi e^2 n_b(r)/m_e$, and $\psi = (\omega/c + k_w)z - \omega t$ denote the ponderomotive phase, $\Delta a = e\Delta A/m_e c^2$; the orbits are given in the idealized approximation (Eqs. (2.103) and (2.104)) subject to the requirement that $|v_w/v_\parallel| << 1$, and the average is over the initial phase as defined in Eq. (5.42). The dynamical equation is solved for an electron beam with a Gaussian density profile:

$$n_b(r)=n_{b0}\exp(-r^2/r_b^2), \tag{6.21}$$

where n_{b0} denotes the maximum density on-axis and r_b is the Gaussian beam radius.

The solution of the dynamical equation may be written in terms of a Green's function which satisfies the equation

$$\left(\frac{1}{r}\frac{\partial}{\partial r}r\frac{\partial}{\partial r}+2i\frac{\omega}{c}\frac{\partial}{\partial z}\right)\delta G(r,z-z')=\frac{\omega}{c}\exp(-r^2/r_b^2)\delta(z-z'). \tag{6.22}$$

Equations (6.20) and (6.22) can be combined to yield

$$\int_0^\infty drr\int_0^{z'+0^+}dz\left[\delta a(r,z)\delta(z-z')-\frac{\omega_{b0}^2}{\omega c}\beta_{z0}\frac{v_w}{v_\parallel}G(r,z'-z)\langle\exp[-i\psi(z)]\rangle\right]$$

$$= \frac{c}{\omega}\int_0^{\infty} dr \int_0^{z'+0^+} dz\left[\frac{\partial}{\partial r}\left(r\delta a(r,z)\frac{\partial}{\partial r}G(r,z'-z)\right) - \frac{\partial}{\partial r}\left(rG(r,z'-z)\frac{\partial}{\partial r}\delta a(r,z)\right)\right]$$

$$-2i\int_0^{\infty} drr \int_0^{z'+0^+} dz\left[\delta a(r,z)\frac{\partial}{\partial z}G(r,z'-z) + G(r,z'-z)\frac{\partial}{\partial z}\delta a(r,z)\right]. \quad (6.23)$$

The first integral on the right-hand side of Eq. (6.23) vanishes subject to the requirements that

$$\delta a(r=\infty,z) = G(r=\infty,z-z') = 0, \quad (6.24)$$

and

$$\frac{\partial}{\partial r}\delta a(r,z)\bigg|_{r=\infty} = \frac{\partial}{\partial r}G(r,z-z')\bigg|_{r=\infty} = 0. \quad (6.25)$$

The second integral on the right-hand side yields the homogeneous solution in the absence of the interaction. Hence, the solution in terms of this Green's function is

$$\delta a(r,z) = \delta a_h(r,z) + \frac{\omega_{b0}^2}{c^2}\beta_{z0}\frac{v_w}{v_\parallel}\frac{c}{\omega}\int_0^z dz' G(r,z'-z)\langle\exp[-i\psi(z')]\rangle, \quad (6.26)$$

where $\omega_{b0}{}^2 = 4\pi e^2 n_{b0}/m_e$ and $\Delta a_h(r,z)$ denote the homogeneous solution which describes the evolution of the vector potential in the absence of the beam.

We now consider the low-gain regime. In order to describe the focusing or defocusing of the radiation due to the interaction, we define an *average* field in the form

$$\langle\delta a(r,z)\rangle_r \equiv \frac{2}{r_b^2}\int_0^{\infty} drr\exp(-r^2/r_b^2)\delta a(r,z). \quad (6.27)$$

This is a weighted average over the cross-section of the electron beam that is a measure of the effective field experienced by the electrons. In the low-gain regime, the ponderomotive phase is determined by the solution of the pendulum Eq. (8.8) in the untrapped regime. Utilization of this average vector potential in the definition of κ^2 in Eq. (8.9), therefore, gives a phase average that varies as

$$\langle\exp[\pm i\psi(z)]\rangle \cong \frac{\kappa^2}{2\Delta k^2}[1 - \exp(\pm i\Delta kz) \pm i\Delta kz\exp(\pm i\Delta kz)], \quad (6.28)$$

where

$$\kappa^2 \cong \langle \delta a(r,z) \rangle_r \frac{v_w}{v_\|} \frac{\omega^2/c^2}{2\gamma_0 \gamma_\|^2 \beta_\|^3}. \tag{6.29}$$

It may be verified by substitution that the Green's function is

$$G(r,z-z') = \frac{1}{2i} \frac{\exp\left[r^2(1+2i(z-z')/z_b)/r_b^2\right]}{1+2i(z-z')/z_b} H(z-z'), \tag{6.30}$$

where $H(z)$ denotes the Heaviside function, and $z_b = \omega {r_b}^2/c$. As a consequence, the average field can be expressed as

$$\langle \delta a(r,z) \rangle_r = \frac{\langle \delta a_h(r,z) \rangle_r}{S(z)}, \tag{6.31}$$

where

$$S(z) \equiv 1 + \frac{v_w^2}{v_\|^2} \frac{\omega_{b0}^2 r_b^2}{16\gamma_0 \gamma_\|^2 v_\|^2} \frac{\omega^2}{c^2 \Delta k^2} \left\langle 1 + \ln\left(1 + i\frac{z}{z_b}\right) - \exp(i\Delta k z) \right.$$

$$\left. - i[1 + \Delta k(z_b + iz)] \exp(-i\Delta k z) \int_0^{z/z_b} dx \frac{\exp(i\Delta k z_b x)}{1+ix} \right\rangle. \tag{6.32}$$

If the homogeneous solution is given in terms of a Gaussian beam which is focused down to the minimum spot size at $z = 0$, then [42]

$$\delta a_h(r,z) = \delta a_0 \frac{\exp\left[r^2(1+2iz/z_R)/r_0^2\right]}{1+2iz/z_R}, \tag{6.33}$$

where Δa_0 is real and denotes the initial amplitude of the field, r_0 denotes the Gaussian radius of the beam, and $z_R = \omega {r_0}^2/c$ denotes the Rayleigh length. As a consequence,

$$\langle \delta a_h(r,z) \rangle_r = \delta a_0 \left(1 + \frac{r_b^2}{r_0^2} + 2i\frac{z}{z_R}\right)^{-1}. \tag{6.34}$$

An effective *focusing factor* may be defined in terms of this average field in the form

$$F(z) \equiv \frac{\left|\langle \delta a(r,z)\rangle_r\right|^2}{\frac{2}{r_b^2}\int\limits_0^\infty drr\left|\delta a(r,z)\right|^2}. \tag{6.35}$$

Since the homogeneous solution for the vector potential describes the propagation of the signal in the absence of the beam, the focusing factor measures the evolution of the radiation spot-size relative to the diffraction of the Gaussian beam in free-space.

The denominator can be determined by energy conservation arguments. Returning to the dynamical equation for the field (6.20), we observe that

$$\delta a^*\left(\nabla_\perp^2 + 2i\frac{\omega}{c}\frac{\partial}{\partial z}\right)\delta a = \frac{\omega_{b0}^2}{c^2}\beta_{z0}\frac{v_w}{v_\parallel}\exp(-r^2/r_b^2)\delta a^*\langle\exp(-i\psi)\rangle, \tag{6.36}$$

and

$$\delta a\left(\nabla_\perp^2 - 2i\frac{\omega}{c}\frac{\partial}{\partial z}\right)\delta a^* = \frac{\omega_{b0}^2}{c^2}\beta_{z0}\frac{v_w}{v_\parallel}\exp(-r^2/r_b^2)\delta a\langle\exp(i\psi)\rangle. \tag{6.37}$$

Subtracting the second equation from the first, we find that

$$2i\frac{\omega}{c}\frac{\partial}{\partial z}|\delta a|^2 = \frac{\omega_{b0}^2}{c^2}\beta_{z0}\frac{v_w}{v_\parallel}\exp(-r^2/r_b^2)[\delta a^*\langle\exp(-i\psi)\rangle - \delta a\langle\exp(i\psi)\rangle]. \tag{6.38}$$

Observe that we have omitted the terms in $\Delta a^*\nabla_\perp{}^2\Delta a - \Delta a\,\nabla_\perp{}^2\Delta a^*$ since they will vanish upon integration over radius. Integration of (6.38) over both axial position and radius, therefore, yields

$$\frac{2}{r_b^2}\int\limits_0^\infty drr|\delta a(r,z)|^2 \cong \delta a_0^2\frac{1+\frac{r_b^2}{r_0^2}+4\frac{z^2}{z_R^2}+G(z)}{1+\frac{r_b^2}{r_0^2}+4\frac{z^2}{z_R^2}}, \tag{6.39}$$

where $G(z)$ is the expression for the power gain in the idealized one-dimensional formulation of the fundamental interaction in the low-gain regime (Eq. 4.170). Combining Eqs. (6.33, 6.34, 6.35 and 6.39), we find that the focusing factor over the total length of the system L_w may be approximated in the form

$$F(L_w) \simeq \frac{1}{|S(L_w)|^2}\left(1+\frac{r_b^2}{r_0^2}+4\frac{L_w^2}{z_R^2}+G(L_w)\right)^{-1}. \tag{6.40}$$

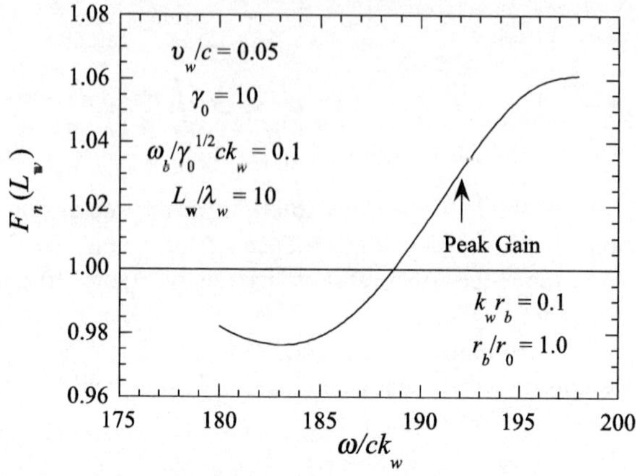

Fig. 6.3 Illustration of the normalized focusing factor for representative parameters in the low-gain regime

Observe that in the limit in which the interaction vanishes (i.e., in the absence of either the beam or the wiggler field), the gain also vanishes, and the focusing factor describes the free-space diffraction of the optical beam

$$F_{fs}(L_w) \cong \left(1 + \frac{r_b^2}{r_0^2} + 4\frac{L_w^2}{z_R^2}\right)^{-1}. \tag{6.41}$$

Hence, the effect of the interaction upon the diffraction of the optical beam can be effectively measured by a normalized focusing factor defined as

$$F_n(L_w) \equiv \frac{F(L_w)}{F_{fs}(L_w)}. \tag{6.42}$$

Hence, when $F_n(L_w) > 1$ the diffraction of the signal is slower than in the free-space limit, and the signal is effectively guided. Conversely, when $F_n(L_w) < 1$, the diffraction is more rapid than in free-space.

The specific example under consideration is that of a system for which $v_w/c = 0.05$, $\omega_{b0}/\gamma_0^{1/2} ck_w = 0.1$, and $\gamma_0 = 10$. Further, it is assumed that there are ten wiggler periods within the interaction length (i.e., $L_w/\lambda_w = 10$), $k_w r_b = 0.1$, and that the optical beam is focused down to the beam radius upon the entry to the wiggler (i.e., $r_b = r_0$). Observe that for this choice of parameters corresponds to that used in the calculation of the relative phase in the idealized one-dimensional regime shown in Fig. 6.2. The peak gain in the idealized one-dimensional limit for this choice of parameters over length L_w is of the order of 20% and occurs for $\omega/ck_w \approx 191.7$. The focusing factor at $z = L_w$ associated with this choice of parameters is shown in Fig. 6.3 for frequencies within the gain band. It is evident that the indicated diffraction of the radiation found on the basis of the calculation of the relative phase and the separable beam limits is qualitatively similar. That is that the diffraction is slower than the free-space value for the low-frequency portion of the gain band but is faster than in free-space at the edges of the gain band.

6.2 Slippage and the Group Velocity

The previous chapter dealt with the nonlinear interaction mechanism when the interaction was between a single-frequency wave with the electron beam. In that case, a nonlinear formalism can be derived based upon the SVEA in which the electromagnetic field amplitude is a slowly varying function of position in which Maxwell's equations are averaged over a wave period to remove the fast-time-scale oscillation from the nonlinear analysis. This greatly simplifies the numerical procedures and is valid for a long pulse electron beam. In order to quantify the meaning of a "long pulse electron beam," we now introduce the concept of *slippage*, which arises because the electromagnetic field propagates with a group velocity near the speed of light in vacuo which is greater than the electron beam velocity; hence, the electromagnetic field *slips* ahead of the electrons during the course of the interaction. While this may not be important in long pulse electron beams generated by pulse line accelerators, modulators, or induction linacs, it can have a significant effect on the interaction for short pulse electron beams produced by RF linacs.

The average axial velocity of the electron beam in the wiggler is given approximately by

$$\frac{v_{\parallel}}{c} = 1 - \frac{1 + K_{rms}^2}{2\gamma^2}, \tag{6.43}$$

where we use K_{rms} ($a_w/\surd 2$ in a planar wiggler and a_w in a helical wiggler) to denote the rms wiggler strength parameter here; hence, $v_{\parallel} = c(\lambda_w - \lambda)/\lambda_w$. The electromagnetic field in vacuo traverses one wiggler period in a time $\tau_{em} = \lambda_w/v_{gr}$. Over this same time, therefore, the electron traverses a distance $(\lambda_w - \lambda)c/v_{gr}$. As a result, when $v_{gr} \approx c$, the electromagnetic field slips ahead of the electron beam by one wavelength per wiggler period, and the slippage time through a wiggler is defined as

$$\tau_{slip} = \frac{N_w \lambda}{c}, \tag{6.44}$$

where N_w is the number of wiggler periods.

However, in the high-gain regime where the power grows exponentially, the group velocity is reduced due to the interaction, and the rate of slippage is also reduced. As shown in Chap. 4, the interaction induces a complex shift in the wavenumber, where the imaginary part describes the gain, and the shift in the real part of the wavenumber describes a change in the refractive index that gives rise to the optical guiding of the electromagnetic wave. This shift in the real part of the wavenumber is accompanied by a shift in the group velocity of the optical mode as well. In order to quantify this, we turn to the one-dimensional dispersion Eq. (4.87). Taking the partial derivative with respect to k_+, we find that

$$\left\{\omega\delta k(\delta k+2\kappa_b)-\frac{\delta k+\kappa_b}{\upsilon_\parallel}\left(\omega^2-c^2k_+^2-\frac{\omega_b^2}{\gamma_0}\right)-\delta k(\delta k+2\kappa_b)(\delta k-\Delta k)\frac{\omega}{\gamma_\parallel^2\beta_\parallel k_w}\right\}\frac{\upsilon_{gr}}{c}$$
$$=c^2k_+\left[(\omega-k\upsilon_\parallel)^2-\kappa_b^2\upsilon_\parallel^2\right]+\upsilon_\parallel(\omega-k\upsilon_\parallel)\left(\omega^2-c^2k_+^2-\frac{\omega_b^2}{\gamma_0}\right), \tag{6.45}$$

where $\upsilon_{gr}=\partial\omega/\partial k_+$, $k=k_++k_w$, $\kappa_b{}^2\upsilon_\parallel{}^2=\omega_b{}^2/\gamma_0\gamma_\parallel{}^2$, and $\gamma_\parallel=\gamma_0/(1+K_{rms}{}^2)^{1/2}$. As in Chap. 4, if we define $\Delta k=k-\omega/\upsilon_\parallel-\kappa_b$, then we may express this as

$$\left\{\omega\delta k(\delta k+2\kappa_b)-\frac{\delta k+\kappa_b}{\upsilon_\parallel}\left(\omega^2-c^2k_+^2-\frac{\omega_b^2}{\gamma_0}\right)-\delta k(\delta k+2\kappa_b)(\delta k-\Delta k)\frac{\omega}{\gamma_\parallel^2\beta_\parallel k_w}\right\}\frac{\upsilon_{gr}}{c}$$
$$=ck_+\delta k(\delta k+2\kappa_b)-\frac{\delta k+\kappa_b}{c}\left(\omega^2-c^2k_+^2-\frac{\omega_b^2}{\gamma_0}\right), \tag{6.46}$$

where

$$\Delta k=\sqrt{\frac{\omega^2}{c^2}-\frac{\omega_b^2}{\gamma_0c^2}}+k_w-\frac{\omega}{\upsilon_\parallel}-\kappa_b. \tag{6.47}$$

If we neglect the backward-propagating wave, then

$$\omega^2-c^2k_+^2-\frac{\omega_b^2}{\gamma_0}\cong-2c^2k_+(\delta k-\Delta k), \tag{6.48}$$

and we have that

$$\left\{\delta k(\delta k+2\kappa_b)+\frac{2}{\beta_\parallel}\frac{ck_+}{\omega}(\delta k+\kappa_b)(\delta k-\Delta k)-(\delta k+2\kappa_b)(\delta k-\Delta k)\frac{\delta k}{\gamma_\parallel^2\beta_\parallel k_w}\right\}\frac{\upsilon_{gr}}{c}$$
$$=\frac{ck_+}{\omega}\left[\delta k(\delta k+2\kappa_b)+2(\delta k+\kappa_b)(\delta k-\Delta k)\right]. \tag{6.49}$$

Finally, if we now assume that space-charge is negligible ($|\Delta k|>>k$), $\Delta k\approx 0$, and $\omega\approx ck_+$, then we find that

$$\frac{\upsilon_{gr}}{c}\cong\left(1+\frac{1}{3\gamma_\parallel^2}\right)^{-1}. \tag{6.50}$$

As a result, the electromagnetic field traverses one wiggler period in the time $\tau_{em}=\lambda_w(1+1/3\gamma_\parallel{}^2)/c$, and the electrons traverse a distance $\lambda_w-\lambda/3$ in this time. As a result, in the high-gain, exponential regime, the slippage time through the wiggler is reduced to

$$\tau_{slip} = \frac{N_w \lambda}{3c}. \tag{6.51}$$

This result is applicable only as long as the interaction is in the exponential regime. As a consequence, slippage occurs at the vacuum rate in the start-up regime prior to exponentiation and increases to the vacuum rate again after saturation of the exponential growth.

6.3 The SVEA, Time Dependence, and the Quasi-Static Assumption

As discussed in the preceding chapter, the SVEA was originally developed to treat long pulse (CW) traveling wave tubes. These devices operated as ultra-narrow bandwidth (essentially singe-frequency) amplifiers. The basic assumption in the SVEA was the quasi-static assumption that all fluctuating quantities, whether the electron velocities or densities as well as the electromagnetic fields were periodic in the underlying wave frequency. As a result, it was possible to decompose the bunch into *beamlets* (or *slices*) each of which was one wavelength long and which interacted with the field in identical ways. As a result, if an electron advanced from one slice to the next while propagating through the interaction region, then a corresponding electron from the preceding slice would advance to replace the advancing electron. This is equivalent to imposing a periodic boundary condition and is applicable when the electron beam properties are constants. However, this is not the case for the electron beams of interest in this chapter where the electron and optical properties are assumed to vary on time scales of picoseconds or shorter.

As a result, while the quasi-static assumption is not rigorously satisfied for these electron beams, we must assume that the variation in the properties of the electron bunch and the optical field vary slowly from *slice* to *slice*. While in the steady-state analysis the electron bunch properties were constant and we imposed the condition that the optical field amplitudes satisfy

$$\frac{\partial}{\partial z}|\delta A| << \frac{1}{\lambda}|\delta A|, \tag{6.52}$$

we now assume that the electron bunch properties vary over the duration of the bunch and the optical field amplitudes vary slowly in both space and time, and we must now impose the additional requirement that

$$\frac{\partial}{\partial t}|\delta A| << \frac{c}{\lambda}|\delta A|, \tag{6.53}$$

and similar requirements must be imposed for the electron bunch properties. In practical terms, this means that the SVEA cannot be used for ultrashort pulses of less than about ten wavelengths long.

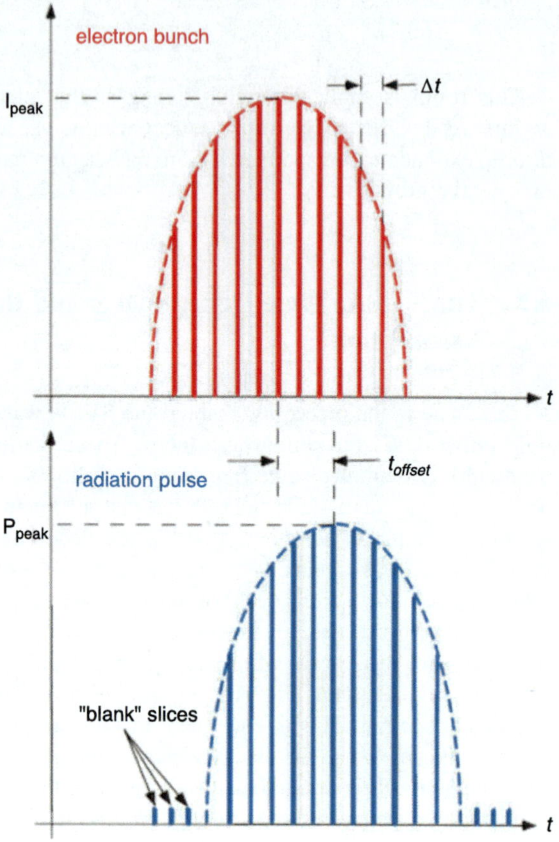

Fig. 6.4 Schematic illustration of the initial electron bunch and radiation pulse

The configuration that is implemented in the time-dependent SVEA formulations is illustrated schematically in Fig. 6.4. Both the electron bunch and the optical pulse consist of a sequence of *slices* each of which is represented by a vertical bar in the figure and corresponds to a section of the beam and the optical pulse that is one wavelength long. The dashed line represents the envelopes that describe the temporal shapes of the electron bunch and the optical pulse. Note that the *slices* need not be contiguous, and fewer *slices* may be used than the total number of wavelengths in the electron bunch or optical pulse. The separation between the slices, Δt, may be varied, but, in practice, we find that the slice spacing must be at least about one every three wavelengths for amplifiers. Each electron *slice* can be characterized by bunch parameters (i.e., current, energy emittance, energy spread, Twiss parameters, etc.) that vary with temporal position in the bunch, although only the current is shown in the figure. The figure shows the power of the optical pulse, but the power, phase, and curvature of the phase front may vary from optical *slice* to optical slice. In addition, the optical pulse shape, and duration, may differ from that of the electron bunch, and the center of the optical pulse may be shifted relative to the center of the electron bunch. Thus, the offset illustrated by t_{offset} is used to describe (1) the injection of a seed pulse into an amplifier that is not synchronized to the electron bunch and (2) for cavity tuning in an oscillator where the roundtrip time for the radiation may be chosen to give arbitrary synchronism between the radiation and the electron bunches from pass to pass. Because the radiation slips ahead of the electrons, allowance must be made in the simulation for radiation *slices* that have no corresponding electron slice, and these are referred to as "blank" *slices* in the figure. Note that the blank *slices* initially are at zero power and grow only due to slippage.

The numerical procedure will be discussed in more detail when specific examples are given for both the one- and three-dimensional analyses. However, in general, the procedure employed is to advance each electron and optical *slice* from $z \rightarrow z + \Delta z$ after which the optical field is allowed to slip relative to the electron bunch.

6.4 The Simulation of Shot Noise

The question of radiation start-up in free-electron lasers is important in the simulation of oscillators and single-pass free-electron lasers based upon self-amplified spontaneous emission (SASE). Both oscillators and SASE free-electron lasers start up from the spontaneous emission due to shot noise on the electron beam, which is essentially due to the random fluctuations in the phase distribution of the electrons. The injected power in a seeded free-electron laser amplifier is usually large enough to overwhelm the shot noise. Of necessity, free-electron laser simulations employ a reduced number of *macro-electrons* to model the electron beam. The phase distributions discussed heretofore were relevant to the simulation of seeded amplifiers and described a *quiet-start* in which the optical field will not grow without a seed pulse. The simulation uses random distribution for the initial phases (i.e., entry times into the wiggler) that satisfies Poisson statistics. The specific manner in which this is implemented depends upon the details of any particular formulation.

At the level of particle in cell (PiC) simulations, no average of any kind is performed, and the step size in time and the grid spacing must be fine enough to resolve the fluctuations in the wave period/wavelength scales. Particles must be loaded into each grid cell. Techniques for including shot noise in PiC simulations have been discussed in the literature [46, 47]. In contrast, SVEA simulations include multiple temporal *slices* that are one wavelength long, and particles must be loaded into each *slice* with an initial phase distribution satisfying Poisson statistics so that $\langle \exp(i\psi_0) \rangle = 1/\sqrt{N_e}$, where N_e is the number of correlated electrons. The rationale for choosing the number of correlated electrons will differ depending upon whether the simulation is performed in the steady-state or with time dependence (i.e., slippage) is included.

The number of electrons per *slice*, n_e, is given by

$$n_e = \frac{I_b \lambda}{e v_b}, \tag{6.54}$$

where I_b is the beam current, λ is the wavelength, e is the electronic charge, and v_b is the bulk axial velocity. Since the number of interacting electrons in steady-state depends upon the distance over which the field exponentiates, the total number of interacting electrons, N_e, will include contributions from multiple slices and will be given approximately by [48]

$$N_e \cong 4.3 \frac{L_G}{\lambda_w} n_e, \tag{6.55}$$

where L_G denotes the power gain length. It should be remarked that steady-state simulations are valid only when the bunch length is much greater than the slippage distance within the wiggler and the inclusion of shot noise in steady-state simulations is equivalent to using an equivalent noise seed power in an amplifier simulation. In time-dependent simulations, one typically chooses an integration window that is many *slices* in length, although the number of *slices* used in simulation is usually less than this total. Thus, each *slice* included in the simulation is separated by some distance $\Delta\lambda_s$, and the number of interacting electrons is

$$N_e = \frac{\Delta\lambda_s}{\lambda} n_e = \frac{I_b \Delta\lambda_s}{e v_b}. \tag{6.56}$$

First consider the particle-loading in the absence of shot noise. This is what has been described in Chap. 5. Particles are loaded into the two-dimensional phase space of the one-dimensional formulation or the six-dimensional phase space of the three-dimensional formulation using Gaussian quadrature in each of the degrees of freedom. This is a deterministic algorithm and the additional degrees of freedom describing the beam (γ_0 in one dimension, and x_0, y_0, p_{x0}, p_{y0}, and γ_0 in three dimensions) do not vary with the initial phase ψ_0. As a result, each choice of ψ_0 is associated with an identical distribution of particles in the additional degrees of freedom. What this means is that if we choose an initial loading in ψ_0 such that

$$\frac{1}{2\pi} \int_0^{2\pi} d\psi_0 \exp(i\psi_0) = 0, \tag{6.57}$$

for one choice of $(x_0, y_0, p_{x0}, p_{y0}, \gamma_0)$, then it will also vanish for all choices of $(x_0, y_0, p_{x0}, p_{y0}, \gamma_0)$, and we will have ensured a *quiet-start* in which some seed power is required for the fields to grow.

In practice, the particles are loaded in ψ_0 (as well as in each of the other degrees of freedom) using an Nth-order Gaussian quadrature in which for some function $f(\psi_0)$

$$\int_0^{2\pi} d\psi_0 f(\psi_0) = \sum_{j=1}^{N} w_j f(\psi_{0j}),, \tag{6.58}$$

where the weights, w_j, and initial phases, ψ_j, are chosen using N^{th}-order Gaussian quadrature. Note that it is not necessary to use the same order N for each of the degrees of freedom. It has been found empirically that Gaussian quadrature ensures that the average $< \exp(i\psi_0) >$ vanishes with the fewest particles in comparison with Simpson's or the trapezoidal rule. We find that this phase average vanishes to within roundoff error with 64-bit machine accuracy with about eight particles in phase (i.e., eighth-order Gaussian quadrature). If it is desired to include up to the hth harmonic in the simulation, then more particles are required in order to resolve the harmonic phase space dynamics, and a rough rule of thumb is that h-times as many particles are needed in the initial phase loading as for simulations including only the fundamental. The beauty of this technique is that if the additional particles are included in the simulation, then $<\exp(ih\psi_0)>$ also vanishes to within machine precision; hence, the *quiet-start* is implicitly obtained for the fundamental and all harmonics.

The starting point in the inclusion of shot noise is the unperturbed *quiet-start* choices for the initial phase $\{\psi_{0j}\}$ and introduces a perturbation to obtain a new distribution $\{\psi_{0j}{}'\}$. The procedure is similar [49], but not identical, to that used by Fawley [50], and we write

$$\psi_{0j}{}' = \psi_{0j} + \sum_{h=1}^{h_{max}} \delta\psi_h \sin\left[h\left(\psi_{0j} - \phi\right)\right], \tag{6.59}$$

where $\Delta\psi_h$ ($<< 1$) is chosen to describe the Poisson statistics and ϕ is chosen randomly over the interval $[0,2\pi]$. If we note that $<\sin(h\psi_0)> = < \cos(h\psi_0)> = 0$ and that $<\sin^2(h\psi_0)> = < \cos^2(h\psi_0)> = ½$, then it follows that

$$\langle \sin(h\psi_0{}')\rangle \cong -\frac{1}{2} h\delta\psi_h \sin(h\phi) \tag{6.60}$$

$$\langle \cos(h\psi_0{}')\rangle \cong \frac{1}{2} h\delta\psi_h \cos(h\phi) \tag{6.61}$$

to lowest order in $\Delta\psi_h$. If we now choose $h\Delta\psi_h = \Delta\psi_1$ and $\Delta\psi_1 = 2/\sqrt{N_e}$, where N_e is the number of correlated electrons as defined in Eq. (6.54) or (6.55) depending on whether the simulation is in the steady-state or time-dependent modes, then

$$|\langle \exp(ih\psi_0{}')\rangle| \cong \frac{1}{2}\delta\psi_1 = \frac{1}{\sqrt{N_e}}, \tag{6.62}$$

and the correct Poisson statistics are recovered.

The shot noise (i.e., phase jitter) is applied as an initial condition which then carries forward self-consistently as the electron beam and radiation propagate through the complete undulator line. This includes any drift spaces, chicanes, quadrupoles, etc., since the particles and fields co-propagate self-consistently throughout since the formulation described herein uses the same integration procedure everywhere.

6.5 Elliptical Wigglers and the *JJ*-Factor

There is strong interest in using free-electron lasers to produce photons with a variety of polarizations ranging from linear, through elliptical, to circular, and novel wiggler configurations have been proposed and implemented for this purpose including the so-called APPLE-II [51] and delta-type wigglers [52].

6.5.1 The APPLE-II Wiggler Representation

An approximate, three-dimensional representation of an APPLE-II wiggler [8] that is valid near the axis of symmetry can be constructed by the superposition of two planar wiggler models that are shifted in phase.

$$\begin{aligned}\mathbf{B}_w(\mathbf{x}) = B_w(z)\left(\sin(k_w z + \phi) - \frac{\cos(k_w z + \phi)}{k_w B_w}\frac{dB_w}{dz}\right)\hat{e}_x \cosh k_w z \\ + B_w(z)\left(\sin k_w z - \frac{\cos k_w z}{k_w B_w}\frac{dB_w}{dz}\right)\hat{e}_y \cosh k_w y \\ + B_w(z)\hat{e}_z \sinh k_w y[\cos(k_w z + \phi) + \cos k_w z], \end{aligned} \tag{6.63}$$

where $B_w(z)$ describes the envelope of the wiggler field that includes the transitions at the entry/exit regions, the uniform field region, and any field taper. The ellipticity of the field is governed by the choice of the phase ϕ.

The ellipticity, u_e, varies over the range $0 \le u_e \le 1$ from planar polarization for $u_e = 0$ to circular polarization for $u_e = 1$ and is given by

$$u_e = \begin{cases} \dfrac{1 - \cos\phi}{1 + \cos\phi} & ; 0 \le \phi \le \pi/2 \\ \dfrac{1 + \cos\phi}{1 - \cos\phi} & ; \pi/2 < \phi \le \pi \end{cases}, \tag{6.64}$$

Fig. 6.5 The on-axis field contours for different choices of the phase ϕ

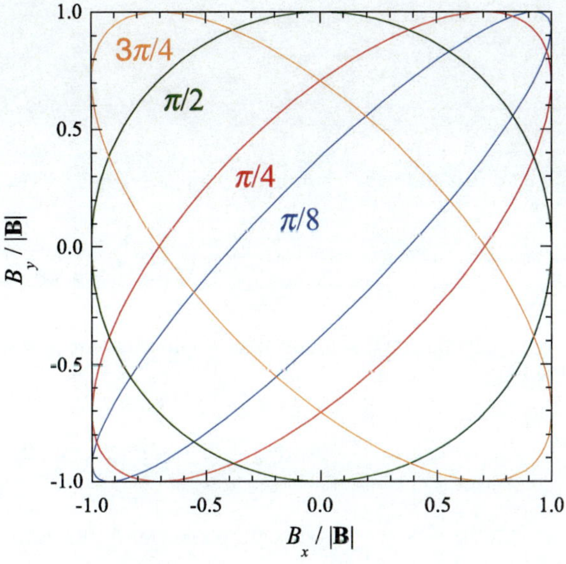

Illustrations of the on-axis field polarization are shown in Fig. 6.5 where the y-component of the field is plotted versus the x-component (normalized for the amplitude) for $\phi = \pi/8$, $\pi/4$, $\pi/2$ (corresponding to helical polarization), and $3\pi/4$.

6.5.2 The Resonance Condition and the JJ-Factor

The resonance condition and the *JJ*-factor are both affected by the ellipticity of the wiggler [53] and are necessary in order to generalize the Ming Xie parameterization for the gain discussed in Chap. 4 to the case of more general polarizations. In order to determine both the resonant frequency and the *JJ*-factor, we start with a one-dimensional representation for the wiggler field

$$\mathbf{B}_w = B_w\left(u_e \hat{\mathbf{e}}_x \cos k_w z + \hat{\mathbf{e}}_y \sin k_w z\right), \tag{6.65}$$

where the ellipticity varies over the range $0 \le u_\varepsilon \le 1$ from planar polarization for $u_e = 0$ to circular polarization for $u_\varepsilon = 1$ as in the case of the APPLE-II wiggler model.

The single-particle trajectories in this field are given by

$$\frac{v_x}{c} = -\frac{K}{\gamma}\cos k_w z, \tag{6.66}$$

$$\frac{v_y}{c} = -u_e \frac{K}{\gamma}\sin k_w z, \tag{6.67}$$

for the transverse motion, and

$$\frac{v_z^2}{c^2} = \frac{v_\parallel^2}{c^2} - \left(1 - u_e^2\right)\frac{K^2}{2\gamma^2}\cos 2k_w z, \tag{6.68}$$

where K is the wiggler strength parameter, and

$$\frac{v_\parallel^2}{c^2} = 1 - \left(1 + u_e^2\right)\frac{K^2}{2\gamma^2}, \tag{6.69}$$

is the average axial velocity. This implies that the resonance condition in an elliptical wiggler is

$$\lambda_{res} = \frac{\lambda_w}{2\gamma^2}\left[1 + \left(1 + u_e^2\right)\frac{K^2}{2}\right], \tag{6.70}$$

so that the rms wiggler strength parameter for the elliptical wiggler becomes

$$K_{rms} = \sqrt{\frac{1 + u_e^2}{2}}K. \tag{6.71}$$

Observe that the resonance and rms wiggler strength parameter reduce to the appropriate values in the limits of a planar ($u_e = 0$) and helical ($u_e = 1$) wiggler.

In order to obtain the generalized JJ-factor, observe that

$$v_z \cong v_\parallel - \left(1 - u_e^2\right)\frac{cK^2}{4\gamma^2}\cos 2k_w v_\parallel t, \tag{6.72}$$

and

$$z \cong z_0 + v_\parallel t - \left(1 - u_e^2\right)\frac{K^2}{8\gamma^2 k_w}\sin 2k_w v_\parallel t, \tag{6.73}$$

which are analogous to Eqs. (4.150) and (4.151). Now consider the interaction with an elliptically polarized electromagnetic wave with wavenumber k and frequency ω,

$$\delta\mathbf{E}(z,t) = \delta E_0\left[\hat{\mathbf{e}}_x \sin(kz - \omega t) + u_e\hat{\mathbf{e}}_y\cos(kz - \omega t)\right], \tag{6.74}$$

so that

$$\mathbf{v}\cdot\delta\mathbf{E} = -\frac{cK}{2\gamma}\delta E_0\left\{\left(1+u_e^2\right)\sin[(k+k_w)z-\omega t]+\left(1-u_e^2\right)\sin[(k-k_w)z-\omega t]\right\}. \tag{6.75}$$

Observe that the oscillation in the axial velocity and the lower beat wave vanishes in the limit of a helical wiggler for which $u_e = 1$, as expected. The development of Eq. (4.154) can be generalized to obtain

$$\mathbf{v}\cdot\delta\mathbf{E} = -\left(1+u_e^2\right)\frac{cK}{2\gamma}\frac{\delta E_0}{2i}\sum_{n=-\infty}^{\infty}(-1)^n\left[J_n(b_e)-\frac{1-u_e^2}{1+u_e^2}J_{n+1}(b_e)\right]$$
$$\times\exp\left\{i\left[kv_\parallel+(1+2n)k_w v_\parallel-\omega\right]t\right\}+c.c., \tag{6.76}$$

where

$$b_e = \frac{\left(1-u_e^2\right)K^2/4}{1+\left(1+u_e^2\right)K^2/2}. \tag{6.77}$$

As a result, the generalized form for the *JJ*-factor is

$$JJ_n(b_e) = J_n(b_e)-\frac{1-u_e^2}{1+u_e^2}J_{n+1}(b_e). \tag{6.78}$$

6.5.3 *The Generalized Pierce Parameter and Ming Xie Parameterization*

The Pierce parameter for an elliptical wiggler and electromagnetic field follow directly from the generalized rms wiggler strength parameter and the *JJ*-factor, and we have

$$\rho^3 \equiv \frac{K_{rms}^2 JJ_0^2(b_e)}{4\gamma_0^3\beta_\parallel^2}\frac{1}{k_w^2R_b^2}\frac{I_b}{I_A}, \tag{6.79}$$

where K_{rms} is given in Eq. (6.71). The parameterizations obtained by Ming Xie for the gain lengths for helical (Eq. 4.374) and planar (Eq. 4.394) can be generalized for the case of elliptical wigglers by using the abovementioned expressions for the resonance, the Pierce parameter, the rms wiggler strength parameter, and the *JJ*-factor. This extension of the parameterization is tested by comparison with three-dimensional, nonlinear simulations of elliptical wigglers using the Apple-II wiggler model that are presented below.

6.6 Quadrupole and Dipole Field Models

Representations have been presented for various wiggler configurations including planar (both flat-pole-face and parabolic-pole-face), helical, and elliptic polarizations. However, many short wavelength free-electron lasers require extremely long wigglers, and it is impractical to construct wigglers of such extreme lengths. As a result, the wiggler line consists of a series of individual wigglers separated by drift spaces. In order to maintain the focus of the electron beam between the wigglers, magnetic quadrupoles are inserted into the drift spaces. In addition, there is often a need for magnetic chicanes in the drift spaces as well in order to shift the phase of the electron beam relative to the optical field and thereby maintain synchronism between the electrons and the light from wiggler to wiggler. These chicanes are composed of three or four magnetic dipoles oriented perpendicularly to the symmetry axis.

The quadrupole field model used in simulation is

$$B_Q(x) = B_Q(z)\left(y\hat{e}_x + x\hat{e}_y\right), \tag{6.80}$$

where $B_Q(z)$ is the field gradient and the quadrupole is defined over a range $z_1 \leq z \leq z_2$. The field gradient can be described by any arbitrary profile; however, hard-edged gradients are used throughout the text, so that this field is both curl- and divergence-free over this range.

The dipole field is modeled using a transversely-oriented magnet of the form

$$B_{dipole} = B_0(z)\left(\cos\phi\hat{e}_x + \cos\phi\hat{e}_y\right), \tag{6.81}$$

where $B_0(z)$ is the field amplitude defined over some range $z_1 \leq z \leq z_2$ and ϕ is the angle defining the orientation of the dipole field. As in the case of the quadrupole field representation, the dipole field is assumed to have a hard edge in all cases under consideration.

6.7 The One-Dimensional Formulation

The one-dimensional, steady-state formulation described in Sects. 5.1 and 5.2 can be generalized to include time dependence for arbitrary polarizations ranging from planar through elliptical to circular polarization and including harmonic generation. Since we are primarily concerned in this chapter with the interaction of an energetic electron beam with optical modes for which $\gamma_0 > > 1$ and $\omega_b < < \omega$, we neglect space-charge waves in the formulation.

6.7.1 The Optical Field Representation

The vector potential of the radiation field that is employed in this chapter is represented in the form of a spectral decomposition

$$\delta\mathbf{A}(z,t) = \sum_{h=1}^{\infty}\left\{\delta A_h^{(1)}(z,t)\left[\hat{\mathbf{e}}_x\cos\varphi_h - u_e\hat{\mathbf{e}}_y\sin\varphi_h\right]\right.$$
$$\left. - \delta A_h^{(2)}(z,t)\left[\hat{\mathbf{e}}_x\sin\varphi_h + u_e\hat{\mathbf{e}}_y\cos\varphi_h\right]\right\}, \tag{6.82}$$

where h denotes the harmonic number, $\varphi_h = h\Delta k(z - ct)$ is the phase of the hth harmonic, and $\Delta A_h{}^{(1,2)}$ are the mode amplitudes that are assumed to vary slowly in z and t. Note that the ellipticity has been chosen to correspond to that used in the wiggler representation.

Observe that this representation differs from that used in Chap. 5, where the vector potential for the SVEA was represented in terms of an overall amplitude and phase, while here we employ two amplitude components. These two representations are equivalent; however, the representation in terms of the amplitude and phase yields an equation for the phase that varies inversely with the amplitude (see Eq. (5.52)). This can lead to numerical problems if the amplitude were vanishingly small as, for example, in the start-up region for self-amplified spontaneous emission (SASE). In comparison with the earlier representation, the amplitude and phase can be expressed as

$$\delta A_h = \delta A_h^{(1)^2} + \delta A_h^{(2)^2}, \tag{6.83}$$

and

$$\delta\varphi_h = \tan^{-1}\left(\frac{\delta A_h^{(2)}}{\delta A_h^{(1)}}\right). \tag{6.84}$$

Time dependence can be simulated using this representation in two quite different ways. If we choose $\Delta k = \omega_{res}/c$, where ω_{res} is the resonant frequency, then this resonant harmonic representation describes the fundamental resonance and its' harmonics as illustrated in Fig. 6.6. Time dependence using this representation proceeds as illustrated in Fig. 6.4 where the field amplitudes in each *slice* vary in both z and t and multiple temporal *slices* are included in the formalism each of which contains the harmonics as indicated in Fig. 6.5. The spectrum about each harmonic is then determined based on the temporal variations in the power and phase of the field from *slice* to *slice*.

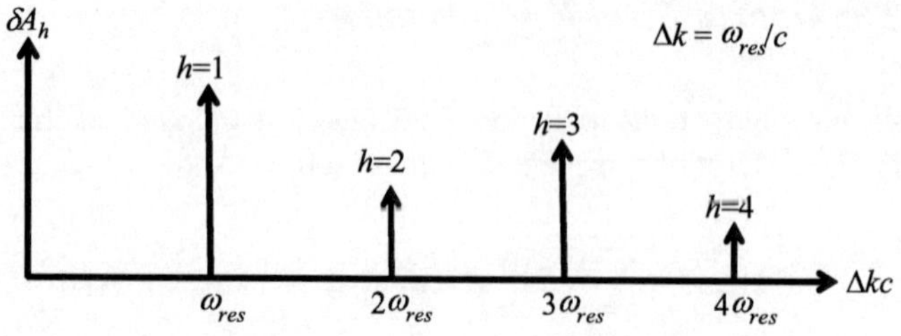

Fig. 6.6 Graphical representation of the resonant harmonic decomposition

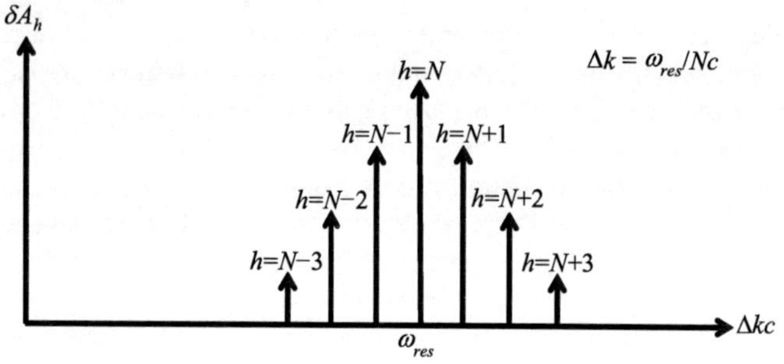

Fig. 6.7 Graphical representation of an explicit spectral decomposition

In contrast, if $\Delta k << \omega_{res}/c$, then the field amplitudes can be assumed to vary only in z, and the entire bandwidth at the fundamental and harmonics by the inclusion of an appropriate number of harmonics of Δk. This representation may be thought of as an explicit spectral decomposition and is illustrated schematically in Fig. 6.7 where $\Delta k = \omega_{res}/Nc$ for $N >> 1$. The frequency band in the vicinity of the fundamental resonance, therefore, can be spanned by choosing $h = N, N \pm 1, N \pm 2, N \pm 3$, etc. Any arbitrary resolution for this spectral decomposition can be obtained by choosing a large enough value for N. Observe also that this representation can span the bandwidth about each harmonic using appropriate choices for h. These two methods of treating time dependence have been formally shown to be equivalent [13].

The electric and magnetic fields associated with the present representation of the vector potential are

$$\delta\mathbf{E}(z,t) = -\sum_{h=1}^{\infty} h\Delta k\Big\{\delta A_h^{(1)}(z,t)\left[\hat{\boldsymbol{e}}_x \sin\varphi_h + u_e\hat{\boldsymbol{e}}_y \cos\varphi_h\right] + \delta A_h^{(2)}(z,t)\left[\hat{\boldsymbol{e}}_x \cos\varphi_h - u_e\hat{\boldsymbol{e}}_y \sin\varphi_h\right]\Big\}, \tag{6.85}$$

and

$$\delta \mathbf{B}(z,t) = \sum_{h=1}^{\infty} h\Delta k \Big\{ \delta A_h^{(1)}(z,t)\left[u_e \hat{\boldsymbol{e}}_x \cos\varphi_h - \hat{\boldsymbol{e}}_y \sin\varphi_h\right]$$
$$- \delta A_h^{(2)}(z,t)\left[u_e \hat{\boldsymbol{e}}_x \sin\varphi_h + \hat{\boldsymbol{e}}_y \cos\varphi_h\right]\Big\}. \tag{6.86}$$

As a result, the time-averaged Poynting flux for each harmonic component is given as

$$P_h = \frac{c}{4\pi}\frac{c\Delta k}{2\pi}\int_0^{2\pi/c\Delta k} dt\, \delta \mathbf{E}_h \times \delta \boldsymbol{B}_h = \frac{ch^2\Delta k^2}{4\pi}\frac{1+u_e^2}{2}\left(\delta A_h^{(1)^2} + \delta A_h^{(2)^2}\right). \tag{6.87}$$

6.7.2 The Dynamical Equations

We now assume that $\Delta kc = \omega$ and $k = \omega/c$ and employ the resonant harmonic decomposition illustrated in Fig. 6.6. Substitution of the field representation into Maxwell's equations yields

$$\left(\frac{\partial}{\partial z} + \frac{\omega}{c^2 k}\frac{\partial}{\partial t}\right)\delta A_h^{(1)} \sin\varphi_h + \left(\frac{\partial}{\partial z} + \frac{\omega}{c^2 k}\frac{\partial}{\partial t}\right)\delta A_h^{(2)} \cos\varphi_h = \frac{2\pi}{hkc}\delta J_x, \tag{6.88}$$

and

$$\left(\frac{\partial}{\partial z} + \frac{\omega}{c^2 k}\frac{\partial}{\partial t}\right)\delta A_h^{(1)} \cos\varphi_h - \left(\frac{\partial}{\partial z} + \frac{\omega}{c^2 k}\frac{\partial}{\partial t}\right)\delta A_h^{(2)} \sin\varphi_h = \frac{2\pi}{u_e hkc}\delta J_y. \tag{6.89}$$

Under the assumption that the field amplitudes vary slowly with respect to the temporal variation within the pulse (6.53) so that the quasi-static assumption is approximately satisfied, these equations are averaged over the wave period to obtain

$$\left(\frac{\partial}{\partial z} + \frac{1}{c}\frac{\partial}{\partial t}\right)\delta A_h^{(1)} \cong \frac{2\pi}{hkc}\frac{2}{1+u_e^2}\frac{\omega}{2\pi}\int_0^{2\pi/\omega} dt\left(\delta J_x \sin\varphi_h + u_e \delta J_y \cos\varphi_h\right), \tag{6.90}$$

and

$$\left(\frac{\partial}{\partial z} + \frac{1}{c}\frac{\partial}{\partial t}\right)\delta A_h^{(2)} \cong \frac{2\pi}{hkc}\frac{2}{1+u_e^2}\frac{\omega}{2\pi}\int_0^{2\pi/\omega} dt\left(\delta J_x \cos\varphi_h - u_e \delta J_y \sin\varphi_h\right). \tag{6.91}$$

The source current given in Eq. (5.23) can be generalized to include both a temporal (i.e., phase) and energy variation over the pulse. Assuming a Gaussian variation in the electron energy, we write

$$\delta\mathbf{J}(z,t) = -en_b c\int_{-T/2}^{T/2} dt_0\sigma(t_0)\int_1^{\infty}\frac{d\gamma_0}{\sqrt{\pi/2}\Delta\gamma(t_0)}\frac{\exp\left[-(\gamma_0-\gamma_b(t_0))^2/2\Delta\gamma^2(t_0)\right]}{1+\mathrm{erf}\left[\gamma_b(t_0)/\sqrt{2}\Delta\gamma(t_0)\right]}$$

$$\times\frac{\delta[t-\tau(z,t_0)]}{|v_z(t,t_0)|},\tag{6.92}$$

where $T\ (>>2\pi/\omega)$ is the pulse time, n_b is the average electron density, and $\sigma(t_0)$, $\gamma_b(t_0)$, and $\Delta\gamma(t_0)$ specify the variations in the charge density, bulk energy, and energy spread over the pulse. As a result, the dynamical equations for the field amplitudes take the form

$$\left(\frac{\partial}{\partial z}+\frac{1}{c}\frac{\partial}{\partial t}\right)\delta a_h^{(1)} \cong -\frac{\omega_b^2}{2h\omega c}\frac{2}{1+u_e^2}\left\langle\frac{v_x\sin\varphi_h+u_e v_y\cos\varphi_h}{|v_z|}\right\rangle,\tag{6.93}$$

and

$$\left(\frac{\partial}{\partial z}+\frac{1}{c}\frac{\partial}{\partial t}\right)\delta a_h^{(2)} \cong -\frac{\omega_b^2}{2h\omega c}\frac{2}{1+u_e^2}\left\langle\frac{v_x\cos\varphi_h-u_e v_y\sin\varphi_h}{|v_z|}\right\rangle,\tag{6.94}$$

where $\Delta a_h{}^{(1,2)} = e\Delta A_h{}^{(1,2)}/m_e c^2$ are the normalized amplitudes, $\omega_b{}^2\ (=4\pi e^2 n_b/m_e)$ is the square of the beam plasma frequency, m_e, and c are the electron rest mass and the speed of light in vacuo, and the averaging operator is defined over a two-dimensional initial phase space as

$$\langle(\cdots)\rangle \equiv \int_0^{2\pi}\frac{d\psi_0}{2\pi}\sigma(t_0)\int_1^{\infty}\frac{d\gamma_0}{\sqrt{\pi/2}\Delta\gamma(t_0)}\frac{\exp\left[-(\gamma_0-\gamma_b(t_0))^2/2\Delta\gamma^2(t_0)\right]}{1+\mathrm{erf}\left[\gamma_b(t_0)/\sqrt{2}\Delta\gamma(t_0)\right]}(\cdots),\tag{6.95}$$

where $\psi_0\ (=-\omega t_0)$, where t_0 is the time the particle crosses the $z=0$ plane defining the start of the wiggler is the initial phase.

The dynamical equations are completed by the specification of the electron orbit equations subject to the one-dimensional elliptical wiggler model (6.65) and the electromagnetic field representation (6.82) and are

$$\frac{d\psi}{dz} = k_w - \frac{\omega}{v_z}\left(1-\frac{v_z}{c}\right),\tag{6.96}$$

for the ponderomotive phase, and

$$\frac{d}{dz}\left(\frac{p_x}{m_e c}\right) = \frac{eB_w}{m_e c^2}\sin k_w z + \sum_h \frac{h\omega}{v_z}\left(1 - \frac{v_z}{c}\right)\left[\delta a_h^{(1)} \sin \varphi_h + \delta a_h^{(2)} \cos \varphi_h\right], \quad (6.97)$$

$$\frac{d}{dz}\left(\frac{p_y}{m_e c}\right) = -u_e \frac{eB_w}{m_e c^2}\cos k_w z - u_e \sum_h \frac{h\omega}{v_z}\left(1 - \frac{v_z}{c}\right)\left[\delta a_h^{(1)} \cos \varphi_h - \delta a_h^{(2)} \sin \varphi_h\right], \quad (6.98)$$

$$v_z \frac{d}{dz}\left(\frac{p_z}{m_e c}\right) = -\frac{eB_w}{m_e c^2}\left(v_x \sin k_w z - u_e v_y \cos k_w z\right) + \sum_h h\omega\left[\delta a_h^{(1)}\left(v_x \sin \varphi_h + u_e v_y \cos \varphi_h\right) + \delta a_h^{(2)}\left(v_x \cos \varphi_h - u_e v_y \sin \varphi_h\right)\right], \quad (6.99)$$

where the sum over h includes those harmonics under consideration.

6.7.3 The Numerical Procedure

The dynamical equations derived using the resonant harmonic decomposition can be used in either steady-state or time-dependent mode. The steady-state mode is obtained by the simple expedient of neglecting the $\partial/\partial t$ term in the dynamical equations for the field amplitudes and including only one temporal *slice* in the simulation. In either case, particles are loaded into the electron *slice(s)* using the Gaussian quadrature in the two degrees of freedom (ψ_0, γ_0) subject to the distribution shown in Eq. (6.86) and as described in Chap. 5. The initial conditions for the radiation slices are chosen to correspond to the injected seed pulse. This centroid of the seed pulse may be shifted relative to that of the electron bunch, and the duration of the seed pulse may be different from that of the electron bunch.

The dynamical equations describe a system of ordinary differential equations including equations for two mode amplitudes per harmonic and four equations per macro-electron, which are integrated simultaneously using a Runge-Kutta algorithm. The numerical procedure is that each slice is first advanced from $z \to z + \Delta z$ by means of the Runge-Kutta algorithm. Note that while the field varies slowly with respect to the wavelength and wiggler period, the step size through the integrator must be small enough to resolve the electrons' wiggler-motion; hence, 20–30 steps per wiggler period are required. Since small integration steps are required in order to resolve the wiggler-motion, it is often possible to use a second-order Runge-Kutta integrator; however, a fourth-order Runge-Kutta integrator is used for all the examples in the work.

It is important to observe that since the dynamical equations involve the solution of the complete Lorentz force equations for the electron trajectories, the effect of the lower beat wave that is present in planar and elliptical wigglers is implicitly included in the formulation. As a result, there is no need to explicitly include the *JJ*-factor in the formulation.

Time dependence is imposed as an additional operation after the integration from $z \rightarrow z + \Delta z$ using the forward time derivative as an additional source term to treat the slippage of the radiation field with respect to the electrons. The simplest way to accomplish this is to use linear interpolation to couple the radiation field in the $(i\text{-}1)^{\text{th}}$ *slice* to that in the ith *slice*. In practice, this slippage of the radiation field may be imposed on every spatial step or after longer intervals. In this way, information is only allowed to propagate in the forward direction. This linear interpolation algorithm implies that the field amplitudes for the ith *slice* are replaced by a linear combination of the fields in the ith and $(i\text{-}1)^{\text{th}}$ *slice* as follows

$$\delta A_i \rightarrow \delta A_i - \frac{\delta z}{\lambda_w} \frac{\lambda}{c\Delta t} (\delta A_i - \delta A_{i-1}), \tag{6.100}$$

where Δz ($\geq \Delta z$) is the interval between applications of slippage and $c\Delta t$ ($\geq \lambda$) is the separation interval between *slices* in the simulation. Observe that if there is one *slice* per wavelength ($c\Delta t = \lambda$) and if slippage is applied once per wiggler period ($\Delta z = \lambda_w$), then the field in the $(i\text{-}1)^{\text{th}}$ *slice* is advanced into the ith *slice* ($\Delta A_i \rightarrow \Delta A_{i\text{-}1}$), and the radiation advances one wavelength per wiggler period as expected.

Energy conservation is preserved between the particles in any given *slice* and the fields in the associated optical *slice* under the integration from $z \rightarrow z + \Delta z$ to within about one part in 10^3 at saturation. The conservation of energy in the optical field associated with the application of slippage depends upon the preservation of the aggregate power in all of the *slices*. The linear interpolation algorithm (6.89) implies that under the application of slippage,

$$\sum_i \delta A_i^2 \rightarrow \sum_i \delta A_i^2 - 2\frac{\delta z}{\lambda_w} \frac{\lambda}{c\Delta t} \left(1 - \frac{\delta z}{\lambda_w} \frac{\lambda}{c\Delta t}\right) \sum_i \delta A_i (\delta A_i - \delta A_{i-1}), \tag{6.101}$$

where the second term on the right-hand side measures the discrepancy in the optical field energy before and after the application of slippage. For example, if there are 30 steps through the integrator per wiggler period and slippage is applied every 30th step ($\Delta z = \lambda_w$) and there is one slice every wavelength ($c\Delta t = \lambda$), then $(\Delta z/\lambda_w)(\lambda/c\Delta t) = 1$ and the energy is conserved under the application of slippage. However, slippage occurs continuously, and this interval between the numerical application of slippage provides a coarse representation of physical reality. In contrast, if slippage is applied on every step through the integrator, then $\Delta z = \lambda_w/30$ and the leading factor is approximately 0.067. In this case, slippage is treated in a continuous fashion, and energy conservation is well-preserved if the field amplitudes vary slowly from *slice* to *slice* which is one of the fundamental assumptions underlying the SVEA and is the case for seeded amplifiers and for oscillators.

The one-dimensional, numerical simulations that are discussed below deal with seeded amplifiers and are performed under the simplifying assumption that the only electron bunch parameter that varies in time is the bunch current, which is characterized by a parabolic temporal profile. It is also assumed that the initial temporal profile of the power in the optical seed pulse synchronized with the electron bunch and is also parabolic with the same full width pulse duration as the electron bunch. As a consequence, the total bunch charge (Q_b) depends upon the peak current (I_b) and the full width bunch duration (τ_b) via

$$Q_b = \frac{2}{3} I_b \tau_b. \tag{6.102}$$

A similar relation holds between the total energy and peak power of the optical seed pulse.

6.7.4 *Simulation of a Seeded Amplifier with a Planar Wiggler*

The first example that is considered is that of a seeded amplifier with a planar wiggler. The wiggler is assumed to have a period of 3.89 cm and an on-axis amplitude of 3.03 kG. The electron beam energy is 100.75 MeV, an energy spread of 0.1%, a peak current of 300 A with an rms beam radius of 300 μm, and a parabolic temporal pulse with a duration of 2.5 psec. This yields a bunch charge of about 500 pC, a Pierce parameter of $\rho \cong 3.9 \times 10^{-3}$, and a normalized plasma frequency of $\omega_b/ck_w \cong 5.48$. The resonant wavelength is about 0.795 μm. The peak power in the seed laser is assumed to be 5000 W over a pulse duration of 6.0 psec, so that the initial pulse energy is about 20 nJ. Observe that the duration of the seed pulse is much greater than that of the electron bunch. The simulation uses a time window of 7.0 psec in duration and 700 *slices*.

A plot showing the growth of the pulse energy along the wiggler is shown in Fig. 6.8. Saturation is found after about 5.6 m at a pulse energy of approximately 270 μJ that corresponds to an efficiency of about 5.4×10^{-3}. The power gain length is approximately 0.41 m, which is in reasonable agreement with the predicted one-dimensional gain length of approximately 0.46 m. Observe also that exponentiation begins approximately 2–3 gain lengths after the start of the wiggler.

The temporal pulse narrows and slips ahead of the electron bunch as the optical field gains energy. This is illustrated in Fig. 6.9 where the temporal pulses at the wiggler entrance (left axis) and at saturation after 5.6 m (right axis) are shown. As is evident from the figure, the FWHM duration of the seed pulse is approximately 4.23 psec, which has narrowed to about 1.36 psec at saturation. Indeed, the full width of the optical pulse at saturation is about 2.4 psec, which corresponds to the full width of the electron bunch thereby illustrating that the wave-particle interaction governs the shape and duration of the optical pulse at saturation. In addition, if we note that the center of the electron bunch is in the center of the time window, then it

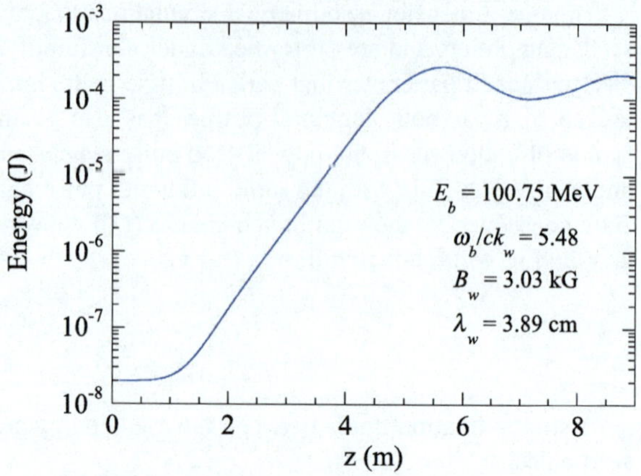

Fig. 6.8 Growth of the pulse energy in a planar wiggler

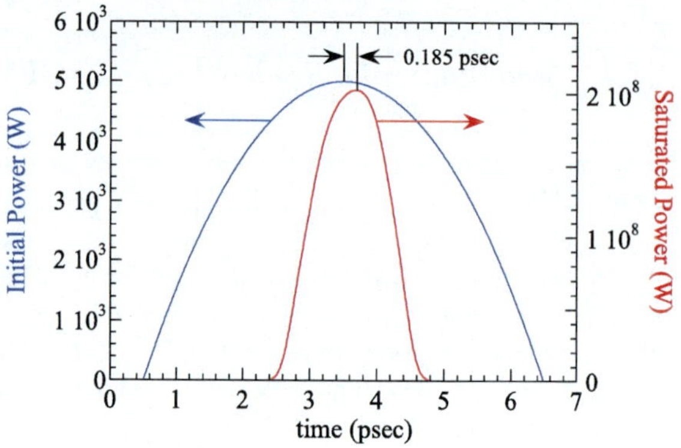

Fig. 6.9 Temporal pulse shapes at the wiggler entrance and at saturation

is clear that the optical pulse has slipped ahead of the electron bunch by about 0.185 psec. Noting that slippage is expected to occur at the normal rate in the initial transient region over the first 1. 5 of wiggler and at one-third of the normal rate in the exponential gain region up to saturation at 5.6 m, then the we expect the optical pulse should slip ahead of the electrons by about 0.195 psec, which is close to that found in simulation.

The narrowing of the temporal pulse at saturation reflects the electron bunch duration and is reflected in a broadening of the spectrum. This is shown in Fig. 6.10 where the seed spectrum and the spectrum at saturation are shown and clearly indicate the spectral broadening associated with the narrowing of the optical pulse. Note also that the central wavelength, however, remains at 795 nm which is the resonant wavelength of the interaction.

The degree of energy conservation over the course of the interaction that includes the balance between the energy lost by the particles and that gained by the optical field in each slice as over the integration $z \rightarrow z + \Delta z$ and that involved in the application of slippage using the linear interpolation algorithm is shown in Fig. 6.11, where the difference between unity and the ratio of the efficiency as determined by

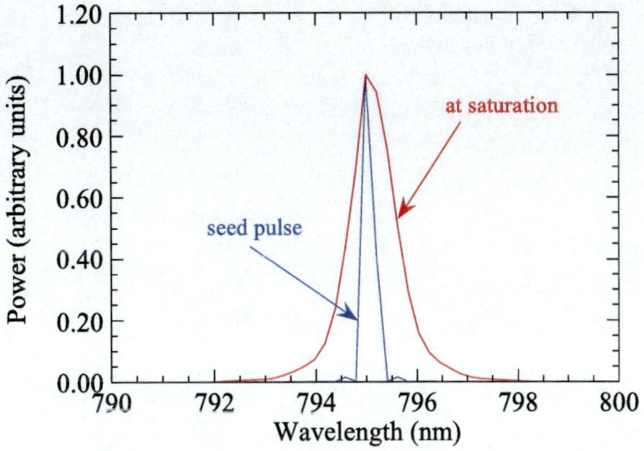

Fig. 6.10 Optical pulse spectra at the wiggler entrance and at saturation

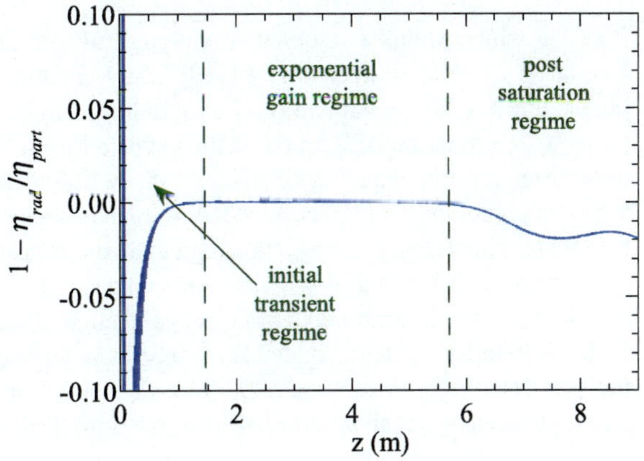

Fig. 6.11 Energy conservation over the course of the interaction

the growth in the energy of the optical field to the efficiency as measured by the loss of energy by the particles is plotted as a function of distance along the wiggler. This measure of energy conservation varies widely in the initial transient region prior to the start of exponentiation where both of these efficiencies are negligibly small. However, once the exponentiation begins, the initial transients are overwhelmed by the interaction, and energy is conserved to within several parts in 10^3 through saturation. The discrepancy in the energy conservation increases after saturation but is still preserved to within about 1–2% over the last 3.4 m of wiggler.

6.7.5 Simulation of a Seeded Amplifier with a Helical Wiggler

The example of a helical wiggler that is considered is that of a seeded amplifier that corresponds to the previously discussed case of a planar wiggler. As such, the wiggler is assumed to have a period of 3.89 cm, but the on-axis amplitude is

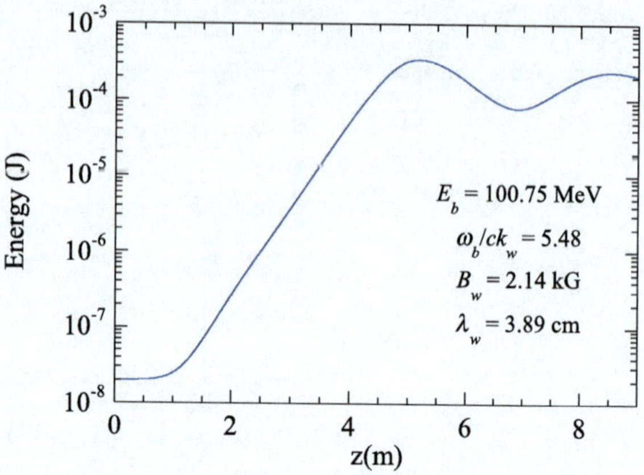

Fig. 6.12 Growth of the pulse energy in a helical wiggler

2.14 kG which yields a resonant wavelength of 795 nm as for the planar wiggler. Similarly, the electron beam energy is 100.75 MeV with an energy spread of 0.1%, a peak current of 300 A with an *rms* beam radius of 300 μm, and a parabolic temporal pulse with a duration of 2.5 psec. This yields a bunch charge of about 500 pC and a normalized plasma frequency of $\omega_b/ck_w \cong 5.48$. The Pierce parameter $\rho \cong 4.6 \times 10^{-3}$, which is higher than for the corresponding planar wiggler example because the lower beat wave is suppressed in a helical wiggler and the *JJ*-factor is unity. As a result of this, it is expected that a helical wiggler that satisfies the same resonance condition as a planar wiggler will exhibit both a shorter gain length and a higher saturation efficiency than that found in a planar wiggler. As in the planar wiggler example, the peak power in the seed laser is assumed to be 5000 W with a pulse duration of 6.0 psec, so that the initial pulse energy is about 20 nJ. The simulation of a helical wiggler also uses a time window of 7.0 psec in duration and 700 *slices*.

The simulation results showing the evolution of the pulse energy with distance along the wiggler is shown in Fig. 6.12. Saturation is found over a distance of 5.23 m at an energy of 341 μJ. As expected, this is both a shorter saturation distance and a higher efficiency than found for the planar wiggler. The power gain length found in simulation is approximately 0.38 m which is shorter than that found for the planar wiggler and in reasonable agreement with 0.39 found from the one-dimensional theory.

6.7.6 Three-Dimensional Extension of the Formulation

It is possible to extend the one-dimensional formulation to treat a limited range of three-dimensional effects using the scale factors developed by Ming Xie [54] that were discussed in Chap. 4 (Eq. (4.394)). In this case, we extend the formulation by scaling the dynamical equation by replacing $\omega_b^2 \rightarrow ff\,\omega_b^2$, where the filling factor is

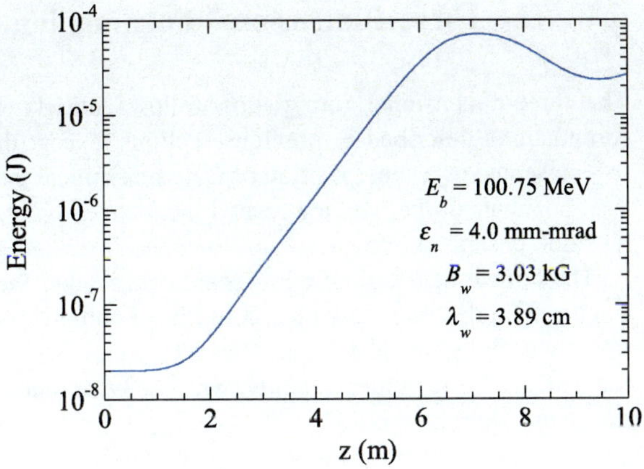

Fig. 6.13 Evolution of the pulse energy subject to the three-dimensional extensions

$$ff = \frac{1 + 0.55\eta_\varepsilon^{1.6} + 3\eta_\gamma^2 + 0.35\eta_\varepsilon^{2.9}\eta_\gamma^{2.4}}{(1 + \eta_{total})^3}, \tag{6.103}$$

where the coefficients η_ε, η_γ, and η_{total} are as used in Eq. (4.395). In addition, the *rms* beam radius is determined by the matched beam condition (see the Beam Optics Appendix) for the wiggler, $\sigma_r = (\varepsilon_n\beta/\gamma)^{1/2}$, where ε_n is the normalized emittance and β is the β-function. Using the appropriate b-function for the wiggler line, this permits the treatment of an arbitrary wiggler line that includes helical, planar, or elliptic wigglers with the possibility of strong focusing by the inclusion of quadrupole focusing.

The example discussed above for the planar wiggler represents the starting point in the simulation of a planar wiggler amplifier experiment performed at Brookhaven National Laboratory [53] which is discussed in more detail in the following section on three-dimensional simulation. The parameters in that experiment were identical to that described above. The normalized emittance was 4.0 mm-mrad, which translated into an rms beam radius of 300 μm for the β-function of the wiggler.

The evolution of the pulse energy using the three-dimensional extension represented by the filling-factor is shown in Fig. 6.13 where the pulse energy is plotted as a function of distance along the wiggler. As is evident from the figure, saturation is found after approximately 7.2 m at a pulse energy of 68 μJ. This compares reasonably well with the observed saturation after about 7.0 m in the experiment with a pulse energy of about 49 μJ.

This three-dimensional extension of the one-dimensional model can be useful in performing rapid parameter scans in the optimization of free-electron laser designs.

6.8 The Three-Dimensional Formulation

The three-dimensional formulation follows closely on that for the steady-state formulations described in previous sections, except that the electromagnetic field is represented as a superposition of Gaussian optical modes [55] in which the x- and y-components of the field are treated independently, and the time-dependent generalization described in Sect. 6.5 for the one-dimensional formulation.

The optical field can be represented using either the Gauss-Hermite and Gauss-Laguerre modes each of which comprise a complete set. However, it is computationally easier to use the Gauss-Hermite modes. Since the x- and y-components of the field are treated independently, we chose, without loss of generality, to focus on the Gauss-Hermite modes.

6.8.1 *The Dynamical Equations for the Gauss-Hermite Modes*

The Gauss-Hermite mode superposition for which the amplitudes are assumed to vary slowly in both position and time can be expressed as follows [10]:

$$\delta\mathbf{A}(\mathbf{x},t) = \hat{\mathbf{e}}_x \sum_{\substack{l,n=0\\h=1}}^{\infty} e_{l,n,h}^{(x)}(x,y)\left[\delta A_{l,n,h}^{(1,x)}(z,t)\sin\varphi_{x,h} + \delta A_{l,n,h}^{(2,x)}(z,t)\cos\varphi_{x,h}\right]$$

$$+\hat{\mathbf{e}}_y \sum_{\substack{l,n=0\\h=1}}^{\infty} e_{l,n,h}^{(y)}(x,y)\left[\delta A_{l,n,h}^{(1,y)}(z,t)\sin\varphi_{y,h} + \delta A_{l,n,h}^{(2,y)}(z,t)\cos\varphi_{y,h}\right], \quad (6.104)$$

where l and n are transverse mode numbers, h is the harmonic number, and

$$e_{l,n,h}^{(j)}(x,y) = \frac{w_{0,j,h}}{w_{j,h}(z)}\exp\left[-r^2/w_{j,h}^2(z)\right]H_l\left(\frac{\sqrt{2}x}{w_{j,h}(z)}\right)H_n\left(\frac{\sqrt{2}y}{w_{j,h}(z)}\right) \quad (6.105)$$

describes the transverse variation of the mode, where H_l is the Hermite polynomial of order l, $w_{0,j,h}$ is the spot size of the mode waist, $wj,_h(z)$ is the spot size and $j = x,y$,

$$\varphi_{j,h} = h(k_0 z - \omega t) + \alpha_{j,h}(z)\frac{r^2}{w_{j,h}^2(z)} \quad (6.106)$$

is the phase for $k_0 = \omega_0/c$, and $\alpha_{j,h}(z)$ describes the curvature of the phase front. We assume that $\delta A_{l,n,h}^{(i,j)}$ $(i = 1,2)$, $w_{j,h}$, and $\alpha_{j,h}$ vary slowly in z and t. Note that this formulation also treats harmonic interactions. The power carried in each mode is given by

$$P_{l,n,h} = \frac{m_e^2 c^5}{8e^2} 2^{l+n-1} l! n! \sum_{j=x,y} k_0^2 w_{0,jh}^2 \left(\delta a_{l,n,h}^{(1,j)^2} + \delta a_{l,n,h}^{(2,j)^2} \right), \tag{6.107}$$

where $\delta a_{l,n,h}^{(i,j)} = e\delta A_{l,n,h}^{(i,j)}/m_e c^2$ and $m_e^2 c^5/8e^2 \cong 1.089$ GW. The spot size and curvature vary in vacuo as

$$w_{j,h}(z) = w_{0,j,h}\left(1 + \frac{(z - z_0)^2}{z_{R,j,h}^2}\right)^{1/2}, \tag{6.108}$$

and

$$\alpha_{j,h}(z) = \frac{z - z_0}{z_{R,j,h}} \tag{6.109}$$

where $z_{Rj,h} = \pi w_{0,j,h}^2/\lambda$ is the Rayleigh range and z_0 represents the position of the mode waist. This describes vacuum diffraction.

Substitution of this field superposition in Maxwell's equations yields

$$\left(\nabla^2 - \frac{1}{c^2}\frac{\partial^2}{\partial t^2}\right)\delta\mathbf{A}(\mathbf{x},t) = -\frac{4\pi}{c}\delta\mathbf{J}(\mathbf{x},t), \tag{6.110}$$

yields

$$\begin{aligned}
\left(\nabla^2 - \frac{1}{c^2}\frac{\partial^2}{\partial t^2}\right)\delta A_x \cong{}& 2k_0 \sum_{l,n,h} h \frac{w_{0,x,h}}{w_{x,h}} \exp(-r^2/w_{x,h}^2) H_l\left(\frac{\sqrt{2}x}{w_{x,h}}\right) H_n\left(\frac{\sqrt{2}y}{w_{x,h}}\right) \\
&\times \left\{ \cos\varphi_{x,h}\left(\frac{d}{dz}\delta A_{l,n,h}^{(1,x)} + K_{l,n,h}^{(x)}\delta A_{l,n,h}^{(2,x)}\right) - \sin\varphi_{x,h}\left(\frac{d}{dz}\delta A_{l,n,h}^{(2,x)} - K_{l,n,h}^{(x)}\delta A_{l,n,h}^{(1,x)}\right)\right\} \\
&+ 2k_0 \sum_{l,n,h} h \frac{w_{0,x,h}}{w_{x,h}} X_{l,n,h}^{(x)}\left(\delta A_{l,n,h}^{(1,x)} \sin\varphi_{x,h} + \delta A_{l,n,h}^{(2,x)} \cos\varphi_{x,h}\right) \\
&\times \left\{ \left[\frac{1}{4} H_{l+2}\left(\frac{\sqrt{2}x}{w_{x,h}}\right) + l(l-1) H_{l-2}\left(\frac{\sqrt{2}x}{w_{x,h}}\right)\right] H_n\left(\frac{\sqrt{2}y}{w_{x,h}}\right) \right. \\
&\left. + \left[\frac{1}{4} H_{n+2}\left(\frac{\sqrt{2}y}{w_{x,h}}\right) + n(n-1) H_{n-2}\left(\frac{\sqrt{2}y}{w_{x,h}}\right)\right] H_l\left(\frac{\sqrt{2}x}{w_{x,h}}\right) \right\}
\end{aligned}$$

$$- 2k_0 \sum_{l,n,h} h \frac{w_{0,x,h}}{w_{x,h}} Y^{(x)}_{l,n,h} \left(\delta A^{(1,x)}_{l,n,h} \cos \varphi_{x,h} - \delta A^{(2,x)}_{l,n,h} \sin \varphi_{x,h} \right)$$

$$\times \left\{ \left[\frac{1}{4} H_{l+2} \left(\frac{\sqrt{2}x}{w_{x,h}} \right) - l(l-1) H_{l-2} \left(\frac{\sqrt{2}x}{wx,h} \right) \right] H_n \left(\frac{\sqrt{2}y}{w_{x,h}} \right) \right.$$

$$\left. + \left[\frac{1}{4} H_{n+2} \left(\frac{\sqrt{2}y}{w_{x,h}} \right) - n(n-1) H_{n-2} \left(\frac{\sqrt{2}y}{w_{x,h}} \right) \right] H_l \left(\frac{\sqrt{2}x}{wx,h} \right) \right\}, \tag{6.111}$$

where the source current is defined as in Eq. (5.89); second-order terms in the derivatives of the field amplitudes, the spot size, and the curvature have been neglected,

$$K^{(x)}_{l,n,h} \equiv (l+n+1) \left(\frac{\alpha_{x,h}}{w_{x,h}} \frac{dw_{x,h}}{dz} - \frac{1}{2} \frac{d\alpha_{x,h}}{dz} - \frac{1+\alpha^2_{x,h}}{hk_0 w^2_{x,h}} \right), \tag{6.112}$$

$$X^{(x)}_h \equiv \frac{1-\alpha^2_{x,h}}{hk_0 w^2_{x,h}} + \frac{\alpha_{x,h}}{w_{x,h}} \frac{dw_{x,h}}{dz} - \frac{1}{2} \frac{d\alpha_{x,h}}{dz}, \tag{6.113}$$

$$Y^{(x)}_h \equiv \frac{2\alpha_{x,h}}{hk_0 w^2_{x,h}} - \frac{1}{w_{x,h}} \frac{dw_{x,h}}{dz}, \tag{6.114}$$

and

$$\frac{d}{dz} = \frac{\partial}{\partial z} + \frac{1}{c} \frac{\partial}{\partial t} \tag{6.115}$$

is the convective derivative. And similarly for the y-components.

Individual harmonic components can be isolated by integrating these equations over

$$\frac{\omega}{2\pi} \int_0^{2\pi/\omega} dt (\sin \varphi_h \cos \varphi_h)$$

to obtain

$$\sum_{l,n} \exp(-r^2/w^2_{x,h}) H_l \left(\frac{\sqrt{2}x}{w_{x,h}} \right) H_n \left(\frac{\sqrt{2}y}{w_{x,h}} \right) \left\{ \left(\frac{d}{dz} \delta A^{(1,x)}_{l,n,h} + K_{l,n,h} \delta A^{(2,x)}_{l,n,h} \right) \right.$$

$$+ X^{(x)}_{l,n,h} \left(\left[\frac{1}{4} \delta A^{(2,x)}_{l-2,n,h} H(l-2) + (l+1)(l+2) \delta A^{(2,x)}_{l+2,n,h} \right] \right.$$

$$\left. + \left[\frac{1}{4} \delta A^{(2,x)}_{l,n-2,h} H(n-2) + (n+1)(n+2) \delta A^{(2,x)}_{l,n+2,h} \right] \right)$$

$$-Y^{(x)}_{l,n,h}\left(\left[\frac{1}{4}\delta A^{(1,x)}_{l-2,n,h}H(l-2)-(l+1)(l+2)\delta A^{(1,x)}_{l+2,n,h}\right]\right.$$
$$\left.\left.+\left[\frac{1}{4}\delta A^{(1,x)}_{l,n-2,h}H(n-2)-(n+1)(n+2)\delta A^{(1,x)}_{l,n+2,h}\right]\right)\right\}$$
$$=-\frac{2w_{x,h}}{w_{0,x,h}}\int\limits_0^{2\pi/\omega} dt\delta J_x\cos\varphi_{x,h}, \tag{6.116}$$

and

$$\sum_{l,n}\exp(-r^2/w^2_{x,h})H_l\left(\frac{\sqrt{2}x}{w_{x,h}}\right)H_n\left(\frac{\sqrt{2}y}{w_{x,h}}\right)\left\{\left(\frac{d}{dz}\delta A^{(2,x)}_{l,n,h}-K^{(x)}_{l,n,h}\delta A^{(1,x)}_{l,n,hh}\right)\right.$$
$$-X^{(x)}_{l,n,h}\left(\left[\frac{1}{4}\delta A^{(2,x)}_{l-2,n,h}H(l-2)+(l+1)(l+2)\delta A^{(2,x)}_{l+2,n,h}\right]\right.$$
$$\left.+\left[\frac{1}{4}\delta A^{(2,x)}_{l,n-2,h}H(n-2)+(n+1)(n+2)\delta A^{(2,x)}_{l,n+2,h}\right]\right)$$
$$-Y^{(x)}_{l,n,h}\left(\left[\frac{1}{4}\delta A^{(1,x)}_{l-2,n,h}H(l-2)-(l+1)(l+2)\delta A^{(1,x)}_{l+2,n,h}\right]\right.$$
$$\left.\left.+\left[\frac{1}{4}\delta A^{(1,x)}_{l,n-2,h}H(n-2)-(n+1)(n+2)\delta A^{(1,x)}_{l,n+2,h}\right]\right)\right\}$$
$$=\frac{2w_{x,h}}{w_{0,x,h}}\int\limits_0^{2\pi/\omega} dt\delta J_x\sin\varphi_{x,h}, \tag{6.117}$$

where $H(l)$ is the Heaviside function. In the derivation of Eqs. (6.117) and (6.117), we have made use of the following recursion relations for the Hermite polynomials:

$$H_l{}'(x)=2lH_{l-1}(x), \tag{6.118}$$

and

$$2xH_l(x)=H_{l+1}(x)+2lH_{l-1}(x). \tag{6.119}$$

Similar equations hold for the y-components.

The transverse modes can be isolated by orthogonalization over x and y. Noting that

$$\int\limits_{-\infty}^{\infty} dx\exp(-a^2x^2)H_m(ax)H_n(ax)=2^n\pi^{1/2}n!\frac{1}{a}\delta_{m,n}, \tag{6.120}$$

we obtain for $j = x,y$

$$
\begin{aligned}
&\frac{d}{dz}\begin{pmatrix} \delta a_{l,n,h}^{(1,j)} \\ \delta a_{l,n,h}^{(2,j)} \end{pmatrix} + K_{l,n,h}^{(j)}\begin{pmatrix} \delta a_{l,n,h}^{(2,j)} \\ -\delta a_{l,n,h}^{(1,j)} \end{pmatrix} + \frac{1}{4}H(l-2)\begin{pmatrix} X_{l,n,h}^{(j)}\delta a_{l-2,n,h}^{(2,j)} - Y_{l,n,h}^{(j)}\delta a_{l-2,n,h}^{(1)} \\ -X_{l,n,h}^{(j)}\delta a_{l-2,n,h}^{(2,j)} - Y_{l,n,h}^{(j)}\delta a_{l-2,n,h}^{(2,j)} \end{pmatrix} \\
&\quad + \frac{1}{4}H(n-2)\begin{pmatrix} X_{l,n,h}^{(j)}\delta a_{l,n-2,h}^{(2,j)} - Y_{l,n,h}^{(j)}\delta a_{l,n-2,h}^{(1,j)} \\ -X_{l,n,h}^{(j)}\delta a_{l,n-2,h}^{(2,j)} - Y_{l,n,h}^{(j)}\delta a_{l,n-2,h}^{(2,j)} \end{pmatrix} \\
&\quad + (l+1)(l+2)\begin{pmatrix} X_{l,n,h}^{(j)}\delta a_{l+2,n,h}^{(2,j)} + Y_{l,n,h}^{(j)}\delta a_{l+2,n,h}^{(1,j)} \\ -X_{l,n,h}^{(j)}\delta a_{l+2,n,h}^{(2,j)} + Y_{l,n,h}^{(j)}\delta a_{l+2,n,h}^{(2,j)} \end{pmatrix} \\
&\quad + (n+1)(n+2)\begin{pmatrix} X_{l,n,h}^{(j)}\delta a_{l,n+2,h}^{(2)} + Y_{l,n,h}^{(j)}\delta a_{l,n+2,h}^{(1)} \\ -X_{l,n,h}^{(j)}\delta a_{l,n+2,h}^{(2)} + Y_{l,n,h}^{(j)}\delta a_{l,n+2,h}^{(2)} \end{pmatrix} = \begin{pmatrix} S_{l,n,h}^{(1,j)} \\ S_{l,n,h}^{(2,j)} \end{pmatrix}, \quad (6.121)
\end{aligned}
$$

where the source terms are

$$
\begin{pmatrix} S_{l,n,h}^{(1,j)} \\ S_{l,n,h}^{(2,j)} \end{pmatrix} = \frac{\omega_b^2}{hk_0c^2}\frac{1}{2^{l+n-1}\pi}\frac{1}{l!n!}\frac{1}{w_{0,j,h}^2}\left\langle e_{l,n,h}^{(j)}\frac{\upsilon_x}{|\upsilon_z|}\begin{pmatrix} -\cos\varphi_{j,h} \\ \sin\varphi j,h \end{pmatrix}\right\rangle, \quad (6.122)
$$

where ω_b is the beam plasma frequency. For a beam with a uniform phase in initial phase and Gaussian distribution in energy, momentum, and beam size,

$$
\begin{aligned}
\langle(\cdots)\rangle &= \int_0^{2\pi}\frac{d\psi_0}{2\pi}\int_1^{\infty}\frac{d\gamma_0}{\sqrt{\pi/2}\Delta\gamma}\frac{\exp\left[-(\gamma_0-\gamma_{avg})^2/2\Delta\gamma^2\right]}{1+\operatorname{erf}\left(\gamma_{avg}/\sqrt{2}\Delta\gamma\right)} \\
&\times \iint\frac{dx_0dy_0}{2\pi\sigma_r^2}\iint\frac{dp_{x0}dp_{y0}}{2\pi\sigma_p^2}\exp\left(-r^2/2\sigma_r^2 - p_{\perp 0}^2/2\sigma_p^2\right)(\cdots). \quad (6.123)
\end{aligned}
$$

Observe that this distribution deals with a beam that is initially injected on-axis with paraxial propagation, although more general distributions can be formulated. In addition, the electron bunch parameters can be varied between *slices* included in time-dependent simulations.

6.8.1.1 Vacuum Diffraction

To show that these equations recover vacuum diffraction, we note that the sources vanish in the limit in which the electron beam is not present; hence, $X_{l,n,h} = Y_{l,n,h} = 0$ for both the x- and y-components. As a result, the spot size and curvature satisfy the following equations:

$$\frac{dw_{j,h}}{dz} = \frac{2\alpha_{j,h}}{hk_0 w_{j,h}}, \tag{6.124}$$

$$\frac{1}{2}\frac{d\alpha_{j,h}}{dz} = \frac{1+\alpha_{j,h}^2}{hk_0 w_{j,h}^2}. \tag{6.125}$$

The solutions to these are given in Eqs. (6.106) and (6.107) for vacuum diffraction.

In addition, substitution of Eqs. (6.122) and (6.123) into Maxwell's equation yields

$$K_{l,n,h}^{(j)} = -\frac{(1+l+n)}{z_{R,j,h}}, \tag{6.126}$$

which is the derivative of the Gouy phase shift $\varphi_{l,n,h}$. If we now express the field components in terms of an amplitude and phase in the form $\delta A_{l,n,h}^{(1)} = \delta A_{l,n,h} \cos\varphi$, and $\delta A_{l,n,h}^{(2)} = \delta A_{l,n,h} \sin\varphi$, then we find that

$$\frac{d}{dz}\delta A_{l,n,h} = 0, \tag{6.127}$$

which shows that the power is constant and

$$\frac{d}{dz}\varphi = \frac{d}{dz}\varphi_{l,n,h}, \tag{6.128}$$

which indicates that the phase variation is described by the Gouy phase. Hence, the dynamical Eq. (6.121) reproduces vacuum diffraction in the absence of an electron beam.

6.8.1.2 The Source-Dependent Expansion

The dynamical equations can be solved under the assumption that the modal superposition is over vacuum modes. In this case, it is assumed that the spot size and curvature are given by Eqs. (6.124) and (6.125) and that $X_{l,n,h} = Y_{l,n,h} = 0$. This superposition is formally correct and can be used in the simulation of the free-electron laser interaction; however, it may require the inclusion of a prohibitively large number of modes to achieve numerical convergence. In order to minimize the number of mode required in any given simulation, an adaptive eigenmode expansion referred to as the source-dependent expansion (SDE) has been developed [56].

The fundamental assumptions of the SDE are that the lowest-order TEM_{00} mode is dominant for any harmonic h and that the mode amplitudes tend to decrease with increasing transverse mode numbers. If the mode amplitudes tend to decrease with mode numbers, then the mode-mode coupling terms can be neglected, and we have that

$$\frac{d}{dz}\begin{pmatrix}\delta a_{l,n,h}^{(1,j)}\\ \delta a_{l,n,h}^{(2,j)}\end{pmatrix}+K_{l,n,h}^{(j)}\begin{pmatrix}\delta a_{l,n,h}^{(2,j)}\\ -\delta a_{l,n,h}^{(1,j)}\end{pmatrix}=\begin{pmatrix}S_{l,n,h}^{(1,j)}\\ S_{l,n,h}^{(2,j)}\end{pmatrix}. \tag{6.129}$$

The SDE describes the diffraction of each harmonic component subject to the interaction with the electron beam. If the TEM_{00} mode is dominant, then from Eq. (6.121), we have that

$$\begin{pmatrix}X_{l.n.h}^{(j)}\delta a_{0,0,h}^{(2,j)}-Y_{l.n.h}^{(j)}\delta a_{0,0,h}^{(1,j)}\\ X_{l.n.h}^{(j)}\delta a_{0,0,h}^{(1,j)}+Y_{l.n.h}^{(j)}\delta a_{0,0,h}^{(2,j)}\\ X_{l.n.h}^{(j)}\delta a_{0,0,h}^{(2,j)}-Y_{l.n.h}^{(j)}\delta a_{0,0,h}^{(1,j)}\\ X_{l.n.h}^{(j)}\delta a_{0,0,h}^{(1,j)}+Y_{l.n.h}^{(j)}\delta a_{0,0,h}^{(2,j)}\end{pmatrix}=4\begin{pmatrix}S_{2,0,h}^{(1,j)}\\ -S_{2,0,h}^{(2,j)}\\ S_{0,2,h}^{(1,j)}\\ -S_{0,2,h}^{(2,j)}\end{pmatrix}. \tag{6.130}$$

The solution of these equations for $X_{l,n,h}$ and $Y_{l,n,h}$ yields

$$X_{l,n,h}^{(j)}=-2\frac{\left(S_{0,2,h}^{(2,j)}+S_{2,0,h}^{(2,j)}\right)\delta a_{0,0,h}^{(2,j)}-\left(S_{0,2,h}^{(1,j)}+S_{2,0,h}^{(1,j)}\right)\delta a_{0,0,h}^{(1,j)}}{\delta a_{0,0,h}^{2}}, \tag{6.131}$$

and

$$Y_{l,n,h}^{(j)}=-2\frac{\left(S_{0,2,h}^{(1,j)}+S_{2,0,h}^{(1,j)}\right)\delta a_{0,0,h}^{(1,j)}+\left(S_{0,2,h}^{(2,j)}+S_{2,0,h}^{(2,j)}\right)\delta a_{0,0,h}^{(2,j)}}{\delta a_{0,0,h}^{2}}, \tag{6.132}$$

where $\delta a_{0,0,h}^2=\delta a_{0,0,h}^{(1)}{}^2+\delta a_{0,0,h}^{(2)}{}^2$. As a result, the diffraction of each harmonic component of the optical field is given by

$$\frac{dw_{j,h}}{dz}=\frac{2\alpha_{j,h}}{hk_0w_{j,h}}-w_{j,h}Y_{l,n,h}^{(j)}, \tag{6.133}$$

and

$$\frac{1}{2}\frac{d\alpha_{j,h}}{dz}=\frac{1+\alpha_{j,h}^2}{hk_0w_{j,h}^2}-\left(X_{l,n,h}^{(j)}+\alpha_{j,h}Y_{l,n,h}^{(j)}\right). \tag{6.134}$$

These equations constitute the SDE, which is a self-consistent adaptive eigenmode representation that tracks the optical guiding of the mode based upon the interaction with the electron beam and which permits a substantial reduction in the number of modes needed to achieve numerical convergence for the field superposition. As is evident, there are two field equations per mode and two further equations per harmonic for any given the number of modes and harmonics included in the simulation. These equations are integrated simultaneously with the three-dimensional orbit equations.

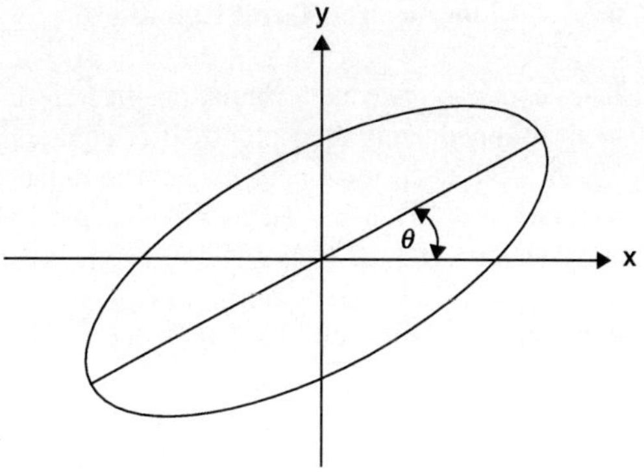

Fig. 6.14 Illustration of the ellipse formed by the polarization of the optical field [10]

6.8.1.3 The Stokes Parameters

We rely on the Stokes parameters (s_0, s_1, s_2, s_3) to characterize the polarization of the optical field [57]. Further, we determine the eccentricity of the polarization ellipse as well as the orientation of the semi-major axis of the ellipse with respect to the x-axis [57] in the case of elliptically polarized light, as illustrated in Fig. 6.14. If we denote the angle between the semi-major axis of the ellipse and the x-axis as θ, then this angle is determined by

$$\theta = \frac{1}{2}\tan^{-1}\left(\frac{s_2}{s_1}\right). \tag{6.135}$$

The eccentricity, ε, of the ellipse is given by

$$\varepsilon = \sqrt{1 - \tan^2\chi}, \tag{6.136}$$

where

$$\chi = \frac{1}{2}\sin^{-1}\left(\frac{s_3}{s_0}\right). \tag{6.137}$$

The eccentricity of a purely circularly polarized optical field is zero, while that for a purely linearly polarized optical field is unity. Elliptical polarizations are characterized by eccentricities between these extrema. Measurements of the polarization in FELs have been characterized by the Stokes parameters [58].

We observe that the s_3 parameter describes the polarization state where s3 $= 0$ denotes plane polarization and $s_3 = 1$ (-1) denotes right-hand (left-hand) circular polarization.

6.8.1.4 The Electron Orbit Equations

In order to complete the formulation, the trajectories of an ensemble of electron must be determined using the Lorentz force equations in the aggregate of all the magnetostatic and electromagnetic fields. The dynamical equations describing these trajectories are given in terms of the particular wiggler model that is under consideration and the Gauss-Hermite modes.

Neglecting derivatives of the slowly-varying amplitudes, spot size, and curvature, the electric and magnetic field associated with the Gauss-Hermite modes are

$$\delta\mathbf{E}(\mathbf{x},t) \cong \hat{\mathbf{e}}_x \sum_{l,n,h} hk_0 e^{(x)}_{l,n,h}(x,y)\left[\delta A^{(1,x)}_{l,n,h}\cos\varphi_{x,h} - \delta A^{(2,x)}_{l,n,h}\sin\varphi_{x,h}\right]$$
$$+\hat{\mathbf{e}}_y \sum_{l,n,h} hk_0 e^{(y)}_{l,n,h}(x,y)\left[\delta A^{(1,y)}_{l,n,h}\cos\varphi_{y,h} - \delta A^{(2,y)}_{l,n,h}\sin\varphi_{y,h}\right], \tag{6.138}$$

and

$$\delta\mathbf{B}(\mathbf{x},t) \cong \hat{\mathbf{e}}_y \sum_{l,n,h}^{\infty} hk_0 e^{(x)}_{l,n,h}(x,y)\left[\delta A^{(1,x)}_{l,n,h}(z,t)\cos\varphi_{x,h} - \delta A^{(2,x)}_{l,n,h}(z,t)\sin\varphi_{x,h}\right]$$
$$+\hat{\mathbf{e}}_x \sum_{l,n,h}^{\infty} hk_0 e^{(y)}_{l,n,h}(x,y)\left[\delta A^{(1,y)}_{l,n,h}(z,t)\cos\varphi_{y,h} - \delta A^{(2,y)}_{l,n,h}(z,t)\sin\varphi_{y,h}\right]. \tag{6.139}$$

As such, the contribution to the Lorentz force equations from the Gauss-Hermite modes is given by

$$\mathbf{F} \cong -m_e c \sum_{l,n,h} h\omega\left\{\left[\hat{\mathbf{e}}_x\left(1-\frac{\upsilon_z}{c}\right)+\frac{\upsilon_x}{c}\hat{\mathbf{e}}_z\right]\left[\delta a^{(1,x)}_{l,n,h}\cos\varphi_{x,h} - \delta a^{(2,x)}_{l,n,h}\sin\varphi_{x,h}\right]\right.$$
$$\left.+\left[\hat{\mathbf{e}}_y\left(1-\frac{\upsilon_z}{c}\right)+\frac{\upsilon_x}{c}\hat{\mathbf{e}}_z\right]\left[\delta a^{(1,y)}_{l,n,h}\cos\varphi_{y,h} - \delta a^{(2,y)}_{l,n,h}\sin\varphi_{y,h}\right]\right\}. \tag{6.140}$$

The Lorentz force equations, therefore, take the form

$$\frac{d}{dz}x = \frac{\upsilon_x}{\upsilon_z}, \tag{6.141}$$

$$\frac{d}{dz}y = \frac{\upsilon_y}{\upsilon_z}, \tag{6.142}$$

$$\frac{d\psi}{dz} = k_0 + k_w - \frac{\omega}{\upsilon_z}, \tag{6.143}$$

and

$$v_z \frac{d}{dz}\mathrm{p} = \mathbf{F} - \frac{e}{c}\mathbf{v} \times \mathbf{B}_{magnetostatic}, \tag{6.144}$$

for an arbitrary choice of the magnetostatic fields. The choice of magnetostatic fields in the formulation is arbitrary and can include, for example, dipoles (i.e., to form magnetic chicanes as phase shifters) and quadrupoles for enhanced focusing. However, we are concerned solely with wigglers in this chapter. This constitutes a set of six equations that must be integrated for each macro-electron included in any given simulation.

6.8.2 The Numerical Procedure

Some discussion of the numerical procedures is appropriate before proceeding with a discussion of simulation results and validation. As in the discussion of the one-dimensional simulation, the field equations are integrated simultaneously with the complete Lorentz force equations for an ensemble of electrons including the optical modes and the wiggler(s) which must now be dealt with using three-dimensional representations. Note that in three-dimensional simulations, there are now six equations per electron, and any additional focusing fields (such as quadrupole magnets) or magnetic dispersive elements (such as a chicane composed of a sequence of magnetic dipoles) must be included. The inclusion of three-dimensional effects in the treatment of the optical field by the use of Gaussian optical modes leads to the use of a two-component amplitude description of each mode which requires the integration of two equations per mode, and the SDE adds two further equations for each harmonic component to track the spot size and curvature. Time dependence is treated in the same way as described for the one-dimensional formulation, so that the electron bunch and the optical pulse are represented by a sequence of temporal *slices* that are one wavelength long as illustrated in Fig. 6.4. Hence, the total number of ordinary differential equations that is integrated is

$$N_{equations} = N_{slices}\left[6N_{particles} + 2N_{harmonics}(N_{modes} + 1)\right]. \tag{6.145}$$

The number of modes required will vary with the specific parameters under consideration and must be determined for each case by adding modes to the simulation until the result converge to within the desired accuracy. In the three-dimensional formulation, all the electron bunch properties may, in principle, be varied from *slice* to *slice*; however, in the examples discussed in this chapter, the only variation under consideration is the beam current, which is assumed to exhibit a parabolic temporal profile.

Because the wiggler-motion must be resolved in the integration of the electron trajectories, the step size in z must be small enough to resolve the wiggle -motion, and the step size for the simulations discussed in the succeeding sections is 1/30th of

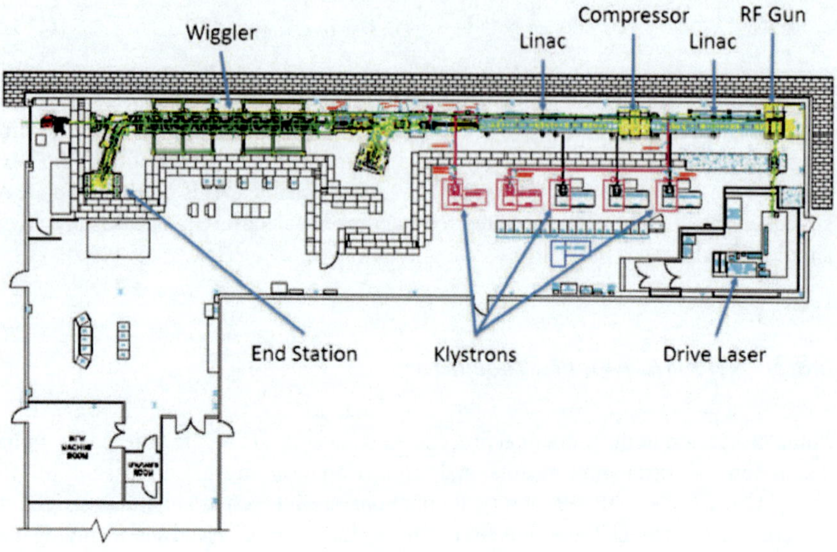

Fig. 6.15 Schematic illustration of the SDL facility. (Courtesy of X.J. Wang)

a wiggler period. While all the examples discussed herein use a fourth-order Runge-Kutta algorithm to integrate these equations, a second-order Runge-Kutta algorithm is often adequate since the step size is small, and this can result in substantial reductions in the computational time required.

As for the one-dimensional formulation, the numerical procedure first advances each *slice* from $z \to z + \Delta z$ using the Runge-Kutta algorithm. Thus far, the procedure follows that employed in steady-state simulations. The time dependence (i.e., slippage) is imposed as an additional operation using the forward time derivative as an additional source term that couples the ith radiation slice to the $(i + 1)^{\text{th}}$ slice using linear interpolation (6.98). In practice the interval between applications of the slippage operation can be varied; however, in all the cases discussed herein, the slippage operation is applied on every spatial step ($\Delta z = \Delta z$). This ensures that the wave-particle interaction is treated in a continuous fashion.

The time-dependent formulation will be both illustrated and validated by comparison with two seeded amplifier experiments at the Source Development Laboratory (SDL) at Brookhaven National Laboratory [59, 60]. The first is a study of the effect of electron beam energy detuning on the efficiency of a free-electron laser amplifier [59], and the second is a tapered wiggler amplifier [60]. Figure 6.15 shows a schematic illustration of the SDL facility, which is an example of the linear arrangement of most short wavelength free-electron lasers (seeded amplifiers or SASE configurations) driven by RF photocathode electron guns. As shown in the figure, the drive laser is injected into the RF gun which feeds a multi-stage linac including bunch compression. The electron beam produced by this process is then injected into the wiggler, and the output light is then directed into an end station for the users.

6.8.2.1 Simulation of Complex Wiggler/Undulator Lines

Unlike the longer wavelength free-electron lasers discussed in the preceding chapter which excite guided modes in waveguide structures, shorter wavelength free-electron lasers generate optical modes that are guided by the interaction. These free-electron lasers often employ high-energy electron beams generated in radio frequency linear accelerators. The gain lengths in these free-electron lasers are often long enough that the wiggler lines needed to reach saturation may be too long to be accommodated by a single long wiggler and a long sequence of relatively short wigglers are employed. In order to maintain the confinement/focusing of the electron beam, quadrupoles are inserted in the drift spaces between the wigglers forming a focusing/defocusing (FODO) lattice which provides for stronger focusing than can be provided by the wigglers alone. In addition, multiple wigglers may be used in optical klystrons which employ a modulating wiggler which introduces some bunching in the electron phase space followed by a magnetic chicane (usually composed of a sequence of magnetic dipoles) which enhances the bunching preparatory to injection into a second wiggler where the radiation grows rapidly to saturation. Free-electron lasers based upon these configurations are discussed in Chaps. 12 and 13.

The formulation described herein is capable of simulation of these complex systems. Since the dynamical equations for the particles and fields are integrated using Runge-Kutta algorithm, the step size may vary from step to step. Thus, while 20–30 steps per wiggler period are needed within the wigglers, once the integrator advances into the drift space beyond the wiggler, the step size is increased in such a way that an integer number of steps are performed until the next magnetic element (either a quadrupole or a dipole) is reached. At that point the step size is adjusted again to resolve the electron trajectories within that element in such a way as to take an integer number of steps. Upon leaving the quadrupole or dipole and entering another field free drift region, the step size is again increased, and an integer number of steps is taken to reach the next wiggler after which the process is repeated. This permits the simulation of arbitrary sequences of wigglers, quadrupoles, and dipoles.

It is important to remark that since the particles and fields are advanced through each of these regions together, the relative phase advance of the fields with respect to the electrons is implicitly included in a self-consistent fashion.

6.8.3 Comparison with an Energy-Detuned Amplifier Experiment

It is well known that the peak saturation efficiency does not coincide with the peak gain. As a result, it may be possible to increase the efficiency of a seeded amplifier by detuning the electron beam energy away from the resonant energy. The purpose of this experiment was to observe and quantify the variation in the saturation efficiency as the electron beam energy is detuned from the resonant energy and to determine whether the central wavelength of the output signal varied with this energy detuning or whether the central wavelength was fixed by the wavelength of the seed pulse. The experimental results were compared with simulations based on the formulation discussed herein, with generally good agreement found [59].

The experiment used the SDL linear accelerator at the National Synchrotron Light Source [59], which uses a Ti:sapphire photocathode drive laser to generate an electron bunch of 350 pC and between 98 and 103 MeV that is compressed to 2–3 psec by a chicane bunch compressor located between two accelerator sections and yielding a peak current of about 230 A. The rms energy spread was about 0.1%, and the normalized emittance is about 4 mm-mrad. The electron beam is propagated through a 10-m-long, weak (two-plane or parabolic-pole-face) focusing wiggler with a period of 3.89 cm and a peak on-axis field of 3 kG [61]. The β-function for this weak, two-plane focusing wiggler is about 2.23 m; hence, the matched beam radius $[= (\varepsilon\beta/\gamma)^{1/2}]$ in the wiggler is about 212 μm, and this is used to initialize the electron beam in the simulation.

The wiggler consists of 16 sections, and the electron beam can be kicked at each section by trim coils. In this way, it is possible to measure the optical pulse energy at fixed distances down the wiggler. It is also possible to introduce a linear down-taper by fixing the gap at the start of any given segment and then opening the jaws of all the downstream segments by a fixed amount, and this is what was done in the tapered wiggler amplifier experiment discussed below.

The photocathode drive laser is also used as the seed laser for the amplifier. After compression to 6 psec with a negative frequency chirp, the seed pulse is directed through a bandpass filter with a bandwidth of 1.5 nm (FWHM) centered at 793.5 nm. The simulation employs a seed pulse with a parabolic temporal profile and a full width of 6 psec. The resonant electron kinetic energy for these parameters is 100.86 MeV.

The simulation imposes a 6 psec time window over which there are 700 *slices*, yielding 1 *slice* approximately every 3 wavelengths. In addition, because a planar wiggler was used in the experiment, the Gauss-Hermite modes are used in the simulation for which it was found that good convergence was achieved with six modes.

For the purposes of the experiment, the peak seed power was fixed at 50 kW yielding an initial pulse energy of 200 nJ. While the resonant energy may be calculated and used in simulation, a determination of the resonant energy in the experiment was more difficult due to uncertainties in the calibration of the accelerator. The experimental procedure used was to run the free-electron laser in SASE (self-amplified spontaneous emission) mode and tune the beam energy produced by the accelerator. The resonant energy will then correspond to the highest output power. When this is found, the energy tune is "locked down" in the accelerator, and the seed power is injected in synchronism with the electron beam. The growth of the seed pulse with distance along the wiggler can then be determined by kicking the electron beam off axis at the various wiggler segments, which has the effect of terminating the interaction at those points. The published comparison between the

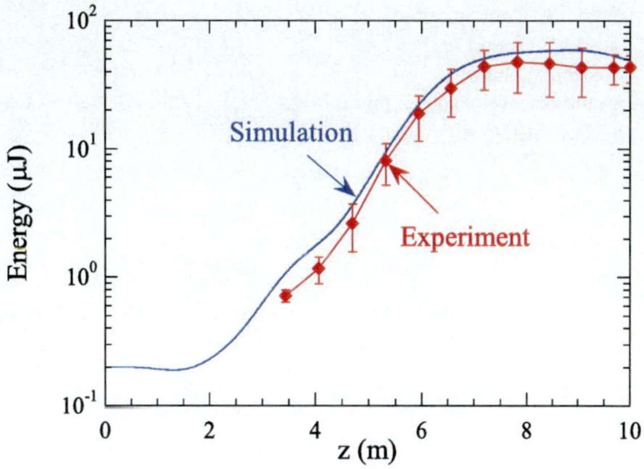

Fig. 6.16 Comparison of the growth of the seed pulse with distance through the wiggler between the simulation and the experiment at the resonant electron energy [59]

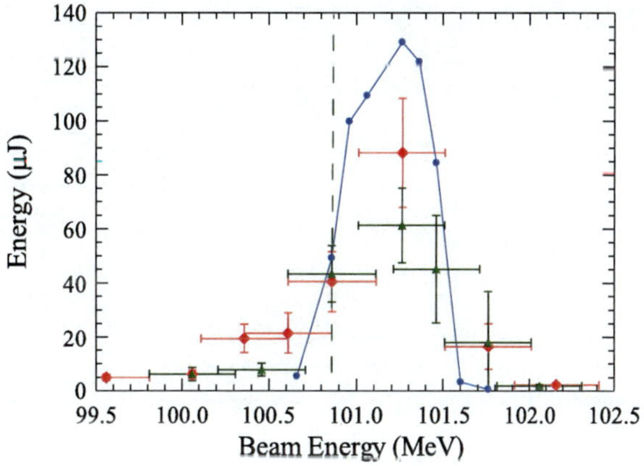

Fig. 6.17 Comparison of the output pulse energy versus electron beam energy from the experiment and simulation [59]

measured pulse energy versus position in the wiggler and that found in simulation is shown in Fig. 6.16 and shows good agreement between the nonlinear formulation and the experimental measurements.

Comparison between the experiment and the simulation of the output pulse energy versus the electron beam energy are shown in Fig. 6.17, where the data are represented in red and green markers that represent data from two series of runs on consecutive days and show the repeatability of the experimental setup. The error bars represent one standard deviation in the data sets. Note that the uncertainty in the energy measurement is of the order of $\pm 0.5\%$. It is also evident that the output pulse energy increases (decreases) as the electron energy increases (decreases) from the resonant energy. It is evident that both experiment and simulation show that the enhancement occurs when the beam energy is increased by about 0.4% from the

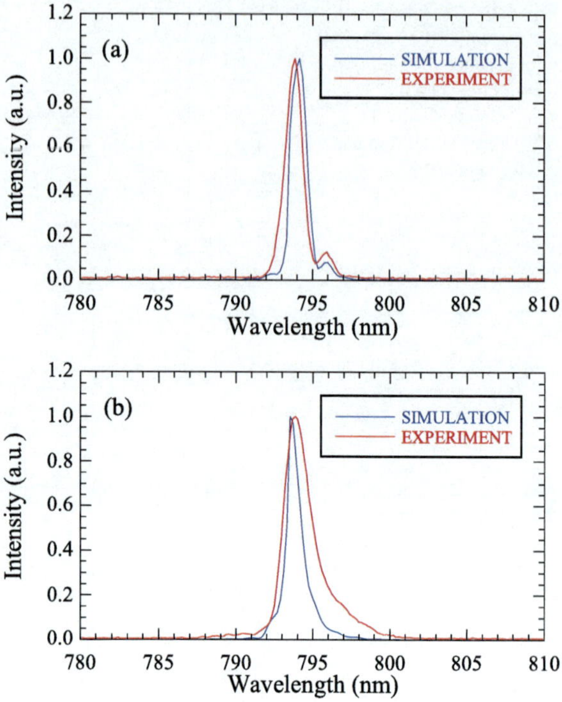

Fig. 6.18 Comparison of the spectra found in the experiment and in simulation on-resonance (**a**) and off-resonance (**b**) at 101.77 MeV [59]

resonance, while a decrease by a similar amount significantly reduces the output energy. It should also be remarked that the maximum enhancement found in simulation is higher than the observed enhancement; however, the data plotted is an average, and the error bars correspond to the standard deviation found over many shots. The maximum energy observed experimentally at the 0.4% detuning was about 134 μJ, which is in substantial agreement with the simulation.

A comparison of the normalized spectra found in experiment and in simulation is shown in Fig. 6.18. Figure 6.18a shows the spectra at the resonant energy. The two spectra are in substantial agreement in regard to (1) the peak in the wavelength, (2) the spectral width, and (3) the secondary maximum (sideband) at a wavelength of about 796 nm. A comparison of the spectra from the experiment and simulation off the resonant energy at 101.77 MeV is shown in Fig. 6.18b. Both experiment and simulation show that the spectrum is determined primarily by the seed laser. However, the experimental spectrum is broader than in the simulation. This might be explained by limitations in the spectrometer resolution and uncertainties in the electron energy. A broadening of the output spectrum is seen in simulation as the electron energy increases beyond 101.77 MeV.

6.8.4 Comparison with a Tapered Wiggler Amplifier Experiment

The tapered wiggler seeded amplifier experiment was designed to study the efficiency and spectrum in a laser-seeded free-electron laser amplifier at a sub-infrared wavelength at both the fundamental and third harmonic using the SDL linear accelerator at the Brookhaven National Laboratory [59]. As such, the apparatus was identical to that described above for the energy-detuned seeded amplifier. As in the preceding energy detuning experiment, the electron beam parameters were unchanged, and the experiment was conducted at the resonant energy of the wiggler. The wiggler taper was implemented by changing the gap in the last four sections of the wiggler, which resulted in a linear amplitude taper that started at about 7 m from the wiggler entrance. Hence the taper was imposed over the last 3 m of the wiggler. As in the energy detuning experiment, the experimental results were compared with simulations using the formulation described herein. The optimal slope was found in both the experiment and simulation to consist of a reduction in the wiggler amplitude of about 4%. Finally, the photocathode drive was also used in this experiment as a seed laser for the amplifier. However, the start-taper point is an important factor in maximizing the efficiency of a tapered wiggler. Since the start-taper point was not a variable, the tapered wiggler interaction was optimized by choosing the proper the seed laser power. It was found that a peak seed power of about 10 kW (corresponding to a pulse energy of 20 nJ) yielded the optimal efficiency enhancement.

A comparison between the experiment and the simulation at the fundamental wavelength for both the uniform and tapered wigglers is shown in Fig. 6.19. The source of the fluctuations in the experiment is attributed laser intensity and timing jitter (both for the photocathode drive and seed laser pulses) and timing jitter of the electron beam relative to the radio frequency drive power of the linear accelerator. The results show that the experiment and the simulation are in substantial agreement.

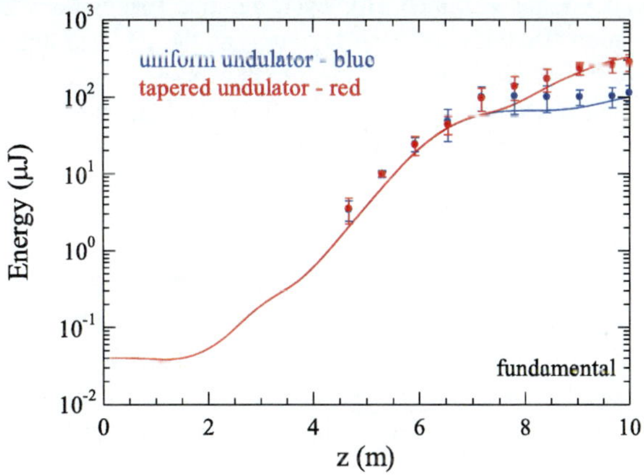

Fig. 6.19 Comparison between the experiment and simulation for the fundamental wavelength for both uniform and tapered wigglers [59]

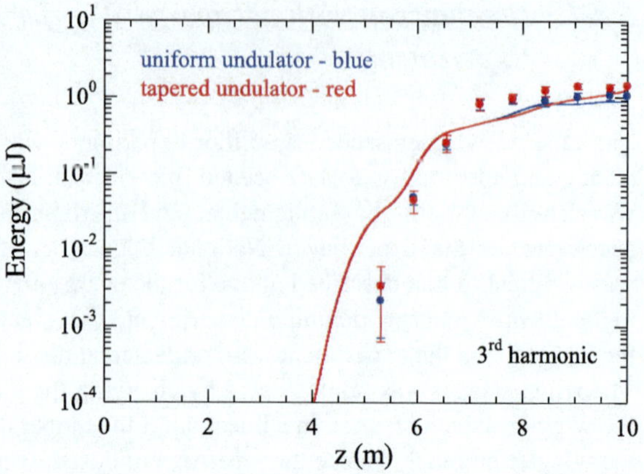

Fig. 6.20 Comparison between the experiment and simulation for the third harmonic for both uniform and tapered wigglers [60]

The output pulse energy for the uniform wiggler was observed to be 113 ± 28 μJ as compared to 100 μJ from the simulation. The measured output pulse energy for the tapered wiggler was 283 ± 68 μJ, which agrees well with the 336 μJ from the simulation.

With the taper optimized for the fundamental, the third harmonic output was also measured in the experiment and determined from the simulation. The evolution of the third harmonic for both the experiment and the simulation are shown in Fig. 6.20 for the uniform and tapered wigglers. The measured third harmonic output pulse energy for the uniform wiggler was 1.06 ± 0.17 μJ, and for the tapered wiggler was 1.42 ± 0.16 μJ. This compares well with the simulation that produced output energies of 0.89 μJ for the uniform wiggler and 1.27 μJ for the tapered wiggler.

Comparison between the spectra for the uniform and tapered wigglers is shown in Fig. 6.21 at 8.5 m, 9.5 m, and 10.0 m from the wiggler entrance. The spectra are in good agreement between the experiment (red) and simulation (blue) at 8.5 m and 9.5 m with respect to the center frequency, linewidth, and presence of sidebands. Note that saturation in a uniform wiggler is found at about 7.5 m downstream from the wiggler entrance. However, the discrepancy between the experiment and simulation becomes increasingly large as the distance increases beyond the saturation point in a uniform wiggler. In the case of a tapered wiggler, the resonance is maintained over an extended distance, and the sidebands are suppressed [60], which is not the case for the uniform wiggler.

6.8.5 Simulation of a Helical Wiggler

In the simulation of a helical wiggler, we employ the same electron beam parameters used for the simulation of the SPARC experiment (see Sect. 12.5.2) but replace the wiggler/quadrupole line with a long helical undulator with the same period as used in

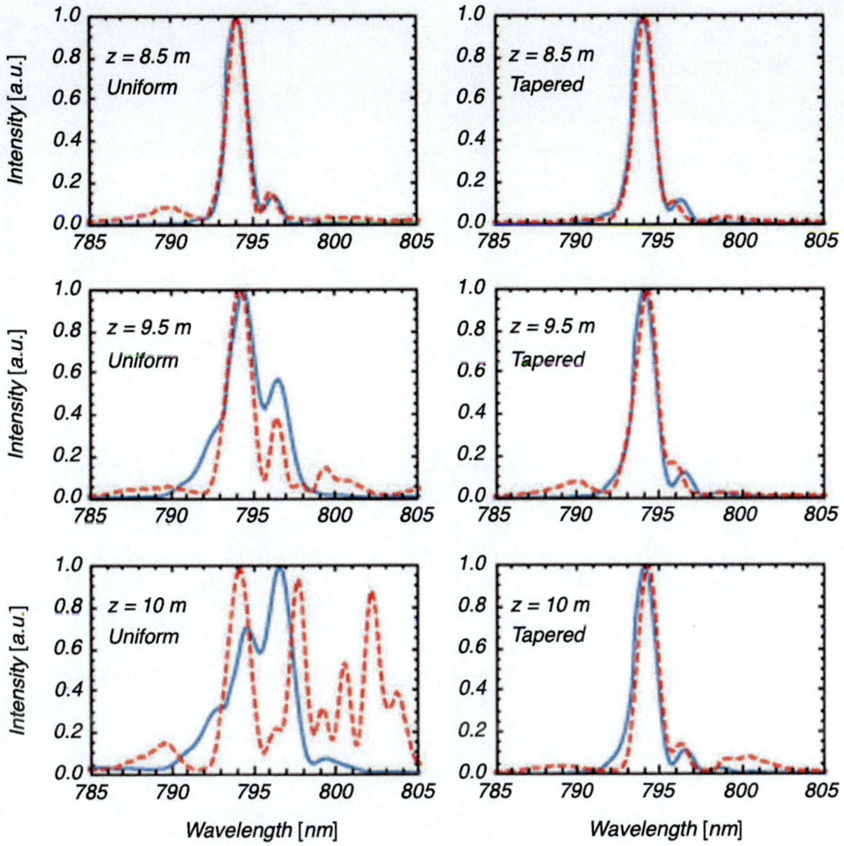

Fig. 6.21 Spectra for the experiment (red) and simulation (blue) for the uniform and tapered wigglers [60]

the SPARC experiment. The wiggler model used in the simulation is that described in Eq. 2.1, and in order to preserve the resonance at 491.5 nm, the peak on-axis field of the helical wiggler is reduced to 5.717 kG. In addition, the Twiss parameters are chosen to match the beam into the weak-focusing helical undulator [8, 10]. As in the case of the SPARC experiment, we simulate SASE for this configuration but only use a single noise seed. This is adequate for the illustration of the formulation.

The evolution of the pulse energy with distance along the helical wiggler is shown in Fig. 6.22 which illustrates the typical start-up regime for SASE and reaches saturation after about 14 m at a pulse energy of 141 μJ. The Stokes s_3 parameter after 15 m is 0.998 which strongly indicates circular polarization. The spectral power density after 15 m of the wiggler is shown in Fig. 6.23.

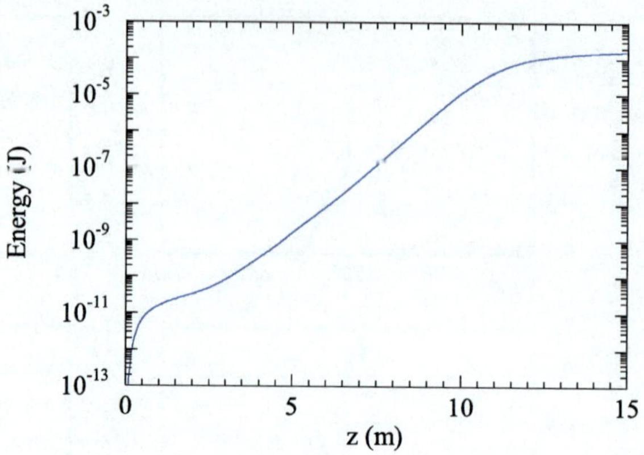

Fig. 6.22 Pulse energy as a function of distance along the helical wiggler

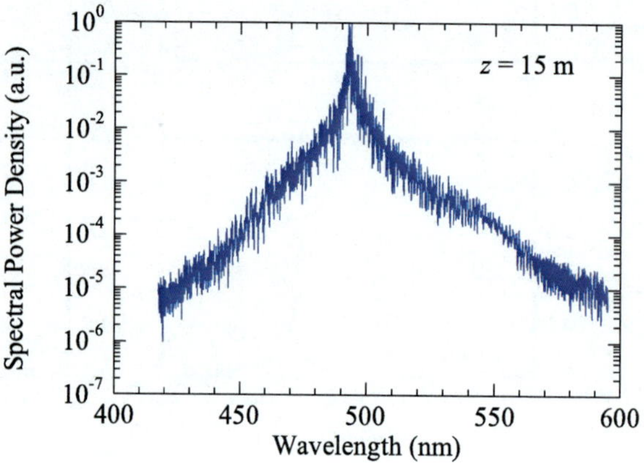

Fig. 6.23 Spectrum after 15 m of helical wiggler

6.8.6 *Simulation of an Elliptic Wiggler/Quadrupole Lattice*

Now consider a wiggler line consisting of variable polarization APPLE-II undulators [51] in conjunction with a focusing lattice composed of quadrupoles in the gaps separating the wigglers. An analytic model for the APPLE-II undulator is described in Sect. 6.5.1. Now consider the basic configuration employed in the SPARC experiment (see Sec. 12.5.2) except that the planar wigglers are replaced with APPLE-II undulators of the same lengths and periods, and consider five different choices for the ellipticity: $u_e = 0.00$, 0.25, 0.50, 0.75, and 1.00. The resonance for a planar wiggler ($u_e = 0$) is at a wavelength of 491.5 nm, and the wiggler field strength must decrease to maintain this resonant wavelength as the ellipticity increases. In order to ensure that the resonance remains at 491.5 nm, the on-axis field amplitudes for these cases were found to be 7.8796 kG, 7.6600 kG, 7.0577 kG, 6.3150 kG, and 5.5817 kG, respectively, as shown in Fig. 6.24 and as determined by Henderson et al. [53].

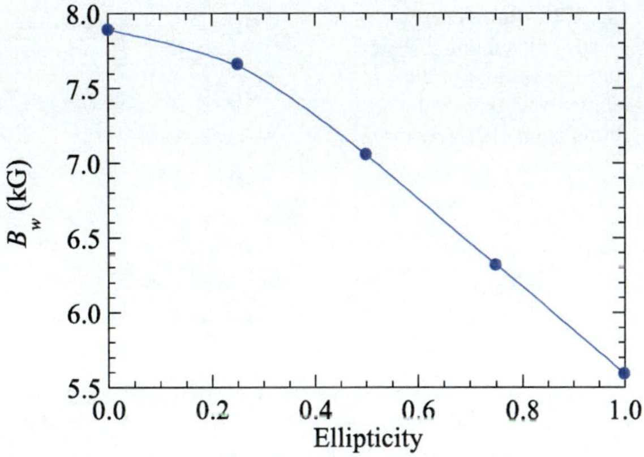

Fig. 6.24 Variation in the on-axis undulator field for various degrees of ellipticity, u_e, required for resonance at 491.5 nm [10]

The electron beam envelope as it propagates through the wiggler/quadrupole line is largely dependent upon the quadrupoles and relatively independent of the polarization and field strength of the wigglers. As such, the matched beam sizes (in the x- and y-directions) are nearly invariant with respect to the wiggler, and the x and y β functions are $\beta_x = 2.08$ m and $\beta_y = 0.58$ m. In order to inject such a matched beam, the simulation initializes the electron beam in the simulation with an rms size in the x-direction of about 132 μm with a Twiss-α_x = 0.938 and an rms size in the y-direction of about 75 μm with a Twiss-α_y = −0.705. The evolution of the beam envelope through the wiggler/quadrupole lattice that results is shown in Fig. 12.20, where the rms beam size in the x-direction is shown in blue on the left axis and the rms beam size in the y-direction is shown in red on the right axis. Observe that this yields an average beam size of about 113 μm.

In order to study the interactions out to saturation, and since the bunch charge was insufficient to reach saturation after 15 m of undulators, we extend the wiggler line to reach 28 m in total length. We also note that the quadrupole positions and field gradients were held fixed as well as the initial Twiss parameters of the electron beam. This means that optimal focusing of the electrons through the undulator line may not be achieved in all cases; however, the transmission of the electron beam is sufficient to illustrate the variations in the polarizations of the optical field.

Now consider SASE using a single, identical noise seed for each of these five choices for the ellipticity. This ensures that the injected electron phase space is also identical for each of these cases. Furthermore, the simulation used 1600 temporal slices, and the optical field was described using 15 Gauss-Hermite modes. The growth of the ensemble average of the pulse energies with propagation distance as found using the new formulation is shown in Fig. 6.25 for three choices of the ellipticity: $u_e = 0.0, 0.50$, and 1.00 which ranges from planar to helical polarizations. We observe that the distance to saturation increases, and, hence, the growth rate decreases, as the ellipticity decreases from a value corresponding to pure linear polarization ($u_e = 0$) to one corresponding to circular polarization ($u_e = 1$). This follows the decrease in the coupling coefficient associated with the increase in the

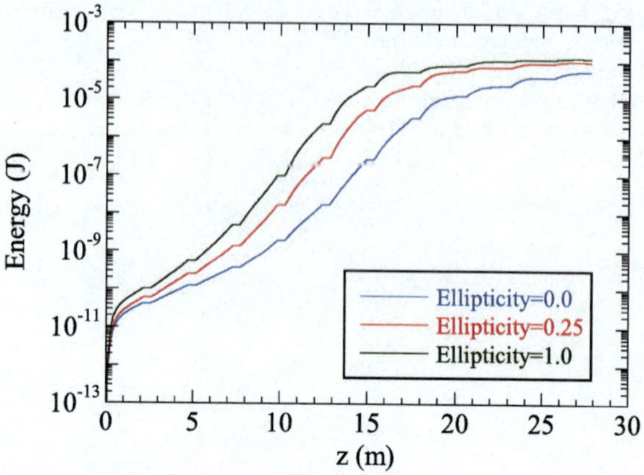

Fig. 6.25 Pulse energy versus propagation distance for three choices of the ellipticity as seen in the new formulation [10]

Table 6.1 Comparison between the gain lengths found in simulation and theory

Ellipticity	L_G (simulation)	L_G (theory)
0.00	0.80 m	0.75 m
0.25	0.79 m	0.72 m
0.50	0.69 m	0.67 m
0.75	0.60 m	0.64 m
1.00	0.58 m	0.63 m

oscillation in the magnitude of the transverse velocity. This is described by the generalized *JJ*-factor (Eq. 6.78). Energy conservation is maintained to within better than one part in 10^3.

An important validation measure is a comparison between the exponentiation (gain) length (L_G) found in simulation and with a theoretical estimate based on the generalized Ming Xie parameterization (see Sect. 6.5.3). Using this generalization, a comparison between the gain lengths found in both simulations and theory for the five ellipticities under consideration is shown in Table 6.1 which shows reasonable agreement.

Using this present diagnostic for the Stokes parameters, the variation in the optical eccentricity versus the ellipticity of the undulator field is shown in Fig. 6.26 in blue, while the ideal polarization is indicated by the red line. Since the optical field is composed of multiple temporal slices, each of which may behave somewhat differently, the average ellipticity shown in the figure is a weighted average over the power in each slice. As is evident in the figure, the eccentricity is unity, and θ corresponds to the direction of wiggle-motion when the Apple-II undulator is configured for linear polarization, as expected, and decreases approximately linearly with increasing ellipticity but more slowly than would be expected from the fixed-phasor representation. We do not find that the eccentricity is zero when the Apple-II undulator is configured for an ellipticity of unity which is close to that of a helical field. It is important to remark here that the Apple-II undulator model is not the same as the helical undulator field used in Eq. 2.1 and, in general, diverges

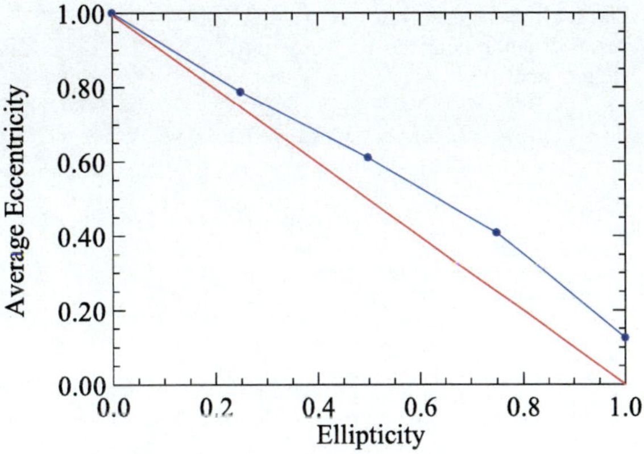

Fig. 6.26 The power-weighted average of the on-axis optical eccentricity for the Apple-II undulator line (blue line and markers). The red line represents the idealized polarization state [10]

from the helical field slowly away from the axis of symmetry. Thus, it may not be possible to achieve pure circular polarization with an Apple-II undulator. The seeded FEL at Fermi Light Source in Trieste uses APPLE-II undulators, and, while this may be due in part to the optical beam transport line from the end of the undulator line, they have indeed observed that the optical polarization is not purely circular [62] at the location of measurement when the ellipticity of the undulators is set to unity.

6.8.7 *Simulation of an APPLE-II Afterburner*

The utility of the formulation described herein where the x- and y-components of the optical field are treated independently is that it can model the variations in the polarization of the optical field along a wiggler line in which the polarizations of the different wigglers may vary in an arbitrary manner. There is interest at the present time in ending a line of wigglers which may, for example, contain plane-polarized wigglers with a final wiggler (afterburner) with a different or variable polarization. In this way, the polarization of the output light may be varied as required for any given experiment.

In order to illustrate the utility of an afterburner, we turn now to the simulation of the SPARC SASE FEL (see Sect. 12.5.2) where we append an APPLE-II afterburner. This necessitates an additional focusing quadrupole that precedes the afterburner. The afterburner is assumed to have the same period as the preceding planar wigglers (2.8 cm) with an overall length of 8.4 m. The magnitude of the field must vary with the choice of the ellipticity as shown in Fig. 6.24 in order to maintain the resonance at 491.5 nm. Simulations describe a parabolic temporal profile with a full width of 12.67 psec using 1600 temporal slices with a time window of 14 psec. We limit consideration to a single noise seed for simplicity as this is adequate to illustrate the utility of the afterburner.

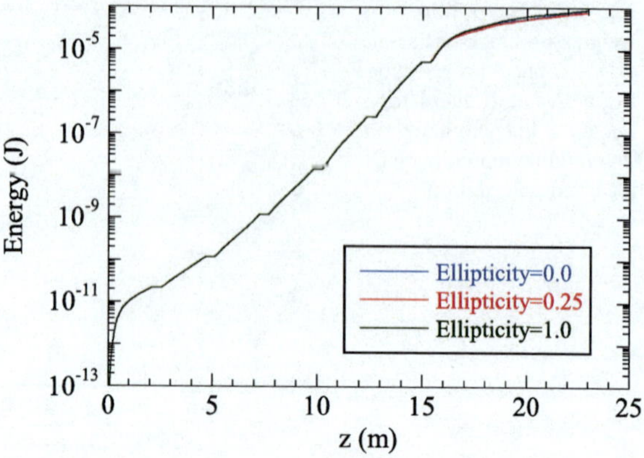

Fig. 6.27 Pulse energy versus distance along the wiggler line

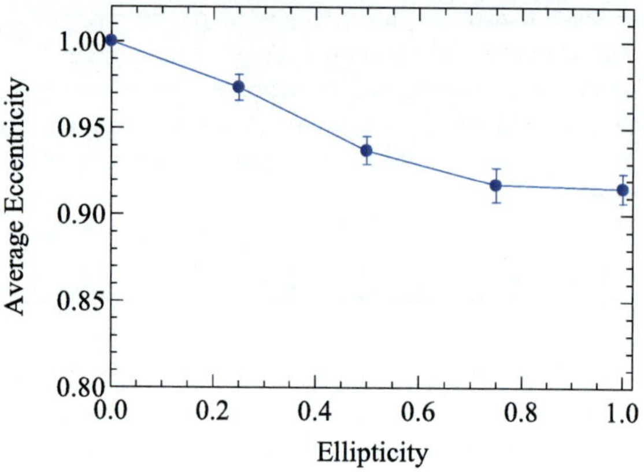

Fig. 6.28 The eccentricity of the optical output from the afterburner averaged over the entire optical pulse

The growth of the optical pulse energy along the wiggler line is shown in Fig. 6.27 for three choices of the ellipticity of the APPLE-II afterburner. Observe that the afterburner extends from about 15.3 m to 23 m. The output energy for these three cases is relatively small and varies between 77 μJ and 90 μJ corresponding to a variation of only about 15%. This is comparable to the variation which can be expected in SASE FELs.

The eccentricity averaged over the entire optical pulse is shown in Fig. 6.28 as a function of the afterburner ellipticity. It is important to emphasize that this represents an average of the pulse and that the error bars in the figure describe the rms variation of the eccentricity within the pulse.

6.8.8 Limitations of the SVEA

In the SVEA, the optical field is represented by a slowly-varying amplitude and phase with respect to the rapid sinusoidal oscillation of the carrier frequency. The field equations are then averaged over the rapid sinusoidal time scale and, thereby, reduced to equations describing the evolution of the slowly-varying amplitude and phase. Although these simulation models have been extremely successful in modeling FELs, the SVEA breaks down for sufficiently short optical/electron pulse interactions. Intensive work is ongoing at the present time into new FEL-based light sources that probe ever shorter wavelengths with a variety of configurations. In particular, there is interest in ultrashort electron bunches that are shorter than the cooperation length and, hence, are shorter than the SASE spike separation. It is important, therefore, to understand the limitations of the SVEA models in the simulation of such short pulse configurations. In contrast to the SVEA formulations, particle-in-cell (PiC) simulations do not rely on an average of Maxwell's equations and can treat arbitrarily short electron/optical pulses. In order to test the pulse length limitations of the SVEA formulations, we compare the operation of the aforementioned numerical formulation with the results of the PUFFIN PiC code [11] in the simulation of the SPARC SASE FEL (see Sect. 12.5.2) [63].

The basic configuration and parameters of the SPARC SASE FEL are described in Sect. 12.5.2 and will not be reproduced here. We first compare the SVEA and PUFFIN simulations with the experimental measurements.

A comparison of the evolution of the pulse energy as found in the SVEA and PUFFIN, and as measured in the experiment, is shown in Fig. 6.29 where the SVEA simulation is indicated by the blue line and the PUFFIN simulation is indicated by the green line and the experimental measurements are indicated by the red markers. Note that this represents an average over 15 runs with different noise seeds in the SVEA and 5 runs with different noise seeds in PUFFIN. The reason for the smaller number of runs with PUFFIN is that each run requires considerably more computer

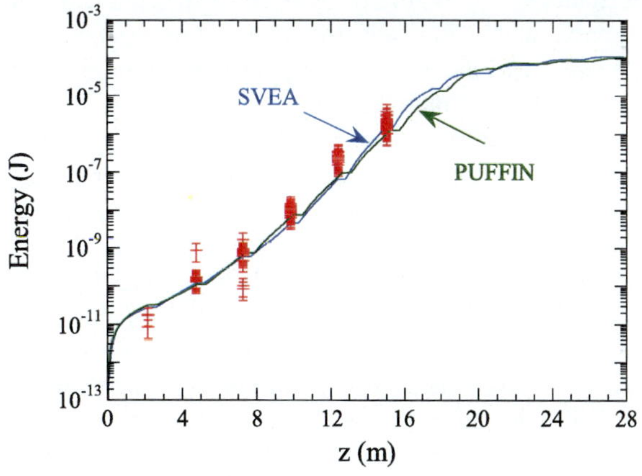

Fig. 6.29 Comparison of simulation results with PUFFIN and MINERVA and the measured pulse energies versus distance through the undulator [63]

time than needed for the SVEA runs. However, convergence is achieved over simulations using multiple noise seeds to within about 5% for both codes. As shown in Fig. 6.27, it is evident that exponential growth starts in the second undulator module and that the start-up region is encompassed in the first module. The experimental measurements indicate that the pulse energy after the first undulator falls into the range of approximately 8.4×10^{-12} J to 1.74×10^{-11} J, while the SVEA yields an average pulse energy of 2.52×10^{-11} J and PUFFIN yields 3.15×10^{-11} J resulting in a discrepancy of only about 24% between the two codes. An experimental value that is somewhat lower than the simulations is to be expected as it is measured at some distance downstream from the first undulator module while the simulations evaluate the pulse energy at every location along the wiggler line. The simulation results are in relatively close agreement with the experiment and with each other. This agreement is an important observation since the particle loading algorithms in the SVEA code [50] and in PUFFIN [46] share no commonality. Another source of the difference in the slightly higher start-up noise in PUFFIN is the fact that PUFFIN naturally includes a wider initial spectral range than the SVEA.

The exponential growth region starts in the second undulator module, and the two codes are in close agreement with each other and with the experimental measurements out to the end of the sixth undulator. These results are in substantial agreement with the parameterization developed by Ming Xie [54]. Using a β-function of approximately 2 m, we find that the Pierce parameter $\rho \approx 2.88 \times 10^{-3}$ and that this parameterization predicts a gain length of 0.67 m, and a saturation distance of 18.1 m (including the additional 3.2 m represented by the gaps between undulator modules). This is in reasonable agreement with the simulations which indicate that saturation occurs after between about 18–20 m of undulator/FODO line.

There is a relatively small difference in the synchronism between the PUFFIN exponentiation and that observed in the experiment and found in the SVEA. This is because the SVEA adjusts the integration step on the fly which permits an integer number of steps through the gaps between the undulators, while PUFFIN uses a fixed step which requires an approximate step length to account for the gap/quad distances. Finally, the predictions of the two codes in the saturation regime after approximately 20 m are also in very good agreement. After 28 m of wiggler/FODO lattice, PUFFIN predicts a pulse energy of 104 μJ, while the SVEA predicts 111 μJ for a difference of 6.3%.

The larger initial spectral linewidth excited in the start-up region exhibited by PUFFIN is shown more clearly in Fig. 6.30 which presents a comparison between the evolution of the relative rms linewidth as determined from PUFFIN and the SVEA and by measurement. It is clear that a PiC code such as PUFFIN predicts the generation of a wider initial bandwidth of incoherent spontaneous emission. Exponential gain due to the resonant FEL interaction starts in the second undulator module, and this is expected to rapidly overcome any incoherent synchrotron radiation from the start-up region in the first undulator module. In view of this, the PUFFIN results converge rapidly to that found by the SVEA and to the measured linewidths after the second wiggler module. Agreement between the simulations and

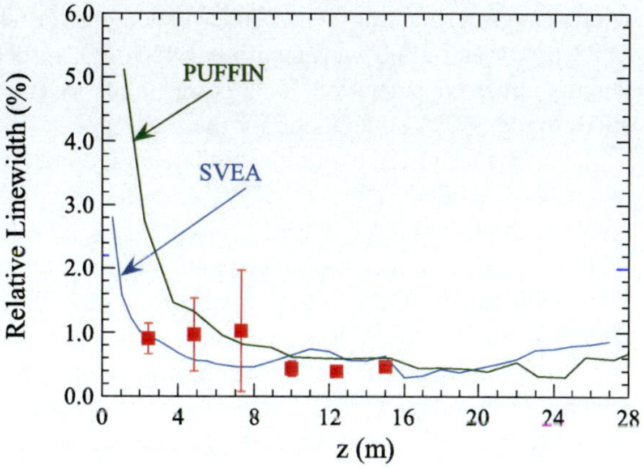

Fig. 6.30 Comparison of the measured relative linewidth in red (with that found in the simulations [63]

the measured linewidth is within about 35% after 15 m. As shown in the figure, the predicted linewidths are in substantial agreement with the experimental measurements, and good agreement between the codes is found over the entire range of integration through the saturated regime.

Having demonstrated the agreement between the SVEA and PUFFIN formulations for the bunch/pulse lengths used in the SPARC SASE FEL, we now consider the case of a short electron bunch and the SASE that is produced using the configuration of the SPARC FEL in which we include only the six undulators used in the experiment for a total length of about 15 m. In this scenario, we study the SASE interaction for a variety of electron bunch lengths in order to determine under what conditions the SVEA model can accurately simulate the physics of the interaction. In particular, we consider three cases corresponding to rms electron bunch lengths/durations of 5.59 μm/18.3 fs, 11.2 μm/36.7 fs, and 16.8 μm/55.0 fs with bunch charges of 2.9 pC, 5.8 pC, and 8.7 pC respectively. The cooperation length is approximately 21 μm for each of these cases. Hence, the three bunch lengths under consideration are all shorter than the separation distance between SASE spikes, and we expect to see single pulses produced in each of these cases rather than the series of spikes usually associated with SASE.

In many FEL simulations using the SVEA, the number of temporal slices used does not always "fill" the entire time window. This means that if the time window is N wavelengths long, then fewer than N temporal slices might be used. However, since we are interested in studying short pulse interactions, it is important to use contiguous temporal slices in all the SVEA simulations so that the number of temporal slices corresponds to the number of wavelengths in the time window. On the other hand, PiC models implicitly simulate a continuous optical and electron pulse.

As there will be substantial slippage between the optical pulse and the electron bunch over the course of six undulators, it is important to choose a temporal simulation window in the SVEA which is of sufficiently broad duration that no optical power exits it. In order to ensure that this doesn't happen, we use time windows corresponding to 1000 wavelengths (1.64 psec), 2000 wavelengths (3.28 psec), and 3000 wavelengths (4.92 psec), respectively, for the three bunch lengths under consideration. This corresponds to 1000, 2000, and 3000 temporal slices in the SVEA simulations for these cases.

Comparisons of the evolution of the average pulse energies along the undulator lattice between PUFFIN and the SVEA for the three bunch lengths under consideration are shown in Figs. 6.31, 6.32, and 6.33. As is evident in Figs. 6.4 and 6.5, while PUFFIN and the SVEA are in substantial agreement at the start up and through the third undulator for the two shorter bunch lengths, they diverge significantly thereafter. We attribute this divergence either, or both, to (1) the breakdown of the SVEA and (2) to the strong coherent synchrotron radiation (CSR) in the short bunches which is captured in PUFFIN but not the SVEA. However, as shown in Fig. 6.33, the two codes are in good agreement for the longest electron bunch which we attribute to the decreased importance of CSR at this bunch length. It is important to remark, though, that this bunch length is still shorter than the separation between SASE spikes.

Now consider the properties of the emission for the longest bunch length of 16.8 μm. Comparisons of the spectra obtained from the two codes after the first, third, and sixth undulators are shown in Figs. 6.34, 6.35, and 6.36 where we plot the normalized spectral power density vs wavelength (normalized to the resonant wavelength). Note that these comparisons are made using a single noise seed in each code.

As shown in Fig. 6.34, the peak in the spectra obtained from the two codes is shifted in wavelength by only about one part in 5×10^3 after the first undulator but shrinks to about one part in 10^4 after the sixth undulator. We attribute this to the larger influence of CSR at the early stage of the interaction which is increasingly overwhelmed by the exponentiation over longer distances. However, PUFFIN exhibits a much broader bandwidth than the SVEA because PUFFIN does not make the resonant wave approximation and implicitly includes a complete spontaneous excitation spectrum. Exponential growth begins in the second undulator, and

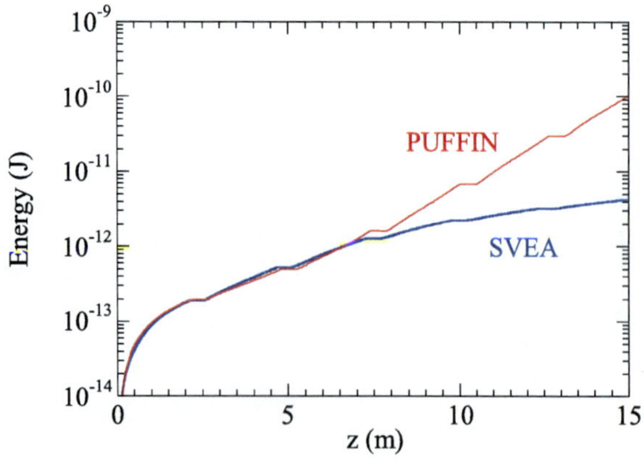

Fig. 6.31 Energy comparison for an rms electron bunch length of 5.59 μm [63]

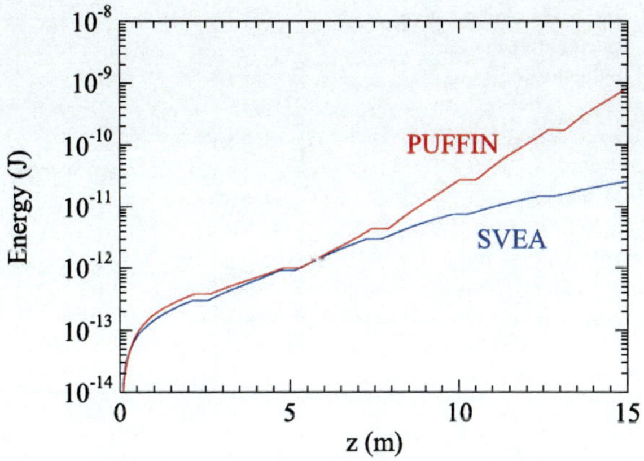

Fig. 6.32 Energy comparison for an rms electron bunch length of 11.2 μm [63]

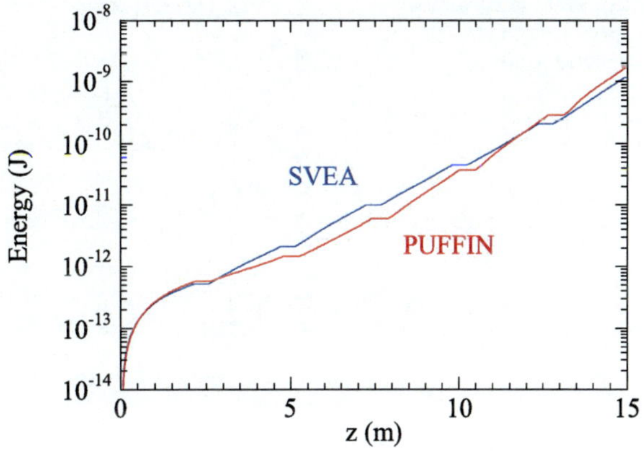

Fig. 6.33 Energy comparison for an rms electron bunch length of 16.8 μm [63]

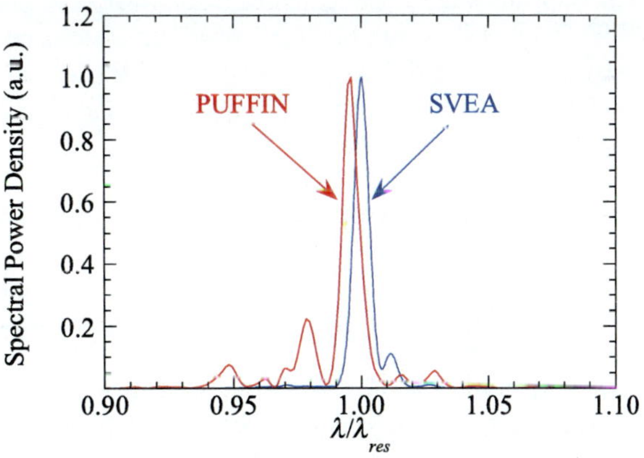

Fig. 6.34 Comparative spectra after the first undulator [63]

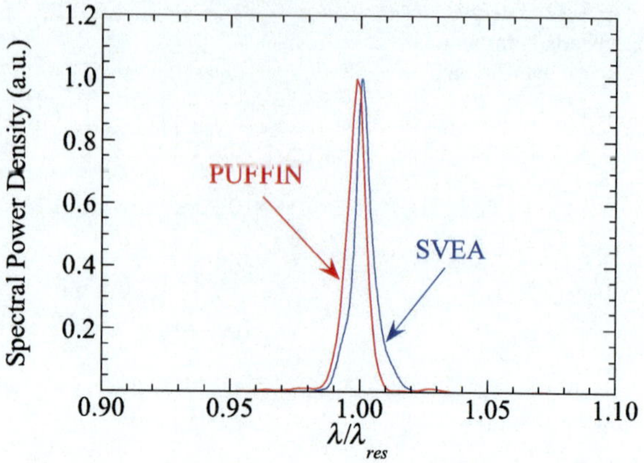

Fig. 6.35 Comparative spectra after the third undulator [63]

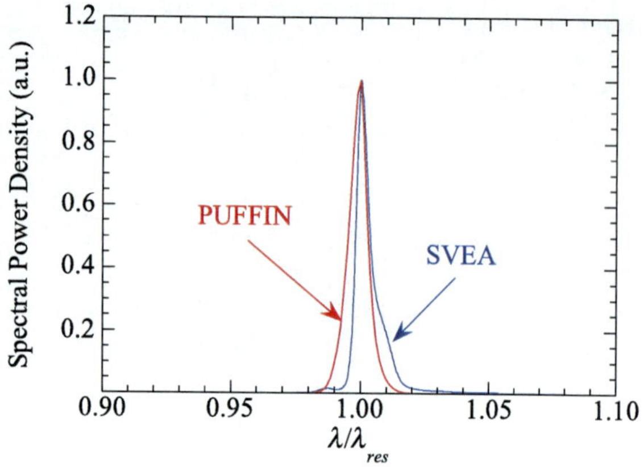

Fig. 6.36 Comparative spectra after the sixth undulator [63]

the resonant wavelength has come to dominate the spectrum by the end of the third undulator, as shown in Fig. 6.35 where the spectra produced by the two codes are very close and PUFFIN now shows a peak near the resonant wavelength. The comparison of the spectra after the sixth undulator follows that after the third undulator as shown in Fig. 6.36.

The temporal pulse shapes after the first and sixth undulators are shown in Figs. 6.37 and 6.38 where the left axis corresponds to the radiation pulse, while the right axis describes the electron bunch. In addition, the horizontal axis corresponds to a portion of the time window (which is 5000 fs in duration) used in the SVEA simulations to better illustrate the characteristics of the pulse. The horizontal axis for the PUFFIN results are mapped onto the same time axis as found in the SVEA such that the peak in the current bunch corresponds to that in the SVEA. It should also be remarked that the pulse shape found in the SVEA is calculated using the Poynting flux averaged over the resonant period, which has averaged out the

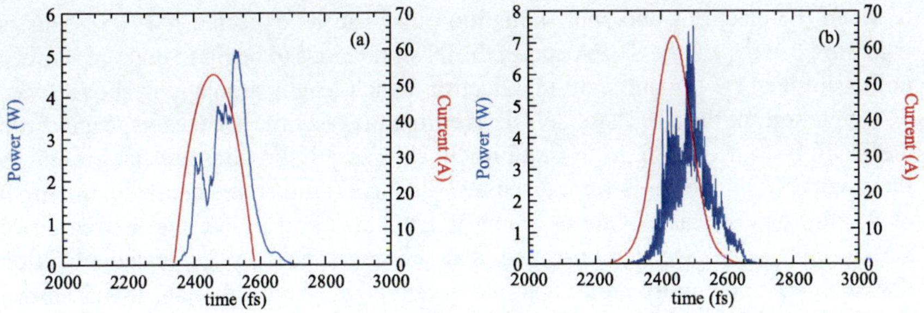

Fig. 6.37 Temporal pulse and current profiles after the first undulator from the SVEA (**a**) and PUFFIN (**b**) [63]

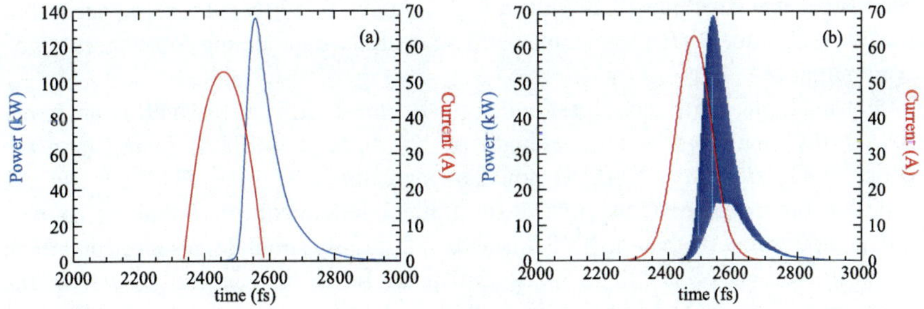

Fig. 6.38 Temporal pulse and current profiles after the sixth undulator from the SVEA (**a**) and PUFFIN (**b**) [63]

fluctuations shown in PUFFIN, while that for PUFFIN is calculated using the instantaneous Poynting flux. We also reiterate that a parabolic current profile was used in the SVEA, while a Gaussian profile was used in PUFFIN, and this is clearly seen in the figure.

The relation between the pulse and current profiles after the first undulator is shown in Fig. 6.37. Note here that these results are for a single noise seed in each code so that the absolute power levels need not agree as closely as the average energy vs z plots would indicate. It is clear from the figure that both codes shown similar levels of slippage of the optical pulse relative to the electron bunch. Similar observations made after the sixth undulator are shown in Fig. 6.38. It should be remarked that such short electron bunch operation produces a single optical pulse rather than the usual sequence of spikes associated with SASE.

The properties of the emission for the longest electron bunch length of 16.8 μm discussed above demonstrate good agreement between the two simulation codes. Here the radiation pulse durations are ~40 wavelengths after the sixth undulator. Going to the shorter electron bunches, this duration can be expected to change in the SVEA and PUFFIN models, and, with the generation of more CSR and a greater stretching of the SVEA approximation, a more complex, dynamic interaction between the electrons and short radiation pulse can be expected [64, 65]. This is where we can expect the SVEA and PUFFIN simulations to begin to diverge, as seen in the shorter 11.2-μm and 5.59-μm electron bunch length simulations above.

Limitations of the SVEA models derive from the fast time scale average and from the inability to treat CSR from first principles. These limitations are not present in PiC codes. The three cases were discussed were shorter than the cooperation length of 21 μm. Hence, each of these bunch lengths are shorter than the separation of SASE spikes, and we demonstrated that the spontaneously generated radiation consists of a single pulse rather than the usual spiky SASE radiation. It was shown that while the SVEA results differ substantially from the PUFFIN results for the two shortest bunch lengths, the two codes are in substantial agreement for the longest bunch which is comparable to, but slightly shorter than, the cooperation length. Since the longest bunch still represents a bunch that is shorter than the SASE spike separation, it is concluded that while a PiC code like PUFFIN is the more general of the two codes, the SVEA model may still have utility in modeling some short bunch configurations.

For example, the bunch duration/length for the LCLS [66], which is an X-ray SASE FEL operating at a wavelength of 1.5 Å, is about 83 fs/25 μm, and the cooperation length is about 20 nm. The agreement between PUFFIN and the SVEA using the SPARC parameters for a bunch length comparable to the cooperation length suggests that the SVEA is able to faithfully simulate rms electron bunch lengths/durations of about 67 as/20 nm in an LCLS-like device. However, the simulation of such short bunches in an LCLS-like device may require the development of a self-consistent model for CSR in the SVEA codes. This may be important because such short bunch operation that results in a single radiation pulse is characterized by near-transform limited pulses, but the computational requirements needed to simulate such an FEL with a PiC code such as PUFFIN may be prohibitively large.

6.9 Code Release (MINERVA)

The three-dimensional, time-dependent simulation code (MINERVA) written to implement the SVEA formulation described in this chapter is available for download from a website hosted at the University of Twente in the Netherlands. The link can be found at

https://gitlab.utwente.nl/tnw/ap/lpno/public-projects/MINERVA/-/releases

In addition, the link to the Optics Propagation Code (see Chap. 10) which can be run in conjunction with MINERVA to simulate FEL oscillators can be downloaded from

https://gitlab.utwente.nl/tnw/ap/lpno/public-projects/Physics-OPC/-/releases

MINERVA is written in generic Fortran 95 and is parallelized using the message passing interface. It has been compiled and run under Windows, Unix, Linux, and the Macintosh OS and runs on laptops, desktops, workstations, and large-scale clusters and supercomputers. Both the source code and selected compiled executables are available for download as well as a user manual which includes examples and the relevant input files.

References

1. H.P. Freund, Nonlinear theory of short wavelength free-electron lasers. Phys. Rev. E **52**, 5401 (1995)
2. S. Reiche, Genesis 1.3: A fully time-dependent FEL simulation code. Nucl. Instrum. Methods Phys. Res. **A429**, 243 (1999)
3. E.L. Saldin, E.A. Schneidmiller, M. Yurkov, FAST: Three-dimensional time-dependent FEL simulation code. Nucl. Instrum. Methods Phys. Res. **A429**, 233 (1999)
4. H.P. Freund, S.G. Biedron, S.V. Milton, Nonlinear harmonic generation in free-electron lasers. IEEE J. Quantum Electron. **36**, 275 (2000)
5. L. Giannessi, In Proceedings of the 2006 FEL Conference, p. MOPPH026 (2006). http://www.jacow.org
6. W. Fawley, *An Informal Manual for GINGER and Its Post-Processor XPLOTGIN* (LBID-2141, CBP Tech Note-104, UC-414, 1995)
7. H.P. Freund, Time-dependent simulation of free-electron laser amplifiers and oscillators. Phys. Rev. ST-AB **8**, 110701 (2005)
8. H.P. Freund, P.J.M. van der Slot, D.L.A.G. Grimminck, I.D. Setya, P. Falgari, Three-dimensional, time-dependent simulation of free-electron lasers with planar, helical, and elliptical undulators. New J. Phys. **19**, 23020 (2017)
9. A. Fisher, P. Musumeci, S.B. van der Geer, Self-consistent numerical approach to track particles in free electron laser interaction with electromagnetic field modes. Phys. Rev. Accel. Beams **23**, 110702 (2020)
10. H.P. Freund, P.J.M. van der Slot, Variable polarization states in free-electron lasers. J. Phys. Commun. **5**, 085011 (2021)
11. L.T. Campbell, B.W.J. McNeil, PUFFIN: A three-dimensional, unaveraged free-electron laser simulation code. Phys. Plasmas **19**, 093119 (2012)
12. C.K. Birdsall, A.B. Langdon, *Plasma Physics Via Computer Simulation* (McGraw-Hill, New York, 1985)
13. N. Piovella, High gain free electron laser amplifiers starting from coherent and incoherent spontaneous emission. Phys. Plasmas **6**, 3358 (1999)
14. E.T. Scharlemann, A.M. Sessler, J.S. Wurtele, Optical guiding in a free-electron laser. Phys. Rev. Lett. **54**, 1925 (1985)
15. G.T. Moore, High-gain and large-diffraction regimes of the free-electron laser. Nucl. Instrum. Methods Phys. Res. **A250**, 381 (1986)
16. J.E. LaSala, D.A.G. Deacon, E.T. Scharlemann, Optical guiding simulations for high-gain short-wavelength free-electron lasers. Nucl. Instrum. Methods Phys. Res. **A250**, 389 (1986)
17. A. Amir, Y. Greenzweig, Three-dimensional free-electron laser gain and evolution of optical modes. Nucl. Instrum. Methods Phys. Res. **A250**, 404 (1986)
18. P. Luchini, S. Solimeno, Optical guiding in a free-electron laser. Nucl. Instrum. Methods Phys. Res. **A250**, 413 (1986)
19. M. Xie, D.A.G. Deacon, Theoretical study of free-electron laser active guiding in the small signal regime. Nucl. Instrum. Methods Phys. Res. **A250**, 426 (1986)

20. J. Gallardo, L.R. Elias, Multimode dynamics in a free-electron laser with energy shift. Nucl. Instrum. Methods Phys. Res. **A250**, 438 (1986)
21. B.D. McVey, Three-dimensional simulations of free-electron laser physics. Nucl. Instrum. Methods Phys. Res. **A250**, 449 (1986)
22. R.H. Pantell, J. Feinstein, Free-electron laser mode propagation at saturation. IEEE J. Quantum Electron. **QE-23**, 1534 (1987)
23. P. Sprangle, A. Ting, C.M. Tang, Analysis of radiation focusing and steering in the free-electron laser by use of a source dependent expansion technique. Phys. Rev. A **36**, 2773 (1987)
24. S.Y. Cai, A. Bhattacharjee, T.C. Marshall, Optical guiding in a Raman free-electron laser. IEEE J. Quantum Electron. **QE-23**, 1651 (1987)
25. P. Luchini, More on optical guiding in a free-electron laser. Nucl. Instr. Meth. **A259**, 150 (1987)
26. R.W. Warren, B.D. McVey, Bending and focusing effects in a free-electron laser oscillator I: Simple models. Nucl. Instrum. Methods Phys. Res. **A259**, 154 (1987)
27. B.D. McVey, R.W. Warren, Bending and focusing effects in a free-electron laser oscillator II: Numerical simulations. Nucl. Instrum. Methods Phys. Res. **A259**, 158 (1987)
28. A. Bhattacharjee, S.Y. Cai, S.P. Chang, J.W. Dodd, T.C. Marshall, Observations of optical guiding in a Raman free-electron laser. Phys. Rev. Lett. **60**, 1254 (1988)
29. G. Bourianoff, B. Moore, M. Rosenbluth, F. Waelbroeck, H. Waelbroeck, H.V. Wong, Adaptive eigenmode expansion for 3-D free-electron laser simulations. Nucl. Instrum. Methods Phys. Res. **A272**, 340 (1988)
30. T.M. Antonsen, B. Levush, Optical guiding in the separable beam limit. Nucl. Instrum. Methods Phys. Res. **A272**, 472 (1988)
31. A. Fruchtman, Optical guiding in a sheet-beam free-electron laser. Phys. Rev. A **37**, 2989 (1988)
32. M.H. Whang, S.P. Kuo, Self-focusing of laser pulses in magnetized relativistic electron beams. Nucl. Instrum. Methods Phys. Res. **A272**, 477 (1988)
33. Y.J. Chen, S. Solimeno, L. Carlomusto, Optical guiding in a free-electron laser with full account of electron wiggling and 3-D propagation. Nucl. Instrum. Methods Phys. Res. **A272**, 490 (1988)
34. J.C. Gallardo, G. Dattoli, A. Renieri, T. Hermsen, Integral equation for the laser field: Multimode description of a free-electron laser oscillator. Nucl. Instrum. Methods Phys. Res. **A272**, 516 (1988)
35. A. Bhowmik, S. Bitterly, R.A. Cover, P. Kennedy, R.H. Labbe, Transverse mode control in high gain free-electron lasers with grazing incidence, unstable ring resonators. Nucl. Instrum. Methods Phys. Res. **A272**, 524 (1988)
36. M. Xie, D.A.G. Deacon, J.M.J. Madey, The guided mode expansion in free-electron lasers. Nucl. Instr. Meth. **A272**, 528 (1988)
37. T.M. Antonsen, G. Laval, Suppression of sidebands by diffraction in a free-electron laser. Phys. Fluids B **1**, 1721 (1989)
38. M. Lontano, A.M. Sergeev, A. Cardinali, Dynamical self-focusing of high-power free-electron laser radiation in a magnetized plasma. Phys. Fluids B **1**, 901 (1989)
39. A. Bhattacharjee, S.Y. Cai, S.P. Chang, J.W. Dodd, A. Fruchtman, T.C. Marshall, Theory and observation of optical guiding in a free-electron laser. Phys. Rev. A **40**, 5081 (1989)
40. N. Metzler, T.M. Antonsen, B. Levush, Nonlinear optical guiding in the separable beam limit. Phys. Fluids B **2**, 1038 (1990)
41. B. Hafizi, A. Ting, P. Sprangle, C.M. Tang, Effect of tapering on optical guiding and sideband growth in a finite-pulse free-electron laser. Nucl. Instrum. Methods Phys. Res. **A296**, 442 (1990)
42. H.P. Freund, C.L. Chang, Effect of the lower beat wave on optical guiding in planar wiggler free-electron lasers. Phys. Rev. A **42**, 6737 (1990)
43. H.P. Freund, T.M. Antonsen Jr., The relationship between optical guiding and the relative phase in free-electron lasers. IEEE J. Quantum Electron. **QE-27**, 2539 (1991)
44. T.J. Orzechowski, E.T. Scharlemann, D.B. Hopkins, Measurement of the phase of the electromagnetic wave in a free-electron laser amplifier. Phys. Rev. A **35**, 2184 (1987)

45. H.P. Freund, Multimode nonlinear analysis of free-electron laser amplifiers in three dimensions. Phys. Rev. A **37**, 3371 (1988)
46. C. Penman, B.W.J. McNeil, Simulation of input electron noise in the free-electron laser. Optics Commun. **90**, 82 (1992)
47. B.W.J. McNeil, M.W. Poole, G.R.M. Robb, Unified model of electron beam shot noise and coherent spontaneous emission in the helical wiggler free-electron laser. Phys. Rev. ST-AB **6**, 070701 (2003)
48. L. Giannessi, *Harmonic Generation and Linewidth Narrowing in Seeded Free-Electron Lasers* (Proceedings of the 26th International Conference on Free-Electron Lasers, 2004), p. 37. www.JACoW.org
49. H.P. Freund, L. Giannessi, W.H. Miner Jr., The effect of shot noise on the startup of the fundamental and harmonics in free-electron lasers. J. Appl. Phys. **104**, 123114 (2008)
50. W.M. Fawley, Algorithm for loading shot noise microbunching in multidimensional free-electron laser simulation codes. Phys. Rev. ST-AB **5**, 070701 (2002)
51. J. Bahrdt, W. Frentrup, A. Gaupp, M. Scheer, W. Gudat, G. Ingold, S. Sasaki, Elliptically polarized insertion devices at BESSY II. Nucl. Instrum. Methods Phys. Res. **A467-468**, 21 (2001)
52. A.A. Lutman et al., Polarization control in an x-ray free-electron laser. Nat. Photon. **10**, 468 (2016)
53. J.R. Henderson, L.T. Campbell, H.P. Freund, B.W.J. McNeil, Modelling elliptically polarized free electron lasers. New J. Phys. **18**, 62003 (2016)
54. M. Xie, *Design Optimization for an X-Ray Free Electron Laser Driven by the SLAC Linac* (Proc. IEEE 1995 Particle Accelerator Conference, Vol. 183, IEEE Cat. No. 95CH35843, 1995)
55. A. Yariv, *Quantum Electronics*, 2nd edn. (Wiley, New York, 1967)
56. P.A. Sprangle, A. Ting, C.M. Tang, Analysis of radiation focusing and steering in the free-electron laser by use of a source-dependent expansion. Phys. Rev. A **36**, 2773 (1987)
57. M. Born, E. Wolf, *Principles of Optics*, 7th edn. (Cambridge University Press, Cambridge, 1999), pp. 25–33
58. J.S. Lee, M. Gensch, K. Hinrichs, W. Seidel, U. Schade, Infrared Phys. & Technol. **51**, 537 (2008)
59. X.J. Wang, T. Watanabe, Y. Shen, R. Li, J.B. Murphy, T. Tsang, H.P. Freund, Efficiency enhancement using electron energy detuning in a laser seeded free electron laser amplifier. Appl. Phys. Lett. **91**, 181115 (2007)
60. X.J. Wang, H.P. Freund, D. Harder, W.H. Miner Jr., J.B. Murphy, H. Qian, Y. Shen, X. Yang, Efficiency and spectrum enhancement in a tapered free electron laser amplifier. Phys. Rev. Lett. **103**, 154801 (2009)
61. D.C. Quimby, S.C. Gottschalk, F.E. James, K.E. Robinson, J.M. Slater, A.S. Valla, Development of a 10-meter, wedged-pole undulator. Nucl. Instrum. Methods Phys. Res. **A285**, 281 (1989)
62. E. Allaria et al., Control of the polarization of a vacuum-ultraviolet, high-gain free-electron laser. Phys. Rev. X **4**, 041040 (2014)
63. L.T. Campbell, H.P. Freund, J.R. Henderson, B.W.J. McNeil, P. Traczykowski, P.J.M. van der Slot, Analysis of ultra-short bunches in free-electron lasers. New J. Phys. **22**, 073031 (2020)
64. B.W.J. McNeil, G.R.M. Robb, D. Jaroszynski, Self-amplification of coherent spontaneous emission in the free electron laser. Optics Comm. **165**, 65 (1999)
65. B.W.J. McNeil, G.R.M. Robb, D. Jaroszynski, SACSE in a FEL amplifier with energy spread. Nucl. Instrum. Methods Phys. Res. **A445**, 72 (2000)
66. P. Emma et al., First lasing and operation of an Angstrom-wavelength free-electron laser. Nat. Photon. **4**, 641 (2010)

Chapter 7
Sideband Instabilities

The growth of sidebands of the primary signal in free-electron lasers can occur after the bulk of the electron beam becomes trapped in the ponderomotive potential formed by the beating of the wiggler and radiation fields [1–31]. Trapped electrons execute an oscillatory bounce motion in the ponderomotive well, and the waves formed by the beating of this oscillation with the primary signal are referred to as the sidebands. The difficulties imposed by the growth of sidebands are that they can compete with and drain energy form the primary signal as well as considerably broaden the output spectrum. Sideband control, therefore, is an important consideration for free-electron laser configurations that operate in the trapped-particle regime. Examples of such systems include (1) tapered wiggler configurations designed to trap the beam at an early stage of the interaction and then extract a great deal more energy from the beam over an extended interaction length and (2) oscillators which run at sufficiently high power over an extended pulse time that the electron beam becomes trapped upon entry to the wiggler.

As a result, a great deal of effort has been expended on techniques of sideband suppression. One method of sideband suppression was employed in a free-electron laser oscillator at Columbia University [12, 14]. This experiment operated at a 2-mm wavelength in which the dispersion due to the waveguide significantly affected the resonance condition. As a consequence, it was found to be possible to shift the sideband frequencies out of resonance with the beam by the proper choice of the size of the waveguide. In order to understand this, observe that the phase of the sideband varies as [13, 14]

$$[(k + \Delta k) + k_w]z - (\omega + \Delta\omega)t = \kappa z, \tag{7.1}$$

where $(\Delta\omega, \Delta k)$ are the shifts in the angular frequency and the wavenumber due to the sideband and κ is the bounce period of the electrons in the ponderomotive potential which is given in Eq. (4.14) in the idealized one-dimensional limit. Noting that $z \approx v_{\parallel}t$ and $\omega \approx (k + k_w)v_{\parallel}$ for a bulk streaming velocity $v_{\parallel}$, it is clear that the shift in the angular frequency due to the sideband is given by

H. P. Freund, T. M. Antonsen, Jr., *Principles of Free Electron Lasers*,
https://doi.org/10.1007/978-3-031-40945-5_7

$$\Delta\omega = \frac{\kappa \upsilon_{\|}}{1 - \upsilon_{\|}/\upsilon_g}, \tag{7.2}$$

where $\upsilon_{gr} = \Delta\omega/\Delta k$ is the group velocity of the wave. The sideband can be suppressed, therefore, when $\Delta\omega = 0$ which requires that the group velocity of the wave matches the streaming velocity of the electron beam. The dispersion equation in a waveguide is given by $\omega^2 = c^2k^2 + \omega_{co}^2$, where ω_{co} is the cutoff frequency. Therefore, the group velocity is given by

$$\upsilon_{gr} = c\sqrt{1 - \frac{\omega_{co}^2}{\omega^2}}. \tag{7.3}$$

Equating this with the axial velocity of an electron in a helical wiggler, we find that the sideband can be suppressed if

$$\omega = \frac{\omega_{co}}{\sqrt{1 + a_w^2}}, \tag{7.4}$$

where $a_w = eA_w/m_ec^2$.

In addition, work on an infrared free-electron laser oscillator at Los Alamos National Laboratory [16, 17, 23, 24] indicate that it is also possible to suppress sidebands by (1) using a Littrow grating to deflect the sidebands out of the optical cavity or (2) changing the cavity length. Hence, it appears that sideband growth due to trapped electrons imposes no essential difficulties for the generation of coherent radiation in free-electron lasers at the present time.

In order to understand the basic nature of the sideband mechanism, we consider the process within the framework of an idealized one-dimensional model [4]. The purpose of the analysis is to examine the gain of the sideband modes that result from the trapped electron motion under a variety of conditions including the effect of a tapered wiggler field and the inclusion of a fluctuating space-charge field. More detailed analyses of the sideband instabilities in free-electron lasers include discussions of the linear kinetic theory of the instability in one [2, 7, 9–11, 30, 31] and two dimensions [15, 21, 22, 26, 27, 29] for a variety of trapped particle distribution functions, as well as nonlinear analyses and simulations of the interaction [1, 3, 6, 19, 28].

7.1 The General Formulation

The physical model under consideration consists of a relativistic electron beam propagating through an idealized one-dimensional helical wiggler represented in Eq. (2.5). The scattered electromagnetic waves are represented as transversely uniform circularly polarized waves in which

$$\delta\mathbb{A}(z,t) = \delta A(z)\left[\hat{e}_x \cos\alpha(z,t) - \hat{e}_y \sin\alpha(z,t)\right] \tag{7.5}$$

represents the large-amplitude primary signal and

$$\delta\mathbb{A}_s(z,t) = \delta A(z)\left[\hat{e}_x \cos\alpha(z,t) - \hat{e}_y \sin\alpha_s(z,t)\right] \tag{7.6}$$

represents the sideband. The phase of each of these waves is represented in the form

$$\alpha(z,t) = \int_0^z dz' k(z') - \omega t, \tag{7.7}$$

and

$$\alpha_s(z,t) = \int_0^z dz' k_s(z') - \omega_s t + \theta_s, \tag{7.8}$$

where (ω,k) and (ω_s,k_s) are the angular frequency and wavenumber of the primary and sideband signals, respectively, and θ_s defines a phase shift of the sideband with respect to the primary. In addition, the fluctuating space-charge fields associated with each of these signals are

$$\delta\Phi(z,t) = \delta\Phi(z) \cos\phi(z,t), \tag{7.9}$$

for the primary signal and

$$\delta\Phi_s(z,t) = \delta\Phi_s(z) \cos\phi_s(z,t), \tag{7.10}$$

for the sideband. The phases for the space-charge waves are defined as

$$\phi(z,t) = \int_0^z dz' k_{sc}(z') - \omega t + \delta, \tag{7.11}$$

and

$$\phi_s(z,t) = \int_0^z dz' k_{sb}(z') - \omega_s t + \delta_s, \tag{7.12}$$

where k_{sc} and k_{sb} denote the wavenumbers and δ and δ_s represent the phase shifts of these waves with respect to the primary. Observe that the amplitudes and wavenumbers of each of these modes are implicitly assumed to vary slowly in comparison with the wavelengths.

Since $\delta\mathbf{A}$ and $\delta\Phi$ represent the primary signal which may either have grown from noise or a low-intensity drive signal or from a large-amplitude external coherent source, we assume that $|\delta\mathbf{A}_s| \ll |\delta\mathbf{A}| \ll |\mathbf{A}_w|$ and $|\delta\Phi_s| \ll |\delta\Phi|$. In addition, since the bounce frequency is typically much less than the resonant frequency of the interaction, we also assume that $|\omega_s - \omega| \ll \omega$.

The fields satisfy the Maxwell-Poisson equations (Eqs. 5.20, 5.21, and 5.22). The analysis of the small-signal gain follows in essentially the same manner as that described in Chap. 5, except that the quasi-static assumption for the multi-wave system may be stated in the form that two electrons which enter the interaction region within an integral multiple of the beat period T [$= 2\pi/|\omega - \omega_s|$] execute identical trajectories. Hence, the dynamical equations that govern the evolution of the amplitudes and phases of the primary and sideband signals are

$$\left[\frac{d^2}{dz^2} + \left(\frac{\omega^2}{c^2} - k^2\right)\right]\delta a \cong -\frac{\omega_b^2}{c^2} \upsilon_{z0} a_w \left\langle \frac{\cos\psi}{\gamma\upsilon_z} \right\rangle, \tag{7.13}$$

$$\left[\frac{d^2}{dz^2} + \left(\frac{\omega_s^2}{c^2} - k_s^2\right)\right]\delta a_s \cong -\frac{\omega_b^2}{c^2} \upsilon_{z0} a_w \left\langle \frac{\cos\psi_s}{\gamma\upsilon_z} \right\rangle, \tag{7.14}$$

$$2k^{1/2}\frac{d}{dz}\left(k^{1/2}\delta a\right) \cong \frac{\omega_b^2}{c^2} \upsilon_{z0} a_w \left\langle \frac{\sin\psi}{\gamma\upsilon_z} \right\rangle, \tag{7.15}$$

$$2k_s^{1/2}\frac{d}{dz}\left(k_s^{1/2}\delta a_s\right) \cong \frac{\omega_b^2}{c^2} \upsilon_{z0} a_w \left\langle \frac{\sin\psi_s}{\gamma\upsilon_z} \right\rangle, \tag{7.16}$$

for the electromagnetic waves and

$$(k + k_w)\begin{pmatrix} \delta\varphi_1 \\ \delta\varphi_2 \end{pmatrix} \cong -2\frac{\omega_b^2}{c^2}\frac{\upsilon_{z0}}{\omega}\begin{pmatrix} \langle\cos\psi\rangle \\ \langle\sin\psi\rangle \end{pmatrix}, \tag{7.17}$$

$$(k_s + k_w)\begin{pmatrix} \delta\varphi_{s1} \\ \delta\varphi_{s2} \end{pmatrix} \cong -2\frac{\omega_b^2}{c^2}\frac{\upsilon_{z0}}{\omega_s}\begin{pmatrix} \langle\cos\psi_s\rangle \\ \langle\sin\psi_s\rangle \end{pmatrix}, \tag{7.18}$$

for the space-charge modes, where $\delta a = e\delta A/m_ec^2$, $\delta\varphi = e\delta\Phi/m_ec^2$, $\delta\varphi_1 = \delta\varphi\cos\delta$, $\delta\varphi_2 = \delta\varphi\sin\delta$, $\delta\varphi_{s1} = \delta\varphi_s\cos\delta_s$, and $\delta\varphi_{s2} = \delta\varphi_s\sin\delta_s$. In addition, the ponderomotive phases are defined as

$$\psi = \psi_0 + \int_0^z dz'\left(k + k_w - \frac{\omega}{\upsilon_z}\right), \tag{7.19}$$

and

$$\psi = \frac{\omega_s}{\omega}\psi_0 + \int_0^z dz'\left(k_s + k_w - \frac{\omega_s}{v_z}\right) + \theta_s, \tag{7.20}$$

where $\psi_0 = -\omega t_0$. The source current and charge density used to derive the dynamical equations are given in Eqs. (5.23) and (5.24), and ω_b denotes the plasma frequency associated with the ambient beam density. Finally, under the quasi-static assumption, Maxwell's equations have been averaged over the beat period to obtain the dynamical equations and the averages that appear are defined as

$$\langle(\cdots)\rangle \equiv \frac{1}{2\pi N}\int_{-N\pi}^{N\pi} d\psi_0\sigma(\psi_0)(\cdots), \tag{7.21}$$

where N [$= \omega/|\omega - \omega_s|$] describes the ratio of the beat period to the period of the primary signal and is assumed to be an integer which is much larger than unity and $\sigma(\psi_0)$ denotes the distribution in the initial phases.

Observe that the time average has been performed over a much longer period than is necessary for a single-frequency amplifier model. The dynamical Eqs. (7.13), (7.14), (7.15), (7.16), (7.17), and (7.18) can be integrated numerically to obtain the complete nonlinear evolution of the sideband instability in a manner analogous to that employed in Chap. 5 for the single-frequency amplifier model and can be extended to include multiple sidebands as well which are separated by the fixed angular frequency $\Delta\omega = |\omega - \omega_s|$. The distinction, however, arises from the time average that is performed, essentially, over a large number of periods of the primary signal. This means that a much larger number of electrons must be included in the description, as the formulation must include N *beamlets* of electrons (where a *beamlet* is defined as the group of electrons which enter the interaction region within one period of the primary signal).

The dynamical Eqs. (7.13), (7.14), (7.15), and (7.16) simplify under the neglect of second-order derivatives of the amplitude and phase. For the purposes of the present analysis of the small-signal gain, therefore, we obtain the more tractable set of equations for the electromagnetic fields

$$\left(\frac{\omega^2}{c^2} - k^2\right)\delta a \cong -\frac{\omega_b^2}{c^2}\frac{v_w}{c}\langle\cos\psi\rangle, \tag{7.22}$$

$$\left(\frac{\omega_s^2}{c^2} - k_s^2\right)\delta a_s \cong -\frac{\omega_b^2}{c^2}\frac{v_w}{c}\langle\cos\psi_s\rangle, \tag{7.23}$$

$$2k\frac{d}{dz}\delta a \cong \frac{\omega_b^2}{c^2}\frac{v_w}{c}\langle \sin\psi\rangle, \tag{7.24}$$

and

$$2k_s\frac{d}{dz}\delta a_s \cong \frac{\omega_b^2}{c^2}\frac{v_w}{c}\langle \sin\psi_s\rangle. \tag{7.25}$$

The phases which appear in the dynamical equations are invariant under the transformation $\psi \to \psi + 2\pi$, $\theta_s \to \theta_s + 2\pi(\omega - \omega s)/\omega$. As a consequence, $\psi(\psi_0 + 2\pi, \theta + 2\pi(\omega - \omega s)/\omega, z) = \psi(\psi_0, \theta, z) + 2\pi$. This symmetry is also possessed by ψ_s. Hence, if it is also required that $\sigma(\psi_0 + 2\pi) = \sigma(\psi_0)$, then the averages in the dynamical equations can be decomposed into

$$\langle\langle\cdots\rangle\rangle \equiv \frac{1}{(2\pi)^2}\int_0^{2\pi} d\theta \int_{-\pi}^{\pi} d\psi_0\sigma(\psi_0)(\cdots), \tag{7.26}$$

for sufficiently large N.

7.2 Trapped Electron Trajectories

Since the vector and scalar potentials are independent of the transverse coordinates, these coordinates are cyclic, and the corresponding canonical momenta are constants of the motion. In view of this, the equation for the axial momentum takes the form

$$\begin{aligned}\frac{d}{dz}p_z = &-m_e c\left[(\delta\varphi_1 \sin\psi - \delta\varphi_2 \cos\psi)\frac{d\psi}{dz} + (\delta\varphi_{s1}\sin\psi_s - \delta\varphi_{s2}\cos\psi_s)\frac{d\psi_s}{dz}\right]\\ &-\frac{m_e c}{2\gamma}\frac{d\mu^2}{dz} - \frac{m_e c}{\gamma}a_w[(k+k_w)\delta a \sin\psi + (k_s + k_w)\delta a_s \sin\psi_s],\end{aligned} \tag{7.27}$$

where $\mu = 1 + a_w^2$ and terms of second order in the fluctuating fields have been neglected. The variation in the total electron energy is given by

$$\begin{aligned}\frac{d\gamma}{dt} = &-v_z\left[(\delta\varphi_1 \sin\psi - \delta\varphi_2 \cos\psi)\frac{d\psi}{dz} + (\delta\varphi_{s1}\sin\psi_s - \delta\varphi_{s2}\cos\psi_s)\frac{d\psi_s}{dz}\right]\\ &-\frac{a_w}{\gamma}\left[\omega\delta a \sin\psi + \omega_s \delta a_s \sin\psi_s\right].\end{aligned} \tag{7.28}$$

The combination of these equations yields the equation for the axial velocity:

$$\frac{d}{dt}v_z = -\frac{c^2\mu^2}{\gamma^3}[(k+k_w)(\delta\varphi_1 \sin\psi - \delta\varphi_2 \cos\psi) + (k_s+k_w)(\delta\varphi_{s1} \sin\psi_s - \delta\varphi_{s2} \cos\psi_s)]$$
$$-\frac{c^2}{2\gamma^2}\frac{d\mu^2}{dz} - \frac{c^2}{\gamma^2}a_w\left[\left(k+k_w-\omega\frac{v_z}{c^2}\right)\delta a \sin\psi + \left(k_s+k_w-\omega_s\frac{v_z}{c^2}\right)\delta a_s \sin\psi_s\right], \tag{7.29}$$

where the terms of order $(\omega - \omega_s)/\omega$ have been ignored.

Substitution of the axial position for the time as the independent variable and the ponderomotive phase for the axial velocity yields a nonlinear pendulum-like equation of the form

$$\frac{d^2}{dz^2}\psi \cong \frac{d}{dz}(k+k_w) - \frac{c^2(k+k_w)}{2\gamma_{ph}^2 v_{ph}^2}\frac{d\mu^2}{dz} - \frac{\mu^2c^2(k+k_w)^2}{\gamma_{ph}^4 v_{ph}^2}a_w\delta a\left(\sin\psi + \varepsilon\frac{\omega_s}{\omega}\sin\psi_s\right)$$
$$+\frac{2\omega_b^2\mu^2 v_{z0}}{\gamma_{ph}^3 v_{ph}^3}\left[\left(\sin\psi\langle\cos\psi\rangle + \frac{\omega_s}{\omega}\sin\psi_s\langle\cos\psi_s\rangle\right)\right.$$
$$\left.-\left(\cos\psi\langle\sin\psi\rangle + \frac{\omega_s}{\omega}\cos\psi_s\langle\sin\psi_s\rangle\right)\right], \tag{7.30}$$

where $\varepsilon = \delta a_s/\delta a$, $v_{ph} = \omega/(k + k_w)$ is the phase velocity of the ponderomotive wave and $\gamma_{ph} = \mu(1 - v_{ph}^2/c^2)^{-1/2}$ is the relativistic factor corresponding to the ponderomotive frame. The inclusion of space-charge effects is made for the case of a diffuse electron beam in which $\omega_b \ll \omega$ and ω_s, so that we may write $\omega \approx ck$ and $\omega_s \approx ck_s$. Under the additional assumptions that $k_w \ll k$ and k_s, the pendulum equation reduces to

$$\frac{d^2}{dz^2}\psi \cong -\frac{\kappa^2}{\cos\psi_{res}}\left(\sin\psi - \sin\psi_{res} + \varepsilon\frac{\omega_s}{\omega}\sin\psi_s\right)$$
$$+\delta\kappa^2\left[\left(\sin\psi\langle\cos\psi\rangle + \frac{\omega_s}{\omega}\sin\psi_s\langle\cos\psi_s\rangle\right)\right. \tag{7.31}$$
$$\left.-\left(\cos\psi\langle\sin\psi\rangle + \frac{\omega_s}{\omega}\cos\psi_s\langle\sin\psi_s\rangle\right)\right],$$

where $\kappa^2 = 4k_w^2 a_w\delta a \cos\psi_{res}/\mu^2$, $\delta\kappa^2 = 2\mu^2\omega_b^2/\gamma^3c^2$, and the resonant phase is determined by

$$\sin\psi_{res} \equiv -\frac{1}{2a_w\delta a}\left[\frac{a_w}{k_w}\frac{da_w}{dz} - \frac{\mu^2}{2k_w^2}\frac{dk_w}{dz}\right], \tag{7.32}$$

and the phase of the sideband can be represented in the form

$$\psi_s \cong \frac{\omega_s}{\omega}\psi + \Delta kz + \theta_s, \tag{7.33}$$

for $\Delta k = k_w(\omega - \omega_s)/\omega$.

The solution to the pendulum Eq. (6.31) is found by perturbation about the resonant phase $\psi \approx \psi_{res} + \delta\psi$, where $\delta\psi \ll \pi$. As a consequence,

$$\begin{aligned}\frac{d^2}{dz^2}\psi \cong & -\kappa^2\delta\psi + \frac{\kappa^2}{2}\delta\psi^2\tan\psi_{res} + \frac{\kappa^2}{6}\delta\psi^3 \\ & - \frac{\varepsilon\kappa^2}{\cos\psi_{res}}\sin\left(\frac{\omega_s}{\omega}\psi_{res} + \Delta kz + \theta_s\right) + \delta\kappa^2[\delta\psi - \langle\delta\psi\rangle].\end{aligned} \tag{7.34}$$

Observe that under the linearization of this equation $2\pi/\kappa$ is the bounce period in the ponderomotive well, while $2\pi/\delta\kappa$ is the oscillation (i.e., plasma) period in the space-charge wave. The perturbed phase is decomposed into $\delta\psi = \delta\psi^{(0)} + \delta\psi^{(1)}$ to denote the zeroth and first-order effects due to the sideband. As a result

$$\delta\psi^{(0)} \cong \alpha\cos\kappa_a z + \frac{\alpha^2}{4}\tan\psi_{res}\left[1 - \frac{1}{3}\cos 2\kappa_a z\right], \tag{7.35}$$

for small deviations from the resonant phase, where a is a constant fixed by the initial condition $\delta\psi^{(0)}(z=0) = \psi_0$ which implies that $\psi_0 - \psi_{res} = \alpha + \alpha^2\tan\psi_{res}/6$ for $|\alpha| \ll 1$, and

$$\kappa_a \equiv \kappa\left[1 - \frac{\alpha}{16}\left(1 + \frac{5}{3}\tan^2\psi_{res}\right)\right], \tag{7.36}$$

is the anharmonic bounce period in the ponderomotive potential. The first-order correction to this motion due to the sideband signal is given by

$$\begin{aligned}\delta\psi^{(1)} \cong & \frac{\varepsilon}{2\cos\psi_{res}}\frac{\kappa}{\Delta\kappa_\pm}\left[1 \pm \frac{\alpha^2\kappa}{16\Delta\kappa_\pm}\left(1 + \frac{5}{3}\tan^2\psi_{res}\right)\right] \\ & \times\left\{\left[1 \pm \frac{\alpha^2\kappa}{16\Delta\kappa_\pm}\left(1 + \frac{5}{3}\tan^2\psi_{res}\right)\right][\cos(\Delta\kappa_\pm z + \phi_{res}) - \cos\phi_{res}]\sin\kappa z\right. \\ & \left.\mp[\sin(\Delta\kappa_\pm z + \phi_{res}) - \sin\phi_{res}]\cos\kappa z \pm \frac{\alpha^2\kappa}{8\Delta\kappa_\pm}\kappa z\left(1 + \frac{5}{3}\tan^2\psi_{res}\right)\sin\phi_{res}\sin\kappa z\right\} \\ & + \frac{\delta\kappa^2}{2\kappa^2}[\alpha - \langle\alpha\rangle]\kappa z\sin\kappa z,\end{aligned} \tag{7.37}$$

where $\phi_{res} = \omega_s\psi_{res}/\omega + \theta_s$, $\Delta\kappa_\pm = \Delta k \pm \kappa$, subject to the initial conditions $\delta\psi^{(1)}(z=0) = 0$ and $d\delta\psi^{(1)}(z=0)/dz = 0$. It is important to bear in mind that this solution requires that $\alpha^2 < \Delta\kappa_\pm/\kappa$, which implies that all electrons are deeply trapped with phases close to the resonant phase.

7.3 The Small-Signal Gain

The gain per pass through a system of length L for the primary and sideband waves is found by integration of Eqs. (7.24) and (7.25) to be

$$G(L) \cong \frac{\omega_b^2}{2\omega\gamma_0 c}\frac{a_w}{\delta a}\int_0^L dz\langle \sin\psi\rangle, \tag{7.38}$$

and

$$G_s(L) \cong \frac{\omega_b^2}{2\omega_s\gamma_0 c}\frac{a_w}{\delta a_s}\int_0^L dz\langle \sin\psi_s\rangle, \tag{7.39}$$

where γ_0 denotes the bulk energy of the beam, and we have assumed that $\omega \approx ck$ and $\omega_s \approx ck_s$.

In the case of the gain of the primary wave, we note that for the lowest nontrivial order, the phase average is

$$\langle \sin\psi\rangle \cong \sin\psi_{res}\left[1 - \frac{1}{2}\left\langle \delta\psi^{(0)^2}\right\rangle - \left\langle \delta\psi^{(0)}\delta\psi^{(1)}\right\rangle\right]. \tag{7.40}$$

Because of the average over the relative phase between the primary and the sideband signals, only those components of $\delta\psi^{(1)}$ due to the space-charge potential can contribute to the gain to this order. In particular, the coupling between the primary and sideband waves cannot affect the gain of the primary to first order in ε. In addition, the space-charge effect is manifested by means of a beating between the bounce motion of the electrons in the ponderomotive potential and the space-charge wave itself. In the evaluation of these averages, the initial phase distribution is assumed to be of the form

$$\sigma(\psi_0) = \frac{\pi}{\Delta}H(\psi_0 + \Delta - \psi_{res})H(\psi_{res} + \Delta - \psi_0), \tag{7.41}$$

in the range $-\pi \le \psi_0 \le \pi$, where $\Delta \ll 1$ and H denotes the Heaviside function. This distribution describes a tightly bunched, flat-topped beam about the resonant phase, and is depicted schematically in Fig. 7.1. As a consequence, the gain associated with this phase space distribution is

$$G(L) \cong \frac{\omega_b^2 L}{2\omega\gamma_0 c}\frac{a_w}{\delta a}\sin\psi_{res}\left[1 + \frac{\Delta^2}{12}\left(\frac{\sin\kappa L}{\kappa L} + \frac{2\sin 2\kappa L}{3\kappa L}\right) + \frac{\delta\kappa^2}{2\kappa^2}\left(\cos 2\kappa L - \frac{\sin 2\kappa L}{2\kappa L}\right)\right]. \tag{7.42}$$

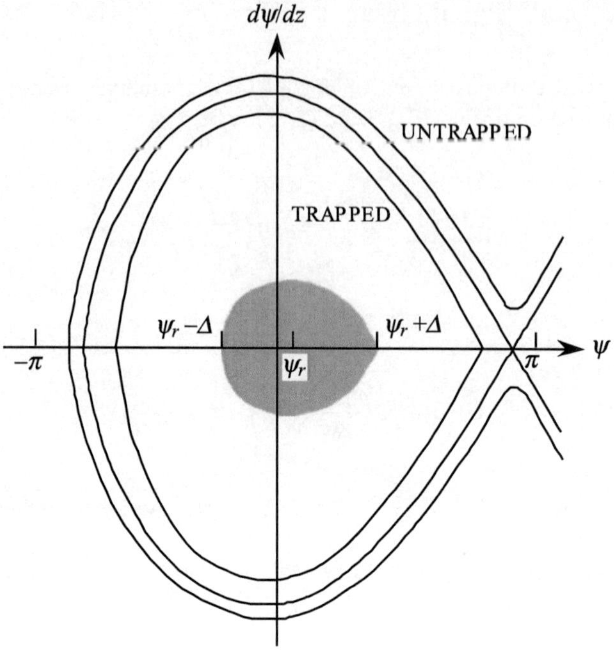

Fig. 7.1 Schematic representation of the phase space distribution of the electron beam about the resonant phase in a tapered wiggler

Since the sideband is not expected to grow prior to the trapping of the beam, the phase space distribution chosen reflects a beam for which the primary signal should be stable in the absence of a tapered wiggler, and this is reflected in the expression for the gain which vanishes when $\sin \psi_{res} = 0$. In addition, the space-charge contribution vanishes in the limit in which $\Delta \rightarrow 0$. This conclusion is quite general and is not dependent upon the particular choice of the phase space distribution. Indeed, it can be shown from Eq. (7.37) that the contribution from the space-charge potential vanishes when the spread in ψ_0 vanishes.

In the case of the sideband signal, the phase average is

$$\langle \sin \psi_s \rangle \cong \frac{\omega_s}{\omega} \left\langle \delta\psi_s^{(1)} \cos \delta\psi_s^{(0)} \right\rangle, \tag{7.43}$$

and the space-charge effect does not contribute to the interaction. Gain, therefore, arises because the presence of the sideband induces an oscillation in the electron bounce motion, and is of the form

$$G_s(L) \cong \frac{\omega_b^2 \kappa L^2}{2\omega_s \gamma_0 c \cos\psi_{res}} \frac{a_w}{\delta a} \times \left[1 \pm \frac{\kappa \Delta^2}{48 \Delta\kappa_\pm}\left(1 + \frac{5}{3}\tan^2\psi_{res}\right)\left(1 - \frac{\Delta\kappa_\pm L}{2}\frac{d}{d\Theta_\pm}\right)\right]\left(\frac{\sin\Theta_\pm}{\Theta_\pm}\right)^2, \tag{7.44}$$

where $\Theta_\pm = \Delta K_\pm L/2$. Observe that it is the lower sideband that is excited.

The preceding analysis applies principally to the case of amplifier configurations. However, it is the relative gain $R = G_s(L)/G(L)$ which determines whether the sideband will grow from noise in an oscillator configuration

$$R \cong \mp \frac{\omega \kappa L}{4\omega_s \sin 2\psi_{res}} \left\{1 \pm \frac{\Delta^2}{24}\left[\frac{\kappa}{2\Delta\kappa_\pm}\left(1 + \frac{5}{3}\tan^2\psi_{res}\right) \mp \frac{\delta\kappa^2}{\kappa^2}\left(\cos 2\kappa L - \frac{\sin 2\kappa L}{2\kappa L}\right)\right]\right\}\left(\frac{\sin\Theta_\pm}{\Theta_\pm}\right)^2, \tag{7.45}$$

If the relative gain is greater than unity for nearly equal loss rates, then the sideband will grow.

References

1. T.J.T. Kwan, Trapped particle dynamics and efficiency optimization in free-electron lasers. Phys. Fluids **23**, 1857 (1980)
2. N.M. Kroll, M.N. Rosenbluth, Sideband instabilities in trapped particle free-electron lasers, in *Physics of Quantum Electronics: Free-Electron Generators of Coherent Radiation*, ed. by S.F. Jacobs, H.S. Pilloff, M. Sargent, M.O. Scully, R. Spitzer, vol. 7, (Addison-Wesley, Reading, 1980), p. 147
3. A.T. Lin, Saturation of sideband instabilities in a free-electron laser, in *Physics of Quantum Electronics: Free-Electron Generators of Coherent Radiation*, ed. by S.F. Jacobs, H.S. Pilloff, M. Sargent, M.O. Scully, R. Spitzer, vol. 9, (Addison-Wesley, Reading, 1981), p. 867
4. H.P. Freund, P. Sprangle, C.M. Tang, Effect of fluctuating space-charge fields on sideband instabilities in free-electron lasers. Phys. Rev. A **25**, 3121 (1982)
5. C.M. Tang, P. Sprangle, Semi-analytic formulation of the two-dimensional pulse propagation in the free-electron laser oscillator, in *Free-Electron Generators of Coherent Radiation*, ed. by C.A. Brau, S.F. Jacobs, M.O. Scully, (Proc. SPIE 453, 1984), pp. 11–24
6. J.C. Goldstein, Evolution of long pulses in a tapered wiggler free-electron laser, in *Free-Electron Generators of Coherent Radiation*, ed. by C.A. Brau, S.F. Jacobs, M.O. Scully, vol. 453, (SPIE, Bellingham, 1984), p. 2
7. R.C. Davidson, W.A. McMullin, Detrapping stochastic particle instability for electron motion in combined longitudinal wiggler and radiation fields. Phys. Rev. A **29**, 791 (1984)
8. D.C. Quimby, J.M. Slater, J.P. Wilcoxon, Sideband suppression in free-electron lasers with multiple synchrotron periods. IEEE J. Quantum Electron. **QE-21**, 979 (1985)
9. R.C. Davidson, Kinetic description of the sideband instability in a helical-wiggler free-electron laser. Phys. Fluids **29**, 2689 (1986)

10. W.B. Colson, The trapped-particle instability in free-electron laser oscillators and amplifiers. Nucl. Instrum. Methods Phys. Res. **A250**, 168 (1986)
11. R.C. Davidson, J.S. Wurtele, Single-particle analysis of the free-electron laser sideband instability for primary electromagnetic waves with constant phase and slowly varying phase. Phys. Fluids **30**, 557 (1987)
12. F.G. Yee, J. Masud, T.C. Marshall, S.P. Schlesinger, Power and sideband studies of a Raman free-electron laser. Nucl. Instrum. Methods Phys. Res. **A259**, 104 (1987)
13. S.S. Yu, W.M. Sharp, W.M. Fawley, E.T. Scharlemann, A.M. Sessler, E.J. Sternbach, Waveguide suppression of the free-electron laser sideband instability. Nucl. Instrum. Methods Phys. Res. **A259**, 219 (1987)
14. J. Masud, T.C. Marshall, S.P. Schlesinger, F.G. Yee, W.M. Fawley, E.T. Scharlemann, S.S. Yu, A.M. Sessler, E.J. Sternbach, Sideband control in a millimeter-wave free-electron laser. Phys. Rev. Lett. **58**, 763 (1987)
15. S. Riyopoulos, C.M. Tang, The free-electron laser sideband instability reconsidered. Nucl. Instrum. Methods Phys. Res. **A259**, 226 (1987)
16. J.C. Goldstein, B.E. Newnam, R.W. Warren, Sideband suppression by an intracavity optical filter in the Los Alamos free-electron laser oscillator. Nucl. Instr. Meth. **A272**, 150 (1988)
17. R.W. Warren, J.C. Goldstein, The generation and suppression of synchrotron sidebands. Nucl. Instrum. Methods Phys. Res. **A272**, 155 (1988)
18. W.M. Sharp, S.S. Yu, Validity of 1-D free-electron laser sideband models. Nucl. Instrum. Methods Phys. Res. **A272**, 397 (1988)
19. H. Takeda, Analysis of sideband spectra from untapered undulator using analytical techniques. Nucl. Instrum. Methods Phys. Res. **A272**, 404 (1988)
20. P.J. Channell, Spectral evolution equation for the sideband instability in a free-electron laser oscillator. Nucl. Instrum. Methods Phys. Res. **A272**, 421 (1988)
21. S. Riyopoulos, C.M. Tang, The structure of the sideband spectrum in the free-electron laser. Phys. Fluids **31**, 1708 (1988)
22. B. Hafizi, A. Ting, P. Sprangle, C.M. Tang, Development of sidebands in tapered and untapered free-electron lasers. Phys. Rev. A **38**, 197 (1988)
23. J.E. Sollid, D.W. Feldman, R.W. Warren, H. Takeda, S.J. Gitomer, W.J. Johnson, W.E. Stein, Sideband suppression in the Los Alamos free-electron laser using a Littrow grating. Nucl. Instrum. Methods Phys. Res. **A285**, 147 (1989)
24. J.E. Sollid, D.W. Feldman, R.W. Warren, Sideband suppression for free-electron lasers. Nucl. Instr. Meth. **A285**, 153 (1989)
25. A. Bhattacharjee, S.Y. Cai, S.P. Chang, J.W. Dodd, T.C. Marshall, Sideband instabilities and optical guiding in a free-electron laser: Experiment and theory. Nucl. Instrum. Methods Phys. Res. **A285**, 158 (1989)
26. S. Riyopoulos, Instability mechanisms in storage ring free-electron oscillators. Nucl. Instrum. Methods Phys. Res. **A296**, 485 (1990)
27. W.M. Sharp, S.S. Yu, Two-dimensional Vlasov treatment of free-electron laser sidebands. Phys. Fluids B **2**, 581 (1990)
28. Y. Kishimoto, H. Oda, M. Shiho, Parasitic wave excitation by multimode coupling in a Raman-regime free-electron laser. Phys. Rev. Lett. **65**, 851 (1990)
29. E.J. Sternbach, Coupled-wave theory of free-electron laser sidebands in a waveguide. IEEE Trans. Plasma Sci **PS-18**, 460 (1990)
30. M.N. Rosenbluth, H.V. Wong, B.N. Moore, Sideband instabilities in free-electron lasers. Phys. Fluids B **2**, 1635 (1990)
31. J.N. Elgin, Analysis of the sideband instability in the free-electron laser. Phys. Rev. A **43**, 2514 (1991)

Chapter 8
Coherent Harmonic Radiation

The electron beam energy required for operation at a given frequency varies inversely as the square root of the harmonic number. For this reason, the mechanisms for the generation of harmonic radiation in free-electron lasers are of interest for many applications in which it is desired to reduce the beam energy requirement in order to generate a given wavelength. Harmonic generation in free-electron lasers occurs by both linear [1–30] and nonlinear mechanisms [31, 32]. The principal difficulty with linear harmonic generation, however, is that the beam quality requirement associated with the growth rate of the linear instability increases with the harmonic number. In general, the free-electron laser resonance condition at the hth harmonic is $\omega \approx (k + hk_w)v_{\|}$; hence, thermal effects have a significant and deleterious impact on the interaction whenever $\Delta v_{\|}/v_{\|} \approx \text{Im}\, k/(\text{Re}\, k + hk_w)$. Nonlinear harmonic generation is driven parasitically off of a high power fundamental and is less sensitive to the electron beam quality than the linear instability.

Harmonics are not commonly seeded in free-electron lasers. The growth of harmonic radiation starts by a combination of shot noise on the electron beam and the initiation of harmonic bunching due to the growth of power at the fundamental. This gives rise to a rapid initial growth of the harmonic power which quickly rolls over to the slower growth associated with the linear instability. The linear instability drives harmonic growth until the fundamental reaches a power level necessary for the nonlinear mechanism to take over, after which the harmonic power grows exponentially with a growth rate that scales as the product of the harmonic number and the growth rate of the fundamental. In this regime, the harmonic grows substantially faster than the fundamental until the harmonic saturates. Saturation of the harmonic typically occurs slightly prior to that of the fundamental due to overbunching of the electrons at the harmonic wavelengths. This is illustrated schematically in Fig. 8.1.

H. P. Freund, T. M. Antonsen, Jr., *Principles of Free Electron Lasers*,
https://doi.org/10.1007/978-3-031-40945-5_8

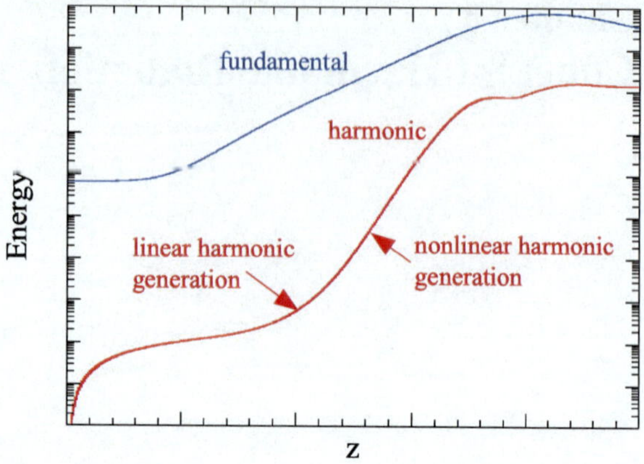

Fig. 8.1 Schematic illustration of harmonic generation showing initial growth driven by the linear instability followed by more rapid growth once the nonlinear harmonic generation mechanism becomes dominant. The harmonics typically saturate at a point prior to that of the fundamental

8.1 Linear Harmonic Generation

The mechanisms for linear harmonic generation differ greatly between planar [1–3, 6–32] and helical [4, 5] wiggler configurations. In its most straightforward interpretation, the planar wiggler causes an oscillation in the axial velocity at a period one-half that of the wiggler period with an amplitude which depends upon the square of the wiggler-induced transverse velocity. This oscillation is the source of a resonant wave-particle interaction at the odd harmonics of the wiggler period. However, the strength of the harmonic interaction increases with the magnitude of the oscillation in the axial velocity that, in turn, is directly proportional to the wiggler strength and inversely proportional to the electron energy. As a result, the harmonic interaction in a planar wiggler is more sensitive to the wiggler strength than is the fundamental and requires extremely large amplitude wigglers to be effective.

In contrast, the linear harmonic interaction in helical wiggler configurations does not depend upon an oscillatory axial velocity. In this case, the electron trajectories describe helices with uniform axial velocities, and the interaction is with a circularly polarized wave. Typically, the electron trajectory in a helical wiggler will vary in azimuth with $\theta \approx k_w z$. As a result, a sinusoidal wave which varies in phase as exp $(ikz + il\theta - i\omega t)$, will resonate at the lth harmonic. This generates waves that propagate at an oblique angle to the axis of symmetry. For the case in which the drift tube acts as a waveguide in controlling the dispersion of the mode, this establishes a selection rule, whereby the TE_{ln} and TM_{ln} modes resonate at the lth harmonic upshift. An important distinction between helical and planar wigglers in this regard is that the strength of the harmonic interaction is no more sensitive, in principal, to the wiggler amplitude than is the fundamental.

8.1.1 Helical Wiggler Configurations

The linear gain in helical wiggler-based free-electron lasers can be treated using the stability analysis derived in Chap. 4 for realizable configurations in the presence of an axial magnetic field. Here, the discussion is confined to the analysis of the harmonic interaction in the low-gain regime [4]. It was shown in Chap. 4 that a selection rule exists, whereby the azimuthal mode number in a cylindrical waveguide determined the harmonic number of the interaction. Specifically, the lth harmonic was resonant with the TE_{ln} or TM_{ln} modes.

8.1.1.1 Harmonic Excitation in the Low-Gain Regime

The small-signal gain for the various waveguide modes is given in Eq. (4.255) and can be used to determine the relative magnitudes of the various harmonics. This equation can be used to determine the ratio of the gain for the $TE_{l'n'}$ mode to that of the TE_{ln} mode for a cylindrical waveguide of radius R_g is given by

$$\frac{G_{l'n'}}{G_{ln}} \cong \frac{l'}{l}\frac{x'^2_{l'n'}}{x'^2_{ln}}\frac{x'^2_{ln}-l^2}{x'^2_{l'n'}-l'^2}\frac{J_l^2(x'_{ln})}{J_{l'}^2(x'_{l'n'})}\frac{J'^2_{l'}(\kappa_{l'n'}r_0)}{J'^2_{l}(\kappa_{ln}r_0)}, \tag{8.1}$$

where $J_l'(x'_{ln}) = 0$, κ_{ln} $(= x'_{ln}/R_g)$ is the cutoff wavenumber of the mode and r_0 is the radius of the steady-state orbit. The ratio of the gain of the $TM_{l'n'}$ mode to that of the TM_{ln} mode is given similarly by

$$\frac{G_{l'n'}}{G_{ln}} \cong \frac{l'^3}{l^3}\frac{x'^2_{l'n'}}{x'^2_{ln}}\frac{J'^2_l(x'_{ln})}{J'^2_{l'}(x'_{l'n'})}\frac{J_{l'}^2(\kappa_{l'n'}r_0)}{J_l^2(\kappa_{ln}r_0)}, \tag{8.2}$$

where $J_l\,(x_{ln}) = 0$, and κ_{ln} $(= x_{ln}/R_g)$ denotes the cutoff wavenumber of the TM mode. Analogous expressions can be derived for the ratios of the TE to TM modes or vice versa.

These expressions simplify considerably in the limit in which $|\upsilon_w/\upsilon_\parallel| << 1$ such that both $\kappa_{ln}r_0 << 1$ and $\kappa_{l'n'}r_0 << 1$. Since $k_w r_0 = |\upsilon_w/\upsilon_\parallel|$ for the steady-state trajectories, the ratio of the gain at the lth harmonic in the $TE_{ln'}$ mode to the gain of the fundamental in the TE_{1n} mode is

$$\frac{G_{l'n'}}{G_{1n}} \cong \frac{l'}{(l'-1)!^2}\left(\frac{x'_{l'n'}}{2k_wR_g}\right)^{2l'-2}\left(\frac{\upsilon_w}{\upsilon_\parallel}\right)^{2l'-2}\frac{x'^2_{l'n'}}{x'^2_{1n}}\frac{x'^2_{1n}-1}{x'^2_{l'n'}-l'^2}\frac{J_1^2(x'_{1n})}{J_{l'}^2(x'_{l'n'})}, \tag{8.3}$$

The corresponding ratio for the TM modes is

$$\frac{G_{l'n'}}{G_{1n}} \cong \frac{l'}{(l'-1)!^2}\left(\frac{x_{l'n'}}{2k_wR_g}\right)^{2l'}\left(\frac{v_w}{v_\parallel}\right)^{2l'}\frac{x_{1n}^2}{x_{l'n'}^2}\frac{J_1'^2(x_{1n})}{J_{l'}'^2(x_{l'n'})}. \tag{8.4}$$

In this limit, it is clear that the gain decreases rapidly with the harmonic number for a sufficiently weak wiggler. However, it is misleading to assume that this conclusion holds in general. The reason for this is that the argument of the Bessel functions, $\kappa_{ln}r_0$, is proportional to the ratio of the beam radius to the waveguide radius, and this can be appreciable even for relatively weak wigglers. Hence, it is possible to design free-electron lasers based upon helical wigglers in which the gain at the harmonics is comparable to that of the fundamental. A nonlinear simulation of such an example follows.

8.1.1.2 Numerical Simulation of Harmonic Excitation

The nonlinear aspects of the harmonic interaction in a helical wiggler/axial magnetic field configuration can be investigated using the nonlinear formalism described in Chap. 5 in cylindrical waveguide geometry. In this case, the dynamical equations for the waveguide modes are given by Eqs. (5.112) and (5.113) for the TE modes and Eqs. (5.114) and (5.115) for the TM modes which are integrated for each waveguide mode in conjunction with the Lorentz force equations (5.128), (5.129), and (5.130) and (5.132), (5.133), and (5.134) for an ensemble of electrons. This formalism is capable of the treatment of an arbitrary harmonic interaction in both the linear and nonlinear regimes.

An example of a second harmonic interaction with the TE_{21} mode is illustrated in Fig. 8.2 in which we plot the efficiency as a function of normalized angular frequency [5]. The parameters of this case are similar to those examined for the fundamental TE_{11} interaction in Fig. 5.17 for Group I orbit parameters. Specifically,

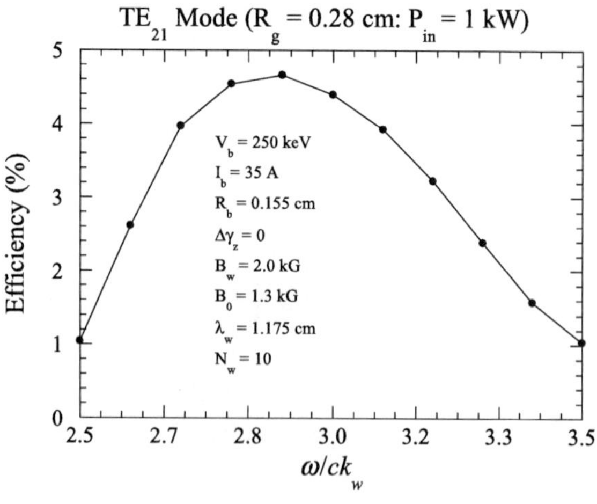

Fig. 8.2 Graph of the efficiency as a function of frequency for a second harmonic TE_{21} mode interaction

this example deals with the propagation of an electron beam with an energy of $V_b = 250$ keV, a current of $I_b = 35$ A, and an initial beam radius of $R_b = 0.155$ cm through a waveguide with a radius of $R_g = 0.28$ cm in the presence of a wiggler field with a magnitude of $B_w = 2.0$ kG and a period of $\lambda_w = 1.175$ cm and an axial solenoidal field of $B_0 = 1.3$ kG. The entry taper region is $N_w = 10$ wiggler periods in length. Note that the wiggler parameter a_w $(= eA_w/m_ec^2) \approx 0.22$ for this example. The only difference with the aforementioned TE_{11} mode analysis is that the waveguide radius has been increased in order to the intersection points between the beam resonance line $[\omega \approx (k + 2k_w)\upsilon_\parallel]$ and the TE_{21} mode dispersion curve close enough together to merge into a single broad-gain band. As shown in the figure, a peak efficiency $\eta \approx 4.6\%$ is found at approximately $\omega/ck_w \approx 2.9$ (74 GHz). While the maximum efficiency of the second harmonic TE_{21} mode interaction is substantially less than that found for the fundamental TE_{11} mode ($\eta \approx 21.4\%$), the growth rate of the TE_{21} mode at this frequency is approximately $\Gamma_{21}/k_w \approx 0.025$ which is comparable to that found for the fundamental TE_{11} mode interaction ($\Gamma_{11}/k_w \approx 0.029$ at $\omega/ck_w \approx 1.3$). In addition, observe that while the bandwidth of the second harmonic interaction $\Delta\omega/\omega \approx 43\%$ is comparable to that of the fundamental, the frequency has been doubled.

8.1.2 *Planar Wiggler Configurations*

The mathematical formalism necessary to treat harmonic excitation in planar wiggler configurations in both the linear and nonlinear regimes has been developed in Chaps. 4 and 5. The linear theory of the interaction is described by the dispersion equation given in Eq. (4.200). This equation was derived subject to an idealized one-dimensional model and is capable, principally, of treating the excitation of odd harmonics. However, beam thermal effects have been included in the analysis, and the effect of an axial energy spread on the linear gain at the fundamental and harmonics can be studied. A nonlinear formulation of the planar wiggler interaction in three-dimensions has also been presented in Chap. 5 (see Eqs. (5.154)–(5.163)) for the case of a beam propagating through a rectangular waveguide in the presence of a planar wiggler generated by a magnet stack with parabolic pole faces for enhanced beam confinement. This formulation is capable of treating both the fundamental and all harmonics and will be applied in this chapter to the study of harmonic emission.

8.1.2.1 Harmonic Excitation in the Linear Regime

The general dispersion equation that describes the linear growth rate for both the fundamental and harmonics [33] in a planar wiggler has been derived in Chap. 4 (see Eq. (4.200)). This dispersion equation can be solved numerically for a case that illustrates the relative growth of the fundamental and the third harmonic. In general,

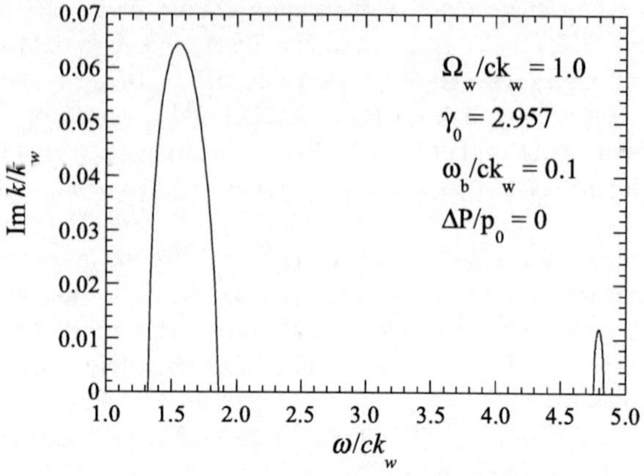

Fig. 8.3 Graph of the linear growth rates of the fundamental and third harmonics in a planar wiggler with $\Omega_w/ck_w = 1.0$, $\gamma_0 = 2.957$, and $\omega_b/\gamma_0^{1/2}ck_w = 0.1$ for an ideal beam

strong harmonic amplification requires a relatively large oscillation in the axial velocity; hence, the growth rate at the harmonics increases rapidly with Ω_w/ck_w. Indeed, the growth rate at the harmonics as predicted by (Eq. (4.200)) can be larger than that at the fundamental when Ω_w/ck_w exceeds unity. However, this is an unjustifiable conclusion based upon the present type of formulation. It is important to bear in mind that the analysis cannot be applied for arbitrarily large values of this parameter because (1) the idealized one-dimensional model breaks down when the displacement of the electrons from the plane of symmetry becomes large, and (2) the Lagrangian time coordinate has been integrated under the assumption that the $\upsilon_w << \upsilon_{\|}$. Therefore, the analysis of cases in which Ω_w/ck_w is greater than unity requires a fully, three-dimensional analysis. This will be discussed within the context of the nonlinear formulation of the interaction in planar wigglers derived in Chap. 5. However, in the analysis of the linear dispersion equation, we restrict the numerical analysis to an example for which $\Omega_w/ck_w = 1$. This is a physically interesting case that is at the fringe of the range of validity of the formulation and will serve to clearly illustrate the relationship of the harmonics to the fundamental. In addition, we shall assume that $\gamma_0 = 2.957$ and $\omega_b/ck_w = 0.1$ as well. It should also be remarked that in order for the thermal effects to result in substantial growth at the even harmonics, $\Delta P/p_0$ must be of the order of unity as well. Since this is unreasonably high for any well-designed experiment, we shall ignore this effect henceforth and concentrate on the emission at the odd harmonics.

The magnitude of the growth rate is plotted as a function of frequency in Fig. 8.3 for the fundamental and the third harmonic. The fundamental exhibits a peak growth rate of $|\mathrm{Im}\, k|/k_w \approx 0.065$ at a normalized frequency of $\omega/ck_w \approx 1.55$. In contrast, the magnitude of the growth rate at the third harmonic is $|\mathrm{Im}\, k|/k_w \approx 0.012$ at a frequency of $\omega/ck_w \approx 4.80$. Observe that both the magnitude and bandwidth of the harmonic is reduced relative to the fundamental.

The effect of the thermal spread on the fundamental and the third harmonic is shown in Fig. 8.4. Here we plot the normalized growth rate defined as the ratio of the maximum growth rate for a specific value of $\Delta P/p_0$ to the maximum growth rate for

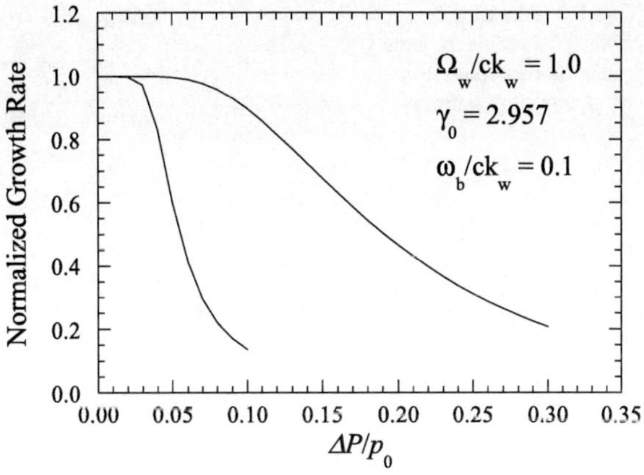

Fig. 8.4 Variation in the normalized growth rate with the axial momentum spread for the fundamental and third harmonic for parameters consistent with Fig. 8.3

$\Delta P/p_0 = 0$ as a function of $\Delta P/p_0$ for the fundamental and third harmonic. Observe that thermal effects are expected to become important on the fundamental when $\Delta v_{\|}/v_{\|} \approx |\mathrm{Im}\, k|/(k_w + \mathrm{Re}\, k) \approx 0.025$, which corresponds to $\Delta P/p_0 \approx 22\%$ (recall that $\Delta v_{\|}/v_{\|} \approx \Delta P^2/2p_0^2$). This is in substantial agreement with the results shown in the figure. For the case of the harmonics, thermal effects are expected to become important at a much reduced thermal spread, specifically, when $\Delta v_{\|}/v_{\|} \approx |\mathrm{Im}\, k|/[(1 + 2\, l)k_w + \mathrm{Re}\, k]$. For the third harmonic in the present example, $\Delta v_{\|}/v_{\|} \approx |\mathrm{Im}\, k|/(3k_w + \mathrm{Re}\, k) \approx 0.0015$. This corresponds to $\Delta P/p_0 \approx 5.5\%$, which is also in good agreement with the calculation.

8.1.2.2 Numerical Simulation of Harmonic Excitation

The nonlinear aspects of the harmonic interaction can be investigated using the formalism described in Chap. 5 for planar wigglers in rectangular waveguide geometry [15, 16, 22]. In this case, the dynamical equations for the waveguide modes are given by Eqs. (5.154)–(5.157) which are integrated for each waveguide mode in conjunction with the Lorentz force equations (5.158)–(5.163) for an ensemble of electrons. The wiggler model employed in this formulation has tapered pole faces (see Eq. (2.3)) to provide for enhanced beam confinement. Observe that no approximations are made in the orbit integrations; hence, unlike the aforementioned linear stability analysis, the formalism is valid for arbitrarily large wiggler amplitudes in which $\Omega_w/ck_w > 1$.

The first example under consideration corresponds to parameters similar to those investigated for the fundamental interaction in Chap. 5. In particular, that of a 3.3-MeV, 100-A electron beam with an initial radius of 0.2 cm propagating through a rectangular waveguide with dimensions $a = 10.0$ cm and $b = 3.0$ cm in the presence of a wiggler with an amplitude $B_w = 4.2$ kG and a period $\lambda_w = 9.8$ cm. The entry taper region into the wiggler is assumed to be $N_w = 10$ wiggler periods in length. This case represents an extremely strong wiggler parameter in which a_w (= $eA_w/$

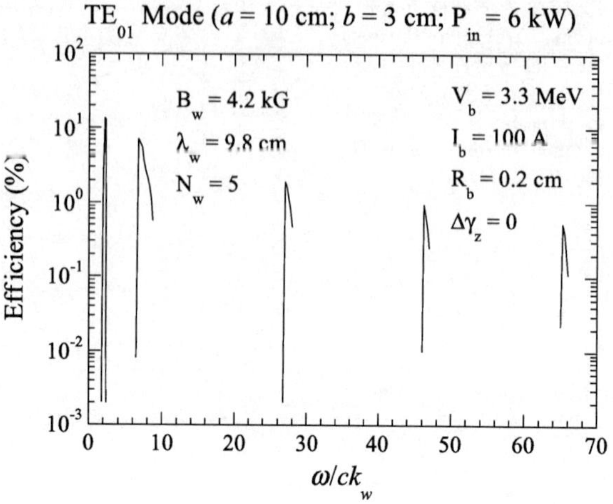

Fig. 8.5 Plot of the efficiency versus frequency at the fundamental, third, fifth, and ninth harmonics

$m_ec^2) \approx 3.84$. Amplification of a signal is found at frequencies in the vicinity of the odd harmonics. A spectrum of the output efficiency, η, of the TE_{01} mode is shown in Fig. 8.5 as a function of frequency for the fundamental through the seventh harmonic for an ideal beam with a vanishing axial energy spread. The doublet indicated in the neighborhood of the fundamental ($\omega/ck_w \approx 7$) corresponds to the upper and lower intersection points between the TE_{01} mode dispersion curve and the beam resonance line [$\omega \approx (k + k_w)v_{\|}$]. It is evident from the figure that there is a sharp decline in the efficiency between the fundamental and the third harmonic. However, the subsequent decrease in the efficiency with harmonic number is rather slow, and substantial power may be found at the higher harmonics. In addition, the decrease in the growth rates with harmonic number is rather slow as well.

It is difficult to estimate the precise growth rates at each harmonic because of (1) variations in the in the launching loss associated with the various harmonics, (2) the oscillations in the power and gain with a period half of $\lambda_w/2$ due to the lower beat wave, and (3) the betatron oscillations on the beam due to the wiggler gradients. However, the distance to saturation, z_{sat}, for a fixed input signal of 6 kW provides a measure of the average growth rate. For example, the peak growth rate is obtained at $\omega/ck_w \approx 7.25$ (22.2 GHz) for the fundamental (upper intersection), and saturation occurs at $k_wz_{sat} \approx 184$ with an efficiency $\eta \approx 5.58\%$. Peak growth is found for $\omega/ck_w \approx 27.25$ (83.4 GHz) at the third harmonic, and saturation is found to occur at $k_wz_{sat} \approx 194$ with an efficiency $\eta \approx 1.61\%$. In view of the combination of the comparable saturation length and lower efficiency at the harmonic, it is clear that the gain per unit length at the harmonic is a substantial fraction of that at the fundamental. A similar situation is found at the fifth harmonic which displays a peak efficiency at $\omega/ck_w \approx 46.5$ (142 GHz), and for which saturation is found for $k_wz_{sat} \approx 205$ at an efficiency $\eta \approx 0.69\%$. The peak growth rate at the seventh harmonic is found for $k_wz_{sat} \approx 172$ at a frequency of $\omega/ck_w \approx 65.5$ (201 GHz) and an efficiency of $\eta \approx 0.39\%$.

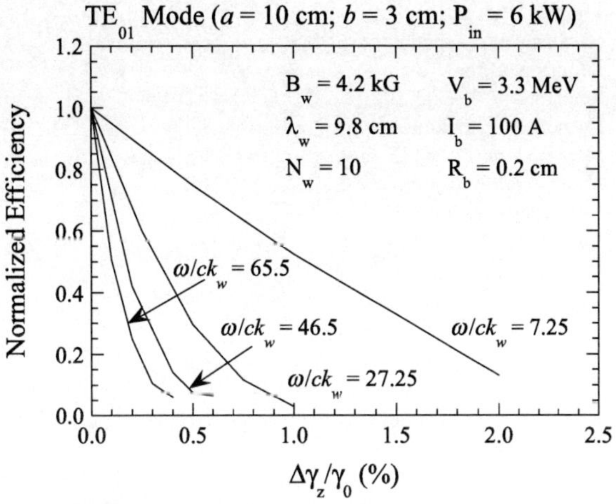

Fig. 8.6 Illustration of the effect of an axial beam energy spread on the efficiency at the fundamental and the harmonics

The effect of an axial energy spread on the beam is illustrated in Fig. 8.6, in which we plot the normalized efficiency η/η_0 (where η_0 is defined as the efficiency in the limit in which $\Delta\gamma_z = 0$) as a function of the axial energy spread $\Delta\gamma_z/\gamma_0$. Recall that in the formulation under consideration, the electron beam distribution is assumed to be monoenergetic but with a pitch angle spread defined in Eq. (5.122). The principal conclusion to be drawn from the figure, as expected, is that the harmonic emission is far more sensitive to the thermal spread than the fundamental. While the efficiency decreases by an order of magnitude for $\Delta\gamma_z/\gamma_0 \approx 2\%$ at the fundamental, a corresponding decrease in the efficiency occurs for $\Delta\gamma_z/\gamma_0 \approx 0.7\%$ at the third harmonic. The decrease is even more rapid at still higher harmonics. The growth rates are also more sensitive to the thermal spread at the harmonics. The saturation length increases form $k_w z_{sat} \approx 184$ to $k_w z_{sat} \approx 194$ at $\omega/ck_w = 7.25$ (fundamental) as the thermal spread increases to $\Delta\gamma_z/\gamma_0 = 1\%$. In contrast, the saturation length increases from $k_w z_{sat} \approx 194$ to $k_w z_{sat} \approx 232$ at $\omega/ck_w = 27.25$ (third harmonic) as the axial energy spread increases to $\Delta\gamma_z/\gamma_0 = 0.5\%$. At $\omega/ck_w = 65.5$ (seventh harmonic), the saturation length increases from $k_w z_{sat} \approx 172$ to $k_w z_{sat} \approx 224$ as the thermal spread increases to $\Delta\gamma_z/\gamma_0 = 0.27\%$. These conclusions dramatically illustrate the importance of beam quality on coherent harmonic emission.

8.1.2.2.1 The Effect of Betatron Oscillations

The next example under to be considered is similar to that shown in Figs. 8.5 and 8.6. The principal difference consists in the thickness of the electron beam upon injection into the wiggler. This beam thickness has significant impact on the betatron oscillations of the beam experienced by the beam that in the case of the wiggler model also has an impact on the harmonic spectrum. The specific parameters deal with the propagation of an electron beam with an energy of 3.5 MeV, a current of 800 A, and

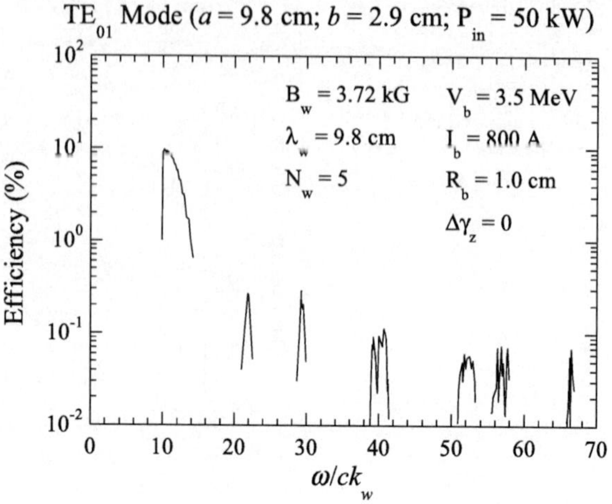

Fig. 8.7 Plot of the efficiency as a function of frequency at the fundamental through fifth harmonics. The doublets centered at $\omega/ck_w \approx 25$ and 55 correspond to emission at the second and fourth harmonics

an initial beam radius of 1.0 cm through a waveguide with dimensions $a = 9.8$ cm and $b = 2.9$ cm in the presence of a wiggler with an amplitude $B_w = 3.72$ kG. The wiggler period and entry taper length are as previously with $\lambda_w = 9.8$ cm and $N_w = 5$. The mode of interest again is the TE_{01} mode but with an input power level of 50 kW. This choice of parameters corresponds closely with the case examined in Chap. 5 for the fundamental interaction.

The spectrum of the saturation efficiency that results is shown in Fig. 8.7 for the fundamental through the fifth harmonics. As expected, emission is found at the fundamental ($\omega/ck_w \approx 11.3$) and the odd harmonics ($\omega/ck_w \approx 40.3$ and 66.5). However, emission is also found in the vicinity of the even harmonics in the form of a frequency doublet. As shown in the figure, emission occurs for $\omega/ck_w \approx 22.0$ and 29.5 corresponding to the second harmonic and at $\omega/ck_w \approx 52.3$ and 56.8 corresponding to the fourth harmonic. The efficiency at the second harmonic ($\eta \approx 0.27\%$) exceeds that found at the third harmonic ($\eta \approx 0.11\%$). As discussed previously, the efficiency decreases sharply between the fundamental and the nearest harmonic, but falls off slowly with harmonic number thereafter.

The source of the doublet emission in the vicinity of the even harmonics is the wiggler-plane focusing due to the parabolic pole face model for the wiggler. The spatial variation of the field in the wiggle-direction induces a modified betatron oscillation in the x-component (i.e., the direction of the wiggler-induced transverse oscillation) of the velocity that is the source of oscillations in the axial velocity with periods $k \pm K_\beta$, where K_β denotes the modified betatron wavenumber. It is this oscillation that drives the even harmonic emission. It should be remarked that emission at the even harmonics is expected to occur for any wiggler configuration that exhibits an inhomogeneity in the direction of the wiggler-induced transverse oscillation.

The modified betatron oscillations in a planar wiggler with parabolic pole faces have been discussed [16] to lowest order in the time domain in Chap. 2. For the purposes of the present discussion, we treat the betatron oscillation in the spatial domain, and write the approximate orbit equations in the xz-plane in the form

$$\frac{d}{dz}v_x = \frac{\Omega_w}{v_z}\cosh\left(\frac{k_w x}{\sqrt{2}}\right)\cos k_w z, \tag{8.5}$$

and

$$\frac{d}{dz}v_z = -\Omega_w\frac{v_x}{v_z}\cosh\left(\frac{k_w x}{\sqrt{2}}\right)\cos k_w z, \tag{8.6}$$

where $\Omega_w = eB_w/\gamma m_e c$ and the y-variations are ignored. Solutions can be found which are of the form

$$x = x_\beta \cos k_\beta z - x_w \cos k_w z, \tag{8.7}$$

and

$$v_x = -v_z\left(k_\beta x_\beta \sin k_\beta z + k_w x_w \sin k_w z\right), \tag{8.8}$$

for which the axial velocity varies as

$$v_z \cong v_\parallel\left(1 + \frac{v_w^2}{4v_\parallel^2}\sin 2k_w z + \frac{v_\beta v_w}{4v_\parallel^2}\left[\cos\left(k_w + k_\beta\right)z - \cos\left(k_w - k_\beta\right)z\right]\right), \tag{8.9}$$

where $v_\parallel$ is the bulk axial velocity of an electron in a planar wiggler, x_w and v_w describe the amplitudes of the fundamental wiggler-induced oscillation, and x_β and v_β ($= -x_\beta k_\beta v_\parallel$) describe the amplitude of the modified betatron oscillations. Under the assumption that $|k_w x| << 1$, substitution of this solution into the orbit equations shows that the modified-betatron wavenumber is

$$k_\beta^2 \cong \frac{\Omega_w^2}{4\gamma^2 v_\parallel^2}\left(1 + \frac{3}{16}\frac{v_w^2}{v_\parallel^2} + \frac{1}{8}k_w^2 x_\beta^2\right). \tag{8.10}$$

The analytic expression also predicts oscillations at $(2k_w \pm k_\beta)$ and $(k_w \pm 2k_\beta)$, although at lower amplitudes. The inhomogeneities in the y-direction, characteristic of the magnet itself, introduce an additional betatron oscillation that does not couple strongly to the axial velocity or the harmonic interaction.

In order to verify the form of this solution for the modified-betatron oscillations, the single-particle equations of motion have been integrated numerically in z under the assumption of the same input taper as for the complete nonlinear analysis. The

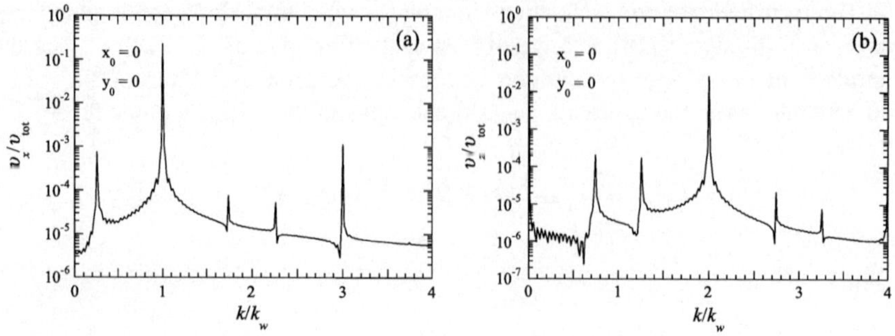

Fig. 8.8 Fourier decomposition of the x- and z-components of the electron velocities (normalized to the magnitude of the total velocity) for an electron which enters the wiggler at $x(z = 0) = 0$ and $y(z = 0) = 0$\

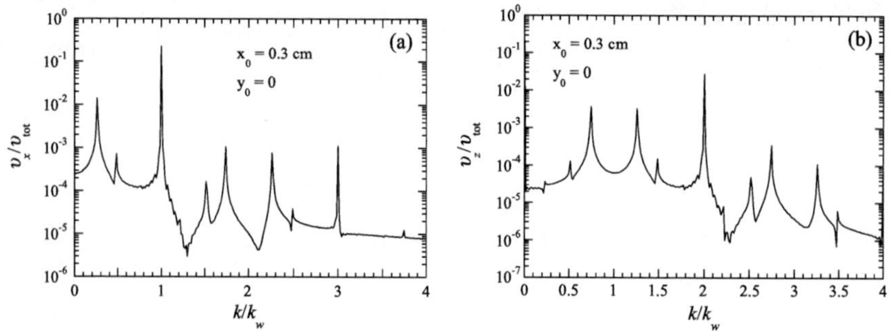

Fig. 8.9 Fourier decomposition of the x- and z-components of the electron velocities (normalized to the magnitude of the total velocity) for an electron which enters the wiggler at $x(z = 0) = 0.3$ cm and $y(z = 0) = 0$

resulting velocities were then Fourier transformed over the uniform wiggler portion of their motion (i.e., $z > N_w\lambda_w$). The results of the Fourier transform of the x-components (a) and z-components of the velocity are shown in Figs. 8.8 and 8.9 for initial conditions in which $y_0 = 0$ and $x_0 = 0$ and 0.3 cm, respectively. The different harmonic structure predicted by the analytic theory is contained in these spectra. Observe that the amplitude of the modified-betatron doublet at $(k_w \pm K_\beta)$ in the axial velocity is relatively independent of the initial y-position of the electron. In addition, the amplitude of the doublet increases with the initial x-position (N.B., the initial x-position is not identical with x_β). This accounts for the lack of emission at the even harmonics in the previous example for which the initial beam radius was only 0.2 cm.

A comparison between the analytic approximation and the numerical results for the wavenumber of the modified oscillation is shown in Fig. 8.10 as a function of x_0 for $y_0 = 0$. It is evident that the analytic approximation holds for $x_0 \leq 0.35$ cm, after which it diverges rapidly from the numerical solution. As shown in Fig. 8.11, the

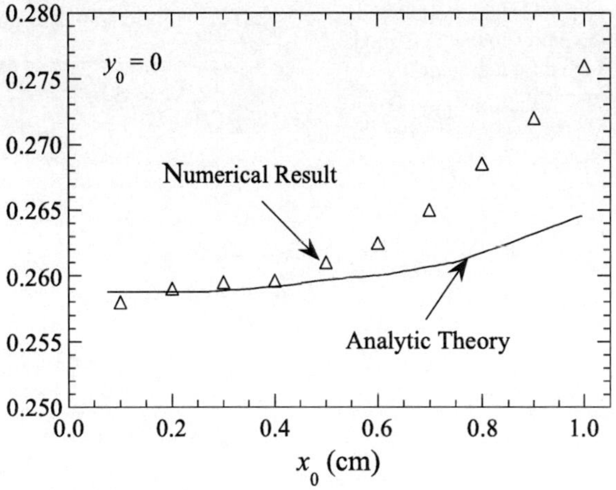

Fig. 8.10 Scaling of the modified betatron wavenumber with the initial *x*-coordinate of the electron

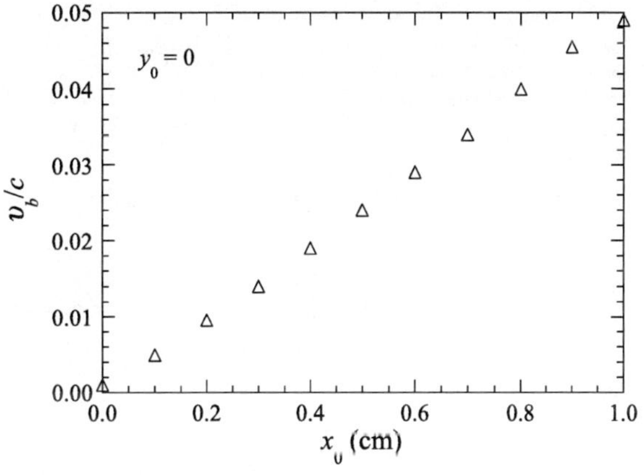

Fig. 8.11 Scaling of the magnitude of the velocity of the modified betatron oscillation with the initial *x*-coordinate of the electron

magnitude of the velocity of the modified-betatron oscillation increases in an approximately linear fashion with x_0.

Two difficulties associated with the design of a harmonic free-electron laser are (1) the suppression of the fundamental and (2) the sensitivity of the harmonic interaction to the axial energy spread of the beam. One method of suppression of the fundamental is applicable when waveguide effects upon the wave dispersion are important is to design the system in such a manner that there is no intersection between the waveguide dispersion curve and the beam resonance line $[\omega \approx (k + k_w)v_\|]$. This allows the harmonic interactions to dominate. The beam quality constraint can be addressed by the use of a tapered wiggler, in order to enhance the efficiency possible for a given axial energy spread.

In order to address these issues, we now consider the case a third harmonic interaction in which the fundamental resonance is below cutoff and, therefore, cannot interact with the beam. Further, this example deals with a single mode in

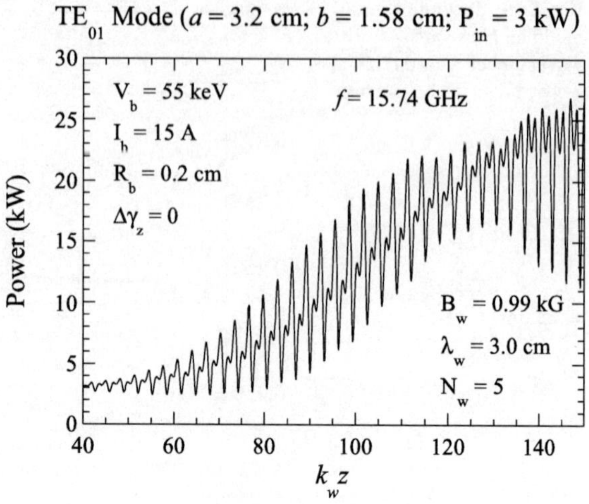

Fig. 8.12 The evolution of the power in the TE_{01} mode for a third harmonic interaction

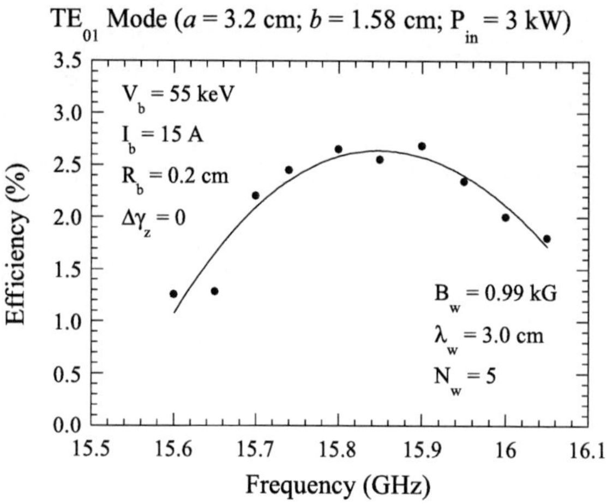

Fig. 8.13 Spectrum showing the efficiency as a function of frequency for an ideal beam

which only the TE_{01} mode is resonant. The electron beam for this example is characterized by an energy of 55 keV, a current of 15 A, and an initial radius of 0.2 cm. The wiggler has an amplitude of 990 G, a period of 3.0 cm, and an entry taper region which is five wiggler periods in length. The waveguide dimensions are $a = 3.2$ cm and $b = 1.58$ cm. We first consider the case of a uniform wiggler and investigate the sensitivity of the interaction to beam thermal effects. The spatial growth of a 3-kW-injected signal at a frequency of 15.74 GHz is shown in Fig. 8.12 for the case of a beam with $\Delta\gamma_z = 0$. Saturation occurs with an efficiency of approximately 2.7%. This gives a net gain of 0.12 dB/cm over a total interaction length (i.e., including the entry taper region) of about 70 cm. This frequency corresponds to the central peak of the interaction spectrum. As shown in Fig. 8.13, resonance occurs from approximately 15.6 GHz through 16.1 GHz with a peak efficiency of the order of 2.8%.

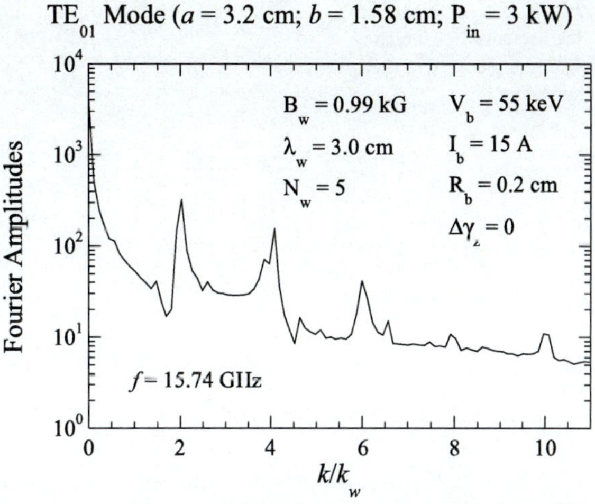

Fig. 8.14 The Fourier spectrum of the spatial evolution of the power

It is evident from Fig. 8.12 that a substantial modulation is superimposed on the bulk growth of the signal. A Fourier transform of this spatial variation is shown in Fig. 8.14, and the modulation is evident with wavenumbers of $2k_w$, $4k_w$, $6k_w$, $8k_w$, $10k_w$, etc. This modulation is due to the effect of the lower beat wave in planar wiggler configurations that results in oscillations with a wavenumber of $2k_w$ at the fundamental. The interaction is similar at the harmonics, and the modulation at $4k_w$, $6k_w$, etc. derives from higher-order oscillations to the wiggler-induced velocity. Observe that such modulation of the signal at the even harmonics of the wiggler period occur, if at all, at a much lower level in the case of a helical wiggler because the symmetry acts to suppress the lower beat wave.

Returning to the question of the sensitivity of the interaction to thermal effects, we plot the decrease in the efficiency with increasing thermal (in this case through a pitch angle) spread in Fig. 8.15. As shown in the figure, the efficiency decreases by a factor of two as $\Delta p_z/p_0$ increases to a value of about 0.045% corresponding to a value of $\Delta\gamma_z/\gamma_0 = 0.01\%$. This is extremely difficult (if not impossible) to achieve given the current state-of-the-art in electron beam technology and motivates our study of the effect of wiggler taper on the harmonic interaction.

8.1.2.2.2 Efficiency Enhancement with a Tapered Wiggler

The effect of a tapered wiggler is shown in Fig. 8.16 in which we plot the maximum efficiency (i.e., if the field is tapered to zero) as a function of the scale length of the taper in the absence of thermal effects. As shown in the figure, the maximum obtainable efficiency is approximately 23% that is almost an order of magnitude greater than that found for the uniform wiggler. In addition, the optimal range for the taper itself is for $0.0005 < |\varepsilon_w| < 0.0009$. The optimal range for the taper is governed by the requirements that the taper be rapid enough to maintain the axial beam

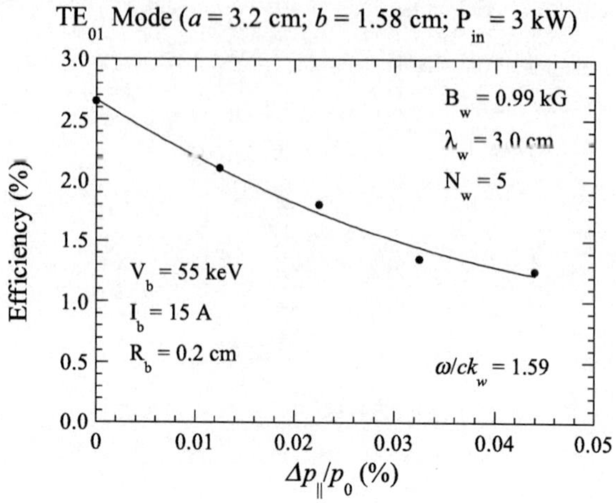

Fig. 8.15 Graph showing the decrease in efficiency with the increase in thermal spread

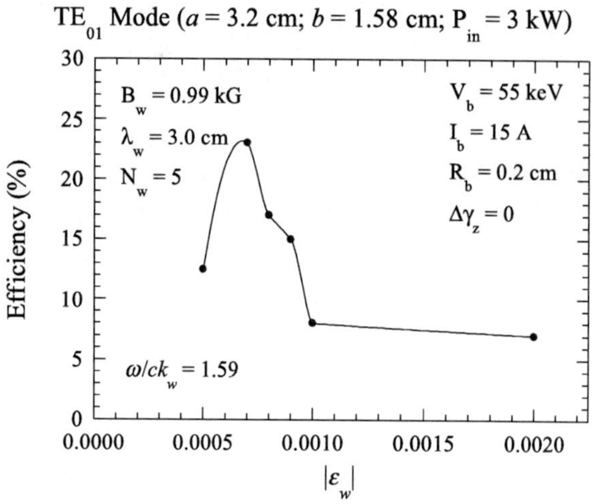

Fig. 8.16 Graph of the maximum efficiency versus the normalized taper

velocity required for the wave particle resonance but not so rapid that the beam becomes detrapped. By way of comparison, $\varepsilon_w = -0.0078$ in the experiment at Lawrence Livermore National Laboratory described in Chap. 5, and simulations of the experiment indicate that the maximum extraction efficiency would occur for $\varepsilon_w \cong -0.002$. It appears, therefore, that harmonic experiments require slower tapers than at the fundamental. The reason for this is that the ponderomotive potential is directly proportional to the normalized wiggler amplitude (Ω_w/k_wc) for the fundamental but varies as the cube of this factor at the third harmonic. As a result, the ponderomotive potential tends to be smaller at the harmonics, and the effect of a given wiggler taper is larger. Indeed, the upper bound on the effective taper shown in the figure for $\varepsilon_w \cong -0.001$ occurs because the effect of the taper has overwhelmed the ponderomotive force, and the particles have become detrapped.

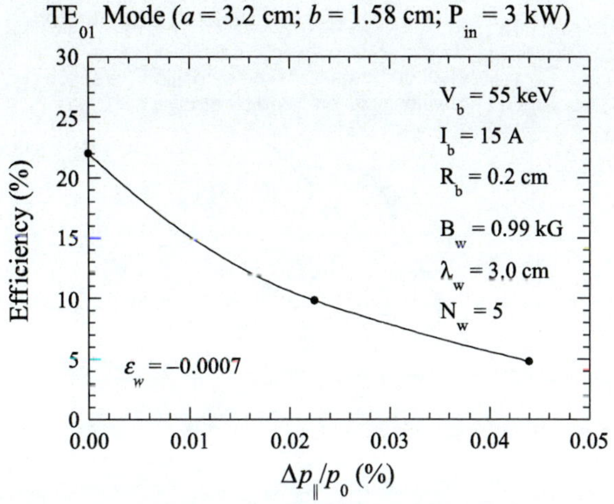

Fig. 8.17 Variation in the maximum efficiency of a third harmonic tapered wiggler interaction with thermal effects

The effect of an axial energy spread upon the third harmonic tapered wiggler interaction is shown in Fig. 8.17, in which we plot the efficiency as a function of the axial momentum spread. As in the case of a uniform wiggler, the efficiency falls by approximately 60% as the axial momentum spread increases to 0.05%. However, the general levels of efficiency to be obtained for a comparable thermal spread are much greater than for the uniform wiggler.

8.1.2.3 The Periodic Position Interaction

The conventional harmonic interaction in planar wiggler free-electron lasers, as described above, occurs at the odd harmonics of the resonance frequency. These velocity harmonics are present even for ideal wiggler with perfect beam injection and give rise to the periodic velocity instability of the free-electron laser. The even harmonic interaction has also been discussed and can result from (1) betatron oscillations and (2) beam energy spread. There is an additional mechanism, however, which results in the excitation of even harmonics which additional periodicity in the electron velocity. Rather, it depends on a synchronism in the electron position with respect to an antisymmetric radiation field. The interaction can occur with either the transverse or axial electric field. The transverse field is required to be odd in the direction of the wiggle motion with respect to the electron beam axis, and the axial field is required to be even for the respective interactions to occur. For a second harmonic interaction, the radiation goes through two cycles as the electron beam traverses one wiggler period. For illustrative purposes, we consider this type of interaction at long wavelengths in which the electron beam interacts with modes supported by a rectangular waveguide. However, the fundamental process is also possible at optical wavelengths.

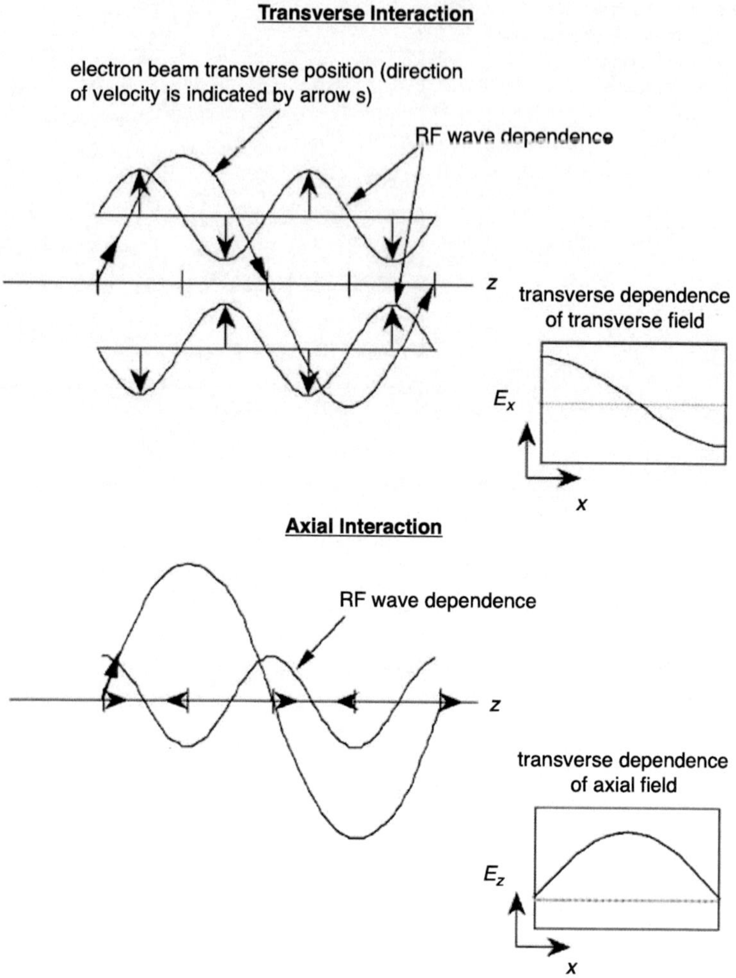

Fig. 8.18 Representation of the physical mechanism underlying the periodic position interaction

A brief discussion of both periodic position interaction types follows. For the transverse interaction, the on-axis electric field of the radiation mode is zero, and the field peaks off-axis. Considering only the central part of the beam, the essentials of the transverse interaction are shown in Fig. 8.18 in which the electron motion is greatly exaggerated and the transverse profile of the field is included (in this case, the TE_{11} rectangular waveguide mode). As seen in the figure, the electron will always be in either a decelerating or a zero electric field. Although a particle displaced from the horizontal center of the beam will be in an accelerating field a portion of the time, the bulk of the beam will be in a decelerating field most of the time, leading to a net amplification of the radiation. The axial interaction is also represented in Fig. 8.18, again for the central part of an on-axis beam. The transverse profile in this case represents the axial field of the TM_{11} mode. Here, even the central particle sees both an accelerating and a decelerating electric field. The particle is in a decelerating field on-axis where the field is at its maximum and the axial electron velocity at a minimum and in an accelerating field off-axis where the field is reduced and the axial velocity is at a maximum. However, the transverse variation of the electric field is greater than the transverse variation of the axial velocity. This results in a stronger interaction on-axis that, again, leads to net amplification.

Although the axial and transverse interactions have been considered separately in the preceding paragraph, it is difficult to completely separate the two interactions. The interaction can be treated using the nonlinear formulation described in Chap. 5. In fact, numerical simulation of the interaction indicates that the overall performance at the second harmonic is improved when the two interactions are combined. Simulation also shows that the second harmonic periodic position interaction can have a stronger growth rate than the fundamental interaction and is a significantly stronger interaction than the third harmonic interaction for the current range of experimental parameters.

The periodic position interaction has recently been demonstrated in the laboratory [34], and we include a comparison of the experimental results with the nonlinear formulation for the interaction. The target frequency of the experiment was between 12 and 18 GHz. The experiment used a cylindrical electron beam tunable in energy from about 30 to 250 keV, with 100 keV as the nominal operating energy for the second harmonic interaction. The wiggler consisted of a permanent magnet-assisted electromagnet with a period of 3 cm and an amplitude which was variable between 670 G and 1300 G, resulting in an a_w of 0.19–0.37. The pole pieces extended partially down the sides of the waveguide to provide wiggle plane focusing. This geometry resulted in a very flat profile near the center of the interaction region with the field rising sharply near the edges. The experiment operated as an amplifier, and the interaction took place in oversized waveguide with dimensions of 3.485×1.58 cm. The second harmonic periodic position interaction occurred with the TE_{11} and TM_{11} modes since these are the lowest-order waveguide modes with the necessary odd transverse symmetry.

Gain due to the second harmonic periodic position interaction was measured at beam energies ranging from 78 keV to 106 keV and currents between 6 A and 10 A. This contrasts with energies in the range of 200–250 keV required for the fundamental interaction. Operation at frequencies between 12.5 GHz and 16.5 GHz was achieved through both energy and wiggler field tuning, and the maximum observed gain was approximately 7 dB. It should be remarked that the experiment was not optimized for this interaction; hence, the interaction was not saturated at this value of gain with the available drive power.

The wiggler model used in the simulation of this experiment differed from the parabolic pole face model described previously and was chosen to describe an inhomogeneity in the direction of the wiggle motion (denoted by the x-axis). The measured field for the experiment was quite uniform about the symmetry axis and rose sharply toward the edges of the interaction region. A quartic profile of the field magnitude with x was found to provide a reasonable fit to this field. As such, the following wiggler model was employed:

$$B_{w,x}(\mathbf{x}) = \left[\left(\sin k_w z - \frac{\cos k_w z}{k_w}\frac{d}{dz}\right)B_w(z)\right]\left[\sinh k_w y - \frac{Y(k_w y)}{2k_w^2}\frac{d^2}{dx^2}\right]\frac{1}{k_w}\frac{d}{dx}X(x), \quad (8.11)$$

$$B_{w,y}(\mathbf{x}) = \left[\left(\sin k_w z - \frac{\cos k_w z}{k_w}\frac{d}{dz}\right)B_w(z)\right]\left[\cosh k_w y - \frac{k_w y \sinh k_w y}{2k_w^2}\frac{d^2}{dx^2}\right]X(x), \quad (8.12)$$

$$B_{w,z}(\mathbf{x}) = B_w(z)\cos k_w z\left[\sinh k_w y - \frac{Y(k_w y)}{2k_w^2}\left(1 + \frac{1}{k_w^2}\frac{d^2}{dx^2}\right)\frac{d^2}{dx^2}\right]X(x), \quad (8.13)$$

where $B_w(z)$ describes the axial variation in the amplitude, $X(x)$ denotes the variation in the wiggle plane, and $Y(k_w y) = k_w y \cosh k_w y - \sinh k_w y$. This field is not self-consistent in that it is divergence-free but not curl-free. However, the approximation is good as long as $B_w(z)$ and $X(x)$ vary slowly compared with λ_w (Fig. 8.19).

The variation in the wiggler-plane is described by a polynomial of the form

$$X(x) = 1 + \frac{1}{2}\left(\frac{x}{\alpha_x}\right)^{2m}, \quad (8.14)$$

where α_x denotes the scale length for variation of the field and m is an integer. When $\alpha_x \to \infty$ this field reduces to that commonly used to describe a wiggler with flat pole faces. A comparison of the actual field with $X(x)$ as used in the code ($m = 2$ and $\alpha_x = 1.4938$) is shown in Fig. 8.19, and it is clear that the approximation gives a reasonable fit to the data.

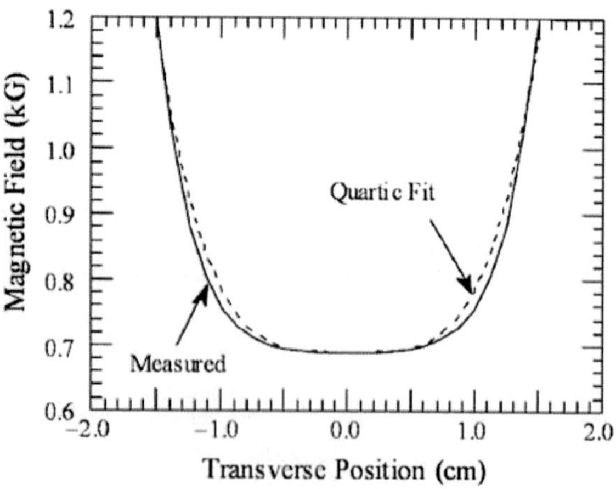

Fig. 8.19 Comparison of the wiggler field as measured in the experiment [33] and as used in simulation

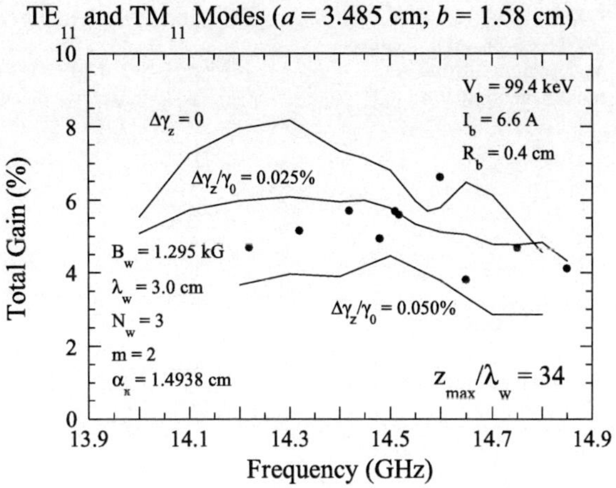

Fig. 8.20 Plot of the gain versus frequency as measured in the experiment (dots) and as determined from numerical simulation [34]

The specific parameters used for comparison are an energy and current of 99.4 keV and 6.6 A with a beam radius of 0.4 cm. The wiggler was characterized by $B_w = 1.295$ kG, $\lambda_w = 3$ cm, an input taper of $N_w = 3$, and a total length of $34\lambda_w$. Both the TE_{11} and TM_{11} modes are included with an initial power split of 9 to 1 (TE to TM) and a total input power of 300 W. Figure 8.20 contains a comparison of the observations with numerical results over the unstable band for three different axial energy spreads: $\Delta\gamma_z/\gamma_0 = 0.0\%$, 0.025%, and 0.05%. The experimental points over the entire frequency band fall, for the most part, between the curves representing an energy spread of 0.025% and 0.05%. This is in good agreement with the estimated energy spread in the experiment based upon electron trajectory calculations of the gun geometry. It is important to bear in mind that the power has not saturated in any of these cases. The evolution of the power for a frequency of 14.4 GHz and an axial energy spread of 0.025% is seen from Fig. 8.21 to result in a gain of about 8 dB with a distance to saturation of approximately $50\lambda_w$. In the absence of an energy spread, the gain for this case is approximately 10 dB with a saturation distance of $40\lambda_w$.

The effect of a wiggler field taper is shown in Fig. 8.22 for the case of an ideal beam ($\Delta\gamma_z = 0$) and a normalized taper of $\varepsilon_w = -0.00083$. The efficiency enhancement is extremely sensitive to the start-taper position, which must be chosen in the vicinity of the point at which the beam becomes trapped in the ponderomotive potential formed by the beating of the wiggler and radiation fields. For this example, the optimal start-taper point is $z_0 \approx 30.2$ cm. Only the total signal and the TE_{11} mode are shown in the figure, and the large oscillations in the total power are caused by the TM_{11} mode. It is evident that the saturated efficiency can be increased relative to that of the uniform-wiggler case for the present parameters by almost threefold through the use of a tapered wiggler.

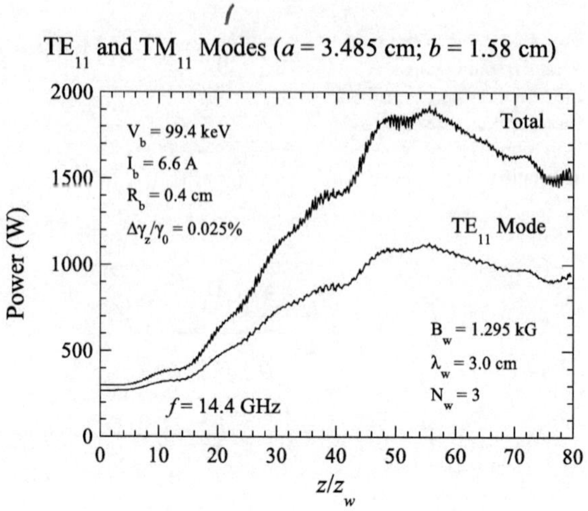

Fig. 8.21 Evolution of the power versus axial distance as determined from numerical simulation [33]

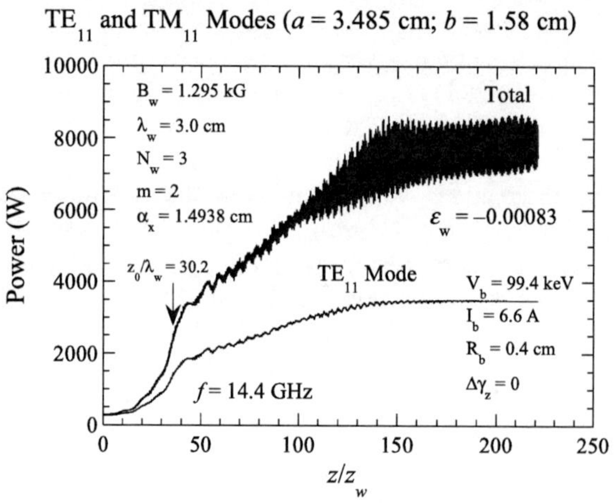

Fig. 8.22 Evolution of the power with axial position for a tapered wiggler interaction [33]

8.2 Nonlinear Harmonic Generation

Harmonics are generated in all vacuum electronic devices due to the bunching of the electron beam as the power at the fundamental resonance grows and becomes extremely strong as the interaction approaches saturation of the fundamental. This nonlinear harmonic generation mechanism is found in the vacuum tubes in audio amplifiers, traveling wave tubes, gyrotrons, klystrons, and free-electron lasers. Indeed, it is the source of the harmonics found in tube-type audio amplifiers that are so prized by audiophiles.

As the power at the fundamental grows, it induces a periodic velocity modulation that is coupled to a periodic density modulation. Harmonics are generated nonlinearly from the beating of these coupled velocity/density modulations. In the initial stages of the interaction where the fundamental power is small, the linear harmonic interaction is dominant; however, the linear harmonic growth rates are typically much less than that of the fundamental, and the harmonic power grows slowly. As the fundamental power grows, the density and velocity modulations reach a point where the nonlinear harmonic generation mechanism becomes dominant. The growth rate of the nonlinear harmonic generation process scales as $h\Gamma_1$, where h is the harmonic number and Γ_1 is the growth rate of the fundamental, so that the harmonics grow rapidly once the nonlinear generation mechanism becomes dominant. Depending upon the electron beam and wiggler parameters, the harmonic power levels can approach 1% that of the fundamental.

8.2.1 The Basis for Nonlinear Harmonic Generation

Both the linear and nonlinear harmonic generation mechanisms are implicitly included in the nonlinear formulations described in Chaps. 5 and 6. This is because the formulations include the harmonic field components and the electron trajectories are integrated using the complete Lorentz force equations. However, it is instructive to consider the mathematical source for the harmonic interactions.

The basis for the nonlinear harmonic generation mechanism can be illustrated by consideration of the coupling between the macroscopic fluid equations for the density and velocity and Maxwell's equations for the electromagnetic field. We now consider a simplified one-dimensional treatment. Assuming that the wiggle-plane corresponds to the x-direction, the wave equation is

$$\left(\frac{\partial^2}{\partial z^2} - \frac{1}{c^2}\frac{\partial^2}{\partial t^2}\right)\delta E_x(z,t) = \frac{4\pi}{c^2}\frac{\partial}{\partial t}\delta J_x(z,t), \tag{8.15}$$

where the electric field is represented as a superposition of harmonic components

$$\delta E_x = \frac{1}{2}\sum_{h=1}^{\infty}\delta E_h(z)\exp[ihk_0(z-ct)] + c.c., \tag{8.16}$$

$k_0 = \omega/c$, and $\delta E_h(z)$ is the slowly varying amplitude. In general, the source current $\delta J_x(z,t)$ describes both the linear and nonlinear interactions. The nonlinear interactions arise in the course of the electron beam bunching process and are due to the beating of the radiation field at various harmonics with various periods over which the electron beam is bunched. To explain this, we consider the fundamental equations describing the response of the electron beam to the radiation field; specifically, the continuity equation

$$\frac{\partial}{\partial t}n+\frac{\partial}{\partial z}(nv_z)=0, \tag{8.17}$$

momentum transfer

$$\frac{d}{dt}\mathrm{v}=-\frac{e}{\gamma m_e}\left[\left(\mathrm{I}-\frac{1}{c^2}\mathrm{vv}\right)\cdot\delta\mathrm{E}+\frac{1}{c}\mathrm{v}\times(\mathrm{B}_w+\delta\mathrm{B})\right], \tag{8.18}$$

and energy balance

$$\frac{d\gamma}{dt}=-\frac{e}{m_ec^2}\mathrm{v}\cdot\delta\mathrm{E}, \tag{8.19}$$

where n and $\mathbf{v}$ are the macroscopic density and velocity of the beam, γ is the relativistic factor corresponding to the bulk energy of the beam, $\mathbf{I}$ is the unit dyadic, $\mathbf{B}_w$ is the wiggler field, and $\delta\mathbf{E}$ and $\delta\mathbf{B}$ are the fluctuating electric and magnetic fields. The wave equation is coupled to the electron dynamics through the source current $\delta J_x=-env_x$.

As the interaction proceeds along the wiggler, the density and velocity modulations grow that describe the bunching of the electron beam. The continuity equation describes the nonlinear coupling between the density and velocity modulations, and the momentum transfer and energy balance equations describe a nonlinear relativistic coupling between the harmonic modulations of the velocity. The nonlinear terms mean that the hth harmonic components of the density and velocity are directly coupled to the $(h-1)^{\text{th}}$ components, which means that it is indirectly coupled to all the lower harmonics. As a result, the hth harmonic components of the density and velocity are affected by all the lower harmonic components. This harmonic dependence carries over to the source current. Assuming that the fundamental harmonic is dominant, the source current can be expressed as

$$\delta J_x=\frac{1}{2}\sum_{h=1}^{\infty}\left[\sigma_h^{(L)}\delta E_h+\sigma_h^{(NL)}\delta E_1^h\right]\exp[ihk_0(z-ct)]+c.c., \tag{8.20}$$

where $\sigma_h^{(L)}$ is the linear conductivity and $\sigma_h^{(NL)}$ is the nonlinear conductivity due to bunching at the fundamental. These conductivities can be found by solution of the electron dynamical equations to each order in the electric field; however, we are not concerned with the detailed solutions because the linear and nonlinear dynamics are implicitly included in the nonlinear formulation, and this discussion is intended merely to illustrate the source of the nonlinear harmonic interaction. As a result, the growth of each harmonic is given by

$$\frac{d}{dz}\delta E_h = -\frac{2\pi}{c}\left[\sigma_h^{(L)}\delta E_h + \sigma_h^{(NL)}\delta E_1^h\right]. \tag{8.21}$$

As illustrated in Fig. 8.1, the harmonic initially grows linearly from noise. However, since the growth rate at the fundamental is much greater than the linear growth rate at the harmonic, the nonlinear interaction will ultimately dominate. If the fundamental grows as $\delta E_1 \propto \exp(\Gamma z)$, then it is clear that nonlinear harmonic growth will scale as $\delta E_h \propto \exp(\Gamma z)$, and the gain length at the harmonics will decrease as h^{-1} from the gain length at the fundamental.

8.2.2 *Planar Wiggler Configurations*

We now consider harmonic generation in a short wavelength free-electron laser with a planar wiggler and employ the nonlinear formulation described in Chap. 6. The parameters under consideration deal with an electron beam with an energy of 219.5 MeV, a peak current of 150 A, a normalized emittance of 5 mm-mrad, and an energy spread of 0.10%. The wiggler is a flat-pole-face design with a peak, on-axis wiggler field strength of 10.06 kG with a wiggler period of 3.3 cm. The wiggler provides weak focusing in both the x- and y-directions. The electron beam is assumed to be matched into the wiggler so that the initial rms beam radius is approximately 183 μm. For these parameters, the fundamental resonance is at a wavelength of 516.8 nm.

It is convenient to illustrate the nonlinear harmonic generation mechanism using steady-state simulations (i.e., using a single temporal *slice*) [31]. To this end, simulations are performed for a seeded amplifier with an initial seed power of 10 W, which is comfortably higher than the equivalent noise power of about 1 W for a SASE configuration. The resulting growth of the power as well as the spot size, $w(z)$, is plotted versus z in Fig. 8.23. Saturation occurs at 110 MW for an efficiency $\eta \approx 0.32\%$ over a length of about 13.3 m. Evidently, the radiation initially undergoes diffraction but experiences rapid focusing at the onset of exponential growth. The gain length found in simulation is approximately 0.62 m, and the exponential growth region extends from 4 to 11 m. Notice that the mode initially exhibits diffraction until the exponentiation begins after approximately 2 m after the start of the wiggler, after which there is some over-focusing and oscillation in the radiation spot size. Finally, the radiation spot size expands rapidly again after saturation.

We consider harmonic generation by including the fundamental through the ninth harmonic in the simulation. Harmonics are included with zero initial power; hence, initial harmonic growth is dependent on bunching at the fundamental. Once the harmonic generation begins, however, further growth is due to both linear and nonlinear mechanisms. While the fundamental and odd harmonic interactions display even symmetry, the interaction with the even harmonics has odd symmetry [34–35], and the Gauss-Hermite superpositions for the even and odd harmonics reflect this symmetry.

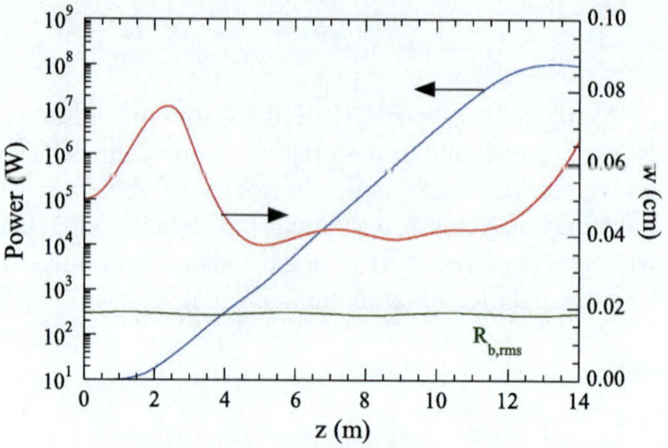

Fig. 8.23 Evolution of the power and radiation spot size for the fundamental [31]

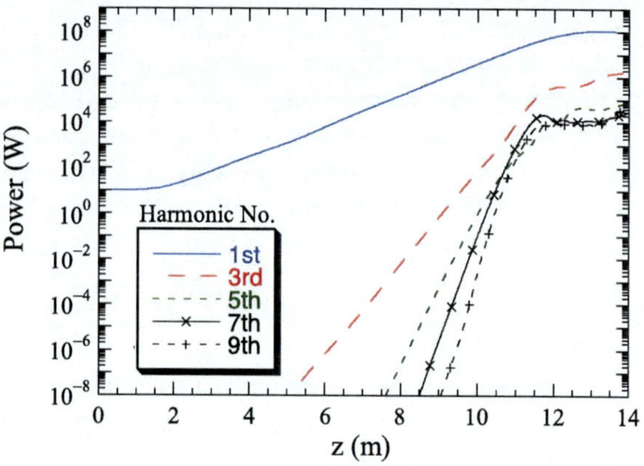

Fig. 8.24 Evolution of the power in the fundamental and odd harmonics [31]

The evolution of the power is shown in Fig. 8.24 for the fundamental and odd harmonics and in Fig. 8.25 for the even harmonics. While not shown in the figures, in the early stages of the interaction, the harmonic growth is linear, but this increases to the nonlinear growth rates once the fundamental power has grown large enough. For the purposes of illustration, we have chosen to include the harmonic power after they have passed 10^{-8} W, which highlights the nonlinear harmonic generation mechanism. Note that the harmonics grow rapidly with growth rates greater than that of the fundamental and which increase with harmonic number. The wavelengths, gain lengths, and peak output power are summarized in Table 8.1. It is well-known that the odd harmonics are favored for planar wigglers, and the powers at the even harmonics are smaller than the adjacent odd harmonics. However, the odd harmonic powers are substantial with 2.67 MW at the third harmonic and decreasing relatively slowly to 50 kW at the ninth harmonic.

It is important to observe that the gain lengths found in simulation vary, approximately, in inverse proportion to the harmonic number as discussed above. This is illustrated in Fig. 8.26 where we plot the growth rate (represented by the inverse gain

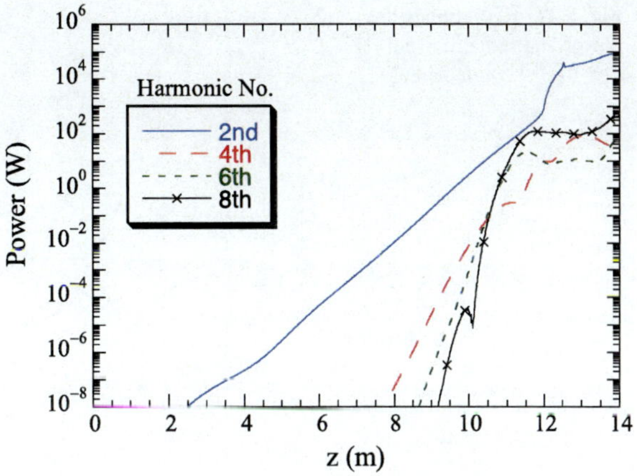

Fig. 8.25 Evolution of the power in the even harmonics [31]

Table 8.1 Variation in the gain length and output power with harmonic number [31]

Harmonic no.	Wavelength (nm)	Gain length (m)	Power
1	518.82	0.619	110 MW
2	259.41	0.364	189 kW
3	172.94	0.217	2.67 MW
4	129.71	0.166	84.3 W
5	103.76	0.128	103 kW
6	86.47	0.115	195 W
7	74.12	0.092	49.5 kW
8	64.85	0.079	478 W
9	57.65	0.072	50.0 kW

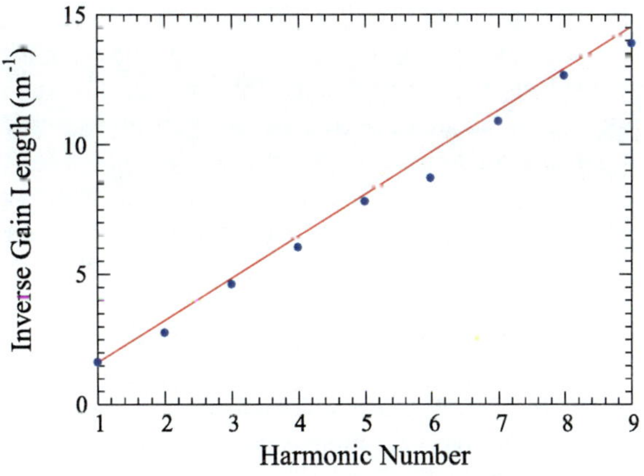

Fig. 8.26 The variation in the inverse gain length with harmonic number [31]

length) versus harmonic number. It is clear from the figure that the harmonic scaling law is well satisfied in the simulation. This cannot be explained by the linear instability and is due to the nonlinear generation mechanism.

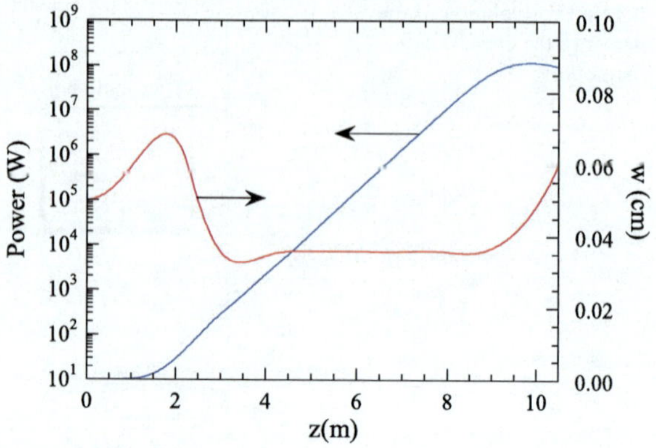

Fig. 8.27 Evolution of the power and radiation spot size for the fundamental

8.2.3 *Helical Wiggler Configurations*

We now consider harmonic generation in a short wavelength free-electron laser with a helical wiggler and employ the nonlinear formulation described in Chap. 6. The parameters under consideration are chosen to be equivalent to the planar wiggler discussed in the preceding section and deal with an electron beam with an energy of 219.5 MeV, a peak current of 150 A, a normalized emittance of 5 mm-mrad, and an energy spread of 0.10%. The helical wiggler has a peak on-axis wiggler field strength of 7.10 kG with a wiggler period of 3.3 cm. The electron beam is assumed to be matched into the wiggler so that the initial rms beam radius is approximately 183 μm. For these parameters, the fundamental resonance is at a wavelength of 516.8 nm.

As in the case of the planar wiggler, we assume that the initial seed power is 10 W. The evolution of the power and spot size are shown in Fig. 8.27. Also, as in the case of the planar wiggler, the optical field initially diffracts but is focused down to an average spot size of about 350 μm once the exponentiation starts. Saturation is found after about 10 m at a power level of 118 MW. We remark that while the resonance condition and saturation efficiency is close to that found for the equivalent planar wiggler example discussed previously, the gain length is shorter for the helical wiggler. This is due to the increase in the planar wiggler gain length due to the *JJ*-factor.

In simulating the harmonic interaction in the helical wiggler, we consider the second through fifth harmonics. The harmonic interaction in a helical wiggler is due to a phase resonance, rather than the oscillation in the axial velocity that occurs in a planar wiggler. As a result, the dynamics of the odd harmonics differs from that of the even harmonics resulting in relatively weaker even harmonic generation. This is not the case for the harmonics in a helical wiggler where the harmonic strength decreases montonically with increasing harmonic number. This is shown in Fig. 8.28 where we plot the evolution of the power versus distance along the wiggler for the fundamental through the fifth harmonic.

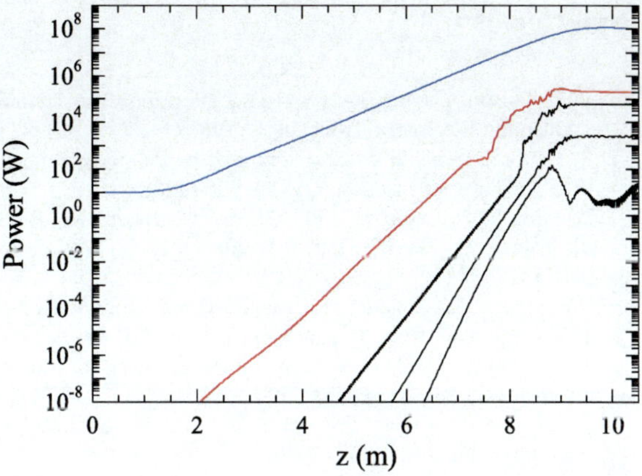

Fig. 8.28 Evolution of the power versus distance for the fundamental through the fifth harmonic

Table 8.2 Variation in the gain length and output power with harmonic number

Harmonic no.	Wavelength (nm)	Gain length (m)	Power
1	518.82	0.465	118 MW
2	259.41	0.206	281 kW
3	172.94	0.133	93.3 kW
4	129.71	0.119	3.62 kW
5	103.76	0.095	139 kW

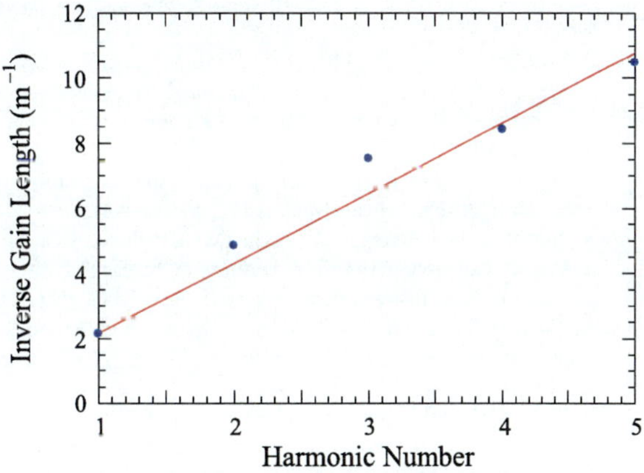

Fig. 8.29 The variation in the inverse gain length with harmonic number

The variation in the gain length and output power with the harmonic number is summarized in Table 8.2 for the helical wiggler. As in the case of the planar wiggler, the gain length of the harmonics scales decreases inversely with the harmonic number as expected for the nonlinear harmonic generation mechanism. This is shown graphically in Fig. 8.29 where we plot the inverse gain length versus the harmonic number.

References

1. W.B. Colson, The nonlinear wave equation for higher harmonics in free-electron lasers. IEEE J. Quantum Electron. **QE-17**, 1417 (1981)
2. R. Coïsson, Generalized description of harmonic generation in a transverse optical klystron. IEEE J. Quantum Electron. **QE-19**, 306 (1983)
3. B. Girard, Y. Lapierre, J.M. Ortega, C. Bazin, M. Billardon, P. Ellaume, M. Bergher, M. Velghe, Y. Petroff, Optical frequency multiplication by an optical klystron. Phys. Rev. Lett. **53**, 2405 (1984)
4. H.P. Freund, S. Johnston, P. Sprangle, Three-dimensional theory of free-electron lasers with an axial guide field. IEEE J. Quantum Electron. **QE-19**, 322 (1983)
5. A.K. Ganguly, H.P. Freund, Nonlinear analysis of free-electron laser amplifiers in three dimensions. Phys. Rev. A **32**, 2275 (1985)
6. R.C. Davidson, Kinetic description of harmonic instabilities in a planar wiggler free-electron laser. Phys. Fluids **29**, 267 (1986)
7. M.J. Schmitt, C.J. Elliot, Even harmonic generation in free-electron lasers. Phys. Rev. A **34**, 4843 (1986)
8. J.M. Ortega, Harmonic generation in the VUV on a storage ring and prospects for Super-ACO at Orsay. Nucl. Instrum. Methods Phys. Res **A250**, 203 (1986)
9. B.M. Kinkaid, Laser harmonic generation using the optical klystron and the effects of wiggler errors. Nucl. Instr. Meth **A250**, 212 (1986)
10. P. Ellaume, Theory of the optical klystron. Nucl. Instrum. Methods Phys. Res **A250**, 220 (1986)
11. M.J. Schmitt, C.J. Elliot, The effects of harmonic wiggler field components on free-electron laser operation. IEEE J. Quantum Electron. **QE-23**, 1552 (1987)
12. A. Gover, A. Friedman, A. Luccio, Three dimensional modelling and numerical analysis of super-radiant harmonic emission in a free-electron laser (optical klystron). Nucl. Instrum. Methods Phys. Res **A259**, 163 (1987)
13. C.J. Elliot, M.J. Schmitt, XUV harmonic enhancement by magnetic fields. Nucl. Instr. Meth **A259**, 177 (1987)
14. R. Prazeres, Y. Lapierre, J.M. Ortega, Monte Carlo simulation of the harmonic generation in an optical klystron or undulator. Nucl. Instrum. Methods Phys. Res **A259**, 184 (1987)
15. H.P. Freund, C.L. Chang, H. Bluem, Harmonic generation in free-electron lasers. Phys. Rev. A **36**, 3218 (1987)
16. H. Bluem, H.P. Freund, C.L. Chang, Harmonic content in a planar wiggler based free-electron laser amplifier. Nucl. Instr. Meth **A272**, 579 (1988)
17. R. Prazeres, J.M. Ortega, C. Bazin, M. Bergher, M. Billardon, M.E. Couprie, M. Velghe, Y. Petroff, Coherent harmonic generation in the vacuum ultraviolet spectral range on the storage ring ACO. Nucl. Instrum. Methods Phys. Res **A272**, 68 (1988)
18. K. Carlson, W. Fann, J.M.J. Madey, Spatial distribution of the visible coherent harmonics generated by the Mark III free-electron laser. Nucl. Instrum. Methods Phys. Res **A272**, 92 (1988)
19. B.A. Hooper, S.V. Benson, A. Cutolo, J.M.J. Madey, Experimental results of two stage harmonic generation with picosecond pulses on the Stanford Mark III free-electron laser. Nucl. Instrum. Methods Phys. Res **A272**, 96 (1988)
20. S.K. Dutt, A. Friedman, A. Gover, C. Pellegrini, Three dimensional simulation of a high harmonic transverse optical klystron. Nucl. Instrum. Methods Phys. Res **A272**, 564 (1988)
21. M.J. Schmitt, C.J. Elliot, B.E. Newnam, Harmonic power implications on free-electron laser mirror design. Nucl. Instrum. Methods Phys. Res **A272**, 586 (1988)
22. H.P. Freund, H. Bluem, R.H. Jackson, Nonlinear theory and design of a harmonic ubitron/free-electron laser. Nucl. Instrum. Methods Phys. Res **A285**, 169 (1989)
23. S.V. Benson, J.M.J. Madey, Demonstration of harmonic lasing in a free-electron laser, in *Proc. Int. Conf. Lasers '88*, ed. by R.C. Sze, F.J. Duarte, (STS Press, McLean, 1989), p. 189

24. D.J. Bamford, D.A.G. Deacon, Measurement of the coherent harmonic emission from a free-electron laser oscillator. Phys. Rev. Lett. **62**, 1106 (1989)
25. S.V. Benson, J.M.J. Madey, Demonstration of harmonic lasing in a free-electron laser. Phys. Rev. A **39**, 1579 (1989)
26. R.W. Warren, L.C. Haynes, D.W. Feldman, W.E. Stein, S.J. Gitomer, Lasing on the third harmonic. Nucl. Instrum. Methods Phys. Res **A296**, 84 (1990)
27. D.J. Bamford, D.A.G. Deacon, Harmonic generation experiments on the Mark III free-electron laser. Nucl. Instrum. Methods Phys. Res **A296**, 89 (1990)
28. W.M. Sharp, E.T. Scharlemann, W.M. Fawley, Three-dimensional simulation of free-electron laser harmonics with FRED. Nucl. Instrum. Methods Phys. Res **A296**, 335 (1990)
29. H. Bluem, R.H. Jackson, D.E. Pershing, J.H. Booske, V.L. Granatstein, Final design and cold tests of a harmonic ubitron amplifier. Nucl. Instrum. Methods Phys. Res **A296**, 37 (1990)
30. P.E. Latham, B. Levush, T.M. Antonsen, N. Metzler, Harmonic operation of a free-electron laser. Phys. Rev. Lett. **66**, 1442 (1991)
31. H.P. Freund, S.G. Biedron, S.V. Milton, Nonlinear harmonic generation in free-electron lasers. IEEE J. Quantum Electron. **36**, 275 (2000)
32. Z. Huang, K.-J. Kim, Three-dimensional analysis of harmonic generation in high-gain free-electron lasers. Phys. Rev. E **62**, 7295 (2000)
33. H.P. Freund, R.C. Davidson, D.A. Kirkpatrick, Thermal effects on the linear gain in free-electron lasers. IEEE J. Quantum Electron. **QE-27**, 2550 (1991)
34. H. Bluem, R.H. Jackson, H.P. Freund, D.E. Pershing, V.L. Granatstein, Demonstration of a new free-electron laser harmonic interaction. Phys. Rev. Lett. **67**, 824 (1991)
35. M.J. Schmitt, C.J. Elliot, Generalized derivation of free-electron laser harmonic radiation from plane-polarized wigglers. Phys. Rev. A **41**, 3853 (1990)

Chapter 9
Oscillator Configurations

In the previous chapters, it has been shown that an energetic electron beam propagating through an oscillatory magnetic field is capable of amplifying electromagnetic radiation. That is, the combination of the beam and the oscillatory field can be regarded as a medium with an inherent gain. The principal focus of the discussion in the previous chapters has been on the amplification of an injected signal with a specified frequency. However, the gain mechanism can also be used to form an oscillator by the feedback of a portion of the output signal. The radiation in an oscillator may be self-excited in the sense that radiation will spontaneously grow from noise in the oscillator if the gain the radiation experiences on traversing the interaction region exceeds the losses the radiation experiences on its return path to the input of the interaction region (including the portion of the radiation that is allowed to leave the oscillator as output). An oscillator may also be mode-locked by the injection of a large-amplitude signal with a frequency within the gain band.

Since the gain in the interaction region increases with beam current, there is a threshold value of current (known as the start current) below which the gain is exceeded by the losses in the cavity and the radiation will not grow. If the beam current exceeds this threshold, spontaneously emitted noise in the electron beam will be amplified and will grow exponentially in time until the radiation in the oscillator reaches a saturation level. It will be shown in this and subsequent chapters that the characteristics of a free-electron laser oscillator are determined to a large degree by the ratio of the operating current to the start oscillation current.

There are a number of practical reasons for constructing an oscillator as opposed to an amplifier. First, it may be the case that no source is available in the frequency range of interest to drive an amplifier. Second, the amplification that the radiation may experience in one pass through the available interaction region in the frequency range of interest may be too small. An amplifier with a small gain is of little practical value. However, if most of the output signal can be fed back to the input, a useful oscillator can be constructed. As a general rule, one finds that shorter wavelength

H. P. Freund, T. M. Antonsen, Jr., *Principles of Free Electron Lasers*,
https://doi.org/10.1007/978-3-031-40945-5_9

devices which rely on relatively high-energy but low-current electron beam sources fall into this category and tend to be oscillators because of their inherent low gain. In contrast, longer wavelength devices that employ lower energy but higher current electron beams can drive amplifiers as well as oscillators.

As previously mentioned, the radiation in an oscillator is self excited, whereas a signal with a specific frequency is injected in an amplifier. Thus, unless the oscillator is operated in a mode-locked manner, the frequency of radiation, which emerges from an oscillator, is determined by the self-consistent action and reaction of the electron beam and the radiation. Consequently, it is more likely to find that the radiation emerging from the device is non-monochromatic (that is multifrequency) in an oscillator than it is in an amplifier. This is particularly the case in the class of oscillators driven by pulsed electron beams, such as those produced by radio frequency accelerators or storage rings, in which the emerging radiation has a temporal structure, which also consists of pulses. In such systems, the radiation and beam pulses overlap within the interaction region. Due to the pulsed nature of the radiation, the emerging signal is necessarily non-monochromatic, and the temporal length of the pulse determines the minimum possible width of the radiation spectrum. For long pulse oscillators in which the electron beam can be regarded as continuous, in contrast, the spectrum may still be broadband due to other effects. In such cases the gain bandwidth (the range of frequencies over which a signal can be amplified) is typically larger than the spacing in frequency between the natural modes of the resonant cavity, which is formed by the interaction region and the feedback loop. This spacing is the inverse of the transit time of radiation through the cavity. At least initially, therefore, one can expect a large number of different frequencies to be excited in the oscillator. Further, as we have seen in Chap. 6 where sidebands were discussed, the presence of a large amplitude signal at one frequency can modify the electron trajectories and generate amplification at a second frequency where previously (i.e., in the absence of the large amplitude radiation signal) there was no amplification. The difference between the frequencies of these two competing signals can be much smaller than the bandwidth of the feedback path. Thus, it may be difficult to discriminate these two signals in order to eliminate the undesired one. All told, the basic conceptual problem which is posed in the description of free-electron laser oscillators is that the radiation field is composed of a spectrum of frequencies which must be determined self-consistently along with the motion of the electrons and which requires the development of a time-dependent formulation of the interaction.

9.1 General Formulation

As discussed in previous chapters, amplification of radiation occurs in a free-electron laser due to the bunching and subsequent coherent radiation of electrons as they move in the combined wiggler and radiation fields. The electrons are bunched by a second-order axial force resulting from the cross product between the first-order wiggle velocities and transverse magnetic fields associated with the radiation and the magnetic wiggler. This *ponderomotive* force appears as a beat wave, which has a phase velocity equal to the radiation frequency divided by the sum of the radiation and wiggler wave numbers [$\omega/(k_w + k)$]. The most important physical processes then are those that cause individual electrons to fall in and out of phase with the beat wave. These processes can be understood within the context of a simple one-dimensional model [1–3] in which the important dynamical variables are the energies and beat wave phases of electrons, and the magnitude and phase of the radiation. Further, since the bunching force is a second-order effect, it is safe to assume that the dynamical variables change slowly on the time scale associated with the traversal of one period of the wiggler. This leads to an additional simplification in that the short time scale associated with the traversal of one wiggler period can be averaged over. As shown in Chap. 5, this wiggler-average removes the effect of the lower beat wave from the analysis of planar wiggler configurations. However, this is adequate for the purposes of determining the average properties of many free-electron lasers. The result is that the bulk energy and beat wave phase of electrons are described by a pendulum equation in which the "effective gravity" is derived from the amplitude of the radiation. Further, the amplitude of the radiation evolves self-consistently in space in time in response to the current generated by the bunched electrons. This simple model has been studied for decades in connection with traveling wave tubes. The first section of this chapter is devoted to a simple derivation of these governing equations.

We begin our consideration of the free-electron laser oscillator equations with a discussion of the representation of the radiation field. For simplicity, we start with a one-dimensional model in which the vector potential that generates the radiation field can be represented in complex phasor notation as

$$\delta\mathbf{A}(z,t) = \delta A(z,t)\widehat{\boldsymbol{e}}_p \exp[ikz - i\omega t] + c.c. \tag{9.1}$$

where $\delta A(z,t)$ is a time- and space-dependent complex amplitude, ω and k are the central frequency and wavenumber of the radiation, $\hat{\boldsymbol{e}}_p$ is a polarization vector, and *c. c.* denotes the complex conjugate. It is assumed that the complex amplitude depends weakly on its arguments in the sense that the amplitude does not change significantly in a distance corresponding to a radiation wavelength or time corresponding to a period of the radiation. The polarization vector describes the relative orientations of the components of the radiation field that will be amplified. For example, a planar wiggler will excite plane-polarized radiation, while a helical wiggler excites circularly polarized radiation. The values of ω and k are chosen to be in the center of the amplification band and are related by the dispersion relation for the propagation of radiation in the absence of an electron beam, for example, $\omega = ck$ for plane waves propagating in vacuum. If the interaction occurs in a waveguide, the dispersion relation is modified to account for the cut off frequency. It should be remarked that the exact values of the frequency and wavenumber are somewhat arbitrary. Different choices of ω and k will lead to differences in the solution for the amplitude but will presumably leave the actual vector potential unaffected. A precise definition of ω and k will be specified as appropriate examples are discussed.

Within the context of the one-dimensional model under consideration, the important coordinates of a particle are its phase relative to the ponderomotive beat wave formed by the beating of the wiggler and radiation fields

$$\psi = (k + k_w)z - \omega t \tag{9.2}$$

and the electron's energy $\gamma m_e c^2$, where $\gamma = (1 - v^2/c^2)^{-1/2}$ is the relativistic factor. As the electrons travel through the combined fields of the wiggler and the radiation, their phases and energies evolve in time according to the pair of equations,

$$\frac{d}{dt}\psi = (k + k_w)v_z - \omega \tag{9.3}$$

and

$$\frac{d}{dt}\gamma = -\frac{e}{m_e c^2}\mathbf{v} \cdot \delta\mathbf{E} \tag{9.4}$$

where δ**E** denotes the radiation electric field and **v** is the electron velocity.

Consider the limit in which the interaction length, L, is many wiggler periods long (i.e., $\lambda_w << L$). One consequence of a long interaction region is that the spectrum of the radiation is narrow. This is evidenced by both the gain bandwidth (see Eq. 4.59 for the low-gain regime), and the bandwidth of spontaneous emission given by (3.25) which both scale as λ_w/L. Thus, the fractional bandwidth scales as the reciprocal of the number of wiggler periods, and the assumption that the radiation amplitude varies slowly in time on the scale of the wave period is justified. A further consequence of this assumption is that the electric field in (9.4) may be evaluated by ignoring the time derivatives of the complex amplitude $\delta A(z,t)$

$$\delta\mathbf{E}(z,t) = \frac{i\omega}{c}\delta A(z,t)\widehat{\mathbf{e}}_p \exp[ikz - i\omega t] + c.c., \tag{9.5}$$

The change in both phase and energy of an electron over one wiggler period is also expected to be small for a long interaction region. Recall that the efficiency will be a maximum and the saturation point will be found when an electron completes half of a synchrotron oscillation in the ponderomotive potential in the time it takes to traverse the interaction region. Alternatively, the ponderomotive phase will change by order unity on traversing the interaction region. Thus, the net change in phase in traversing one period of the wiggler will be small.

Equations. (9.3) and (9.4) can be integrated by perturbation about the single-particle trajectories over a time equal to the travel time in traversing one period of the wiggler. In this case, it is assumed that the velocity (on the right-hand side of the equations) is given solely by its value in the presence of the wiggler field.

Furthermore, the slow variation of the amplitude of the radiation vector potential can be ignored on the space and time scales associated with the traversal of one wiggler period. The average rate of change of phase and energy, therefore, can be expressed as

$$\frac{\Delta\psi}{\Delta t} = (k + k_w)v_{\parallel} - \omega \tag{9.6}$$

and

$$\frac{\Delta\gamma}{\Delta t} = -\frac{i\omega}{c}\delta a \int_{t}^{t+\Delta T} \frac{dt'}{\Delta T}\mathbf{v}(z')\cdot\hat{\boldsymbol{e}}_p \exp[ikz' - i\omega t'] + c.c. \tag{9.7}$$

where $\delta a = e\delta A/m_e c^2$ is the normalized vector potential,

$$\Delta T \equiv \int_{z}^{z+\lambda_w} \frac{dz'}{v_z(\gamma, z')}, \tag{9.8}$$

is the travel time over one wiggler period, $v_{\parallel} = \lambda_w/\Delta T$ is the wiggler-averaged axial velocity, and z' appearing in Eq. (9.7) represents the position of a particle at time t'.

For the sake of simplicity, we will complete this derivation under the assumption of a helical wiggler. The discussion of the case of a planar wiggler will be given later. The helical wiggler case is easier to analyze due to the fact that the axial velocity of the beam in the wiggler is constant. In contrast, the axial velocity of an electron in a planar wiggler varies as it travels through the periodic magnetic field. While this complicates the analysis, the final equations for a wiggler-averaged orbit analysis of a planar wiggler configuration are formally identical to those for a helical wiggler. In the case of the idealized helical wiggler field given by Eq. (2.5), the single-particle electron velocity is given by

$$\mathbf{v}(z) = v_w\left[\hat{\boldsymbol{e}}_x\cos(k_w z) + \hat{\boldsymbol{e}}_y\sin(k_w z)\right] + v_{\parallel}\hat{\boldsymbol{e}}_z, \tag{9.9}$$

where $v_w = -\Omega_w/k_w$ is the wiggler-induced transverse velocity (see Eqs. (2.7) and (2.8) for an idealized beam in which the canonical momenta vanish) and $v_{\parallel}$ is given by solution of Eq. (2.16). The appropriate polarization vector in this case describes a circularly polarized wave for which

$$\hat{\boldsymbol{e}}_p = \frac{1}{2}\left(\hat{\boldsymbol{e}}_x + i\hat{\boldsymbol{e}}_y\right), \tag{9.10}$$

and the time to traverse one wiggler period is $\Delta T = \lambda_w/v_{\parallel}$. Insertion of Eqs. (9.9) and (9.10) into (9.7) allows the integral in Eq. (9.7) to be evaluated yielding the average rate of change of energy,

$$\frac{\Delta\gamma}{\Delta T} = -\frac{\omega}{2\gamma}u_w[i\delta a(z,t)\exp(\iota\psi) + c.c.], \tag{9.11}$$

where $a_w/\gamma = v_w/c$.

Consider the energy dependence of the rate of change of phase given by Eq. (9.6). There exists a resonant energy, $\gamma_r m_e c^2$, for which the phase is constant. At this energy, the average velocity of a particle is equal to the phase velocity of the beat wave, $v_{\parallel} = \omega/(k + k_w)$. Electrons with energies far from this resonance value will not interact strongly with the radiation and, hence, will not contribute to amplification. The width of the resonance can be estimated by consideration of the range of energies for which the total change in phase in traversing the interaction region will be of order unity; specifically,

$$|\gamma - \gamma_r| = \left[L(k + k_w)\frac{\gamma_r}{v_{\parallel}}\frac{dv_{\parallel}}{d\gamma_r}\right]^{-1}. \tag{9.12}$$

For a long interaction region, the relative deviation of an electron's energy from the resonant value will be small. As a consequence, a perturbation analysis of the dynamical equations which govern the phase and energy about the resonant energy may be performed in which $\gamma = \gamma_r + \delta\gamma$, where it is assumed that $|\delta\gamma| << \gamma_r$. The average rate of change of phase and energy are evaluated along the trajectory of a beam electron, and the time derivatives are expressed in Eulerian form as

$$\frac{\Delta\psi}{\Delta T} = \left(\frac{\partial}{\partial t} + v_{\parallel}\frac{\partial}{\partial z}\right)\psi = \omega_\gamma\delta\gamma, \tag{9.13}$$

and

$$\frac{\Delta\gamma}{\Delta T} = \left(\frac{\partial}{\partial t} + v_{\parallel}\frac{\partial}{\partial z}\right)\delta\gamma = -\frac{\omega}{2\gamma_r}a_w[i\delta a(z,y)\exp(i\psi) + c.c.], \tag{9.14}$$

where

$$\omega_\gamma \equiv (k + k_w)\frac{dv_{\parallel}}{d\gamma_r} \simeq \frac{\omega}{v_{\parallel}}\frac{dv_{\parallel}}{d\gamma_r}, \tag{9.15}$$

is related to the derivative of the average velocity with respect to the relativistic factor. Observe that the Eulerian treatment describes the convective derivatives of the phase and energy along the electron trajectories. The importance of this will be seen in the later treatment of beam slippage.

Equations (9.13) and (9.14) determine the trajectory of a single electron in the combined wiggler and radiation fields. These two coupled first-order differential equations require two initial conditions for solution, specifically, the values of $\delta\gamma$ and ψ for an electron when it enters the interaction region. With multiple electrons, this information must be specified for all of them. This is accomplished by definition of a distribution function $f_0(\delta\gamma,\psi,t)$ where $f_0(\delta\gamma,\psi,t)d\delta\gamma d\psi$ denotes the fraction of electrons which enter the interaction region at time t with energies and phases in the interval $\delta\gamma$ through $\delta\gamma + d\delta\gamma$ and ψ and $\psi + d\psi$, respectively. In addition, the normalization requirement on this distribution is chosen as

$$\int_{-\infty}^{\infty} d\delta\gamma \int_{0}^{2\pi} d\psi f_0(\delta\gamma,\psi,t) = 1. \tag{9.16}$$

Observe that if electrons enter the interaction region with entrance times distributed uniformly over a wave period, then it is appropriate to assume that the distribution function is independent of the phase.

Equations (9.13) and (9.14) are equivalent to the pendulum equation discussed in Chap. 4. These equations of motion can be derived from a Hamiltonian, which has the form

$$H(\delta\gamma,\omega,z,t) = \frac{\omega_\gamma}{2}\delta\gamma^2 + \frac{\omega}{2\gamma_r}a_w[\delta a(z,t)\exp(i\psi) + c.c.], \tag{9.17}$$

Hamilton's equations,

$$\frac{d}{dt}\psi = \frac{\partial H}{\partial \delta\gamma} \text{and} \frac{d}{dt}\delta\gamma = -\frac{\partial H}{\partial \psi}, \tag{9.18}$$

then yield the dynamical Eqs. (9.13) and (9.14), where the total time derivative follows the trajectory of an electron moving with the resonant velocity. If the radiation amplitude is independent of space and time, then the Hamiltonian is conserved, and one may draw the phase space trajectories and separatrices as was done in Chap. 4. In the general case however, the field amplitude depends on both space and time leading to extremely complicated electron motion.

The electron equations of motion can be replaced by the Vlasov equation which describes the evolution of the local distribution function $f(\delta\gamma,\psi,z,t)$

$$\left(\frac{\partial}{\partial t} + v_\parallel \frac{\partial}{\partial z}\right) f + \frac{\partial H}{\partial \delta\gamma}\frac{\partial f}{\partial \psi} - \frac{\partial H}{\partial \psi}\frac{\partial f}{\partial \delta\gamma} = 0. \tag{9.19}$$

The solution of the Vlasov equation must satisfy the boundary condition $f(\delta\gamma,\psi, z = 0,t) = f_0(\delta\gamma,\psi,t)$ representing the distribution of $\delta\gamma$ and ψ values for the injected beam. The particle equations of motion and the Vlasov equation describe exactly the same physics, and the particular choice of formulation is largely a matter of convenience.

It remains to determine the equations that describe the evolution of the radiation field. To do this, the specific form of the radiation vector potential is inserted into the one-dimensional wave Eq. (5.19). Observe that by writing the electric field in terms of the vector potential of an electromagnetic wave and not including a scalar potential, it is implicitly assumed that collective (i.e., Raman) effects may be neglected. For the specific choice of the vector potential in Eq. (9.1), the wave equation takes the form

$$\left(\frac{\partial}{\partial t}+v_g\frac{\partial}{\partial z}\right)\delta A(z,t)=\frac{2\pi ic}{\omega}\widehat{e}^*_{\mathbf{p}}\cdot\mathbf{J}(z,t)\exp[-ikz+i\omega t], \tag{9.20}$$

where $\mathbf{J}(z,t)$ is the current density and $v_g = kc^2/\omega$ denotes the group velocity of the radiation. Note that only first derivative terms of the complex amplitude have been retained due to the assumed slow space and time variation of the amplitude. For the plane wave case developed here, $\omega = kc$ and the group velocity equal the speed of light. In a waveguide, ω and k satisfy a different dispersion relation that results in a group velocity, which is below the speed of light. The distinction between the group velocity and the speed of light will be maintained here in the interest of generality even though they are identical for this particular example. The components of the current density which resonantly drive the radiation field can be computed by forming the product of the local beam charge density and the single-particle electron velocity vector, $\mathbf{J}(z,t) = \rho(z,t)\mathbf{v}(z)$, where $\mathbf{v}(z)$ is given by Eq. (9.9) for the case under consideration of a helical wiggler.

The beam charge density $\rho(z,t)$ evolves into a nonuniform state in space and time owing to the bunching in phase of the beam particles as they interact with the radiation. The charge density can be visualized as consisting of clumps of charge moving at the resonant velocity $v_\parallel$, with a spatial wavenumber $k_b = \omega/v_\parallel$. This corresponds to a frequency of oscillation equal to that of the radiation. The current density, which is the product of the charge density and the electron velocity, will have a frequency equal to that of the radiation and a spatial variation with wavenumbers, which are the sum and difference of the wiggler wavenumber and the wave number of the charge density k_b. According to the resonance condition $[\omega = (k + k_w)v_\parallel]$, only the portion of the current density whose spatial wave number is the difference of the wiggler, and charge density wavenumbers will resonantly excite the radiation. This portion of the current density which resonantly interacts with the radiation can then be expressed in terms of an average of the factor $\exp(-i\psi)$ as follows:

$$\mathbf{J}(z,t)\exp[ikz+i\omega t]\rightarrow J(t_e)\frac{v_w}{v_\parallel}\widehat{e}_p\langle\exp(-i\psi)\rangle, \tag{9.21}$$

where $J(t_e)$ is the current density entering the interaction region at time t_e. It is important to observe that the transverse current for the choice of a helical wiggler is precisely that required to excite the circularly polarized wave. The function $t_e(z,t)$ specifies the entrance time of the group of electrons which contribute to the high-frequency current at the point z and time t:

$$t_e(z,t) = t - z/v_{\|}, \tag{9.22}$$

This entrance time function will appear later in the discussion of slippage.

The average appearing in Eq. (9.21) can be represented as an average over the initial energies and entrance phases of beam particles injected at time t_e as follows:

$$\langle \exp(-i\psi) \rangle = \int_{-\infty}^{\infty} d\delta\gamma_i \int_0^{2\pi} d\psi f_o(\delta\gamma_i, \psi_i, t_e) \exp[-i\psi(\delta\gamma_i, \psi_i, t_e)], \tag{9.23}$$

where the subscript i refers to the initial condition of the electron upon entry into the interaction region and $\psi(\delta\gamma_i,\psi_i,t_e)$ is the phase of the particle launched at time $t_e(z,t)$ with initial values of phase and energy deviation given by ψ_i and $\delta\gamma_i$, respectively. Equivalently, if the Vlasov equation has been solved, then the average can be taken over the distribution of energies and phases at the specific point (z,t):

$$\langle \exp(-i\psi) \rangle = \int_{-\infty}^{\infty} d\delta\gamma \int_0^{2\pi} d\psi f(\delta\gamma, \psi, z, t_e) \exp(-i\psi), \tag{9.24}$$

Once again, we observe that these two averages are equivalent and which is used depending largely on convenience. In terms of these averages, the final wave equation can now be written in the form

$$\left(\frac{\partial}{\partial t} + v_g \frac{\partial}{\partial z}\right) \delta A(z,t) = \frac{2\pi i}{\omega} \frac{c^2 a_w}{\gamma_r v_{\|}} J(t_e) \langle \exp(-i\psi) \rangle, \tag{9.25}$$

The solution of this equation requires specifying the complex radiation amplitude at the entrance of the interaction region, $\delta A(z = 0,t)$.

In order to complete this formulation, an equation must be introduced which describes the feedback loop. The simplest model for this process is one in which the radiation propagates without dispersion through an optical system from the end of the interaction region at $z = L$ to the entrance at $z = 0$, taking a time T_r and suffering an amplitude loss factor μ. The field amplitude at $z = 0$, therefore, can be expressed as

$$\delta A(z=0,t) = \mu \exp[i(kL + \omega T_r)] \delta A(z=L, t - T_r), \tag{9.26}$$

where μ may be complex. The central frequency and wavenumber can be specified by the requirement that the product $R = \mu \exp[i(kL + \omega T_r)]$ is real. In this case, the natural frequencies of the cavity formed by the feedback loop and the interaction region are given by

$$\omega_n = \omega + \frac{2\pi n + i \ln R}{T}, \tag{9.27}$$

where n is an integer (positive or negative) that labels the different longitudinal modes of the oscillator cavity and

$$T \equiv T_r + \frac{L}{v_g}, \tag{9.28}$$

is the total round-trip time for radiation to circulate through the cavity. The imaginary part of the natural frequency ω_n describes the damping of the radiation in the cavity due to losses and output coupling. The power will decay exponentially with a characteristic time T_d given by

$$T_d \equiv \frac{T}{|\ln R^2|}. \tag{9.29}$$

The quality factor for the cavity is defined based upon this decay rate as

$$Q = \omega T_d. \tag{9.30}$$

Due to the assumed narrowness of the spectrum and the fact that all frequencies are attenuated by the same factor R, the quality factor is essentially the same for all modes. Thus, in the present model, there is no mechanism to control the frequency of oscillation other than the natural gain bandwidth of the free-electron laser amplification mechanism. Solutions of this governing system of equations (Eqs. (9.13), (9.14), and (9.25) and the associated boundary conditions) describe a variety of different time-dependent phenomena which can occur in a free-electron laser oscillator. A discussion of these solutions for both continuous and pulsed electron beam systems forms the bulk of the remaining discussion of free-electron laser oscillators.

9.2 Planar Wiggler Equations

The derivations of Eqs. (9.13), (9.14), and (9.25) need to be modified if a planar wiggler is used. The important difference that arises in the consideration of planar wigglers is that the instantaneous axial velocity is no longer a constant as an electron propagates through one period of the wiggler field. For example, in the planar wiggler field given by Eq. (2.99), the electrons will have a velocity $\mathbf{v}$ given by

$$\mathbf{v} = v_w \hat{\mathbf{e}}_x \cos k_w z + v_z \hat{\mathbf{e}}_z, \tag{9.31}$$

where the instantaneous z-dependent axial velocity is determined from energy conservation to be

$$v_z = c\left(1 - \frac{1}{\gamma^2} - \frac{v_w^2}{c^2}\cos^2 k_w z\right)^{1/2}, \tag{9.32}$$

and it has been assumed that the electrons are injected into the wiggler with no transverse canonical momentum. The wiggle velocity for this case is in the x-direction, and, as a result, the radiation will be plane polarized with $\widehat{\boldsymbol{e}}_p = \widehat{\boldsymbol{e}}_x$.

The periodic modulations in the axial velocity cause the ponderomotive phase ψ given in (9.3) to have periodic modulations as well. Equation (9.6), which is obtained by averaging over the travel time through one period of the wiggler, represents the rate of change of the average value of the ponderomotive phase with the periodic modulations removed. Thus, it is convenient to write ψ as the sum of its slowly varying average part ψ_0 and the periodic modulations $\delta\psi$ with period equal to that of the wiggler. The periodic modulation is expressed in terms of the difference in time to reach a particular point in the wiggler between an electron traveling at the instantaneous velocity and at the average velocity; hence

$$\delta\psi = -\omega \int_{z_o}^{z} dz' \left(\frac{1}{v_z} - \frac{1}{v_{\|}} \right), \tag{9.33}$$

where $v_{\|}$ denotes the average axial velocity defined in Eq. (2.105). The time integral appearing in Eq. (9.7) can now be evaluated as an integral over z',

$$\int_{t}^{t+\Delta T} \frac{dt'}{\Delta T} \mathbf{v}(z') \cdot \widehat{\boldsymbol{e}}_p \exp[ikz' - i\omega t'] = \frac{1}{2} \mathrm{v}_w C_1 \exp(i\psi_0), \tag{9.34}$$

where C_1 is defined by the integral

$$C_1 \equiv \int_{z}^{z+\lambda_w} \frac{dz'}{v_z \Delta T} [1 - \exp(-2ik_w z')] \exp[i\delta\psi(z')]. \tag{9.35}$$

The modification to the equations of motion for the planar wiggler case, therefore, can be accounted for by replacing a_w by the product $a_w C_1$.

Similar considerations apply to the derivation of the wave Eq. (9.25). The periodic modulation of the electron phase must be accounted for in the excitation of the radiation by replacing the factor a_w appearing in (9.25) by the product $a_w C_1{}^*$.

Finally, it is a straightforward procedure to incorporate the effects of higher harmonics in the radiation field in the present model. Let δa_l be the complex amplitude of the lth harmonic of the radiation having a frequency $l\omega$ and a wavenumber lk. These harmonics contribute to the rate of change of particle energy in Eq. (9.14) according to the substitution

$$\delta a \exp(i\psi) \rightarrow \sum_l \delta a_l \exp(il\psi_0), \tag{9.36}$$

where a coefficient C_l is defined in analogy to Eq. (9.35),

$$C_l \equiv \int_{z}^{z+\lambda_w} \frac{dz'}{\upsilon_z \Delta T}[1 - \exp(-2ik_w z')] \exp[il\delta\psi(z')]. \tag{9.37}$$

The dependence of these coefficients on harmonic number has been discussed in Chap. 8. In particular, C_l is nonzero only for odd values of the harmonic number l and decreases with harmonic number. The wave equation must also be modified in the sense that a separate equation must be written for each harmonic amplitude, δA_l, which takes the form

$$\left(\frac{\partial}{\partial t} + \upsilon_g \frac{\partial}{\partial z}\right)\delta A_l = \frac{2\pi i}{l\omega} \frac{c^2 a_w}{\gamma_r \upsilon_\parallel} C_l^* \exp(il\psi_0). \tag{9.38}$$

The question of harmonic generation is important in free-electron laser oscillators because it is possible, by control of the optical feedback path, to cause the device to lase in a particular harmonic of the fundamental resonant frequency. In this way, shorter wavelengths can be generated without either increasing beam energy or decreasing the wiggler period.

9.3 Characteristics: Slippage

In the preceding sections, a system of equations describing the interaction of the radiation and beam particles in a free-electron laser oscillator has been derived. The system consists of an ensemble of first-order differential equations where the time and space derivatives represent a convective time derivative following the trajectory of the beam electrons, Eqs. (9.13), (9.14), and (9.19), or following the trajectory of the radiation, Eq. (9.25). It is usually the case that the radiation propagates faster than the beam. Thus, radiation, which enters the interaction region at a particular time, will overtake some group of electrons which have entered earlier. This is known as slippage, and its effects can be most easily understood by plotting the characteristics of the first-order partial differential equations in the z versus t plane [4].

Figure 9.1 shows a plot of the relevant characteristics. The orbit equations can be integrated to find either the distribution function or the phases and energies of particles at a particular space-time point (z_1,t_1). The group of electrons which arrive at the point z_1 at time t_1 have moved along a trajectory $t = t_1 + (z - z_1)/\upsilon_\parallel$. This trajectory is known as a characteristic for the equation of motion since the convective derivative in Eqs. (9.13), (9.14), and (9.19) can be written as a total derivative along this trajectory (shown as a dashed line in the figure). Electrons arriving at the space-time point (z_1,t_1) are only affected by the complex amplitude of the radiation, $\delta a(z,t)$, evaluated for points on the characteristic trajectory. For example, one can write the solution of Eq. (9.14) as an integral along this characteristic:

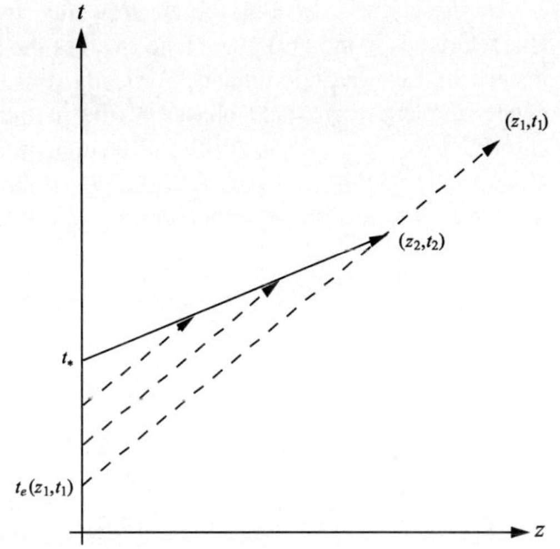

Fig. 9.1 Illustration of the characteristics in an oscillator configuration

$$\delta\gamma = \delta\gamma_i + \frac{\omega a_w}{2\gamma_r}\int_0^{z_1}\frac{dz}{v_\parallel}[i\delta a(z,t)\exp(i\psi) + c.c.], \tag{9.39}$$

where $\delta\gamma_i$ is the injected value of $\delta\gamma$ and z and t are related by the condition $t = t_1 + (z - z_1)/v_\parallel$.

To evaluate the integral in (9.39), the value of the radiation amplitude for points (z,t) on the characteristic must be determined. Let the space-time point (z_2,t_2) be a point on this characteristic where the complex amplitude of the radiation field is to be evaluated. To determine the field at this point, the wave equation Eq. (9.25) is integrated along its characteristics, that is, along a trajectory given by $t = t_2 + (z - z_2)/v_g$,

$$\delta A(z_2, t_2) = \delta A_e(t_2 - z_2/v_g) + \frac{2\pi i c^2 a_w}{\omega\gamma_r v_\parallel}\int_0^{z_2}\frac{dz}{v_g}J[t_e(z,t)]\langle\exp(-i\psi)\rangle, \tag{9.40}$$

where $\delta A_e(t)$ denotes the radiation amplitude upon entering the interaction region at time t. The characteristic for the wave equation is shown as a solid line in Fig. 9.1.

In order to perform the integral in (9.40), it will be necessary to know the high-frequency current density at all points along the wave equation characteristic ending at (z_2,t_2). To determine this current density, it is necessary to integrate the equations of motion along a number of different characteristics of the type shown as dashed lines in Fig. 9.1. In this way, the value of the complex amplitude of the radiation field at (z_2,t_2) becomes dependent on the values of the radiation amplitude at the entrance of the interaction region for a range of times given by $t_* = t_2 - z_2/v_g > t > t_e(z_2, t_2) = t_2 - z_2/v_\parallel$. Different portions of the radiation pulse can influence each other in this way.

An alternative to writing the electron trajectories in the form of integrals along characteristics in the (z,t) plane is to express the equations of motion in Lagrangian instead of Eulerian coordinates. Basically, this amounts to expressing the dependence of the energies and phases of the particles as functions of position z and entrance time $t_e = t - z/v_{\|}$. With this change of variables, the convective derivative along the trajectory of a particle in the (z,t) plane becomes an ordinary derivative with respect to position z, with entrance time t_e held fixed,

$$v_{\|}\frac{d}{dz}\psi(z,t_e) = \omega'\delta\gamma(z,t_e), \tag{9.41}$$

and

$$v_{\|}\frac{d}{dz}\delta\gamma(z,t_e) = -\frac{\omega}{2\gamma_r}a_w\left[i\delta a\left(z,t_e + z/v_{\|}\right)\exp(i\psi) + c.c.\right]. \tag{9.42}$$

The space and time dependence of the radiation field amplitude in Eqs. (9.41) and (9.42) have been left in terms of the Eulerian variables z and t. In order to continue to write the wave equation in Eulerian variables, we must still express the high-frequency current density appearing in Eqs. (9.25) and (9.40) in Eulerian variables:

$$J(t_e)\langle\exp(-i\psi)\rangle = J[t_e(z,t)]\langle\exp[-i\psi(z,t_e(z,t))]\rangle, \tag{9.43}$$

where $\psi(z,t_e)$ is the solution for the phase which satisfies the Lagrangian Eqs. (9.41) and (9.42).

A particularly simple case is the one in which the group velocity of the radiation is equal to the beam velocity. In this case, the characteristics of the wave equation and the equations of motion coincide. Since the electrons and radiation enter the interaction region and propagate together, the complex amplitude of the radiation at any space-time point (z_2,t_2) is, therefore, dependent only on the radiation which entered the interaction region at the time $t_e = t_2 - z/v_{\|}$ and the distribution function and current for the beam particles entering at that time as well. In this case different portions of the radiation pulse never communicate with one and another. While it seems that the case of equal beam and radiation speeds is rather special, it turns out that for highly relativistic electron beams, the beam velocity is close enough to the speed of light that in some cases the electron slippage can be neglected.

Figure 9.2 shows the system of characteristics for the determination of the amplitude of the radiation at the entrance of the interaction region $z = 0$ at the time t_3. The wavy line in the figure represents the return path of the radiation through the feedback loop. From the figure, it can be seen that the amplitude of the radiation entering the interaction region at time t_3 depends on the radiation that entered in the past between times $t_3 - T$ and $t_3 - T(1 + \varepsilon)$ where

$$\varepsilon = \frac{L}{T}\left(\frac{1}{v_{\|}} - \frac{1}{v_g}\right) = \frac{L}{L_c}\left(\frac{v_g}{v_{\|}} - 1\right), \tag{9.44}$$

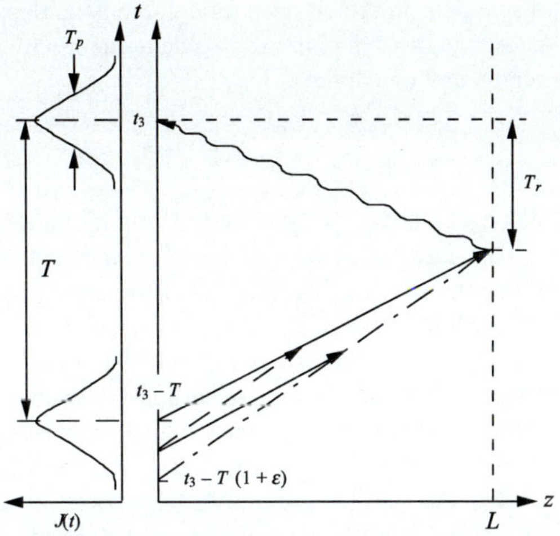

Fig. 9.2 Schematic illustration of the synchronism and slippage conditions between the electron beam pulses and the round-trip time of the radiation within the cavity

is the slippage parameter which measures how much the electrons slip behind the radiation during a time equal to the round-trip travel time of radiation through the optical length L_c of the cavity, where $T = L_c/v_g$.

Also shown on the left-hand side of Fig. 9.2 are plots, which characterize the injected electron current in a free-electron laser, which is driven by a pulsed electron beam. In the plot, the electron pulses, known as micropulses, are shown arriving at times separated by the round-trip travel time of radiation in the cavity so that the returning radiation pulse overlaps with the arrival of the next beam pulse. It will be seen at a later stage of the discussion that due to an effect known as *laser lethargy* [5], the electron pulses actually should have a separation in arrival times which is slightly larger than the round-trip travel time. The width of the electron pulse is characterized by a time T_p. A measure of the importance of slippage is the parameter $\varepsilon T/T_p$, which is the ratio of the distance the electron pulse slips behind the radiation pulse over the interaction length to the spatial width of the electron pulse. This parameter is, typically, smaller than unity. In a free-electron laser oscillator, which is driven by a continuous electron beam the relevant parameter is simply ε, which can be extremely small. Since it is through the process of slippage that different portions of the radiation pulse communicate with one another, the size of the slippage parameter determines the rate at which temporal coherence is established in the radiation pulse.

The integration in Eqs. (9.39) and (9.40) can be performed for a relatively simple analytic formulation, known as the klystron model [4]. The coupling between the radiation field and the electrons is mediated by the wiggler field, as evidenced by the appearance of the constant a_w in both the equation of motion (9.14) and the wave Eq. (9.25). If the amplitude of the wiggler field were a function of position z, this would result in the coupling coefficient having a z-dependence as well. In principle, this would lead to an axial dependence of the resonant velocity $v_{\parallel}(\gamma_r)$ if the wiggler parameter were sufficiently large. This possibility will be neglected in the following discussion. In the klystron model, the axial dependence of the wiggler parameter is chosen to be of the form of two delta functions, one at the entrance of the interaction region and one at the exit:

$$a_w(z) = a_{w0}L[\delta(z) + \delta(z - L)], \tag{9.45}$$

where a_{w0} is a constant amplitude giving the strength of the delta function.

The interaction in this model is identical to that in a conventional klystron. Electrons are impulsively kicked in energy by the radiation upon entry to the interaction region (i.e., at $z = 0$) and then ballistically bunch in phase while traversing the interaction region. The radiation is amplified by the coherently bunched particles as they leave the interaction region (i.e., $z = L$). While a delta function seems to be a rather radical dependence for the wiggler field, all that is required is that the distance over which the wiggler parameter is nonzero be short enough that the phase of individual particles remains relatively constant in the region of nonzero wiggler field. Further, the klystron model is only considered here for the insight, it provides into the behavior of the more complicated system.

The equations of motion and the wave equation for the given axial profile of the wiggler field strength can now be solved. An electron entering the interaction region with relativistic factor $\delta\gamma_i$ and phase ψ_i immediately receives a kick in energy. The relativistic factor for such an electron just after the first delta function is given by

$$\delta\gamma = \delta\gamma_i + \frac{\omega a_{w0}L}{\gamma_r v_{\|}}|\delta a_e(t)|\sin(\psi_i + \theta_e(t)), \tag{9.46}$$

where $\delta a_e(t) = |\delta a_e| \exp(i\theta_e)$ is the complex amplitude of the radiation entering the interaction region at time t. The beam electron then travels the length of the interaction region with a constant energy and a phase which increases linearly with time according to Eq. (9.13). When this electron arrives at the end of the interaction region, it has a phase given by

$$\psi(L, t) = \psi_i + p_i + |X(t_e)|\sin(\psi_i + \theta_e(t_e)), \tag{9.47}$$

where

$$p_i = \frac{L\omega_\gamma}{v_{\|}}\delta\gamma_i, \tag{9.48}$$

is a normalized energy deviation,

$$X(t_e) = \frac{\omega\omega_\gamma a_{w0}L^2}{\gamma_r v_{\|}}\delta a_e(t_e), \tag{9.49}$$

is a normalized radiation amplitude, and $t_e(L,t)$ is the entrance time of the electron arriving at the exit at time t. Here, $X(t_e)$ plays the role of the bunching parameter in a klystron.

The complex amplitude of the radiation, which satisfies Eq. (9.25), will be constant as the radiation propagates through the interaction region except that it will exhibit discontinuous jumps at the points where the wiggler field is nonzero. If the electron beam has a uniform phase distribution upon entry to the interaction region, then there will be no jump at the entrance due to the fact that the phase average vanishes for a uniform distribution. In other words, the beam has yet to be bunched by the radiation. However, the radiation is amplified as it passes through the exit of the interaction region. Using Eq. (9.40), the complex amplitude of the radiation leaving the interaction region at time t can be expressed in terms of an average over the phases of particles at that point:

$$\delta A(L,t) = \delta A_e\left(t - L/\mathrm{v}_g\right) + \frac{2\pi i c^2 a_{w0} L}{\omega \gamma_r v_\parallel v_g} J(t_e) \langle \exp(-i\psi(L,t)) \rangle, \tag{9.50}$$

The average over injected phases and energies can now be written as

$$\langle \exp(-i\psi(L,t)) \rangle = iX(t_e) g(|X(t_e)|), \tag{9.51}$$

where

$$g(|X|) = -i \iint d\delta\gamma_i d\psi_i \frac{f_0(\delta\gamma_i, t_e)}{|X|} \exp[-i\psi_i - i\theta_e - ip_i - i|X| \sin(\psi_i + \theta_e)], \tag{9.52}$$

can be thought of as a complex, nonlinear gain function for the interaction region. The integral over ψ_i is readily carried out yielding

$$g(|X|) = 2\pi \frac{J_1(|X|)}{|X|} \int d\delta\gamma_i f_0(\delta\gamma_i, t_e) \exp[-i(p_i - \pi/2)], \tag{9.53}$$

where J_1 is a regular Bessel function of the first kind. The remaining integral over energy can be carried out once the distribution of injected energies is specified. The important feature of the nonlinear gain function is that it decreases as the amplitude of the radiation X increases. This provides for the saturation of growth in the oscillator.

The derivation is completed by application of the feedback boundary condition (9.26). It is reasonable to express the amplitude of the radiation in terms of the bunching parameter X since it appears in the argument of a Bessel function. When this is done, one obtains a two-time delay equation which describes the nonlinear, multifrequency behavior of radiation in a free-electron laser oscillator

$$X(t) = R[X(t-T) + I_b(t')X(t')g(|X(t')|)]_{t'=t-T(1+\varepsilon)}, \tag{9.54}$$

The normalized beam current $I_b(t')$ appearing in Eq. (9.54) is given by

$$I_b(t) = -\frac{2\pi eJ(t)a_{w0}^2 L^3 \omega_\gamma}{m_e v_\parallel^3 v_g \gamma_r^2}, \tag{9.55}$$

The various terms in Eq. (9.54) can be understood with the aid of Fig. 9.2. The first term on the right-hand side represents the contribution to the radiation arriving at the entrance of the interaction region which comes from the propagation of radiation along the characteristic indicated with a solid line in Fig. 9.2, that is, the direct path of radiation through the cavity. The second term represents the contribution from the path, which is indicated by the dashed-dot line in the figure. This path represents the portion of the signal, which is carried by the modulated beam in the interaction region. In the klystron model, it is only necessary to consider one such path since the electrons are affected by the radiation only just as they enter the interaction region. Observe that in the general case of a distributed interaction region, the contributions of the beam will depend on the radiation amplitude for all entrance times between $t - T$ and $t - T(1 + \varepsilon)$.

It is interesting to note that by introduction of normalized variables, an equation with very few parameters has been obtained. These parameters are the reflection coefficient R, the slippage parameter ε, the distribution of injected p_i values, and a normalized current I_b. The time can be normalized to T for a continuous beam or T_p for a pulsed beam. The same parameters can be used to describe the more general system from which the klystron model was derived [2, 6–9], and this will be shown in the next section where oscillator gain is discussed.

This concludes the general discussion of slippage in free-electron laser oscillators. However, extensive use of the basic ideas developed in this section (as well as the klystron model) will be used in subsequent sections and chapters.

9.4 Oscillator Gain

A free electron laser oscillator may be characterized according to its single-pass gain. Low-gain oscillators are those in which the radiation amplitude grows by a small factor (e.g., 10%) over a single pass through the interaction region. High-gain oscillators are those in which the radiation grows by a factor of order unity or larger in traversing the interaction length. It is important to realize that the amplitude of the radiation in a low-gain oscillator can eventually grow large and may be comparable to the power found in a high-gain oscillator. If almost all the radiation power is fed back to the input of the interaction region, the power can grow to large amplitude after many passes through the interaction region. What is important in both high- and low-gain oscillators is that the radiation losses over a single pass are lower than the gain.

Whether or not an oscillator with a particular set of parameters is a high-gain or low-gain device can be determined by calculating the maximum spatial growth rate using, for example, the theory of Chap. 5. For the sake of clarity and continuity of discussion, we shall briefly re-derive the gain in the Compton regime by an alternate method than that used in Chap. 5. If the spatial growth length in the infinite interaction region limit is much shorter than the actual length of the interaction region, then the device must be of the high-gain type. To determine the gain for the Compton regime model described by Eqs. (9.13), (9.14), and (9.25), consider the spatial dependence of the amplitude of the vector potential to be of the form

$$\delta A(z,t) = \delta\hat{A}\exp(i\delta kz), \tag{9.56}$$

where δk is a complex wave number whose imaginary part determines the spatial rate of growth of the radiation signal and whose real part determines the shift in wave number of the radiation from the vacuum case. It has been assumed here that the frequency of the radiation which is being amplified is ω and, thus, the complex amplitude $\delta A(z,t)$ has no explicit time dependence.

To determine the linearized spatial growth rate, it is convenient to use the Vlasov Eq. (9.19) rather than the individual particle equations of motion (9.13) and (9.14). Thus, the distribution function is expressed as a sum of the value at injection plus a small perturbation:

$$f = f_0 + [\delta f\exp(i\delta kz) + c.c.], \tag{9.57}$$

where the amplitude of the small perturbation satisfies the linearized version of (9.25)

$$\left[i\delta kv_\parallel + \omega_\gamma\delta\gamma\frac{\partial}{\partial\psi}\right]\delta f - \frac{i\omega}{2\gamma_r}\delta\hat{a}\exp(i\psi)\frac{\partial}{\partial\delta\gamma}f_0 = 0, \tag{9.58}$$

where $\delta\hat{a} = e\delta\hat{A}/m_ec^2$ is the normalized amplitude of the vector potential. Equation (9.58) can be readily solved for the perturbed distribution function in the form

$$\delta f = \frac{\omega a_w\delta\hat{a}}{2\gamma_r(\delta kv_\parallel + \omega_\gamma\delta\gamma)}\exp(i\psi)\frac{\partial}{\partial\delta\gamma}f_0, \tag{9.59}$$

The perturbed distribution function, which is proportional to the radiation amplitude, is then inserted into the wave Eq. (9.25) yielding the following dispersion relation:

$$v_g\delta k = -\frac{\pi a_w^2\omega_b^2}{\gamma_r}\int\frac{d\delta\gamma}{\delta kv_\parallel + \omega_\gamma\delta\gamma}\frac{\partial}{\partial\delta\gamma}f_0, \tag{9.60}$$

where $\omega_b{}^2 = -4\pi eJ/(m_e v_\|)$ is the nonrelativistic beam plasma frequency. Solutions of Eq. (9.60) for δk give the local spatial growth rate. If this growth rate satisfies

$$\mathrm{Im}(\delta kL) << 1 \tag{9.61}$$

then the device will operate in the low-gain regime.

To assess the gain for a particular device, it is necessary to specify the distribution function for the injected beam. All injected beam electrons have the same energy in the idealized limit, and the distribution is of the form

$$f_0 = \frac{1}{2\pi}\delta(\delta\gamma - \delta\gamma_0), \tag{9.62}$$

In this case, the integral over the energy in the dispersion relation can be readily performed to obtain the standard dispersion relation for traveling wave amplifiers [10, 11]:

$$v_g \delta k = -\frac{a_w^2 \omega_b^2 \omega_\gamma}{2\gamma_r^2 \left(v_\| \delta k + \omega_\gamma \delta\gamma_0\right)^2}. \tag{9.63}$$

The denominator on the right-hand side of (9.63) can be cast into a more familiar form using the definition of ω_γ in Eq. (9.15)

$$v_\| \delta k + \omega_\gamma \delta\gamma_0 = (k_w + k + \delta k)v_\| + (k + k_w)\delta v_\| - \omega, \tag{9.64}$$

where $(k + k_w)\delta v_\| = \omega_\gamma \delta\gamma$. The cubic dispersion relation that results is identical to Eq. (4.96) discussed in Chap. 4. The spatial growth rate is a maximum when the frequency is such that the average velocity of the beam matches the phase velocity of the ponderomotive wave. This occurs when $\delta\gamma_0 = 0$, in which case

$$(L\delta k)^3 = -\frac{a_w^2 \omega_b^2 \omega_\gamma L^3}{2\gamma_r^3 v_g v_\|^2} = -I_b, \tag{9.65}$$

where I_b is the normalized current density introduced in Eq. (9.52). As a consequence, the oscillator is of the low-gain type if the normalized current density satisfies [2, 6–8]

$$I_b << 1, \tag{9.66}$$

The maximum spatial growth rate appearing in Eq. (9.65) was derived for an ideal, monoenergetic beam in an ideal wiggler. It is often the case that these assumptions are not satisfied, and, as a result, the gain is substantially reduced from the ideal value. A discussion of the various causes for the reduction of gain by thermal effects can be found in Chap. 4 and will not be repeated here. However, it should be emphasized that these effects are important and often are the crucial factors limiting the gain in a particular device.

9.5 The Low-Gain Regime

It is possible to further reduce the governing system of equations, (9.13), (9.14), and (9.25) for devices which operate in the low-gain regime [2, 4, 8]. In this case, the change in the amplitude of the radiation during each circuit of the cavity will be small. Hence, the time dependence of the radiation amplitude will be nearly periodic with a period equal to the round-trip time of radiation in the cavity. However, it is important to recognize that the radiation amplitude can still have large variations on the time scale associated with the round-trip travel time of radiation through the cavity. For example, in the case of an oscillator driven by a pulsed electron beam as illustrated in Fig. 9.2, it can be expected that the radiation entering the interaction region will develop into pulses which match the pulses of the beam current. Because the gain and losses are both small, the radiation pulses will vary little from pulse to pulse. That is, the radiation entering the interaction region will be nearly periodic in time with a period given by the separation in arrival times of successive beam pulses, which must be nearly the same as the round-trip travel time T. The shape of the radiation pulses can change substantially but only after a large number of circuits of the cavity.

Consider one round-trip of the radiation in the cavity by beginning with the wave equation integrated along its characteristic as is done in Eq. (9.40). In the low-gain limit, the second term on the right-hand side of (9.40), which represents the increase in amplitude of the radiation due to the beam, will be small compared to the first term. Thus, to a first approximation, the radiation amplitude in the interaction region can be expressed in terms of the amplitude of the radiation entering the interaction region:

$$\delta A(z,t) \cong \delta A_e\left(t - z/v_g\right), \tag{9.67}$$

This space-time dependence can then be inserted into the equations of motion, which in Lagrangian coordinates become

$$v_\parallel \frac{d}{dz}\psi = \omega_\gamma \delta\gamma, \tag{9.68}$$

and

$$v_\parallel \frac{d}{dz}\delta\gamma = -\frac{\omega}{2\gamma_r} a_w \left[i\delta e_e\left(t + z\left(v_\parallel^{-1} - v_g^{-1}\right)\right) \exp(i\psi) + c.c.\right], \tag{9.69}$$

The solutions for the phases of particles may now be inserted in the wave Eq. (9.40) to determine the small gain that the radiation experienced on one transit of the interaction region:

$$\delta A(L,t_2) = \delta A_e(t_2 - L/v_g) + \frac{2\pi i c^2 a_w}{\omega\gamma_r v_{\|}} \int_0^L \frac{dz}{v_g} J(t_e)\langle\exp[-i\psi(z,t_e)]\rangle, \tag{9.70}$$

where $t_e = t_e(z,t_2 + (z - L)/v_g) = t_2 - L/v_g - z(v_{\|}^{-1} - v_g^{-1})$. Note that the phase variable has been written in terms of its Lagrangian coordinates as discussed in (9.43).

The feedback boundary condition (9.26) can now be used to relate the radiation amplitude entering the cavity at time $t + T$ to the amplitude entering earlier

$$\delta A_e(t+T) = R\left[\delta A_e(t) - \frac{2\pi i c^2 a_w}{\omega\gamma_r v_{\|}} \int_0^L \frac{dz}{v_g} J(t_e)\langle\exp[-i\psi(z,t_e)]\rangle\right], \tag{9.71}$$

where it has been assumed that $t_2 = t + T - T_r$; hence, $t_e = t - z(v_{\|}^{-1} - v_g^{-1})$.

The similarity of Eq. (9.71) with the klystron model (9.54) is evident. In particular, the first term on the right-hand side of Eq. (9.71) represents the direct propagation of radiation through the cavity, while the second term represents the contributions of the beam. This term depends on the complex amplitude of the radiation entering the cavity only for times $t > t' > t - \varepsilon T$.

As previously discussed, it is expected that the solutions of (9.71) will yield a radiation amplitude which is nearly periodic in time with a period equal to the arrival time of electron pulses which must be close to the round-trip time for the radiation. To describe this expected time dependence of the radiation pulse, it is convenient to introduce a new time variable t' defined by

$$t' = t - nT_a, \tag{9.72}$$

where T_a is the separation in arrival times of the beam pulses (known as micropulses) and n is the integer that puts t' in the range $^-T_a/2 < t' < T_a/2$. Thus, the dependence of the radiation amplitude on the variable t' will determine the shape of the pulse and the dependence on the integer n will describe how the pulse evolves after many circuits of the resonator. In terms of a multiple time scale perturbation theory, t' is described as a fast time variable and the index n as a slow time variable. The amplitude of the radiation entering the cavity as a function of time is written in the form

$$\delta A_e(t) = \delta A_e(n,t'), \tag{9.73}$$

and the amplitude of radiation entering a period T later is given by

$$\delta A_e(t+T) = \delta A_e(n+1, t' + T_\delta), \tag{9.74}$$

where $T_\delta = T - T_a$ is the difference in round-trip time of the radiation and the separation in the arrival times of successive beam pulses. The time T_δ measures what is known as cavity detuning, and it will be found that it must be negative in order that arbitrarily small initial radiation amplitudes build up to saturation in a pulsed beam oscillator [5].

The fact that the gain is small and the reflection coefficient R is close to unity allows considerable simplifications to be made. First, the low-gain assumption is that the radiation amplitude changes slowly with n. Second, T_δ, may be assumed to be small in comparison with the temporal width of the electron pulse T_p. If T_δ is not small, the radiation and beam pulses drift out of step before the interaction between the beam and the radiation can act to force synchronism. Therefore, a Taylor expansion in the dependence of the two arguments of δA_e yields a partial differential-integral equation for the radiation amplitude

$$\left(\frac{\partial}{\partial n} + T_\delta \frac{\partial}{\partial t'}\right)\delta A_e = (R-1)\delta A_e + \frac{2\pi i c^2 a_w}{\omega \gamma_r \upsilon_\parallel} \int_0^L \frac{dz}{\upsilon_g} J(t_e) \langle \exp[-i\psi(z, t_e)] \rangle, \tag{9.75}$$

where we have assumed that $R \approx 1$ in the term containing the response of the beam.

At this point, it is convenient to introduce normalizations as in the case of the klystron model. First, axial distance is normalized to the length of the interaction region, with the dimensionless distance given by $\xi = z/L$. Second, the energy deviation $\delta\gamma$ is normalized as in Eq. (9.48)

$$p = \frac{L\omega_\gamma}{\upsilon_\parallel}\delta\gamma, \tag{9.76}$$

where p is the normalized energy deviation. Note that the variable p can also be thought of as a normalized detuning parameter. In particular, p can also be expressed in terms of the average parallel velocity as

$$p = \frac{L}{\upsilon_\parallel}\left[(k + k_w)\left(\upsilon_\parallel + \delta\upsilon_\parallel\right) - \omega\right], \tag{9.77}$$

Thus, p is the frequency detuning of a particle with energy $(\gamma + \delta\gamma)m_e c^2$ normalized to the time of flight through the interaction region. The choice of the letter p is motivated by the fact that this variable plays the role of a momentum in terms of the pendulum Hamiltonian.

The radiation amplitude is now normalized in the same way as the bunching parameter introduced in the klystron model; specifically,

$$X(n, t') = \frac{\omega\omega_\gamma |a_w L^2}{\gamma_r \upsilon_\parallel}\delta a_e(n, t'), \tag{9.78}$$

where $\delta a_e = e\delta A_e/m_ec^2$. With these normalizations, the Lagrangian equations of motion can be expressed as

$$\frac{\partial \psi}{\partial \xi} = p, \tag{9.79}$$

and

$$\frac{\partial p}{\partial \xi} = -\frac{1}{2}[iX(n, t_e + \varepsilon\xi T)\exp(i\psi) + c.c.], \tag{9.80}$$

where the slippage parameter ε has been introduced as defined by Eq. (9.44). The normalization for the radiation amplitude is understood by consideration of the discussion of the pendulum equation in Chap. 4. In particular, the normalized amplitude can be expressed as $X = (k_sL)^2$ where k_s is the synchrotron wavenumber for a particle at the bottom of the ponderomotive potential well of the beat wave.

Application of the preceding normalizations to Eq. (9.75) results in

$$\left(\frac{\partial}{\partial n} + T_\delta \frac{\partial}{\partial t'}\right)X(n, t') = (R-1)X(n, t') - i\int_0^1 d\xi I_b(t_e)\langle \exp[-i\psi(\xi, t_e)]\rangle, \tag{9.81}$$

where $t_e = t' - \varepsilon\xi T$ and $I_b(t_e)$ is defined in Eq. (9.55).

The various terms on the left-hand side of Eq. (9.81) describe the slow evolution of the radiation pulse due to cavity detuning, and the terms on the right-hand side describe the effects of cavity losses and the gain due to interaction of the radiation with the electron beam. The real-time dependence of the amplitude of the radiation field entering the cavity can be obtained from the solution for $X(n,t')$ by replacing n by $t/T_a \approx t/T$ and t' by t. For example, in the absence of an electron beam, the solution of (9.81) is

$$X(n, t') = X(0, t' - nT_\delta)\exp[-n(1-R)], \tag{9.82}$$

where $X(0,t')$ is periodic in t' with period T_a (the electron pulse arrival time) and represents the initial pulse shape. In the multiple time scale variables n and t', the solution (9.82) represents a pulse which drifts to increasing values of t' (for $T_\delta > 0$) as the round-trip circuit index n increases. Furthermore, the amplitude of the pulse decreases with n due to cavity losses. The solution is illustrated schematically in Fig. 9.3. The decay of the radiation field is easily understood in that the power in the pulse decays exponentially in real time with characteristic time constant $T_d = T/[2(1-R)]$, which for R close to unity is identical to the decay rate predicted by Eq. (9.29). The drifting of the pulse in t' due to the cavity detuning is a consequence of having defined t' (9.72) in terms of the arrival time T_a of beam pulses which is not necessarily the same as the round-trip time for radiation T. Equation (9.82) describes

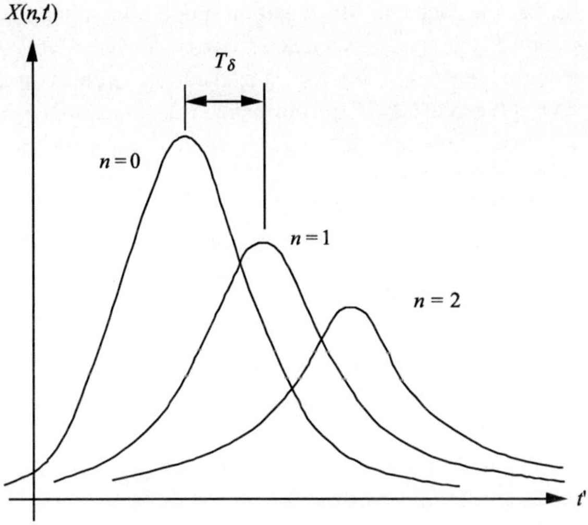

Fig. 9.3 Schematic illustration of a decaying radiation pulse

the solution in the absence of the electron beam. In this case, the radiation pulses arrive separated by the cavity round-trip time T. Thus, the radiation pulse appears to slip in the *beam pulse frame* by an amount $T_\delta = T - T_a$ on every round-trip of the cavity. An alternate way to represent the solution (9.82) is to write the periodic function $X(0,t')$ in terms of a Fourier series with Fourier coefficients $X_m(0)$. Upon replacement of t' by t and n by t/T, Eq. (9.82) becomes

$$X(t/T,t) = \sum_{m=-\infty}^{\infty} X_m(0)\exp\{-i[2\pi m - i(1-R)t/T]\}, \tag{9.83}$$

which is seen to be a superposition of signals whose frequencies are those of the modes of the empty cavity given by Eq. (9.27).

Equation (9.81) does not represent the most compact form for the equation describing a low-gain oscillator. In principle, one could normalize the two-time variables n and t'. The round trip index could be normalized to the number of round trips of radiation occurring in the cavity decay time T_d. This would be accomplished by dividing (9.81) through by $2(1-R)$. The normalized current would then enter only in combination with the quantity $2(1-R)$. Physically, this is a consequence of the fact that the important parameter measuring the current in a low-gain oscillator is the ratio of the current to the losses. Equivalently, the ratio of the operating current to the start current could be specified which also depends only on the ratio of the current to the losses.

Finally, observe that the time variable t' that gives the shape of the radiation pulse has not been normalized. This is because the most appropriate normalization depends on whether a pulsed beam or continuous beam is driving the oscillator. In the case of a pulsed beam, the appropriate choice is to normalize t' to the width of the electron pulse T_p. Typically, this time is much shorter than the repetition time T_a;

hence, the fact that the radiation amplitude is supposed to be periodic in t' with period T_a can be neglected and t' can be considered to run over all time. In the case of a continuous beam, the normalized current is constant and one can take $T_a = T$. The appropriate choice for the normalization of t' is, therefore, T. These two cases, that of a pulsed beam and that of a continuous beam, will be considered in more detail in subsequent sections.

9.6 Long Pulse Oscillators

Oscillators driven by continuous electron beams are referred to as *long pulse oscillators* and are capable in principle of oscillating at a single frequency. This is in contrast to devices driven by pulsed electron beams where the radiation field takes on a pulsed nature that, necessarily, consists of spectrum of frequencies. What is meant by *single frequency* is that solutions of Eqs. (9.13), (9.14), and (9.25) can be found in which none of the quantities depend on the variable t. The fluctuating fields and current densities in a real device would have a more complicated time dependence which, nevertheless, is periodic with a period corresponding to the frequency ω. Hence, the real spectrum would consist of the frequency ω and all its harmonics. In the simple formulation described by (9.13), (9.14), and (9.25), the small components of the fields and current densities which are oscillating at harmonics of the basic frequency ω have been neglected. Therefore, these solutions are effectively single-frequency solutions.

The problem of finding the single-frequency states can be thought of as a nonlinear eigenvalue problem. Solutions of the governing equations, including the feedback boundary condition (9.26), exist only for specific values of the frequency ω and the amplitude of the radiation entering the interaction region. The solution of this nonlinear eigenvalue problem can be obtained by specification of ω and the initial amplitude of the radiation. The equations in the interaction region can then be solved as if the device were an amplifier with a specified injected signal. Using the feedback boundary condition, the magnitude and phase of the returning radiation can be found and compared with the assumed injected radiation. To find an eigensolution, the frequency and magnitude of the injected radiation must be adjusted to match the magnitudes and phases of the injected and returning radiation. The phase of the injected radiation is arbitrary in that changing the phase of the injected radiation simply changes the phase of the radiation amplitude at any other point in the device by the same amount. In general, there are many different possible eigenfrequencies for a given set of beam, wiggler, and feedback path parameters. The separations between the frequencies in the present model are approximately uniform and separated by the inverse of the travel time of the radiation through the cavity. In the low-gain limit, the frequencies are nearly the same as those of the empty cavity given by Eq. (9.27). Thus, it is sometimes useful to think of these eigenfrequencies as corresponding to modes of the cavity.

While it may be clear from the above discussion that single-frequency states always exist in principle, it is not obvious whether it is possible to access these states in practice. For example, the single-frequency states may be unstable in the sense that radiation at other frequencies will spontaneously grow from noise even in the presence of a large-amplitude saturated signal. The sideband instability of Chap. 7 is an example of just such a situation. The problem of the stability of single-frequency states in an oscillator will be considered in detail in the next section.

A second reason preventing single the attainment of true single-frequency states is the inherent noise present in any electron beam. There is always a minimum level of noise associated with discreteness of the charge of an electron. This is the same source of noise responsible for the spontaneous emission discussed in Chap. 3 and can cause spectral broadening in two ways. First, the radiation field may be essentially one of the single-frequency nonlinear eigenstates described in the preceding paragraphs. However, as was discussed, the precise phase of one of these states is an arbitrary constant. In the presence of noise, the phase becomes a time dependent, random variable which gives rise to a finite spectral width for radiation. This width is dependent on the amplitude of the noise. A second, and distinct possibility, is that the noise can excite other modes of the cavity which can achieve large amplitudes (depending on parameters). These two separate limits on the achievable spectral width will be determined later in this chapter.

A third impediment to the achievement of a true single-frequency state of the cavity is the finite time duration of any practical electron beam. One limit is the spectral width associated with the reciprocal of the temporal duration of the electron beam, although this width can be made very small. Further, even if the beam is of extremely long duration, its parameters may vary with time. This can lead to a situation in which a stable single-frequency state becomes unstable and is replaced by a state (or states) with different frequencies. A more important limitation, however, is the slow rate at which coherence is established in devices with small slippage parameters. This effect was alluded to in Sect. 9.3 and will be considered in more detail in the following sections. The net result is that a device with a small slippage parameter will tend to initially oscillate over a relatively broad range of frequencies, and this range will narrow slowly but progressively over time. The rate of narrowing can be increased by various techniques that will be discussed.

Single-frequency states are difficult, if not impossible to achieve in practice. However, their study for long pulse oscillators is important since the interaction of these states (or their modification due to the temporal variation of a real beam pulse) provides a way of understanding the temporal behavior of the radiation in a long pulse oscillator.

9.6.1 Single-Frequency States

The previous discussion regarding the problems of the existence, stability, and sensitivity of single-frequency states to noise applied to oscillators of both the high- and low-gain type. While the physical effects that were discussed are general and can be expected to occur independent of the oscillator gain, they are most easily analyzed, and understood, if one adopts the low-gain approximation. Therefore, in the next several sections, we will concentrate by studying the solutions of the low-gain equations, in particular the normalized Eqs. (9.79), (9.80), and (9.81).

For the case of a continuous electron beam, the normalized current appearing in (9.81) can be assumed to be a constant. Further, while the beam pulse arrival time T_a has no specific meaning, the radiation amplitude will be nearly periodic with period equal to the round-trip time for radiation in the cavity T. Accordingly, we set $T_a = T$. The fast time dependence of the normalized radiation amplitude can then be expressed in terms of a Fourier series:

$$X(n,t') = \sum_{m=-\infty}^{\infty} X_m(n)\exp(-i\Lambda\omega_m t') \tag{9.84}$$

where

$$\omega_m = \frac{2\pi m}{T}, \tag{9.85}$$

and the Fourier index m labels the mth mode of the cavity relative to the mode with frequency ω. That is, the Fourier amplitudes $X_m(n)$ are the amplitudes of the modes of the empty cavity whose real frequencies are approximately $\omega + \Delta\omega_m$. These mode amplitudes will change slowly with time (n denotes the round-trip index) due to the interaction of the modes with the electron beam and due to the output coupling and cavity losses. These effects are described by the two terms on the right-hand side of Eq. (9.81).

The modal representation for the normalized field amplitude can then be inserted in the particle equations of motion (9.79) and (9.80). It is then necessary to integrate the equations of motion for an ensemble of initial conditions corresponding to a uniform distribution of phases, a distribution of momentum values p reflecting the energy dependence of the injected beam, and, finally, a distribution of entrance times t_e. Since the radiation amplitude is periodic in the fast time variable t' with period T, the solutions of (9.79) and (9.80) for the orbits will have the same periodicity. Therefore, it is only necessary to integrate the equations of motion for a distribution of entrance times t_e on an interval of length T.

As has been discussed in Sect. 9.1, it is always possible to introduce a Vlasov equation to replace the individual particle equations:

$$\left[\frac{\partial}{\partial\xi} + p\frac{\partial}{\partial\psi} - \frac{1}{2}(iX(n,t_e+\varepsilon\xi T)\exp(i\psi) + c.c.)\frac{\partial}{\partial p}\right]F(p,\psi,\xi,t_e) = 0, \tag{9.86}$$

where $F(p,\xi,\psi,t_e)$ is the distribution function which satisfies the boundary condition

$$F(p,\psi,\xi=0,t_e) = F_0(p), \tag{9.87}$$

at the entrance of the cavity. Here $F_0(p)$ is the distribution function in normalized momentum p for the incoming beam, which satisfies a normalization condition

$$\int_0^{2\pi} d\psi \int_{-\infty}^{\infty} dp F_0(p) = 1, \tag{9.88}$$

Comparison of the preceding with (9.16) along with the definition of normalized momentum p, (9.76), reveals that f_0 and F_0 differ only by a multiplicative constant, $L\omega_\gamma / v_\parallel$.

Once the orbits or the distribution function are calculated, the slow rate of change of the mode amplitudes with round-trip number is obtained from the mth Fourier component of Eq. (9.81):

$$\left[\frac{\partial}{\partial n} + (1-R)\right] X_m(n) = -iI_b \int_0^T \frac{dt'}{T} \int_0^1 d\xi \langle \exp(i\Delta\omega_m t' - i\psi(\xi, t_e))\rangle, \tag{9.89}$$

Since the orbits are periodic in the entrance time variable t_e, the integral over t' in Eq. (9.89) can be transferred to an integral over t_e resulting in the mode equation:

$$\left[\frac{\partial}{\partial n} + (1-R)\right] X_m(n) = -iI_b \int_0^T \frac{dt_e}{T} \int_0^1 d\xi \langle \exp(i\Delta\omega_m (t_e + \varepsilon\xi T) - i\psi(\xi, t_e))\rangle, \tag{9.90}$$

where the angular brackets imply an average over the distribution function of the incoming beam and can be evaluated as an average over particles as in (9.23), or as an average weighted by the local distribution function as (9.24).

The mode amplitude evolution Eq. (9.90), along with the equations of motion, describes the nonlinear competition of the modes of a low-gain oscillator cavity. Before analyzing the nonlinear saturated single-frequency solutions of this system, it is useful to examine the linear regime in which the radiation amplitude is small. It is more convenient in the linear regime to work with the Vlasov Eq. (9.86), which we then linearize by writing the distribution function as the sum of the injected distribution function plus a small perturbation:

$$F = F_0(p) + [F_+(p, \xi, t_e)\exp(i\psi) + c.c.], \tag{9.91}$$

The perturbation satisfies the linearized Vlasov equation:

$$\left[\frac{\partial}{\partial \xi} + ip\right] F_+ = \frac{i}{2} \sum_{m=-\infty}^{\infty} X_m \exp[-i\Delta\omega_m (t_e + \varepsilon\xi T)] \frac{\partial F_0}{\partial p}, \tag{9.92}$$

subject to the boundary condition that the perturbed distribution function vanishes at the entrance of the interaction region. Equation (9.92) can be integrated to obtain the perturbed distribution function:

$$F_+ = \frac{i}{2}\int_0^{\xi} d\xi' \sum_{m=-\infty}^{\infty} X_m \exp[-ip(\xi-\xi') - i\Delta\omega_m(t_e + \varepsilon\xi T)]\frac{\partial F_0}{\partial p}, \tag{9.93}$$

Substitution of the perturbed distribution function into the equation for the slow evolution of the mth mode amplitude (9.90) results in

$$\left[\frac{\partial}{\partial n} + (1-R)\right]X_m = I_b G_m X_m, \tag{9.94}$$

where G_m is the linear, complex, normalized gain for the mth mode

$$G_m = \pi^2 \int_{-\infty}^{\infty} dp\Delta(p - \varepsilon\Delta\omega_m T)\frac{\partial F_0}{\partial p}, \tag{9.95}$$

where the resonance function $\Delta(p)$ is given by

$$\Delta(p) = -\frac{i}{\pi}\int_0^1 d\xi\frac{1-\exp(-ip\xi)}{p} = \frac{1-\cos p + i(\sin p - p)}{\pi p^2}, \tag{9.96}$$

A mode will grow if the product of the real part of the gain and the normalized current exceeds the cavity losses:

$$\operatorname{Re} G_m > \frac{1-R}{I_b}, \tag{9.97}$$

For a cold beam with a monoenergetic distribution $F_0(p) = \delta(p - p_0)/2\pi$, the real part of the gain function reduces to that derived in Chap. 4 and displayed in Eq. (4.60):

$$G_m = -\frac{1}{2}\frac{\partial}{\partial p}\frac{1-\cos p + i(\sin p - p)}{p^2}\Bigg|_{p=p_0-\varepsilon\Delta\omega_m T}, \tag{9.98}$$

That is, with $\Theta = -p/2$, Re $G_m = F(\Theta)/8$ where $F(\Theta)$ is defined in Eq. (4.60). The maximum value of G_m is approximately 0.068 and occurs for a value of its argument equal to approximately 2.6.

The solutions of Eq. (9.94) describe the initial growth of the cavity modes. The modes will grow with a temporal rate γ_m and have a frequency shift $\delta\omega_m$ where

$$\gamma_m = \frac{I_b}{T}\left[\mathrm{Re}\, G_m - \frac{(1-R)}{I_b}\right], \tag{9.99}$$

and

$$\delta\omega_m = -\frac{I_b}{T}\mathrm{Im}G_m, \tag{9.100}$$

The effective density of modes can be determined by the dependence of the gain function on the mode index *m*. In particular, for a monoenergetic beam, the argument of the gain function is the detuning

$$p_m = p_0 - 2\pi m\varepsilon \cong \frac{L}{v_\parallel}\left[(k_m + k_w)\left(v_\parallel + \delta v_\parallel\right) - (\omega + \Delta\omega_m)\right], \tag{9.101}$$

where $k_m = (\omega + \Delta\omega_m)/c$ is the spatial wave number of the mth mode and $(k_m + k_w) L\delta v_\parallel/v_\parallel = p_0$ measures the deviation in velocity of the incoming beam from exact resonance with the $m = 0$ mode. As has been indicated in Chap. 4, the first positive peak of the gain function corresponds to values of the argument of G between approximately 0 and π. The number of modes, which fall into this range, is $N \approx (2\varepsilon)^{-1}$. Thus, the slippage parameter ε defined in Eq. (9.44) determines the mode density in a continuous beam oscillator. With a small value of the slippage parameter, a large number of cavity modes will have positive growth rates and can be expected to compete for the free energy of the beam. Further, each mode is also characterized by a frequency shift, which is induced by the beam. That is, the frequency of oscillation of the mode is not exactly equal to the oscillation frequency of a mode in the empty cavity, but rather is shifted relative to the corresponding empty cavity frequency by a small amount (in the low-gain regime) which is proportional to the beam current and depends on detuning.

The gain function for arbitrary distribution functions is defined in Eq. (9.95). In general, as the spread in p values (parallel energy in (9.76) and (9.77)) increases, the gain is reduced, and the range of frequencies (or detunings defined by (9.98) with p_0 being the mean value of p) for which the gain positively increases. This transition occurs when the spread in p values is comparable to unity. In the case of a beam with an extremely large spread in p values, the integral in (9.95) can be performed by considering the derivative of the unperturbed distribution function to be constant over the range of p values for which the other factor Δ in the integrand is large; specifically, for $|p - \Delta\omega_m T| < \pi$. In this case, the factor $\Delta(p - \varepsilon\Delta\omega_m T)$ acts as a delta-function, and the gain is proportional to the slope of the distribution function at the value of p which is resonant with the mode in question:

$$G_m \cong \frac{\pi}{2} \frac{\partial F_0}{\partial p}\bigg|_{p=\varepsilon\Delta\omega_m T}, \tag{9.102}$$

Gain will be positive or negative depending upon the slope of the distribution function. A positive slope indicates that there are more electrons with energies greater than the resonant energy than with energies less than the resonant energy. As a result, energy is extracted from the beam and the radiation is amplified. A situation where the distribution function might be broad enough that (9.102) applies is the case of a storage ring free-electron laser where the same group of electrons transits the interaction region many times acquiring a large spread in parallel energies.

Equation (9.94) describes the small signal growth of modes in a low-gain continuous beam oscillator. If one of these modes grows and becomes dominant (i.e., it is the only mode with a substantial amplitude), then it can be regarded as a single-frequency state. Without loss of generality, we can label this mode the $m = 0$ mode. The problem, therefore, is to analyze the characteristics of an oscillator in which the mode amplitude X_0 is sufficiently large so that the gain function depends on the mode amplitude. Saturation will be achieved when the mode amplitude reaches a sufficient size that the power extracted from the beam balances the power lost from the cavity. The saturation process can be understood in the following way. In the low-gain single-mode regime, the Hamiltonian

$$H = \frac{p^2}{2} + \mathrm{Re}[X_0 \exp(i\psi)], \tag{9.103}$$

is conserved by electrons as they transit the interaction region. This is in contrast to the case of a high-gain amplifier or oscillator where the dependence of the radiation field amplitude on axial distance destroys the constancy of the equivalent pendulum Hamiltonian. In spite of this difference, Fig. 4.1 also applies qualitatively to a low-gain oscillator. The efficiency at which energy is extracted from the beam is measured by the mean change in the momentum variable for the ensemble of electrons:

$$\Delta p = \langle p_0 - p(\xi = 1) \rangle. \tag{9.104}$$

Recall that the momentum variable p is proportional to the energy deviation $\Delta\gamma$ according to Eq. (9.76). In terms of the dimensionless momentum change Δp, the average energy extracted per particle is

$$\Delta\gamma = \Delta p \frac{v_\parallel}{L\omega_\gamma} \cong \frac{\gamma \Delta p}{2\pi N_w} \frac{\beta_\parallel^2}{1 + \beta_\parallel}, \tag{9.105}$$

where in the second equality, N_w is the number of wiggler periods and we have evaluated the factor ω_γ defined in (9.15) for the case of a helical or a weak planar

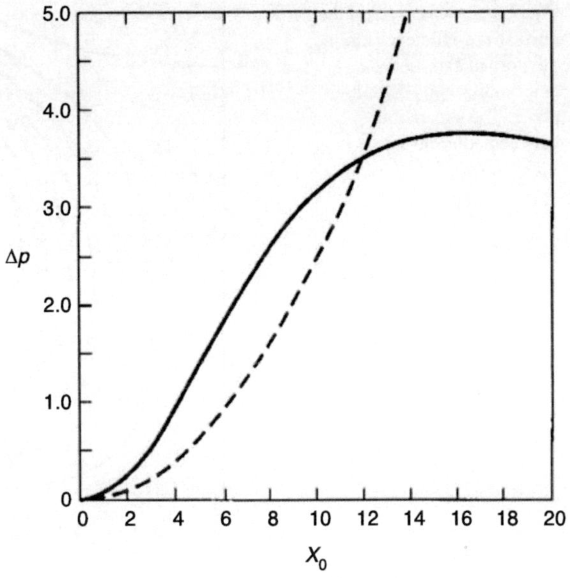

Fig. 9.4 Normalized energy extracted per electron as a function of radiation amplitude (solid line) and normalized energy radiated per electron (dashed line)

($\gamma >> 1 + a_w^2$) wiggler. In the relativistic case, this gives the standard estimate that the fractional change in energy scales as the reciprocal of the number of wiggler periods.

The dependence of Δp on the radiation amplitude for a specific value of p_0 (= 2.6) is shown in Fig. 9.4. For small values of radiation amplitude, the dimensionless efficiency is a quadratic function of the radiation amplitude. As the radiation amplitude increases, the dimensionless efficiency increases (but at a rate which is slower than predicted by the quadratic dependence in the linear regime) and reaches a maximum corresponding to the value of the radiation amplitude in Fig. 4.1c. This maximum is seen to correspond to the case in which the bulk of the electrons has made half a synchrotron oscillation in the ponderomotive well.

The process by which saturation is achieved is illustrated by the addition to the figure of the dashed line which represents the *per electron* losses. The rate at which energy is lost from the cavity is quadratic in the field amplitude. Thus, the amount of energy per electron that must be extracted from the beam to maintain a given amplitude will appear as a parabola with a coefficient of proportionality which depends on the beam current and reflection coefficient. For currents above the start current, the energy extracted per electron exceeds that which is lost, and the mode grows. As the mode amplitude increases, the losses increase faster than the energy extracted per electron. Saturation of growth occurs when these two effects balance, as represented by the intersection of the solid and dashed curves in Fig. 9.4.

Curves similar to those of Fig. 9.4 could be plotted for different values of the detuning p_0. Instead, we show in Fig. 9.5 the level curves of Δp in the field amplitude versus detuning plane. For the range of parameters shown, the maximum dimensionless efficiency is $\Delta p \approx 5.5$ and occurs for a detuning 5.2 and a field amplitude

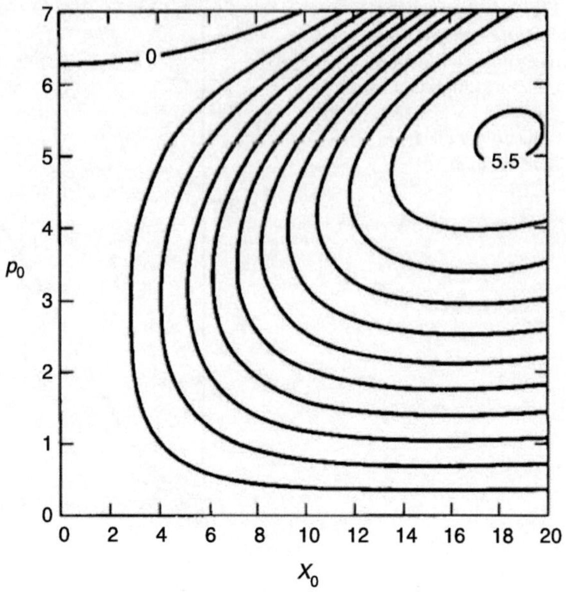

Fig. 9.5 Level curves of the normalized efficiency Δp in the normalized mode amplitude versus detuning plane. The separation between adjacent levels is $\Delta p = 0.5$

$X_0 = 18.1$. As the range of detunings and field amplitudes is increased, other local maxima appear. However, as we shall soon see, these maxima occur for parameters for which one would not expect to find single-mode operation in an oscillator.

The efficiency of energy extraction is closely related to the single-mode nonlinear gain of the oscillator [12]. When a single mode is present ($m = 0$), Eq. (9.94) can be generalized by the introduction of the nonlinear gain function G_{nl} defined by

$$G_{nl}(|X_0|, p_0) = -\frac{i}{X_0}\int_0^1 d\xi\langle \exp(-i\psi)\rangle, \tag{9.106}$$

That the gain depends only on the magnitude of the radiation amplitude can be seen by noting that a change in the phase of X_0 can be absorbed in the definition of the particle phase ψ both in (9.103) and (9.106). Since the average in Eq. (9.106) is taken over an ensemble of phases distributed over an interval of 2π, this change in phase leaves the nonlinear gain unaffected. The real part of the gain can now be related to the dimensionless efficiency by means of energy conservation arguments, and we find using (9.80), (9.84), and (9.106) that

$$\Delta p = -\int_0^1 d\xi\left\langle\frac{dp}{d\xi}\right\rangle = |X_0|^2\,\mathrm{Re}[G_{nl}(|X_0, p_0|)], \tag{9.107}$$

The plot in Fig. 9.5 shows the efficiency as a function of mode amplitude and detuning. While the detuning is determined directly by the incoming beam, the mode amplitude is determined by power balance considerations as illustrated in Fig. 9.4. Specifically, the mode amplitude adjusts itself until the power extracted from the beam equals that dissipated in and radiated from the cavity. This power balance constraint can be expressed by searching for steady-state solutions of Eq. (9.94) for the $m = 0$ mode using the nonlinear gain function defined in Eq. (9.106). In these steady-state solutions, the amplitude of the radiation is constant on the slow time scale (round-trip index n) and has a phase which decreases steadily with time representing the frequency shift $X_0 = |X_0| \exp(-i\delta\omega_0 Tn)$. The real and imaginary parts of Eq. (9.94) then give

$$(1-R) = I_b \,\mathrm{Re}[G_{nl}(|X_0|, p_0)], \tag{9.108}$$

and

$$\delta\omega_0 T = -I_b \mathrm{Im}[G_{nl}(|X_0|, p_0)], \tag{9.109}$$

The current required to maintain a particular mode amplitude can be obtained from the power balance relation (9.108). It is useful to compare this current with the minimum current necessary to initiate oscillations in a cavity with a given loss factor $1 - R$. This minimum current, known as the start current, is obtained by inserting the expression for the linear, small signal gain evaluated at the detuning $p_0 = 2.6$ in Eq. (9.98). Hence,

$$I_{st} = \frac{1-R}{G_0(2.6)} \cong 15(1-R). \tag{9.110}$$

The level curves of the current required to maintain a given mode amplitude given by Eq. (9.108) can be plotted versus mode amplitude and detuning. Such a plot appears in Fig. 9.6 where we have normalized the current to the start current. It is evident that a current about four times the start current is required to reach the maximum efficiency equilibrium. Further, for a given value of current in excess of the start current, there is a range of detuning values for which single-frequency equilibrium states exist. If the slippage parameter is small, then according to Eq. (9.101), this range in detuning parameters translates into a large number of modes. Thus, with a given current in excess of the start current, there are typically a large number of possible single-frequency equilibria in a long pulse oscillator. Which of these equilibria are stable and which can be accessed will be discussed in the subsequent sections.

Finally, we plot the level curves of the dimensionless frequency shift $\delta\Omega$ [$= \delta\omega_0 T/2(1 - R)$] in the field amplitude versus detuning plane in Fig. 9.7. The frequency shift is normalized to the decay time of the radiation in the empty cavity T_d (i.e., the Q resonance width of the cavity). This frequency shift will be important when the effects of diffraction on mode competition are discussed in a later section.

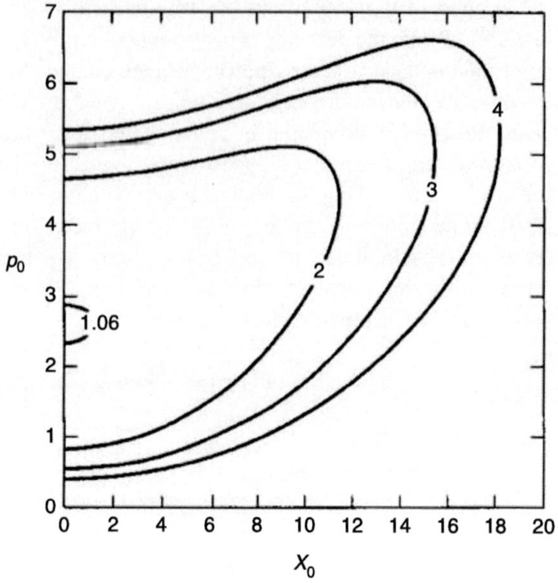

Fig. 9.6 Level curves of current ($\chi \equiv I_b/I_{st}$) required to maintain power balance, plotted in the normalized mode amplitude vs detuning plane [15]

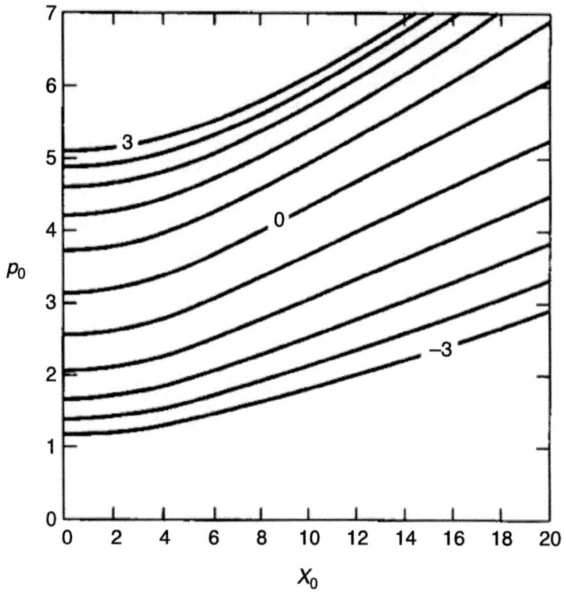

Fig. 9.7 Level curves of normalized frequency shift [$\delta\Omega = \delta\omega_0 T/2(1\text{-R})$] in the normalized mode amplitude vs detuning plane. The separation between adjacent levels is 0.6

9.6.2 Stability of Single-Frequency States

The single-frequency states of the previous section were found upon requiring that the radiation have a single spectral component. It was found that these states represented valid equilibrium solutions of the basic governing equations. In this section, the stability of these equilibrium states to perturbations will be examined. Clearly, due to the ubiquitous presence of noise, stability of a single-frequency state is necessary for the state to ever be reached in practice.

There are two principal considerations in the stability of single-frequency equilibria in the low-gain regime. The first is the stability of the equilibrium mode against perturbations in its amplitude and phase. The second consideration is the stability of the equilibrium mode against the introduction of other modes. The first question can be answered relatively easily based upon information from the previous section regarding the characteristics of single-frequency equilibria. Since a free running oscillator has no preferred phase of oscillation, the equilibrium state is neutrally stable to perturbations in its phase. Therefore, a small perturbation in the phase of the complex amplitude of the single cavity mode will neither grow nor decay in time. This is manifested mathematically in the fact that the nonlinear gain function defined in (9.106) is independent of the phase of the radiation. Perturbations to the magnitude of the mode will decay if the rate at which power is radiated increases faster with field amplitude than the rate at which power is extracted from the beam. For example (see Fig. 9.4), if the radiation amplitude is increased (due to the perturbation) to a point above the intersection of the power extracted per electron and losses curves, then the power lost from the cavity exceeds that extracted from the beam. As a result, the mode amplitude will decay back to the equilibrium point. We will see that this can, in general, be determined from the level curves of current plotted in Fig. 9.6.

The second stability consideration is whether the equilibrium state is stable with respect to the introduction of other modes. Specifically, can a mode other than the equilibrium state (which is excited by noise) grow exponentially in time? The answer to this question depends upon both the parameters of the equilibrium state and the nearness in frequency to the equilibrium mode of the perturbing mode(s). In particular, there is a wide range of equilibrium states which are capable of suppressing competing modes [2, 13–15]. This mode suppression can be envisioned in the following manner. Assume that the current is several times the start current defined in Eq. (9.110), and the slippage parameter ε is small. In this case, a large number of modes have positive growth rates when all modes are small. However, if one mode is able to reach large amplitude, then the same nonlinear mechanism that reduces the gain of this mode at saturation also reduces the gain of the competing modes. The gain reduction of the competing modes is sufficient to cause the losses to exceed the gain for these modes, with the result that these other modes decay away with time.

It is important to bear in mind that the mode suppression mechanism just described does not work for all the equilibrium states in the mode amplitude versus

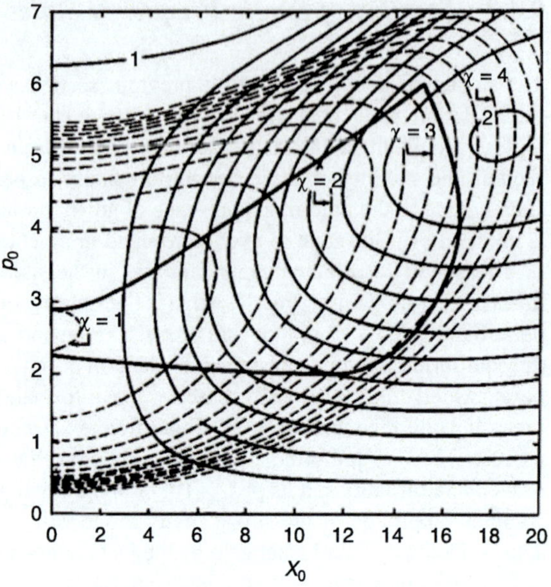

Fig. 9.8 Region of stable single-mode operation in the normalized mode amplitude vs detuning plane [15]. Only inside the triangular-shaped region is a single-frequency state stable to the introduction of other modes

detuning plane of Figs. 9.5, 9.6, and 9.7. Only those equilibria, which fall within the triangular-shaped region of Fig. 9.8, are robustly stable. The triangular-shaped region has three fundamental boundaries, and equilibria outside these boundaries are unstable to the growth of other modes. There are basically two distinct types of instability that determine the three boundaries. The equilibria just outside the upper and lower boundaries are unstable to the growth of nearby-in-frequency modes. This instability is called a phase instability for reasons that will be given subsequently. The basic cause of the instability is the fact that if the equilibrium mode has a detuning p_0 which is too far from resonance (at which the linear growth is maximum), then a mode with a detuning which is closer to the resonance (and has a higher linear gain) will be able to grow. Equilibria, which are just across the third boundary defining the right side of the triangle, are unstable to the growth of modes which are predominantly lower frequency than the equilibrium mode. This instability is called variously the sideband instability [3, 6, 16], the synchrotron instability [13], the trapped particle instability [14], the overbunch instability [4, 15], or the spiking mode [17, 18] and can be explained as follows. Due to the phase-trapping nature of the saturation mechanism, the equilibrium mode is not able to extract any more beam energy once the bulk of the particles have completed about half of a synchrotron oscillation in the ponderomotive well. A lower-frequency mode with a resonant detuning below that of the equilibrium mode can still extract energy from these particles. In fact, it is now in a better position to trap the electrons in its ponderomotive well than it was in the absence of the equilibrium mode. This situation is illustrated schematically in Fig. 9.9 where the separatrices corresponding to the equilibrium mode and the lower frequency satellite mode are superimposed.

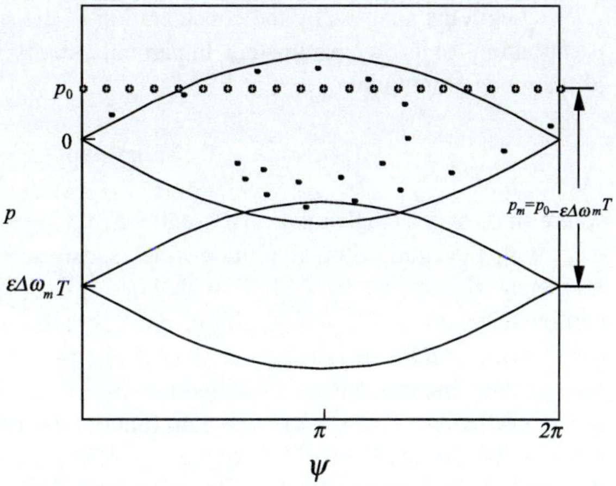

Fig. 9.9 Schematic phase space plot illustrating the mechanism by which a lower frequency sideband extracts energy from electrons trapped in the ponderomotive well of the carrier

Based on Fig. 9.9, a mode with a resonant detuning lower than that of the equilibrium mode by about half the width of the trapped region of phase space for the mode at saturation would be most unstable. From the definition of detuning (9.101), this gives

$$\varepsilon\Delta\omega_m T \cong -\sqrt{2|X_0|} \cong -5.7, \tag{9.111}$$

where we have substituted $|X_0| \approx 16$ as an estimate of the saturation amplitude. Interestingly, this estimate is identical to that of the frequency of the most unstable sideband mode obtained by the often-used theoretical argument that the sideband should be down shifted by the synchrotron frequency. It is clear, however, that the threshold of the instability occurs for mode amplitudes such that particles execute only half a synchrotron oscillation on one transit of the interaction length, and the conventional argument is only qualitative. A third explanation of this instability based on the klystron model introduced in Sect. 9.3 will be given in this section. This explanation is based on time domain considerations and gives a different insight into the overbunch-sideband-spiking mode instability.

A final point that will be brought out in this section is that for equilibria within the stable triangle in parameter space, nearby modes decay slowly in time [15]. The slow temporal decay is a consequence of the fact that coherence in the radiation pulse is established slowly for small slippage parameters. Hence, the smaller the slippage parameter, the closer in detuning neighboring modes are to the equilibrium mode (see Eq. (9.101)), and the slower these modes decay. This important dependence of the damping rate on mode number is a consequence of the equal spacing in frequency of the cavity modes and, as a result, leads the nonlinear interaction of modes to be of the four-wave type [19]. Modifications to the frequencies of cavity modes, which cause the separation in frequency of adjacent modes to become nonuniform (e.g., by including a dispersive element in the feedback loop), therefore, can increase the rate at which coherence is established [20].

We begin the analysis by the consideration of the stability of the equilibrium to perturbations of its own parameters. In particular, we write the slow time dependence of the mode amplitude $X_0(n)$:

$$X_0(n) = [X_{00} + \delta X_0(n)] \exp(-i\delta\omega_0 nT), \tag{9.112}$$

where since only a single mode is present, we may label it the $m = 0$ mode, X_{00} is the magnitude the equilibrium amplitude which satisfies (9.108), $\delta\omega_0$ is the equilibrium frequency shift given by (9.109) with $|X_0|$ replaced by X_{00}, and δX_0 is a small complex perturbation. The real part of δX_0 gives the perturbation of the amplitude of the mode, and the imaginary part gives the perturbation of the phase. This form of the radiation amplitude is then inserted into the equation for the slow time evolution (9.94) for the $m = 0$ mode with the gain function G_0 replaced by the nonlinear gain function defined in (9.106). The arguments of the gain function are the magnitude of the radiation amplitude $|X_0|$ and the detuning p_0. Expansion of the gain function to first order in the perturbation yields

$$\begin{aligned} G_{nl}(|X_0|, p_0) &\cong G_{nl}(X_{00}, p_0) + \delta X_{0,r} \frac{\partial G_{nl}(X_{00}, p_0)}{\partial X_{00}}, \\ &\equiv G_{nl}(X_{00}, p_0) + \frac{\delta X_{0,r}}{X_{00}} G'_{nl} \end{aligned} \tag{9.113}$$

where $\delta X_{0,r}$ is the real part of δX_0 and G_{nl} is the normalized derivative of the nonlinear gain. The real and imaginary parts of (9.94) then give equations for the evolution of the real and imaginary parts of the perturbation:

$$\frac{\partial}{\partial n} \delta X_{0,i} \cong I_b \delta X_{0,r} \operatorname{Re} G'_{nl}, \tag{9.114}$$

and

$$\frac{\partial}{\partial n} \delta X_{0,i} \cong I_b \delta X_{0,r} \operatorname{Im} G'_{nl}, \tag{9.115}$$

where $\delta X_{0,i}$ is the imaginary part of the perturbation to the mode amplitude.

There are two independent solutions to Eqs. (9.114) and (9.115). The first solution has $\delta X_{0,r} = 0$ and $\delta X_{0,i} = \delta X_{0,i}(0)$ which is a constant. This corresponds to a neutrally stable perturbation of the radiation phase. The second solution is given by

$$\delta X_{0,r}(n) \cong \delta X_{0,r}(0) \exp(-\eta\gamma_a T), \tag{9.116}$$

and

$$\delta X_{0,i}(n) \cong \delta X_{0,i}(0) + I_b \operatorname{Im}(G'_{nl}) \int_0^n dn' \delta X_{0,r}(n'), \tag{9.117}$$

and represents stable or unstable perturbations in the amplitude of the radiation field with a damping rate $\gamma_a = -(I_b/T)\,\mathrm{Re}\,G'_{nl}$. The solution will be stable as long as the gain decreases with radiation amplitude:

$$\mathrm{Re}\left[\frac{\partial G_{nl}(X_{00}, p_0)}{\partial X_{00}}\right] \leq 0. \tag{9.118}$$

This is precisely the requirement that the rate at which power is lost from the cavity increases faster with radiation amplitude than the rate at which power is extracted from the beam.

Some of the equilibria represented by points in the amplitude versus detuning plane of Figs. 9.5, 9.6, 9.7 and 9.8 are not stable according to the requirement (9.118). The unstable equilibria can be found by consideration of the level curves of current shown in Fig. 9.6. The functional dependence of the level curves of current in the mode amplitude versus detuning plane is given by Eq. (9.108). Differentiation of (9.108) with respect to mode amplitude yields

$$\frac{\partial I_b}{\partial X_{00}} G_{nl}(X_{00}, p_0) + I_b\,\mathrm{Re}\left[\frac{\partial G_{nl}(X_{00}, p_0)}{\partial X_{00}}\right] = 0, \tag{9.119}$$

Thus, if the current required to maintain an equilibrium increases with mode amplitude, then the gain at that mode amplitude will decrease with mode amplitude and the equilibrium will be a stable operating point. Examination of Fig. 9.6 shows that equilibria with relatively high detunings but low-field amplitudes where the level curves of current have a positive slope correspond to currents which decrease with mode amplitude and, therefore, are unstable.

The stability of single-mode equilibria to perturbations in the amplitude of the equilibrium mode can be analyzed in terms of derivatives of the gain function. In general, the stability of a nonlinear, single mode equilibrium to perturbations by other modes must be analyzed numerically. This is not the case, however, if one adopts the klystron model introduced in Sect. 9.3. Consideration of the klystron model leads to the two-time delay Eq. (9.54), and the same approximations leading to the low-gain equations of Sect. 9.5 can be carried out on the klystron model. In particular, we introduce the two time-scale variables t', and n defined in Eq. (9.72) and assume the current and losses are small. This enables us to cast Eq. (9.54) into the form of (9.81):

$$\left(\frac{\partial}{\partial n} + T_\delta \frac{\partial}{\partial t'}\right) X(n, t') = (R-1)X(n, t') + [I_b(t'')X(n, t'')g(|X(n, t'')|)]_{t''=t'-\varepsilon T}, \tag{9.120}$$

where g denotes the gain in the klystron model defined by Eq. (9.53).

Equations (9.81) and (9.120) are similar except for the terms that describe the effect of the beam. In the klystron model, the interaction with the beam is dependent on the radiation amplitude at the time a particular group of electrons enters the interaction region, $t' - \varepsilon T$. In the more general model described by Eq. (9.81), the interaction with the beam is dependent on the amplitude of the radiation entering the cavity for all times between t' and $t' - \varepsilon T$. A slight improvement to the klystron model is obtained by replacement of the slippage parameter ε by $\varepsilon/2$ to account for the fact that in the more general case, some weighted average of the entering radiation between the times t' and $t' - \varepsilon T$ determines the response of the beam.

To study the stability of single-frequency states using the klystron model, we again take the normalized current I_b to be constant and set T_δ equal to zero. Further, the fast time scale dependence of the radiation field is expressed as the sum of the large amplitude equilibrium mode with $m = 0$ and an ensemble of small perturbing modes with $m \neq 0$:

$$X(n,t') = \exp(-i\delta\omega_0 nT)\left[X_{00} + \sum_{m\neq 0}\delta X_m(n)\exp(-i\Delta\omega_m t')\right], \tag{9.121}$$

where $\Delta\omega_m$ is defined in Eq. (9.85). The argument of the gain function in Eq. (9.120) is the magnitude of the radiation amplitude evaluated at the delayed time $t' - \varepsilon T$. For small amplitude of the perturbing modes, this becomes

$$\begin{aligned} |X(n',t'-\varepsilon T)| &\cong X_{00} + \mathrm{Re}\left[\sum_{m\neq 0}\delta X_m(n)\exp(-i\Delta\omega_m(t'-\varepsilon T))\right] \\ &\cong X_{00} + \frac{1}{2}\sum_{m\neq 0}\left[\delta X_m(n) + \delta X^*_{-m}(n)\right]\exp(-i\Delta\omega_m(t'-\varepsilon T)), \end{aligned} \tag{9.122}$$

where in the second equality we have used the fact that $\Delta\omega_m = -\Delta\omega_{-m}$. It is important to note that the usual relation between the Fourier amplitudes of a real function, $\delta X_{-m} = \delta X_m{}^*$, is not satisfied in the present case since the time-dependent radiation amplitude $X(n,t')$ is complex. Physically, this is a result of the fact that the $\pm m$ modes are distinct, and there is no a priori reason why their amplitudes should be related.

An evolution equation for the amplitude of the mth perturbing mode can be obtained by the following procedure. We substitute the expression for the radiation amplitude (9.122) into the nonlinear gain function g. This expression is expanded in powers of the field amplitude and subsequently inserted into the dynamical Eq. (9.120). In addition, the equilibrium relations (9.108) and (9.109) (with G_{nl} replaced by g) are used to eliminate the equilibrium terms. Finally, we multiply by $\exp(i\Delta\omega_m t')$ and integrate over t' to select the mth Fourier component:

$$\begin{aligned} &\left(\frac{\partial}{\partial n} - i\delta\omega_0 T + (1-R)\right)\delta X_m(n) = \\ &= \left(g(X_{00})\delta X_m(n) + \frac{1}{2}g'(X_{00})\left[\delta X_m(n) + \delta X^*_{-m}(n)\right]\right)\exp(-i\varepsilon\Delta\omega_m T), \end{aligned} \tag{9.123}$$

An important consequence of Eq. (9.123) is that modes with equal and opposite mode numbers (i.e., modes which are equally spaced in frequency about the equilibrium mode) couple together in the presence of the equilibrium mode. This coupling vanishes as the amplitude of the equilibrium mode goes to zero, and [15, 19]

$$g'(X_{00}) \equiv X_{00}\frac{\partial g(X_{00})}{\partial X_{00}} \approx X_{00}^2, \tag{9.124}$$

The coupling can be thought of as the result of a nonlinear four wave mixing process [19]. In such a case, four modes with frequencies ω_1, ω_2, ω_3, and ω_4 will couple nonlinearly if the frequency matching criterion

$$\omega_1 + \omega_2 = \omega_3 + \omega_4, \tag{9.125}$$

is satisfied. In the present case, one can take the frequencies ω_1 and ω_2 to be those of the perturbing modes ($\omega + \Delta\omega_{\pm m}$) and the frequencies ω_3 and ω_4 to be that of the equilibrium mode ω. Because of this coupling, it is not proper to speak of the gain of a single perturbing mode in the presence of an equilibrium mode. Rather, one must always treat coupled pairs of modes. This coupling has important consequences regarding the damping rate of stable modes.

Since modes with equal and opposite mode numbers are coupled, it is convenient to write the evolution equations for both the mode amplitudes δX_m and $\delta X_{-m}{}^*$, and treat these mode amplitudes as two independent variables. The result is two coupled equations of the form of Eq. (9.123):

$$\frac{\partial}{\partial n}\delta X_m = I_b\left[g\delta X_m(\exp(i\varepsilon\Delta\omega_m T) - 1) + \frac{g'}{2}\left(\delta X_m + \delta X^*_{-m}\right)\exp(i\varepsilon T\Delta\omega_m)\right], \tag{9.126}$$

and

$$\frac{\partial}{\partial n}\delta X^*_{-m} = I_b\left[g^*\delta X^*{}_m(\exp(i\varepsilon\Delta\omega_m T) - 1) + \frac{g'^*}{2}\left(\delta X_m + \delta X^*_{-m}\right)\exp(i\varepsilon T\Delta\omega_m)\right], \tag{9.127}$$

where we have used the equilibrium condition, $-i\delta\omega_0 T + (1 - R) = I_b g$ to simplify the coefficients in Eqs. (9.126 and 9.127).

One limit in which the two coupled Eqs. (9.126) and (9.127) are easily solved is the limit where $\varepsilon\Delta\omega_m T \rightarrow 0$, which corresponds to the limit in which the perturbing modes are close in frequency to the equilibrium mode. In this case, *close in frequency* means that the spacing between the perturbing modes and the equilibrium mode is a small fraction of the gain bandwidth. In this limit, the sum and difference of Eqs. (9.126) and (9.127) yields

$$\frac{\partial}{\partial n}\left(\delta X_m + \delta X^*_{-m}\right) = I_b\,\mathrm{Re}(g')\left(\delta X_m + \delta X^*_{-m}\right), \tag{9.128}$$

and

$$\frac{\partial}{\partial n}\left(\delta X_m - \delta X^*_{-m}\right) = iI_b\mathrm{Im}(g')\left(\delta X_m + \delta X^*_{-m}\right), \tag{9.129}$$

Observe that these equations are formally identical to the pair of Eqs. (9.114) and (9.115) which describe the stability of the equilibrium state against perturbations of its magnitude and phase. In particular, there are two solutions: one which is damped corresponding to perturbations of the magnitude of the radiation $\delta X_m + \delta X_{-m}{}^* \neq 0$ and one which is neutrally stable corresponding to a perturbation of the phase of the radiation $\delta X_m + \delta X_{-m}{}^* = 0$ and $\delta X_m - \delta X_{-m}{}^* \neq 0$. Thus, to lowest order, nearby modes which couple together in such a way that only the phase of the radiation field is perturbed are undamped and not suppressed. Observe that if the coupling between the $\pm m$ modes had not been included in Eqs. (9.126) and (9.127); then we would have erroneously predicted that all nearby modes were strongly damped in the presence of the equilibrium mode at a rate exactly half that at which the magnitude of the equilibrium mode decays when perturbed (as in (9.116)).

While the preceding result was derived for the klystron model, it is in fact more general. In particular, the lowest-order stability of nearby in frequency modes was determined by taking the limit of vanishingly small slippage parameter. If the slippage parameter vanishes in Eqs. (9.79)–(9.81), then both the electron equations of motion and the source term representing the effect of the electron beam in (9.81) become identical to the single-frequency equations leading to (9.103) and (9.106), except that the mode amplitude $X_0(n)$ is replaced by the fast time-dependent amplitude $X(n,t_e = t')$. This is precisely the situation discussed in Sect. 9.3 when the case of equal beam and radiation speeds was considered. In the absence of slippage, electrons and photons move in synchronism through the interaction region. Therefore, the amplitude of the radiation entering the cavity at time $t + T$ depends only on the amplitude of the radiation entering at time t, and there is no communication between portions of the radiation pulse having different entrance times. The same situation effectively results with finite slippage if it is assumed that the radiation amplitude consists of a superposition of nearby in frequency modes. The variation in the field amplitude on the times scale on which slippage is important (i.e., εT) is small for closely spaced modes, and slippage can be neglected to lowest order. Thus, in the limit of vanishing slippage, the radiation amplitude in the general case evolves according to (9.120) with T_δ set equal to zero and the klystron gain g replaced by the general nonlinear gain G_{nl}:

$$\frac{\partial}{\partial n}X(n,t') = (R-1)X(n,t') + I_b G_{nl}(|X(n,t')|,p)X(n,t'), \tag{9.130}$$

The nonlinear evolution of the field amplitude which satisfies an equation of the form of (9.120) with slippage parameter ε and detuning time T_δ set equal to zero can be described as follows. An arbitrary small initial radiation amplitude $X(0,t')$ will grow to saturation (with an equilibrium magnitude satisfying Eqs. (9.108) and (9.109)). This magnitude at saturation is independent of the fast time variable t'. The phase of the complex amplitude at saturation, however, will be determined by the initial conditions and will depend on the fast time variable t'. In the absence of slippage, there is no mechanism to bring the phases of different portions of the radiation pulse into coherence. Therefore, because the phase is time dependent, the radiation pulse will consist of a superposition of cavity modes expressed in the form of a Fourier series as in (9.84) [21].

The neutrally stable solution of (9.128) and (9.129) for the linear growth of nearby frequency satellite modes is modified when the effects of nonzero frequency separation are included. The growth or damping of the satellite modes in this case can be determined by looking for solutions of Eqs. (9.126) and (9.127) in which the slow time dependence of the mode amplitudes is expressed in the form $X_m(n) = X_{m0}\exp(\gamma_m nT)$ where γ_m is the complex growth rate. Substitution of this solution into Eqs. (9.127) and (9.128) results in a quadratic dispersion relation for the growth rate:

$$\begin{aligned}&\left\{\gamma_m T - I_b\left(g[\exp(i\varepsilon t\Delta\omega_m)-1] + \frac{g'}{2}\exp(i\varepsilon T\Delta\omega_m)\right)\right\}\\ &\times\left\{\gamma_m T - I_b\left(g^*[\exp(i\varepsilon t\Delta\omega_m)-1] + \frac{g'^*}{2}\exp(i\varepsilon T\Delta\omega_m)\right)\right\}\\ &= \frac{I_b^2}{4}\exp(2i\varepsilon T\Delta\omega_m)|g'|^2,\end{aligned} \tag{9.131}$$

Observe that in the limit in which $\varepsilon\Delta\omega_m T \to 0$, the previously discussed solutions for amplitude and phase perturbations are obtained.

Certain properties of the solutions of Eq. (9.130) can be determined from symmetry arguments. In particular, the equation is invariant under the combined operations of replacement of m by $-m$ and complex conjugation. Hence, $\gamma_{-m} = \gamma_m^*$, or if we write γ_m in terms of its real and imaginary parts, $\mathrm{Re}(\gamma_{-m}) = \mathrm{Re}(\gamma_m)$ and $\mathrm{Im}(\gamma_{-m}) = -\,\mathrm{Im}(\gamma_m)$. Thus, the growth or damping rate of the satellite pair is an even function of mode number, and the frequency shift is an odd function of mode number. This implies that the complex growth rate of the solution (which in the limit of vanishing $\varepsilon\Delta\omega_m T$ corresponds to a perturbation of the phase of the radiation) must have the following Taylor expansion for small $\varepsilon\Delta\omega_m T$:

$$\gamma_m \cong i\varepsilon\Delta\omega_m T\omega' - D(\varepsilon\Delta\omega_m T)^2, \tag{9.132}$$

to second order in $\varepsilon T\Delta\omega_m$, where the coefficients are obtained from the solution of (9.131) in ascending powers of $\varepsilon T\Delta\omega_m$. The stability of the satellite pair is determined by the sign of the coefficient D. If D is positive, nearby satellite pairs are weakly stable, and if D is negative, the pair are weakly unstable [15]. The situation in which the nearby pairs are weakly unstable is termed a phase instability. For the klystron model, the coefficients D and ω' can be determined analytically with some effort [15]. For the general case, it can be shown that the frequency shift coefficient ω' can be related to the equilibrium frequency shift (9.109) by

$$\omega' = \frac{d}{dp_0}\delta\omega_0(X_{00}, p_0), \tag{9.133}$$

where the derivative in (9.133) is performed by allowing X_{00} to vary in accordance with maintaining I_b constant in Eq. (9.108). The steps required to obtain this result will be presented when nonlinear gain narrowing is discussed. The important coefficient D, which determines the stability of the satellite pair, must be determined numerically in general. An order of magnitude estimate of D based on (9.131) indicates that it scales as I_b/T, which scales as the inverse of the cavity time T_d in equilibrium.

So far only the growth and damping of nearby frequency modes has been discussed. Recall that the term nearby frequency describes pairs of modes whose separation in frequency from the equilibrium mode is small compared with the gain bandwidth. Pairs of satellites that are displaced from the equilibrium mode by approximately the synchrotron frequency can also become unstable at large field amplitudes. To find this instability in the klystron model [4], we set $\varepsilon\Delta\omega_m T = \pi$ and impose the simplifying assumption that the nonlinear gain is real. From the definition of the klystron gain in Eq. (9.53), it is seen that the gain is real if the distribution of energies of the injected beam is centered about a detuning value $p_i = \pi/2$. This corresponds to the maximum gain in the klystron model. With these assumptions, the dispersion Eq. (9.131) yields two possible solutions:

$$\gamma_m = -\frac{2gI_b}{T}, \tag{9.134}$$

and

$$\gamma_m = -\frac{I_b[2g + g']}{T}, \tag{9.135}$$

The solution given in (9.134) is always stable, while that in (9.135) may be either stable or unstable. The latter of these corresponds to the case in which $\delta X_m = \delta X_{-m}{}^*$, that is, when the phases of the two satellites are such that the magnitude of radiation is perturbed. The satellite pair will be unstable if

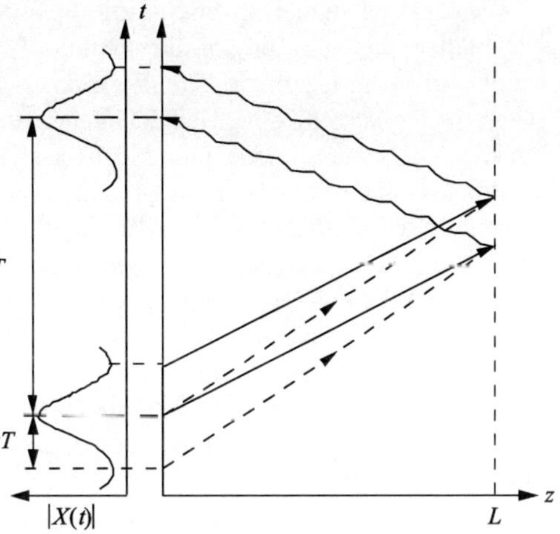

Fig. 9.10 Characteristic plots illustrating the mechanism by which nonlinear saturation of gain combined with slippage gives rise to the spiking mode

$$2g + X_{00}\frac{\partial g}{\partial X_{00}} < 0. \tag{9.136}$$

Hence, if the gain decreases too rapidly with mode amplitude, then sidebands with $\varepsilon\Delta\omega_m T = \pi$ become unstable. Furthermore, these sidebands produce perturbations in the magnitude of the radiation. Recalling the relationship between nonlinear gain and efficiency from Eq. (9.107), the condition (9.136) is seen to correspond to the field amplitude for maximum efficiency that can be expressed as

$$\frac{\partial}{\partial X_{00}}\left(gX_{00}^2\right) = \frac{\partial}{\partial X_{00}}\Delta p < 0, \tag{9.137}$$

Therefore, the sidebands become unstable when the equilibrium mode amplitude exceeds that value at which the efficiency begins to decrease with field amplitude.

A physical picture of the instability [4] can be drawn using characteristic plots such as Fig. 9.10. The radiation amplitude shown on the left side of the figure has developed modulations with a period equal to twice the slippage time εT. In terms of modes, this is described by a superposition of the equilibrium mode and two satellites with frequencies displaced from the equilibrium mode by $\varepsilon\Delta\omega_m T = \pi$. The modulations grow because they are self-reinforcing if the gain decreases too rapidly with radiation amplitude. Electrons, which enter the interaction region when the radiation amplitude is small, produce a significantly larger gain than those entering when the radiation amplitude is large. In traversing the interaction region, the electrons slip behind the radiation by half a period of the modulation. Hence, the high-gain electrons then contribute to the high amplitude of the radiation on the next pass, and the low-gain electrons contribute to the low-field amplitude, thereby reinforcing the perturbation. This instability occurs when the gain decreases rapidly with field amplitude and has been named the overbunch instability [22].

The klystron model affords us the opportunity to obtain a qualitative, if not quantitative, understanding of the stability of single-frequency states in a low-gain, continuous beam oscillator. The determination of the stability of single-frequency states for the case of a distributed interaction region, as opposed to the klystron model, must proceed along numerical lines [15]. However, certain aspects of the klystron calculation can be employed. In particular, a radiation amplitude expressed in the form of Eq. (9.121) will turn Eq. (9.90) into an equation of the form of (9.123). Thus, a satellite mode with index m is driven by beam currents, which are the responses to the presence of not only the mth satellite mode but also the $-m^{\text{th}}$ mode. The coefficients of proportionality must be determined numerically by integrating the equations of motion in the presence of each satellite. Once the coefficients are determined, the stability of the satellites is found by solving a quadratic equation of the form (9.131). This procedure has been carried out for a 50 ∞ 50 grid of equilibria in the amplitude versus detuning plane [15]. For each equilibrium, the growth rates $\gamma_m(X_{00},p_0)$ were calculated for ten pairs of satellites and slippage parameter $\varepsilon = 0.1$. From this data, the boundary separating stable and unstable equilibria was interpolated as shown in Fig. 9.8. Examples of phase unstable and overbunch unstable equilibria are shown in Figs. 9.11 and 9.12, where the growth rate of the most unstable of the two solutions of the quadratic equation corresponding to Eq. (9.131) is plotted versus detuning difference $\varepsilon\Delta\omega_m T$. As can be seen for the phase unstable equilibrium, the most unstable modes are near to the equilibrium mode in frequency with growth rates that depend quadratically on detuning difference for small detuning differences. On the other hand, for the overbunch unstable equilibrium, the unstable modes are displaced in detuning difference from the equilibrium mode.

In conclusion, the single-frequency states found in the previous section will be stable for equilibrium parameters that place them within the triangular shaped region

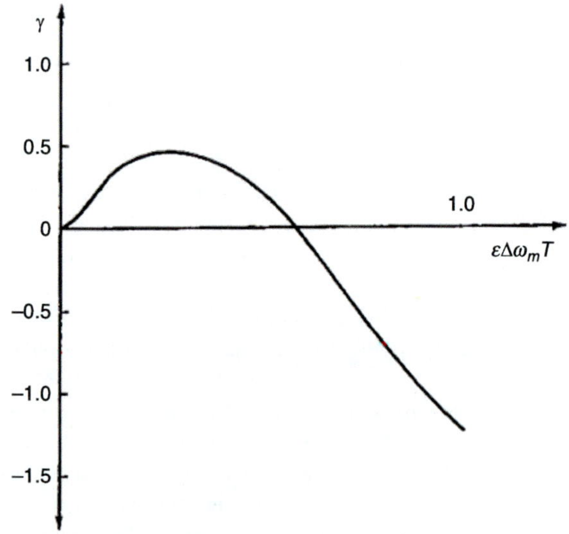

Fig. 9.11 Satellite growth rate versus detuning difference illustrating the phase instability. Equilibrium parameters are $X_0 = 6.0$ and $p_0 = 5.14$ [15]

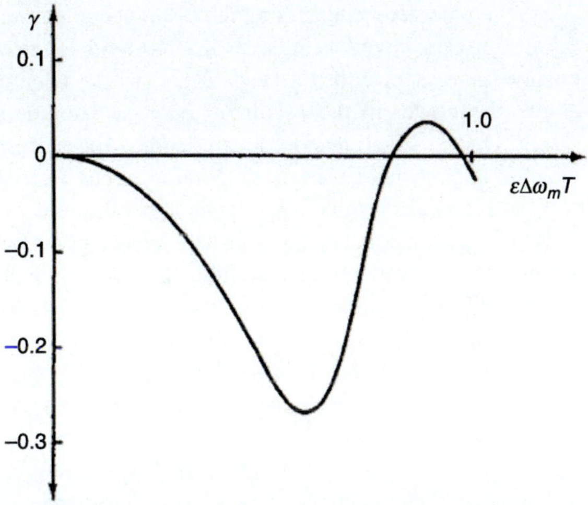

Fig. 9.12 Satellite growth rate versus detuning difference illustrating the overbunch instability. Equilibrium parameters are $X_0 = 14.8$ and $p_0 = 2.6$ [15]

of Fig. 9.8. These stable equilibria are capable of suppressing parasitic modes, although this suppression becomes weak for pairs of parasitic modes equally spaced close in frequency to the equilibrium mode. The decay rate of these modes scales as the square of the detuning difference between the equilibrium mode and the satellites. The fact that these satellites decay slowly has two consequences. First, the slow decay implies that a single-frequency equilibrium will only be reached after a time sufficiently long so that all satellites decay [15, 21]. Second, noise fluctuations can excite the weakly damped satellites to an unexpectedly large amplitude which places a more stringent condition on the noise level required to reach a minimum bandwidth. Finally, other modes will not be suppressed for equilibrium parameters outside the stable region. In the case of a phase instability, the tendency is for nearby frequency modes to grow, and the unstable equilibrium mode is eventually replaced by one that is stable. In the case of the overbunch instability, the radiation amplitude develops modulations, and single-frequency equilibria are not possible. This is known as the spiking mode, and it is expected to occur when the beam current exceeds the start current by a factor of between three and four (depending on the detuning of the equilibrium). For higher currents, the time dependence of the radiation field becomes progressively more complicated and eventually chaotic as will be discussed in Chap. 11.

9.6.3 *The Effects of Shot Noise*

So far, our treatment of oscillators has neglected the effects of noise on the incoming electron beam. For example, in our discussion of long pulse oscillators, the beam is characterized by a current and a distribution function that are independent of time.

When the parameters describing the beam become time dependent in a random way this can be considered as noise. Noise has both adverse and positive effects. On the positive side, it provides a seed signal, which becomes amplified with time and allows the power in the oscillator to reach saturation without the need to inject power. On the negative side, it can degrade the performance of the oscillator from which it is expected based on time-independent beam parameters.

Classical noise results from the random variations of the injected beam over time. In the low-gain oscillator, the time dependence of the beam current enters the system of equations in three places with three distinct time scales. The shortest time scale is the reciprocal of the radiation frequency. Variations in beam current on this time scale result from the fact that the beam is composed of discrete electrons, which have random entrance times. This source of noise is responsible for the spontaneously emitted radiation that was discussed in Chap. 3 and will be discussed most extensively here. This noise may be included in the present theory by allowing the injected distribution function to be composed of a superposition of delta functions representing the discrete charges with random values of entrance phase ψ_i [23–26].

The next shortest time scale in a low-gain oscillator is the radiation round-trip time. Variation of the beam parameters on this time scale is particularly important for oscillators driven by pulsed electron beams [27]. If the beam pulses do not arrive at uniformly separated times, synchronism with the radiation pulses will be spoiled. If the pulse-to-pulse variations in the arriving beam are correlated over a time, which is short compared with the cavity decay time, then they can be modeled by appropriate modification of the parameters of the incoming beam. For example, pulse-to-pulse variations in beam energy can be treated as an additional energy spread for a beam whose pulses are identical. Pulse-to-pulse variations in the arrival time of the beam effectively broaden the temporal width and lower the peak current of the beam micropulse.

Finally, the longest time scale is the cavity decay time. Variations of the beam parameters over this time scale can lead to mode-hopping [28]. In this case, a stable single-frequency state may have established itself for a given set of beam parameters. However, if the beam parameters change, then this mode may become unstable and be replaced by another mode or modes. This will be the case if the beam energy varies, since the free-electron laser resonance is particularly sensitive to energy. Fractional changes in beam energy of order $1/N_w$ (where N_w is the length of the interaction region measured in wiggler periods) are sufficient to change the detuning of a mode and move it out of the gain bandwidth.

The shot noise associated with the discreteness of the electron charge is the most straightforward to calculate and probably represents a minimum to the noise that would be encountered in a realistic beam. For this reason, we will focus our attention on it here. It will be seen that shot noise can be modeled by adding a random source term to the right-hand side of Eq. (9.90). The random source term can be taken to be a Gaussian white noise process with respect to the round-trip index n. It is characterized by a matrix of correlation functions between the amplitudes of the source terms driving modes with different indices m [23]. If the mode amplitudes are all small (as they would be during the start-up phase of the oscillator), this matrix will be diagonal; that is, the source terms driving different modes are uncorrelated. The diagonal terms will be found to be proportional to the rate of spontaneous emission of radiation into each mode of the cavity. A consequence of the independence of the source terms is that each mode will grow independently during the initial start-up phase of the oscillator and, at any instant of time before saturation, the mode amplitudes will have a Gaussian probability distribution characterized by a mode number-dependent temperature. Thus, there will be large statistical fluctuations in the amplitudes of individual modes even though the expected amplitude of a mode (the mode temperature) depends smoothly on mode numbers [29].

In the nonlinear regime, shot noise will cause the phase of the signal to diffuse in round-trip time n if conditions of the previous section are satisfied for the existence of stable single-frequency equilibria. This phase diffusion with time leads to a broadened spectrum in the form of a Lorentzian distribution with a width proportional to the spontaneous emission rate [25].

It is convenient to begin the analysis with Eq. (9.90), which describes the evolution of the amplitude of the mth cavity mode in a low-gain oscillator. The time integral and average on the right-hand side of this equation represent the contribution to the growth of the amplitude of the mth mode of an ensemble of electrons with a uniform distribution of entrance times t_e and phases ψ_i. Such a uniform distribution approximates the situation when a large number of electrons enter the device at random times. The average in (9.90) is more properly thought of as a sum over all electrons that entered the device during a time period T. Given that the electrons enter at random times, the right-hand side will have an average component and a random component which fluctuate with iteration number n. Let us define the exponent inside the average of (9.90) to be some phase:

$$\theta_m(t_e,\psi_i) = \Delta\omega_m(t_e + \varepsilon\xi T) - \psi(\xi, t_e, \psi_i), \tag{9.138}$$

where we have expressed explicitly the dependence of θ_m on the initial entrance phase and entrance time of the particle and suppressed writing the dependence of θ_m on the axial distance, initial energy, and the amplitudes of the modes $X_m(n)$.

The average over entrance phases and entrance times in (9.90) are now replaced by a sum over entrance times and entrance phases of N_T individual particles:

$$Z_m = \frac{1}{N_T}\sum_{j=1}^{N_T} d\xi \exp\left[i\theta_m\left(t_{ej}, \psi_{ij}\right)\right], \tag{9.139}$$

The quantity Z_m is a complex random variable, which is the sum of a large number N_T of individual random variables. If we assume that the entrance times and entrance phases for the individual electrons are random variables which are independent and identically distributed, then by the mean value theorem, Z_m will be a Gaussian random variable characterized only by a mean and a variance which describes the random nature of the electron beam. The mean value of Z_m, denoted by, is found by averaging over the entrance times and entrance phases of all N_T

particles under the assumption that the entrance times and entrance phases of different particles are independent and identically distributed and that the distribution is uniform with respect to entrance time and entrance phase. Hence,

$$\overline{Z}_m = \frac{1}{(2\pi T)^{N_T}} \int_0^{2\pi} d\psi_{i1} \cdots d\psi_{iN_T} \int_0^T dt_{e1} \cdots dt_{eN_T} Z_m(\psi_{i1}, \cdots, \psi_{iN_T}, t_{e1}, \cdots, t_{eN_T}), \tag{9.140}$$

Evaluation of these integrals yields a mean value for Z_m of

$$\overline{Z}_m = \frac{1}{(2\pi T)} \int_0^{2\pi} d\psi_i \int_0^T dt_e \int_0^1 d\xi \exp[i\theta_m(t_e, \psi_i)], \tag{9.141}$$

As expected, this gives precisely the right-hand side of (9.90). The variance of $|Z_m|^2$ is obtained by forming the square of the magnitude of Z_m given by (9.139)

$$|Z_m|^2 = \frac{1}{N_T^2} \sum_{j,j'=1}^{N_T} \int_0^1 d\xi \exp\left[i\theta_m\left(t_{ej}, \psi_{ij}\right)\right] \int_0^1 d\xi \exp\left[-i\theta_m\left(t_{ej'}, \psi_{ij'}\right)\right], \tag{9.142}$$

and averaging over the entrance times and entrance phases of all the electrons. The double sum in Eq. (9.142) results in two types of terms when averaged. There are N_T diagonal terms for which $j = j'$, and $N_T(N_T - 1)$ terms where particles $j \neq j'$. Under the assumption that N_T is large, this results in

$$\overline{|\delta Z_m|^2} = \overline{|Z_m|^2} - |\overline{Z}_m|^2 = \frac{1}{N_T} \int_0^{2\pi} \frac{d\psi_i}{2\pi} \int_0^T \frac{dt_e}{T} \left| \int_0^1 d\xi \exp[i\theta_m(t_e, \psi_i)] \right|^2, \tag{9.143}$$

Thus, the root-mean-square value of the random fluctuation of Z_m from its mean value scales as $N_T^{-1/2}$, which is small when the number of electrons, is large.

The random component of the source term in (9.90) will fluctuate with round-trip index n. Each time the round-trip index increases by one, a new group of N_T electrons will have passed through the interaction region. This group will be uncorrelated with the previous groups of electrons. Thus, the effective correlation time in round-trip index n is unity. Since field amplitudes in the low-gain limit can change only after many round-trip times, a correlation time of one round-trip is essentially the same as a correlation time of zero. In other words, the noise term can be taken to represent white noise:

$$\overline{\delta Z_m(n) \delta Z_m^*(n')} = \delta(n - n') \overline{|\delta Z_m|^2}. \tag{9.144}$$

In principle, there can be correlations between the source terms (i.e., on the right-hand side of Eq. (9.90)) driving different modes. That is the expectation value of the cross term $\delta Z_n{}^*\delta Z_m$ may be nonzero [23]. The cross-correlation function between the sources for two modes labeled m and n can be defined in analogy to (9.144)

$$\overline{\delta Z_n^*\delta Z_m} = \frac{1}{N_T}\int\limits_0^{2\pi}\frac{d\psi_i}{2\pi}\int\limits_0^{T}\frac{dt_e}{T}\int\limits_0^{1}d\xi\exp[-i\theta_n(t_e,\psi_i)]\int\limits_0^{1}d\xi\exp[i\theta_m(t_e,\psi_i)], \tag{9.145}$$

Examining Eq. (9.138) for the phase θ_n, we see that the cross-correlation between the noise signal driving the two modes will be nonzero if the phase ψ depends periodically on the entrance time t_e with a frequency equal to the difference of the frequencies of the two modes. This requires that a large amplitude mode with this difference frequency be present in the cavity to modify the electron trajectories. If the amplitudes of all modes are small, as they are during the linear start-up phase of the oscillator, then the cross-correlations vanish, and each mode is excited by its own independent white noise source. On the other hand, if the mode amplitudes are large, then there will be correlations between the noise sources driving each mode. Another instance in which the noise sources are cross-correlated is the case of a pulsed beam oscillator. In this case the current in (9.90) is time dependent and should be taken inside the integral over entrance time. This same factor would then appear, but squared, inside the integral in (9.145). The time dependence of the beam current would then induce a correlation between the noise signals driving two different modes even when the mode amplitudes are small.

The simplest case to analyze is the case in which all modes are assumed to be small. This was done in Chap. 3 in which spontaneous emission was studied in detail. The amplitude of the fluctuations expressed in (9.143) can be calculated by substitution of the expression for the free-streaming phase of an electron in the interaction region in terms of its initial phase and entrance time

$$\psi(\xi,\psi_i,t_e) = \psi_i + p_0\xi, \tag{9.146}$$

Evaluation of (9.143) then gives the spontaneous emission noise amplitude in terms of the dimensionless detuning $p_m = p_0 - \varepsilon\Delta\omega_m T = p_0 - 2\pi\varepsilon m$, for which

$$\overline{|\delta Z_m|^2} = \frac{1}{N_T}\left|\frac{\exp(ip_m)-1}{ip_m}\right|^2 = \frac{1}{N_T}\frac{\sin^2\left(\frac{p_m}{2}\right)}{p_m^2}, \tag{9.147}$$

Examination of the effect of noise on the evolution of individual mode amplitudes requires solution of the set of equations represented by (9.90) and either the particle Eqs. (9.79) and (9.80) or the Vlasov Eq. (9.86) supplemented by the addition of the random source representing the fluctuations in the variables Z_m. These equations can be solved analytically in two limits: (1) the limit of small mode amplitudes (i.e., the start-up phase) and (2) the limit of a single mode. In both cases, correlations between different modes can be neglected. In the general case, computer simulation of the system is required with specific realizations for the random variables. A technique for performing such a simulation, which correctly models the noise without requiring the solution for the trajectories of an unduly large number of electrons, will be given at the end of this section.

We now consider the case of the evolution of a single mode in the presence of a white noise source. Because of the presence of a random noise source, the complex amplitude of the mode X_m itself becomes a random variable. At a particular time, labeled by the round-trip index n, there is a probability distribution function, which gives the probability of observing the mode to have a given amplitude. Since the amplitude is a complex variable, we must introduce the joint probability distribution function for either the real and imaginary parts of the amplitude, or its magnitude and phase. It is convenient at this point to introduce the two-dimensional vector $\mathbf{X}$ whose components are the real and imaginary parts of the mode amplitude X_m. This permits the use of vector notation to write the probability of observing the vector mode amplitude $\mathbf{X}$ in the small two-dimensional area dX^2 centered at $\mathbf{X}$

$$dP = P(\mathbf{X}, n) d^2X. \tag{9.148}$$

Because the mode amplitudes change slowly with time and because the noise is white noise, the evolution of the probability distribution function is the governed by the Fokker-Planck equation:

$$\frac{\partial}{\partial n} P(\mathbf{X}, n) = \frac{\partial}{\partial \mathbf{X}} \cdot \left[-\frac{\overline{\Delta \mathbf{X}}}{\Delta n} P(\mathbf{X}, n) + \frac{\partial}{\partial \mathbf{X}} \cdot \left(\frac{1}{2} \frac{\overline{\Delta \mathbf{X} \Delta \mathbf{X}}}{\Delta n} P(\mathbf{X}, n) \right) \right], \tag{9.149}$$

where the coefficients in (9.149) give the expected rates of drift and diffusion of the mode amplitude vector $\mathbf{X}$. Derivations of the Fokker-Planck equation can be found in standard text books on kinetic equations [30]. The coefficients in (9.149) are most easily written when the vector $\mathbf{X}$ is represented in polar coordinates (X,φ):

$$\frac{\partial}{\partial n} P(\mathbf{X}, n) = \frac{1}{X} \left[\frac{\partial}{\partial X} (X\Gamma_X) + \frac{\partial}{\partial \varphi} \Gamma_\varphi \right], \tag{9.150}$$

where the fluxes are given by

$$\begin{aligned} \Gamma_X &= X[(1-R) - I_b \operatorname{Re} G_{nl}]P + \frac{1}{X} \left[\frac{\partial}{\partial X} (XPD_{XX}) + \frac{\partial}{\partial \varphi} (PD_{\varphi X}) - PD_{\varphi\varphi} \right], \\ \Gamma_\varphi &= -X\delta\omega_0 P + \frac{1}{X} \left[\frac{\partial}{\partial X} (XPD_{X\varphi}) + \frac{\partial}{\partial \varphi} (PD_{\varphi\varphi}) + PD_{\varphi X} \right] \end{aligned} \tag{9.151}$$

and the components of the diffusion tensor in polar coordinates are

$$D_{XX} = \frac{I_b^2}{2N_T} \int_0^{2\pi} \frac{d\psi_i}{2\pi} \int_0^T \frac{dt_e}{T} \left[\int_0^1 d\xi \sin \theta_m(\psi_i, t_e) \right]^2$$
$$D_{\varphi\varphi} = \frac{I_b^2}{2N_T} \int_0^{2\pi} \frac{d\psi_i}{2\pi} \int_0^T \frac{dt_e}{T} \left[\int_0^1 d\xi \cos \theta_m(\psi_i, t_e) \right]^2 \quad (9.152)$$
$$D_{X\varphi} = D_{\varphi X} = \frac{I_b^2}{2N_T} \int_0^{2\pi} \frac{d\psi_i}{2\pi} \int_0^T \frac{dt_e}{T} \int_0^1 d\xi \sin \theta_m(\psi_i, t_e) \int_0^1 d\xi \cos \theta_m(\psi_i, t_e)$$

In evaluating the average drift term (the first terms in Eq. (9.151)), we have used the result that the expected value of the variable Z_m given by (9.141) can be expressed in terms of the nonlinear gain function defined in (9.106). This introduces the real part of the gain and the frequency shift through Eq. (9.109). The diffusive terms (in particular the elements of the diffusion tensor) are evaluated by solution for the trajectories of electrons in the presence of a real mode amplitude $X_m = X$. Equation (9.150) is expected to apply if many modes are present (but all are small), or if only the mth mode is present. Observe that when the mode amplitude is small in the linear regime, the diffusion tensor is diagonal and can be expressed in terms of the normalized spontaneous emission rate (9.147):

$$D_d = \frac{I_b^2}{4N_T} \frac{\sin^2(p_m/2)}{p_m^2} \equiv \frac{1}{2} S_m, \quad (9.153)$$

where D_d denotes the diagonal elements of the diffusion tensor and S_m is used to denote the source term which is the normalized rate of spontaneous emission.

In the linear regime, we can look for solutions of Eq. (9.150) of the form of a Gaussian probability distribution function with a time-varying-mode temperature:

$$P(\mathbf{X}, n) = \frac{1}{2\pi T_m(n)} \exp\left(- \frac{X^2}{2T_m(n)} \right), \quad (9.154)$$

Insertion of the assumed form of the distribution function into Eq. (9.150) results in a consistency relation determining the growth with time of the mode temperature:

$$\left(\frac{d}{dn} + \gamma_m T \right) T_m = S_m, \quad (9.155)$$

where γ_m is the growth rate in the linear regime given by (9.99). If the macroscopic beam parameters such as current and voltage are constant in time, Eq. (9.155) is readily integrated to obtain

$$T_m(n) = \frac{S_m}{2\gamma_m T}[\exp(2\gamma_m nT) - 1]. \tag{9.156}$$

For times shorter than the growth time $\gamma_n nT < < 1$, the mode temperature increases linearly with time reflecting the initial buildup of energy in the mode due to the constant rate of spontaneous emission (the exponential growth is unimportant on this time scale). Initially, the temperature increases by an amount S_m in the time it takes the radiation to circuit the cavity once. For times greater than the growth time $\gamma_n nT > > 1$, the temperature grows exponentially due to gain with an effective initial amplitude proportional to the spontaneous emission rate. This effective initial amplitude corresponds to a temperature $T_m(0) = S_m/(2\gamma_m T)$ which is roughly the temperature achieved at the end of the initial phase after a number $(2\gamma_m T)^{-1}$ round trips of the radiation through the cavity. If the macroscopic beam parameters vary with time, Eq. (9.155) can be solved numerically to determine the mode temperature in the linear regime. An example of this will be given in the section on nonlinear gain narrowing.

In addition to providing a seed signal to start the oscillator, shot noise also causes fluctuations in the amplitude and phase of a mode at saturation, which may be determined quantitatively from Eq. (9.150). We assume that a single mode is present and calculate the time asymptotic probability distribution function for the amplitude of this mode. When expressed in polar coordinates, Eq. (9.150) describes the drift and diffusion of the amplitude and phase of the cavity mode. Due to the diffusion of the phase, we can assume that the probability distribution function becomes independent of the phase of the mode asymptotically in time. The dependence of the distribution function on the magnitude of the mode is obtained by averaging Eq. (9.150) over phase (the polar angle of $\mathbf{X}$) and integrating once with respect to mode amplitude under the assumption that a steady state has been reached:

$$\Gamma_X = X[(1-R) - I_b \operatorname{Re} G_{nl}]P + \frac{1}{X}\left[\frac{\partial}{\partial X}\left(XPD_{XX} - PD_{\varphi\varphi}\right)\right] = 0, \tag{9.157}$$

where we have set the constant of integration to zero by virtue of the requirement that the distribution function must vanish as $X \to \infty$. Equation (9.157) can be integrated once more with respect to the magnitude of the mode amplitude to find the time asymptotic probability distribution function:

$$P = \frac{C}{D_{XX}} \exp\left[-\int_0^X dX' \frac{X'[(1-R) - I_b \operatorname{Re} G_{nl}] + \left(D_{XX} - D_{\varphi\varphi}\right)/X'}{D_{XX}}\right], \tag{9.158}$$

where C is a constant chosen to normalize the probability distribution function to unity. The probability distribution function described by (9.158) will be peaked at values of mode amplitude X where the integrand in the exponent of P vanishes. If the noise is weak (i.e., $X^2(D_{\varphi\varphi} - D_{\psi\psi}) < < 1 - R$ and I_b Re G_{nl}), then the peak will occur at the value of the mode amplitude that gives saturation in the absence of noise (9.108) (i.e., $1 - R = I_b \operatorname{Re} G_{nl}(X_0)$). The nonlinear coefficients appearing in (9.158) can be evaluated by Taylor expansion around this value. Specifically, the probability distribution becomes a Gaussian peaked at the saturated mode amplitude X_0:

$$P(X, n \to \infty) = \frac{1}{2\pi^{3/2} X_0 \Delta X} \exp\left[- \frac{(X - X_0)^2}{(\Delta X)^2} \right], \tag{9.159}$$

where the width ΔX is proportional to the spontaneous emission rate:

$$(\Delta X)^2 = \frac{2D_{XX}}{\gamma_a T} = \frac{\Phi_A(X_0, P_0)}{N_*}, \tag{9.160}$$

and the quantity γ_a appearing in (9.160) is defined in (9.116) and represents the damping rate for perturbations of the mode amplitude away from equilibrium. It must be positive for the saturated equilibrium to be stable. The noise-induced fluctuations in the mode amplitude are inversely proportional to this rate. Further, Eq. (9.160), along with the expression for the diffusion tensor (9.152), shows that the relative fluctuations in the amplitude of a saturated mode are inversely proportional to the square root of the number of electrons which pass through the interaction region during this relaxation time. This dependence is more concisely expressed in the second Eq. (9.160) where Eq. (9.152) has been used to eliminate D_{XX} and the equilibrium relation (9.108) (i.e., $1 - R = I_b$ Re G_{nl} $(X_0, p)0$ has been used to eliminate the current. This allows the expression for the noise-induced amplitude fluctuations to be expressed in terms of $N_* = N_T/[2(1 - R)]$ which is the number of electrons passing through the device in a cavity decay time, $T_d = T/[2(1 - R)]$, and a form factor

$$\Phi_A(X_0, P_0) = \frac{\int_0^{2\pi} \frac{d\psi_i}{2\pi} \int_0^T \frac{dt_e}{T} \left[\int_0^1 d\xi \sin\theta_0(\psi_i, t_e) \right]^2}{2\operatorname{Re}[G(X_0, p_0)] \operatorname{Re}[X_0 \partial G(X_0, p_0)/\partial X_0]}, \tag{9.161}$$

which depends only on the normalized mode amplitude and the detuning for the nonlinear saturated mode. Level curves of the form factor Φ_A are plotted in Fig. 9.13. Only those values inside the stable triangle should be considered since is single-mode operation is possible only for these parameters. The dominant dependence of the amplitude fluctuations is, therefore, determined by N_* (the number of electrons passing through the device in a cavity decay time). Since this is typically a large number, the relative fluctuation in the amplitude of a large saturated mode due to shot noise is expected to be quite small and will have a negligible effect on the spectrum of the radiation.

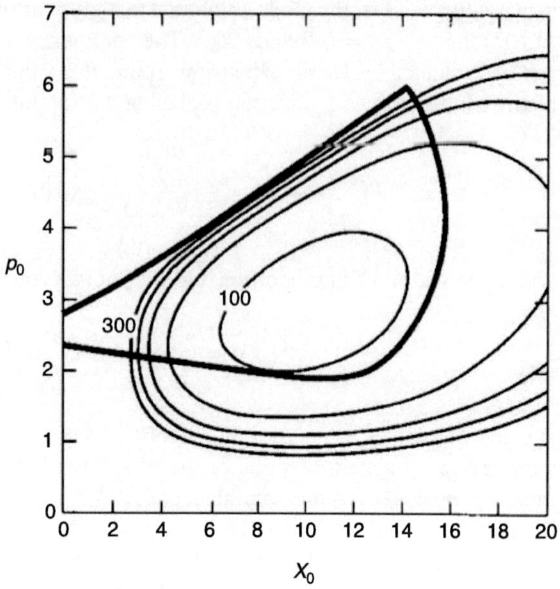

Fig. 9.13 Level curves of the form factor Φ_A determining the noise-induced amplitude fluctuations for a nonlinearly saturated mode. The separation between adjacent levels is 50

The dominant effect of shot noise on the spectral width of a saturated mode is due to the diffusion in time of the phase of the mode. To calculate this effect, we introduce the power spectrum, which is the time Fourier transform of the two-time correlation function of the complex mode amplitude:

$$S(\nu) = \int_{-\infty}^{\infty} dnT \exp(i\nu nT)\overline{X_m(n)X_m^*(0)}, \tag{9.162}$$

We must find the expected value of the product of the complex mode amplitude and its complex conjugate (i.e., the two-time correlation function) evaluated at two different times 0 and nT in order to evaluate the power spectrum. Under the assumption that the fluctuations in the magnitude of the complex amplitude are small, the time dependence of the correlation function is dominated by the diffusion of the phase of the mode with time. The correlation function may be determined by assuming the magnitude of the amplitudes of the mode at the two times are the same and equal to the expected value of the amplitude at saturation. The expected dependence on the phase difference between the complex amplitudes of the mode at the two different times can be determined by solving the Fokker-Planck Eq. (9.150) subject to the initial condition that the phase of the mode at time $n = 0$ is known. Since the probability distribution will be narrowly peaked around the saturation amplitude X_0, Eq. (9.150) can be integrated over the magnitude of $\mathbf{X}$ to obtain an equation for the phase dependence of the amplitude-integrated probability distribution function:

$$\frac{\partial}{\partial n}\overline{P}(\varphi,n)=\frac{\partial}{\partial\varphi}\left[-\delta\omega_0 T(X_0,p_m)\overline{P}+\Delta\nu T\frac{\partial}{\partial\varphi}\overline{P}\right], \tag{9.163}$$

where

$$\overline{P}(\varphi,n)=\int_0^\infty dXXP(X,\varphi,n), \tag{9.164}$$

and

$$\Delta\nu=\frac{D_{\varphi\varphi}}{TX_0^2}=\frac{\Phi_B}{T_dN*}, \tag{9.165}$$

is the spontaneous emission-induced diffusion coefficient for the phase. The diffusion tensor is evaluated at the saturated mode amplitude X_0.

The diffusion rate $\Delta\nu$ will be shown to determine the width of the spectrum for the saturated mode. The second equality in (9.165) shows that the width scales inversely with the cavity decay time and with the number of electrons transiting the device in the cavity decay time. Again, there is a form factor:

$$\Phi_B(X_0,P_0)=\frac{\int_0^{2\pi}\frac{d\psi_i}{2\pi}\int_0^T\frac{dt_e}{T}\left[\int_0^1 d\xi\cos\theta_0(\psi_i,t_e)\right]^2}{8X_0^2(\mathrm{Re}[G(X_0,p_0)])^2}, \tag{9.166}$$

which depends only on the normalized parameters of the saturated equilibrium. The level curves of Φ_B are shown in Fig. 9.14.

The analysis continues by solution of the diffusion Eq. (9.163) subject to the condition that the initial phase is specified as φ_0; hence,

$$\overline{P}=\frac{1}{\sqrt{2\pi|n|\Delta\nu T}}\exp\left[-\frac{(\varphi-\varphi_0+\delta\omega_0 nT)^2}{4|n|\Delta\nu T}\right], \tag{9.167}$$

which shows the phase to be both drifting in time with an average rate equal to the frequency shift defined for the saturated mode amplitude and to be diffusing due to the shot noise. The correlation function that is the integrand in (9.163) is then obtained by integration of the amplitude averaged probability distribution function over all angles weighted by the initial and final mode amplitudes. Fourier transformation in the time domain then yields the power spectrum:

$$S(\nu)=\frac{2\Delta\nu X_0^2}{(\nu-\delta\omega_0)^2+\Delta\nu^2}, \tag{9.168}$$

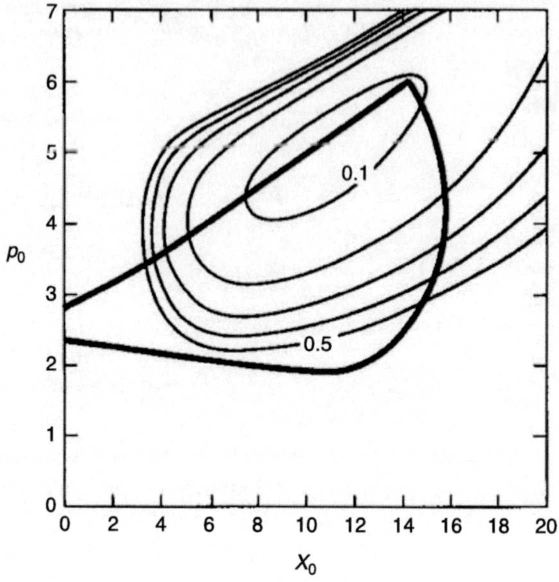

Fig. 9.14 Level curves of the form factor Φ_B determining the noise-induced spectral width for a single nonlinearly saturated mode. The separation between adjacent levels is 0.1

Thus, the power spectrum is peaked at a frequency $\delta\omega_0$ corresponding to the shifted frequency of a saturated mode given by (9.109) and has a Lorentzian shape with a width given by $\Delta\nu$. This power spectrum was defined for the time dependence of the mode amplitudes on the slow time scale. The actual spectrum will, of course, be shifted by the empty cavity frequency of the mode in question $\omega_m = \omega + \Delta\omega_m$.

The spectral width defined in Eq. (9.165) assumes that only one mode is present in the time asymptotic state. However, it was shown in the last section that nearby modes are weakly damped with a damping rate that scales as the square of the frequency difference between the mode in question and the large amplitude equilibrium mode. Equation (9.132) expresses this damping rate in terms of a coefficient D (not to be confused with the elements of the Fokker-Planck diffusion tensor defined in (9.152)), which is roughly the order of the inverse of the cavity decay time T_d. Using the previous calculations as a guide (in particular Eq. (9.160)) which states that the RMS level of fluctuations in a mode scale as the normalized spontaneous emission rate divided by the damping rate, we arrive at the estimate:

$$|X_m|^2 = \frac{S_m}{DT(\varepsilon\Delta\omega_m T)^2} \approx \frac{\Phi_C(X_0, p_0)}{N_*(2\pi\varepsilon m)^2}, \tag{9.169}$$

where Φ_C is a form factor analogous to Φ_A and Φ_B and we have used $\Delta\omega_m T = 2\pi m$. Recall that Eq. (9.132) corresponds to the damping rate of a coupled pair of modes; however, this will not affect the order of magnitude estimate given here. If the index m corresponds to the nearest neighbor pair of modes [$\Delta\omega_m T = 2\pi$], then this pair of modes can be excited to large amplitude even if the number of electrons transiting in a cavity decay time is large. A noise level consistent with

$$\varepsilon^2 N_* < 1, \tag{9.170}$$

is required in order that the single-mode regime be achieved. Thus, if the slippage parameter is too small, then the single-mode spectral width calculated in (9.165) will not apply. Instead, the final asymptotic state of the oscillator will be one in which the shot noise excites a number of modes to large amplitude. This number is estimated by using Eq. (9.169) to determine the mode number for which the energy in the noise-excited satellites is comparable to that equilibrium mode. That is if a large number of modes are excited but the spectrum is still narrower than the gain bandwidth we expect $\mathfrak{R}_m|X_m|^2 = |X_0|^2$, where X_0 is the saturated amplitude for a single mode with the corresponding central detuning. Thus, the number of excited modes can be estimated from (9.169) by finding the value of m such that $m|X_m|^2 \approx |X_0|^2$. The resulting spectral width is given by

$$\Delta\omega_m T \cong \frac{1}{\varepsilon^2 N_*}. \tag{9.171}$$

To summarize, shot noise can be responsible for the initial excitation of modes in the cavity which then grow due to their positive gain. Modes are excited with a Gaussian probability distribution function prior to saturation characterized by a time-dependent mode temperature. Due to the statistical nature of the noise excitation process, a given mode will have 100% fluctuations in its amplitude from shot to shot. Equivalently, nearby modes in the same shot will have large fluctuations in their amplitudes. If the beam parameters are such that a stable single-frequency equilibrium is possible, the spectrum will tend to approach the single-frequency state asymptotically in time. However, it will ultimately be limited by noise. This limit may be either of the form of a broadening of a single mode (9.165) [25] or the excitation of a number of modes (9.171). The process of saturation and spectral narrowing which leads to the time asymptotic single-mode state will be discussed in the next section.

Finally, we conclude this section with the description of a numerical technique for incorporating noise realistically into particle simulations of free-electron laser oscillators [29]. As previously noted, electrons do not enter the interaction region with phases and entrance times uniformly distributed over the intervals [0,2π] and [0, T]. Rather, a large number of electrons enter with random entrance times and phases that are independent and identically distributed. This is the source of the noise. To account for the noise, it would be impractical to distribute the phases and entrance times randomly and independently since this would require simulating the trajectories of too many electrons in order to achieve a low enough level of noise to be comparable to that in an experiment. Instead, one can ascribe a weight $W_j = 1 + \delta W_j$ to each particle in the sum in (9.139) and simulate the trajectories of N_s electrons where $N_s << N_T$. The electrons included in the simulation can then be uniformly distributed on a grid of entrance phases and entrance times. If the δW_j were zero ($W_j = 1$), then there would be no noise. To adjust the level of noise in the simulation to be equal to that in an experiment, one can allow the δW_j to be random variables that are independent and identically distributed with a zero mean and a variance given by

$$\overline{\delta W_j^2} = \Delta n \frac{N_S}{N_T}, \tag{9.172}$$

where Δn is the time step used in the integration of the mode amplitude equation. The noise in the simulation will be the result of the addition of a large number N_s of random variables. Hence, the noise will be Gaussian white noise with identical statistics to that expected from experimental shot noise. The mode amplitudes calculated in this way will represent a single shot, and a number of simulations will be required to obtain statistical averages. Examples of the start-up phase of a long pulse oscillator using this technique will be presented in the next section.

9.6.4 Linear and Nonlinear Spectral Narrowing

In the previous sections, it has been shown that both the spectrum of spontaneous emission and the intrinsic gain have frequency bandwidths that correspond to modes whose dimensionless detunings fall in the range $0 \le p_m \le \pi$. This is evidenced by the dependence of both the rate of spontaneous emission (9.147) and the gain (9.98) on the dimensionless detuning parameter. This range of dimensionless detunings translates, using the definitions appearing in Eqs. (9.101) and (9.44), into a fractional frequency bandwidth given by

$$\frac{\omega_m}{\omega} \cong \frac{\pi}{\varepsilon \omega T} = \frac{1}{2N_w}, \tag{9.173}$$

where we use N_w here to denote the number of wiggler periods in the interaction region. Thus, in an oscillator that is initiated by shot noise, modes in this frequency range will be excited. If the level of noise is small, then the radiation in the cavity will have to be greatly amplified over many passes before saturation is achieved. During this time, the modes with the largest growth rates (i.e., the highest gains) will outgrow modes with smaller gain, so that the spectrum of modes by the time saturation is achieved will be narrower than that which was initially excited by noise. This process, which is common to lasers, is known as gain narrowing.

The amount of gain narrowing that is expected to occur can be quantified using Eq. (9.156) for the mode temperature (the expected squared amplitude of the mode X_m) in the linear regime. For large times nT, the dominant dependence of the mode temperature on the mode number is through the dependence of the growth rate appearing in the exponent. If the growth rate is expanded about the detuning value

p_{max}, which gives the maximum growth rate (i.e., $p_{max} \approx 2.6$ for a monoenergetic beam), then the dependence of the mode temperature on detuning is of the form of a Gaussian:

$$T_m \cong T_{\max} \exp\left[-\frac{(p_m - p_{\max})^2}{2\delta p^2}\right], \tag{9.174}$$

where T_{max} is the amplitude of the largest mode and δp is the width in the dimensionless detuning of the spectrum:

$$\delta p^2 = \frac{1}{nT}\left|\frac{d^2\gamma_m}{dp_m^2}\right|^{-1}. \tag{9.175}$$

Equation (9.175) shows that the spectral width starts approximately at the gain bandwidth and decreases steadily with time. This form of linear gain narrowing is expected to proceed at least until saturation occurs at which time the factor nT can be estimated from (9.156) as

$$nT \cong \frac{1}{2\gamma_m} \ln\left|\frac{\gamma_m T T_{sat}}{S_m}\right|. \tag{9.176}$$

where T_{sat} is the mode temperature corresponding to a saturated amplitude and S_m is the spontaneous emission rate. Therefore, the larger the saturated amplitude compared with the emission rate, the more growth is required to reach saturation, and the narrower the spectrum as saturation is approached.

Once the saturation process begins, the mode amplitudes interact nonlinearly and the formulae describing linear gain narrowing are no longer valid. It will be shown, however, that formula (9.175) is approximately valid with the spectral width decreasing with time but with the second derivative of the linear growth rate replaced by the coefficient D, which appears in the formula for the damping rate of nearby frequency modes in the presence of a large amplitude equilibrium mode. Note that the coefficient D is the second derivative with respect to detuning of the damping rate of nearby satellite modes. Since the coefficient D is a frequency which scales as the inverse of the cavity decay time T_d (see the discussion following (9.133)), we can write the expression for the time-dependent spectral width as

$$\delta p^2 \cong \frac{T_d}{nT}, \tag{9.177}$$

which when expressed as a fractional bandwidth (as in (9.173)) becomes [15, 21]

$$\frac{\Delta\omega}{\omega} \cong \frac{1}{2N_w}\left(\frac{T_d}{nT}\right)^{1/2}, \tag{9.178}$$

Therefore, many cavity decay times are required for the spectrum to become significantly narrower than the gain bandwidth.

Recall that the slow decay of these coupled pairs of modes was a consequence of the fact that coherence in the radiation pulse is established only by the slippage of the beam electrons with respect to the radiation. Thus, if the slippage is small, a long time is required before all temporal portions of the radiation pulse are oscillating in phase. The rate at which coherence is established can be increased if some other mechanism can be found to provide communication between different portions of the radiation pulse. An example is the addition of a dispersive element to the radiation feedback path [20]. The effect of a dispersive element on the rate of spectral narrowing will be discussed at the end of this section.

A second method to achieve a narrow bandwidth in a relatively short time is to *seed* the oscillator with a signal at the desired frequency. This process is often referred to as *mode-locking*. That is, if radiation is injected into the cavity at a level significantly larger than the noise level, and if the gain at the frequency of the injected radiation is near the peak of the gain spectrum, then the radiation spectrum at saturation will be essentially single frequency [31].

An example of a free-electron laser where gain narrowing and mode competition has been important is the experiment at the University of California at Santa Barbara [28, 31–33]. The slippage parameter in this experiment was of the order of $\varepsilon \approx 1.7 \infty 10^{-3}$, the cavity decay time was 0.2 μsec, and the duration of the electron pulse was about 2.75 μsec. Thus, a large number of modes were initially excited. These modes grew and saturated, but there was insufficient time for coherence to be established. Indeed, measurements have shown that in a large number of cases, the spectrum was multi-moded at the end of the shot [33]. In a smaller number of cases (roughly 29% of the discharges), a single mode appeared to dominate. The criterion for single-mode domination was that of the 20 or so modes whose frequencies fell in the range of the detector, the power in one mode exceeded by a factor of two the power in any other mode. This is most likely due to the statistical fluctuations in the spontaneous emission process discussed in the previous section.

A complicating factor in the Santa Barbara experiment was that the beam energy decreased in time. Thus, the modes that dominated at the end of the pulse were not necessarily the ones that grew initially [28]. It will be shown in this section that the time dependence of the beam energy has a dramatic effect on the efficiency of energy extraction. Specifically, the peak efficiency can vary by over a factor of five depending on whether the beam voltage increases or decreases over time. An increasing beam voltage is preferred because it allows the high-efficiency single-mode equilibrium occurring at high-detuning parameter [$p_0 \approx 5.14$] to be accessed.

We now describe the results of a numerical simulation of the Santa Barbara experiment [29]. The equations which were solved numerically were the electron equations of motion (9.79) and (9.80):

$$\frac{d\psi}{d\xi} = p, \tag{9.79}$$

and

$$\frac{dp}{d\xi} = -\frac{1}{2}[iX(n, t_e + \varepsilon\xi T)\exp(i\psi) + c.c.], \tag{9.80}$$

for an ensemble of electrons, with the dimensionless field amplitude represented as a superposition of cavity modes as in (9.84):

$$X(n, T') = \sum_{m=-\infty}^{\infty} X_m(n)\exp(-i\Lambda\omega_m t'), \tag{9.84}$$

The equations of motion were solved at each time step in round-trip index n for an ensemble of particles launched on a uniform grid of initial phases $(0,2\pi)$ and entrance times $(0,T)$ with an initial momentum (detuning parameter for the central mode) p_0 appropriate for a monoenergetic beam. The expression for the detuning parameter is given in (9.101)

$$p_m = p_0 - 2\pi m\varepsilon = \frac{L}{v_\parallel}\left[(k_m + k_w)\left(v_\parallel + \delta v_\parallel\right) - (\omega + \Delta\omega_m)\right], \tag{9.101}$$

The evolution of the mode amplitudes with slow time (round-trip index n) is governed by Eq. (9.90):

$$\begin{aligned}\left[\frac{\partial}{\partial n} + (1-R)\right]X_m(n) &= -iI_b Z_m \\ &= -\frac{iI_b}{T}\int_0^T dt_e \int_0^1 d\xi\langle\exp[i\Delta\omega_m(t_e + \varepsilon\xi T) - i\psi(\xi, t_e)]\rangle,\end{aligned} \tag{9.90}$$

To accurately model spontaneous noise in the simulation, the entrance phase and entrance time averages in Eq. (9.90) are evaluated as sums over randomly weighted particles as given by (9.172)

$$Z_m = \frac{1}{N_S}\sum_{j=1}^{N_S} W_j \int_0^1 d\xi \exp\left[i\theta_m\left(t_{ej}, \psi_{ij}\right)\right], \tag{9.139}$$

where N_s is the number of simulation particles traversing the interaction region per slow time step and W_j denotes the particle weights.

Simulation parameters were chosen to match the conditions of the experiment as closely as possible. Recall that the slippage parameter in the experiment was very small with $\varepsilon \approx 1.7 \times 10^{-3}$. Using this value in a simulation would require keeping approximately 750 modes (to span the gain bandwidth). Instead, a value of $\varepsilon = 5.0 \times 10^{-3}$ was chosen, and 251 modes were retained in Eqs. (9.90), (9.79), and (9.80). Thus, each mode in the simulation can be thought of as representing three modes in the experiment. With this number of modes, it was necessary to solve the equations of motion for 300 different entrance times and 15 different entrance phases, yielding $N_s = 15 \times 300 = 4500$ electrons per step.

The beam voltage in the Santa Barbara experiment decreased linearly with time at a rate of approximately 2 kV/μsec. This is modeled by allowing the detuning parameter p_0 to be a function of the slow time variable n through the dependence of the injected parallel velocity on beam energy (c.f. Eq. (9.101)). Evaluation of the formula for the normalized current for the parameters of the experiment requires making assumptions on the structure of the modes in the cavity. In the low-gain limit, these can be taken to be the same as the vacuum modes in the cavity. The result is that the actual current is found to be about three times the start current. This is consistent, as will be seen, with the observed rate of increase of power in the experiment. The final important parameter is the noise level, which is modeled using the technique described in the last section. The relevant experimental number that must be supplied for the simulation is the number of electrons traversing the interaction region in a given time that is determined by the beam current (≈ 1 A).

The normalized efficiencies Δp (defined in Eq. (9.104)) versus slow time $\tau_s = n\ T/T_d$ (where T_d is the decay time of the empty cavity from (9.29)) are shown for three different runs in Fig. 9.15. Using Eq. (9.105), it is seen that a value of $\Delta p = 2\pi$ corresponds (for a relativistic beam) to the standard estimate $\Delta\gamma/\gamma \approx 1/2N_w$. The three curves of Fig. 9.15 correspond to different dependences of beam energy on time. The curve labelled a corresponds to a falling beam voltage as in the experiment, the curve labelled b corresponds to a constant beam voltage, and the curve labelled c corresponds to a beam voltage that rises with time at rate opposite to that of

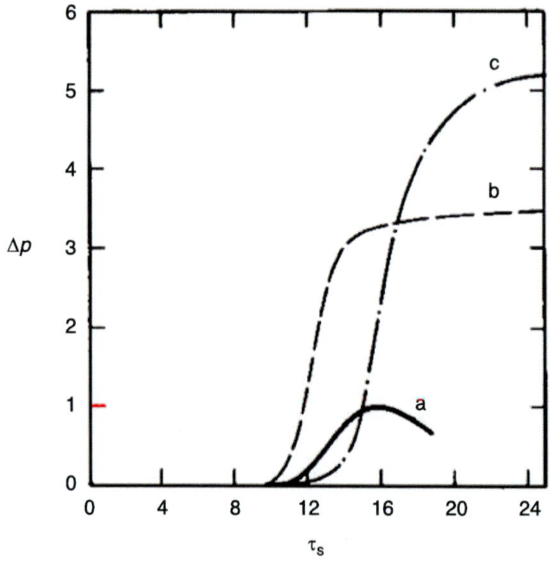

Fig. 9.15 Normalized efficiency versus time for three simulations appropriate to the parameters of the USCB continuous beam experiment. The three cases correspond to (**a**) a falling beam voltage, (**b**) a constant beam voltage, and (**c**) a rising beam voltage [29].

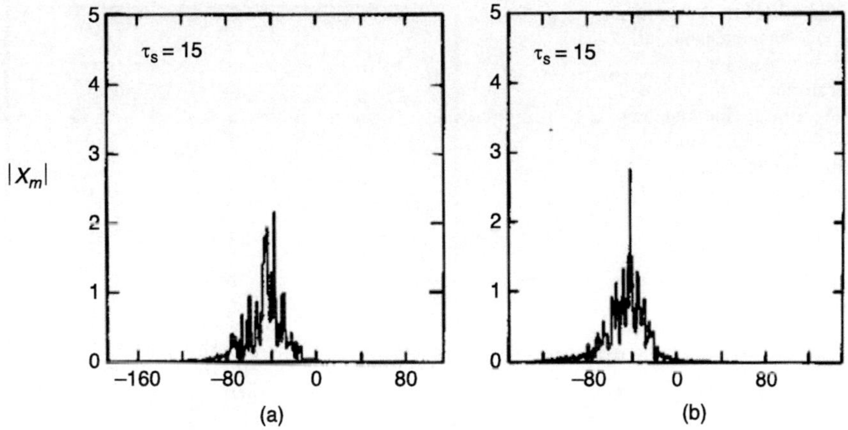

Fig. 9.16 Spectra at saturation for two different realizations of the UCSB experiment. The difference is in the random number *seed* used to generate the spontaneous noise [29]

case *a*. It is evident that the time history of the beam voltage has a large effect on the efficiency of energy extraction. Note that for the conditions of the experiment shown in curve *a*, the efficiency is about 16% of the much-quoted standard value.

Figure 9.16 shows histograms of the mode amplitudes at various times during the simulations illustrated Fig. 9.15. In particular, Fig. 9.16a and b show the spectrum at $\tau_s = 15$ of the case *a* corresponding to a falling beam voltage. The realizations differ in the random numbers that generated the incoherent noise and illustrate the statistical fluctuations predicted to occur when modes are excited by incoherent noise. In particular, there are 100% fluctuations in the magnitude of a given mode from realization to realization as well as 100% fluctuations in the magnitude of neighboring modes. Plots of the efficiency versus time for these two realizations are virtually identical. The statistical fluctuations in mode amplitudes are further illustrated by a comparison of the expected amplitudes of the modes just prior to saturation given by the solution of Eq. (9.155) including the appropriate time dependence of the beam energy. The simulation results for one realization of the incoherent noise is shown in Fig. 9.17. The spectrum is approximately Gaussian as predicted by Eq. (9.174).

For the three cases of increasing, uniform, and decreasing voltage, the spectra at saturation are peaked at different mode numbers. The central detunings are given by $p_m = 5.67$, 2.60, and 1.16 for these cases, respectively. If these spectra were single-moded with the same detunings, Figs. 9.5 and 9.6 could be used to predict the efficiency. With the given value of the normalized current (about three times the start current) and the three values of the detuning from the simulation, the predicted efficiencies at saturation are $\Delta p = 5.0$, 3.4, and 1.0, which is in agreement with the simulations. Thus, as far as efficiency is concerned, the simulations are behaving as if the spectra were single-moded. This is to be expected since a significant amount of gain narrowing has occurred during the exponential growth phase.

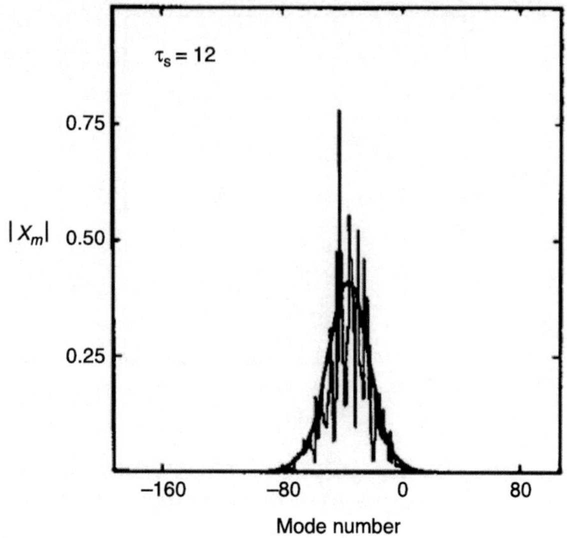

Fig. 9.17 Spectrum of mode amplitudes just prior to saturation for one realization of the UCSB experiment. Superimposed is the expected value of the mode amplitudes as predicted by (9.174) [29]

The reason that the spectrum is peaked at different detunings depending on the time history of the beam voltage can be understood by examination of Eq. (9.101) for the dimensionless detuning. Due to the time dependence of the beam voltage, the detuning of the central mode p_0 and hence the detunings of all modes becomes a function of time. The mode which will experience the greatest amount of linear gain will be that mode whose detuning spends the longest time in the vicinity of the maximum instantaneous growth rate $p_{max} = 2.6$. If the voltage and detunings are rising, then the detuning of the mode with the maximum total gain will be higher than p_{max} when it reaches saturation, while if the voltage is falling the detuning will be lower than p_{max}. Once saturation is achieved, the detuning of the dominant mode will still continue to change with beam voltage. As a result, the detuning of the dominant mode will be pulled out of the stable triangle of Fig. 9.8, and other modes whose frequencies are more in resonance with the beam will grow. This process of mode hopping [28, 29] will continue as long as the beam voltage is changing, and a monochromatic spectrum will never be achieved.

A continuation of the simulation with constant beam parameters results in a spectrum which progressively narrows with time [34]. Figure 9.18 shows spectra from a simulation in which the beam energy becomes constant in time after an initial transient. Spectra are shown at two different times after the power has saturated. It is evident that the spectrum narrows very slowly with time. In particular, it is still multi-moded after about 1000 cavity decay times. It has been verified [34] that the width of the spectrum was governed by (9.178). Also shown are reconstructions of the fast time dependence of the magnitude of the radiation field corresponding to the two spectra. Note the absence of modulations in the magnitudes in accordance with the zero-slippage prediction. A quantitative analysis of nonlinear spectral narrowing proceeds by assuming that as the radiation grows and approaches saturation, the

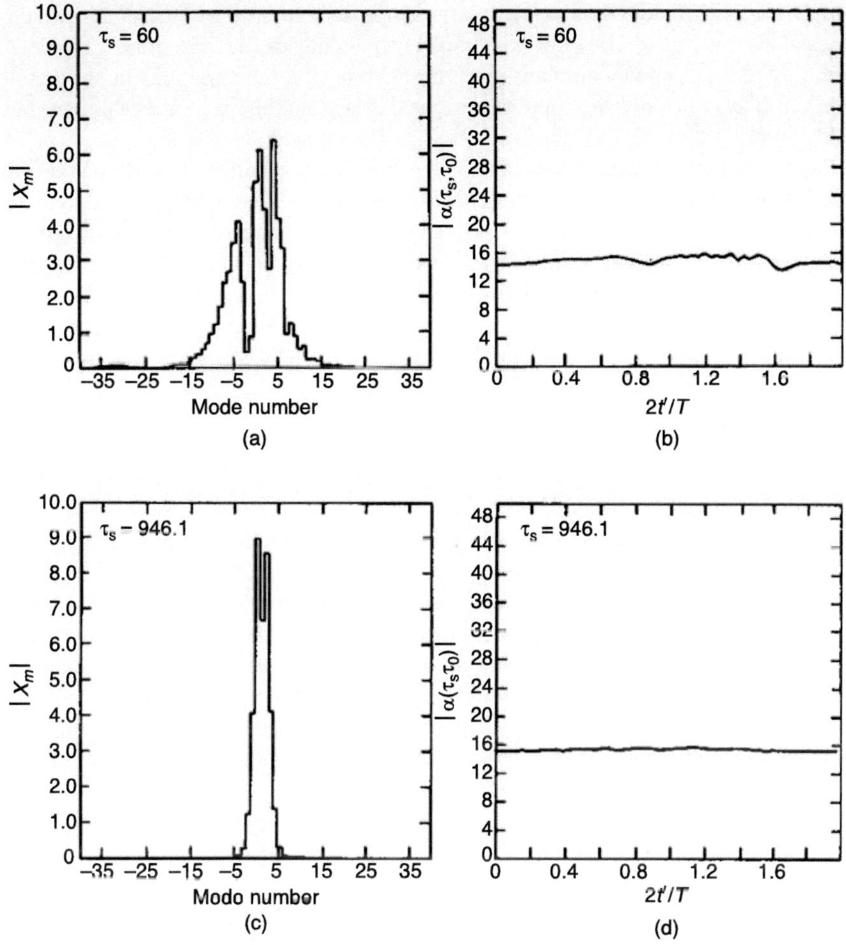

Fig. 9.18 Spectrum of modes at constant voltage showing nonlinear spectral narrowing [34]

spectrum has already narrowed linearly to some degree so that the width of the spectrum is considerably less than the gain bandwidth when nonlinear effects come into play. Thus, the radiation spectrum will be peaked around a mode with a dimensionless detuning parameter, which we label p_0. In the simplest case of a beam whose parameters are constant in time, this will correspond to the detuning which gives maximum linear gain (i.e., $p_0 \approx 2.6$ for a monoenergetic beam). As has been seen, if the beam parameters vary with time, the detuning will correspond to the value which maximizes the growth of the modes over the time history of the beam. The nonlinear evolution of the radiation pulse which consists of a narrow spectrum of modes peaked about a mode detuning p_0 is governed by Eq. (9.130) when the effects of slippage are ignored. As discussed, the evolution of the complex amplitude of the radiation field in this case is to asymptotically approach a constant magnitude, described by Eq. (9.108) but with phase depending on the fast time variable t' leading to a non-monochromatic spectrum. Ultimately, the variations in phase will diminish as coherence in different portions of the radiation pulse is established, but this involves the action of slippage, which is not included in (9.130).

To examine the way in which slippage eventually establishes coherence in the radiation pulse, we return to the consideration of the basic low-gain system of equations ((9.79), (9.80), and (9.81)) and treat these equations by expansion in powers of the slippage parameter ε. Equation (9.130) is the result of the lowest-order expansion if we set the slippage parameter to zero. This equation leads to an asymptotic solution in which the magnitude of the radiation is independent of the fast time variable t', reflecting local power balance, and the phase advances with round-trip index at a constant rate:

$$\frac{1}{T}\frac{\partial \phi(n,t')}{\partial n} = -\,\delta\omega_0, \tag{9.179}$$

where $X(n,t') = X_{00}\exp[i\phi(n,t')]$ is the complex amplitude of the radiation. The value of the amplitude X_{00} and the frequency shift $\delta\omega_0$ are determined by the nonlinear equilibrium Eqs. (9.108) and (9.109) for the given values of the beam current and central detuning. The lack of coherence in the absence of slippage is reflected in the fact that Eq. (9.179) does not require that the phase be independent of the fast time variable t' but only that it advance at the same rate everywhere in the radiation pulse. Keeping higher-order terms in the slippage parameter will lead to an evolution equation for the phase which couples together the values of the phase at different times t' in the radiation pulse:

$$\frac{1}{T}\frac{\partial \phi}{\partial n} = -\,\delta\omega_0 - C_1\varepsilon T\frac{\partial \phi}{\partial t'} - \frac{C_2}{2}(\varepsilon T)^2\left(\frac{\partial \phi}{\partial t'}\right)^2 + C_3(\varepsilon T)^2\frac{\partial^2 \phi}{\partial t'^2}, \tag{9.180}$$

where the coefficients C_1, C_2, and C_3 depend on the magnitude of the radiation X_{00} and the central detuning p_0. That the evolution equation which must be of this form can be deduced from the way in which the slippage parameter enters Eqs. (9.79), (9.80), and (9.81). In particular, it always appears multiplied by the product of the round-trip time T and the dimensionless axial distance ξ, where $0 \le \xi \le 1$. Further, this product always appears additively in the fast time argument of the radiation amplitude and beat wave phase. Thus, in an expansion in ε such as (9.180), each power of ε must be accompanied by a derivative with respect to the fast time variable. Physically, the slippage induces a coupling between neighboring in time phases of the radiation amplitude. Equation (9.180) represents all possible terms in an expansion up to and including the ε^2 term. Direct calculation of the coefficients appearing in Eq. (9.180) would be tedious and, in the end, requires numerical evaluation. It is possible, however, to deduce the coefficients on the basis of some general considerations.

The coefficients in Eq. (9.180) can be determined based on our knowledge of the equilibrium and stability of single-frequency states. For example, suppose the mode phase depends linearly on the fast time variable:

$$\phi = \phi_0(n) - \Delta\omega_m t', \tag{9.181}$$

This variation of the phase corresponds to a single-frequency state with a detuning parameter $p_m = p_0 - \varepsilon\Delta\omega_m T$. Thus, the rate at which the phase advances with respect to the slow time index n must be of the form

$$\frac{1}{T}\frac{\partial\phi_0}{\partial n} = -\delta\omega_0(I_b, p_0 - \varepsilon\Delta\omega_m T), \tag{9.182}$$

where we have written the frequency shift explicitly in terms of the beam current as opposed to the mode amplitude since the current is held fixed and the mode amplitude adjusts itself according to Eq. (9.108) to maintain power balance. Insertion of Eqs. (9.181) into (9.180) allows the coefficients C_1 and C_2 to be identified in terms of the derivatives of the frequency shift with respect to detuning at constant current by expansion of the frequency shift to second order in the slippage parameter:

$$\begin{aligned} C_1 &= -\frac{d}{dp_0}\delta\omega_0 = -\delta\omega_0', \\ C_2 &= -\frac{d^2}{dp_0^2}\delta\omega_0 = -\delta\omega_0''. \end{aligned} \tag{9.183}$$

To determine the final coefficient, consider the case in which the phase is constant in the fast time variable except for a small sinusoidal perturbation:

$$\phi = \phi_0(n) + \phi_m \cos(\Delta\omega_m t' + \alpha_m), \tag{9.184}$$

Since only the phase of the radiation is perturbed, this corresponds to a large equilibrium mode perturbed by two small satellites having $X_m + X_{-m}{}^* = 0$. Based on the results of the previous section concerning the stability of single -frequency equilibria, the perturbation ϕ_m must decay with round-trip time n at a rate given by (9.132). Therefore, the coefficient C_3 must be the same as the coefficient D in (9.132). Further, using (9.183) we can deduce relation (9.133). As before, the coefficient C_3 must be determined numerically for each equilibrium and will only be positive within the triangular shaped region of Fig. 9.18 corresponding to stable single-frequency equilibria.

Having deduced the coefficients in the nonlinear evolution Eq. (9.180), we now turn to its solution. This equation is a form of Burger's equation, and, consequently, we look for a solution which is expressed in the form:

$$\phi(n, t') = \delta\omega_0 nT - \frac{2C_3}{C_2}\ln[1 + u(n, t')], \tag{9.185}$$

where insertion of (9.185) in (9.180) yields a linear differential equation for the function $u(n,t')$

$$\frac{1}{T}\frac{\partial u}{\partial n} = -C_1\varepsilon T\frac{\partial u}{\partial t'} - C_3(\varepsilon T)^2\frac{\partial^2 u}{\partial t'^2}. \tag{9.186}$$

Thus, we have converted a nonlinear equation for the phase into a linear equation for the function u. Representing the function $u(n,t')$ in the form of a Fourier series in the fast time variable t', the solution of Eq. (9.186) can be written

$$u(n, t') = \sum_{m=-\infty}^{\infty} u_m(0)\exp[-i\Delta\omega_m t' + \gamma_m nT], \tag{9.187}$$

where $u_m(0)$ are the initial Fourier amplitudes of the function $u(0,t')$ and the complex growth rate γ_m is given by (9.132). Interestingly, the time dependence of the Fourier components of $u(n,t')$ is precisely the same as that predicted by linear theory for the small perturbing amplitudes in the presence of a large equilibrium mode (Eq. (9.132)). Complete solution of the problem thus requires specifying the initial fast time dependence of function $u(0,t')$ which is determined by the spontaneous emission. However, it is clear that the function u will eventually decay to a constant independent of t' at a rate determined by the damping coefficient C_3. Consequently, the phase given by Eq. (9.185) becomes independent of the fast time variable, indicating that all temporal portions of the radiation pulse ultimately oscillate coherently.

The slow rate of spectral narrowing predicted by the previous analysis is a consequence of the fact that radiation propagates in this model without dispersion, and only the slippage of the beam provides communication between different temporal portions of the radiation pulse. The communication between different portions can be increased, along with the rate of spectral narrowing, by adding a dispersive element to the path of the radiation [20]. The effect of weak dispersion is to cause the frequencies of the longitudinal modes of the cavity to become nonuniformly spaced. That is, the natural frequencies will have a dependence on mode number of the form

$$\omega_m = \omega + \frac{m}{T}(2\pi + \alpha m), \tag{9.188}$$

where the coefficient α measures the amount of dispersion and may be either positive or negative. If we assume the dispersion is small, then its effect will only be apparent after a time corresponding to many circuits of the radiation through the cavity. Therefore, the dispersion can be included by modifying the slow time evolution of the radiation amplitude as given by (9.81) [4]:

$$\left[\frac{\partial}{\partial n} - i\alpha\left(\frac{T}{2\pi}\right)^2 \frac{\partial^2}{\partial t'^2} + (1-R)\right] X(n,t') = -iI_b \int_0^1 d\xi \langle \exp(-i\psi) \rangle, \qquad (9.189)$$

It is clear that in the absence of a beam, the solutions of (9.189) give precisely the dispersion-modified frequencies indicated in (9.188).

The effect of dispersion on the rate of spectral narrowing can now be assessed by calculating the rate of decay of small perturbing satellites in the presence of a large equilibrium mode. In order to accomplish this, the effect of slippage is neglected so that the beam response can be modeled as in Eq. (9.130). Specifically, the beam provides a nonlinear gain dependent on both the amplitude of the radiation field and the detuning of the equilibrium mode. The radiation field is then written as a superposition of a large amplitude equilibrium mode and small perturbing satellites as in (9.121). The evolution of the small amplitude satellites is governed by an equation of the form of (9.123) except if that slippage is set to zero and dispersion is added. This gives the equation

$$\left[\frac{\partial}{\partial n} - i(\delta\omega_0 T + \alpha m^2) + (1-R)\right] \delta X_m = I_b \left[G\delta X_m + \frac{1}{2} G' \left(\delta X_m + \delta X^*_{-m}\right)\right], \qquad (9.190)$$

Again, it is seen that the beam couples modes, which are equally spaced in mode number about the equilibrium mode. Writing the corresponding evolution equation for the mode amplitude δX^*_{-m} as in (9.127) and assuming an exponential slow time dependence, we arrive at a quadratic dispersion relation replacing (9.131):

$$\left[\gamma_m T + i\alpha m^2 - \frac{1}{2} G'\right]\left[\gamma_m T - i\alpha m^2 - \frac{1}{2} G'^*\right] = \frac{I_b^2}{4} |G'|^2, \qquad (9.191)$$

The solutions for the complex growth rate of the coupled pairs of satellites is found to be

$$\gamma_m T = \frac{I_b}{2} \,\mathrm{Re}\, G' \pm \frac{1}{2}\left[(I_b \mathrm{Re}\, G')^2 - 4\alpha m^2 \left(1 - \frac{I_b \mathrm{Im} G'}{\alpha m^2}\right)\right]^{1/2}, \qquad (9.192)$$

and both solutions are damped provided

$$1 > \frac{I_b}{\alpha m^2} \mathrm{Im} G'. \qquad (9.193)$$

Notice that if the dispersion vanishes, then one of the two solutions is strongly damped (corresponding to a perturbation of the magnitude of the radiation field), and the other solution is marginally stable (corresponding to perturbations of the phase). Dispersion causes a coupling of amplitude and phase perturbations, which result in both solutions of the quadratic equation being strongly damped. Thus, a dispersive element can be used to hasten the spectral narrowing. In the limit in which (9.193) is strongly satisfied,

$$\gamma_m T = \frac{I_b}{2} \operatorname{Re} G' \pm i\alpha m^2, \tag{9.194}$$

which is the damping rate one would obtain had one neglected the coupling between modes with opposite signs of the mode index m [35] in Eq. (9.190). In this case, dispersion has modified the frequencies of the cavity to the point that the modes are no longer effectively coupled. Specifically, they no longer satisfy the four-wave matching condition given by Eq. (9.125). Dispersive elements have been used to hasten the coherence in the radiation pulse in storage ring free-electron lasers [36] and have the advantage over filters in that the tunability of the free-electron laser with beam energy is not compromised by the dispersive element.

9.7 Repetitively Pulsed Oscillators

The basic operating wavelength of a free-electron laser scales inversely with the energy of the electron beam. In order to operate at shorter wavelengths, one approach is to increase the energy of the beam. The electron beams in most high-energy accelerators, however, are not continuous in time but rather consist of a series of micropulses of a short duration with a fixed time between pulses. The series of micropulses is known as a macropulse. As a result, it is of fundamental importance to understand the effect of the pulsed nature of the electron beam on the operation of the oscillator in which it is being used [5, 17, 18, 37–41]. When pulsed beams are used to drive an oscillator, the radiation develops a temporal structure that more or less matches that of the electron beam. It is necessary, therefore, to synchronize the time of flight of radiation in the cavity with the arrival of successive beam micropulses, thus providing for temporal overlap of the radiation with the beam pulses. The degree of synchronism that is required, as well as the temporal structure of the resulting radiation pulse, will depend on the amount of electron slippage, the duration of the micropulse, and the gain and losses.

Since the radiation emerging from an oscillator driven by a pulsed electron beam also consists of pulses, the spectrum will be broader than that which could be achieved with a continuous electron beam. If the duration of the radiation pulses is the same as the beam micropulse T_p, then the minimum spectral width will be determined by the Fourier transform limit, $2\pi/T_p$. By comparison, the gain bandwidth is approximately $2\pi/\varepsilon T$. Thus, pulses which are of a duration T_p which is longer than the slippage time εT can have spectra which are narrower than the gain bandwidth. This can be visualized in the time domain using Fig. 9.2 and imagining that the micropulse duration T_p is much greater than the slippage time εT. In this case, the behavior of the radiation in the center of the pulse is to some degree independent of the shape of the pulse and only dependent on the instantaneous beam current and energy. As a result, in a limited sense, the center of the pulse can be treated as if the electron beam were continuous. Thus, concepts that have been developed in the treatment of continuous beam oscillators can be applied to the understanding of the more complex pulsed beam case if the duration of the micropulse is greater than the slippage time. In particular, the conditions for starting oscillations from noise, the mechanism for saturation of the radiation by phase trapping of electrons, and the criteria for the onset of sidebands (overbunch instability) in pulsed beam oscillators may be inferred from the continuous beam result.

9.7.1 Cavity Detuning

There are important differences, however, in the behavior of pulsed and continuous beam oscillators even when the micropulse duration is long compared with the slippage time. These differences relate to the interaction of slippage and cavity detuning in determining the dynamics of the radiation pulse [5]. Cavity detuning relates to the difference between the round-trip time of the radiation in the cavity T and the arrival time of electron beam pulses T_a and is characterized by the difference of these two times $T_\delta = T - T_a$. It will be shown in this section that for normal slippage $\varepsilon > 0$, it is necessary to have a negative value of the cavity detuning time T_δ in order for coherent oscillations to grow from arbitrarily small noise levels [5, 38–42]. This requirement is attributed to an effect known as *laser lethargy* [5] and is analogous to a similar requirement on the conditions for the existence of absolute instability in a medium that supports two waves which when coupled together are unstable [43, 44]. The requirement is that if the medium is of finite size, then the group velocities of the two waves must be oppositely directed. If the group velocities are not oppositely directed, then a small initial disturbance will grow in time due to the instability, but at the same time, it will convectively escape from the medium, so that the perturbations ultimately decay to zero. If noise is continually present, then perturbations will grow to a size, which is determined by the noise level and the gain a perturbation experiences before being carried out of the unstable region. On the other hand, if the group velocities are oppositely directed, an inherent feedback loop exits, which allows an arbitrarily small initial perturbation to grow to saturation. This is known as an absolute instability.

To see how these ideas apply to the interplay of slippage and cavity detuning, we redraw Fig. 9.2 in the variables n and t', introduced in Eq. (9.72), corresponding to the round-trip index and the time within the radiation pulse. Figure 9.19, therefore, attempts to illustrate the way in which the field amplitude at time $t' = t_*$ and round-trip index $n + 1$ is determined by the field amplitude at the earlier round-trip index n and fast time variable t'. The two cases shown in the figure correspond to positive cavity detuning and negative cavity detuning, respectively. The various lines in the figure correspond to terms in Eq. (9.81) which determine the evolution of the complex radiation amplitude $X(n,t')$. The dashed lines represent the contribution to

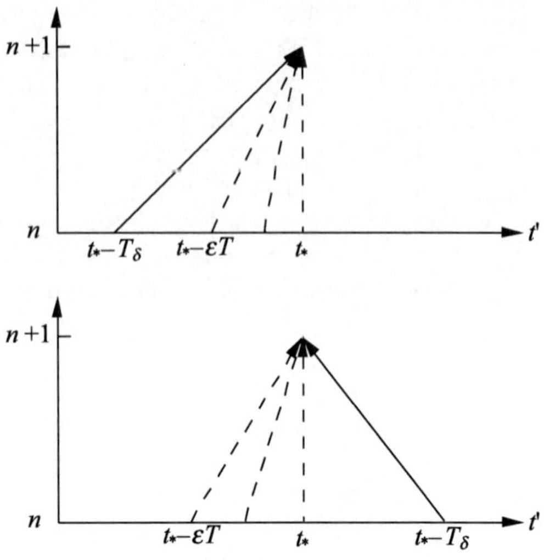

Fig. 9.19 Characteristics illustrating the dependence of the radiation amplitude entering the cavity at a time t_* on the radiation amplitude entering the cavity with the previous micropulse. For case (**a**) cavity detuning is positive and no absolute instability is possible, and for case (**b**) cavity detuning is negative and absolute instability is possible

the rate of change of the radiation amplitude with round trip index due to the interaction with the beam. Due to electron slippage, the effect of the beam on the radiation amplitude at fast time $t' = t_*$ is determined by the radiation amplitude the previous round-trip for fast time in the range $t_* > t' > t_* - \varepsilon T$. Recall that t' is defined such that events occurring at the same value of t' and successive values of the round-trip index n are actually separated in time by T_a, the time between arrival of successive beam pulses. The solid lines show the contribution to the radiation amplitude from the direct propagation of radiation through the cavity. Due to cavity detuning, the amplitude of the radiation at time t_* is determined by the radiation entering one round-trip earlier at a time $t' - T_\delta$. This is also illustrated in Fig. 9.3. As is evident, the effect of slippage is such that information from the interaction with the beam always propagates to larger values of t'. If the cavity detuning is positive, information resulting from the direct path of the radiation also propagates to larger values of t'. In this case, both signal paths have positive *group velocities*, if one thinks of t' as a space-like variable and n as time-like variable. If the electron pulse is of a finite duration in t', disturbances will grow but also convect out of the electron pulse. In this case, no self-sustaining oscillations can develop. On the other hand, if the cavity detuning is negative, information from the direct path of the radiation is propagated toward negative values of t', and the *group velocities* of the two signal paths are oppositely directed. As a result, self-sustaining oscillations can grow to saturation from arbitrarily small noise levels.

The effect of cavity detuning on the start current of a pulsed beam oscillator is illustrated in Fig. 9.20 which plots the start current for a monoenergetic beam normalized to the minimum start current for the corresponding continuous beam (cf. Eq. (9.110)) versus normalized cavity detuning parameter. The calculation applies to the case in which the slippage time is much shorter than the micropulse

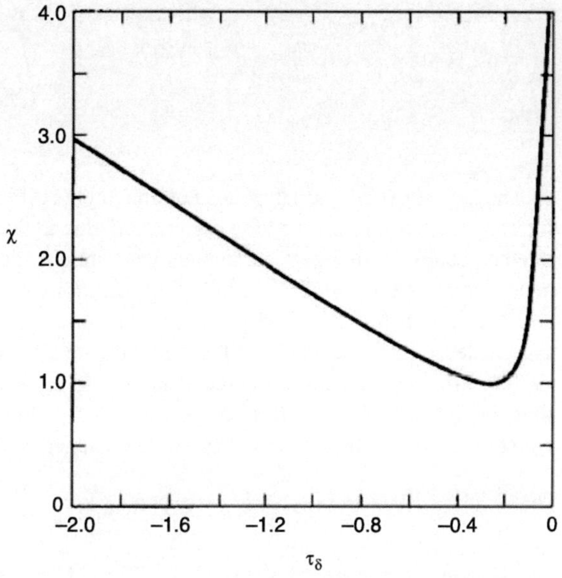

Fig. 9.20 The peak current required to start oscillations in a monoenergetic pulsed beam normalized to the corresponding start current for a continuous beam as a function of normalized cavity detuning time $\tau_\delta = T_\delta/[\varepsilon T(1-R)]$. The calculation has assumed that the duration of the micropulse T_p is much longer than the slippage time εT

pulse duration, and, as a result, the shape of the beam micropulse is not important. The steps required to obtain this plot will be detailed subsequently. The plot shows that there is an optimum cavity detuning time T_δ (which is negative) for which the start current is the same as in a continuous beam. Actually, the start current for the pulsed beam will be slightly higher than the corresponding continuous beam if corrections due to the finite value of the pulse duration time are retained. Further, the start current is infinite for positive values of the cavity detuning and becomes large for large negative values of cavity detuning.

9.7.2 Supermodes

A theoretical treatment of the small signal growth of radiation in pulsed oscillators as described in the previous paragraph proceeds as follows. To determine the conditions for the existence of absolute instability, one linearizes the equations of motion and evaluates the last term on the right-hand side of Eq. (9.81) describing the interaction of the radiation with the beam. The result can be represented in the form of a convolution of the radiation amplitude with a kernel and is written [37, 39, 40]

$$\left(\frac{\partial}{\partial n}+T_\delta\frac{\partial}{\partial t'}\right)X(n,t')=(R-1)X(n,t')+\int\limits_{t'-\varepsilon T}^{t'}dt''K(t'',t'-t'')X(n,t''). \tag{9.195}$$

The kernel $K(t'',t'-t'')$ depends not only on the difference time $t'-t''$ accounting for slippage but also on the time argument t'' due to the temporal dependence of the injected beam. If the beam parameters were independent of time, the kernel would depend only on the difference time, and solutions of the form $X(n,t')\exp(-i\Delta\omega t')$ could be found (as discussed in the sections on continuous beam oscillators). The time scale for the dependence of the kernel on the difference time is the slippage time εT. The time scale for the explicit dependence of the kernel on t'' is the characteristic time of duration of the electron micropulse T_p. Solutions of this linear integral equation corresponding to exponential dependence on the round-trip index n

$$X(n,t')=\widehat{X}(t')\exp(\gamma_{sm}nT), \tag{9.196}$$

are known as supermodes [37, 39, 40] and in general must be determined numerically.

Analytic progress can be made when the duration of the micropulse is much longer than the slippage time. In this case the explicit dependence of the kernel on the time variable t'' is relatively weak,; and one can expect the radiation field to be locally of the form of a complex exponential

$$\widehat{X}(t')\cong\widehat{X}\exp\left[-i\int^{t'}dt''\Delta\omega(t'')\right]. \tag{9.197}$$

That is, one can use a WKB representation for the radiation amplitude. Equation (9.197) is then inserted in the integral equation with the parameters of the beam evaluated at time t'. The contribution of the beam is then formally identical to that which would be found for a monochromatic radiation signal interacting with a continuous electron beam with the instantaneous parameters of the micropulse. We can thus express the response of the beam in terms of a linear gain function of the form derived in (9.95). The resulting dispersion equation determining the instantaneous frequency of the radiation including the effect of cavity detuning follows from (9.195):

$$\gamma_{sm}T-iT_\delta\Delta\omega(t')=R-1+I_b(t')G[t',-\varepsilon T\Delta\omega(t')], \tag{9.198}$$

where $I_b(t')$ denotes the time-dependent electron beam current and $G[t', -\varepsilon T\Delta\omega(t')]$ is the instantaneous gain defined in analogy to (9.95)

$$G[t',-\varepsilon T\Delta\omega(t')]=\pi^2\int dp\Delta(p-\varepsilon T\Delta\omega)\frac{\partial}{\partial p}F_0(p,t'), \tag{9.199}$$

where $\Delta(p)$ is defined in Eq. (9.96) and G is time dependent due to the fact that the beam distribution function may have a time dependence within the micropulse. For example, the gain would be given by Eq. (9.98) with the parameter p_0 a function of the fast time variable t' for a beam which at any instant is monoenergetic but whose energy depends on time.

Equation (9.198) is insufficient by itself to determine the complex growth rate of the radiation on the round-trip time scale. In fact, for the given values of t' and the fast time frequency shift $\Delta\omega$, there will always be solutions for the slow time growth rate γ_{sm}. What is required is some way of enforcing the boundary condition that the field amplitude vanishes far outside of the beam micropulse as $|t'| \to \infty$. Within the context of WKB theory, the problem of enforcing the boundary conditions at infinity reduces to the problem of finding *two turning points*, specifically, two values of t' (t_1 and t_2) where two distinct solutions for the fast time frequency ($\Delta\omega_1$ and $\Delta\omega_2$) coalesce. The correct slow time growth rate is then determined by the quantization condition that the integral of the difference of the two fast time frequencies between the two turning points is given by

$$\left(m + \frac{1}{2}\right)\pi = \int_{t_1}^{t_2} dt' [\Delta\omega_1(t') - \Delta\omega_2(t')], \tag{9.200}$$

where m is a positive integer labeling the various normal modes (supermodes). The problem is further complicated by the fact that, in general, the two turning points may occur for complex values of t'. Thus, even though the solution can be written in a compact form, implementation of this solution still requires computation.

The lowest-order solution (i.e., the ground state) with $m = 0$ can be found by noting that the quantization condition requires that the two turning points be close to each other on the time scale of the duration of the electron micropulse. Thus, not only must there be a coalescence of the two solutions for the fast time frequency $\Delta\omega$ but also the two turning points must coalesce. Thus, to lowest order, the complex growth rate can be determined by solving Eq. (9.198) along with the subsidiary conditions that the time t' and frequency shift $\Delta\omega$ are such that derivatives of (9.198) with respect to time and frequency shift vanish:

$$\frac{dI_b(t')}{dt'} G(t', -\varepsilon T\Delta\omega) + I_b(t') \frac{\partial}{\partial t'} G(t', -\varepsilon T\Delta\omega) = 0, \tag{9.201}$$

and

$$\varepsilon T I_b(t') \frac{\partial G}{\partial \tilde{p}}\bigg|_{\tilde{p} = -\varepsilon T\Delta\omega} = iT_\delta. \tag{9.202}$$

These latter two conditions ensure that the dispersion equation is solved for parameters that give simultaneous coalescence of both the turning points and the fast time frequencies. It is now a good idea to count the number of equations and unknowns. There are three complex Eqs. (9.198, 9.201, and 9.202) along with three complex unknowns: the location of the coalesced turning points t', the complex frequency at the turning points $\Delta\omega$, and the complex growth rate on the round-trip time scale γ_{sm}. Thus, we have the right number of knowns and unknowns to calculate the round-trip time growth rate.

To simplify matters further, assume that the distribution function for the injected beam is independent of time and only the beam current varies within the micropulse. Thus, the gain function is independent of time, and Eq. (9.201) requires that the coalesced turning points occur for real values of t' at the maximum of the beam current. Since the coalesced turning points occur for real t', the normalized beam current appearing in (9.198) and (9.202) is real and equal to the maximum beam current. As a consequence, Eq. (9.202) now implicitly determines the value of the complex frequency shift $\Delta\omega$. This value may then be inserted into (9.198) to determine the complex supermode growth rate γ_{sm}. This procedure was followed for the case of a monoenergetic beam and the value of the beam current required to initiate absolute instability (Re $\gamma_{sm} = 0$) is plotted versus cavity detuning in Fig. 9.20.

As pointed out previously, the minimum starting current predicted by this method is the same as the minimum starting current in a continuous beam oscillator. However, this minimum occurs only for a specific value of cavity detuning. This optimum detuning can be found from (9.198) and (9.202) by the following argument. The real part of Eq. (9.202) requires that the derivative of the real part of the gain with respect to frequency shift vanishes. This certainly occurs for the real frequency shift that maximizes the gain. At this frequency shift, the start current (Re γ_{sm} =0) can be obtained from the real part of (9.198):

$$I_b(t') = \frac{1-R}{\mathrm{Re}[G(t', -\varepsilon T\Delta\omega)]}, \tag{9.203}$$

where t' and $\Delta\omega$ give the maximum current and gain. This expression is identical to the corresponding condition for starting oscillations in a continuous beam oscillator, (9.97). The optimum value of the cavity detuning which produces the minimum start current is then obtained from the imaginary part of Eq. (9.202).

The structure of the supermode on the fast time scale variable t' predicted by this theory will tend to be Gaussian as will be seen. At first one might expect that the radiation pulse predicted by this theory would be peaked on the maximum current since this is the location of the coalesced turning points in t'. It must be remembered, however, that the coalesced frequency shift $\Delta\omega_c$ is complex. Thus, depending on the sign of the imaginary part of $\Delta\omega_c$ at the coalescence points, the radiation peak will be either advanced or retarded with respect to the peak of the beam current. Generally, the more negative the detuning, the more retarded the peak is. The special case of optimum detuning predicts a pulse whose peak coincides with that of the beam current.

At this point it must be recalled that the present method of solution has relied on the approximation that the micropulse duration is much longer than the slippage time. When this is not the case, the full integral Eq. (9.195) must be solved, and one can expect the start current to be higher when higher-order finite pulse duration effects are included. This can be seen when one calculates the first-order corrections to the starting current or growth rate implied by the quantization condition (9.200). Rather than perform the integral in (9.200), we will convert the local dispersion relation (9.198) to the corresponding second-order differential equation by expanding about the fast time $t' = t_{max}$ and frequency shift $\Delta\omega = \Delta\omega_c$ implied by simultaneous solution of (9.201) and (9.202). We again assume the beam distribution function is time independent (only the beam current varies), and we define $\tau = t' - t_{max}$ where $I_b(t_{max})$ is the maximum beam current. To obtain a differential equation, we expand the frequency shift about the coalescence frequency shift by substitution of

$$\Delta\omega = \Delta\omega_c - i\frac{\partial}{\partial\tau}, \tag{9.204}$$

and expand Eq. (9.198) to second order in τ and its derivative. The result is a harmonic oscillator equation of the form [36, 42]

$$\left[T_1^2\frac{d^2}{d\tau^2} - \frac{\tau^2}{T_2^2} - 2\Lambda\right]\hat{X}(\tau) = 0, \tag{9.205}$$

where the coefficients appearing in (9.205) are defined by

$$T_1^2 = -(\varepsilon T)^2 \frac{1}{G}\frac{\partial^2 G}{\partial p^2}\bigg|_{p=-\varepsilon T\Delta\omega_c} \approx (\varepsilon T)^2, \tag{9.206}$$

$$T_2^{-1} = -\frac{1}{I_b}\frac{\partial^2 I_b}{\partial t'^2}\bigg|_{t'=t_{max}} \approx T_p^{-2}, \tag{9.207}$$

and

$$\Lambda = \frac{\gamma_{sm}T - iT_\delta\Delta\omega_c - (1-R)}{I_b(t_{max})G(t_{max}, -\varepsilon T\Delta\omega_c)} - 1. \tag{9.208}$$

Note that we have indicated the order of magnitude of the coefficients T_1 and T_2. In particular, T_1 scales as the slippage time εT and T_2 scales as the pulse duration T_p.

The solution of the harmonic oscillator equation is given in terms of Hermite polynomials:

$$\widehat{X}_m(\tau) = H_m\left(\frac{\tau}{\sqrt{T_1 T_2}}\right)\exp\left[-\frac{\tau^2}{2T_1T_2}\right], \tag{9.209}$$

and

$$\Lambda_m = -\frac{T_1}{T_2}\left(m + \frac{1}{2}\right), \tag{9.210}$$

where the index m labels various supermodes by the number of their temporal nodes. The lowest-order solution has a start current which is elevated slightly above that predicted on the basis of (9.198), (9.201), and (9.202) from $\Lambda_0 = 0$. As can be seen, this difference is small when pulse duration exceeds the slippage time. Also of interest is the temporal width of the solution $\sqrt{(T_1 T_2)}$, which is the geometric mean of the slippage time and the pulse duration. For the case of the optimum detuning, this is the width of the supermode. For cavity detunings which are not optimum, the radiation pulse is peaked away from the maximum of the beam current. The width of the radiation peak is determined by the WKB representation of the solution (9.197) expanded about the fast time t', where the imaginary part of the local frequency shift vanishes. This width also scales as the geometric mean of the slippage and pulse duration times. As the current is raised above the start value, an increasing number of supermodes are driven unstable. The fastest-growing mode is the lowest-order one which has the narrowest temporal width. Thus, as radiation approaches saturation, it will be somewhat narrower than the pulse duration and will correspond to a spectral width that is the geometric mean of the gain bandwidth and the pulse duration Fourier transform limit. The shape of the saturated radiation pulse depends on the nature of the saturation mechanism. In storage ring free-electron lasers, where the electron beam is continually recirculated through the interaction region, saturation is achieved at low-field amplitudes by a process in which the quality of the beam is continually degraded until the linear gain exactly balances losses [45, 46]. In this case, the supermode structure is likely to remain intact [36, 42, 45, 46]. In free-electron lasers driven by linear accelerators, saturation is achieved by phase trapping at large field amplitude, and the radiation pulse shape is significantly altered from that in the small signal phase [14, 17, 18, 27, 41, 42].

It is important to emphasize at this point that the previous discussion of supermodes focuses on the question of whether the conditions of cavity detuning, slippage, and gain are such that an absolute instability is present. That is, will radiation grow to saturation if the noise level is arbitrarily small? In practice, the noise level is not arbitrarily small, and radiation can grow to a substantial level even if it is only convectively unstable. Noise that is excited at some point in the micropulse will grow as it convects out of the pulse. If the gain during the time that the disturbance is in the pulse is large, then a small but nonzero noise signal can amplify to a significant level. The gain can be estimated for various parameters from (9.198)

$$N_e \cong T_p \mathrm{Im}\Delta\omega, \tag{9.211}$$

where N_e is the number of exponentiations and T_p is the pulse duration. Under the assumption that the detuning time T_δ is large compared with the slippage time εT, the imaginary part of the fast time frequency is determined by gain and losses;

$$\mathrm{Im}\Delta\omega = \frac{1}{T_\delta}\left(I_b \mathrm{Re}\left[G(t_{max}, -\varepsilon T\Delta\omega)\right] - (1-R)\right), \tag{9.212}$$

where the gain is evaluated with a real frequency shift. Thus, the number of exponentiations N_e scales as

$$N_e \cong \frac{T_p}{T_\delta}\left(I_b \mathrm{Re}\left[G(t_{max}, -\varepsilon T\Delta\omega)\right] - (1-R)\right), \tag{9.213}$$

which for long pulses can be a large number even if the conditions for absolute instability are not satisfied. In this limit one can think of cavity detuning as providing an additional loss mechanism where the fraction of radiation which is lost each round trip is the ratio of the cavity detuning time to the micropulse duration time. We caution the reader that the simple estimate give above applies only if the cavity detuning time is much longer than the slippage time and much shorter than the pulse duration. In cases where both these inequalities cannot be satisfied, it is necessary to resort to numerical solution to determine more precise estimates of the amount of radiation growth.

9.7.3 Spiking Mode and Cavity Detuning

If the maximum beam current in the micropulse exceeds either of the thresholds discussed in the previous section, then radiation can grow to large amplitude and saturate. The resulting radiation pulse will have a time dependence determined by the shape of the beam micropulse as well as other parameters such as cavity detuning. Due to the time dependence, the radiation must always have a nonzero spectral width.

The characteristics of the spectrum depend sensitively on whether the conditions for the excitation of the overbunch instability are satisfied. The overbunch instability was discussed in Sect. 9.6 in the context of continuous beam oscillators. It occurs when the radiation amplitude becomes sufficiently large that electrons are trapped in the ponderomotive well and execute about half a synchrotron oscillation on transit through the interaction region. The instability manifests itself by the appearance of sidebands in the spectrum that correspond to temporal oscillations of the magnitude of the radiation amplitude whose period is roughly the slippage time. Thus, the sidebands are displaced from the original signal by roughly the gain bandwidth. If the overbunch instability is not present, then the spectrum can be much narrower than the gain bandwidth, and its width can approach the Fourier transform limit determined by the duration of the beam micropulse.

In the case of a continuous beam oscillator, the conditions for the excitation of the overbunch instability can be determined with great precision and are displayed in Fig. 9.8. A similarly detailed calculation of the conditions for the onset of this instability in the case of a pulsed beam oscillator is not available due to the added complexity introduced by the time dependence of the beam within a micropulse and the important effect of cavity detuning. There has been, however, extensive numerical simulation of pulsed beam oscillators that tend to confirm that the threshold for the onset of the overbunch stability is correlated with the peak radiation amplitude exceeding a value for which electrons can complete half a synchrotron oscillation [14, 17, 18, 27, 41, 47]. This rule is not expected to be hard and fast due to the complicated interaction of slippage and cavity detuning. Recall that the instantaneous linear gain given by Eq. (9.198) can be positive yet if the cavity detuning is not optimized, small signals will not grow significantly before being convected out of the beam pulse. Similarly, the conditions for the onset of the overbunch instability may be locally satisfied at some point in the radiation pulse, but, due to a large value of cavity detuning, the instability can be suppressed. Indeed, it has been found that cavity detuning is effective in suppressing the overbunch instability [17, 18].

The features of the onset of the overbunch instability can be illustrated with the use of the klystron model introduced previously. In particular, Figs. 9.21, 9.22 and 9.23 show examples of the radiation profile obtained by numerical solution of Eq. (9.120) that may be rewritten as

$$\left(\frac{\partial}{\partial n}+T_\delta\frac{\partial}{\partial t'}+(1-R)\right)X(n,t')=I_b(t'')X(n,t'')g(|X(n,t'')|)|_{t''=t'-\varepsilon T}, \quad (9.120)$$

For the particular simulations shown, the time dependence of the beam current was chosen to be a Gaussian with a width T_p

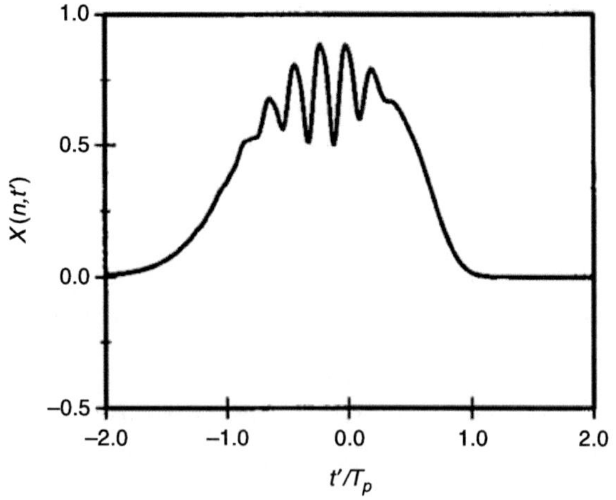

Fig. 9.21 The time dependence of the radiation pulse showing the spiking mode. The magnitude of the radiation pulse is plotted versus t_0' for a specific value of the cavity detuning τ_δ

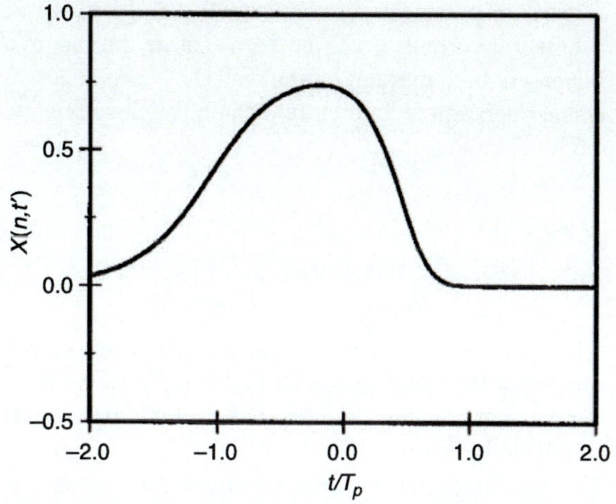

Fig. 9.22 The time-dependence of the radiation pulse showing the absence of the spiking mode. The magnitude of the radiation pulse is plotted versus t_0' for the same parameters as Fig. 9.21 except that the magnitude of the cavity detuning τ_δ has been increased

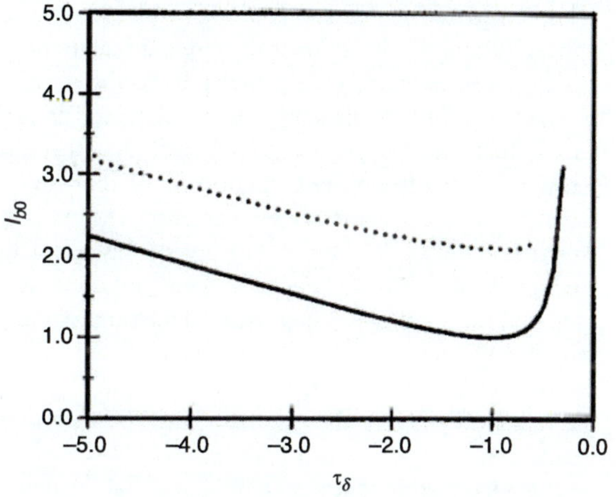

Fig. 9.23 The peak current required to initiate the spiking mode in the klystron as a function of the normalized cavity detuning time $\tau_\delta = T_\delta/[\varepsilon T(1 - R)]$. Also shown is the start current versus normalized detuning

$$I_b(t'') = I_{b0} \exp\left(-\frac{t''^2}{T_p^2}\right), \tag{9.214}$$

and the gain function g defined in Eq. (9.53) was taken to be simply

$$g = 1 - |X|^2, \tag{9.215}$$

which has the required feature to produce the overbunch instability; specifically, the gain decreases with radiation amplitude. This requirement is discussed in Sect. 9.6 and illustrated in Fig. 9.10. The boundary in current versus cavity detuning for the excitation of the overbunch instability obtained by solution of Eq. (9.120) for a large number of parameters is shown in Fig. 9.20, along with the curve giving the start oscillation current. It can be seen that as the magnitude of the cavity detuning is increased, both the start current and the overbunch threshold current increase. Thus, stable operation at high current can be achieved by increasing the magnitude of the cavity detuning.

9.8 Multidimensional Effects

Our treatment of oscillators has thus far been strictly one-dimensional in that all quantities have been assumed to depend only on the axial coordinate and time. Of course, in a real device, the radiation field, the wiggler field, and the electron beam will have significant variations in the directions transverse to the axis of the beam. The resulting multidimensional effects have been discussed in extensive detail for the case of amplifiers in the early chapters of this book. In particular, the linear and nonlinear interaction of a beam with multiple transverse modes in a waveguide has been considered in Chaps. 4 and 5, while the related problem of diffraction and optical guiding for the case of unconfined radiation was discussed in Chap. 8.

The basic physical effects that govern the transverse dependence of radiation profiles in amplifiers also apply to oscillators. The problem is more complicated in an oscillator, however, due to the fact that the radiation circulates continually through the interaction region. In an amplifier, the transverse distribution of the radiation field that enters the interaction region is determined by the launching structure more or less independently of the interaction of the radiation with the electron beam. The radiation is then amplified and the transverse distribution modified as it propagates down the interaction region. In an oscillator, the amplified radiation that leaves the interaction region is fed back through an optical system and returns to the entrance of the interaction region. Thus, the transverse dependence of the radiation entering the oscillator must be determined self consistently with the interaction.

If the interaction is strong in the sense that the device is of the high-gain type, the modification of the transverse dependence of the radiation profile by the beam can be significant and must be accounted for in the design of experiments. Sophisticated computer codes which self-consistently determine the transverse and axial dependences of the radiation field have been written for this task [48–52]. If the device is of the low-gain type, then the transverse dependence of the radiation field can be assumed to be the same as that of the eigenmodes of the empty cavity, and the one-dimensional theory presented here can be adopted with certain modifications. The remainder of this section will address the issue of incorporation of multidimensional effects into the low-gain model of an oscillator.

The first modification to the one-dimensional theory introduced in Sect. 9.1 is that the full three-dimensional variation of the vector potential must be included. Thus, Eq. (9.1) for the vector potential is rewritten:

$$\delta\mathbf{A}(\mathbf{x},t) = \delta A(x,t)\widehat{\mathbf{e}}_p \exp(ikz - \omega t) + c.c. \tag{9.216}$$

The equations for the rate of change of the beat wave phase and the electron energy (Eqs. (9.3) and (9.4)) are formally the same, except that the full three-dimensional dependence of the radiation electric field must be included. As a consequence, it is necessary also to introduce equations describing the transverse location of the beam electrons since electrons at different transverse locations experience different radiation fields. Within the approximations leading to the wiggler period averaged equations (Eqs. (9.6) and (9.7)), it is now assumed that the average transverse location of an electron does not change much during the time it takes the electron to travel one wiggler period. Thus, in Eq. (9.14), the normalized radiation vector potential is evaluated at the average transverse location of an electron:

$$\left(\frac{\partial}{\partial t} + v_{\parallel}\frac{\partial}{\partial z}\right)\delta\gamma = -\frac{\omega}{2\gamma_r}a_w\left[i\delta a(\mathbf{x}_{0\perp}, z, t)\exp(i\psi) + c.c.\right]. \tag{9.217}$$

This position which we label $\mathbf{x}_{0\perp}$ is sometimes termed the *oscillation center* since the motion of the electron in the wiggler consists of periodic oscillations about the average trajectory $\mathbf{x}_{0\perp}(t)$. By replacement of the exact transverse position of an electron with the average transverse position, we implicitly assume that variations of quantities such as the wiggler field strength and the radiation field strength are small over a distance corresponding to the displacement of the electrons in the wiggler field. Note that this assumption eliminates the possibility of including periodic position interactions of the type described in Chap. 7. The equations for energy and phase must now be supplemented by an equation for the transverse location of the oscillation center. The simplest case is the one in which the transverse motion is determined solely by the strong focusing of the wiggler field and corresponds to betatron oscillations with betatron wave number k_β:

$$\left(\frac{\partial}{\partial t} + v_{\parallel}\frac{\partial}{\partial z}\right)^2 \mathbf{x}_{0\perp} = -k_\beta^2 v_{\parallel}^2 \mathbf{x}_{0\perp}. \tag{9.218}$$

A further simplification is the case in which the amplitude of the betatron oscillations is so small that the transverse dimension of the beam is much smaller that the scale length over which the radiation vector potential varies. In this case, all electrons sample only the radiation field on-axis, and it is not necessary to follow the transverse motion of the electrons at all.

Inclusion of transverse dependence in the radiation field results in a modification to the wave Eq. (9.25) to account for diffraction; specifically, the transverse Laplacian must be added to the left-hand side of Eq. (9.25), and the source term on the right-hand-side now acquires transverse dependence due to the fact that the beam is of finite size:

$$\left(\frac{\partial}{\partial t}+c\frac{\partial}{\partial z}-\frac{ic}{2k}\nabla_{\perp}^{2}\right)\delta A(\mathbf{x}_{\perp},z,t)=\frac{2\pi i}{\omega}\frac{c^{2}a_{w}}{\gamma_{r}v_{\parallel}}\langle J(t_{e})\exp(-i\psi)\rangle. \tag{9.219}$$

Here we have replaced v_g by c because we are explicitly treating the transverse dependence, and Eq. (9.219) determines self-consistently the group velocity. The new system of equations is now capable of describing all the effects of optical guiding discussed in Chap. 8. Finally, to treat an oscillator, the equation describing the optical feedback path must be modified. In the one-dimensional case, this equation was a simple algebraic relation between the radiation leaving the interaction region at time $t - T_r$ and the radiation entering the interaction region at time t (Eq. (9.26)). When multidimensional effects are included, this algebraic relation is replaced by a complicated integral relation describing the transformation of the transverse profile of the radiation by the system of mirrors (and/or waveguides) composing the feedback loop. We will not discuss this problem in detail here; however, there have been several publications describing numerical methods for treating these effects [48–52]. It will be seen that in the low-gain limit, the effects of the optical feedback path enter in determining the empty cavity modes, and that once these are found the effect of the interaction can be treated by perturbation.

We proceed with a discussion of the low-gain limit by consideration of solutions of (9.219) under the assumption that the right-hand side is small. The development parallels that given in Sect. 9.5 for the one-dimensional case. In the absence of a beam, the solution of the wave equation should yield a radiation field whose structure can be described by a superposition of empty cavity modes. Inclusion of the beam in the low-gain regime will cause the amplitudes of the various cavity modes to evolve in time as in the one-dimensional limit described by Eq. (9.90). We assume for simplicity that only the lowest-order transverse eigenmodes of the cavity are excited and that these modes have a transverse dependence which is Gaussian with a minimum spot size σ_0 occurring at an axial location z_0. The radiation field in the interaction region to which the electrons respond can then be written in analogy to (9.67):

$$\delta A(\mathbf{x}_{\perp},z,t)=\delta A_{e}(t-z/c)u(\mathbf{x}_{\perp},\sigma), \tag{9.220}$$

where the functions $u(\mathbf{x}_{\perp},\sigma)$ and $\sigma(z)$ describe the transverse dependence of the radiation field

$$u(\mathbf{x}_{\perp},\sigma)=\frac{\sigma_{0}}{\sigma(z)}\exp\left[-x_{\perp}^{2}/2\sigma(z)\right], \tag{9.221}$$

and

$$\sigma(z)=\sigma_{0}+\frac{i(z-z_{0})}{k}, \tag{9.222}$$

determines the local spot size. The field given by Eq. (9.221) is then inserted into the electron equations of motion to calculate the interaction with the electron beam. This leads to equations of the form of (9.68) and (9.69) except that the factor $u(\mathbf{x}_{0\perp},\sigma)$ multiplies the radiation amplitude. These equations are subsequently normalized to yield the multidimensional counterparts of (9.79) and (9.80):

$$\frac{\partial \psi}{\partial \xi} = p, \tag{9.223}$$

and

$$\frac{\partial p}{\partial \xi} = -\frac{1}{2}[iu(x_{0\perp},\sigma)X(n, t_e + \varepsilon\xi T)\exp(i\psi) + c.c.], \tag{9.224}$$

where $\xi = z/L$ is substituted in Eq. (9.223). These equations are to be solved for an ensemble of electrons with initial conditions appropriate to the incoming beam. In addition, the equations for the trajectories of the oscillation centers (Eq. (9.219)) must also be solved with a distribution of initial conditions describing the transverse positions and angular divergences of the electrons in the beam.

The solutions for the phases and transverse positions of the electrons must now be inserted in the wave Eq. (9.220) to determine the gain that the radiation experiences on one transit of the interaction region. The interaction with the beam produces a small increase in the amplitude of the radiation whose transverse dependence is described by the function $u(\mathbf{x}_\perp,\sigma)$. To find this small increase, we multiply the wave equation by u^*, integrate over transverse coordinates and along the radiation characteristics in the (z,t) plane, and obtain an expression analogous to (9.70) of the one-dimensional case:

$$\delta A(L,t) = \delta A_e(t - L/c) + \frac{2\pi i}{\omega}\frac{ca_w}{\gamma_r v_\parallel}\frac{\int_0^L dz \int d^2x_\perp u^* \langle J\exp(-i\psi)\rangle}{\int d^2x_\perp |u|^2}, \tag{9.225}$$

The transverse average appearing on the right-hand side of (9.225) can be expressed in terms of an average over the initial coordinates of the beam electrons entering the interaction region and an effective current density equal to the total beam current divided by the minimum spot area:

$$\frac{\int d^2x_\perp u^* \langle J\exp(-i\psi)\rangle}{\int d^2x_\perp |u|^2} = J_{eff}\langle u^*\exp(-i\psi)\rangle, \tag{9.226}$$

where $J_{eff} = I_{total}/(\pi\sigma_0)$ and the average is over the initial distribution of the injected electrons.

Thus, it is seen that multidimensional effects can be included in the one-dimensional formulation in the low-gain regime by introducing transverse profile functions $u(\mathbf{x}_\perp,\sigma)$ in the equations of motion and the wave equation and by defining an effective one-dimensional current density. This leads to the following general conclusions in regard to the choice of experimental parameters based on the requirement that the profile function remain of order unity for all electrons in the interaction region. First, the transverse size of the beam must be kept smaller than the minimum spot size in order that all electrons in the beam interact strongly with the radiation field. Second, the interaction length L must not be much greater than the Rayleigh length so that radiation does not diffract significantly over a distance corresponding to the interaction length. However, if the gain is high, this latter condition can be relaxed due to the optical guiding effect discussed in Chap. 8.

9.9 Storage Ring Free-Electron Lasers

The high-energy electrons circulating in storage rings are suitable for driving free-electron laser oscillators. In these devices, the same groups of electrons circulate continually around the ring with their energy determined by a balancing of gain due to an accelerating RF field and loss due to synchrotron radiation. Thus, when a wiggler is added to the ring, the properties of the beam entering the interaction region vary on each pass around the storage ring in response to the interaction of the beam with the radiation on the previous pass. This situation for the beam is now analogous to that of the radiation in an oscillator and introduces another degree of complexity in the design and analysis of these devices.

The interaction with the radiation acts primarily to alter the energy of the electrons in the beam, and this leads to a modification of the interaction of the electrons with the radiation on its subsequent passes through the interaction region. The modifications of the interaction can be divided into two basic classes: (1) modifications due to the sensitivity of the performance of the storage ring to changes in beam energy and (2) modifications due to the sensitivity of the basic free-electron laser interaction to the energy of the entering beam. Examples of the first class relate to the dispersion and energy acceptance of the ring. As the electron energy changes, the path followed through the ring also changes, which is known as dispersion. If this change is comparable to the spot size of the radiation, then the interaction of the electron with the radiation will be weakened if the radiation profile remains fixed. Accompanying the change in path for the electron is a change in travel time around the ring. If this change in travel time becomes comparable to the period of the accelerating field, then the electron will fall out of phase with the accelerating field and no longer be accelerated. This determines the energy acceptance. Both these effects serve to remove an electron from interaction with the radiation once its energy has changed by a specific amount. The importance of these effects depends on details of the storage ring and will not be pursued here.

The second and perhaps more fundamental class of modifications relates to the effect of changes in the energy of electrons on the free-electron laser interaction itself. The interaction with the radiation lowers, on average, the energy of electrons and thus produces an amplification of the radiation. However, the energies of different electrons are changed by different amounts. Thus, the interaction also leads to an increase in the spread of energies; at the same time, the average is lowered. The spread in energy of the beam degrades the interaction on its subsequent passes around the ring. Energy spread degrades the interaction in two ways. First, it lowers the gain directly since with a spread all electrons cannot have the optimum detuning. Second, a spread in energy leads to an increase in the length of the micropulse and a lowering of the peak beam current [46, 53].

The rates at which energy is extracted from the beam and the rate at which the energy spread of the beam is increased are related directly in the small radiation amplitude limit by Madey's second theorem [54]:

$$\langle \gamma_f - \gamma_i \rangle = \frac{1}{2} \frac{\partial}{\partial \gamma_i} \left\langle (\gamma_f - \gamma_i)^2 \right\rangle, \tag{9.227}$$

where γ_f and γ_i are the final and initial relativistic factors for electrons transiting the interaction region and the average is over an ensemble of electrons entering the interaction region with a relativistic factor γ_i and a uniform distribution of entrance phases. Relation (9.227) is a consequence of the fact that the average rate of energy extraction is second order in the radiation field strength, whereas the magnitude in the change in energy for an electron is first order. That is, in a group of electrons entering the interaction region with a uniform distribution of entrance phases, half of the electrons are accelerated initially by the radiation field, and half are decelerated leading to a spread in energies. To the first order in the radiation field, the net change in energy averages to zero. Due to the first-order acceleration and deceleration, electrons become bunched in phase as they transit the interaction region. The slightly higher energy electrons catch up to the lower energy ones. This leads to a net change in energy which is second order in the field strength and which satisfies Madey's theorem. Madey's theorem will be derived in this section by solution of the Vlasov Eq. (9.86) in powers of the radiation field. The development leads to what is known as a quasilinear theory [55, 56] of the evolution of the energy distribution function for the beam electrons [46, 57, 58].

The increase in energy spread of the beam due to its interaction with the radiation leads to a limit, known as the Renieri limit [45, 46], on the average amount of power that can be extracted from the beam. In the absence of the free-electron laser interaction, the beam energy spread is damped by the combined effects of synchrotron radiation and RF acceleration. Since the rate of synchrotron radiation increases strongly with energy while the rate of acceleration is approximately independent of energy (in the typical range of interest), the competition of the two causes all electrons to tend to the energy at which the rate of increase of energy due to acceleration and the rate of decrease due to radiation balance. Synchrotron radiation scales as the second power of energy and magnetic field strength; therefore, the rate at which deviations in energy decay is given by [59]

$$\nu_s = \frac{P_s(2+D_b)}{\gamma m_e c^2}, \tag{9.228}$$

where P_s is the power radiated by an electron in the ring, the factor of 2 comes from the energy dependence of the synchrotron radiation, and the factor D_b accounts for the variation of magnetic field strength with the variation of the path (dispersion) implied by the variation in energy. The Renieri limit states that the maximum average power extracted from the beam by the free-electron laser interaction is given by [45, 46]

$$\overline{P} = \frac{P_s}{2N_w}, \tag{9.229}$$

where N_w denotes the number of wiggler periods in the interaction region. This power, while a fraction of the synchrotron radiation power, is concentrated in a narrow range of frequencies and in a well-collimated beam and is thus still of practical interest.

Finally, with the available current in storage rings, it is frequently difficult to obtain a significant gain with an interaction region whose length is not too long as to be incompatible with operation of the storage ring. As a result, a wiggler configuration has been developed known as the optical klystron [60, 61], which is capable of achieving enhancements in gain above that predicted for the conventional uniform wiggler. The basic operation of the optical klystron can be explained in terms of the klystron model introduced in Sect. 9.3. This will be discussed briefly at the end of this section.

We begin our analysis of the dynamics of the radiation and the electron beam in a storage ring oscillator by making a number of simplifying assumptions. First, we assume that a one-dimensional, low-gain theory, as developed in Sect. 9.5. This imposes the requirement that the size of the electron beam must be less than the minimum spot size of the radiation and that the interaction length must be shorter than the Rayleigh length. Further, we assume that the duration of the micropulses in the storage ring is much longer than the slippage time and neglect the details of the pulsed nature of the electron beam. Recall from (9.203) the start oscillation current for a pulsed beam device is the same as that in a continuous beam device in this limit at the optimum cavity detuning. Thus, the theory will parallel that of a continuous beam oscillator as developed in Sect. 9.6, except that we now must account for the recirculation of the beam electrons through the interaction region.

The basic equations governing the dynamics of the beam particles are the Vlasov Eq. (9.86) along with boundary condition (9.87), which must now be modified to describe the recirculation of the electron beam. The distribution function for injected electrons will vary in time as it is affected by the radiation. This variation will be ascribed to the slow time variable n, which measures the number of round trips of the radiation through the cavity. We shall assume (1) that the radiation field in the interaction region is weak in the sense that electrons will suffer only small changes in energy on each pass through the device, (2) that the beam electrons return to the entrance of the interaction region after a time T_{sr} which depends weakly on the energy of the electrons, and (3) that during this time the energy of each electron is constant. Actually, electrons are reaccelerated and lose energy to synchrotron radiation during their trip around the ring, but this effect can be easily added later in the analysis. The normalized energies and phases of electrons entering the interaction region at round-trip index n can now be expressed in terms of the same quantities for electrons leaving the interaction region at a round trip index N_{sr} earlier, where $N_{sr} = T_{sr}/T$. In particular, the energies are unchanged, and the phases have advanced by an amount $\Delta\psi$ where

$$\Delta\psi = \omega T_{sr}(\gamma) \cong \frac{2\pi L_{sr}(\gamma)}{\lambda}, \tag{9.230}$$

is the change in phase accumulated in the time T_{sr} it takes for electrons to return to the interaction region. The second equality in (9.230) expresses the phase change in terms of the path length L_{sr} for the returning electrons assuming they travel at the speed of light and λ is the wavelength of the radiation.

Equation (9.230) allows for the possibility that returning electrons maintain phase coherence with the radiation. If the variations in the phase change from electron to electron are small compared with unity, then the entering beam is effectively *prebunched* by its previous encounters with the radiation. This could lead, in principle, to large enhancements in gain [62]. The requirements for phase coherence are quite stringent in that variations of the return time with energy are large enough that $\Delta\psi$ varies considerably over the range of energies of interest [63]. In this case it is appropriate to assume that phase coherence is lost by the returning electrons and the distribution function for electrons entering the interaction region can be taken to be independent of entrance phase with an energy dependence given by the phase average of the distribution function for electrons leaving the interaction region on the previous circuit of the ring. Further, consistent with our neglect of the pulsed nature of the electron beam, we assume that the entering distribution function is independent of entrance time during the time interval T_p (i.e., the pulse duration) and equal to the time average of the distribution function for electrons leaving the interaction region.

An evolution equation for the phase and entrance time-averaged distribution function entering the interaction region at round-trip index n is then obtained by integrating the Vlasov Eq. (9.86) over all phases and entrance times within a micropulse, and over axial distance through the interaction region:

$$\begin{aligned}\overline{F}(p, n + N_{sr}) = \overline{F}(p, n) + \frac{1}{2}\frac{\partial}{\partial p}\int_0^1 d\xi \int_0^{2\pi} \frac{d\psi}{2\pi} \int_0^{T_p} \frac{dt_e}{T_p} \\ \times [iX(n, t_e + \varepsilon\xi T)\exp(i\psi) + c.c.]F(p, \psi, \xi, t_e),\end{aligned} \tag{9.231}$$

To solve (9.231) we must still supply on the right-hand side the detailed, phase-dependent distribution function for electrons in the interaction region. To do this, we make what is known as the quasilinear approximation [55, 56], which assumes that the distribution function is perturbed only slightly by the radiation on each pass of the electrons through the interaction region. This is valid if the amplitude of the radiation is small enough that the usual phase trapping effects that cause nonlinear saturation are not operative. Instead, the energies of electrons, as measured by the dimensionless momentum p, change only by small amounts on each pass through the interaction region. After many passes, a large change in the distribution function can occur, and the nature of this change is diffusive. On each pass through the interaction region, an electron receives a small kick in normalized momentum. Kicks on subsequent passes are uncorrelated due to the loss of phase coherency by the electrons during their trip around the ring. Thus, the normalized momentum of each electron executes a random walk leading to a diffusion equation. The quasilinear diffusion equation is obtained by calculating the first-order correction to the distribution function, as one would do in linear theory, and inserting it in (9.231). Specifically, we represent the radiation field during the time interval T_p in terms of a Fourier series (as in Eq. (9.84)) except that the frequency separation defined in (9.85) will now be based on the pulse duration time T_p instead of the round-trip travel time of the radiation T. The distribution function is then written as the sum of the injected distribution and a small perturbation proportional to the radiation field as in (9.91). The perturbation satisfies the linearized Vlasov Eq. (9.92) and can be written as an integral as in (9.93). Substitution of the perturbed distribution function into (9.231) and performing the averages over phase and entrance time then yields the quasilinear diffusion equation:

$$\overline{F}(p,n+N_{sr})=\overline{F}(p,n)+\frac{\partial}{\partial p}D_{QL}(p)\frac{\partial}{\partial p}\overline{F}(p,n), \tag{9.232}$$

where the diffusion coefficient D_{QL} obtained after a number of integrations is second order in the radiation amplitude

$$D_{QL}(p)=\frac{\pi}{2}\sum_{m=-\infty}^{\infty}|X_m(n)|^2\,\mathrm{Re}[\Delta(p-\varepsilon\Delta\omega_m T)]. \tag{9.233}$$

The resonance function $\Delta(p)$ appearing in Eq. (9.233) is defined in (9.96), and it appears in the expressions for the gain of the radiation (9.95) as well. It is referred to as a resonance function since it is positive, peaked at zero argument, and has unit area. Equation (9.233) states that the contribution of each component of the radiation spectrum to the diffusion coefficient is peaked at a normalized energy corresponding to the resonance with the beat wave for that frequency. The width of the resonance, as measured in normalized energy, is unity corresponding to the range of energies for which the Doppler shifted beat wave frequency is of the order of the inverse of the transit time through the interaction region. If the dependence of the spectral amplitudes on the mode index m is weak or, equivalently, if the spectrum is broader than the gain bandwidth, then the sum in (9.233) can be replaced by an integral and the resonance function treated as a delta function.

Madey's second theorem [54] can be derived by comparing the quasilinear Eq. (9.232) with the Fokker-Planck equation [30] for the distribution function so that

$$\overline{F}(p, n+N_{sr}) = \overline{F}(p,n) + \frac{\partial}{\partial p}\left[\frac{\partial}{\partial p}\left(\frac{\Delta p^2}{2}\overline{F}(p,n)\right) - \Delta p \overline{F}(p,n)\right], \qquad (9.234)$$

where Δp and Δp^2 are the average rate of drift and diffusion of the normalized energy on each pass through the interaction region. In order that the Fokker-Planck equation reduce to the quasilinear equation, it is required that

$$\Delta \dot{p} = \frac{1}{2}\frac{\partial}{\partial p}\Delta p^2, \qquad (9.235)$$

Expressing the normalized energy in terms of the initial and final relativistic factors then gives (9.227). The above relationship is a consequence of the underlying Hamiltonian dynamics as has been discussed by Manheimer and Dupree [64] and by Krinsky et al. [65].

Madey's theorem, as derived from the quasilinear equation, does not depend on the detailed form of the diffusion coefficient, but only on the order of the diffusion coefficient and the derivatives with respect to normalized energy appearing in the quasilinear equation. In particular, the diffusion coefficient is placed between the two p derivatives, not outside. That this must be the case that follows from conservation of particles and phase space area. Specifically, in order that the total number of electrons is conserved, the change in the distribution function must be expressed as the derivative of a flux. This fixes the first derivative on the left of the diffusion coefficient. In addition, the flow in phase space implied by the Vlasov equation is incompressible. Thus, if the injected distribution function were independent of normalized energy in a particular range of energies, then there could be no resulting change in the distribution function in this range of energies. This fixes the second derivative with respect to p in that it must act directly on the distribution function. Finally, there can be only two derivatives with respect to p since the quasilinear equation is second order in the radiation field strength and the derivative with respect to normalized energy in the original Vlasov equation is multiplied by the radiation field strength.

The evolution of the distribution function due to the interaction of the electrons and the radiation is described as quasilinear flattening. Specifically, the diffusion in energy of the electrons causes the distribution function to become independent of energy in the range of normalized energies which are resonant with the radiation. This leads to a progressive reduction in the gain given by (9.95) since the gain is proportional to the average of the slope of the distribution function weighted by the resonance function. Suppose that only a single spectral component is present. The distribution function satisfying the quasilinear equation will then tend monotonically toward a constant for values of normalized momentum p between the first two zeros of the resonance function $p^{-} \varepsilon \Delta\omega_m T = \pm 2\pi$. Thus, the beam acquires a spread in energies corresponding to $\delta\gamma \approx \gamma/2N_w$ (where N_w denotes the number of wiggler periods in the interaction region). Further, in the absence of other effects, specifically, reacceleration and synchrotron cooling, the gain falls below the level required to overcome losses, and the radiation decays away.

The basic time scale over which quasilinear flattening occurs is governed by the rate of buildup of the radiation field; hence, the flattening will occur over a time which scales as the inverse of the initial exponentiation rate of the radiation. Typically, this rate is much greater than the rate at which synchrotron radiation and acceleration combine to restore the beam to its initial state (9.228). Thus, on the initial flattening times scale, the effects of acceleration and synchrotron radiation can be neglected. Ultimately, the distribution function is determined by the balancing of the diffusive effects of the radiation and the damping effect of the synchrotron radiation and acceleration. In addition, the balance of these effects determines the average rate at which power can be extracted from the beam by the free electron laser interaction.

An order of magnitude estimate of the average radiation power extracted from the beam going to the free-electron laser interaction can be obtained as follows. In equilibrium, the average rate of increase of the normalized energy spread due to the free-electron laser interaction is balanced by synchrotron cooling which occurs at a rate given by (9.228). Thus, for a typical electron we have

$$\nu_s T_{sr} p^2 \cong \left|\Delta\dot{p}^2\right| \cong \left|\Delta\dot{p}\right|, \tag{9.236}$$

where the second approximation follows from Madey's theorem along with the knowledge that the gain is saturated when the distribution function acquires a spread in p values of order unity. Finally, since the spread in p values is of order unity, we have $|\Delta p| = \nu_s T_{sr}$, where $|\Delta p|$ is the typical energy lost to the free electron laser interaction per trip around the ring. Restoring units to the normalized energy, we arrive at Renieri limit (9.229).

The above calculation applies in steady state. However, a weakly damped relaxation oscillation cycle is often excited due to the large disparity in time scales between growth of radiation and the damping of energy deviations by synchrotron radiation [53, 66, 67]. This cycle is characterized by bursts of radiation that increase the spread in energies of the electron beam beyond that which is necessary to suppress growth of the radiation. The radiation then decays to a low level, while the beam cools due to synchrotron radiation. Beam cooling continues for a time even after the gain of the radiation becomes positive since modes do not grow to finite amplitude immediately. As a result, radiation then grows again to a large amplitude, and the bursting process is repeated. The power level during one of the bursts can be many times the average, thus temporarily exceeding the Renieri limit. Further, it has been demonstrated that the timing of the bursts has been controlled by periodically changing the overlap of the radiation and beam pulses [53].

9.10 Optical Klystrons

In the low-gain regime, the one-dimensional theory predicts a single pass gain for the radiation that scales as the third power of the length of the interaction region. In particular, in the small signal limit the growth of the radiation with each pass through the interaction is given by (9.94). The gain is thus determined by the product I_b and G_m, where I_b is a normalized current given by Eq. (9.55) and G_m is a normalized gain factor which accounts for the energy distribution of the incoming beam. The dependence on interaction length enters through the definition of the normalized current. Physically, the third power of the length enters due to the fact that there are three spatial derivatives in the governing equations: one in the wave equation and two in the equations of motion. Thus, if one doubles the length, keeping the radiation amplitude fixed, the first-order change in the electron energy will also double since the radiation electric field acts to accelerate electrons over a longer distance. This causes the degree of phase bunching to quadruple since both the variation of particle energy and the length of the interaction region have doubled. Finally, the amplification of the radiation will increase by a factor of eight due to the fourfold increase in the bunched current, and the factor of two increase in the length over which amplification occurs.

In a short wavelength device, the third power of the length is reduced to only the second power by multidimensional effects. Recall from Eq. (9.226) that when the finite transverse sizes of the beam and radiation envelopes are included in the analysis, the effective current density is the total current divided by the minimum spot size of the radiation. The minimum spot size is constrained by the requirement that the Rayleigh length be of the order of or less than the interaction length $L/k\sigma_0$. This effectively reduces by one power the length dependence of the gain. It is still advantageous to design a wiggler, which is long to increase gain if the energy spread of the available beam permits. However, there may be physical limitations in the case of a storage ring simply due to the available space between bending magnets [53].

A method of increasing the gain without increasing the overall length of the interaction region is to use the optical klystron configuration [60, 61, 68–70]. The optical klystron works by effectively increasing the bunching length for the electrons. To see this, we note that the normalized current is proportional to the factor:

$$\frac{\omega_\gamma L}{v_\parallel} = -\frac{d}{d\gamma}\left(\frac{\omega L}{v_\parallel}\right), \tag{9.237}$$

where ω_γ is defined in Eq. (9.15). Hence, the degree of bunching (and, hence, the gain) is proportional to the derivative with respect to energy of the phase change of an electron in traversing the interaction region. For electrons traveling along the same path at nearly the speed of light, this derivative is small. However, it can be enhanced greatly if high-energy electrons travel on a shorter path than low energy electrons. This is accomplished by constructing an interaction region consisting of two sections of undulator separated by a drift region with a dispersive magnetic field. This is the optical klystron. The dispersive magnetic field imparts an impulse to the electron in the direction transverse to the axis of the beam and causes the path of the electron to veer off axis and then to return. Due to the relativistic energy dependence of the electron mass, the path followed by a high-energy electron will be shorter than that followed by a low energy electron. For small deviations from the axis, the energy dependence of the time of flight through the drift region can be expressed as [53, 70]

$$T_d(\gamma) = \frac{L_d}{\upsilon_\parallel} + \frac{1}{2}\int_0^{L_d} \frac{dz}{\upsilon_\parallel}\left[\int_0^z \frac{dz'}{\upsilon_\parallel}\frac{eB(z')}{\gamma m_e c}\right]^2, \tag{9.238}$$

where L_d is the length of the drift region and $B(z)$ is the magnitude of the dispersive magnetic field. Differentiating this expression with respect to energy and assuming highly relativistic motion one finds that the effective bunching length of the drift region is now given by

$$L_{eff} = L_d + \int_0^{L_d} dz\left[\int_0^z dz'\frac{eB(z')}{\gamma m_e c}\right]^2, \tag{9.239}$$

The operating characteristics of the optical klystron which can be understood at a basic level by studying the klystron model of Sect. 9.3 that provided a number of parameters are replaced by values which are appropriate to the use of a dispersive drift section. In particular, the normalized momentum p_i (9.48) is now defined in terms of the energy dependence of the phase change in transiting the dispersive section:

$$p_i = \frac{L_{eff}\omega_\gamma}{\upsilon_\parallel}\delta\gamma_i, \tag{9.240}$$

The normalized radiation amplitude $X(t_e)$ (see Eq. (9.49)) measures the strength of the kick in energy that an electron receives in the bunching section. Thus, it is normalized by the strength (a_{w1}) and length (L_1) of the wiggler in the buncher, the spot size in the buncher (σ_0/σ_1), and the energy dependence of the phase change through the drift region:

$$X(t_e) = \frac{\sigma_0 a_{w1} L_1 L_{eff}\omega\omega_\gamma}{\sigma_1\gamma_r\upsilon_\parallel}\delta a_e, \tag{9.241}$$

The change in amplitude of the radiation as expressed in (9.50) is proportional to the strength (a_{w2}) and length (L_2) of the wiggler as well as the spot size of the radiation in the output section (σ_0/σ_2). This leads again to the klystron Eq. (9.54) except that the normalized current is now given by

$$I(t) = \frac{2e\sigma_0 a_{w1} a_{w2} L_1 L_2 L_{eff} \omega_\gamma}{\sigma_1 \sigma_2^* \gamma_r^2 m_e c v_\parallel^3} I_T(t), \tag{9.242}$$

Finally, the slippage parameter is defined such that εT is the difference in time for the propagation of radiation and electrons through the interaction region:

$$\varepsilon = \frac{L}{L_c}\left[\frac{T_d(\gamma)c}{L} - 1\right], \tag{9.243}$$

where $T_d(\gamma_r)$ is the mean travel time for electrons through the drift section.

The small signal gain for an arbitrary distribution of injected electrons is obtained from (9.53) and (9.54). For a monoenergetic beam interacting with a mode with frequency shifted from exact resonance by an amount $\Delta\omega$, the gain has a purely sinusoidal dependence on frequency shift:

$$\left|\frac{X(t+T)}{X(t)}\right| = \left|1 + \frac{I_b}{2}\exp\left(i\frac{\pi}{2} + i\varepsilon\Delta\omega T\right)\right|, \tag{9.244}$$

The klystron model has vanishingly small buncher and output sections, and this leads artificially to the appearance of gain over in infinite range of frequencies. If the nonzero lengths of the buncher and output sections are accounted for, then the frequency dependence of the gain is modulated by an envelope function with width scaling as the gain bandwidth of the individual wiggler sections [70].

The effect of energy spread on gain in the optical klystron is particularly easy to evaluate since (9.53) reveals that the gain is proportional to the Fourier transform of the distribution function. Thus, for a Gaussian spread in energies with a width $\Delta\gamma$

$$f(\delta\gamma_i) - \frac{1}{2\pi^{3/2}\Delta\gamma}\exp\left[-\left(\frac{\delta\gamma_i}{\Delta\gamma}\right)^2\right], \tag{9.245}$$

the gain function g is reduced by a factor h from its value for a monoenergetic beam, where

$$h = \exp\left[-\left(\frac{L_{eff}\omega_\gamma\Delta\gamma}{v_\parallel}\right)^2\right]. \tag{9.246}$$

The requirement that this factor not be too small sets an upper limit on the effective length of the dispersive section for a given beam quality. In addition, saturation of the gain occurs when the energy spread of the electron beam acquires a width $\Delta\gamma$ such that the reduced gain implied by (9.246) balance losses. In conclusion, the use of the optical klystron is an effective way to increase gain in cases which physical space limitations as opposed to beam quality limit the usable length of a conventional undulator.

References

1. W.B. Colson, S.K. Ride, The free-electron laser, Maxwell's equations driven by single particle currents, in *Physics of Quantum Electronics: Free-Electron Generators of Coherent Radiation*, ed. by S.F. Jacobs, H.S. Pilloff, M. Sargent, M.O. Scully, R. Spitzer, vol. 7, (Addison-Wesley, Reading, Massachusetts, 1980), p. 377
2. Y.L. Bogomolov, V.L. Bratman, N.S. Ginzburg, M.I. Petelin, A.D. Yunakovsky, Nonstationary generation in free-electron lasers. Opt.Commun. **36**, 209 (1981)
3. N.M. Kroll, P.L. Morton, M.N. Rosenbluth, Free-electron lasers with variable parameter wigglers. IEEE J. Quantum Electron. **QE-17 1436** (1981)
4. N.S. Ginzburg, M.I. Petelin, Multifrequency generation in free-electron lasers with quasi-optical resonantors. Int. J. Electronics **59**, 291 (1985)
5. H. Al-Abawi, F.A. Hopf, G.T. Moore, M.O. Scully, Coherent transients in the free-electron laser: Laser lethargy and coherence brightening. Optics Commun. **30**, 235 (1979)
6. S.S. Yu, W.M. Sharp, W.M. Fawley, E.T. Scharlemann, A.M. Sessler, E.J. Sternbach, Wave-guide suppression of the free-electron laser sideband instability. Nucl. Instrum. Methods Phys. Res. **A259**, 219 (1987)
7. W.B. Colson, J. Blau, Parameterizing physical effects in free-electron lasers. Nucl. Instrum. Methods Phys. Res. **A272**, 386 (1988)
8. B. Levush, T.M. Antonsen Jr., Regions of stability of free-electron laser oscillators. Nucl. Instrum. Methods Phys. Res. **A272**, 375 (1988)
9. W.B. Colson, Classical free-electron laser theory, in *The Laser Handbook: Free-Electron Lasers*, ed. by W.B. Colson, C. Pellegrini, A. Renieri, vol. 6, (North Holland, Amsterdam, 1990), p. 115
10. J.R. Pierce, *Traveling Wave Tubes* (Van Nostrand, New York, New York, 1950)
11. R.E. Collin, *Foundations for Microwave Engineering* (McGraw-Hill, New York, 1966)
12. F.A. Hopf, P. Meystre, G.T Moore, and M.O. Scully, Nonlinear theory of free-electron devices, in Physics of Quantum Electronics: Novel Sources of Coherent Radiation, eds. S.F. Jacobs, M. Sargent, and M.O. Scully (Addison-Wesley, Reading, Massachusetts, 1978) Vol.5, p. 41
13. W.B. Colson, R.A. Freedman, Synchrotron instability for long pulses in free-electron lasers. Opt. Commun. **46**, 37 (1983)
14. W.B. Colson, The trapped particle instability in free-electron laser oscillators and amplifiers. Nucl. Instrum. Methods Phys. Res. **A250**, 168 (1986)
15. T.M. Antonsen Jr., B. Levush, Mode competition and suppression in free-electron laser oscillators. Phys. Fluids B **1**, 1097 (1989)
16. J. Masud, T.C. Marshall, S.P. Schlesinger, F.G. Yee, W.M. Fawley, E.T. Scharlemann, S.S. Yu, A.M. Sessler, E.J. Sternbach, Sideband control in a millimeter-wave free-electron laser. Phys. Rev. Lett. **58**, 763 (1987)
17. R.W. Warren, J.C. Goldstein, B.E. Newnam, Spiking mode operation for a uniform-period wiggler. Nucl. Instr. Meth. **A250**, 19 (1986)
18. R.W. Warren, J.E. Sollid, D.W. Feldman, W.E. Stein, W.J. Johnson, A.H. Lumpkin, J.C. Goldstein, Near-ideal lasing with a uniform wiggler. Nucl. Instrum. Methods Phys. Res. **A285**, 1 (1989)
19. G.S. Nusinovich, The mode interaction in free-electron lasers. Sov. Phys. Tech. Phys. **6**, 848 (1980)
20. E.R. Stanford, T.M. Antonsen Jr., The effect of dispersion on mode competition in free-electron laser oscillators. Nucl. Instr. Meth. **A304**, 659 (1991)
21. T.M. Antonsen Jr., B. Levush, Mode competition and control in free-electron laser oscillators. Phys. Rev. Lett. **62**, 1488 (1989)
22. N.S. Ginzburg, S.P. Kuznetsov, T.M. Fedoseeva, Theory of transients in backward wave tubes. Radiofiz. **21**, 1037 (1978)
23. P. Sprangle, C.M. Tang, I.B. Bernstein, Initiation of a pulsed beam free-electron laser. Phys. Rev. Lett. **50**, 1775 (1983)

24. K.J. Kim, An analysis of self amplified spontaneous emission. Nucl. Instrum. Methods Phys. Res. **A250**, 396 (1986)
25. W. Becker, J. Gea-Banacloche, M.O. Scully, Intrinsic linewidth of a free-electron laser. Phys. Rev. A **33**, 2174 (1986)
26. A. Friedman, A. Gover, G. Kurizki, S. Ruschin, A. Yariv, Spontaneous and stimulated emission from quasi-free electrons. Rev. Mod. Phys. **60**, 471 (1988)
27. R.W. Warren, J.C. Goldstein, The generation and suppression of synchrotron sidebands. Nucl. Instrum. Methods Phys. Res. **A272**, 155 (1988)
28. L.R. Elias, R.J. Hu, G.J. Ramian, The UCSB electrostatic accelerator free-electron laser: First operation. Nucl. Instrum. Methods Phys. Res. **A237**, 203 (1985)
29. T.M. Antonsen Jr., B. Levush, Spectral characteristics of a free-electron laser with time-dependent beam energy. Phys Fluids B **2**, 2791 (1990)
30. R.L. Liboff, *Introduction to the Theory of Kinetic Equations* (Wiley, New York, 1969)
31. A. Amir, R.J. Hu, F. Kielmann, J. Mertz, L.R. Elias, Injection locking experiment at the UCSB free-electron laser. Nucl. Instrum. Methods Phys. Res. **A272**, 174 (1988)
32. L.R. Elias, G.J. Ramian, J. Hu, A. Amir, Observation of single mode operation in a free-electron laser. Phys. Rev. Lett. **57**, 424 (1986)
33. B.G. Danly, S.G. Evangelides, T.S. Chu, R.J. Tempkin, G.J. Ramian, J. Hu, Direct spectral measurements of a quasi-cw free-electron laser. Phys. Rev. Lett. **65**, 2251 (1990)
34. B. Levush, T.M. Antonsen Jr., Nonlinear mode competition and coherence in low gain free-electron laser oscillators. Nucl. Instrum. Methods Phys. Res. **A285**, 136 (1989)
35. I. Kimmel, L.R. Elias, Long-pulse free-electron lasers as sources of monochromatic radiation. Nucl. Instrum. Methods Phys. Res. **A272**, 368 (1988)
36. V.N. Litvenenko, N.A. Vinokurov, Lasing spectrum and temporal structure in storage ring free-electron lasers: Theory and experiment. Nucl. Instrum. Methods Phys. Res. **A304**, 66 (1991)
37. G. Dattoli, A. Renieri, Classical multimode theory of the free-electron laser. Lett. Nuovo Cimento **59B**, 1 (1979)
38. W.B. Colson, Optical pulse evolution in the Stanford free-electron laser and in a tapered undulator, in *Physics of Quantum Electronics:Free-Electron Generators of Coherent Radiation*, ed. by S.F. Jacobs, G.T. Moore, H.S. Pilloff, M. Sargent, M.O. Scully, R. Spitzer, vol. 8, (Addison-Wesley, Reading, Massachusetts, 1982), p. 457
39. G. Dattoli, T. Hermsen, A. Renieri, A. Torre, J.C. Gallardo, Lethargy of laser oscillations and supermodes in free-electron lasers: I. Phys. Rev. A **37**, 4326 (1988)
40. G. Dattoli, T. Hermsen, L. Mezi, A. Renieri, A. Torre, Lethargy of laser oscillations and supermodes in free-electron lasers: II-quantitative analysis. Phys. Rev. A **37**, 4334 (1988)
41. J.C. Goldstein, B.E. Newnam, R.W. Warren, R.L. Sheffield, Comparison of the results of theoretical calculations with experimental measurements from the Los Alamos free-electron laser oscillator experiment. Nucl. Instrum. Methods Phys. Res. **A250**, 4 (1986)
42. P. Elleaume, Storage ring free-electron laser theory. Nucl. Instrum. Methods Phys. Res. **A237**, 28 (1985)
43. N.M. Kroll, Excitation of hypersonic vibrations by means of photoelastic coupling of high intensity light waves to elastic waves. J. Appl. Phys. **36**, 34 (1965)
44. D. Pesme, G. Laval, R. Pellat, Parametric instabilities in bounded plasmas. Phys. Rev. Lett. **31**, 203 (1973)
45. A. Renieri, Storage ring operation of a free-electron laser: The amplifier. Nuovo Cimento **53B**, 160 (1979)
46. G. Dattoli, A. Renieri, Storage ring operation of a free-electron laser: The oscillator. Nuovo Cimento **59B**, 1 (1980)
47. J.C. Goldstein, Evolution of long pulses in a tapered wiggler free-electron laser, in *Free-Electron Generators of Coherent Radiation*, ed. by C.A. Brau, S.F. Jacobs, M.O. Scully, (Proc. SPIE 453, Bellingham, Washington, 1984), p. 2
48. W.B. Colson, J.L. Richardson, Multimode theory of free-electron laser oscillators. Phys. Rev. Lett. **50**, 1050 (1983)
49. J.C. Goldstein, B.D. McVey, B.E. Carlsten, L.E. Thode, Integrated numerical modeling of free-electron laser oscillators. Nucl. Instrum. Methods Phys. Res. **A285**, 192 (1989)

50. J.C. Goldstein, B.D. McVey, R.L. Tokar, C.J. Elliot, M.J. Schmidt, B.E. Carlsten, L.E. Thode, Simulation codes for modeling free-electron laser oscillators, in *Modeling and Simulation of Laser Systems*, ed. by D.L. Bullock, vol. 1042, (Proc. SPIE, Bellingham, Washington, 1989), p. 28
51. S. Riyopoulos, P. Sprangle, C.M. Tang, A. Ting, Reflection matrix for optical resonators in free-electron laser oscillators. Nucl. Instrum. Methods Phys. Res. **A272**, 543 (1988)
52. D. Iracane, J.L. Ferrer, An optimal basis equation for solving the time-dependent Schrödinger equation: Simulation of guiding and multifrequency mechanisms. Nucl. Instr. Meth. **A296**, 417 (1990)
53. D.A.G. Deacon, J.M. Ortega, The storage ring free-electron laser, in *The Laser Handbook: Free-Electron Lasers*, ed. by W.B. Colson, C. Pellegrini, A. Renieri, vol. 6, (North Holland, Amsterdam, 1990), p. 345
54. J.M.J. Madey, Relationship between mean radiated energy, mean squared radiated energy, and spontaneous power spectrum in a power series expansion of the equation of motion in a free-electron laser. Nuovo Cimento **50B**, 64 (1979)
55. A.A. Vedenov, E.P. Velikov, R.Z. Sagdeev, Nonlinear oscillations of rarified plasma. Nucl. Fusion **1**, 82 (1961)
56. W.E. Drummond, D. Pines, Nonlinear stability of plasma oscillations. Nucl. Fusion Suppl. **3**, 1049 (1962)
57. T. Taguchi, K. Mima, T. Mochizuki, Saturation mechanism and improvement of conversion efficiency of the free-electron laser. Phys. Rev. Lett. **46**, 824 (1981)
58. N.S. Ginzburg, M.A. Shapiro, Quasilinear theory of multimode free-electron lasers with an inhomogeneous frequency broadening. Opt. Commun. **40**, 215 (1982)
59. D.A. Edwards, M.J. Syphers, An introduction to the physics of particle accelerators, in *Physics of Particle Accelerators*, ed. by M. Month, M. Dienes, vol. 1, (American Institute of Physics Conference Proceedings #184, New York, 1989), p. 2
60. N.A. Vinokurov, A.N. Skrinsky, Optical range klystron oscillator using ultrarelativistic electrons, Preprint 77–59 of the Institute of Nuclear Physics, Novosibirsk, 1977
61. N.A. Vinokurov, A.N. Skrinsky, On ultimate power of the optical klystron installed on electron storage ring, Preprint 77–67 of the Institute of Nuclear Physics, Novosibirsk, 1977
62. D.A.G. Deacon, J.M.J. Madey, Isochronous storage ring laser: A possible solution to the electron heating problem in recirculating free-electron lasers. Phys. Rev. Lett. **44**, 449 (1980)
63. A. Van Steenbergen, Accelerators and storage rings for free-electron lasers, in *The Laser Handbook: Free-Electron Lasers*, ed. by W.B. Colson, C. Pellegrini, A. Renieri, vol. 6, (North Holland, Amsterdam, 1990), p. 417
64. W.M. Manheimer, T.H. Dupree, Weak turbulence theory of velocity space diffusion and nonlinear Landau damping of waves. Phys. Fluids **11**, 2709 (1968)
65. S. Krinsky, J.M. Wang, P. Luchini, Madey's gain spread theorem for the free-electron laser and the theory of stochastic processes. J. Appl. Phys. **53**, 5453 (1982)
66. M. Billardon, P. Elleaume, J.M. Ortega, C. Bazin, M. Bergher, M. Velghe, Y. Petroff, D.A.G. Deacon, K.E. Robinson, J.M.J. Madey, First operation of a storage ring free-electron laser. Phys. Rev. Lett. **51**, 1652 (1983)
67. P. Elleaume, Macrotemporal structure of free-electron lasers. J. Phys. **45**, 997 (1984)
68. P. Elleaume, Optical klystron spontaneous emission and gain, in *Physics of Quantum Electronics: Free-Electron Generators of Coherent Radiation*, ed. by S.F. Jacobs, G.T. Moore, H.S. Pilloff, M. Sargent, M.O. Scully, R. Spitzer, vol. 8, (Addison-Wesley, Reading, Massachusetts, 1982), p. 119
69. C.C. Shih, M.Z. Caponi, An optimized multicomponent wiggler design for a free-electron laser. IEEE J. Quantum Electron. **QE-19, 369** (1983)
70. P. Elleaume, Free-electron laser undulators, electron trajectories and spontaneous emission, in *The Laser Handbook: Free-Electron Lasers*, ed. by W.B. Colson, C. Pellegrini, A. Renieri, vol. 6, (North Holland, Amsterdam, 1990), p. 91

Chapter 10
Oscillator Simulation

Free-electron laser oscillators have demonstrated high average power operation at infrared wavelengths using energy recovery linacs [1, 2] and using racetrack microtrons at terahertz wavelengths [3]. For example, the 10-kW Upgrade experiment at the Thomas Jefferson National Accelerator Facility (henceforth referred to as JLab) [1] showed a saturated power level of 14.3 kW at a wavelength of 1.6 μm using a concentric resonator with transmissive output coupling. As a result, free-electron laser configurations using energy recovery linacs have the potential for achieving sustained high-power operation.

The interaction in an oscillator is governed by a balance between amplification of the optical field in the gain medium and losses due to out-coupling, mirror heating and distortion, resistive wall effects, etc. A steady-state is achieved when the nonlinear, saturated gain, G, is balanced by the total losses, L. If the power at the wiggler entrance after the nth pass is denoted by P_n, then the power after the $(n + 1)^{\text{th}}$ pass is given by $P_{n+1} = (1 - L)(1 + G)P_n$. Once the oscillator reaches equilibrium, $P_{n+1} = P_n$; hence,

$$G = \frac{L}{1 - L}. \tag{10.1}$$

In a free-electron laser, the gain tends to increase with the length of the wiggler. The low gain regime refers to configurations where the wiggler is not long enough to reach the exponential gain regime. Typically, in the low gain regime $G \propto N_w^3$ [Eq. 1.9], where N_w is the number of periods in the wiggler. However, the saturation efficiency is estimated as $\eta \approx (2.4N_w)^{-1}$ [Eq. 1.14]. Hence, the performance of a low gain free-electron laser oscillator must strike a balance between maximizing both the gain and the efficiency. Often, this is determined by the properties of the mirrors. For example, if we neglect other losses and if the mirror out-coupling is 20%, then the saturated gain must be 25%. This low gain regime implies that most of the power is recirculating through the resonator; hence, this type of oscillator has low gain but high Q. The alternative is a high-gain/low-Q oscillator design that makes use of a

H. P. Freund, T. M. Antonsen, Jr., *Principles of Free Electron Lasers*,
https://doi.org/10.1007/978-3-031-40945-5_10

long wiggler to achieve the high gain. For example, if the losses are as high as 95%, then the required gain would be 1900%, and this would necessitate a long wiggler with a correspondingly long cavity length. Of course, the saturation efficiency for such a configuration is $\eta \approx \rho$, which is lower than that typically achieved in low-gain/high-Q oscillators; however, this may be the only choice for an oscillator at wavelengths where there are no high reflectivity mirrors. This is important, in particular, at x-ray wavelengths [4–6]. This configuration is often referred to as a *regenerative amplifier*, or RAFEL. Both low-gain/high-Q and high-gain/low-Q configurations will be discussed in this chapter.

10.1 The General Simulation Procedure

Simulations of free-electron laser oscillators have appeared in the literature using a variety of approaches [7–11]. The approach used in this chapter has been previously discussed in Refs. [9, 10]. There are two basic elements of the simulation procedure discussed. The first is the gain medium represented by the propagation of the electrons through a wiggler. The second is the propagation of the electromagnetic field through some kind of optical resonator. This chapter is concerned with short wavelength free-electron laser oscillators driven by a sequence of short electron bunches that might be produced in RF linacs or storage rings. The general procedure that is employed makes use of the Gaussian optical mode formulation discussed in Chap. 6 to deal with the interaction of the electrons and the optical field.

A schematic illustration of the simulation procedure is shown in Fig. 10.1 for a generic concentric resonator although arbitrary resonator configurations may be used. In order to simulate an oscillator, the interaction between the electrons and the optical field is first simulated in the wiggler. This is done here using the nonlinear formulation described in Chap. 6. The optical field at the exit from the wiggler is then mapped onto a grid, which is then imported into the OPC for transport through the resonator and back to the wiggler entrance. At that point, the field on the grid is decomposed into the appropriate Gaussian mode representation for importation into the simulation of the interaction with the electrons in the next pass through the wiggler. This process is repeated as many times as necessary for the oscillator to attain the steady-state or beyond.

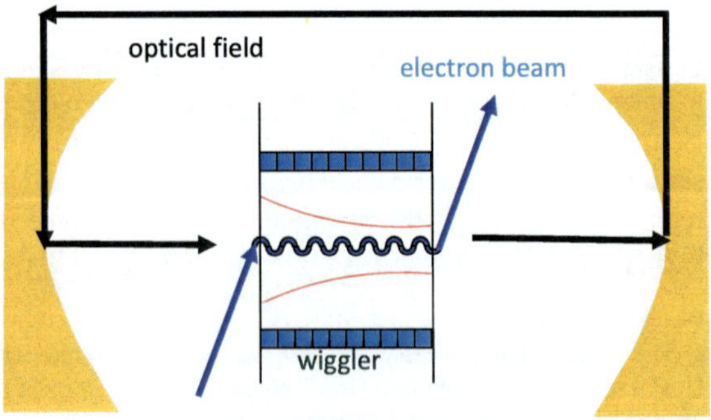

Fig. 10.1 Schematic illustration of the numerical simulation procedure

The simulations can be performed either in the steady-state or with time dependence. In the latter case, the synchronism between the entrance of individual electron bunches into the wiggler and the returning optical pulses must be considered. This synchronism depends upon the separation distance between adjacent electron bunches and the roundtrip travel time of the optical field through the resonator. Since the roundtrip time depends upon the path length through the resonator, this is often referred to as *cavity detuning*. This detuning is accomplished in the simulation by shifting the temporal *slices* of the optical field either forward or back relative to those of the electrons at the wiggler entrance.

The detailed properties of the optical resonator must be included in the transport of the optical field. These include the detailed geometrical configuration of the resonator including, but not limited to, concentric and ring resonators. Examples of a low-gain/high-Q oscillator using a concentric resonator and a high-gain/low-Q oscillator using a ring resonator are described in this chapter. Transmissive, hole, or edge out-coupling of the optical field may be described. In addition, mirror distortion by heating or the formation of color centers due to harmonic generation can also be included.

10.2 The Optics Propagation Code (OPC)

The optics propagation code (OPC) [12, 13] can be used for the simulation of oscillators or propagating an optical field beyond the end of the wiggler line to a point of interest. OPC propagates the optical field using either the Fresnel diffraction integral or the spectral method in the paraxial approximation using fast discrete Fourier transforms (FFT). A modified Fresnel diffraction integral [14, 15] is also available and allows the use of FFTs in combination with an expanding grid on which the optical field is defined. This method is often used when diffraction of the optical field is large. Propagation can be done either in the time or frequency domain. The latter allows for the inclusion of dispersion and wavelength dependent properties of optical components. Currently, OPC includes mirrors, lenses, phase and amplitude masks, and round and rectangular diaphragms. Several optical elements can be combined to form more complex optical component, for example, by combining a mirror with a hole element, extraction of radiation from a resonator through a hole in one of the mirrors can be modeled. Phase masks can be used, for example, to model mirror distortions or to create nonstandard optical components like a cylindrical lens.

In a typical resonator configuration, OPC handles the propagation from the end of the gain medium to the first optical element, applies the action of the optical element to the optical field, and propagates it to the next optical element and so on until it reaches the entrance of the gain medium. Diagnostics can be performed at the planes where the optical field is evaluated. Some optical elements, specifically diaphragms and mirrors, allow forking of the optical path. For example, the reflected beam of a partial transmitting output mirror forms the main intracavity optical path, while the transmitted beam is extracted from the resonator. When the intracavity propagation reaches the output mirror, this optical propagation can be temporarily suspended, and the extracted beam can be propagated to a diagnostic point for evaluation. Then the intracavity propagation (main path) is resumed.

10.3 Cavity Detuning

Synchronism between the incoming electron bunches and the returning optical pulses is an important consideration in designing free-electron laser oscillators employing pulsed electron beams because if the there is no synchronism, then the optical field cannot grow beyond what is generated from shot noise in a single pass of the electrons through the wiggler.

If we consider a concentric resonator, then the separation between optical pulses in the resonator is given by $\tau_{optics} = 2L_{cav}/(Mv_{gr})$, where L_{cav} is the separation distance between the mirrors, v_{gr} is the group velocity of the light, and M is the number of pulses in the cavity at any given time. For perfect synchronism, this separation time must coincide with the separation time between electron bunches, which is $\tau_{electrons} = 1/f_{rep}$, where f_{rep} is the electron bunch repetition frequency. The *zero-detuning* cavity length, L_0, is found by equating these separation times; hence,

$$L_0 = \frac{Mv_{gr}}{2f_{rep}}. \tag{10.2}$$

The peak performance of a low-gain/high-Q oscillator is typically found near *zero-detuning*. The shift in the peak performance from the *zero-detuning* cavity length is due to the slippage of the optical field relative to the electron bunch in the wiggler. The detuning range depends upon the gain in the wiggler, the amount of slippage through the wiggler, the length of the wiggler, and the length of the electron bunch and is often in the range of about 10 wavelengths.

We remark that for high-average-power free-electron laser oscillators, it can be necessary to a large cavity length in order for the optical field to expand far enough that the mirror loading is reduced below the damage threshold, and this may require a large number of pulses in the cavity (i.e., $M > > 1$).

10.4 The Stability of Concentric Resonators

All of the free-electron laser oscillator experiments to date have employed stable resonators. This means that if the optical mode departs slightly from the symmetry axis due, for example, to bunch-to-bunch misalignments of the electron beam, then the mode does not "walk" out of the cavity.

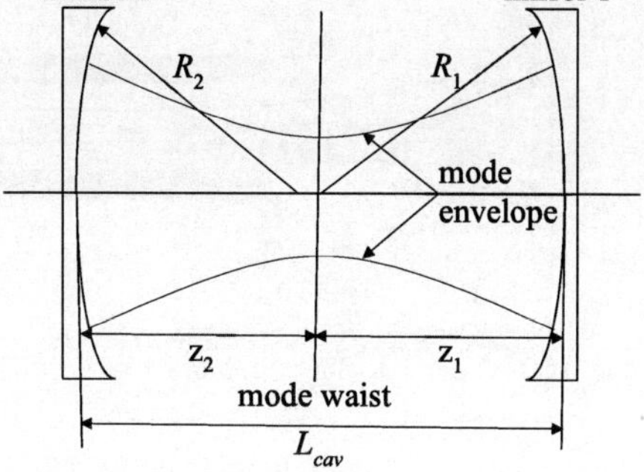

Fig. 10.2 Schematic illustration of a concentric resonator

Now consider the stability of a concentric resonator with a separation L_{cav}, as illustrated in Fig. 10.2, which shows the radii of curvature $R_{1,2}$ for the two mirrors as well as the distance from the mode waist to each mirror, $z_{1,2}$. We now define ($i = 1,2$)

$$g_i \equiv 1 - \frac{L_{cav}}{R_i}, \tag{10.3}$$

where the Rayleigh range in the cavity is given by

$$\frac{z_R^2}{L_{cav}^2} = \frac{g_1 g_2 (1 - g_1 g_2)}{(g_1 + g_2 - 2g_1 g_2)^2}. \tag{10.4}$$

Since the Rayleigh range must be positive, the condition for the stability of the concentric resonator is that $g_1 g_2 < 1$, which implies that

$$L_{cav} < R_1 + R_2. \tag{10.5}$$

Finally, we note that

$$\frac{z_1}{L_{cav}} = \frac{g_2(1 - g_1)}{g_1 + g_2 - 2g_1 g_2}, \tag{10.6}$$

and that $z_2 = L_{cav} - z_1$.

10.5 Low-Gain/High-Q Oscillators

The majority of low-gain/high-Q free-electron laser oscillators employ concentric resonators, as shown schematically in Fig. 10.3. As shown in the figure, the electron bunches are directed into the wiggler/resonator. The cavity length must be chosen so that the returning optical pulses are in near-synchronism with the electron bunches.

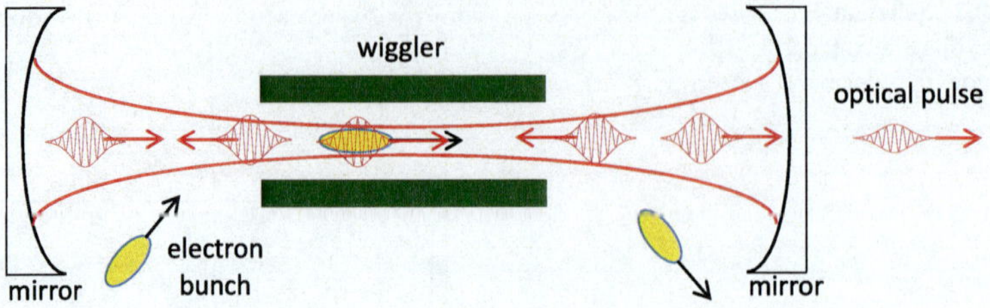

Fig. 10.3 Schematic illustration of a low-gain/high-Q oscillator using a concentric resonator

This synchronism can be accomplished in such a way that there are more than one optical pulse in the cavity at any given time. In addition, as described by Eq. 10.1, some fraction of the recirculating pulse energy is out-coupled from the resonator.

10.5.1 The Efficiency in the Low-Gain Regime

The phase trapping efficiency is determined subject to the electron beam losing the energy associated with twice the difference between the bulk axial electron velocity, υ_b and the phase velocity, υ_{ph}, of the ponderomotive wave, as given in Eq. 5.2. Using Eq. 4.65, the phase velocity of the ponderomotive wave at the resonant frequency can be expressed as

$$\upsilon_{ph} = \upsilon_b \left(1 - \frac{1}{2.4N_w}\right)\left(1 - \frac{\upsilon_b/c}{2.4N_w}\right)^{-1}, \tag{10.7}$$

where N_w is the number of periods in the wiggler; hence,

$$\upsilon_b - \upsilon_{ph} = \frac{\upsilon_b(1 - \upsilon_b/c)}{2.4N_w}\left(1 - \frac{\upsilon_b/c}{2.4N_w}\right)^{-1}. \tag{10.8}$$

As a result, the expression for the efficiency [Eq. 5.2] in the limit in which $\gamma >> 1$ and $N_w >> 1$ becomes

$$\eta \cong \frac{1}{2.4N_w}. \tag{10.9}$$

This is often expressed in the literature simply as $\eta \approx 1/2N_w$.

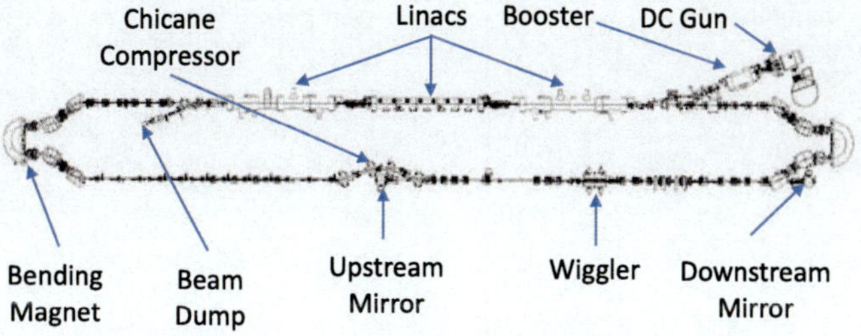

Fig. 10.4 Schematic drawing of the JLab free-electron laser oscillator (Courtesy S.V. Benson)

10.5.2 The JLab 10-kW Upgrade Experiment

The configuration of the JLab free-electron laser oscillator is that of an energy recovery linac (ERL) and is illustrated in Fig. 10.4 where the principal elements of the system have been labeled. A 9 MeV electron beam composed of bunches with 2–5 psec durations at a repetition rate of 74.85 MHz is produced by a DC photocathode gun followed by a booster linac, which is then directed into the first leg of the circuit consisting primarily of a linac composed of three superconducting cryomodules operating at a frequency of 1500 MHz that is capable of boosting the energy up to a maximum of about 160 MeV. The first and third cryomodules use an eight-cavity design with five cells per cavity, while the second cryomodule uses an eight-cavity design with seven cells per cavity. A bending magnet in a Bates bend configuration is used to take the energetic beam into the second leg of the circuit that contains the wiggler, which is preceded by a dipole chicane to compress the electron bunches. The overall bunch compression in the Bates bend [16] and the bunch compressor produces bunch with durations of about 500 fsec, after which the bunches enter the wiggler. A second Bates bend directs the spent beam from the wiggler back into the linac for same-cell energy recovery, after which the electrons are directed into the beam dump. In typical operation, all of the energy is recovered less than about 10–12 MeV, which is dissipated in the beam dump. Average currents of up to about 9 mA have been achieved in normal operation.

A concentric resonator was employed in the experiment with a cavity length of about 32 m and transmissive out-coupling of about 21% of the power through the downstream mirror. The radii and radii of curvatures of the mirrors were 3.5 cm and 16.056 m respectively for both the upstream and downstream mirrors. This was a stable concentric resonator with $g_1g_2 = 0.993$. The Rayleigh range for this resonator was approximately 0.75 m in vacuo. The experiment achieved a high average output power of 14.3 ± 0.72 kW [1], which (1) drove the requirement for the 32-m cavity length ($M = 16$) to keep mirror loading small, but (2) still necessitated the use of cooled mirrors. In this case, the upstream mirror was back-cooled, while the downstream mirror, through which the power was out-coupled, was edge-cooled.

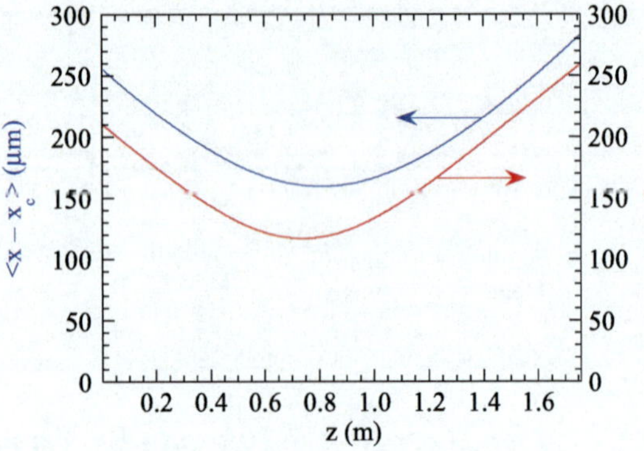

Fig. 10.5 Beam transport though the wiggler

The 10-kW upgrade experiment employed an electron beam with a kinetic energy of 115 MeV and a bunch charge of 115 pC with an rms bunch duration of 390 fsec and an rms energy spread of 0.3%. The normalized emittance was 9 mm-mrad in the wiggler-plane and 7 mm-mrad in the direction normal to the wiggle-plane. In contrast to amplifier and SASE free-electron lasers in which the electron beam is matched into a long wiggler line, the electron beam in a low-gain oscillator is matched into the optical mode, which comes to a focus near the center of the resonator. A flat-pole-face wiggler was employed, which was 30 periods in length with a period of 5.5 cm and an on-axis amplitude of 3.75 kG. As a result, the free-electron laser was tuned to a wavelength of 1.6 μm. The electron beam was matched to the optical mode whose waist, in vacuo, is located at the center of the resonator. However, the electron beam was focused to a spot slightly off-center from that of the wiggler and resonator. This was accomplished in the experiment by the settings of quadrupole and sextupole steering magnets and is determined in the simulation by choosing the proper initial Twiss parameters. For this purpose, the initial rms beam sizes were 257 μm and 212 μm in the x- and y-directions, respectively, with the initial Twiss-α parameters of 1.25 in both directions. The focusing of the electron beam in the simulation is shown in Fig. 10.5, where x_c and y_c denote the beam centroids in the x- and y-directions and the rms beam sizes for the x- and y-directions are shown on the left and right axes, respectively.

Observe that the energy contained in each electron bunch is approximately 132 mJ. Given that the average measured output power was 14.3 kW and the repetition frequency was 74.85 MHz, the optical energy per pulse that was out-coupled from the resonator was $E_{pulse} = 14.3\ \text{kW}/f_{rep} = 0.191$ mJ corresponding to an efficiency of 1.44%. This is close to the theoretical prediction of the efficiency [Eq. 10.9] of 1.39%.

Simulations were performed using six Gauss-Hermite modes with an initial mode waist in the center of the wiggler and a Rayleigh range of 0.75 m [9]. Simulation scans over cavity lengths in the vicinity of 32 m show that the optimal cavity length was found to be 32.0417850 m. A plot of the evolution of the recirculating energy in each pulse (left axis) and the average output power (right axis) versus pass in

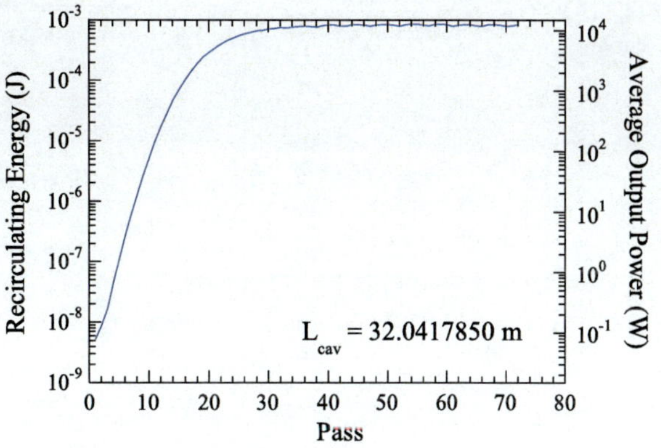

Fig. 10.6 Peak recirculating pulse energy and average output power for the optimal cavity length [9]

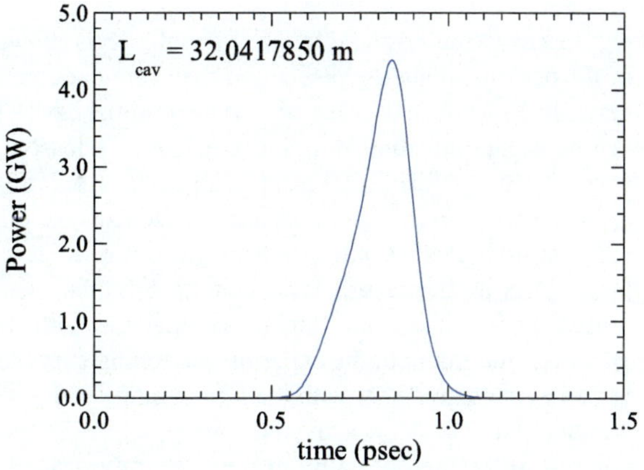

Fig. 10.7 Power in the optical pulse versus time in the steady-state [9]

Fig. 10.6. An equilibrium state is achieved after about 40 passes with a recirculating pulse energy of about 0.8 mJ. Since 21% of the pulse energy is out-coupled per pass at a repetition rate of 74.85 MHz, this translates into an average output power of 12.3 kW, so that the simulation result is only approximately 9% lower than the experimental observation corresponding to an efficiency of about 1.31%.

The optical pulse interacts with a fresh electron bunch on each pass, and the electron bunch is assumed in simulation to initially have a symmetric parabolic shape. However, due to slippage within the wiggler and different synchronism conditions between the optical pulse and the electron bunch due to the cavity tuning, the optical pulse is distorted relative to the electron bunch. This is shown in Fig. 10.7 where the power versus time of the optical pulse in the steady-state is plotted for the optimal cavity length. Note that within the context of the limits on the time axis, the electron bunch would be centered at 0.75 psec. As a result, the optical pulse has slipped ahead of the electron bunch by about 90 fsec, which is comparable to the slippage time in the wiggler. The pulse shape is also distorted and exhibits a sharp drop-off at the head of the pulse and a more gradual rise from the tail.

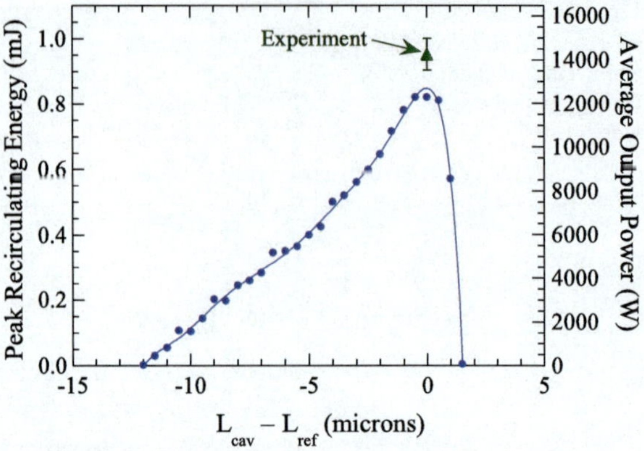

Fig. 10.8 The cavity tuning curve [9]

The cavity tuning curve found in simulation is shown in Fig. 10.8, which is a plot of the peak recirculating power and the average output power versus the difference between a specific cavity length and the optimal cavity length. The dots in the figure denote each simulation run, and the line is a least-squares fit to the points. The triangle is the experimental result with an error bar of ±5%. Only one experimental point is shown in the figure because of the extreme difficulty in obtaining a tuning curve at high average power arising from mirror steering caused by absorption of laser power in the mirror and its mounts. For this reason, tuning curves are taken in pulsed mode and are not directly comparable with the high average power runs. However, the extent of the experimental tuning curve taken in pulsed mode is found to be in good agreement with the tuning range of 12–13 μm found in simulation. In addition, Fig. 10.8 shows that the tuning curve exhibits a roughly triangular shape in contrast to the more usually observed sharply peaked behavior. This is due to the relatively large out-coupling and is also observed in the pulsed mode experiment.

The mode sizes found in simulation on the downstream and upstream mirrors versus pass are shown in Figs. 10.9 and 10.10, respectively. The average rms mode size on both mirrors in the steady state is approximately 10–11 mm, and this is in good agreement with observations. The fluctuation occurs on a period of about five passes and has a magnitude of about ±15%. There was no unambiguous diagnostic to either confirm or refute this behavior. We note that there is a corresponding oscillation in the output power (which is difficult to observe on the logarithmic scale in Fig. 10.6) of about ±3%, which is 180° out of phase with the oscillations in the mode size.

The optical mode quality observed in the experiment is near diffraction limited, and similar results are found in simulation. The optical mode pattern for the optimal cavity length at the wiggler exit is shown in Fig. 10.11 corresponding to the peak of the pulse in steady-state. The mode radius is about 1 mm, and the shape is a nearly perfect Gaussian exhibiting a low higher order mode content. As indicated in Figs. 10.9 and 10.10, the mode expands as it propagates to the downstream mirror, and the mode pattern at the downstream mirror is shown in Fig. 10.12 with a mode radius of 10–15 mm as expected.

Fig. 10.9 Variation in the rms mode size on the downstream mirror versus pass [9]

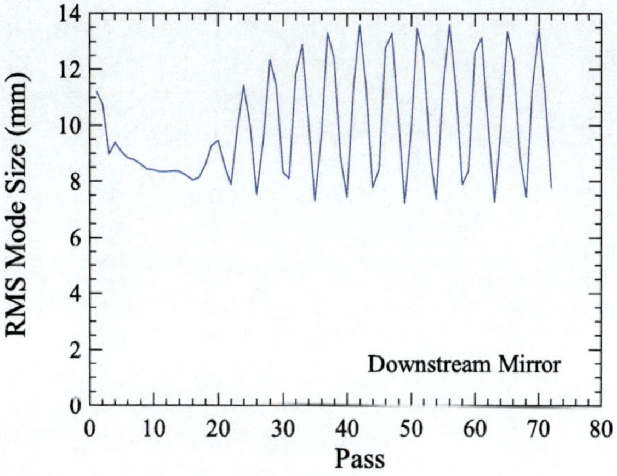

Fig. 10.10 Variation in the rms mode size on the upstream mirror versus pass [9]

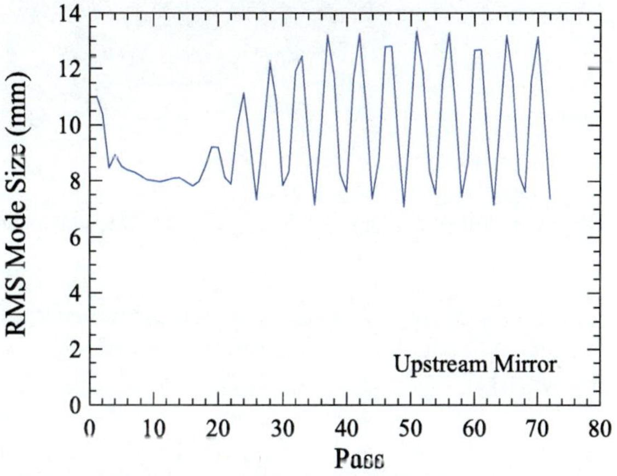

10.6 High-Gain/Low-Q Oscillators (RAFELs)

High-gain/low-Q oscillators are often referred to as *regenerative amplifiers* (or RAFELs). Equation 6.1 relating the gain at saturation to the resonator losses can be inverted to relate the losses in equilibrium to the gain, that is, $L = G/(1 + G)$, which indicates that if the gain is low, then the losses must also be low. However, low-loss optics may not exist at short wavelengths, and this necessitates using a high-gain wiggler where the radiation grows exponentially on a single pass through the wiggler with a high loss resonator where a substantial fraction of the pulse energy is out-coupled on each pass. This is particularly the case at x-ray wavelengths.

Because exponential gain is reached in the wiggler, a RAFEL exhibits many of the properties of a SASE free-electron laser; for example, the efficiency $\eta \cong \rho$ as in the case of the high-gain Compton regime. In this chapter, we discuss the general

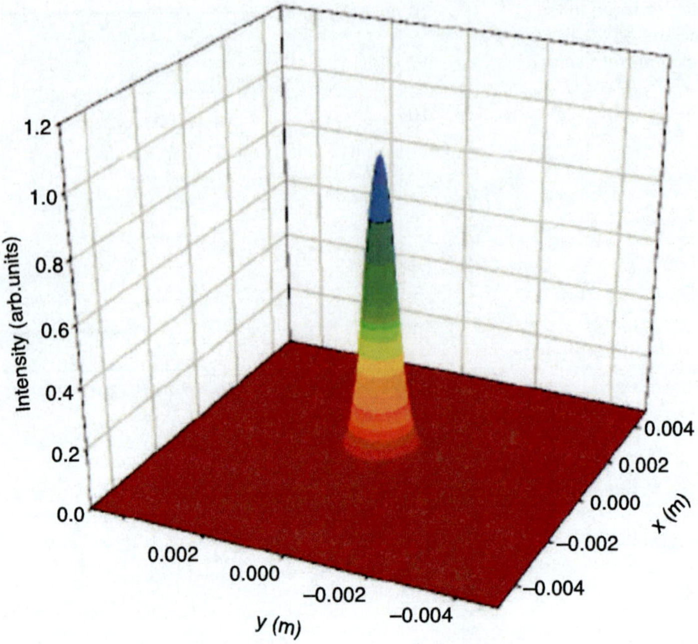

Fig. 10.11 The optical mode at the wiggler exit corresponding to the peak in the output pulse [9]

properties of a RAFEL using simulations for a configuration based on a concentric resonator. This is followed by simulations of an infrared RAFEL that used a ring resonator [17].

We now consider the simulation of a 2.2-μm RAFEL based upon a concentric resonator with hole out-coupling [10]. The electron beam is assumed to have a kinetic energy of 55 MeV and a bunch charge of 800 pC with a parabolic temporal profile having a full width of 1.2 psec and a repetition rate of 87.5 MHz. The normalized emittance is 15 mm-mrad in both the *x*- and *y*-directions, and the energy spread is 0.25%. A planar wiggler with two-plane focusing (parabolic-pole-faces) is assumed with a period of 2.4 cm, a peak on-axis field of 6.5–7.0 kG (K_{rms} = 1.01–1.11), and a length of 100 periods with one period entry and exit tapers (i.e., 98 periods of uniform field). The electron beam is injected into this long wiggler with the matched beam radius for the two-plane-focusing wiggler of 392 μm.

The resonator is concentric with the power coupled out through a 5.0 mm hole in the downstream mirror, which provides for a typical average out-coupling of about 97%. Given the repetition rate, f_{rep} = 87.5 MHz, the nominal *zero-detuning* cavity length, L_0 (= $Mc/2f_{rep}$, where M is the number of pulses in the resonator), is 6.85239904 m when $M = 4$. Both upstream and downstream mirrors have radii of curvature of 3.5 m, so that the Rayleigh range of the vacuum resonator is 0.5 m.

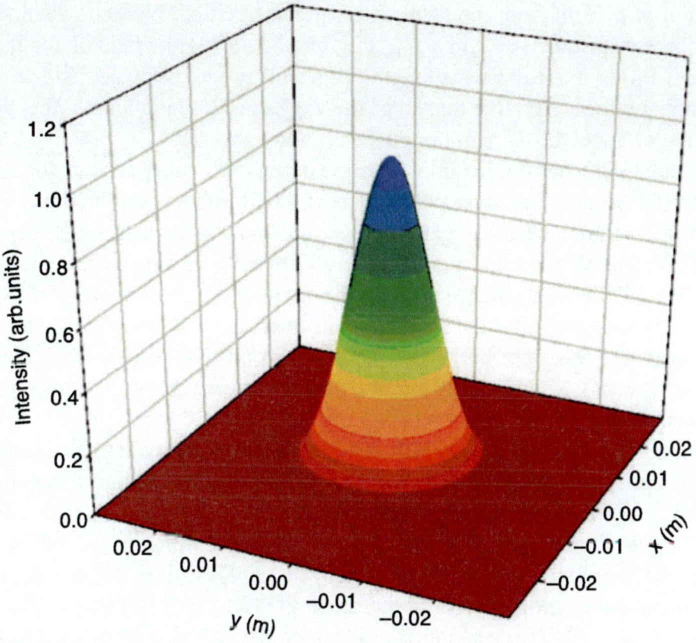

Fig. 10.12 The optical mode at the downstream mirror corresponding to the peak in the optical pulse [9]

Simulations are conducted assuming startup from noise on each pass; however, the shot noise is an important contributor only on the first several passes. In addition, the temporal window is an important numerical consideration and must be chosen to be large enough to accommodate the maximum cavity detuning length that is consistent with pass-to-pass amplification so that the optical pulse remains within the time window for all the usable choices of cavity length. In practice for this example, we choose a temporal window of 4.0 psec and include 182 temporal slices, which corresponds to the inclusion of one temporal slice every three wavelengths.

10.6.1 The Single-Pass Gain

Since the RAFEL wiggler is long enough to achieve exponential growth, it is instructive to determine the exponentiation (or gain) length, L_G, found in a single pass simulation and to compare that with the predictions based upon the empirical formula developed by Ming Xie [18]. The variation in the gain length with on-axis wiggler field as found in simulation is shown in Fig. 10.13, where the minimum gain length of about 0.176 m is found for a wiggler field of about 6.7 kG ($K_{rms} = 1.06$) that nominally corresponds to the closest resonance. This compares well with the prediction of 0.16 m from the Ming Xie parameterization. Note, however, that the well-known resonance $\lambda = \lambda_w(1 + K_{rms}^2)/2\gamma^2$ predicts a wiggler field of about 6.81 kG ($K_{rms} = 1.08$) at a 2.2 μm wavelength. This shift in the resonance is due to three-dimensional effects. Observe that since the wiggler is 2.4 m in length, this permits a maximum of about 12–14 gain lengths within the wiggler.

The gain length has implications over the permissible range of K_{rms} for which the RAFEL will operate (i.e., over which there is pass-to-pass amplification). Since the RAFEL will saturate when the gain balances the loss and the loss for the resonator is about 97%, this implies that the RAFEL will operate as long as the gain exceeds about 3200–3300%. In order to identify this range more closely, we perform multi-pass simulations and take the average gain over the first ten passes. We take an average because there are fluctuations in the gain on pass-to-pass basis (i.e., limit-cycle oscillations), which will be discussed in more detail below. The average gain is shown as a function of the on-axis wiggler field under the assumption of a cavity length of 8.65238144 m in Fig. 10.14. This represents a cavity detuning with respect to the *zero-detuning* length of $\Delta L_{cav} = -8\lambda$. It is clear from the figure that the gain is relatively constant over the range of about 6.65–6.85 kG ($K_{rms} = 1.054$–1.085) and falls off rapidly as the field diverges outside this range, which is consistent with the behavior of the gain length shown in Fig. 10.13. The cutoff for a gain of about 3300% occurs for field levels of about 6.518 kG ($K_{rms} = 1.03$) at the low end and 6.878 kG ($K_{rms} = 1.09$) at the high end, and the RAFEL is not expected to function outside of this range of wiggler fields.

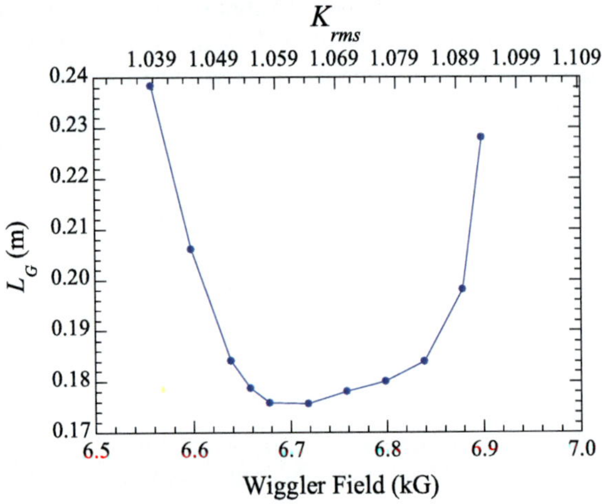

Fig. 10.13 The gain length as a function of the on-axis wiggler field [10]

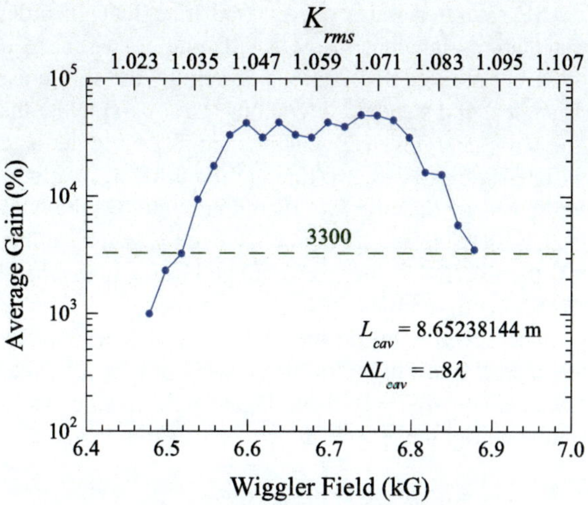

Fig. 10.14 Average gain over the first ten passes versus the on-axis wiggler field [10]

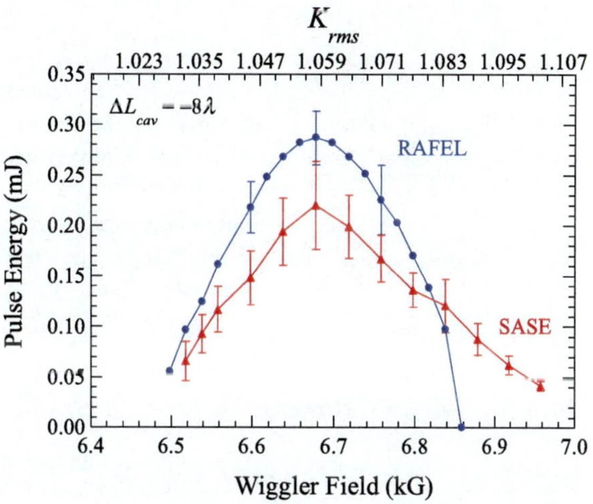

Fig. 10.15 Optical pulse energy for the RAFEL and for the equivalent SASE as a function of on-axis wiggler field [10]

10.6.2 Comparison with a SASE Free-Electron Laser

Since the RAFEL starts from shot noise and employs a high-gain wiggler, it is instructive to compare the performance of a RAFEL with a similar SASE free-electron laser. The performance of the RAFEL is shown in Fig. 10.15, where we plot the average pulse energy (blue circles) in the steady-state as a function of the on-axis wiggler field for the same cavity detuning ($\Delta L_{cav} = -8\lambda$) as used for Fig. 10.14. The error bars indicate the level of so-called limit-cycle oscillations. Observe that the RAFEL reaches its peak pulse energy for a wiggler field of 6.678 kG ($K_{rms} = 1.058$) and falls to zero outside the range predicted in Fig. 10.2. We also plot the equivalent

SASE saturated pulse energy (red triangles). In order to deal with the statistical fluctuations inherent in the SASE output, we made a large number of runs with different random distributions describing the shot noise and found the average (red triangles) and standard deviations (error bars). Note that the SASE results represent the pulse energies over whatever length of wiggler is required to reach saturation. It is interesting to observe that (1) the RAFEL configuration saturates with a higher pulse energy than the SASE configuration, (2) the fluctuations in the RAFEL in the steady-state regime are comparable to the statistical fluctuations found in SASE, and (3) the FWHM of the tuning range in K_{rms} is comparable for both the RAFEL and SASE configurations.

The spectral linewidth of the RAFEL also differs from that of a low-gain oscillator. The full width of the spectrum for a typical low-gain oscillator is given by $\Delta\omega/\omega = 1/N_w = 0.01$ for the example under consideration. This can be translated into a tuning range over the wiggler field as follows:

$$\left|\frac{\Delta B_w}{B_w}\right| = \frac{1 + K_{rms}^2}{2K_{rms}^2}\left|\frac{\Delta\omega}{\omega}\right|. \tag{10.10}$$

This implies a full width tuning range of $\Delta B_w \approx 0.063$ kG ($\Delta K_{rms} = 0.001$), which is much narrower than what we find in simulation. The SASE linewidth is given by $(\Delta\omega/\omega)_{rms} \approx \rho$, where ρ denotes the Pierce parameter. For this example, $\rho \approx 0.0097$ so that $(\Delta\omega/\omega)_{rms} \approx 0.0097$. Converting this to a tuning range in the wiggler field and going from the rms width to a FWHM tuning range, we obtain $(\Delta B_w/B_w)_{\text{FWHM}} \approx 0.022$, which compares well with the simulation results that give $(\Delta B_w/B_w)_{\text{FWHM}} \approx 0.019$. As a result, in regard to the spectral linewidth, the RAFEL behaves more like a SASE configuration than a typical low-gain oscillator.

10.6.3 Cavity Detuning

Another way in which the RAFEL differs from a low-gain oscillator is in the cavity detuning. The *zero-detuning* length is defined as the synchronous length for a group velocity, v_{gr}, equal to the speed of light in vacuo, c, throughout the resonator. However, v_{gr} is reduced in a RAFEL by the interaction in the wiggler and results in synchronism for cavity lengths less than the typical *zero-detuning* length. The cavity detuning depends on the group velocity reduction in the wiggler. In a low-gain oscillator, the group velocity reduction is small, and the slippage is one wavelength per wiggler period; hence, the slippage distance is $l_{slip} = N_w\lambda$, where N_w is the number of periods in the wiggler. However, the slippage per wiggler period is reduced in a high-gain RAFEL, or in any FEL where there is exponential growth because the gain medium reduces both the phase and group velocities. The reduced phase velocity results in the optical guiding of the radiation, while the reduced group velocity results in less slippage. It has been shown that $l_{slip} = N_w\lambda/3$ at the resonant wavelength [see Eq. 6.51]. For the example under consideration, this yields $l_{slip} \approx 72$ μm, which is much less than the slippage length of 220 μm if the RAFEL behaved as a low-gain oscillator.

In order to estimate the effect of this on the detuning length, we note that the zero detuning length is found by equating the roundtrip time of the radiation through the cavity with the spacing between electron bunches ($1/f_{rep}$). If v_{gr} is reduced as the radiation traverses a wiggler of length L_w, then

$$\frac{1}{f_{rep}} = \frac{2L_{cav} - L_w}{c} + \frac{L_w}{v_{gr}}, \tag{10.11}$$

so that

$$L_{cav} = \frac{c}{2f_{rep}} - \frac{L_w}{2}\frac{c}{v_{gr}}\left(1 - \frac{v_{gr}}{c}\right). \tag{10.12}$$

This implies a shift in the cavity length of

$$\Delta L_{cav} = -\frac{L_w}{6\gamma^2}\left(1 + K_{rms}^2\right) = -\frac{N_w \lambda}{3} \tag{10.13}$$

from the expected zero-detuning length. This lower value for the slippage is comparable to what is seen in simulation.

The detuning curve found in simulation is shown in Fig. 10.16, where the output pulse energy is plotted versus cavity detuning for a wiggler field of 6.658 kG ($K_{rms} = 1.055$). Here, the cavity detuning is defined relative to the nominal *zero-detuning* length so that $\Delta L_{cav} = L_{cav} - L_0$. As shown in the figure, we find a full width detuning range of about $50\lambda = 110$ μm and a FWHM detuning range of about 40λ, which are in reasonable agreement with the estimate based on the one-dimensional analysis of slippage.

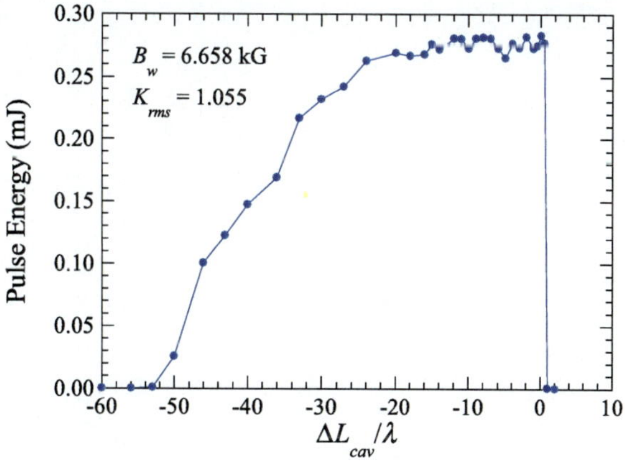

Fig. 10.16 Cavity detuning curve for an on-axis wiggler field of 6.658 kG [10]

10.6.4 The Temporal Evolution of the Pulse: Limit-Cycle Oscillations

The temporal evolution of the output pulse energy is shown for a wiggler field of 6.658 kG ($K_{rms} = 1.055$) in Fig. 10.17, where the pulse energy is plotted versus pass number through the wiggler for the choice of several cavity detunings. It is clear that significant fluctuations are found over a large range of detunings and that both the magnitude and period of the fluctuations decrease as the magnitude of the detuning increases, although the magnitude of the fluctuations decreases as well near the *zero-detuning* length.

The fluctuations seen in simulation of the RAFEL can be rapid and irregular. There are two possible explanations for these characteristics. One explanation is that due to the high gain and high out-coupling in the RAFEL, small changes in the mode structure from pass to pass can result in relatively large changes in the gain and, hence, the pulse energy. These "small" changes can include variations in the transverse mode structure (both in terms of the modal decomposition and spot size) and the temporal pulse shape. The second explanation, related to the first, is that since we have employed hole out-coupling, these relatively small changes in the transverse mode structure at the mirror can give rise to large differences in the out-coupling of the optical mode. It is not surprising, therefore, to expect that the magnitude of the fluctuations will vary depending on the cavity detuning. In Fig. 10.6, we show the variation in the rms magnitude of the fluctuations in the out-coupled pulse energy as a function of the cavity detuning. It is clear from the figure that the fluctuation level is relatively constant at about the 0.03 mJ level over most of the detuning range but with rapid declines at the ends of the detuning range. Also, the oscillation period is of the order of a few passes through the resonator (Fig. 10.18).

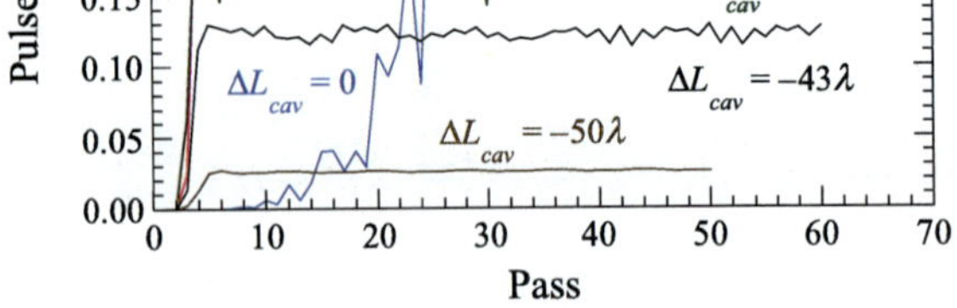

Fig. 10.17 Temporal evolution of the pulse energy for various cavity detunings [10]

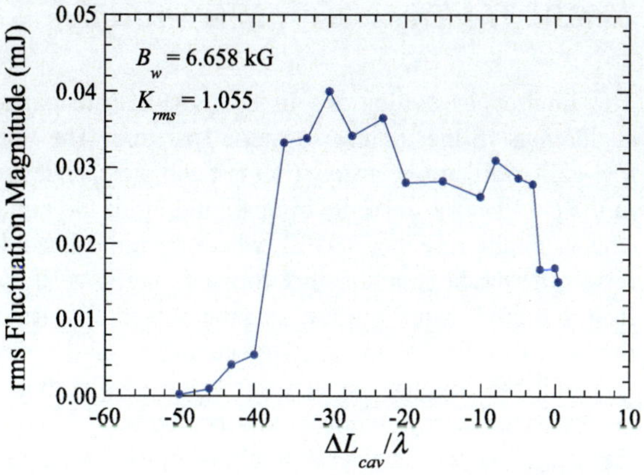

Fig. 10.18 The variation in the rms fluctuation level with cavity detuning [10]

Fluctuations/oscillations have been observed in low-gain oscillators and are referred to as limit-cycle oscillations. The observation of limit-cycle behavior in the low-gain FELIX free-electron laser oscillator corresponds to an oscillation period of [19].

$$\Delta\tau = -\tau_{slip}\frac{L_{cav}}{\Delta L_{cav}}, \qquad (10.14)$$

where τ_{slip} ($= l_{slip}/c$) is the slippage time. For the case of FELIX, $L_{cav} \approx 6$ m, $\lambda = 40$ mm, and $N_w = 38$, and the cavity detuning ranges over about 160 μm. As a result, $\tau_{slip} \approx 5.1$ psec and $\tau_{roundtrip}$ ($= 2L_{cav}/c$) ≈ 40 nsec is the nominal roundtrip time; hence, this implies that the limit cycle oscillation occurs over a period of about 3 μsec or 75 passes for a cavity detuning of -100 μm, which is consistent with observations. In contrast, if we apply the slippage time for the high gain RAFEL, under consideration

$$\Delta\tau = -\frac{N_w\lambda}{3\Delta L_{cav}}\frac{L_{cav}}{c} = -\frac{\tau_{roundtrip}}{2}\frac{N_w\lambda}{3\Delta L_{cav}}. \qquad (10.15)$$

As such, we expect the oscillation period to occur on the scale of a small number of passes for the indicated cavity detuning range. This is indeed what is observed in Fig. 10.17. For example, the oscillations occur approximately every two to four passes for $\Delta L_{cav}/\lambda = -8$, which is consistent with Eq. 10.15. However, there is not a great deal of variation with detuning possible when the oscillations occur on such a fast time scale, and we must take Eqs. 10.14 and 10.15 as approximate measures of the oscillation period. Still, the observed oscillation period is well described by the formula for the oscillation period for limit-cycle oscillations found in in low-gain oscillators when the appropriate slippage is taken into account. For a high-gain RAFEL, the much lower slippage results in very short oscillation periods.

10.6.5 The Transverse Mode Structure

The limit-cycle oscillations in the RAFEL are correlated with the fluctuations/oscillations in the transverse mode structure. The transverse mode structure in a low-gain oscillator is largely (but not completely) determined by the mode structure in the cold cavity since the optical guiding of the radiation in the wiggler is weak. This is not the case in a RAFEL where the mode is guided through the wiggler. As a result, the mode structure that forms as the RAFEL saturates differs substantially from the cold cavity modes, and the choice of a Rayleigh range of 0.5 m serves mainly to determine the radii of curvature of the mirrors. Since the radiation is guided in the wiggler, the spot size at the wiggler exit may be smaller than it would be in the cold cavity, which means that the Rayleigh range of the radiation as it exits the wiggler is smaller than it would be in the cold cavity. This implies, in turn, that the optical mode will expand more rapidly as it propagates to the downstream mirror. Alternatively, decomposing the smaller spot size at the wiggler exit in cold cavity modes necessarily leads to higher-order transverse modes in the optical field. After propagating to the outcoupling mirror, the superposition of these modes determines the fraction of the optical field coupled out through the hole, and similarly, after propagation to the wiggler entrance, the superposition sets the field profile at wiggler entrance. Small variations in the exponential growth rate, for example, due to changing coupling of the electrons to the optical field at the wiggler entrance, lead to relatively larger effects on the optical guiding of the radiation. This in turn changes the spot size at the wiggler exit and, hence, the energy coupled out of the resonator and the spot size at the wiggler entrance.

This is illustrated in Fig. 10.19 where the pass-to-pass variation in the width of the optical mode is plotted on the downstream and upstream mirrors for $B_w = 6.585$ kG ($K_{rms} = 1.043$) and $\Delta L_{cav} = -8\lambda$. It is clear from the figure that both the spot size and the fluctuations of the spot size on the upstream mirror are greater than those on the downstream mirror due to the optical properties of the resonator. At saturation, the

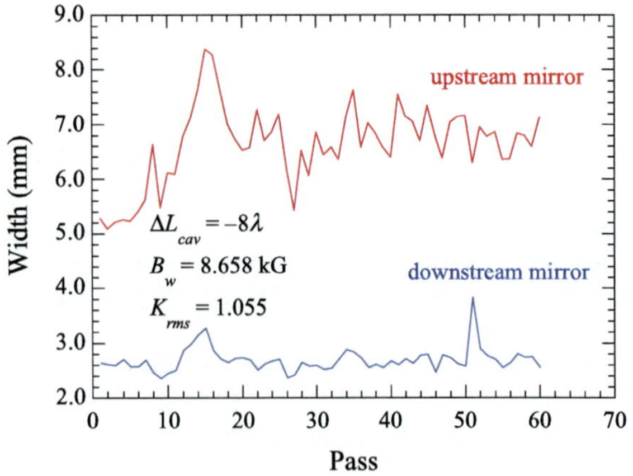

Fig. 10.19 Evolution of the optical mode widths on the upstream and downstream mirrors [10]

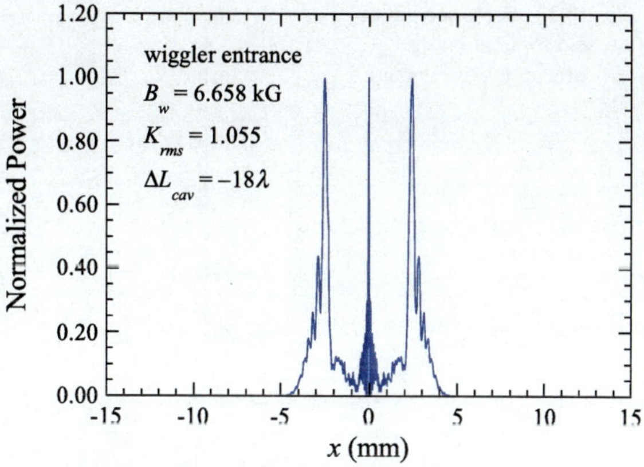

Fig. 10.20 Cross section of the field at the wiggler entrance on pass 60 [10]

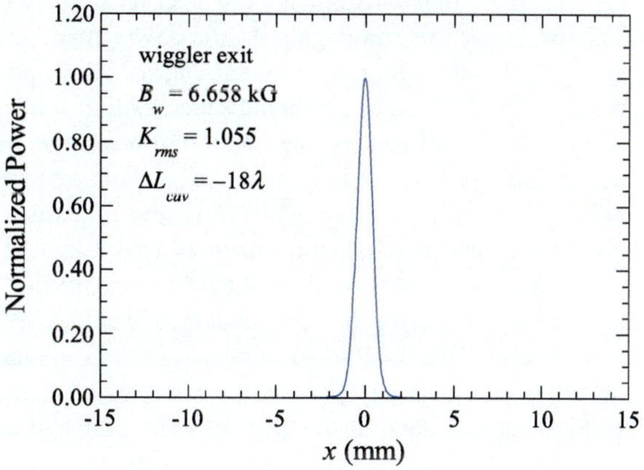

Fig. 10.21 Cross section of the field incident on the downstream mirror on pass 60 [10]

location of the smallest optical beam size moves over the axis of the wiggler and, consequently, due to the change in optical magnification. The optical field changes at both mirrors and the entrance of the wiggler changes as well. The resulting oscillations in size of the optical beam on the mirrors can be observed in Fig. 10.19.

The transverse mode structure is not only a result of optical guiding, it is also affected by the hole out-coupling. Consider the case of $B_w = 6.658$ kG ($K_{rms} = 1.055$) and $\Delta L_{cav} = -18\lambda$. The cross section of the field delivered to the wiggler entrance on pass 60 is shown in Fig. 10.20 where the normalized power in the x-direction (i.e., the wiggle plane) is plotted. Observe that the bulk of the power is at the edge of the optical field, but there is a spike at the center due to the presence of high-order modes. Despite the multiple peaks in the cross section at the wiggler entrance, the strength of the interaction in the wiggler yields a near-Gaussian mode peaked on-axis at the wiggler exit, as shown in Fig. 10.21. What has happened is that the interaction with the electron beam, which has a diameter of about 0.784 mm, essentially amplifies and guides the central peak shown in Fig. 10.20, while the

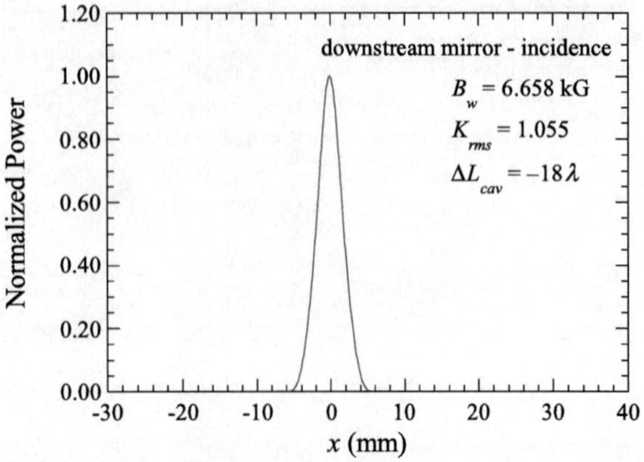

Fig. 10.22 Cross section of the field incident on the downstream mirror on pass 60 [10]

power in the wings falls outside the electron beam and is not amplified. This near-Gaussian mode then propagates to the downstream mirror during which it expands by about a factor of four, as shown in Fig. 10.22, where the FWHM is about 3.9 mm in width. The FWHM of the modal superposition at the wiggler exit is about 1.2 mm.

Ignoring the hole in the out-coupling mirror, the waist (0.59 mm) of the fundamental cold cavity mode is designed to be about $\sqrt{2}$ times the matched electron beam radius in the wiggler. The FWHM of the fundamental cold cavity mode at the wiggler exit and downstream mirror is 1.84 and 4.82 mm, respectively. We thus observe that the optical mode in the RAFEL expands faster from the wiggler exit to the downstream mirror than the fundamental cold cavity mode (factor 3.25 and 2.62, respectively). That the RAFEL mode size is still smaller at the downstream mirror is the result of a balance between the faster expansion and smaller spot size of the RAFEL optical mode at the wiggler exit compared to the cold cavity mode. The smaller spot size at the wiggler exit is due to gain guiding as described above. Both the smaller spot size at the wiggler exit and faster expansion of the RAFEL optical beam again indicate that at the wiggler exit, the optical field consists of fundamental and higher-order cold-cavity modes. Note a fundamental Gaussian beam having a waist at the wiggler exit with the same size as the RAFEL optical beam would have a Rayleigh range of 1.49 m. Finally, the cross section of the mode incident on the upstream mirror is shown in Fig. 10.23. After reflection from the upstream mirror and propagation through the resonator to the wiggler entrance, this field results in a modal pattern similar to that shown Fig. 10.20.

10.6.6 Temporal Coherence

The temporal coherence produced in a RAFEL is also an important property of the output light. The evolution of the temporal pulse on the first pass through the wiggler for $B_w = 6.658$ kG ($K_{rms} = 1.055$) is shown in Fig. 10.24. Since the RAFEL starts

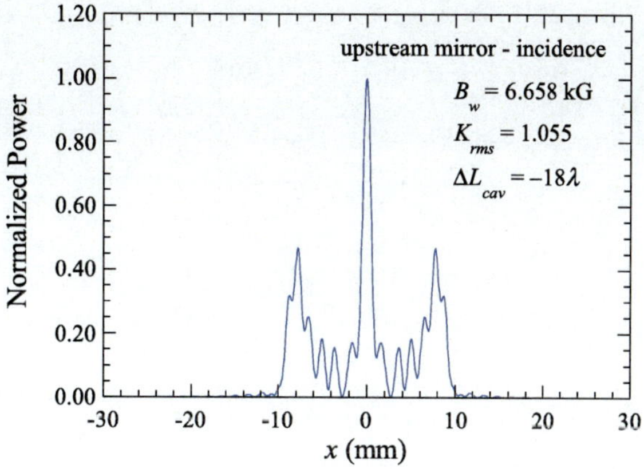

Fig. 10.23 Cross section of the field incident on the upstream mirror on pass 60 [10]

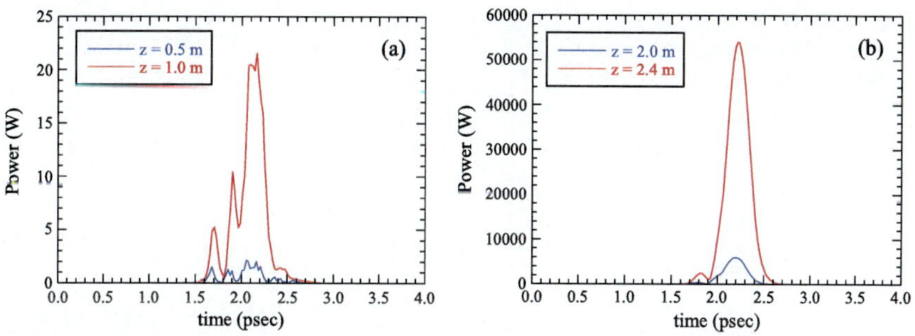

Fig. 10.24 Evolution of temporal coherence on the first pass for $B_w = 6.658$ kG [10]

from shot noise on the beam, the initial growth of the mode starts from spiky noise and, just as in a SASE FEL, develops coherence as it propagates through the wiggler. However, the wiggler is not long enough to reach saturation on a single pass, so that the evolution to temporal coherence develops over multiple passes. In Fig. 10.24, we plot the temporal pulse shapes found on the first pass (a) at $z = 0.5$ m and 1.0 m and (b) at $z = 2.0$ m and 2.4 m (wiggler exit). The figure shows the full-time window used in the simulation, and it should be noted that the electron beam is centered in the time window with a full (parabolic) width of 1.2 psec. It is evident that the pulse shows many spikes at 0.5 m but that it has coalesced into about five spikes after 1.0 m. The development of temporal coherence continues until at the wiggler exit at 2.4 m only two spikes remain.

The multi-pass development of temporal coherence after the first pass depends strongly on the cavity detuning. The temporal pulse shapes on the 60th pass at the wiggler exit for $B_w = 6.658$ kG ($K_{rms} - 1.055$) and for cavity detunings of $\Delta L_{cav}/\lambda = 0, -8, -18$, and -43 are shown in Figs. 10.25, 10.26, 10.27, and 10.28, respectively. As demonstrated previously, the synchronized cavity length for a RAFEL is shorter than the synchronized cavity length of a low-gain FEL oscillator.

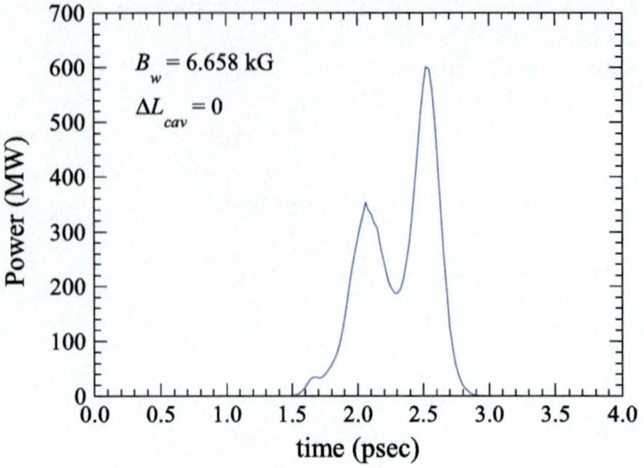

Fig. 10.25 Pulse shapes at the wiggler exit in the steady-state regime at zero detuning [10]

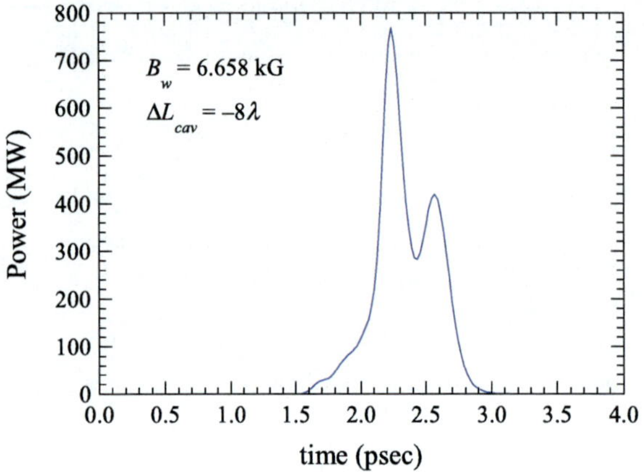

Fig. 10.26 Pulse shapes at the wiggler exit in the steady-state regime for $\Delta L_{cav}/\lambda = -8$ [10]

Therefore, as we have used the speed of light in vacuo instead of the actual group velocity to define the cavity detuning, we note that $\Delta L_{cav} = 0$ actually corresponds to a cavity length that is larger than the synchronized length for the high-gain RAFEL. Consequently, the returning optical pulse will lag behind the center of the electron bunch. The pulse will be amplified as it propagates through the wiggler, but as shown in Fig. 10.25, the two spikes formed over the first pass remain. This behavior is also found for small detunings as shown in Fig. 10.26 for $\Delta L_{cav}/\lambda = -8$.

If the detuning is closer to the center of the detuning range, then the synchronism between the returning optical pulse and the electrons is a better match, and the multiple spikes are "washed out." This is shown in Fig. 10.27, where $\Delta L_{cav}/\lambda = -18$, and we see that a broader pulse has formed. As the cavity length is decreased further, the optical pulse arrives increasingly near the head of the electron bunch. In this case, there is not enough gain to wash out the multi-spike character of the signal. This is shown in Fig. 10.28 for $\Delta L_{cav}/\lambda = -43$ where the total power (or pulse energy is much reduced).

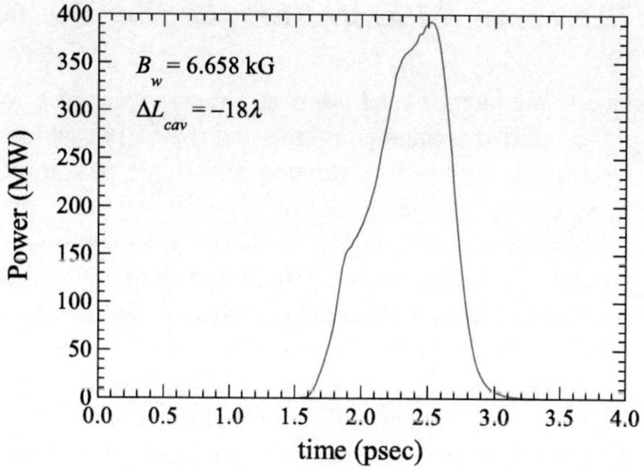

Fig. 10.27 Pulse shapes at the wiggler exit in the steady-state regime for $\Delta L_{cav}/\lambda = -18$ [10]

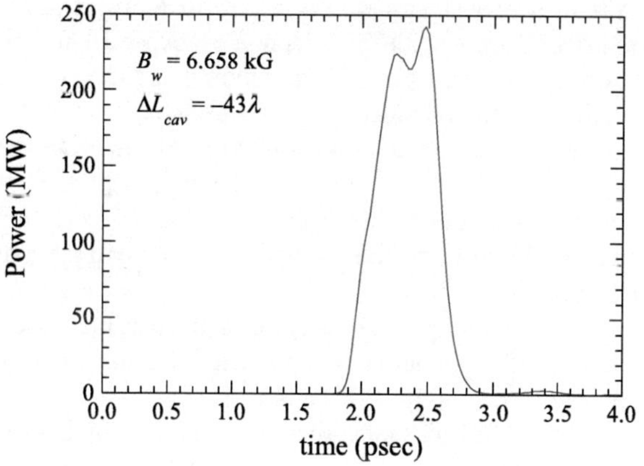

Fig. 10.28 Pulse shapes at the wiggler exit in the steady-state regime for $\Delta L_{cav}/\lambda = -43$ [10]

To summarize, while the RAFEL can be thought of as a low-Q oscillator, the high exponential gain in the wiggler leads to significant differences between the RAFEL and typical low-gain oscillators. The extraction efficiency and the linewidth in the RAFEL showed how they are more accurately specified by the efficiency and linewidth of a SASE free-electron laser than by the expectations for a low gain oscillator. In addition, the slippage in the RAFEL is reduced in comparison with the slippage in a low-gain oscillator due to the interaction in the high-gain regime, and this has a significant impact on both the detuning of the cavity and the period of the limit-cycle oscillations. However, it is important to observe that the shot-to-shot fluctuations in the RAFEL are comparable to those for an equivalent SASE configuration.

10.6.7 An X-Ray RAFEL with Hole Out-Coupling

Due to the lack of seed lasers at x-ray wavelengths, x-ray FEL facilities rely upon self-amplified spontaneous emission (SASE) in which the optical field grows from electron shot noise to saturation in a single pass through a long undulator. While pulse energies of the order of 2 milliJoules have been achieved at Ångstrom to sub-Ångstrom wavelengths, SASE exhibits shot-to-shot fluctuations in the output spectra and power of about 10–20%. For many applications, these fluctuations are undesirable, and efforts are underway to find alternatives. The utility of an x-ray FEL oscillator (XFELO) has been under study for a decade [4–6, 20–24] making use of resonators based upon Bragg scattering from high-reflectivity diamond crystals [25–29]. The development of these crystals is a major breakthrough in the path toward an XFELO. Estimates indicate that using a superconducting rf linac producing 8 GeV electrons at a 1 MHz repetition rate is capable of producing 10^{10} photons per pulse at a 0.86 Å wavelength with a FWHM bandwidth of about 2.1×10^{-7}. As a consequence, such an XFELO facility is expected to result in a decrease in SASE fluctuations in the power and spectrum and to narrow the spectral linewidth.

As with the majority of FELOs to date [1, 30, 31], the aforementioned XFELOs use low-gain/high-Q resonators with transmissive out-coupling through thin diamond crystals [24]. Potential difficulties with low-gain/high-Q resonators derive from sensitivities to electron beam characteristics and mirror loading and alignments. While experiments show that diamond crystals can sustain relatively high thermal and radiation loads [27, 29], transmissive out-coupling cannot be easily achieved at the photon energies of interest here. Hence, we consider an XFELO design with high out-coupling efficiency using a pinhole diamond mirror based on a regenerative amplifier (RAFEL).

The RAFEL described here is based on a high-gain/low-Q resonator [32] where the majority of the power is out-coupled on each pass through a pinhole in the diamond mirrors [33, 34]. Typically, this ranges from 90% to 95% of the power. Hence, the mirror loading can be significantly reduced. Since the interaction in a RAFEL optically guides the light, the optical mode is characterized by high purity with $M^2 \approx 1$ whether hole or transmissive out-coupling is used [10]. This might include an unstable resonator; however, it was shown by Siegman [14] that gain guiding, such as in an FEL, will stabilize a resonator that is otherwise (i.e., in vacuo) unstable.

Consider a six-crystal, tunable, compact cavity [32] as illustrated in Fig. 10.29 [the top view is shown in (a) and the side view is shown in (b)]. The crystals are arranged in a non-coplanar (3-D) scattering geometry. There are two *backscattering* units comprising three crystals (C_1, C_2, and C_3) on one side of the undulator and three crystals (C_4, C_5, and C_6) on the other side. Collimating and focusing elements are shown as $CRL_{1,2}$, which could be grazing-incidence mirrors but are represented in the figure by another possible alternative – compound refractive lenses [35]. In each backscattering unit, three successive Bragg reflections take place from three individual crystals to reverse the direction of the beam from the undulator. Assuming

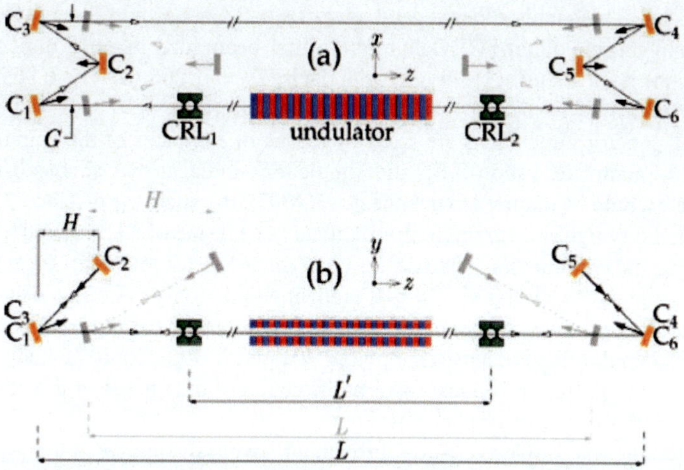

Fig. 10.29 A tunable s-x-crystal resonator [32]

that all the crystals and Bragg reflections are the same, the Bragg angles can be chosen within the range $30° < \theta < 90°$; however, Bragg angles close to $\theta = 45°$ should be avoided to ensure high reflectivity for both linear polarization components, as the reflection plane orientations for each crystal change. The cavity allows for tuning the photon energy in a large spectral range by synchronously changing all Bragg angles. In addition, to ensure constant time of flight, the distance L (which brackets the undulator), and the distance between crystals as characterized by H must be changed with θ. The lateral size G is kept constant as the resonator is tuned.

Because the C_1C_6 and C_3C_4 lines are fixed, intracavity radiation can be out-coupled simultaneously for several users at different places in the cavity, only out-coupling through C_1 is considered here. Out-coupling through crystals C_1 and C_4 is most favorable, since the direction of the out-coupled beams does not change with photon energy, but out-coupling for more users through crystals C_3 and C_6 is also possible. Such multiuser capability is in stark contrast with present SASE beamlines, which support one user at a time.

We consider that the electron beam propagates from right to left through the undulator and the out-coupling is accomplished through a pinhole in the first downstream mirror (C_1). The electron beam has an energy of 4.0 GeV, a bunch charge in the range of 10–30 pC with an rms bunch duration (length) at the undulator of 2.0–173 fs (0.6–52 μm) and a repetition rate of 1 MHz. The peak current at the undulator is 1000 A with a normalized emittance of 0.45 mm-mrad, and an rms energy spread of about 1.25×10^{-4}. The undulators are plane-polarized with a period of 2.6 cm. A fundamental resonance at 3.05 keV ($\approx$ 4.07 Å) yields an undulator field of 5.61 kG. Each undulator is has 130 periods, and the first and last periods describe entry/exit tapers. There is a total of 32 segments. The break sections between the undulators are 1.0 m in length and contain quadrupoles, which are located in the center of the breaks. The quadrupoles are 7.4 cm in length with a field gradient of

1.71 kG/cm. This yields a Pierce parameter of $\rho \approx 5.4 \times 10^{-4}$. In order to match this beam into the undulator/FODO line, the initial beam size in the x-dimension (y-dimension) is 37.87 μm (31.99 μm) with Twiss $\alpha_x = 1.205$ ($\alpha_y = -0.8656$). Note that this yields Twiss $\beta_x = 24.95$ m and $\beta_y = 17.80$ m.

The resonator dimensions are fixed by means of estimates of the gain using the Ming Xie parameterization [18], and simulations indicate that about 40–60 m of undulator would be required to operate as a RAFEL. As such, the distance, L, is fixed between the two mirrors framing the undulator as 130 m, which is also the distance separating the two mirrors on the back side of the resonator (elements C_3 and C_4). In studying the cavity tuning via time-dependent simulations, these two distances are allowed to vary while holding fixed the configurations of the *backscattering* units. The compound refractive lenses are placed symmetrically around the undulator and are designed to place the optical focus at the center of the undulator in vacuo. Here, the focal length is approximately 94.5 m.

In order to out-couple the x-rays through a transmissive mirror at the wavelength of interest, the diamond crystal would need to be impractically thin (about 5 μm); hence, an out-coupling through a hole in the first downstream mirror is used. All the mirrors are 100 μm thick, and the out-coupling of the fundamental is through a hole in the first downstream mirror (C_1). We begin with an optimization of the RAFEL with respect to the hole radius and the undulator length using steady-state (i.e., time-independent) simulations.

The choice of hole radius is important because if the hole is too small, then the bulk of the power remains within the resonator, while if the hole is too large, then the losses become too great, and the RAFEL cannot lase. The results for the optimization of the hole radius indicate that the optimum hole radius is 135 μm, which allows for 90% out-coupling, where the undulator line consists of 11 undulator segments. This is shown in Fig. 10.30 where the output power is plotted as a function of pass number for the optimum hole radius and the variation in the saturated power with the hole radius (inset) based upon time-independent simulations.

A local optimization on the undulator length for a hole radius of 135 μm is shown in Fig. 10.31 where the peak recirculating power (left axis) and the average output power (right axis) are plotted and which is also based upon time-independent simulations. The error bars in the figure indicate the level of pass-to-pass fluctuations in the power, which is generally smaller than the level of shot-to-shot fluctuations in SASE. Note that while this represents steady-state simulations, the average power is calculated under the assumption of an electron bunch with a flat-top temporal profile having a duration of 24 fs, which yields a duty factor of 2.4×10^{-8}. Each point in the figure refers to a given number of undulators ranging from 9 to 13 segments. It is evident from the figure that the optimum length is 47.18 m corresponding to 11 segments, as this length both maximizes the average output power and is close to the minimum in the pass-to-pass power fluctuations. Note that while the output power drops quickly when the hole size increases beyond the optimum value (cf. Fig. 10.30), the output power changes are more gradual and almost symmetric when the number of undulator segments deviate from the optimum number (cf. Fig. 10.31).

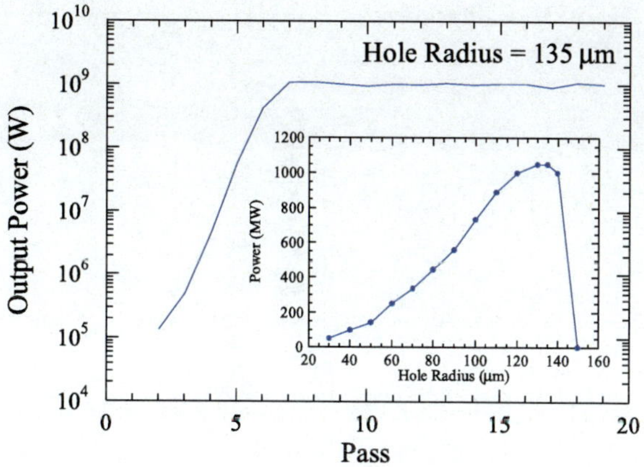

Fig. 10.30 Output power versus pass for a hole radius of 135 μm and the output power as a function of hole radius (inset) as obtained from time-independent simulations [33]

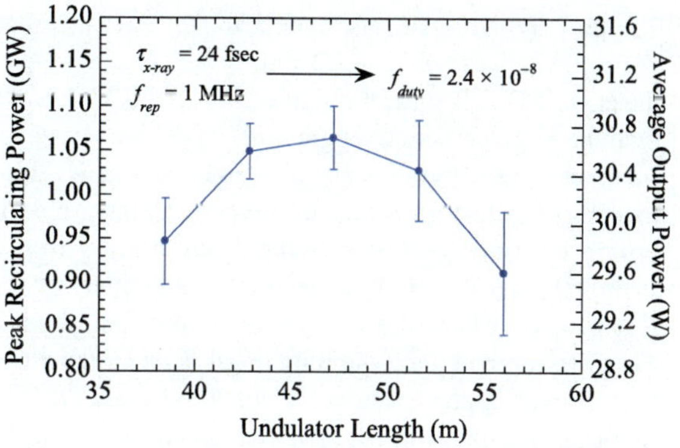

Fig. 10.31 Optimization on undulator length as obtained from time-independent simulations [33]

Having optimized the hole radius and the undulator length, consider now time-dependent simulations of the RAFEL under the abovementioned assumption of electron bunches with a flat-top temporal profile having a full width duration of 24 fs and a peak current of 1000 A. This corresponds to a bunch charge of 24 pC. The hole radius in the first downstream mirror (C_1) is taken to be 135 μ m. The simulation is set up to have a bandwidth of 0.8%. We consider startup from noise on the electron beam on the first pass (with noise included in the simulations for each successive pass as well), and since the RAFEL employs a high-gain undulator line, the pulse energy after the first pass reaches about 5 nJ, and subsequent growth is rapid despite an out-coupling from the first downstream mirror of about 90% of the incident pulse. Typically, saturation is achieved after about 15–25 passes.

The detuning curve defining which cavity lengths are synchronized with the repetition rate of the electrons is shown in Fig. 10.32. The synchronous cavity length (the so-called zero-detuning length) is $L_{vac} = c/f_{rep}$, where $L_{\rm vac}$ denotes the synchronous, roundtrip cavity length for the vacuum resonator, f_{rep} is the repetition rate, and c is the speed of light in vacuo. Here, $L_{vac} = 299.7924580$ m. As the flat-top temporal profile of an electron bunch corresponds to a bunch length of about

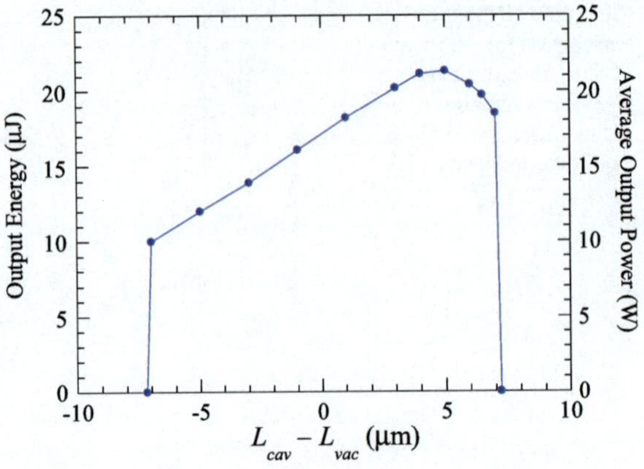

Fig. 10.32 Cavity detuning curve [33]

7.2 μm, synchronism is expected for cavity lengths in the range of L_{vac} − 7.2 μm < L_{cav} < L_{vac} + 7.2 μm, where L_{cav} denotes the total roundtrip length of the cavity. This is indeed observed in Fig. 10.32. Once the cavity is detuned by more than the electron bunch length (either positive or negative), the RAFEL fails to lase. As shown in the figure, since the single-pass gain is high in a RAFEL, this transition to complete desynchronization occurs rapidly. It is important to remark that the "sharp transitions" shown in the figure derives from the assumption of a flat-top temporal profile. Note that the output pulse energy at the peak of the detuning curve is about 21.2 μJ and the average output power is about 21.2 W for 3.05 keV photons. For an assumed bunch duration of 24 fs and a repetition rate of 1 MHz, this implies that the output power per pulse would be about 880 MW and the long-term average output power is about 21.2 W.

The evolution of the output energy at the fundamental and the third harmonic and the spectral linewidth of the fundamental versus pass are shown in Fig. 10.33 for a detuning of 5 μm, which is close to the peak in the detuning curve (Fig. 10.32). While it is not evident in the figure, the rms fluctuation in the energy from pass to pass is about 0.3 μ J (< 2%), as derived from 14 post saturation passes through the RAFEL. At least as important as the output power is that the linewidth contracts substantially during the exponential growth phase and remains constant through saturation. Starting with a relative linewidth of about 3.7×10^{-4} after the first pass, corresponding to non-saturated SASE, the linewidth contracts to about 6.0×10^{-5} at saturation with a pass-to-pass rms fluctuation of about ±2%. The relative linewidth after the first pass through the undulator is somewhat smaller than the predicted saturated SASE linewidth based on 1-D theory [36], which is approximately 5×10^{-4}. Hence, the RAFEL is expected to have both high average power and a stable narrow linewidth.

The third harmonic grows parasitically from high powers/pulse energies at the fundamental in a single pass through the undulator [37] and has been shown to reach output intensities of 0.1% that of the fundamental in a variety of FEL configurations, and this is what is found in the RAFEL simulations. As shown in Fig. 10.33, the third

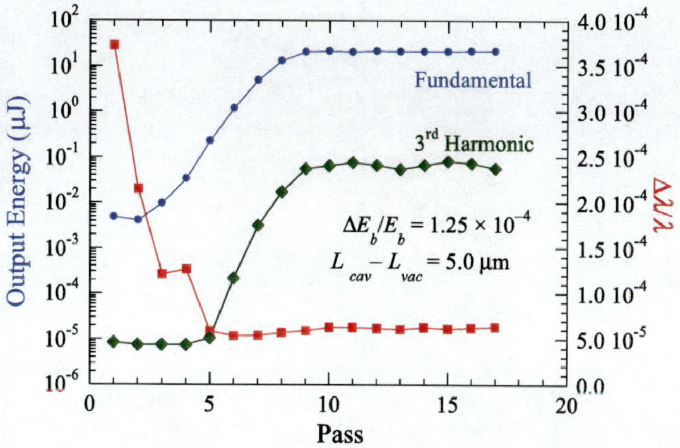

Fig. 10.33 Evolution of the fundamental pulse energy (blue, left) and the third harmonic (green), as well as the linewidth (red, right). Note, only about half of the post-saturation passes are shown in this figure [33]

harmonic intensity remains small until the fundamental pulse energy reaches about 1 μJ after which it grows rapidly and saturates after about 12 passes. This is close to the point at which the fundamental saturates as well. The saturated pulse energies at the third harmonic reach about 0.067 μJ. Given a repetition rate of 1 MHz, this corresponds to a long-term average power of 67 mW.

A comparison of the performance of the RAFEL with that of an equivalent SASE FEL based upon the same electron beam/undulator/FODO line, which is long enough to reach saturation, shows that the RAFEL exhibits comparable pulse energy and narrower linewidth than the SASE FEL with smaller rms fluctuations. SASE simulations using 15 different noise seeds, which achieves convergence of the average pulse energy to within a few percent, indicates that the average pulse energy is about 22.8 μJ with an rms fluctuation of ±4%, which is comparable to the 21.2 μJ found for the RAFEL. However, the average relative linewidth in SASE is about 4.3×10^{-4} with a fluctuation of ±16%, which is larger than that found for the RAFEL.

The reduction in the linewidth after saturation shown in Fig. 10.33 indicates that a substantial level of longitudinal coherence has been achieved in the saturated regime. The RAFEL starts from shot noise on the beam during the first pass through the undulator. Despite the large roundtrip loss of more than 98%, the optical energy returned to the undulator is still dominant over the noise and longitudinal coherence develops over the subsequent passes. This is depicted in Fig. 10.34, which shows the evolution of the normalized optical spectrum with increasing pass number. The optical spectrum is normalized to the maximum spectral power density for each pass to allow easy comparison of the shape of the spectrum. Figure 10.34 is derived from a second simulation using exactly the same parameters as for Fig. 10.33 except for shot noise in the electron bunches. Both simulations show that the optical field

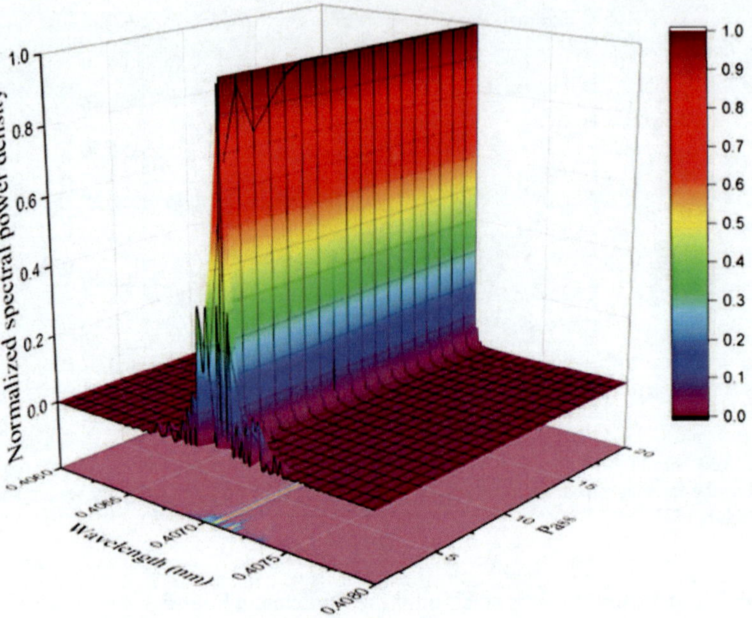

Fig. 10.34 Evolution of the normalized optical spectrum with increasing pass number. Same parameters as for Fig. 10.33 except for shot-noise in the electron bunches [33]

becomes independent of the shot noise in the electron bunches in less than five passes through the RAFEL. Figure 10.34 shows that only about three roundtrips are needed before the optical spectrum condensates into its final shape. That only such a small number of roundtrips are needed is due to the additional spectral filtering by the Bragg mirrors.

Hence, the temporal profile of the optical field after the first pass can be expected to exhibit the typical spiky structure associated with SASE and will strongly depend on the shot noise in the first electron bunch. Indeed, this is depicted in Figs. 10.35 and 10.36, which shows the temporal profile and associated optical spectrum at the undulator exit after the first pass for the simulation that was used to produce Fig. 10.33. Comparing the spectrum shown in Fig. 10.36 with the spectrum shown in Fig. 10.34 for pass one of the other simulation run, we observe the fluctuations expected from the different shot-noise in the electron bunches. The number of temporal spikes expected for saturated SASE, N_{spikes}, is given approximately by $N_{spikes} \approx l_b/(2\pi l_c)$, where l_b is the rms bunch length and l_c is the coherence length. For the present case, $l_b \approx 7.2$ μm and $l_c \approx 60$ nm, which implies that $N_{spikes} \approx 19$. Observe about 14 spikes are evident in Fig. 10.35, which is in reasonable agreement with the expectation. Note that the time axis encompasses the time window used in the simulation.

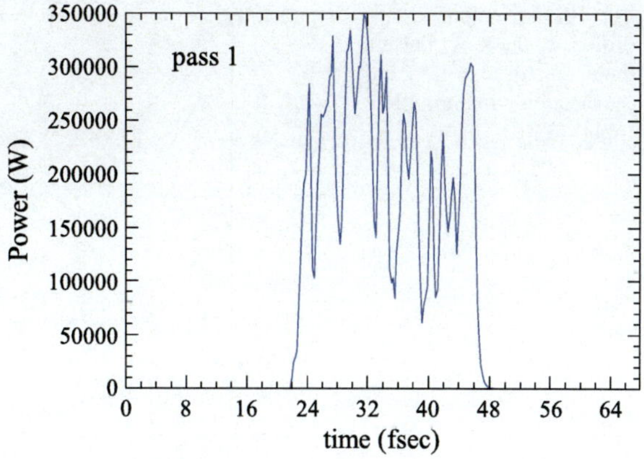

Fig. 10.35 Temporal profile of the optical pulse at the undulator exit after the first pass [33]

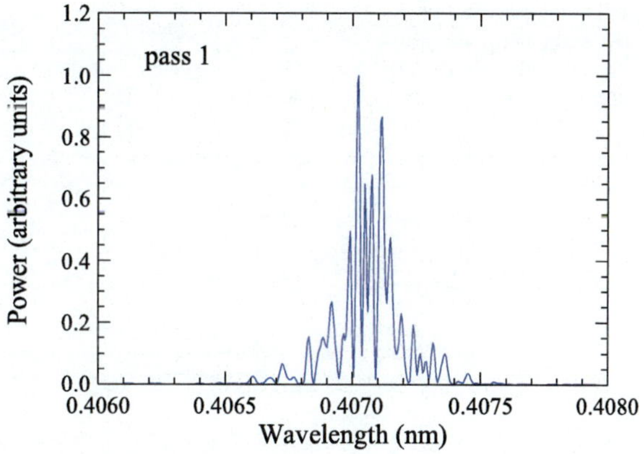

Fig. 10.36 Spectrum at the undulator exit after the first pass for the temporal profile shown in Fig. 10.35 [33]

As indicated in Fig. 10.33, the linewidth after the first pass is of the order of 3.7×10^{-4}, which is relatively broad and corresponds to the interaction due to SASE. The spectral narrowing that is associated with the development of longitudinal coherence as the interaction approaches saturation (see Fig. 10.34) results in a smoothing of the temporal profile. This is illustrated in Fig. 10.37, where the temporal profiles of the optical field are plotted at the undulator exit corresponding to passes 12–16, which are after saturation has been achieved (left axis). As shown in the figure, the temporal pulse shapes from pass to pass are relatively stable and exhibit a smooth plateau with a width of about 23–24 fs, which corresponds to, and overlaps, the flat-top profile of the electron bunches, which is shown on the right axis. Significantly, the smoothness of the profiles corresponds with the narrow linewidth that results from both having an oscillator configuration as well as the spectral filtering caused by the Bragg mirrors. Both the pass-to-pass stability and smoothness of the output pulses contrast markedly with the large shot-to-shot fluctuations and the spikiness expected from the output pulses in pure SASE.

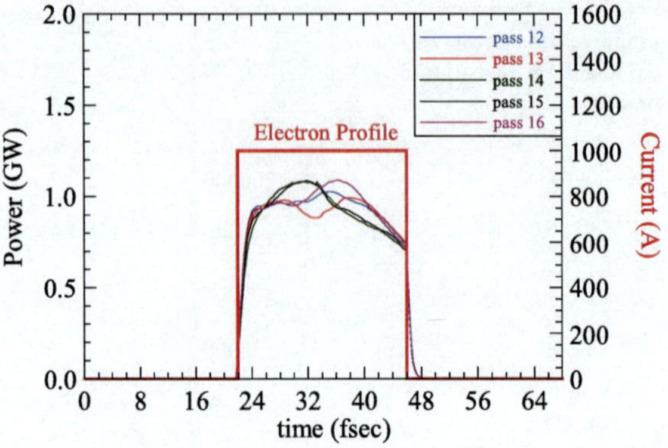

Fig. 10.37 Temporal profiles of the x-ray pulse at the undulator exit (left) and the electron bunch profile (right) [33]

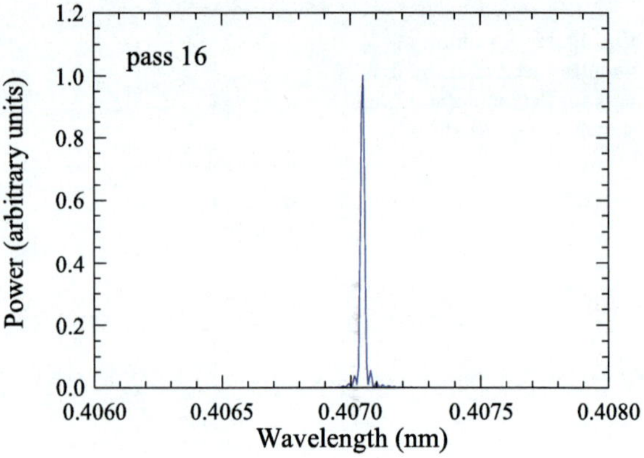

Fig. 10.38 Spectrum at the undulator exit after pass 16 for the temporal profile shown in Fig. 10.37 [33]

For clarity, the narrow relative linewidth in this regime of about 7.3×10^{-5} at the undulator exit is shown again in Fig. 10.38 after pass 16 and corresponds to the temporal profile shown in Fig. 10.37. Such a narrow linewidth as well as the smooth temporal profiles are associated with longitudinal coherence after saturation is achieved.

Similar performance is expected for photon energies other than 3.05 keV as the resonator comprised of the Bragg mirrors can be tuned over a large rang of photon energies by appropriate rotation of the Bragg mirrors while retaining its reflective properties. Using a different temporal profile for the electron bunches would not fundamentally change the characteristics of the x-ray RAFEL. For example, using a parabolic temporal profile in an infrared RAFEL still produces temporally smooth optical pulses [10]. As such, we conclude that an x-ray RAFEL may constitute an important alternative to SASE XFELs.

References

1. G.R. Neil, C. Behre, S.V. Benson, M. Bevins, G. Biallas, J. Boyce, J. Coleman, L.A. Dillon-Townes, D. Douglas, H.F. Dylla, R. Evans, A. Grippo, D. Gruber, J. Gubeli, D. Hardy, C. Hernandez-Garcia, K. Jordan, M.J. Kelley, L. Merminga, J. Mammosser, W. Moore, N. Nishimori, E. Pozdeyev, J. Preble, R. Rimmer, M. Shinn, T. Siggins, C. Tennant, R. Walker, G.P. Williams, S. Zhang, The JLab high power ERL light source. Nucl. Instrum. Methods Phys. Res. **A557**, 9 (2006)
2. R. Hajima, T. Shizuma, M. Sawamura, R. Nagai, N. Nishimori, N. Kikuzawa, E.J. Minehara, First demonstration of energy-recovery operation in the JAERI superconducting linac for a high-power free-electron laser. Nucl. Instrum. Methods Phys. Res. **A507**, 115 (2003)
3. N.G. Gavrilov, B.A. Knyazev, E.I. Kolobanov, V.V. Kotenkov, V.V. Kubarev, G.N. Kulipanov, A.N. Matveenko, L.E. Medvedev, S.V. Miginsky, L.A. Mironenko, A.D. Oreshkov, V.K. Ovchar, V.M. Popik, T.V. Salikova, M.A. Scheglov, S.S. Serednyakov, O.A. Shevchenko, A.N. Skrinsky, V.G. Tcheskidov, N.A. Vinokurov, Status of the Novosibirsk high-power terahertz FEL. Nucl. Instrum. Methods Phys. Res. **A575**, 54 (2007)
4. K.-J. Kim, Y. Shvyd'ko, S. Reiche, A proposal for an x-ray free-electron laser oscillator with an energy-recovery linac. Phys. Rev. Lett. **100**, 244802 (2008)
5. K.-J. Kim, Y.V. Schvyd'ko, Tunable optical cavity for an x-ray free-electron laser oscillator. Phys. Rev. ST AB **12**, 030703 (2009)
6. R.R. Lindberg, K.-J. Kim, Y.V. Schvyd'ko, W.M. Fawley, Performance of the x-ray free-electron laser with crystal cavity. Phys. Rev. ST-AB **14**, 010701 (2011)
7. W.B. Colson, J.L. Richardson, Multimode theory of free-electron laser oscillators. IEEE J. Quantum Electron. **QE-19**, 272 (1983)
8. B.D. McVey, Three-dimensional simulations of free-electron laser physics. Nucl. Instrum. Methods Phys. Res. **A250**, 449 (1986)
9. P.J.M. van der Slot, H.P. Freund, W.H. Miner Jr., S.V. Benson, M. Shinn, K.-J. Boller, Time-dependent, three-dimensional simulation of free-electron laser oscillators. Phys. Rev. Lett. **102**, 244802 (2009)
10. H.P. Freund, D.C. Nguyen, P.A. Sprangle, P.J.M. van der Slot, Three-dimensional, time-dependent simulation of a regenerative amplifier free-electron laser. Phys. Rev. ST-AB **16**, 010707 (2013)
11. J. Blau, K. Cohn, W.B. Colson, *Four-Dimensional Models of Free-Electron Laser Amplifiers and Oscillators* (Proceedings of the 37th International Free-Electron Laser Conference, Daejon, 2015) http://www.jacow.org
12. J.G. Karssenberg, P.J.M. van der Slot, I.V. Volokhine, J.W.J. Vershuur, K.-J. Boller, Modeling paraxial wave propagation in free-electron laser oscillators. J. Appl. Phys. **100**, 093106 (2006)
13. http://lpno.tnw.utwente.nl/opc.html
14. A. Siegman, *Lasers* (University Science Books, Mill Valley, 1986)
15. M. Born, E. Wolf, *Principles of Optics* (Pergamon Press, Oxford, 1984)
16. J.B. Flanz, S. Kowalski, C.P. Sargent, An isochronous beam recirculation magnet system. IEEE Trans. Nuclear Sci. **NS-28**, 2847 (1981)
17. D.C. Nguyen, R.L. Sheffield, C.M. Fortgang, J.C. Goldstein, J.M. Kinross-Wright, N.A. Ebrahim, First lasing of a regenerative amplifier free-electron laser. Nucl. Instrum. Methods Phys. Res. **A429**, 125 (1999)
18. M. Xie, *Design Optimization for an X-Ray Free Electron Laser Driven by the SLAC Linac* (Proc. IEEE 1995 Particle Accelerator Conference, Vol. 183, IEEE Cat. No. 95CH35843, 1995)
19. D.A. Jaroszynski, R.J. Bakker, D. Oepts, A.F.G. van der Meer, P.W. van Amersfoort, Limit cycle behavior in FELIX. Nucl. Instrum. Methods Phys. Res. **A331**, 52 (1993)
20. R.R. Lindberg, K.-J. Kim, Mode growth and competition in the x-ray free-electron laser oscillator start-up from noise. Phys. Rev. ST-AB **12**, 070702 (2009)
21. K.-J. Kim, Y. Shvyd'ko, R.R. Lindberg, An x-ray free-electron laser oscillator for record high spectral purity, brightness, and stability. Synchrotron Rad. News **25**, 25 (2012)

22. T.J. Maxwell, J. Arthur, Y. Ding, W.M. Fawley, J. Frisch, J. Hastings, Z. Huang, J. Krzywinski, G. Marcus, K.-J. Kim, R.R. Lindberg, D. Shu, Yu Shvyd'ko, S. Stoupin, *Feasibility Study for an X-Ray FEL Oscillator at the LCLS-II* (Proc. of the International Particle Accelerator Conference, Richmond, 2015), p. 8197. 3–8 May 2015
23. K.-J. Kim, *A Harmonic X-Ray FEL Oscillator* (Paper Presented at the High-Brightness Sources and Light-Driven Interactions Meeting, Long Beach, California, 20–22 March 2016, OSA Technical Digest paper ES2A.2, 2016)
24. W. Qin, S. Huang, K.X. Liu, K.-J. Kim, R.R. Lindberg, Y. Ding, Z. Huang, T. Maxwell, K. Bane, G. Marcus, *Start-to-End Simulations for an X-Ray FEL Oscillator at the LCLS-II* (Proc. of the 38th International Free-Electron Laser Conference, Santa Fe., 20–25 August 2017, 2017), p. 247
25. Y. Shvyd'ko, S. Stoupin, A. Cunsolo, A.H. Said, X. Huang, High-reflectivity high resolution x-ray crystal optics with diamonds. Nat. Phys. **6**, 196 (2010)
26. Y. Shvyd'ko, S. Stoupin, V. Blank, S. Terentyev, Near-100% Bragg reflectivity of x-rays. Nature Phot. **5**, 539 (2011)
27. S. Stoupin, A.M. March, H. Wen, D.A. Walko, Y. Li, E.M. Dufresne, S.A. Stepanov, K.-J. Kim, Y. Shvyd'ko, V.D. Blank, S. Terentyev, Direct observation of dynamics of thermal expansion using pump-probe high-energy-resolution x-ray diffraction. Phys. Rev. B **86**, 054301 (2012)
28. Yu. Shvyd'ko, *Feasibility of X-Ray Cavities for Free-Electron Laser Oscillators* (International Committee for Future Accelerators: Beam Dynamics Newsletter, No. 60, Issue Editor, G.R. Neil, 2013), p. 68. https://icfa-usa.jlab.org/archive/newsletter/icfa_bd_nl_60.pdf
29. T. Kolodziej, P. Vodnala, S. Terentyev, V. Blank, Y. Shvd'ko, Diamond drunhead crystals fo x-ray optics applications. J. Appl. Crystallogr. **49**, 1240 (2016)
30. G.R. Neil, C.L. Bohn, S.V. Benson, G. Biallas, D. Douglas, H.F. Dyllla, R. Evans, J. Fugitt, A. Grippo, J. Gubeli, R. Hill, K. Jordan, R. Li, L. Merminga, P. Piot, J. Preble, M. Shinn, T. Siggins, R. Walker, B. Yunn, Sustained kilowatt lasing in a free-electron laser with same-cell energy recovery. Phys. Rev. Lett. **84**(662) (2000)
31. G.R. Neil, C. Behre, S.V. Benson, M. Bevins, G. Biallas, J. Botce, J. Coleman, L.A. Dillon-Townes, D. Douglas, H.F. Dylla, R. Evans, A. Grippo, D. Gruber, J. Gubeli, D. Hardy, C. Hernandez-Garcia, K. Jordan, M.J. Kelly, L. Merminga, J. Mammossser, W. Moore, N. Nishimori, E. Posdeyev, J. Preble, R. Rimmer, M. Shinn, T. Siggins, C. Tennant, R. Walker, G.P. Williams, S. Zhang, The Jlab high power ERL light source. Nucl. Instrum. Methods Phys. Res. A **557**, 9 (2006)
32. B.W.J. McNeil, A simple model of the free electron laser oscillator from low to high gain. IEEE J. Quantum Electron. **26**, 1124 (1990)
33. H.P. Freund, P.J.M. van der Slot, Yu Shvyd'ko, An x-ray regenerative amplifier free-electron laser using diamond pinhole mirrors. New J. Phys. **21**, 93028 (2019)
34. G. Marcus, A. Halavanau, Z. Huang, J. Krzywinski, J. MacArthur, R. Margraf, T. Raubenheimer, D. Zhu, Refractive guide switching a regenerative amplifier free-electron laser for high peak and average power hard x-rays. Phys. Rev. Lett. **125**, 254801 (2020)
35. A. Snigerev, V. Kohn, I. Snigereva, B. Lengeler, A compound refractive lens for focusing high-energy x-rays. Nature **384**, 49 (1996)
36. Z. Huang, K.-J. Kim, Review of free-electron laser theory. Phys. Rev. ST-AB **10**, 034801 (2007)
37. H.P. Freund, S.G. Biedron, S.V. Milton, Nonlinear harmonic generation in free-electron lasers. J. Quantum Electron. **36**, 275 (2000)

Chapter 11
Wiggler Imperfections

The free-electron laser operates by the coherent axial bunching of electrons in the ponderomotive wave formed by the beating of the wiggler and radiation fields. The interaction is extremely sensitive to the axial energy spread of the electron beam, and an energy spread of a percent or less is sufficient to cause substantial reductions in the efficiency due to the detuning of the wave-particle resonance. A related effect is caused by random imperfections in the wiggler field. Planar wigglers can easily exhibit a random rms fluctuation of 0.5% from pole to pole [1]. This yields a velocity fluctuation that causes a phase jitter that also detunes the wave-particle resonance. In this chapter, we explore the effects of wiggler imperfections on free-electron laser performance and compare the effects of wiggler imperfections with those of an axial energy spread.

The effects of random wiggler imperfections have been studied using a random walk model for the electron trajectories and their effects upon both spontaneous emission [2] and the linear gain [3, 4]. Nonlinear modeling of wiggler field imperfections has been based [4–7] upon the inclusion of an analytic model of the random walk in a wiggler-period averaged formalism of the electron trajectories. In contrast to these approaches, we adapt the nonlinear formalism described in Chap. 5 to treat the effect of wiggler imperfections [8, 9]. No average over a wiggler period is performed in this approach, and no explicit assumption of the random walk is included. Instead, this formalism relies upon a model of the imperfections in the wiggler field, and the evolution of the electron trajectories, as well as the growth of the radiation field, is then determined self-consistently by integration of the coupled nonlinear differential equations for the electrons and the fields.

Consideration of the effects of wiggler imperfections shows that any perturbation induced in one of the pole pieces of a planar wiggler will induce a series of correlated changes in the field over several adjacent wiggler periods. This effect has been measured in the laboratory on a prototype planar wiggler design [10]. Here, an error was introduced by reducing the gap spacing between one set of pole pieces. An axial scan of the on-axis field showed that the error propagated through ± 1 wiggler period (± 2 pole pieces for this design) with an increase in amplitude at the adjacent poles of

H. P. Freund, T. M. Antonsen, Jr., *Principles of Free Electron Lasers*,
https://doi.org/10.1007/978-3-031-40945-5_11

approximately 55% and at the next poles of approximately 10%. The amplitude and extent of these correlations are dependent upon the detailed design of any given wiggler and can be substantial. Thus, the question of the nature of "random" imperfections in wiggler magnets requires further study. As a first step, a continuous mapping of random field variations from pole-to-pole has been included.

Two specific examples are discussed. The first is at a relatively long wavelength of approximately 8 mm and corresponds to the 35 GHz example discussed in Sect. 5.4. As such, the discussion makes use of the nonlinear formalism developed in Chap. 5 directly. The second example corresponds to a short wavelength free-electron laser. For this purpose, the nonlinear formalism developed in Chap. 5 is extended to treat an ensemble of optical modes.

11.1 The Wiggler Model

The wiggler configuration we employ for both the long and short wavelength examples is that of the parabolic-pole-face planar wiggler introduced in Eq. 2.3. This wiggler model provides enhanced focusing of the electron beam with respect to a planar wiggler with flat pole faces and was shown to provide good agreement with experiment (see Sect. 5.4.4). The variation in the wiggler amplitude in the axial direction will be assumed to contain both systematic and random components, and we shall write $B_w(z) = B_{w0}(z) + \Delta B_w(z)$, where $B_{w0}(z)$ denotes the systematic variation and $\Delta B_w(z)$ denotes the random variation.

The systematic variation in the wiggler amplitude describes both the adiabatic entry taper as well as any amplitude tapering for efficiency enhancement. In this chapter, however, we shall ignore any tapering for efficiency enhancement and assume that the systematic variation in the wiggler amplitude solely includes the adiabatic entry taper as

$$B_{w0}(z) = \begin{cases} B_w \sin^2\left(\dfrac{k_w z}{4N_w}\right) & ; 0 \le z \le N_w\lambda_w \\ B_w & ; N_w\lambda_w < z \end{cases}, \tag{11.1}$$

where B_w is the systematic amplitude of the uniform wiggler, λ_w is the wiggler period, k_w ($= 2\pi/\lambda_w$) is the wiggler wavenumber, and the adiabatic entry taper is over N_w wiggler periods.

The random component of the amplitude is chosen at regular intervals using a random number generator, and a continuous map is used between these points. Since a particular wiggler may have several sets of pole faces per wiggler period, the interval is chosen to be $\Delta z = \lambda_w/N_p$, where N_p is the number of pole faces per wiggler period. Hence, a random sequence of amplitudes $\{\Delta B_n\}$ is generated, where $\Delta B_n = \Delta B_w(n\Delta z)$. The only restriction is that $\Delta B_w = 0$ over the entry taper region (i.e.,

$\Delta B_n = 0$ for $0 \le n \le 1 + N_pN_w$) to ensure a positive amplitude. The variation in $\Delta B_w(z)$ between these points is given by

$$\Delta B_w(n\Delta z + \delta z) = \Delta B_n + (\Delta B_{n+1} - \Delta B_n)\sin^2\left(\frac{\pi\delta z}{2\Delta z}\right), \tag{11.2}$$

where $0 \le \delta z \le \Delta z$. It is important to note that it is possible to model the effects of pole-to-pole variations in specific wiggler magnets with this formulation.

Before turning to a detailed numerical analysis of the effect of wiggler imperfections, we consider the effect of the wiggler fluctuations on the variations in the axial energy along the electron trajectory. It is clear that the effect of axial variations in the wiggler field will be to cause an oscillation in the axial velocity of the electrons. It is possible to show that the fluctuation in the axial electron velocity Δv_z caused by some given fluctuation in the wiggler field ΔB_w is given by

$$\frac{\Delta v_z}{v_z} = -\left(\frac{v_w}{v_z}\right)^2 \frac{\Delta B_w}{B_w}, \tag{11.3}$$

where v_z is the bulk axial velocity, v_w is the wiggler-induced transverse velocity, and

$$\left(\frac{v_w}{v_z}\right)^2 = \frac{K_{\rm rms}^2}{\gamma_0^2}, \tag{11.4}$$

for a planar wiggler. However, in order to neglect the effects of the axial velocity spread, we must require that

$$\frac{\Delta v_z}{v_z} \ll \frac{{\rm Im}(k)}{{\rm Re}(k)} = \frac{1}{4\pi}\frac{\lambda}{L_G}, \tag{11.5}$$

where k is the complex wavevector, λ is the wavelength, and L_G is the power gain length. As a result, the effect of wiggler imperfections will be small when

$$\frac{\Delta B_w}{B_w} \ll \frac{1}{4\pi}\frac{\gamma_0^2}{K_{\rm rms}^2}\frac{\lambda}{L_G}. \tag{11.6}$$

11.2 The Long Wavelength Regime

Recall that the nonlinear formulation developed in Chap. 5 includes the simultaneous integration of a slow time-scale formulation of Maxwell's equations for an ensemble of TE and TM modes as well as the complete Lorentz force equations for an ensemble of electrons. The wiggler model includes an adiabatic entry taper that describes the injection of the beam into the wiggler. As a result, the initial conditions on the electron beam are specified at the entrance to the wiggler, and the subsequent evolution of the electromagnetic field and the electron beam are integrated in a self-consistent manner. Thermal effects are included under the assumption that the electron beam is initially monoenergetic but with a pitch angle spread to describe the axial energy spread.

The waveguide configuration employed is that of the rectangular geometry described in Chap. 5 in which the electron beam propagates through a loss-free rectangular waveguide with the dimensions $-a/2 \leq x \leq a/2$ and $-b/2 \leq y \leq b/2$. The equations that govern the evolution of the individual TE and TM modes of this structure are given in Eqs. 5.154, 5.155, 5.156, and 5.157.

The specific example under consideration is one in which a 3.5 MeV/850 A electron beam with an initial radius of 1.0 cm propagates through a rectangular waveguide ($a = 9.8$ cm, $b = 2.9$ cm) in the presence of a wiggler with $B_w = 3.72$ kG, $\lambda_w = 9.8$ cm, and $N_w = 5$. Hence, the wiggler parameter $a_w = 3.404$. Resonant interaction occurs with the TE_{01}, TE_{21}, and TM_{21} modes at frequencies of 30–40 GHz, and the efficiency decreases with increasing frequency across this band. For an ideal beam (with a vanishing thermal energy spread for which $\Delta\gamma_z = 0$) and wiggler ($\Delta B_w = 0$), the efficiency falls off from a maximum $\eta = 12.4\%$ at 30 GHz to a minimum of $\eta = 3.6\%$ at 40 GHz. A frequency of 34.6 GHz is selected for the comparison since this was the operational frequency used in the experiment.

The variation in the wiggler imperfection-induced axial energy fluctuation versus the wiggler fluctuation given in Eqs. 11.3, 11.4, and 11.5 is shown in Fig. 11.1 for this case. It is evident that the axial energy fluctuation increases rapidly with the magnitude of the wiggler fluctuation. In particular, a wiggler fluctuation of 5% results in a corresponding fluctuation in the axial energy of the beam of approximately 22%. One important question that we shall address in the numerical analysis of wiggler imperfections is whether such wiggler-induced fluctuations in the axial energy of the beam will have a comparable effect on the free-electron laser interaction as a thermal axial energy spread.

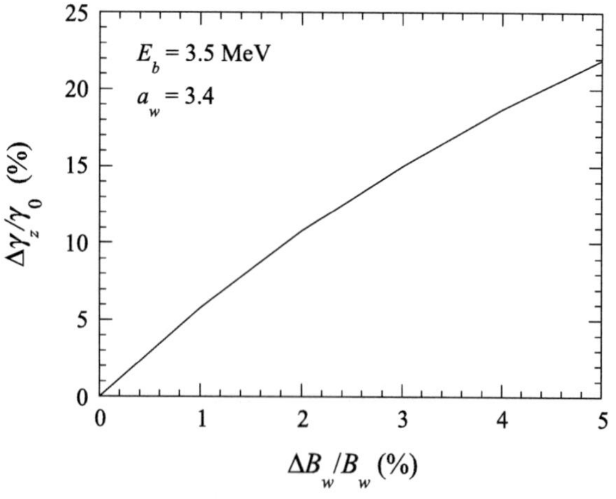

Fig. 11.1 Plot of the variation in the axial electron energy as a function of the amplitude of the wiggler fluctuation

In the case of an ideal wiggler (i.e., $\Delta B_w = 0$) discussed in Sect. 5.5.3, the initial drive powers in the modes are chosen to be 50 kW in the TE_{01} mode, 500 W in the TE_{21} mode, and 100 W in the TM_{21} mode. The evolution of the power as a function of axial distance is shown in Fig. 5.53 for the choice of an initial axial energy spread of $\Delta\gamma_z/\gamma_0 = 0$. Observe that this refers to the thermal energy spread and not the induced energy fluctuation due to wiggler imperfections. We find that saturation occurs after a distance of 153 cm at a total power level in all modes of 260 MW for an efficiency of 8.6%. The effect of the axial energy spread is illustrated in Fig. 5.57 in which the extraction efficiency is plotted as a function of $\Delta\gamma_z/\gamma_0$. As shown, the efficiency decreases gradually with the axial energy spread (due to the relatively high a_w) for $\Delta\gamma_z/\gamma_0 \leq 2.5\%$, at which point the efficiency has fallen to approximately 5.9%.

Random wiggler fluctuations can take many different forms for a fixed rms value. It is most natural to consider a random fluctuation that is relatively uniform over the interaction region (i.e., $<\Delta B_w> = 0$); however, other configurations are possible. For example, fluctuations where the wiggler field is always greater (or less) than the systematic value for B_w are possible, as is one in which ΔB_w is very large over a small range and zero elsewhere. These are only limited examples, and a thorough analysis necessitates a large number of simulations with different random wiggler fluctuations to obtain adequate statistics. Typically, we find that a choice of approximately 35 different wiggler fluctuation distributions is required in order for the mean efficiency to converge to within 1%.

The effect of random wiggler errors is shown in Fig. 11.2 where the efficiency is plotted versus the rms wiggler variation (for $\Delta\gamma_z = 0$) for $N_p = 1$. This describes random wiggler variations at intervals of the wiggler period. The dots represent the average efficiency over the *ensemble* of random fluctuations, and the error bars denote the standard deviation. As shown, the average efficiency is relatively insensitive to wiggler errors for $(\Delta B_w/B_w)_{rms} \leq 5\%$, although the standard deviation

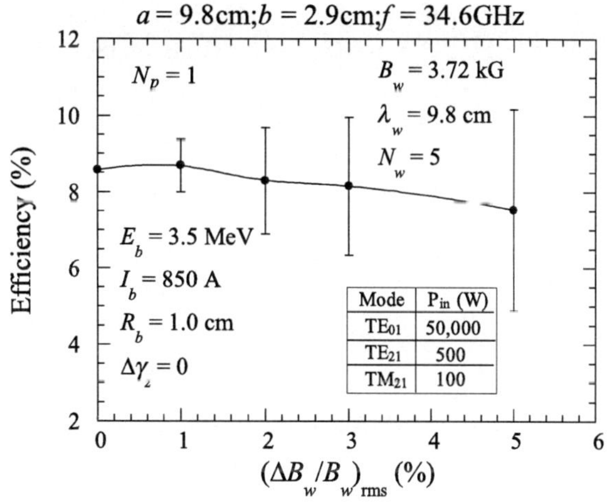

Fig. 11.2 Plot of the variation in the ensemble-averaged efficiency as a function of the rms wiggler fluctuation for $N_p = 1$. The error bars denote the standard deviation

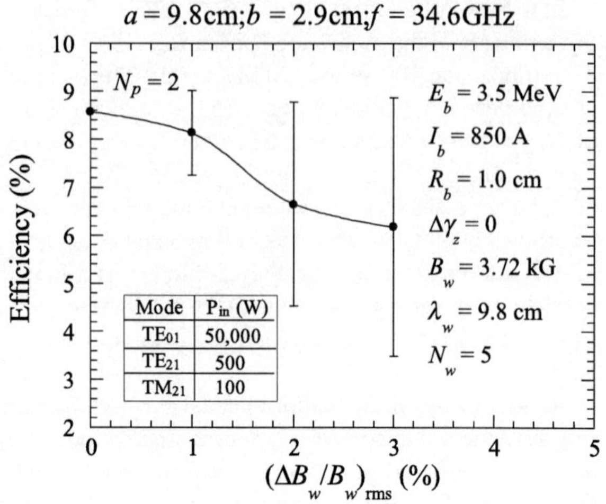

Fig. 11.3 Plot of the variation in the ensemble-averaged efficiency as a function of the rms wiggler fluctuation for $N_p = 2$. The error bars denote the standard deviation

increases with the rms error. For this example, the effect of a given $(\Delta B_w/B_w)_{\mathrm{rms}}$ is much more benign than for a comparable $\Delta\gamma_z/\gamma_0$. We note from Fig. 11.1 that a wiggler fluctuation of 5% would correspond to a fluctuation in the axial energy of approximately 22%.

The interval over which the random wiggler imperfections occur is dependent upon a particular wiggler design; specifically, upon the number of pole pieces per wiggler period. This can be important because the effect of wiggler imperfections varies somewhat with the period over which the imperfections occur. This is illustrated in Fig. 11.3 in which the average efficiency is plotted versus the rms wiggler variation for $N_p = 2$, which describes random wiggler fluctuations at half-wiggler-period intervals. In comparison with Fig. 11.2, the efficiency falls somewhat more rapidly in this case and drops from 8.6% to 6.2% as the rms wiggler fluctuation increases to 3%. However, the effect of decreasing the interval between the random fluctuations does not result in a monotonic variation in the average efficiency. This is shown in Fig. 11.4 in which the variation in the average efficiency is plotted versus N_p for $(\Delta B_w/B_w)_{\mathrm{rms}} = 3\%$.

Observe that the efficiency increases relative to the ideal wiggler case for particular wiggler fluctuations. In order to understand this, recall that the efficiency varies across the frequency band. This tuning can also be accomplished by variations in the wiggler magnitude, and an increase (decrease) in the mean B_w can result in an increase (decrease) in the efficiency as long as the chosen frequency remains in the resonant bandwidth of the interaction. Another way in which the wiggler fluctuation can affect the efficiency is if the field exhibits a bulk taper either upward or downward over the interaction region. A downward (upward) taper increases (decreases) the efficiency. In fact, it is this average up- or down-taper, which accounts for the extrema in simulation.

Fig. 11.4 Variation in the average efficiency versus the interval between random wiggler fluctuations

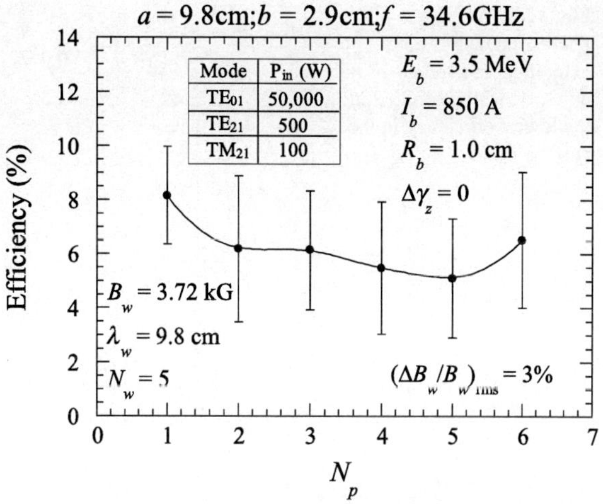

Fig. 11.5 Wiggler fluctuation distribution exhibiting a down-taper resulting in the highest efficiency in the ensemble

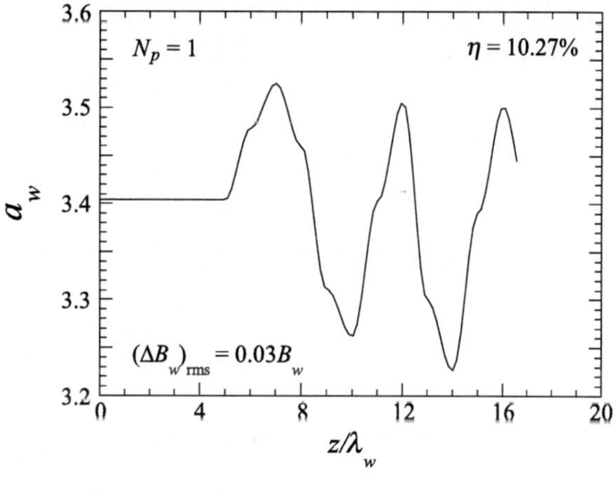

In order to illustrate this, consider the case for which $(\Delta B_w/B_w)_{\mathrm{rms}} = 3\%$ and $N_p = 1$. Wiggler error distributions, which give rise to $\eta = 10.3\%$ and 5.9% (compared to $\eta = 8.6\%$ for an ideal wiggler), respectively, are shown in Figs. 11.5 and 12.6. The average a_w for each of these cases is close to the systematic value of 3.404; however, the field exhibits a downward taper in Fig. 11.5 and an upward taper in Fig. 11.6.

In general, the statistical distribution of the efficiency differs from the normal distribution, and the standard deviation must be used with some caution. For example, 35 runs were generally required to obtain adequate statistics, and the probability histogram is shown in Fig. 11.7 for $(\Delta B_w/B_w)_{\mathrm{rms}} = 3\%$ and $N_p = 1$. Here, the skewness ≈ -0.41 and the kurtosis ≈ 0.92 indicating a distribution skewed below the mean and more peaked than the normal distribution.

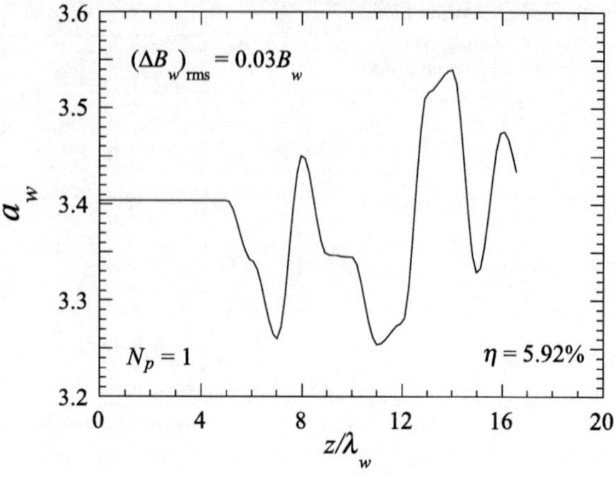

Fig. 11.6 Wiggler fluctuation distribution exhibiting an average up-taper, which resulted in the lowest efficiency in the ensemble

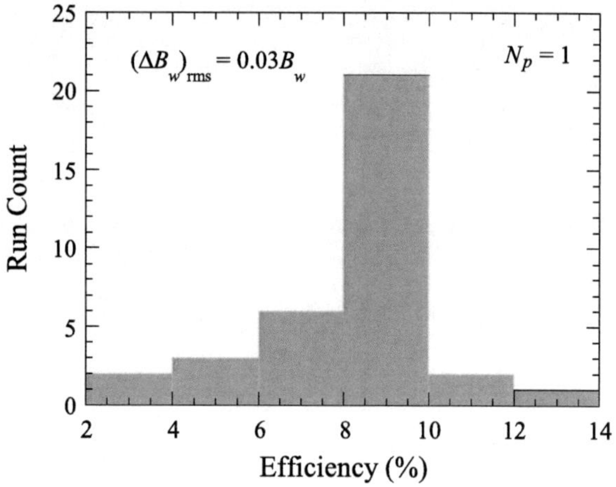

Fig. 11.7 Probability histogram including 35 different wiggler fluctuation distributions

11.3 The Short Wavelength Regime

In treating the effect of wiggler imperfections in short wavelength free-electron lasers, we turn to the nonlinear formulation based upon Gaussian optical modes [11] described in Chap. 6. In this section, we consider an example of a planar wiggler seeded amplifier, and we use a Gauss-Hermite modal superposition to describe the optical field. As in the preceding nonlinear analyses, the electron dynamics are treated using the complete three-dimensional Lorentz force equations for the magnetostatic and electromagnetic fields; however, since the wavelength of interest here is less than or of the order of several microns, the collective Raman effects due to the beam space-charge waves are neglected.

The example under consideration here is the energy-detuned amplifier experiment discussed in Chap. 6 that was performed at Brookhaven National Laboratory [12]. It is convenient to restate the basic configuration used in the experiment. The experiment used an rf linac with an injector that employed a Ti/sapphire photocathode drive laser to generate an electron bunch of 350 pC and between 98 and 103 MeV that is compressed to 2–3 psec by a chicane bunch compressor located between two accelerator sections and yielding a peak current of about 230 A. The rms energy spread was about 0.1%, and the normalized emittance is about 4 mm-mrad. The electron beam propagated through a 10-m long, weak (two-plane or parabolic-pole-face) focusing wiggler with a period of 3.89 cm and a peak on-axis field of 3 kG [13]. The β-function for this weak, two-plane focusing wiggler is about 2.23 m; hence, the matched beam radius $[= (\varepsilon\beta/\gamma)^{1/2}]$ in the wiggler is about 212 μm, and this is used to initialize the electron beam in the simulation. The wiggler consists of 16 sections, and the electron beam can be kicked at each section by trim coils. In this way, it is possible to measure the optical pulse energy at fixed distances down the wiggler. It is also possible to introduce a linear down-taper by fixing the gap at the start of any given segment and then opening the jaws of all the downstream segments by a fixed amount, and this is what was done in the tapered wiggler amplifier experiment discussed below.

The photocathode drive laser is also used as the seed laser for the amplifier. After compression to 6 psec with a negative frequency chirp, the seed pulse is directed through a bandpass filter with a bandwidth of 1.5 nm (FWHM) centered at 793.5 nm. The simulation employs a seed pulse with a parabolic temporal profile and a full width of 6 psec. The resonant electron kinetic energy for these parameters is 100.86 MeV.

We base the analysis of wiggler imperfections on the interaction at the resonant energy with a peak seed power of 50 kW yielding an initial pulse energy of 200 nJ. Comparison of the simulation (in the absence of wiggler imperfections) with the measured average pulse energies along the wiggler is shown in Fig. 6.15. The average is over a large number of randomly generated wiggler imperfection sequences, and the error bars represent one standard deviation about the average values. We remark that the power gain length for this experiment was approximately 0.54 m, so that we expect that on the basis of the inequality expressed in (11.6) that the interaction will be relatively insensitive to wiggler imperfections as long as $(\Delta B_w/B_w)_{\mathrm{rms}} \ll 0.76\%$.

The simulations of wiggler imperfections were performed using $N_p = 2$, and the results comparing the evolution of the pulse energy along the wiggler are shown in Fig. 11.8 for $(\Delta B_w/B_w)_{\mathrm{rms}} = 0$, 0.25%, and 0.50%. It is clear from the figure that the peak pulse energy has decreased by almost 30% as $(\Delta B_w/B_w)_{\mathrm{rms}}$ increases to 050%, which is in accord with the prediction based upon the inequality (11.6).

In order to illustrate the variation of the performance with increasing levels of wiggler imperfections, we plot the average peak pulse energy (≈ 9.0 m) versus the level of wiggler imperfections in Fig. 11.9. We remark that the performance is relatively independent of the wiggler imperfections as long as $(\Delta B_w/B_w)_{\mathrm{rms}}$ is less than about 0.20–0.25%, after which the degradation in performance is rapid. In view of the agreement between the simulation for a perfect wiggler and the measurements, this implies that $(\Delta B_w/B_w)_{\mathrm{rms}}$ was less than this critical value.

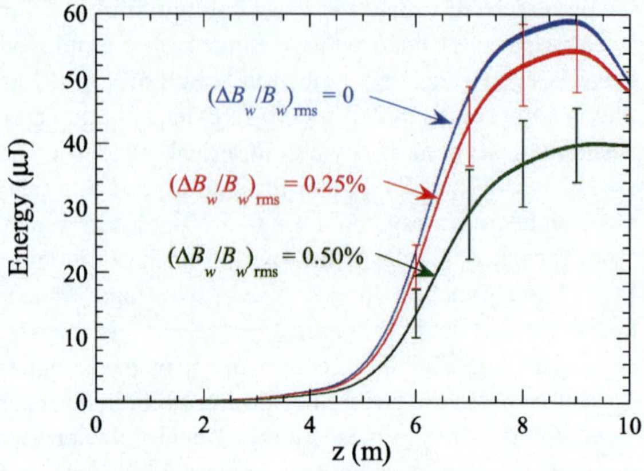

Fig. 11.8 Comparison of the evolution of the pulse energy for different choices of $(\Delta B_w/B_w)_{rms}$

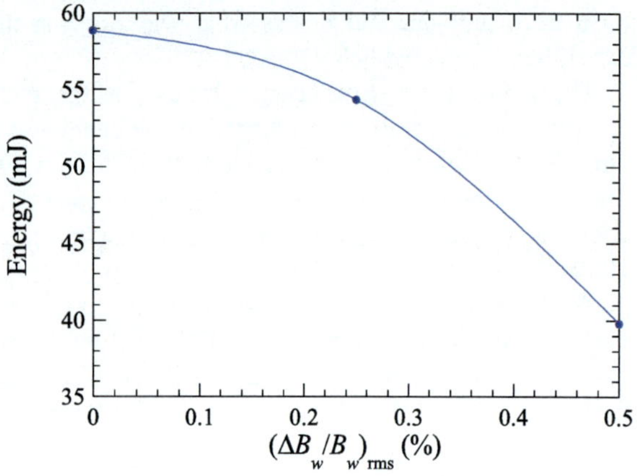

Fig. 11.9 Variation in the peak pulse energy versus the level of wiggler imperfections

11.4 Summary

In summary, a self-consistent analysis of the effect of random wiggler errors on the saturation efficiency of the free-electron laser has been presented in which no a priori assumption of a random walk of the electron orbits has been imposed. For the specific parameters under study, the results indicate that the effects of random wiggler errors are relatively more benign than the effects of beam energy spread, and some error configurations chosen at random were found to result in efficiency enhancements due to effective increases.

One final issue that merits discussion is the relevance of a statistical analysis of wiggler errors to any specific free-electron laser experiment. Consider the construction of a planar wiggler from an assembly of permanent magnets. Such a collection of permanent magnets can be expected to exhibit a fluctuation in magnetization of as much as ±5%. If the wiggler is constructed using a random selection process, then the present analysis may be expected to characterize the statistical properties of *a large number of wigglers constructed in this way*, and the performance of any specific wiggler can be expected to fall within the range determined by the statistical analysis. Wigglers are not constructed using a random selection process, however, but by sorting the permanent magnets to either minimize or optimize the field fluctuations. In either case, however, the performance of a free-electron laser is a deterministic process that is governed by the electron trajectories in the specific wiggler employed in the device. The formulation described herein is capable of modeling specific free-electron lasers by the relatively simple expedient of using measured values for the pole-to-pole fluctuations $\{\Delta B_w\}$ of any given wiggler.

References

1. E. Hoyer, T. Chan, J.Y.G. Chin, K. Halbach, K.J. Kim, H. Winick, J. Yang, The beam line VI rec-steel hybrid wiggler for SSRL. IEEE Trans. Nucl. Sci. **NS-30**, 3118 (1983)
2. B.M. Kincaid, Random errors in undulators and their effects on the radiation spectrum. J. Opt. Soc. Am. B **2**, 1294 (1985)
3. W.P. Marable, E. Esarey, C.M. Tang, Vlasov theory of wiggler field errors and radiation growth in a free-electron laser. Phys. Rev. A **42**, 3006 (1990)
4. L.H. Yu, S. Krinsky, R.L. Gluckstern, J.B.J. van Zeijts, Effect of wiggler errors on free-electron laser gain. Phys. Rev. A **45**, 1163 (1992)
5. C.J. Elliot, B. McVey, Analysis of undulator field errors for XUV free-electron lasers, in *Undulator Magnets for Synchrotron Radiation and Free Electron Lasers*, ed. by R. Bonifacio, L. Fonda, C. Pellegrini, (World Scientific, Singapore, 1987), p. 142
6. H.D. Shay, E.T. Scharlemann, Reducing sensitivity to errors in free-electron laser amplifiers. Nucl. Instrum. Methods Phys. Res. Sect. A **A272**, 601 (1988)
7. W.P. Marable, C.M. Tang, E. Esarey, Simulation of free-electron lasers in the presence of correlated magnetic field errors. IEEE J. Quantum Electron. **QE-27**, 2693 (1991)
8. H.P. Freund, R.H. Jackson, Self-consistent analysis of wiggler field errors in free-electron lasers. Phys. Rev. A **45**, 7488 (1992)
9. H.P. Freund, R.H. Jackson, Self-consistent analysis of the effect of wiggler field errors in free-electron lasers. Nucl. Instrum. Methods Phys. Res. Sect. A **A331**, 461 (1993)
10. H. Bluem, Generation of harmonic radiation from a ubitron/FEL configuration, Ph.D. Thesis, University of Maryland (1990)
11. H.P. Freund, P.J.M. van der Slot, D.L.A.G. Grimminck, I.D. Setya, P. Falgari, Three-dimensional, time dependent simulation of free electron lasers with planar, helical, and elliptical undulators. New J. Phys. **19**, 023020 (2017)
12. X.J. Wang, T. Watanabe, Y. Shen, R. Li, J.B. Murphy, T. Tsang, H.P. Freund, Efficiency enhancement using electron energy detuning in a laser seeded free electron laser amplifier. Appl. Phys. Lett. **91**, 181115 (2007)
13. D.C. Quimby, S.C. Gottschalk, F.E. James, K.E. Robinson, J.M. Slater, A.S. Valla, Development of a 10-meter, wedged-pole undulator. Nucl. Instrum. Methods Phys. Res. Sect. A **A285**, 281 (1989)

Chapter 12
X-Ray Free-Electron Lasers and Self-Amplified Spontaneous Emission (SASE)

Early experiments on Raman free-electron masers [1–23] were often based upon self-amplified spontaneous emission (SASE) where the electromagnetic field grew from shot noise in a single pass through the wiggler; however, the term often used for this configuration at the time was *super-radiant amplifier*. These experiments often made use of pulse line accelerators, modulators, or induction linacs that produced relatively long-pulse (10 nsec through 1 μsec), intense (1–10 kA) electron beams with relatively low energies (less than about 1–2 MeV) and operated at frequencies below about 100 GHz; hence, these experiments operated at millimeter wavelengths and longer. While these electron sources were typically low repetition rate accelerators, they were capable of producing bunch charges exceeding 300 μC; hence, the electromagnetic field could grow from noise to saturation in wigglers of 1–2 m in length.

At the present time, SASE free-electron lasers are commonly used to generate short wavelengths where either lasers that can provide a seed pulse do not exist or where there are no high-reflectivity mirrors to create an optical resonator. In particular, x-ray free-electron lasers based on SASE configurations are becoming common. These free-electron lasers are driven by radio frequency (rf) linacs that produce high energy/short pulse electron beams. An intermediate SASE free-electron laser, which illuminated the principal characteristics of SASE, namely, large fluctuations in the pulse energy and spectrum from shot to shot, was conducted at the Massachusetts Institute of Technology [24]. This experiment was discussed in Chap. 5 and operated at wavelengths in the neighborhood of 500–600 μm, so that the wavelengths excited in this experiment were much shorter than the transverse extent of the drift tube. Hence, the electromagnetic fields that were generated were transitional between the guided modes that existed in the long wavelength Raman free-electron masers and the optical modes that arise in the more recent shot wavelength rf linac-based SASE free-electron lasers [25–32].

The development of laser-driven photocathodes [33–35] and their application in free-electron lasers [36] for use in either rf or dc electron guns to produce high brightness electron bunches combined with rf linacs to accelerate the electrons to

H. P. Freund, T. M. Antonsen, Jr., *Principles of Free Electron Lasers*,
https://doi.org/10.1007/978-3-031-40945-5_12

high energies has permitted free-electron lasers to operate at wavelengths from the infrared through x-rays. This development has been crucial to the application of SASE free-electron lasers to the short wavelength range of the spectrum where high-reflectivity mirrors or seed lasers do not exist; in particular, to the development of UV and x-ray free-electron lasers.

12.1 The Ming Xie Parameterization and the Equivalent Noise Power

The Ming Xie parameterization for the exponential gain [37] has been adapted to describe SASE free-electron lasers by the application of an equivalent noise power derived from the derivation of the incoherent undulator radiation [see Chap. 3]. The equivalent noise power is the power that would be required for a seeded amplifier to match the shot noise and is given approximately by Ming Xie as

$$P_{\text{noise}} \approx \frac{\rho^2 c E_b}{\lambda_{\text{res}}}, \tag{12.1}$$

where ρ is the Pierce parameter, E_b is the electron beam energy, and λ_{res} is the resonant wavelength. In order to understand this, we return to Eq. 3.39 where the power radiated into the forward cone from the electrons within a volume V is given by

$$P_{\text{noise}} \cong 10.9\pi^3\rho^3 E_b \frac{c}{\lambda_{\text{res}}} \frac{\gamma_\parallel^2 V}{\lambda_w^3}, \tag{12.2}$$

and we have made use of the expression for the resonant frequency, $\omega_{\text{res}} = 2\gamma_\parallel^2 k_w c$. We note that several gain lengths are required before the start of exponential gain. This *lethargy* region is typically 2–3 gain lengths in extent. We now choose a cylindrical volume element with a radius equal to the matched beam radius, R_b, and a length three times the gain length, so that the volume element is $V = 3\pi R_b^2 L_G$, where $L_G = \lambda_w / 4\sqrt{3}\rho$. The matched beam radius is typically

$$R_b = \sqrt{\frac{\varepsilon\widehat{\beta}}{\gamma_b}}, \tag{12.3}$$

where ε is the normalized emittance, $\widehat{\beta}$ is the betatron period, and γ_b is the relativistic factor corresponding to the beam energy. For diffraction limited propagation

$$\varepsilon \leq \frac{\lambda_{res}}{4\pi} \cong \frac{\lambda_w}{8\pi\gamma_\parallel^2}. \tag{12.4}$$

As a result, the spontaneous power emitted from this volume element is

$$P_{noise} \cong 18.3\rho^2 E_b \frac{c}{\lambda_{res}} \frac{\widehat{\beta}}{\lambda_w \gamma_b}. \tag{12.5}$$

As an example, if we consider the Linac Coherent Light Source (LCLS) at the Stanford Linear Accelerator Center [30], which is a 1.5 Å SASE free-electron laser and which uses a strong focusing FODO lattice (see Appendix) with a betatron period of $\widehat{\beta} \approx 30$ m, and an energy such that $\gamma_b \approx 2.6 \times 10^4$, with a wiggler period $\lambda_w = 3$ cm; hence, $\widehat{\beta}/\lambda_w\gamma_b \approx 0.038$ and

$$P_{\text{noise}} \cong 0.7\rho^2 E_b \frac{c}{\lambda_{\text{res}}}, \tag{12.6}$$

which is comparable to the Ming Xie formula.

The power gain length, L_G, in this parameterization is given in Eqs. 4.374 and 4.394. It is important to note in this regard that the matched beam radius is also used in the calculation of the Pierce parameter so that this formulation can treat both strong and weak focusing wigglers. As discussed in Chap. 4, there are three solutions to the cubic dispersion equation in the high-gain Compton regime: a growing root, a damping root, and a root with a neutral growth rate. Since the free-electron laser acts as a power amplifier of the spontaneous emission, the length of the lethargy region is the distance it takes for the growing mode to dominate, and the exponential growth of the power is given by

$$P = \frac{1}{9} P_{\text{noise}} \exp(z/L_G). \tag{12.7}$$

The distance to saturation therefore is given approximately by

$$L_{\text{sat}} = L_G \ln\left(\frac{9P_{\text{sat}}}{P_{\text{noise}}}\right), \tag{12.8}$$

where P_{sat} is the saturated power. In an idealized one-dimensional model, the saturated power is given approximately by $P_{\text{sat}} \approx \rho P_{\text{ebeam}}$, where P_{ebeam} is the electron beam power. However, it was found by comparison with simulation results [38] that

$$P_{\text{sat}} \approx 1.6\rho \left(\frac{L_{1D}}{L_G}\right)^2 P_{\text{ebeam}}, \tag{12.9}$$

where L_{1D} is the one-dimensional gain length. These relatively simple formulae have been found to give reasonable approximations to both experimental and simulation results.

12.2 Electron Bunch Compression

Many of the short wavelength free-electron lasers are driven by high energy rf linacs. Because the growth rate is inversely proportional to the beam energy and because rf linacs produce electron bunches with relatively low charges, in the range of nano-Coulombs or less, compression of the bunches to achieve high peak currents is required in order to obtain reasonably short gain lengths and saturation distances. This bunch compression is done prior to the wiggler, and so the dynamics of bunch compression is outside the intended scope of this work; however, some discussion of the basic physics of the compression process is appropriate.

Electron bunch compression is typically done in magnetic chicanes formed by a sequence of dipole magnets. A schematic illustration of the process is shown in Fig. 12.1. The chicane illustrated in the figure consists of four magnetic dipoles of equal lengths and magnetic fields in which the fields in the outer dipoles are directed in the opposite direction to those of the inner dipoles. Note that this is not the only possible configuration. For example, the two center dipoles could be replaced by a single dipole (1) of the same length but twice the field of the outer dipoles or (2) twice the length but the same field as the outer dipoles. In addition, more complex chicanes can be designed, and some compression can also be done in the bending magnets used in recirculating beam/wiggler lines such as in the energy recovery linac used in the free-electron laser oscillator at JLab (see Chap. 10).

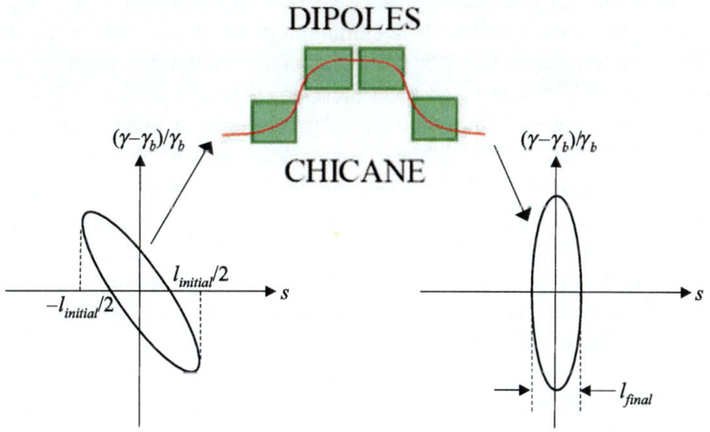

Fig. 12.1 Schematic illustration of bunch compression in a dipole chicane

Bunch compression in the chicane occurs because the orbital deflection due to the dipole fields decreases with increasing electron energy; hence, the path length through the chicane decreases with increasing energy. Thus, higher energy electrons in the tail of the bunch will overtake lower energy electrons in the head of the bunch. This is illustrated in Fig. 12.1 where the electron bunch is assumed to enclosed within an ellipse in the phase space of energy (γ) versus position in the bunch (s). The convention used here is that the trailing edge of the bunch is to the left and that γ_b represents the average electron energy. The phase space prior to injection into the chicane is characterized by an ellipse that is tilted such that higher energy decreases from the tail to the head of the bunch. The initial bunch length is l_{initial}. If the dipole fields are chosen correctly, then the ellipse can be rotated so that it stands "on end" and the bunch length is compressed to $l_{\mathrm{final}} < l_{\mathrm{initial}}$. The field strength required for this degree of compression is determined based upon the slope of the ellipse, which determines the R_{56} factor of the chicane required to give the proper rotation (see Appendix).

Of course, the process in the laboratory is complicated by effects such as coherent synchrotron radiation (CSR). This is discussed in the Appendix and refers to the emission of radiation in the dipole magnets with wavelengths longer than the bunch length. This CSR has the effect of increasing the emittance and energy spread of the electrons, which limits the compression process and can degrade to interaction in the wiggler. Thus, the optimal compression factor is found by balancing the peak current desired against the deleterious effects of increased emittance and energy spread.

12.3 SASE and MOPA Comparison

The interaction in a tapered wiggler has been discussed for both guided and optical modes for master oscillator power amplifiers (MOPAs), otherwise called simply seeded amplifiers, for which enhancements in the efficiency of from three to five times have been demonstrated both in simulation and in the laboratory. The efficient use of a tapered wiggler requires the optimization of both the start-taper point and the slope of the taper. The start-taper point must be chosen to correspond to the point where the electrons begin to cross the phase space separatrix between free streaming and trapped trajectories. While this is easily done in a MOPA, the shot-to-shot fluctuations associated with SASE imply that the optimal start-taper point will also fluctuate. In addition, SASE free-electron lasers have a reduced longitudinal coherence relative to MOPAs, and this too may affect the overall enhancement that can be achieved with the taper. In this section, we compare the performance of SASE and MOPA configurations with a tapered wiggler using the nonlinear optical mode formulation discussed in Chap. 6.

12.3.1 The Case of a Uniform Wiggler

The example under consideration is that of an infrared free-electron laser [39]. The electron beam kinetic energy is 100.75 MeV, and the peak current is 270 A. The bunch shape is assumed to be parabolic with a full width of 2.5 psec, and this yields a bunch charge of 450 pC. The normalized emittance is 4 mm-mrad, and the energy spread is 0.1%. The initial electron beam radius is chosen to be properly matched into a weak-focusing (i.e., parabolic pole face) wiggler with a peak on-axis amplitude of 3.03 kG and a period of 3.89 cm. This results in an rms wiggler strength parameter of 0.78. These beam and wiggler parameters yield a resonance at a wavelength of 795 nm. In the MOPA simulations, we assume that the initial seed laser pulse is synchronized with the electron bunch and has the same temporal profile (i.e., parabolic) and the same pulse duration. Finally, the simulation window is taken to be 7.0 psec in duration to allow for possible slippage of the light relative to the electron beam, and 700 slices of both the electron beam and the electromagnetic field are used.

The fluctuations in the SASE simulations are assumed to derive solely from the phase noise in the initial electron distribution. The shot-to-shot fluctuations in the macroscopic beam parameters (such as the energy, current, emittance, or energy spread) are not included, but it should be noted that any variation in these parameters will result in larger fluctuations than is predicted by the phase noise alone. However, these macroscopic fluctuations can also be expected to affect MOPAs, so that we can limit the SASE simulations to fluctuations due to phase noise without invalidating the conclusions.

We start by conducting SASE simulations for the uniform wiggler. To this end, it was necessary to perform many individual runs each with a different seed for the random number generator in order to reach numerical convergence to the average saturation efficiency and saturation distance. In practice, 24 runs were needed to reach convergence to within better than 1% accuracy. The evolution of the pulse energy versus distance as averaged over these 24 simulation runs is shown in Fig. 12.2. The average pulse energy in the SASE simulations is 96 ± 5.4 μJ. The average saturation distance is 14.1 ± 0.61 m and corresponds to an average power gain length of about 0.57 m. Hence, the phase noise introduces fluctuations of about 5% that of the average values for these parameters.

The predicted equivalent seed power for these parameters is about 1 W. Also shown in Fig. 12.2 are the results of a MOPA simulation in which the equivalent seed power was found to be about 1.2 W, which is reasonable agreement with the prediction. It is clear from the figure that the two curves are very close and that the MOPA and SASE results with regard to the total pulse energy are nearly identical for the uniform wiggler. While the overall efficiencies of the MOPA and SASE FELs are very similar, the details of the pulse shapes and spectra are very different.

These results are in reasonable agreement with the prediction based upon the Ming Xie parameterization [37]. This predicts a power gain length of 0.54 m and a saturation power (P_{sat}) of 100 MW over a saturation distance of 12.2 m. The electron

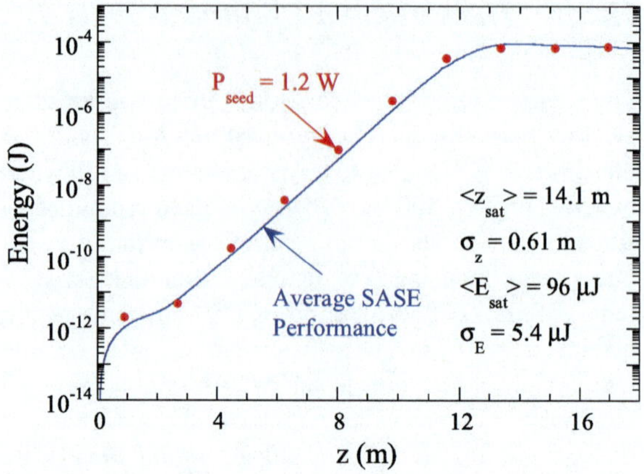

Fig. 12.2 Evolution of the pulse energy versus distance for the average ASASE performance and for a MOPA with the equivalent seed power of 1.2 W [39]

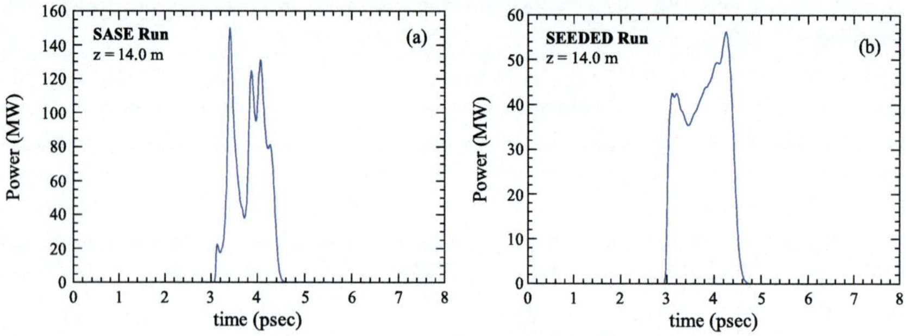

Fig. 12.3 Temporal pulse shapes of the (**a**) SASE, and (**b**) MOPA simulations at the output [39]

bunch was assumed to have a parabolic bunch shape over a 2.5 psec bunch duration (τ_b). If the optical pulse shape mirrored that of the electron bunch, then this implies a saturation energy of about $E_{sat} = (2/3)P_{sat}\tau_b$. However, the pulse shape for SASE exhibits a, possibly, large number of spikes, which result in a reduced output pulse energy with respect to the parabolic pulse. It is found empirically in simulations that the saturated pulse energy is reduced by half, so that the pulse energy that is expected to be produced by a parabolic electron bunch is $E_{sat} \approx (1/3)P_{sat}\tau_b$. For the case under consideration, this produces a predicted output pulse energy of about 84 μJ, which is close to what is found in simulation.

A comparison of the saturation pulse shapes for a particular SASE run with performance close to the average and with the MOPA simulation is shown in Fig. 12.3a, b respectively. It is clear from Fig. 12.3a that full temporal coherence has not been achieved in the SASE simulation as the pulse exhibits sharp spikes. This contrasts with the MOPA pulse output pulse shape shown in Fig. 12.3b. Here, the initial pulse shape was parabolic, and while local maxima are found at the head and tail of the output pulse, this pulse is relatively uniform. Note, however, that the total energy in each case is nearly identical.

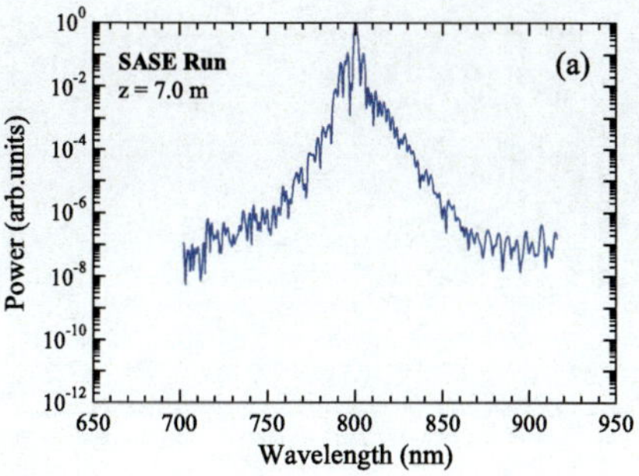

Fig. 12.4 Spectra from SASE simulations at (**a**) 7.0 m prior to saturation, and (**b**) 14.0 m at saturation [39]

The number of spikes to be expected in the SASE output pulse has been derived theoretically by S. Krinsky [40] for a Gaussian bunch shape. Observing from Fig. 12.3a that the radiation pulse duration is about 1.5 psec and that the electron bunch duration is 2.5 psec, this theory predicts that there should be two to three spikes in the SASE output pulse. Although the simulation assumes a parabolic (rather than Gaussian) electron bunch shape, this is in reasonable agreement with the SASE simulations.

More information about the temporal coherence can be obtained from the radiation spectra. In Fig. 12.4, we show the SASE spectra at (a) $z = 7.0$ m in the exponential gain region and (b) at 14.0 m at saturation. It is clear from the figure that the spectrum has narrowed substantially at saturation and the noise floor has dropped by an order of magnitude as well. Thus, while full temporal coherence has not been achieved, the coherence level of the SASE radiation has increased markedly near saturation from the initial shot noise because the wavelengths with the highest growth rates come to dominate the SASE spectrum as the interaction proceeds.

The corresponding spectra for the MOPA simulations are shown in Fig. 12.5 where we show the spectra at (a) 7.0 m in the exponential gain regime and at (b) 14.0 m at saturation. In contrast to the SASE simulation results, the MOPA spectra show broadening as the interaction proceeds. This is because the initial seed has a very narrow spectrum and spectral broadening occurs over the course of the interaction due to the amplification of wavelengths outside the seed laser linewidth. If we had taken the interaction past saturation, where substantial sideband growth is expected to occur, then the spectral broadening would be still more pronounced. Nevertheless, the spectra are considerably narrower in the MOPA simulation than in the SASE case, and the overall noise floor is several orders of magnitude lower in the MOPA than in the SASE free-electron laser.

To summarize thus far, the uniform wiggler results indicate that while the total pulse energy in MOPA and SASE free-electron lasers are comparable, the longitudinal coherence is much poorer in the SASE free-electron laser. The issue now is what effect this difference in the longitudinal coherence has on the tapered wiggler interaction.

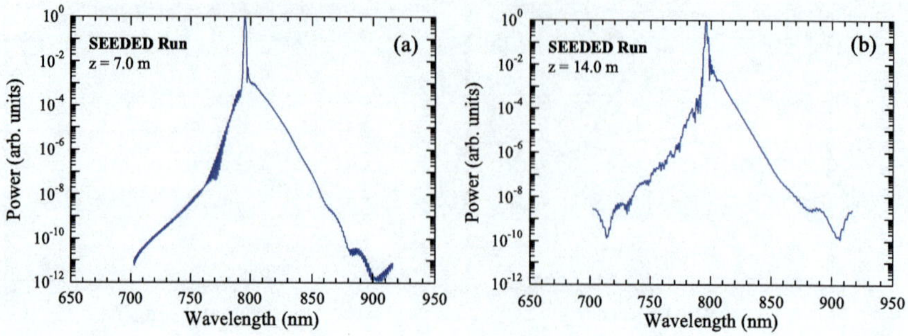

Fig. 12.5 Spectra from MOPA simulations at (**a**) 7.0 m prior to saturation, and (**b**) 14.0 m at saturation [39]

12.3.2 The Case of a Tapered Wiggler

The purpose of a tapered wiggler is to maximize the efficiency, and this is sensitive to both the start taper point and the slope of the taper. Typically, the start taper point must be chosen to be a point prior to saturation. Since the phase noise introduces fluctuations in the saturation point, it seems reasonable to optimize the taper parameters to correspond to the particular SASE simulation that is closest to the average performance. In practice, this corresponds to a specific noise seed in the random number generator. After making this choice, we then made a sequence of tapered wiggler simulations where we varied both the start taper point and the taper slope and fixed the total wiggler length to be about 18 m. The results of these simulations are shown in Fig. 12.6 where we plot the output energy versus both the start taper point and the taper slope. Optimal performance is found when the start taper point is located 12.8 m after the wiggler entrance and for a decreasing slope of 44 G/m. The output pulse energy for this case is 300 μJ. Higher efficiencies would be possible for still longer wigglers.

The optimum shown in Fig. 12.6 is fairly broad, and the overall efficiency enhancement over the uniform wiggler performance is found to be more than 100% over a relatively broad parameter range. Because of the breadth of this peak, the efficiency enhancement due to a tapered wiggler will still be effective for SASE free-electron lasers. Using the optimal start taper point and taper slope described above, we made a sufficient number of simulations using different phase noise distributions to reach numerical convergence. As a result, we find that the output pulse energy is 276 ± 42 μJ. Hence, the phase noise introduces a 15% fluctuation in the shot-to-shot output pulse energy for these parameters.

We conclude that the phase noise fluctuations are not great enough to render the tapered wiggler interaction ineffective. Significant efficiency enhancements are possible in SASE free-electron lasers with tapered wigglers, although the shot-to-shot variation might have an impact on specific applications.

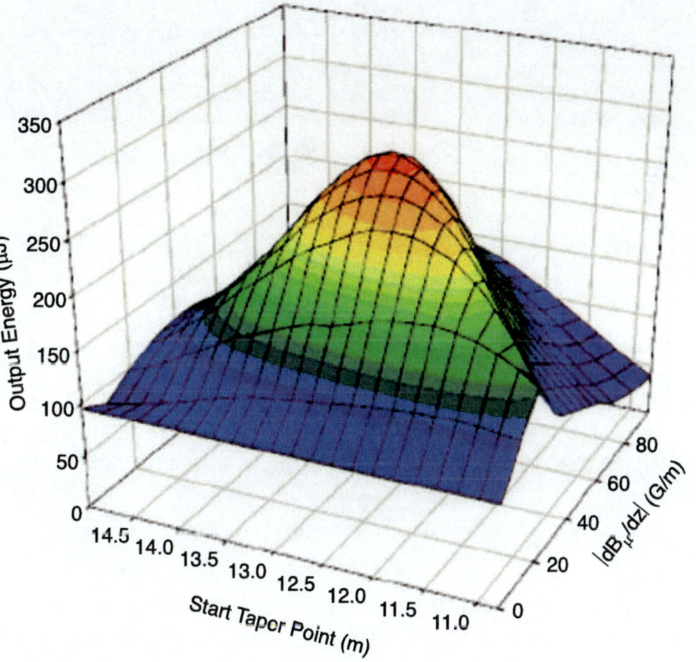

Fig. 12.6 Variation in the output pulse energy with variations in the start taper point and the taper slope for an average SASE run [39]

In order to compare the tapered wiggler performance of the SASE free-electron laser with that of a MOPA with the equivalent seed power, we optimized the tapered wiggler interaction of the MOPA with respect to the start taper point and the taper slope. The results are shown in Fig. 12.7. The optimal performance for a MOPA is found for a start taper point of 12.0 m and a decreasing slope of 48.9 G/m. This is close, but not identical, to the optimal parameters for the tapered wiggler SASE performance. However, the optimal output energy for the MOPA is much higher than that found for the optimal SASE configuration, and we find that the optimal output energy for the MOPA is 616 μJ, which is more than twice that for the SASE example. This is illustrated in Fig. 12.8, where we plot the evolution of the energy versus distance for the optimal parameters for the MOPA and SASE examples.

Therefore, while a tapered wiggler can substantially enhance the efficiency in a SASE free-electron laser, the overall performance of an equivalent MOPA is substantially greater.

The reason for the difference between the tapered wiggler interaction in the MOPA and SASE free-electron lasers derives from the longitudinal coherence of the interaction in each configuration. As shown in Fig. 12.4, full longitudinal coherence has not formed at saturation in the SASE example, which is close to the start taper point as shown by the multiple sharp spikes in the temporal pulse shape and the broad spectral

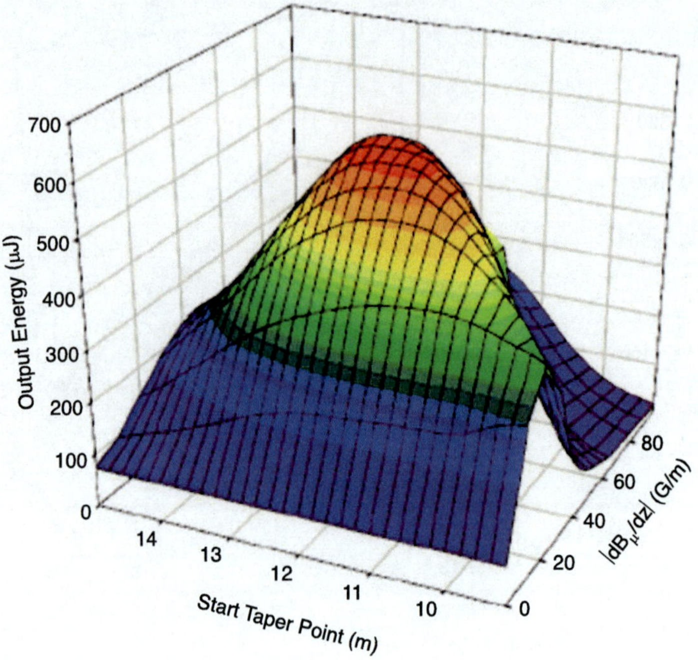

Fig. 12.7 Variation in the output pulse energy with variations in the start taper point and the taper slope for a MOPA [39]

Fig. 12.8 Evolution of the pulse energy versus distance for the optimal tapered wiggler parameters for the average SASE performance and for a MOPA with the equivalent seed power of 1.2 W [39]

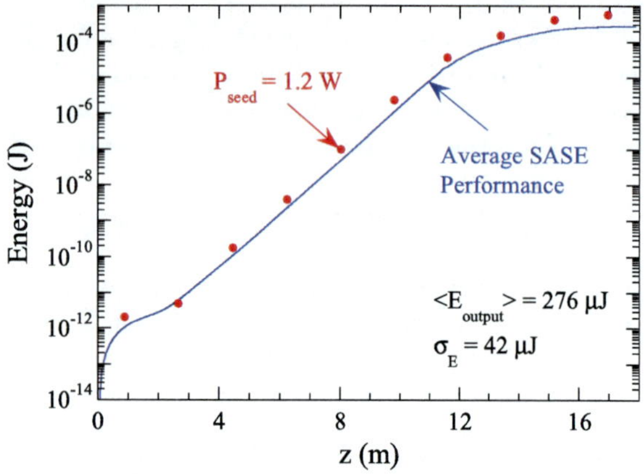

width. The temporal pulse shape at the output of the tapered wiggler for the SASE example is shown in Fig. 12.9a. It is evident from the figure that while a single sharp spike is still present, full longitudinal coherence has not formed. The output spectrum for the SASE example is shown in Fig. 12.9b where it is clear that while the noise floor is comparable to that found for the uniform wiggler, the tapered wiggler spectrum is substantially broader than that for the uniform wiggler.

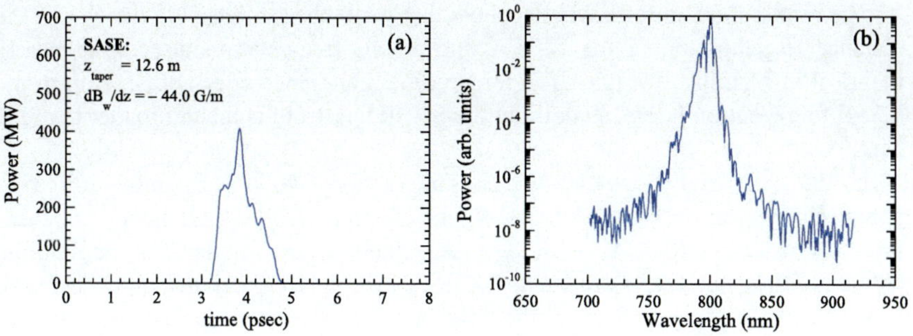

Fig. 12.9 Temporal shape (**a**) and spectrum (**b**) of the optimal SASE output pulse [39]

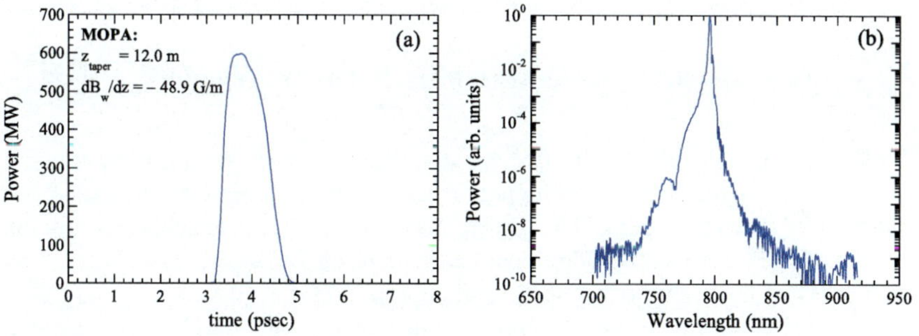

Fig. 12.10 Temporal shape (**a**) and spectrum (**b**) of the optimal MOPA output pulse [39]

The temporal pulse shape and the spectrum for the MOPA example are shown in Fig. 12.10. In contrast to the SASE example, the output pulse MOPA shows good longitudinal coherence, and the spectrum is much narrower, and the noise floor is an order of magnitude lower.

12.3.3 Summary

We have addressed three questions regarding the difference between SASE and MOPA free-electron lasers. First, how do the shot-to-shot fluctuations in SASE free-electron lasers affect the efficiency enhancement in tapered wigglers, and how does this compare with the efficiency enhancement in MOPAs? Second, what are the coherence properties in MOPAs and SASE free-electron lasers? Third, do the differences in the spectral and coherence properties of MOPAs and SASE free-electron lasers result in differences in the total efficiency enhancement that can be achieved using these two different configurations?

The results for a uniform wiggler show that while the saturated efficiency and the distance to saturation in the MOPA and SASE free-electron laser were nearly identical in a uniform wiggler, full longitudinal coherence does not develop in the SASE free-electron laser. In addition, the SASE spectrum is much broader than the MOPA.

In the case of the tapered wiggler interaction, while substantial efficiency enhancements are possible for a SASE free-electron laser, the efficiency enhancement is greater in a MOPA. As in the case of the uniform wiggler, full longitudinal coherence does not develop in the SASE free-electron laser, and the output spectrum is substantially broader than in the MOPA. These differences between the two configurations may have profound implications for specific applications where a choice between the MOPA and SASE free-electron lasers is possible.

12.4 Slippage and Phase Matching Between Wigglers

Most short wavelength SASE free-electron lasers require high energy electron beams. The growth rate in the high-gain Compton regime scales inversely with beam energy, hence, extremely long wigglers is often required to achieve saturation. Because of (1) the need for additional beam focusing and steering over these long distances and the multiple diagnostic ports required to control the beam and (2) the difficulty in producing extremely long wiggler, many of these SASE free-electron lasers employ multiple wiggler segments with focusing/steering/diagnostics located in the gaps between segments. In many of these SASE free-electron lasers, the gain length is comparable to the length of the wiggler segments so that it is necessary to ensure that the phase shift of the optical field relative to the electrons in the gaps between the wiggler is close to a multiple of 2π (i.e., the resonant wavelength) in order to ensure that there is coherent amplification of the optical field from segment to segment. This can be accomplished by the proper choice of the gap length between the wigglers or through the use of weak magnetic chicanes to increase the effective path length of the electron trajectories.

12.4.1 The Phase Match in a Uniform Wiggler Line

A schematic illustration of the typical configuration showing two wigglers with a quadrupole in the gap is shown in Fig. 12.11. The field in each wiggler is assumed to be constant except for transition regions at the ends of the wiggler needed to match the electron beam into and out of the wiggler. These transition regions are indicated by the solid bars at the ends of the wiggler that are an integer number of wiggler periods in length (i.e., $N_{trans}\lambda_w$). The length of the gap, L_{gap}, includes the drift spaces separating the wigglers from the quadrupole as well as the length of the quadrupole.

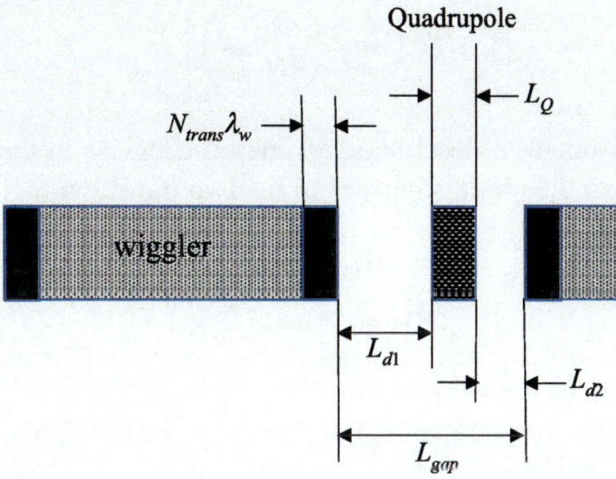

Fig. 12.11 Schematic illustration of two wiggler segments with a quadrupole magnet in the gap

Observe that the quadrupole need to be located in the center of the gap. Note that the quadrupole is used for enhanced focusing of the electron beam and has no effect on the path length of the electrons.

A general analytic formula for the phase slippage of the optical field relative to the electrons over a distance L is [41]

$$\frac{\varphi}{2\pi} = -\frac{L}{\lambda}\frac{c}{v_b}\left(1 - \frac{v_b}{c}\right), \tag{12.10}$$

where λ is the wavelength, v_b is the axial electron velocity, and c is the speed of light in vacuo. In gaps between the wigglers $v_b/c \cong 1–1/2\gamma_b{}^2$, where $\gamma_b = 1 + E_b/m_ec2$, E_b is the kinetic energy of the electron beam, and e and m_e are the electronic charge and rest mass. The phase slippage in the gaps at the resonant wavelength is

$$\frac{\varphi_{\text{gap}}}{2\pi} \cong -\frac{L_{\text{gap}}}{\lambda_w\left(1 + K_{\text{rms}}^2\right)}, \tag{12.11}$$

The phase slippage in the uniform wiggler region will vary with the wavelength as

$$\frac{\varphi_w}{2\pi} \cong -\frac{L_{\text{gap}}}{\lambda_w}\left(1 - \frac{\lambda_{\text{res}} \quad \lambda}{\lambda_{\text{res}}}\right), \tag{12.12}$$

where L_w is the length of the uniform wiggler region and the resonant wavelength $\lambda_{\text{res}} = \lambda_w(1 + K_{\text{rms}}^2)/2\gamma_b{}^2$. This describes the slippage of the optical field relative to the electrons as one wavelength per wiggler period at the resonant wavelength. Using the transition model described in Eq. 5.14 for the entry taper, the wiggler amplitude is given by $B_w(z) = B_{w0}\sin^2(k_wz/4N_{\text{trans}})$, the axial velocity is

$$\frac{v_b}{c} \cong 1 - \frac{1}{\gamma_b^2}\left[1 + K_{rms}^2 \sin^4\left(\frac{k_w z}{4N_{trans}}\right)\right]. \quad (12.13)$$

The exit transition is assumed to describe a symmetric decrease in the wiggler amplitude, so that the sum of the phase slippages in the two transition regions is

$$\frac{\varphi_{trans}}{2\pi} \cong -\frac{2L_{eff}}{\lambda_w\left(1 + K_{rms}^2\right)}, \quad (12.14)$$

where the effective length of the transition region is $L_{eff} = N_{trans}\lambda_w(1 + 3K_{rms}^2/8)$. As a consequence, the total phase advance through the wiggler (including the uniform and transition regions) and the gap is

$$\frac{\varphi_{total}}{2\pi} \cong -\frac{L_{gap} + 2L_{eff}}{\lambda_w\left(1 + K_{rms}^2\right)} - \frac{L_{gap}}{\lambda_w}\left(1 - \frac{\lambda_{res} - \lambda}{\lambda_{res}}\right). \quad (12.15)$$

The variation of φ_{total} with wavelength in the gaps and the transition regions over the full amplification band is small and may be neglected. However, this is not necessarily the case in the uniform wiggler regions. The range of wavelengths in a typical free-electron laser that can be amplified is small; in particular, $|\lambda - \lambda_{res}| << \lambda$. However, when $L_w >> \lambda_w$, the phase shift in the uniform sections of the wigglers can vary by an appreciable fraction of a wavelength across the free-electron laser amplification band.

In the example under consideration, the electron beam is Gaussian with a 14.36 GeV energy and a 3272 A peak current. The beam is not symmetric, and the normalized emittances are 0.8869 mm-mrad and 0.8198 mm-mrad in the x- and y-directions, respectively. The beam is elliptical with rms beam dimensions of 19.32 (24.22) microns in the x (y)-direction. Since many simulation runs are required to fully examine the phase slippage, the beam model is simplified by assuming that the energy spread vanishes. This does not change any of the essential conclusions regarding the phase slippage while reducing the computational load significantly. The transport line consists of 33 wigglers and quadrupoles (located in the centers of the gaps). The undulators have a peak on-axis field (B_w) of 13.25 kG and a period (λ_w) of 3.0 cm. An analytic field model of a flat-pole-face undulator field is used with a uniform field section of 112 periods in length and one-period entry and exit transitions. The field is oriented so that the wiggle motion is in the x-direction. A transition region of one period in length is used ($N_{trans} = 1$). The quadrupoles are configured as a FODO lattice (see Appendix). A hard-edged field model is used with a field gradient of 10.6 kG/cm over a length of 0.05 m. The Twiss α-parameters are assumed to be zero, so that the transport line is configured with a half-quadrupole before the first undulator to rotate the phase space so that the beam is matched into the FODO lattice. It should be remarked that while the wigglers provide weak focusing in the y-direction, beam focusing is dominated by the quadrupoles in the FODO lattice. This ensures that the beam is focused in both the x- and y-directions.

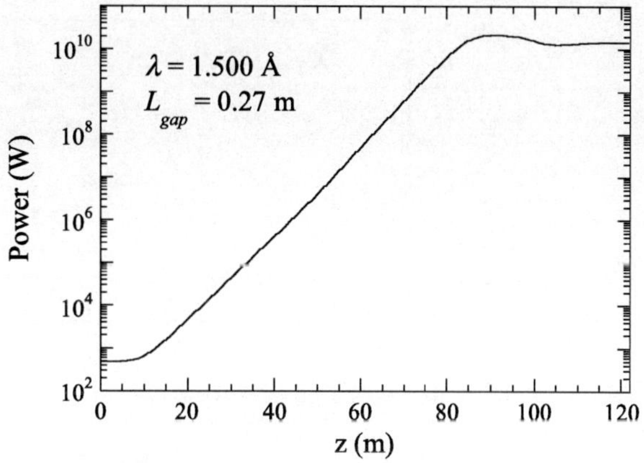

Fig. 12.12 Evolution of the power versus distance for a wavelength of 1.5 Å and a gap length of 0.27 m [41]

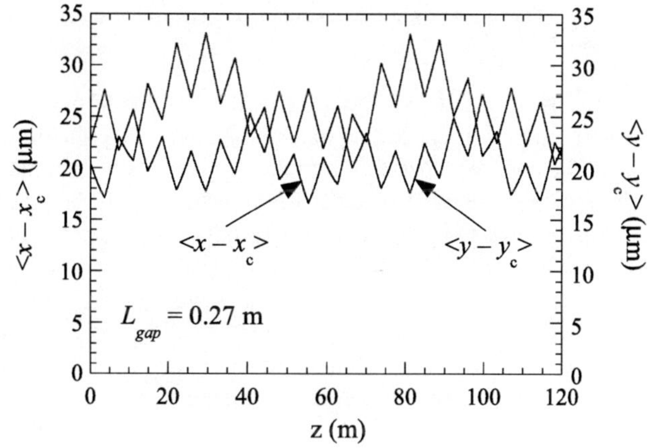

Fig. 12.13 Evolution of the beam envelope as a function of distance [41]

Simulations are performed that correspond to a seeded amplifier at fixed wavelengths. An example of the results is shown in Fig. 12.12 where the power is plotted versus distance for a gap length of 0.27 m at a wavelength of 1.500 Å with an initial drive power of about 500 W corresponding to the equivalent noise power. After the initial transient region, the power grows exponentially until shortly before saturation at about 91 m with a power of 21.5 GW. This gap length corresponds to an ideal phase match between the optical field and the electrons. The evolution of the transverse beam dimensions versus distance for this case is shown in Fig. 12.13. Observe that the FODO lattice initially focuses the beam in the *x*-direction and defocuses in the *y*-direction. While the beam match can be improved, this is not an important consideration for the relative phase shifts between the optical field and the electrons.

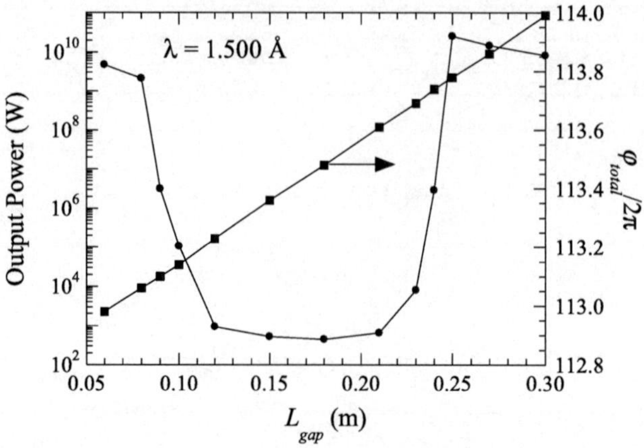

Fig. 12.14 Variation in the output power (circles) and phase slippage with gap length [41]

The phase slippage in simulation differs from the analytic formula because there is an axial velocity spread corresponding to the emittance. Nevertheless, comparison between simulation and the analytic formula is good. For example, the phase slippage in the transition regions is about 0.91λ using Eq. 12.14. This compares well with the average phase slippage of 0.92λ for all of the electrons in simulation. The phase slippage across the unit cell consisting of the wiggler and gap is about 114λ for a gap length of 0.27 m at the resonant wavelength of 1.4975 Å. At the wavelength of 1.500 Å used in Fig. 12.12, this phase slippage is 113.86λ. The full width of the amplification band found in simulation extends from about 1.490–1.506 Å, but this bandwidth would be smaller had an energy spread been used in simulation. Since $\Delta\lambda/\lambda_{res} \approx \pm 0.008$ and the uniform undulator region is $112\lambda_w$ long, the phase slippage in the undulator will vary by as much as ±80% of a wavelength across the amplification band. However, coherent amplification does not occur over the entire amplification band at this, or any other, choice of gap length.

The variation in performance with gap length is shown in Fig. 12.14, where the output power at the end of the undulator line (circles) and phase slippage through a unit cell (squares) is plotted versus gap length. Note that no effort to retune the FODO lattice has been made as the gap length changes. Since the wiggler line is longer than required to reach saturation, the points shown in the figure correspond to a range of cases from where the power has saturated prior to the end of the undulator line to cases where there is no saturation at all. The phase slippage increases linearly with gap length, and it is clear that the free-electron laser will reach saturation when the phase slippage through the unit cell is close to an integer number of wavelengths, but this can vary by as much as about $\pm 0.2\lambda$ without seriously degrading performance. Since this is smaller than the total variation in phase slippage with wavelength across the amplification band in the undulators ($\approx \pm\ 0.8\lambda$), the phase mismatch associated with the segmented undulator will result in a narrowing of the spectrum. Note also that the phase slippage varies by an integer number of wavelengths as the gap length increases by $\lambda_w\,(1 + K_{rms}^2) \approx 0.24$ m.

It should be remarked that if there were high gain in each wiggler, then the system would not be sensitive to the gap length. For wigglers much longer than an e-folding length, the phases of the electrons self-adjust over a distance comparable to an e-folding length after which exponential growth resumes [42]. However, the wigglers for most SASE x-ray free-electron laser designs are comparable to the e-folding length. For this case, the e-folding length is approximately 3.3 m, which is comparable to the undulator length of 3.42 m. As a result, the undulators are not long enough to overcome the initial transient response due to a mismatch in the phase slippage, and care is needed to ensure that the phase slippage is properly matched for coherent amplification in each undulator.

As noted above, the phase slippage in the uniform wiggler region can vary by a substantial fraction of a wavelength over the amplification band. This implies that the proper phase match for coherent amplification over the wiggler/FODO line will occur at different wavelengths as the gap length changes. The change in gap length required to maintain a constant phase slippage as the wavelength changes is

$$\Delta L_{\text{gap}} = L_w \frac{\Delta\lambda}{\lambda_{\text{res}}}\left(1 + K_{\text{rms}}^2\right), \tag{12.16}$$

so that ΔL_{gap} varies across the amplification band by about 0.27 m for the parameters under consideration. Because $\Delta L_{\text{gap}} > \lambda_w(1 + K_{\text{rms}}^2)$, a range of wavelengths in the amplification band can be found that reaches saturation for any desired gap length. This is summarized in Fig. 12.15 where the range of wavelengths that reach saturation is plotted versus gap length. The "error bars" in the figure represent the range of wavelengths over which saturation is reached and the central circle corresponds to the wavelength with a slippage that is an integer number of wavelengths. The vertical dashed lines mark a change in gap length of 0.24 m.

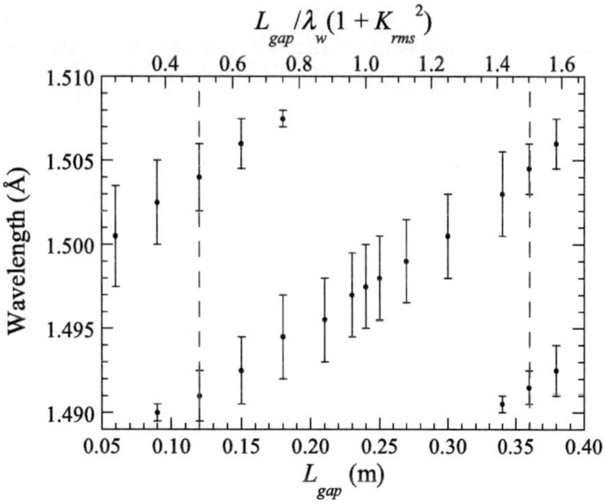

Fig. 12.15 Variation in the range of wavelengths that reach saturation versus gap length [41]

There are several characteristics shown in Fig. 12.15 that deserve mention. First, the range of wavelengths that reach saturation increases linearly with gap length in accord with Eq. 12.16. Second, the expected periodicity at 0.24 m is found in simulation. Third, while the full amplification band extends over 1.490–1.506 Å, the requirement that the phase slippage must be close to an integer number of wavelengths results in a narrowing of this band, and in some cases, two distinct amplification bands are found that are subsets of the full amplification band.

To summarize, the phase slippage observed in simulation is in good agreement with the analytic model described for this slippage. It is found that the phase slippage through a wiggler and gap must be within about 20% of a wavelength for optimal performance for the parameters studied. Since the variation in phase slippage through the wigglers with wavelength across the amplification band exceeds this value, there is a narrowing of the amplification spectrum. This means that (1) saturation can be expected for any choice of gap length, and (2) it may be possible to tune a seeded free-electron laser using both energy and gap length.

These conclusions were obtained for a seeded free-electron laser, but the implications for SASE free-electron lasers may be more complex. In a SASE free-electron laser, the noise power at each frequency is independently amplified in the exponential regime prior to saturation. In this regime, the conclusions, such as the frequency band(s) that can be coherently amplified, will be similar to those found for a seeded amplifier. However, the wavelength preferentially excited in a SASE free-electron laser will correspond to that wavelength in the amplification band that is characterized by a good phase match; hence, the gap length will tune the interaction. In addition, when a SASE free-electron laser nears saturation, there is strong competition between the different frequency components that results in a substantial spectral narrowing. While the effects of this mode competition are not included in the present analysis, it is possible that the gap lengths can be adjusted for detailed spectral control or to reduce spiking modes.

12.4.2 Optimizing the Phase Match in a Tapered Wiggler Line

Enhancements in the efficiency can be obtained by imposing a step taper in a segmented wiggler line. By this technique, the downward step taper from wiggler to wiggler would be imposed starting with a wiggler near the saturation point. However, since the wiggler field is changing from segment to segment, the phase matching condition must also vary to compensate for the taper.

Under the assumption that the resonant wavelength remains fixed, the phase advance depends only upon the phase advances in the transitions and the gap. As such, this phase advance is

$$\frac{\varphi_{tg}}{2\pi} = \frac{L_{gap} + 2N_{trans}\lambda_w\left(1 + 3K_{rms}^2/8\right)}{\lambda_w\left(1 + K_{rms}^2\right)}. \tag{12.17}$$

As a result, the change in the phased advance due to changes in the gap length and wiggler field is

$$\frac{\Delta\varphi_{tg}}{2\pi} = -\frac{\Delta L_{gap}}{\lambda_w\left(1 + K_{rms}^2\right)} + \left(\frac{L_{gap}}{\lambda_w} - \frac{5}{4}N_{trans}\right)\frac{2K_{rms}\Delta K_{rms}}{\left(1 + K_{rms}^2\right)^2}. \tag{12.18}$$

If it is now required that $\Delta\varphi_{tg} = 0$, then

$$\Delta L_{gap} = \left(L_{gap} + \frac{5}{4}N_{trans}\right)\frac{2K_{rms}\Delta K_{rms}}{\left(1 + K_{rms}^2\right)}. \tag{12.19}$$

Denoting the K_{rms} and gap length prior to the ith wiggler as K_i and L_i, respectively, then the gap length after the ith wiggler must be adjusted in such a way that

$$L_i = \left(1 + \frac{2K_{i-1}\Delta K_{rms}}{1 + K_{i-1}^2}\right)L_{i-1} + \frac{5}{4}N_{trans}\lambda_w\frac{2K_{i-1}\Delta K_{rms}}{1 + K_{i-1}^2}. \tag{12.20}$$

in order to maintain the proper phase slippage.

12.4.3 Phase Shifters

It is often either difficult or impractical to adjust the gaps between wigglers to maintain the proper phase shifts when the operating parameters of the free-electron laser are changed. As a result, weak dipole chicanes may be used to shift the phase advance between the optical field and the electrons in the gaps between the wigglers. This applies to both uniform and tapered wiggler lines. In such cases, the chicane is not designed to compress the electron bunch but, rather, to increase the effective path length of the electrons in the gaps between the wigglers.

The electron trajectories through a chicane are described in the Appendix. Under the assumption that the electrons are propagating along the axis of symmetry as they enter the dipole having a field strength of B_0 and a length of L_d, the path length, s_d, can be written as

$$s_d = \frac{\gamma_b v_b}{\Omega_0}\sin^{-1}\left(\frac{L_d\Omega_0}{\gamma_b v_b}\right), \tag{12.21}$$

where $\gamma_b = \left(1 - v_b^2/c^2\right)^{-1/2}$, $\Omega_0 = eB_0/m_ec$ is the gyrofrequency, and it has been assumed that the dipole length is shorter than the radius of curvature (i.e., $L_d\Omega_0 << $

$\gamma_b v_b$). Expanding the inverse sine function in powers of the argument, the path length is given approximately by

$$s_d \cong L_d + \frac{1}{6}\frac{L_d^3\Omega_0^2}{\gamma_b^2 v_b^2}. \qquad (12.22)$$

As a result, the effective increase in the path length due to the dipole is

$$\Delta L_d = \frac{1}{6}\frac{L_d^3\Omega_0^2}{\gamma_b^2 v_b^2}. \qquad (12.23)$$

A dipole chicane may be configured in various ways. For example, it is common for the chicane to consist of four dipoles of equal lengths and magnetic fields where the outer dipoles are oriented oppositely to the two center dipoles. In this case, the effective path length is simply the sum of the increases due to each dipole; hence,

$$\Delta L_{\text{chicane}} = \frac{2}{3}\frac{L_d^3\Omega_0^2}{\gamma_b^2 v_b^2}. \qquad (12.24)$$

This also holds if there are three dipoles of equal field strengths but where the two center dipoles are replaced by a single dipole having twice the length of the outer dipoles. However, if the single center dipole is the same length as the outer dipoles but has twice the field strength, then the effective path length is increased by

$$\Delta L_{\text{chicane}} = \frac{L_d^3\Omega_0^2}{\gamma_b^2 v_b^2}, \qquad (12.25)$$

where Ω_0 denotes the gyrofrequency in the outer dipoles.

As a result, dipole chicane phase shifters may be used to adjust the path length of the electrons through the gaps between the wigglers in either uniform or tapered wiggler lines.

12.5 Comparison Between Simulation and Experiments

In this section, simulations using the formulation described in Chap. 6 are compared with two SASE free-electron lasers: the Linac Coherent Light Source [29] and at a wavelength of 1.5 Å and the "Sorgente Pulsata ed Amplificata di Radiazione Coerente" (SPARC) experiment, which is a SASE FEL located at ENEA Frascati [30] at a wavelength of 492 nm.

12.5.1 The Linac Coherent Light Source (LCLS)

The linac coherent light source (LCLS) at the Stanford Linear Accelerator Center (SLAC) [29] is a SASE free-electron laser user facility that became operational in 2009 based upon one-third of the SLAC linac. The LCLS employs a 13.64 GeV/250 pC electron beam with a flat-top temporal pulse shape of 83 fsec duration. The normalized emittance (x and y) is 0.4 mm-mrad, and the rms energy spread is 0.01%. The wiggler line consists of 33 segments with a period of 3.0 cm and a length of 113 periods including one period each in entry and exit tapers. A mild down-taper in field amplitude of −0.0016 kG/segment starting with the first segment (with an amplitude of 12.4947 kG and $K_{rms} = 2.4748$) and continuing from segment to segment was used. This is the so-called *gain taper*. The electron beam was matched into a FODO lattice consisting of 32 quadrupoles each having a field gradient of 4.054 kG/cm and a length of 7.4 cm. Each quadrupole is placed a distance of 3.96 cm downstream from the end of the preceding wiggler segment. The gap lengths between the wiggler segments are not uniform but occur in a repeating sequence of 0.48 m, 0.48 m, and 0.908 m.

The initial Twiss parameters for this FODO lattice include an rms beam size of 21.5 μm in the x-direction and 19.5 μm in the y-direction with Twiss-$\alpha_x = 1.1$ indicating focusing in the x-direction and Twiss-α_y in the y-direction of −0.82 indicating defocusing. The initial Twiss $\widehat{\beta}$ are 30.85 m and 25.38 m in the x- and y-directions, respectively. The propagation of the beam through the wiggler/quadrupole lattice as found in simulation is shown in Fig. 12.16 [43], where the beam envelope is plotted in x (blue, left axis) and y (red, right axis) versus position. Observe that the beam is well confined over the 130 meters of the extended lattice with an average beam size of about 21 μm that conforms to the expected value of $\sqrt{\varepsilon_n\widehat{\beta}/\gamma_b}$.

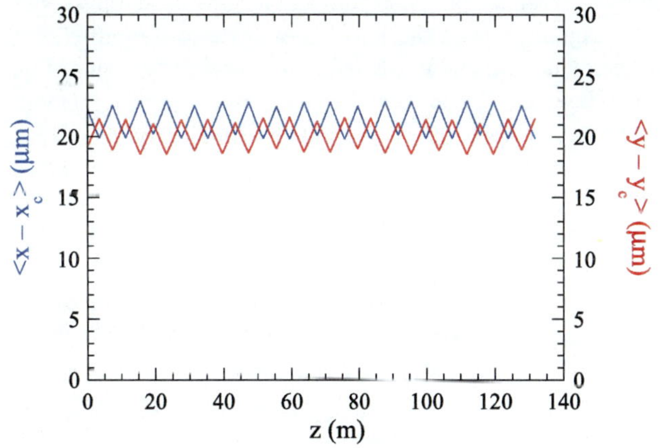

Fig. 12.16 Simulated propagation of the LCLS beam [43]

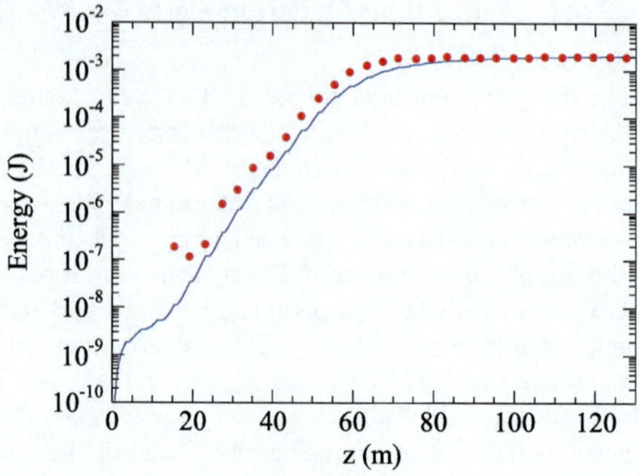

Fig. 12.17 Comparison between the average pulse energy found in simulation and the experimental data (circles) [43]

The LCLS produces pulses of about 1.89 mJ at the end of the wiggler line [29], and saturation is found after about 65–75 m. A comparison between the measured pulse energies (red circles), as obtained by giving the electrons a kick to disrupt the free-electron laser interaction, and the simulation (blue) is shown in Fig. 12.17. The experimental data is courtesy of P. Emma and H.-D. Nuhn at SLAC, and the simulation results represent an average over an ensemble of 25 runs performed with different noise seeds. As shown in the figure, the simulations are in good agreement with the measurements in the exponential growth region with close agreement for the gain length. The simulation exhibits saturation at the same distance as the experiment in the range of 65–75 m at a pulse energy of 1.5 mJ. After saturation, in view of the gain taper, the pulse energy grows more slowly to about 2.02 mJ at the end of the undulator line, which is approximately 8% higher than the observed pulse energy. The agreement between simulation and experiment for the pulse energy is poorer during the early stages of the interaction. This may be due to a variety of reasons. On the experimental side, as the pulse energy grows by 5–6 orders of magnitude from the initial shot noise to saturation, it is difficult to calibrate the detectors for the low pulse energies at the early stages of the interaction. Also, while kicking the electrons provides a fast measurement method, it is accompanied by a larger background signal and the possibility of restarting the free-electron laser interaction downstream along the wiggler line when the kick is performed at the beginning of the undulator line [44]. On the simulation side, there may be some inaccuracies in the shot noise algorithm that underestimates the initial noise level.

Experiments have also been performed at the LCLS [44] to investigate enhancing the efficiency using a more sharply tapered wiggler. The LCLS configured with a stronger taper for the last segments has demonstrated enhancements in the efficiency. This experiment employed an undulator in which the aforementioned mild linear down-taper is enhanced by the addition of a more rapid down-taper starting at the 14th wiggler segment. This is referred to as the *saturation taper* profile and is shown

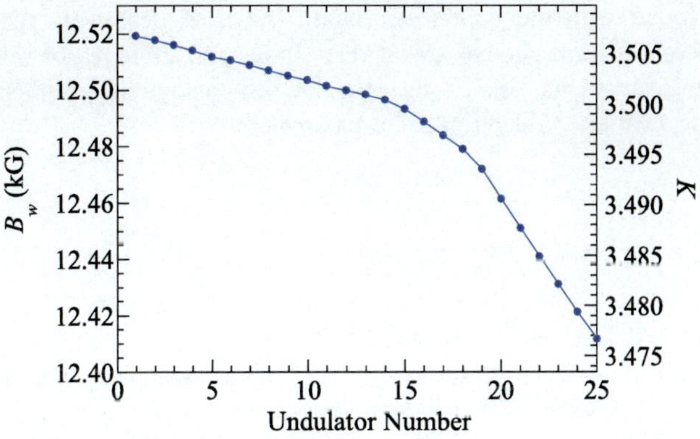

Fig. 12.18 The experimentally applied saturation taper profile [43, 44]

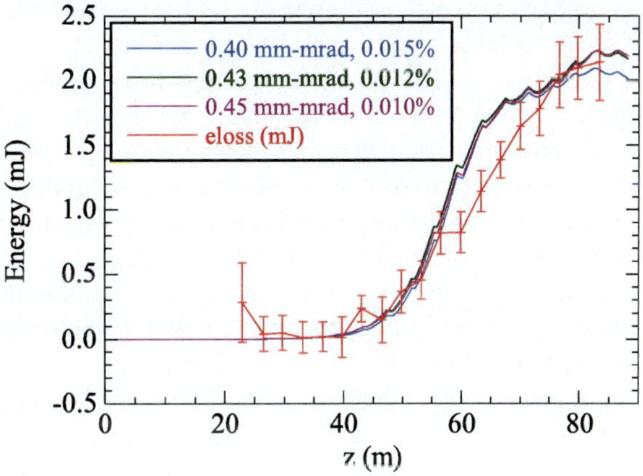

Fig. 12.19 Comparison between the experimental data and simulations for a variety of emittances and energy spreads [43]

in Fig. 12.18 (data courtesy of D. Ratner). No effort was made to adjust the phase match to account for the enhanced taper profile.

In comparison with the wiggler and electron beam properties employed in the first lasing experiments, the tapered wiggler experiment employed wigglers tuned to somewhat different field strengths and electron beam parameters that may have varied from the first lasing experiment represented in Fig. 12.17. The pulse energies in the experiment were obtained by measuring the energy loss in the electron beam. We present the results of simulations conducted over a parameter range including emittances of 0.40–0.45 mm-mrad and energy spreads of 0.010–0.015% that are thought to characterize the electron beam.

A comparison between the measured pulse energies and simulations over the parameter range that most closely agree with the experiment is shown in Fig. 12.19, where the experimental results are shown in with error bars (courtesy of (D. Ratner) and have been obtained by measuring the electron energy loss. The maximum pulse energy shown represents an enhancement of the efficiency by a factor of 2–3 over what was found with the *gain taper* alone. As is evident from the figure, the simulations for the three choices are all very similar and are in reasonable agreement with the measurements, indicating that the efficiency enhancement could be achieved for a variety of electron beam parameters.

12.5.2 The SPARC Experiment

The "Sorgente Pulsata ed Amplificata di Radiazione Coerente" (SPARC) experiment is a SASE free-electron laser located at ENEA Frascati [30]. The electron beam energy was 151.9 MeV, with a bunch charge of 450 pC and a bunch width of 12.67 psec. The peak current was approximately 53 A for a parabolic temporal bunch profile. The x and y emittances were 2.5 and 2.9 mm-mrad, respectively, and the rms energy spread was 0.02%. There were six undulators each of which was 77 periods in length (with one period for the entrance up-taper and another for the exit down-taper) with a period of 2.8 cm and an amplitude of 7.88 kG. The bunch charge was not high enough for the experiment to reach saturation after six wigglers; hence, for the purposes of the simulation, we shall add two extra wigglers to bring the interaction to saturation. The gap between the undulators was 0.4 m in length and the quadrupoles (0.053 m in length with a field gradient of 0.9 kG/cm) forming a strong focusing lattice were located 0.105 m downstream from the exit of the previous undulator. Note that the quadrupole orientations were fixed and did not alternate. The electron beam was matched into the undulator/focusing lattice. The resonance occurred at a wavelength of 491.5 nm. The pulse energies were measured in the gaps between the undulator segments.

In order to match the beam into this strong focusing lattice, the initial rms beam sizes in the x- and y-directions are 132 and 75 μm, respectively. The initial Twiss-α parameters are 0.938 and −0.705 in the x- and y-directions indicating that the beam was initially focused in the x-direction and defocused in the y-direction with Twiss-β parameters $\widehat{\beta}_x = 2.08$ m and $\widehat{\beta}_y = 0.58$ m. The propagation of the beam through the wiggler/quadrupole lattice as found in simulation is shown in Fig. 12.20, where the beam envelope is plotted in x (blue, left axis) and y (red, right axis) versus position. Observe that the beam is well confined over the 20 m of the extended lattice with an average beam size of about 115 microns.

A comparison of the evolution of the pulse energy as found in simulation and as measured in the experiment is shown in Fig. 12.21 where the simulation is indicated by the line and is an average taken over 20 simulation runs with different noise seeds. The pulse energy was measured in the gaps between the wigglers, and the results for a sequence of shots are indicated by the markers (data courtesy of L. Giannessi). Observe that the agreement between the simulation and the measured performance is excellent over the entire range of the experiment. In addition, the simulation shows that saturation could have been reached after about 18–20 m with two additional undulator segments.

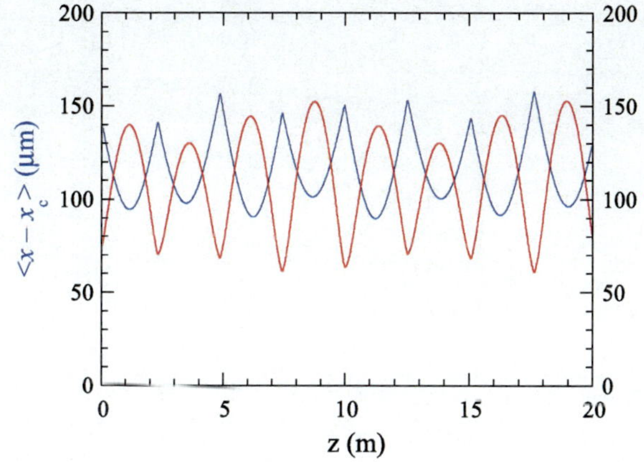

Fig. 12.20 Simulated propagation of the SPARC electron beam [43]

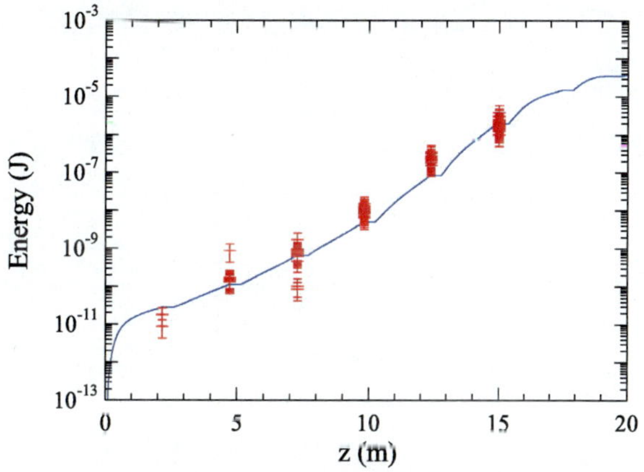

Fig. 12.21 Comparison of simulation results and the measured pulse energies versus distance [43]

A comparison between the evolution of the relative linewidth as determined from simulation and by measurement (data courtesy of L. Giannessi) is shown in Fig. 12.22 over the range of the installed undulators and agreement between the simulation and the measured linewidth is within about 35% after 15 m. As shown in the figure, the predicted linewidths are in substantial agreement.

The initial decrease in linewidth shown in Fig. 12.23 results from the development of temporal coherence as can be seen in Figs. 12.23, 12.24, and 12.25, where we plot the power versus time within the optical pulse. The time window used in the simulation was chosen to be 14 psec in order to allow for slippage across the 12.67 psec electron bunch. The optical pulse at the startup of the SASE interaction is expected to contain a large number of "spikes." This is indeed what is found in simulation as shown in Fig. 12.23, where we plot the power in the pulse over the entire time window. This pulse is near the start of the undulator line and exhibits a broad distribution of spikes coinciding roughly with the center of the electron bunch,

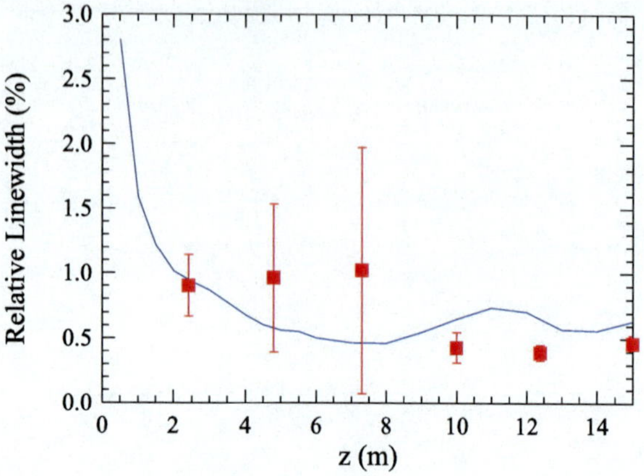

Fig. 12.22 Comparison of the measured relative linewidth with that found in simulation [43]

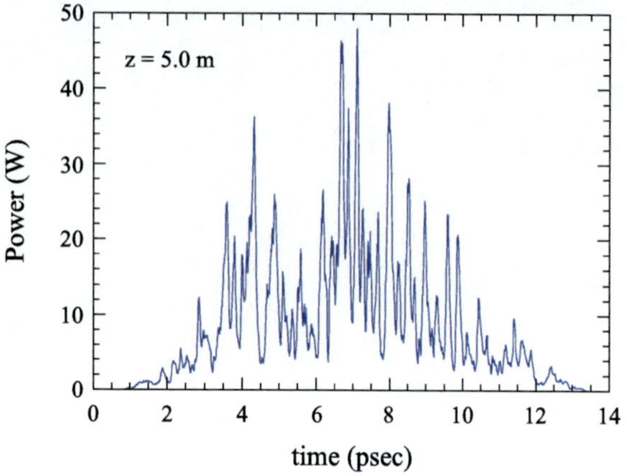

Fig. 12.23 Temporal pulse shape at $z = 5.0$ m [43]

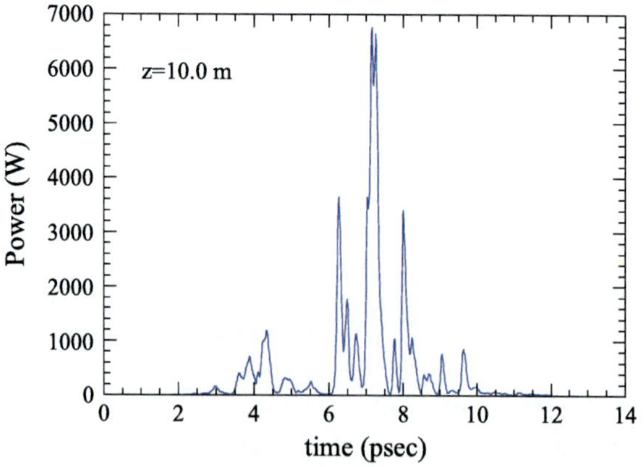

Fig. 12.24 Temporal pulse shape at $z = 10.0$ m [43]

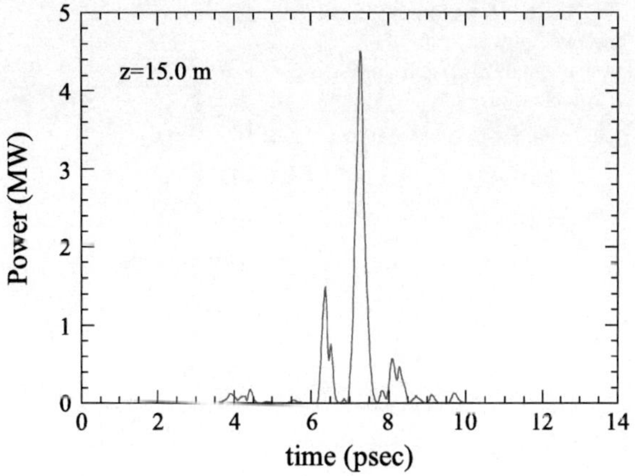

Fig. 12.25 Temporal pulse shape at $z = 15.0$ m [43]

which is located at the center of the time window. As shown in Fig. 12.22, the linewidth narrows as the interaction proceeds, and this corresponds to the development of temporal coherence.

This evolution of temporal coherence is illustrated in Figs. 12.24 and 12.25, by comparison with Fig. 12.23, where the temporal pulses are shown at $z = 10.0$ m and 15.0 m, respectively. These two figures correspond to the exponential gain region prior to saturation. It is clear in these figures that the early collection of a large number of spikes has coalesced into a more sharply peaked distribution containing a smaller number of spikes. This corresponds to the narrowing of the linewidth due to the development of coherence in the exponential gain region. Comparing Figs. 12.21 and 12.22, we observe that the measured pulse energy is somewhat higher and the relative linewidth somewhat lower than the simulation results, which may be due to a small difference in the beam match.

12.6 Enhanced Harmonic Radiation

In the interests of generating shorter and shorter wavelengths without the necessity of increasing the electron beam energy, a method was proposed for enhancing the generation of harmonics by suppressing the fundamental [45, 46]. This method relies on shifting the phase of the fundamental in the gaps between the wiggler segments to a value that suppresses the fundamental but optimizes that of a harmonic. For the third harmonic, this phase shift should be $2\pi/3$. Adding to this method, it is suggested that enhanced harmonic generation could be obtained by optimizing the β-function for operation at any given harmonic [45, 46]. Simulations have shown these suggestions to be feasible [47].

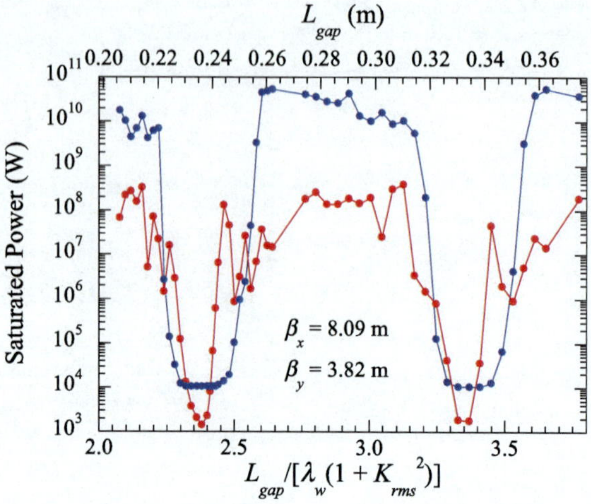

Fig. 12.26 Variation in the saturated power for the fundamental (blue) and third harmonic (red) [47]

As discussed above, a change in the phase shifts in the gaps between the wigglers can be compensated for in a SASE free-electron laser by a shift in the wavelength excited within the wigglers. Thus, a change in the gap length by ΔL_{gap} will result in a change in the wavelength excited by

$$\delta\lambda = -\frac{L_w}{2\gamma_b^2\lambda_w}\Delta L_{\text{gap}}, \tag{12.26}$$

where L_w denotes the length of the uniform section of the wigglers. As a consequence, introducing the same phase shift in each of the gaps will effectively retune the wavelength that is excited. In order to suppress the fundamental, a variety of different phase shifts (i.e., gap lengths) must be imposed between different wiggler sections. Note that this same detuning can be accomplished using phase shifters.

We now consider single-slice, time-independent simulations for convenience. The example under consideration [47] employs a beam with a kinetic energy of 12 GeV, a peak current of 3400 A, a normalized emittance of 0.2 mm-mrad, and an energy spread of 0.01%. Flat-pole-face wigglers are assumed, which have a period of 2.49 cm, with a peak on-axis field strength of 10.5134 kG. This yields a fundamental resonance at a wavelength of 0.9 Å. Each wiggler segment is composed of 103 periods having a uniform field with transitions that are one period in length for a total of 105 periods in all per wiggler. A strong focusing FODO lattice is composed of quadrupoles of 0.744 m in length located in the center of the gaps.

We first consider the effect of detuning the gap length for $\widehat{\beta}_x = 8.09$ m and $\widehat{\beta}_y = 3.82$ m and plot the saturated power at the fundamental (blue) and third harmonic (red) in Fig. 12.26 as determined in simulation as functions of the gap length, L_{gap}, and the normalized gap length, $L_{\text{gap}}/[\lambda_w(1 + K_{\text{rms}}^2)]$. These values of $\widehat{\beta}$ correspond to the optimal performance of the fundamental. Note that a change in the normalized

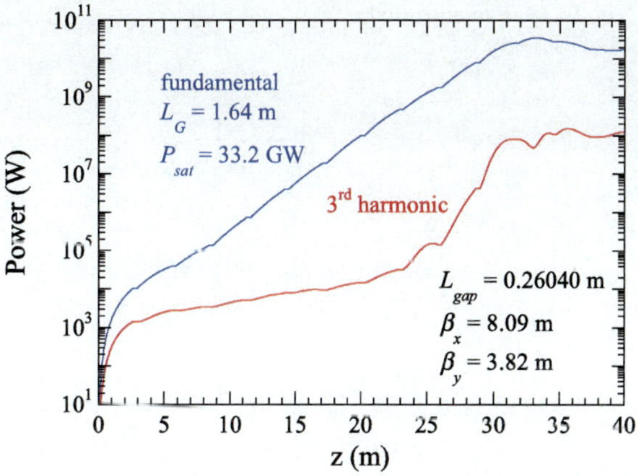

Fig. 12.27 Evolution if the fundamental and third harmonic for $L_{gap} = 0.2784$ m [47]

gap length of unity corresponds to a phase shift of 2π. It is evident that there are peaks in the harmonic emission corresponding to phase shifts of approximately every $2\pi/3$ as indicated by intervals in the normalized gap length of 1/3. These peaks in the harmonic emission correspond to nonlinear harmonic generation when the fundamental is not detuned and to linear harmonic generation when the fundamental is detuned. In either case, however, the peaks in the harmonic emission reach about the same power levels. These harmonic peaks are reduced relative to the fundamental by about 2–3 orders of magnitude, which is what is typically expected of the harmonic emission. However, this choice of the $\widehat{\beta}$-functions is near the optimal values for fundamental emission. As will be discussed, the harmonic emission is substantially enhanced for increased values of the $\widehat{\beta}$-functions.

The optimal gap length for fundamental emission, as determined by the minimum gain length, occurs for $L_{gap}/[\lambda_w(1 + K_{rms}^2)] \approx 2.6228$ ($L_{gap} \approx 0.26040$ m) over the range of gap lengths shown. At this gap length, the nonlinear harmonic generation mechanism is dominant, and the evolution of the fundamental and third harmonic is shown in Fig. 12.27. The evolution of the harmonic exhibits the expected behavior of the nonlinear harmonic generation mechanism; specifically, there is linear harmonic growth following the initial startup region followed by nonlinear harmonic growth once the fundamental reaches high powers. For this case, linear harmonic generation is observed over the range of about 5 m $< z <$ 26 m after which nonlinear harmonic generation is dominant. Saturation of the third harmonic is found slightly before the fundamental saturates at a distance of about 34 m. Note that since the phase shift is periodic, this performance for the fundamental will be replicated for integer increments of this normalized phase shift.

The third harmonic is found to grow exponentially for a gap length where the fundamental is suppressed. This is linear harmonic generation and is shown in Fig. 12.28 where we plot the evolution of the third harmonic power as a function of position along the wiggler line. Observe that the saturation distance is now over 100 m, which results from the longer gain length for linear harmonic generation of the harmonic.

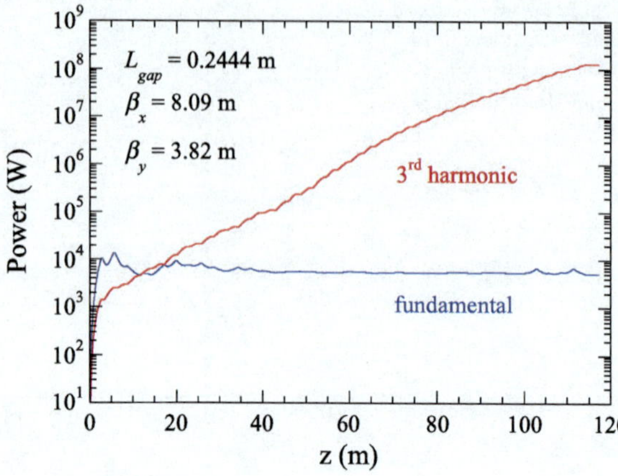

Fig. 12.28 Evolution of the third harmonic for $L_{\text{gap}} = 0.244$ m [47]

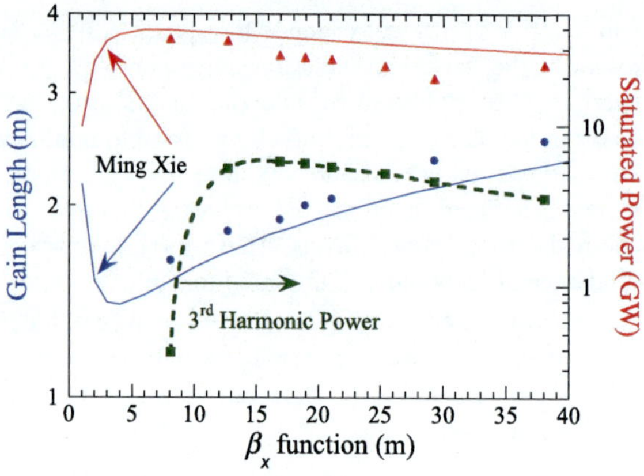

Fig. 12.29 Comparison of simulations and the Ming Xie parameterization [47]

As previously mentioned, the choice of $\widehat{\beta}_x = 8.09$ m and $\widehat{\beta}_y = 3.82$ m is near the optimal values for performance at the fundamental, and this is in good agreement with predictions based on the parametric scaling developed by Ming Xie [37]. A comparison of the fundamental gain length (blue circles) and saturated power found in simulation (red triangles) and the parameterization due to Ming Xie (blue and red lines) is shown in Fig. 12.29 versus $\widehat{\beta}_x$.

Also shown in Fig. 12.29 is the maximum saturated power of the third harmonic as found in simulation (green squares), where the dashed line connecting the points is simply a fit to the data and not an analytic extension of the Ming Xie parameterization. Note that the peak powers as found with nonlinear harmonic generation and linear harmonic generation are comparable. Observe that the qualitative behavior of the third harmonic is similar to that of the fundamental in that there is a relatively rapid increase in power as $\widehat{\beta}_x$ increases up to the optimal value of $\widehat{\beta}_x$, followed by a relatively slow decline thereafter. For the third harmonic, the optimal value for $\widehat{\beta}_x$ is approximately 18 m for these beam, wiggler, and quadrupole parameters. There is a

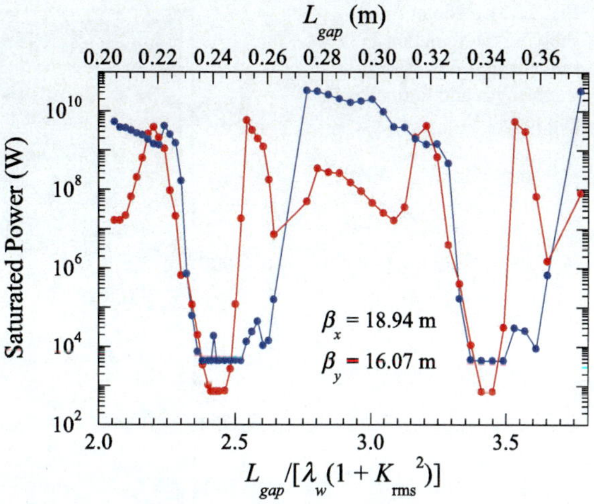

Fig. 12.30 Variation in the saturated power at the fundamental and third harmonic for the optimal $\widehat{\beta}_x$ for the third harmonic [47]

substantial enhancement in harmonic power with respect to that found at the $\widehat{\beta}_x$ that yields the maximum power at the fundamental. The third harmonic power when $\widehat{\beta}_x=$ 8.09 m, corresponding to the optimal fundamental power, is about 380 MW but rises to about 6.0 GW when $\widehat{\beta}_x = 18$ m. This is an impressive increase of more than an order of magnitude in the third harmonic power.

The variation of the saturated power at the fundamental (blue) and the third harmonic (red) as functions of gap length found in simulation for the near-optimal $\widehat{\beta}_x$ for the third harmonic is shown in Fig. 12.30. Observe that the periodicities indicated in Fig. 12.26 for smaller values of $\widehat{\beta}_x$ are more pronounced here where the third harmonic is more prominent.

The simulations showing the variation in the tuning with gap length were done at a fixed wavelength. However, since the phase shift from the start of one wiggler to the next varies substantially with wavelength over the amplification band, changes in the gap will result in changes in the wavelength excited within the resonant bandwidth. This is illustrated for the fundamental in Fig. 12.31, which is a contour plot of the fundamental output power versus both wavelength and (normalized) gap length. Observe that the peak powers are found for wavelengths close to 0.9 (±0.004) Å and that the output is periodic in gap length.

A corresponding contour plot of the third harmonic power versus both wavelength and (normalized) gap length is shown in Fig. 12.32. In this case, we note that there are five peaks corresponding to the nonlinear harmonic generation and linear harmonic generation harmonic peaks over the range shown in Figs. 12.26 and 12.32, although the central peak is much reduced relative to the outer peaks.

Figures 12.31 and 12.32 provide a measure of the bandwidth that would be excited at any given gap length, as well as the relative magnitude of the radiated powers.

Fig. 12.31 Contour plot showing the variation in fundamental power with wavelength and normalized gap length [47]

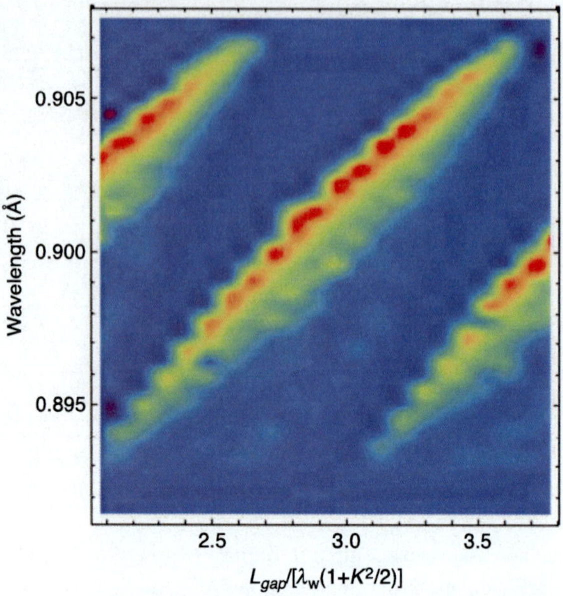

Fig. 12.32 Contour plot showing the variation in the third harmonic power with wavelength and normalized gap length [47]

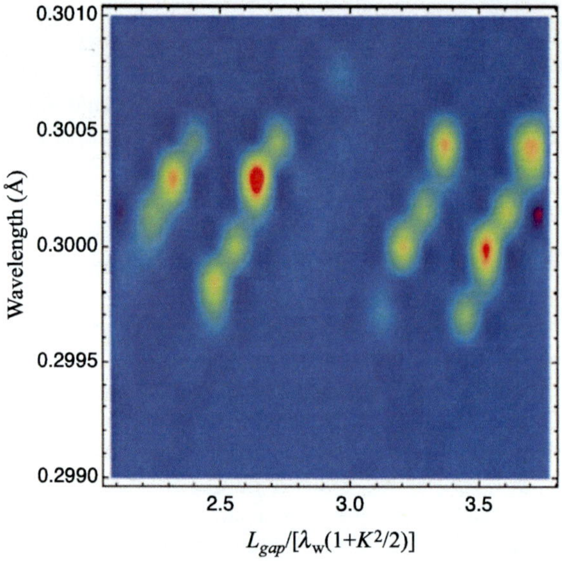

It is interesting to compare the performance of the third harmonic at a wavelength of 0.3 Å for a wiggler tuned to a fundamental wavelength of 0.9 Å as described previously with the performance at 0.3 Å when the wiggler is tuned to a fundamental at that wavelength. To tune the free-electron laser to lase at 0.3 Å at the fundamental, we use the same beam parameters as used previously, but we reduce the wiggler

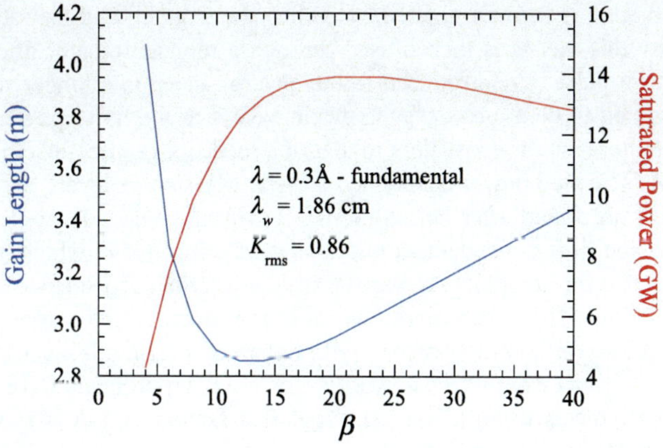

Fig. 12.33 Gain length and saturation power for wiggler parameters consistent with operation at a fundamental wavelength of 0.9 Å found using the Ming Xie parameterization [47]

period and amplitude. In particular, the wiggler period is 1.86 cm, and the peak on-axis amplitude is 7.0 kG yielding an rms wiggler strength parameter of $K_{rms} = 0.86$. It should be remarked that a wiggler producing this field would have the same gap as the field used in the harmonic simulations. As shown previously in Fig. 12.29, the performance at the fundamental found in simulation is in close agreement with the results of the parameterization developed by Ming Xie [37]. In view of this, we show the predicted performance at the fundamental of 0.3 Å using the parameterization in Fig. 12.33 where we plot the gain length (blue) and the saturated power (red) versus $\widehat{\beta}$. As shown in the figure, the saturated power reaches a maximum value of about 14 GW when $\widehat{\beta} = 20$ m. This is only a factor of about 2.3 greater than the highest power found when the wiggler is tuned to a third harmonic at 0.3 Å and $\widehat{\beta} = 18$ m.

It is clear both linear and nonlinear harmonic generation can be substantially enhanced by the optimal choice of $\widehat{\beta}$ for the FODO lattice in an x-ray free-electron laser. Moreover, at a given wavelength, substantial power at harmonics can be generated by the linear harmonic generation mechanism without any noticeable radiation at the fundamental wavelength.

12.7 Terawatt X-Ray Free-Electron Lasers

Recent simulations [48, 49] using a novel superconducting helical undulator design and a quadrupole FODO lattice indicate that a terawatt (TW) x-ray free-electron laser is possible. The configuration studied was based upon a self-seeding scheme whereby the electron beam is propagated through a wiggler that is long enough to reach the exponential regime via SASE but not long enough to reach saturation. The

SASE interaction is then terminated by ending the wiggler, and the radiation pulse produced by this means is then passed through a monochromator after which the filtered optical pulse is reintroduced to the electron beam in a longer wiggler. This seed is assumed to be at power levels far in excess of the noise power; hence, the subsequent interaction is equivalent to that of a seeded x-ray free-electron laser, that is, a MOPA. The seed power is assumed to be at MW power levels, and TW output power levels are found after an additional 100 m of a tapered wiggler line. Both steady-state and time-dependent simulations were described in which it was found, as expected, that the power levels were reduced when time dependence was included in the simulations. This result represents an enhancement of the output power in the tapered wiggler configuration of nearly 2 orders of magnitude over the saturated power in a uniform wiggler. Historically, however, experiments have shown efficiency enhancements using a tapered wiggler of factors of 3–5 [44, 50, 51]. It is important, therefore, to understand what gives rise to such extreme efficiency enhancements with tapered wigglers.

In this section, we discuss simulations for a variety of configurations that reach TW or near-TW power levels. Unless otherwise specified, the simulations that we discuss are in the steady-state (i.e., time-independent) regime since that is sufficient to illustrate the essential physics of the process. However, we shall also include some discussion of time-dependent simulations for comparison purposes.

12.7.1 The Case of a Helical Wiggler (MOPA)

The configuration that we consider is based upon a self-seeding scenario in which the interaction in a SASE free-electron laser is halted at an early stage by passing the optical field through a monochromator to extract a narrow band of the SASE radiation after which this narrow bandwidth light is then re-injected in synchronism with the electron beam. Hence, the light acts as the seed in a master oscillator power amplifier (MOPA).

The electron beam is assumed to be characterized by an energy of 13.22 GeV, a peak current of 4000 A, an rms energy spread of 0.01% and normalized emittances of 0.3 mm-mrad in both the x- and y-directions. The transverse profile of the electron beam is assumed to be characterized by a Gaussian distribution, and the beam is matched into either the natural focusing of a helical wiggler or the FODO lattice/helical wiggler system. We initially study the case of a single, long wiggler with a period of 2.0 cm and a peak on-axis amplitude of 16.1 kG. This is equivalent to a wiggler parameter $K = 3.01$ and yields a resonance at a wavelength of 1.5 Å. The simulations do not include the initial SASE region prior to the monochromator, and the wiggler in the MOPA region is assumed to be 100 m in length. While it is not practical to construct a wiggler that is 100 m in length, this model is useful to study the essential physics of the interaction in a tapered wiggler with strong focusing applied. Since optical guiding ceases in the gaps between undulator segments, it is expected that the interaction will be less efficient in a segmented wiggler, and an equivalent simulation of a segmented wiggler will be described for comparison.

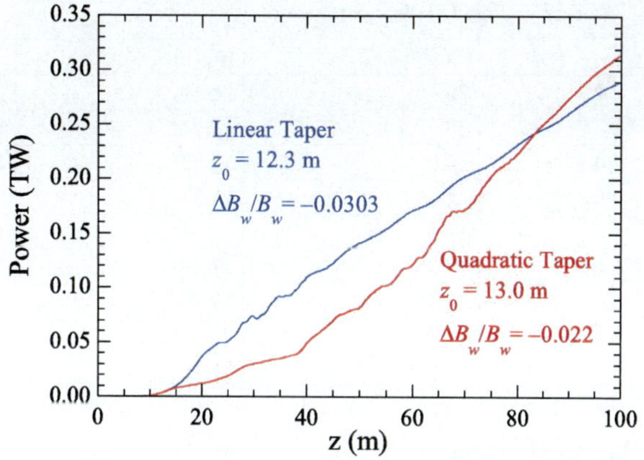

Fig. 12.34 The growth of the power for optimized linear and quadratic taper profiles for a weak-focusing helical wiggler

We first consider the case of weak, natural focusing. In this case, the β-function is 38.7 m, and the matched rms beam radius is 29.9 μm. Assuming that the seed power is 5 MW, saturation in the uniform wiggler is found after about 12.5 m at a power level of about 40 GW. Two taper profiles have been considered: linear and quadratic. The growth of the power for each profile is shown in Fig. 12.34, where the optimal values of the start-taper point and the total decrease in the wiggler field are indicated. As shown in the figure, although the quadratic taper produces somewhat more power than the linear taper, there is little difference for the two taper profiles after a distance of 100 m. As a result, we will limit subsequent discussions to the case of a linear down-taper for simplicity. Observe that the maximum power achieved after 100 m is about 0.3 TW in either case, which represents an enhancement of the efficiency over the saturated power in the uniform undulator by about a factor of about 7.5. Note, however, that the quadratic taper would result in increasingly higher powers relative to the linear taper for longer undulators.

In order to study the effect of increasingly strong focusing, we consider nine different FODO lattices as shown in Table 12.1. In each case, the quadrupole length is assumed to be 0.074 m. The leftmost column in the table represents the length of the FODO cell, while the second column is the field gradient. The third column is a measure of the average β-function, while the two rightmost columns describe the rms beam radius and the initial beam sizes in the x- and y-directions. The Twiss-α parameters are not shown but are $\alpha_x \approx 1.3$ and $\alpha_y \approx -0.77$.

A summary plot showing the optimized output power for a linear taper profile after 100 m for the six different FODO lattices and the weak-focusing wiggler is shown in Fig. 12.35 for an assumed seed power of 5 MW. As expected, strong-focusing wiggler enhances the interaction with respect to the weak-focusing wiggler. However, there appear to be two regimes associated with strong-focusing wiggler. At the longer FODO cell lengths ($L_{FODO} \geq 4.4$ m), the simulation predicts output powers within the range of approximately 0.4–0.8 TW. However, there is a more dramatic increase in the output powers as the FODO cell length decreases below

Table 12.1 FODO lattice parameters

L_{FOD} (m)	Gradient (kG/cm)	$(\beta_x^2 + \beta_y^2)^{1/2}$ (m)	Radius (μm)	x_{rms}/y_{rms} (μm)
2.2	26.40	6.49	10.3	8.03/6.22
3.3	18.69	9.61	12.2	9.68/7.47
4.4	14.02	12.32	14.0	11.31/8.72
5.0	12.34	14.57	15.2	12.06/9.29
5.5	11.22	16.04	16.2	12.65/9.74
6.0	10.28	17.49	16.5	13.0/10.2
6.6	9.35	19.26	16.9	13.9/10.7
8.8	7.01	25.69	19.2	16.0/12.3
11.1	5.56	32.41	21.0	18.0/13.8

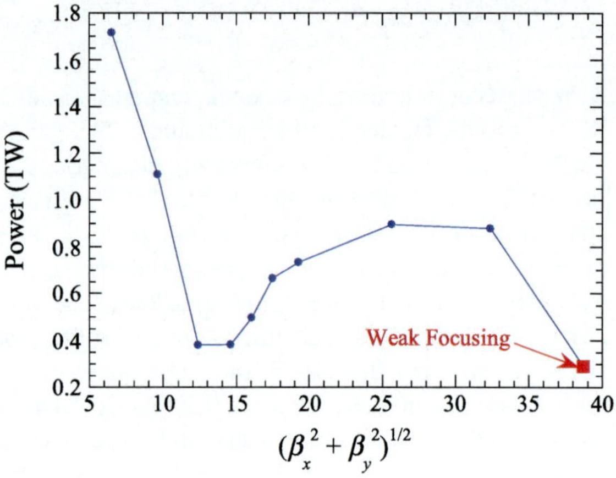

Fig. 12.35 Optimized output power after 100 m for the six FODO lattices and the weak-focusing case

4.4 m. In this regime, we observe an approximately linear increase in the output power with decreasing β-functions and where the maximum output power exceeds 1.7 TW for the cases under consideration.

The current density increases as the FODO cell length decreases so that the Pierce parameter (Fig. 12.36) increases with decreasing β-functions, and this leads to increasing interaction strengths. The reason for the dramatic increase in the output power when the FODO cell length decreases below 4.4 m is twofold. In the first place, the stronger interaction strength yields optical guiding even in the tapered regime. In the second place, the smaller beam size results in a more coherent interaction with the optical field, and this results in a higher trapping fraction.

The evolution of the output power (left axis, blue) and the spot size of the optical field (right axis, red) versus position in the wiggler is shown in Fig. 12.37 for each of the six choices of the FODO lattice. The green line in the figures represents the rms electron beam radius. In all of these cases, saturation in the uniform undulator occurs after between 10 and 15 m, and the optimal start-taper points range from about 9 to 12 m. Figure 12.37a, b, d, and f correspond to the longest FODO cells with the lower power levels (0.4–0.8 TW). Diffraction and optical guiding have been discussed

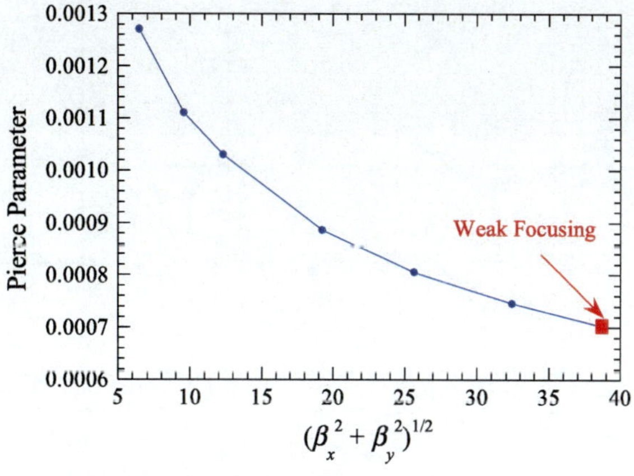

Fig. 12.36 The variation in the Pierce parameter with the β-function

frequently in the literature [9–11]. It is evident from these figures that substantial diffraction occurs after the start of the taper where the optical field expands from about 10 to 20 μm at the start-taper point to between 250 and 350 μm after 100 m. This is in stark contrast with what is found when the FODO cell is shorter than 4.4 m (shown in Fig. 12.37e). As shown in the figure, the optical mode experiences substantial guiding in the tapered regime and the maximum expansion found after 100 m of undulator is to a mode radius of between 40 and 60 μm.

It was pointed out by Jiao et al. [52], that the "decreasing of refractive guiding is the major cause of the efficiency reduction, particle detrapping, and then saturation of the radiation power" in a tapered wiggler. Hence, the continuation of optical guiding, and the associated increase in the trapping fraction, in the case of extreme transverse compression of the electron beam is an important factor in reaching TW power levels. It was also pointed out that the interaction might be further optimized by varying the electron beam size within the tapered wiggler. Simulation of this requires a multiparameter optimization of the FODO lattice including both the quadrupole field gradients and positions.

Comparable behavior in the case of a weak-focusing helical wiggler is shown in Fig. 12.38 where we plot the power (left axis, blue) and spot size (right axis, red) versus distance along the wiggler and where the green line represents the rms electron beam radius. Observe that there is very little optical guiding of the field in the tapered region after about 15 m but that the spot size of the optical mode remains below about 100 μm due to the relatively large transverse size of the electron beam.

Optical guiding is thought to be composed of two effects: gain guiding and refractive guiding. Gain guiding occurs because the amplification of the field occurs only in the presence of the electron beam. In the exponential regime, the theory shows that the interaction leads to a complex solution for the wavenumber, which has a shift in the real part of the solution that is proportional to the growth rate and which shifts the refractive index and causes the refractive guiding. However, these two effects cannot be separated and optical guiding occurs when the characteristic growth length of the optical field is shorter than the Rayleigh range. This is usually

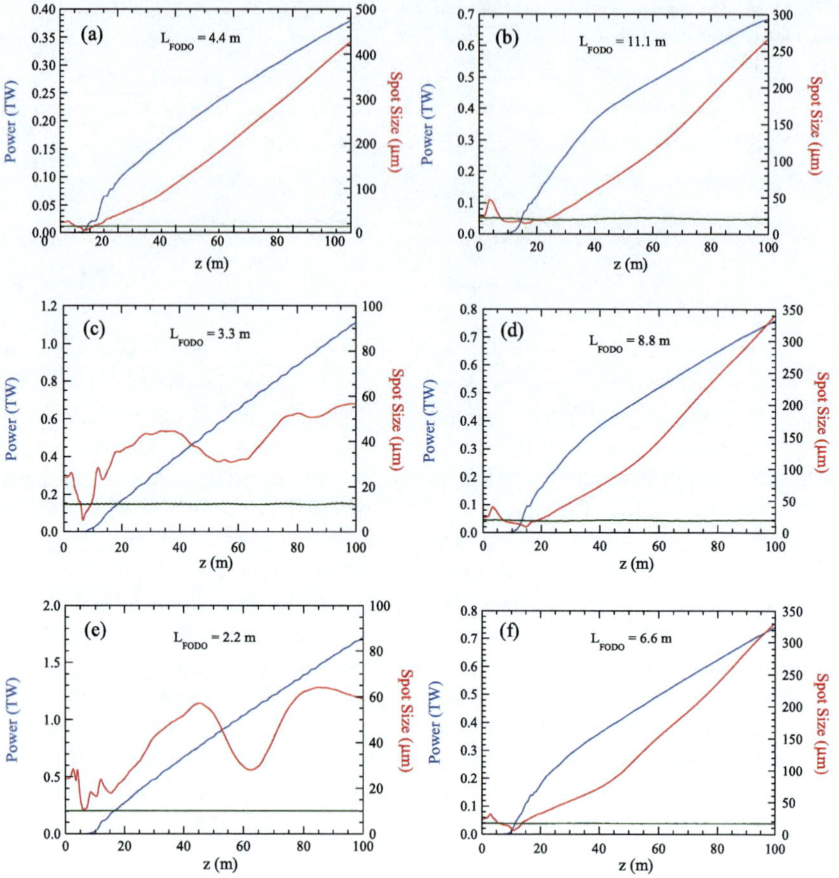

Fig. 12.37 Plots of the power (left axis, blue) and spot size (right axis, red) versus distance along the undulator for different levels of focusing. The green line represents the rms electron beam radius. Observe that the optical field is guided when the length of the FODO cell is less than about 4.4 m

the case in the exponential growth regime, but it is not typically found in a tapered undulator where the power grows more slowly than exponential. However, the extreme focusing that we find when $L_{FODO} < 4.4$ m leads to sufficiently rapid growth that optical guiding can occur through gain guiding. This is seen by noting that the power grows in the linear tapered region as $P(z) = P_0[1 + (z - z_0)/L_G]$, where P_0 is the power at the start of the tapered region, z_0 is the start-taper point, and L_G is the characteristic growth length. The Rayleigh range is about 15 m at the start of the tapered region. When $L_{FODO} < 4.4$ m, the characteristic growth length is $L_G \approx 13$ m, which is comparable to but shorter than the Rayleigh range indicating that optical guiding is possible. The characteristic growth length decreases as L_{FODO} decreases further, thereby enhancing the optical guiding. Observe that the spot size exhibits

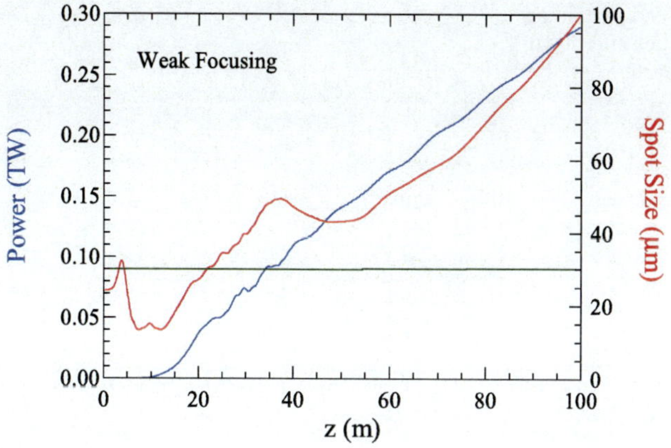

Fig. 12.38 Plot of the evolution of the optimized power and the spot size versus distance for the case of a weak-focusing helical wiggler

bounded oscillations when $L_{FODO} < 4.4$ m indicating that the guiding is modulated by the detailed phase space evolution of the electrons, which undergo synchrotron oscillations in the ponderomotive potential with a period in the range of 20–30 m. In contrast, L_G is approximately 21 m or greater when $L_{FODO} \geq 4.4$ m indicating that optical guiding cannot occur and the spot size is found to grow monotonically without bound in these cases.

The physics of optical guiding in the tapered undulator has been studied analytically [53] in which it was shown that refractive guiding varies inversely with the on-axis field strength, which typically reaches an asymptotic limit. Once that limit is achieved, refractive guiding ceases, and diffraction takes hold leading to expansion of the optical field. At the same time, the optical power continues to increase linearly. This is, effectively, what is shown in Figs. 12.37a, b, d, and f. The results shown in Figs. 12.37c, e represent a new regime where, while the power still increases linearly with distance, it grows rapidly enough to overcome diffraction; hence, gain guiding becomes an important effect.

In all cases, the enhanced focusing due to the FODO lattice enhances the performance with respect to the case of weak focusing. As the focusing increases with decreasing FODO cell lengths, the optimized output power initially increases with the corresponding increase in the Pierce parameter. However, diffraction becomes increasingly important as the transverse beam size decreases since the optical field expands beyond the electron beam more rapidly. As shown in Fig. 12.35, diffraction overcomes the effect of increasing Pierce parameter for β-functions in the range 12–25 m (FODO cell lengths between 4.4 and 8.8 m), after which the output power decreases with decreasing β-functions. Diffraction, therefore, provides a limiting factor with respect to the increasing focusing strength of the FODO lattice until the transverse beam size becomes small enough that the growth rate of the field in the tapered region is fast enough to overcome diffraction and

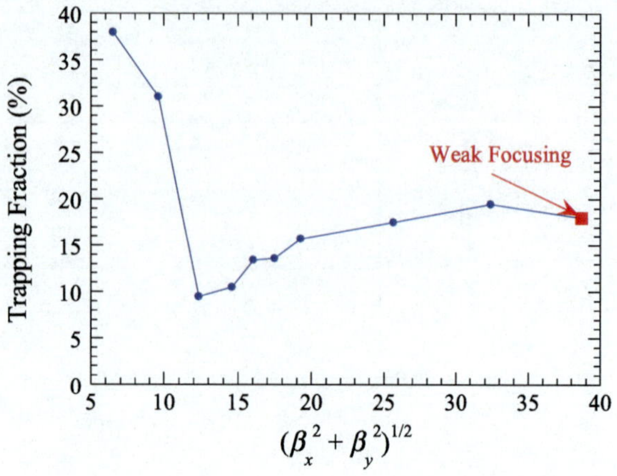

Fig. 12.39 Variation in the trapping fraction after 100 m with the β-function

enable optical guiding. For the configuration under study, this transition point is reached at when the FODO cell length is approximately 4.4 m corresponding to a β-function of 12 m.

The second major factor in achieving such extreme enhancements of the efficiency is how strong-focusing affects the trapping fraction. In this regard, we find that the trapping fraction is highest at about 38% of the beam for the shortest FODO cell length ($L_{FODO} = 2.2$ m) and generally mirrors the behavior of the power versus average β function. This is shown in Fig. 12.39 where we plot variation in the trapping fraction after 100 m of undulator versus the β-function. This is related to the optical guiding where the mode size remains closest to that of the transverse extent of the electron beam, and this enhances the coupling of the field to the electrons.

Examples of the spent beam distributions for (a) strong focusing with the highest power case ($L_{FODO} = 2.2$ m) and (b) for the weak-focusing case are shown in Fig. 12.40. It is clear from the figures that the extreme focusing has a significant impact om the trapping fraction, which is 38% for the strong-focusing lattice but only 18% for the weak-focusing wiggler. In addition, a greater amount of energy has been extracted from the trapped portion of the beam when the strong focusing lattice is used.

We now consider the case of a segmented helical wiggler with the strong-focusing FODO lattice with a cell length of 2.2 m. In order to configure this wiggler/FODO lattice properly, the wigglers are 0.96 m in length with 46 periods (for a 2.0 cm period) over with one period each in the entry and exit transition, and the gaps between the wigglers are 0.16 m in length. Because the simulation procedure does not automatically select the optimum phase shift between the wiggler sections, we have adjusted the wiggler field strength slightly to 16.135 kG in order to optimize the phase shift between undulator segments in the uniform wiggler section for a resonance at 1.5 Å.

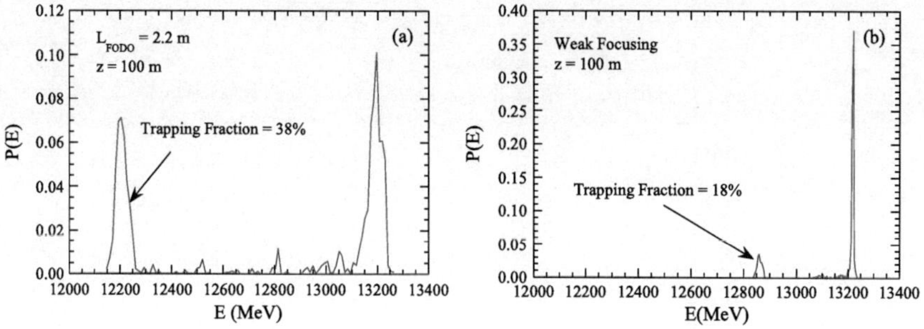

Fig. 12.40 The spent beam distributions for (**a**) strong focusing with $L_{FODO} = 2.2$ m, and (**b**) weak focusing

It should be remarked that since the wiggler field strength decreases in the tapered section, the factors controlling the optimal phase shift will also vary. The optimal phase shift can be selected by inserting phase shifters between the wigglers or by varying the gap lengths. However, the gaps are quite short given the short length of the FODO cell, and this would make it difficult to insert phase shifters in the configuration under consideration here. In addition, changing the gap lengths could, in principle, affect the lengths of the FODO cells. Of course, it may be that the interaction can be further optimized by changing the focus of the electron beam along the wiggler, which implies varying the parameters of the FODO lattice. As a result, it is clear that a more complete optimization of the segmented wiggler configuration is a complicated procedure and is beyond the scope of the present discussion. As such, therefore, we shall restrict the present analysis to an optimization over the start-taper segment and the (linear) taper profile, but it should be recognized that this is not a complete optimization.

The evolution of the power for an optimized taper profile and the corresponding spot size are shown in Fig. 12.41. The start-taper point is found to be the 16th wiggler for a seed power of 5 MW, and the optimal (linear) down taper is 6%. It is expected that the effect of the segmented wiggler is to degrade the interaction because the optical field is not guided in the gaps and the phase shift may not be optimized either, and this is found to be the case where the output power reaches approximately 0.73 TW, which is reduced relative to the 1.7 TW found for the single, long wiggler. Nevertheless, this still represents an enhancement by a factor of more 15 over the saturated power in a uniform wiggler.

It should be noted that the output power and spot size shown in Fig.12.41 are comparable to that found for the single long wiggler when optical guiding is absent (see Figs. 12.37a, b, d, and f). This is because (1) the power growth in the tapered region is reduced relative to that obtained in the case of a single, long wiggler due to phase mis-matches in the gaps between the wigglers, and (2) there is diffraction in the gaps between the wigglers. In regard to (1), the optimal phase match between wigglers in the tapered region will vary from gap to gap due to decreasing wiggler amplitudes, and no effort has been made to find the (different) optimal gap lengths in the tapered section.

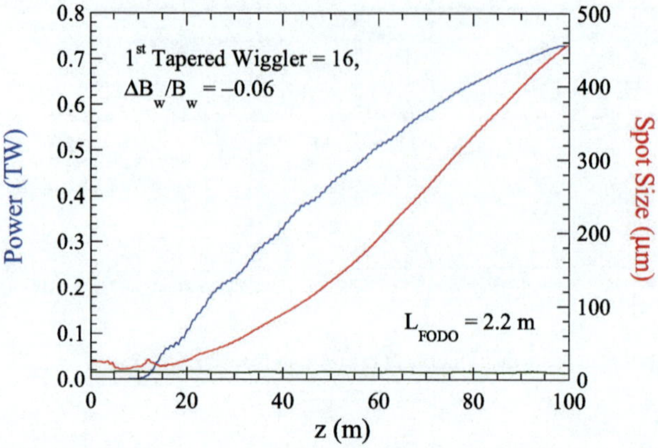

Fig. 12.41 Evolution of the power and spot size for an optimized tapered wiggler line

12.7.2 *The Case of a Planar Wiggler (MOPA)*

We now consider a flat-pole-face planar wiggler with a 3.0 cm period and an on-axis field magnitude of 12.49 kG. These parameters correspond to the period and magnitude of the wiggler in the LCLS; however, we consider the case of a single, long wiggler for the present study. The electron beam has an energy of 13.64 GeV, a peak current of 4000 A, a normalized emittance of 0.4 mm-mrad, and an rms energy spread of 0.01%. This configuration is resonant at a wavelength of 1.5 Å. The performance of the self-seeded MOPA configuration is studied using two FODO lattices. The FODO lattice used in the LCLS had a FODO cell length of approximately 7.3 m and a field gradient of 4.05 kG/cm. Note that the quadrupole length that we have been using (0.074 m) also corresponds to the quadrupoles used in the LCLS. We compare the performance of the self-seeded MOPA based upon this FODO lattice with the extreme-focusing lattice shown in Table 12.1 with a FODO cell length of 2.2 cm. A seed power of 5 MW is used to determine the performance of the self-seeded MOPA for both FODO lattices.

The Twiss parameters used to match the electron beam into the 7.3 m long FODO cell correspond to initial rms sizes of 22 μm in the *x*-direction and 19 μm in the *y*-direction with Twiss-α parameters of $\alpha_x = 1.1$ and $\alpha_y = -0.82$. Saturation for this lattice is found after about 25 m at a power level of close to 10 GW. The optimized linear down-taper is found to correspond to a start-taper point of 20 m with a total down-taper of 1.9% over the additional 80 m of tapered wiggler. The evolution of the power and spot size for this optimized taper is shown in Fig. 12.42. The output power reaches 0.12 TW after 100 m of wiggler, which represents an enhancement over the saturation power for the uniform wiggler by a factor of 12.

Electron beam propagation is determined largely by the FODO lattice rather than the wiggler, and the initial Twiss parameters specified in Table 12.1 for the 2.2 m FODO cell length are applicable for this planar wiggler as well. Saturation in the

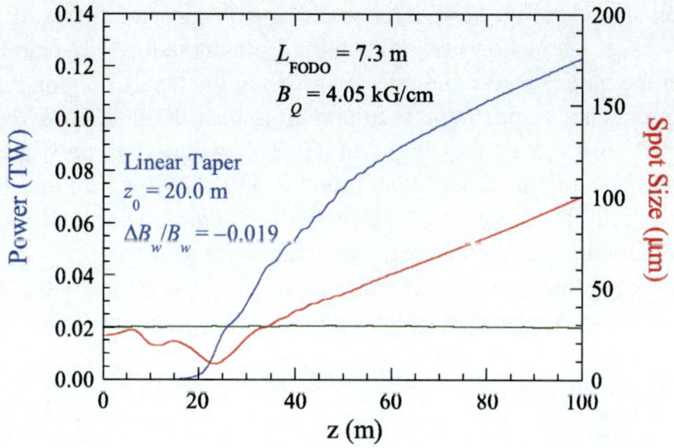

Fig. 12.42 The evolution of the power (left axis, blue) and spot size (right axis, red) for the optimized taper using the lattice with the 7.3 m long FODO cell. The green line represents the rms electron beam radius

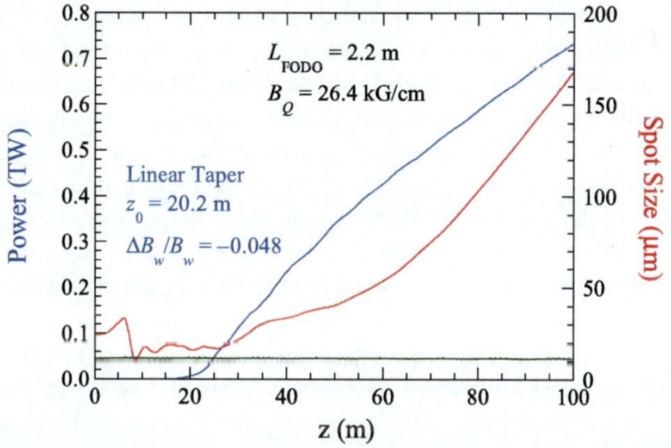

Fig. 12.43 The evolution of the power (left axis, blue) and spot size (right axis, red) for the optimized taper using the lattice with the 2.2 m long FODO cell. The green line represents the rms electron beam radius

uniform wiggler for this FODO lattice is found after about 23.5 m at a power level of 15 GW. The optimized linear taper profile is characterized by a start-taper point of 20.2 m and a total down-taper of 4.8% over the length of the taper region. The evolution of the power and spot size for the optimized taper profile is shown in Fig. 12.43, where the output power reaches 0.73 TW. This corresponds to an enhancement by a factor of about 50 over the saturated power in the uniform wiggler and a factor of six greater than the output power in the FODO lattice with the 7.3 m cell length.

The result shown in Fig. 12.43 refers to a planar wiggler where we might expect a somewhat weaker interaction strength coming from the reduced *JJ*-factor; hence, the reduction in the output power and rate of growth in the tapered region relative to the helical wiggler is not surprising. It is important to bear in mind, however, that while the spot size grows over the 80 m of tapered wiggler, it only reaches about 170 microns, which is much less than found in Fig. 12.41 or in Figs. 12.37a, b, d, and f. As a result, this represents an intermediate regime where diffraction occurs more slowly and does not completely overwhelm the growth.

It is clear, therefore, that an extreme-focusing FODO lattice will also bring the performance using a planar wiggler to near-TW power levels in 100 m long wigglers.

12.7.3 The Case of Pure SASE

The utility of extreme focusing by means of a FODO lattice to achieve near-TW or TW output powers for the self-seeded MOPA configuration suggests that such extreme focusing will also improve the performance of pure SASE free-electron lasers. Our formulation is capable of simulating the startup from shot noise on the electron beam within the context of steady-state simulations, and we now consider SASE using the long, single, helical wiggler described previously in conjunction with the extreme FODO lattice with a cell length of 2.2 m. As in the previous discussion, the resonance at 1.5 Å is found using the electron beam energy of 13.22 GeV, a peak current of 4000 A, an rms energy spread of 0.01%, and normalized emittances of 0.3 mm-mrad in both the *x*- and *y*-directions. The transverse profile of the electron beam is characterized by a Gaussian distribution, and the beam is matched into the FODO lattice as described in Table 12.1.

The evolution of the average power along a uniform wiggler/FODO lattice is shown in Fig. 12.44. The average is over an ensemble of 30 different initial random phase distributions. As shown in the figure, saturation is found after about 16 m at a power of about 30 GW.

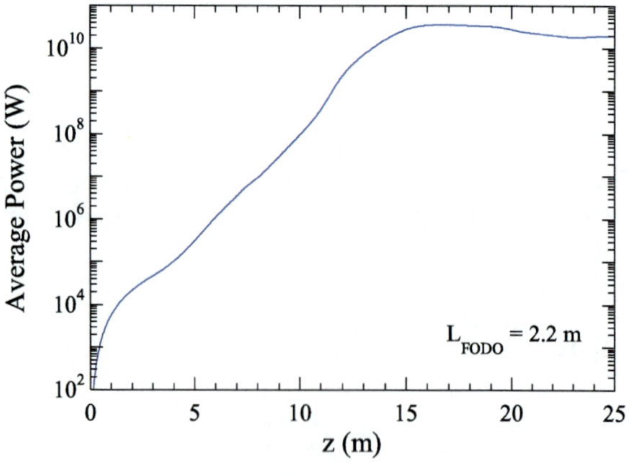

Fig. 12.44 Evolution of the average power along the wiggler for SASE simulations using 30 initial random phase distributions

A (linearly) tapered wiggler has been optimized based upon the saturation of the average power shown in Fig. 12.44 for the extreme-focusing FODO lattice, and the result is compared with that found for the self-seeded MOPA configuration (see Fig. 12.37e). The self-seeded MOPA simulation is based upon a seed power of 5 MW. As indicated in Fig. 12.44, the average SASE performance indicates that the 5 MW power level is reached after about 7 m of wiggler. As a result, it is appropriate to consider a 107 m wiggler/FODO lattice for purposes of the comparison with the self-seeded MOPA performance.

The optimal performance for the tapered wiggler is found for a start-taper point of 14.0 m and a total down-taper of about 9.1% over the length of the tapered region. The optimization was performed over an ensemble of 30 different noise realizations for the initial phase distribution, and the evolution of the average power and spot size is shown in Fig. 12.45. The average output power at the end of the wiggler is 1.5 TW with a standard deviation over the 30 different noise seeds of about 3.5%. It is expected that the tapered wiggler interaction will be less effective for SASE than for a seeded MOPA [39] due to the less coherent nature of the SASE interaction; however, this represents only a small penalty in comparison with the 1.7 TW found for the self-seeded MOPA. We also point out that, as in the case of the self-seeded MOPA, substantial optical guiding is also found here.

On the basis of this result, it appears that extreme focusing in a wiggler/FODO lattice can be used to reach TW power levels in pure SASE x-ray FELs.

12.7.4 Time-Dependent Simulations

We now consider time-dependent simulations based upon the single, long helical wiggler configuration described above. This is important for SASE in particular where the interaction can be degraded because the slippage of the optical spikes relative to the electrons can result in particle detrapping. In this section, we also consider the strongest FODO lattice with a FODO cell length $L_{FODO} = 2.2$ m. The basic electron beam parameters for these time-dependent simulations are the same as used in the steady-state simulations except that we assume a top-hat temporal profile for the electron bunch with a full-width duration of 24 fsec.

A comparison of the output pulse energy found in simulation from the self-seeded MOPA for uniform and tapered wigglers is shown in Fig. 12.46. As shown in the figure, the saturated pulse energy in a uniform wiggler is about 0.96 mJ. This pulse energy can be increased to about 28 mJ using a taper, which starts at 9.2 m and is tapered downward by 8.6% over the remaining length of wiggler. Estimation of the average output power can be obtained by dividing the pulse energy by the bunch duration and yields an output power of about 1.2 TW.

Turning to a time-dependent SASE simulation, a comparison of the evolution of the pulse energies for uniform and tapered wigglers is shown in Fig. 12.47. As seen in the figure, the saturated pulse energy in the uniform wiggler is about 1.17 mJ. This

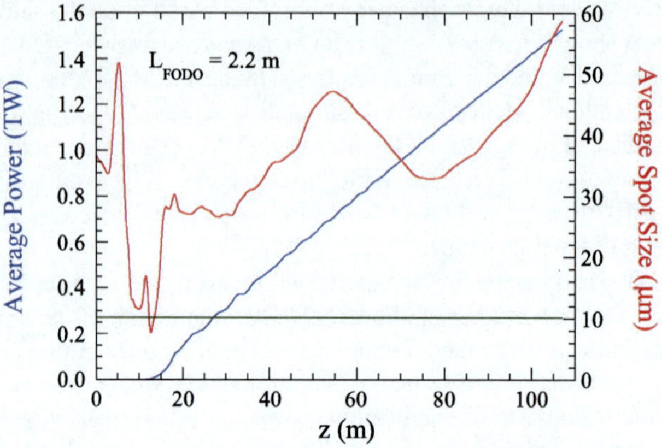

Fig. 12.45 The evolution of the average power (left axis, blue) and spot size (right axis, red) for the optimized taper using the lattice with the 2.2 m long FODO cell. The green line represents the rms electron beam radius

Fig. 12.46 Output pulse energies for the self-seeded MOPA configuration using uniform and tapered wigglers

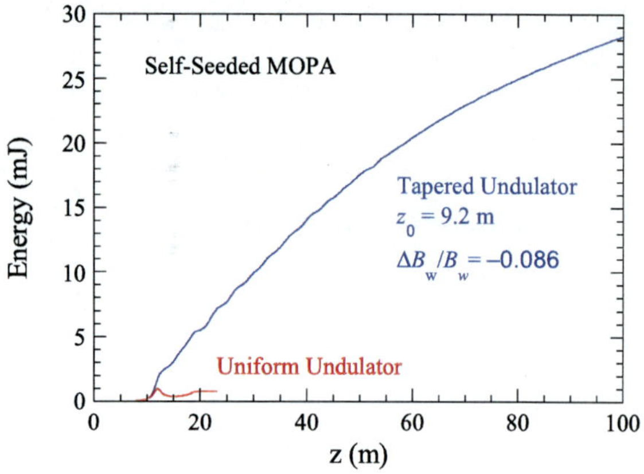

is increased to about 10.0 mJ using a downward taper of 8.2% starting at 13.0 m and extending over a total length of 100 m. Estimation of the average power over the pulse yields a figure of about 0.42 TW, which, while less than that found for the self-seeded MOPA, is still substantial.

The observation that the tapered wiggler is less effective for the SASE configuration than the self-seeded MOPA was also found in the steady-state simulation, although the relative decrease associated with SASE was less in the steady-state simulation. Additionally, it should be remarked that while the pulse energy is still increasing after 100 m for the self-seeded MOPA, the interaction appears to asymptote after about 80 m for the SASE configuration. It is expected that this is due, at least in part, to the spiky nature of the SASE pulse, which can result in detrapping of the electrons as the spikes in the optical pulse slip relative to the electron bunch.

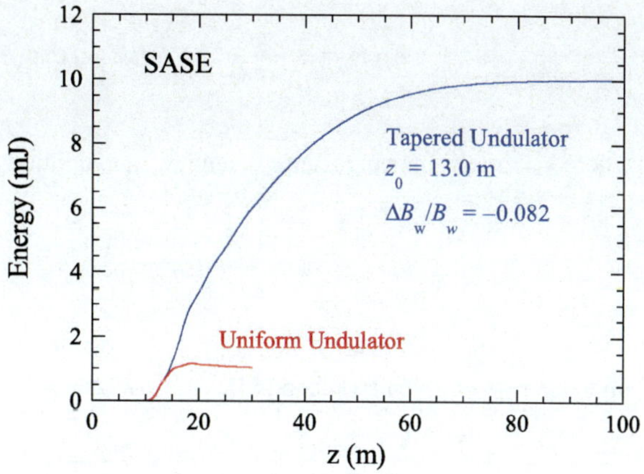

Fig. 12.47 Output pulse energies for the SASE configuration using uniform and tapered undulators

12.8 Resistive Wall Wakefields

The need to maximize the growth rate in extremely short wavelength free-electron lasers leads to extreme bunch compression and tight focusing of the electron beam both of which lead to very compact (both longitudinally and transversely), high intensity electron beams. In addition, high on-axis wiggler field strengths are also necessary, and this means that the magnet poles that comprise the wiggler must be placed very close together. These two requirements mean that the intense electron beams are directed through a very narrow beam pipe. As a result, relatively strong wakefields may be generated due to the resistance [54–56] and roughness [57] of the walls of the beam pipe. These wakefields can result in perturbations of the electron beam that cause a degradation in the performance of the free-electron laser; hence, it is important to be able to characterize these wakefields.

12.8.1 The Wakefields in a Cylindrical Beam Pipe

The analysis presented here follows the derivation of the wakefields given by Bane and Sands [54], where the longitudinal dc impedance per unit length is expressed as

$$Z(k) = -\frac{1}{qc} \int_{-\infty}^{\infty} dsE_z(s)\exp(iks), \tag{12.27}$$

where q is the charge per unit length, and this is $-q^{1}$ times the Fourier transform of the axial electric field E_z. The transform can be inverted to obtain the wakefield in terms of the impedance, so that

$$E_z(s) = -\frac{qc}{2\pi s_0}\int_{-\infty}^{\infty} d\kappa Z(\kappa)\exp(-i\kappa s/s_0), \tag{12.28}$$

where $\kappa = s_0 k/s$, the characteristic length, s_0 is defined as

$$s_0 = \left(\frac{ca^2}{2\pi\sigma_{dc}}\right)^{1/3}, \tag{12.29}$$

where a is the drift tube radius and σ_{dc} is the dc conductivity of the wall. The impedance was given by Chao [57].

$$Z(k) = \frac{2}{ca}\left[\left(\frac{\lambda}{k}+\frac{k}{\lambda}\right)\left(1+\frac{i}{2\lambda a}\right) - \frac{ika}{2}\right]^{-1}, \tag{12.30}$$

where

$$\lambda = \frac{4\pi i\sigma_{dc}k}{c}. \tag{12.31}$$

Taking the root in the upper half-plane under the assumption that k is real, we write

$$\lambda s_0 = \sqrt{\frac{2\pi\sigma_{dc}s_0}{c}}|\kappa|[i+\operatorname{sgn}(\kappa)]. \tag{12.32}$$

Neglecting the high-frequency (k/λ) and low-frequency ($i/2\lambda a$) terms, we have that

$$Z(\kappa) = \frac{2s_0}{ca^2}\left[\frac{i+1}{\kappa^{1/2}} - \frac{i\kappa}{2}\right]^{-1} \tag{12.33}$$

for $\kappa > 0$. The electric field is obtained by integrating

$$E_z(s) = -\frac{q}{\pi a^2}\int_{-\infty}^{\infty} d\kappa\left[\frac{i+1}{\kappa^{1/2}} - \frac{i\kappa}{2}\right]^{-1}\exp(-i\kappa s/s_0). \tag{12.34}$$

The integrand has poles for $\kappa = 2i$ and $\pm\sqrt{3} - i$. Integration of Eq. 12.34 over the contour shown in Fig. 12.48 yields

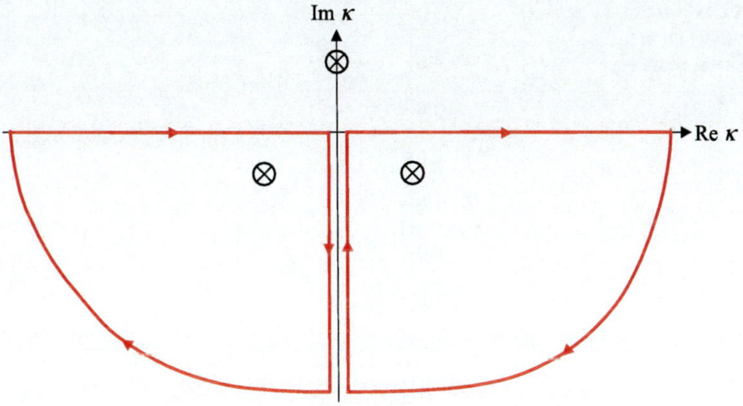

Fig. 12.48 The integration contour for the wakefield

$$E_z(s) = -\frac{q}{a^2} F_{cyl}^{(dc)}(s, s_0), \tag{12.35}$$

where the characteristic function for the dc wakefields in a cylindrical beam pipe is

$$F_{cyl}^{(dc)}(s, s_0) = 16\left[\frac{1}{3}\exp(-s/s_0)\cos\left(\frac{\sqrt{3}s}{s_0}\right) - \frac{\sqrt{2}}{\pi}\int_0^\infty dx \frac{x^2\exp(-x^2 s/s_0)}{x^6+8}\right]. \tag{12.36}$$

A plot of the wake function versus position is shown in Fig. 12.49.

The ac wakefields are obtained by replacing the replacing the dc conductivity by the ac conductivity via

$$\sigma_{dc} \rightarrow \sigma_{ac} = \frac{\sigma_{dc}}{1 - ikc\tau}, \tag{12.37}$$

where $k = \omega/c$ and τ is the relaxation time. The relaxation time for copper (aluminum) at room temperature is $c\tau \cong 8.1$ μm (2.4 μm). If we now define $\Gamma = c\tau/s_0$ and write

$$\lambda = \frac{a}{s_0}|\kappa|^{1/2}\left(1+\kappa^2\Gamma^2\right)^{-1/4}\left(i\sqrt{1+t_\lambda} + \operatorname{sgn}(\kappa)\sqrt{1-t_\lambda}\right), \tag{12.38}$$

where

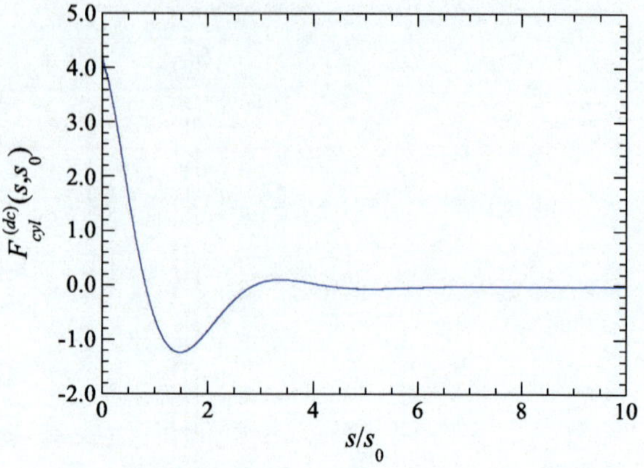

Fig. 12.49 Plot of the dc wake function in a cylindrical beam pipe

$$t_\lambda = \frac{|\kappa|\Gamma}{\sqrt{1+\kappa^2\Gamma^2}}. \tag{12.39}$$

Substitution of this value of λ in the expression for the impedance, we obtain

$$Z(\kappa) = \frac{2s_0}{a^2c}\left[\frac{\left(i\sqrt{1+t_\lambda} + \mathrm{sgn}(\kappa)\sqrt{1-t_\lambda}\right)}{|\kappa|^{1/2}\left(1+\kappa^2\Gamma^2\right)^{1/4}} - \frac{i\kappa}{2}\right]^{-1}. \tag{12.40}$$

Following the procedure described for the dc wakefields, the ac wakefields are of the form

$$E_z(s) = -\frac{q}{a^2}F^{(ac)}_{cyl}(s, s_0), \tag{12.41}$$

where the characteristic wakefield function is

$$F^{(ac)}_{cyl}(s, s_0) = \frac{1}{\pi} \times \int_{-\infty}^{\infty} d\kappa \exp(-i\kappa s/s_0)\left[\frac{\left(i\sqrt{1+t_\lambda} + \mathrm{sgn}(\kappa)\sqrt{1-t_\lambda}\right)}{|\kappa|^{1/2}\left(1+\kappa^2\Gamma^2\right)^{1/4}} - \frac{i\kappa}{2}\right]^{-1}. \tag{12.42}$$

A plot of the wake function is plotted versus position in Fig. 12.50.

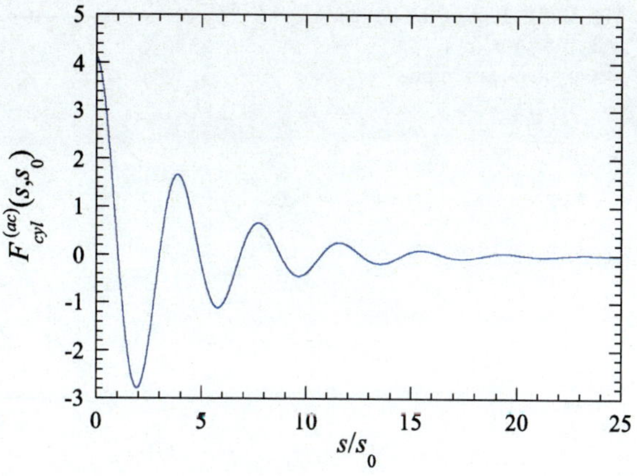

Fig. 12.50 Plot of the ac wake function in a cylindrical beam pipe

12.8.2 The Wakefields in a Rectangular Beam Pipe

The dc impedance of a rectangular wall, or parallel plate, beam pipe of separation $2a$ is [55]

$$Z(k) = \frac{1}{ac} \int_{-\infty}^{\infty} dx \left[\frac{\lambda}{k} \cosh^2 x - \frac{ika}{x} \cosh x \sinh x \right]^{-1}, \tag{12.43}$$

where

$$\lambda = \sqrt{\frac{2\pi\sigma_{dc}}{c} |k|} [i + \mathrm{sgn}(k)]. \tag{12.44}$$

As in the case of a cylindrical beam pipe, we define $\kappa = ks_0$, and write the impedance as

$$Z(k) = \frac{2s_0}{a^2 c} \int_0^{\infty} dx \left[\frac{1+i}{|\kappa|^{1/2}} \cosh^2 x - \frac{i\kappa}{x} \cosh x \sinh x \right]^{-1}. \tag{12.45}$$

As a consequence, the dc wakefields can be shown as

$$E_z(s) = -\frac{q}{a^2} F^{(dc)}_{rec}(s, s_0), \tag{12.46}$$

where the characteristic function is

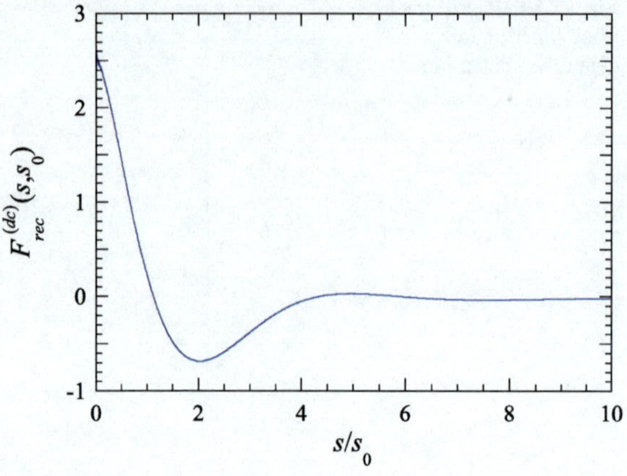

Fig. 12.51 Plot of the dc wake function in a rectangular beam pipe

$$F_{rec}^{(dc)}(s,s_0) = \frac{1}{\pi}\int_{-\infty}^{\infty} d\kappa \exp(-i\kappa s/s_0)\int_0^{\infty} dx\left[\frac{\cosh^2 x}{|\kappa|^{1/2}} - \frac{i\kappa}{x}\cosh x \sinh x\right]^{-1}. \tag{12.47}$$

The wake function for the dc wakefields in a rectangular beam pipe is shown in Fig. 12.51.

Following the procedure used to obtain the ac wakefields in a cylindrical drift tube, the ac impedance in a rectangular drift tube can be written as [55].

$$Z(\kappa) = \frac{2s_0}{a^2 c}\int_0^{\infty} dx\left[\frac{\left(i\sqrt{1+t_\lambda} + \operatorname{sgn}(\kappa)\sqrt{1-t_\lambda}\right)}{|\kappa|^{1/2}\left(1+\kappa^2\Gamma^2\right)^{1/4}}\cosh^2 x - \frac{i\kappa}{2}\cosh x \sinh x\right]^{-1}. \tag{12.48}$$

so that the ac wakefield in a rectangular beam pipe is

$$E_z(s) = -\frac{q}{a^2}F_{rec}^{(ac)}(s,s_0), \tag{12.49}$$

where the wake function is

$$F_{rec}^{(dc)}(s,s_0) = \frac{1}{\pi}\int_{-\infty}^{\infty} d\kappa \exp(-i\kappa s/s_0)$$

$$\times \int_0^{\infty} dx\left[\frac{\left(i\sqrt{1+t_\lambda} + \operatorname{sgn}(\kappa)\sqrt{1-t_\lambda}\right)}{|\kappa|^{1/2}\left(1+\kappa^2\Gamma^2\right)^{1/4}}\cosh^2 x - \frac{i\kappa}{x}\cosh x \sinh x\right]^{-1}. \tag{12.50}$$

The wake function is plotted versus distance in Fig. 12.52.

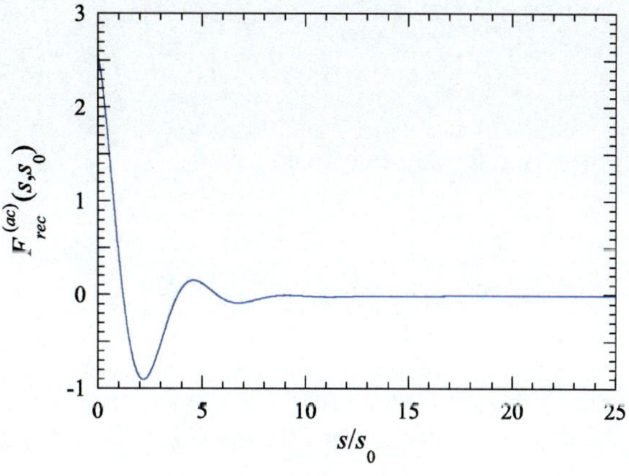

Fig. 12.52 The ac wake function in a rectangular beam pipe

12.8.3 The Energy Variation Within the Bunch

The electric field at some position s within the bunch is due to the wakes generated by those electrons ahead of that point; hence, the electric field is given by

$$E_z(s) = -\frac{1}{a^2}\int_s^{s_b} ds' q_b(s')F(s'), \tag{12.51}$$

where s_b is the bunch length, $F(s)$ is the wake function, and $q_b(s)$ describes the longitudinal (i.e., temporal) charge profile and is related to the total bunch charge

$$Q_b = -\int_0^{s_b} ds\, q_b(s). \tag{12.52}$$

This can be expressed in terms of the peak current, I_b, across the bunch and the bunch duration via

$$Q_b = \frac{I_b s_b}{c}\int_0^{s_b} ds\, \sigma_b(s), \tag{12.53}$$

where $\sigma_b(s)$ denotes the current profile. As a result, the cumulative energy change in over some distance L of the electrons at a position s within the bunch is

$$\frac{\Delta\gamma_L(s)}{\gamma_b} \cong -\frac{I_b}{I_A}\frac{Ls_b}{a^2}\int_s^{s_b} ds'\sigma_b(s')F(s'), \tag{12.54}$$

where I_A is the Alfvén current.

12.8.4 An Example: The LCLS

The simulations discussed previously in this chapter of the LCLS were performed without regard to the wakes generated by the electron beam; nevertheless, good agreement was found between the simulations and the measured pulse energies. Because of this, it is important to consider the wakes expected for the LCLS. Consider parameters similar to the reported properties of the first lasing experiment on the LCLS [29]. The copper beam pipe in the LCLS was approximately cylindrical with a radius of 5 mm, through which a flat top electron bunch propagated with an energy of about 13.64 GeV, a bunch charge of 250 pC, and a peak current of 3000 A. The electron bunch was approximately 25 μm in length and $s_0 \cong 12.7$ μm or about half the bunch length. Saturation was found after approximately 60 m. The Peirce parameter for the LCLS is $\rho \approx 5.8 \times 10^{-4}$, and it is expected that the wakes will not be important if they induce energy changes along the bunch of $\Delta\gamma/\gamma_b < \rho$. The energy changes induced on the bunch by the dc and ac wakes are shown in Fig. 12.53. It is clear from the figure that the magnitude of the induced energy changes is smaller than the Pierce parameter; hence, the wakefields should not be expected to have a large impact on the LCLS.

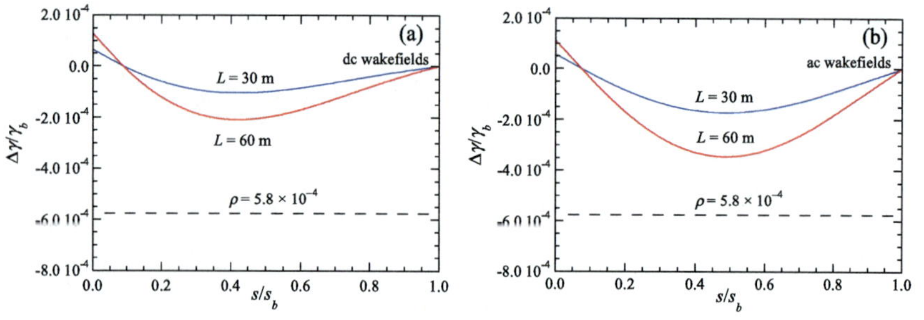

Fig. 12.53 The energy changes in the LCLS bunch due to the dc and ac wakefields

References

1. V.L. Granatstein, S.P. Schlesinger, M. Herndon, R.K. Parker, J.A. Pasour, Production of megawatt submillimeter pulses by stimulated magneto-Raman scattering. Appl. Phys. Lett. **30**, 384 (1977)
2. D.B. McDermott, T.C. Marshall, S.P. Schlesinger, R.K. Parker, V.L. Granatstein, High-power free-electron laser based on stimulated Raman backscattering. Phys. Rev. Lett. **41**, 1368 (1978)
3. R.K. Parker, R.H. Jackson, S.H. Gold, H.P. Freund, V.L. Granatstein, P.C. Efthimion, M. Herndon, A.K. Kinkead, Axial magnetic field effects in a collective-interaction free-electron laser at millimeter wavelengths. Phys. Rev. Lett. **48**, 238 (1982)
4. R.H. Jackson, S.H. Gold, R.K. Parker, H.P. Freund, P.C. Efthimion, V.L. Granatstein, M. Herndon, A.K. Kinkead, J.E. Kosakowski, T.J.T. Kwan, Design and operation of a collective millimeter-wave free-electron laser. IEEE J. Quantum Electron. **QE-19**, 346 (1983)
5. S.H. Gold, W.M. Black, H.P. Freund, V.L. Granatstein, R.H. Jackson, P.C. Efthimion, A.K. Kinkead, Study of gain, bandwidth, and tunability of a millimeter-wave free-electron laser operating in the collective regime. Phys. Fluids **26**, 2683 (1983)
6. S.H. Gold, W.M. Black, H.P. Freund, V.L. Granatstein, A.K. Kinkead, Radiation growth in a millimeter-wave free-electron laser operating in the collective regime. Phys. Fluids **27**, 746 (1984)
7. J.A. Pasour, R.F. Lucey, C.A. Kapetanakos, Long-pulse, high-power free electron laser with no external beam focusing. Phys. Rev. Lett. **53**, 1728 (1984)
8. S.H. Gold, D.L. Hardesty, A.K. Kinkead, L.R. Barnett, V.L. Granatstein, High-gain 35 GHz free-electron laser amplifier experiment. Phys. Rev. Lett. **52**, 1218 (1984)
9. J.A. Pasour, R.F. Lucey, C.W. Roberson, *Long pulse free-electron laser driven by a linear induction accelerator*, in *Free-Electron Generators of Coherent Radiation*, eds. C.A. Brau, S.F. Jacobs, and M.O. Scully, Proc. SPIE 453, p. 328 (1984)
10. J.A. Pasour, S.H. Gold, Free-electron laser experiments with and without a guide magnetic field: A review of the millimeter-wave free-electron laser research at the Naval Research Laboratory. IEEE J. Quantum Electron. **QE-21 845** (1985)
11. S.H. Gold, A.K. Ganguly, H.P. Freund, A.W. Fliflet, V.L. Granatstein, D.L. Hardesty, A.K. Kinkead, Parametric behavior of a high-gain 35 GHz free-electron laser amplifier with guide magnetic field. Nucl. Instrum. Methods Phys. Res. **A250**, 366 (1986)
12. J. Mathew, J.A. Pasour, High-gain, long-pulse free-electron laser oscillator. Phys. Rev. Lett. **56**, 1805 (1986)
13. J.A. Pasour, J. Mathew, C. Kapetanakos, Recent results from the Naval Research Laboratory experimental free-electron laser program. Nucl. Instr. Meth. **A259**, 94 (1987)
14. D.S. Birkett, T.C. Marshall, S.P. Schlesinger, D.B. McDermott, A submillimeter free-electron laser experiment. IEEE J. Quantum Electron. **QE-17**, 1348 (1981)
15. J. Masud, T.C. Marshall, S.P. Schlesinger, F.G. Yee, Gain measurements from start-up and spectrum of a Raman free-electron laser oscillator. Phys. Rev. Lett. **56**, 1567 (1986)
16. J. Masud, T.C. Marshall, S.P. Schlesinger, F.G. Yee, W.M. Fawley, E.T. Scharlemann, S.S. Yu, A.M. Sessler, E.J. Sternbach, Sideband control in a millimeter-wave free-electron laser. Phys. Rev. Lett. **58**, 763 (1987)
17. J. Masud, T.C. Marshall, S.P. Schlesinger, F.G. Yee, Regenerative gain in a Raman free-electron laser oscillator. IEEE J. Quantum Electron **QE-23 1594** (1987)
18. F.G. Yee, J. Masud, T.C. Marshall, S.P. Schlesinger, Power and sideband studies of a Raman free-electron laser. Nucl. Instrum. Methods Phys. Res. **A259**, 104 (1987)
19. F.G. Yee, T.C. Marshall, S.P. Schlesinger, Efficiency and sideband observations of a Raman free-electron laser oscillator with a tapered undulator. IEEE Trans. Plasma Sci. **PS-16**, 162 (1988)
20. S.Y. Cai, S.P. Chang, J.W. Dodd, T.C. Marshall, H. Tang, Optical guiding in a Raman free-electron laser: Computation and experiment. Nucl. Instrum. Methods Phys. Res. **A272**, 136 (1988)

21. J.W. Dodd, T.C. Marshall, Spiking radiation in the Columbia free-electron laser. Nucl. Instrum. Methods Phys. Res. **A296**, 4 (1990)
22. J. Fajans, G. Bekefi, Y.Z. Yin, B. Lax, Spectral measurements from a tunable, Raman free-electron laser. Phys. Rev. Lett. **53**, 246 (1984)
23. J. Fajans, G. Bekefi, Y.Z. Yin, B. Lax, Microwave studies of a tunable free-electron laser in combined axial and wiggler magnetic fields. Phys. Fluids **28**, 1995 (1985)
24. D.A. Kirkpatrick, G. Bekefi, A.C. DiRienzo, H.P. Freund, A.K. Ganguly, A millimeter and submillimeter wavelength free-electron laser. Phys. Fluids B **1**, 1511 (1989)
25. D.C. Nguyen, R.L. Sheffield, C.M. Fortgang, J.C. Goldstein, J.M. Kinross-Wright, N.A. Ebrahim, Self-amplified spontaneous emission driven by a high-brightness electron beam. Phys. Rev. Lett. **81**, 810 (1998)
26. S.V. Milton et al., Exponential gain and saturation of a self-amplified spontaneous emission free-electron laser. Science **292**, 2037 (2001)
27. P. Frigola et al., Initial gain measurements of an 800 nm SASE FEL, VISA. Nucl. Instrum. Methods Phys. Res. **A475**, 339 (2001)
28. V. Ayvazyan et al., First operation of a free-electron laser generating GW power radiation at 32 nm wavelength. Eur. Phys. J. D **37**, 297 (2006)
29. P. Emma et al., First lasing and operation of an Ångstrom-wavelength free-electron laser. Nature Phot. **4**, 641 (2009)
30. L. Giannessi et al., Self-amplified spontaneous emission for a single pass free-electron laser. Phys. Rev. ST-AB **14**, 060712 (2011)
31. T. Tanaka, S. Goto, T. Hara, T. Hatsui, H. Ohashi, K. Togawa, M. Yabashi, H. Tanaka, Undulator commissioning by characterization of radiation in x-ray free-electron lasers. Phys. Rev. ST-AB **15**, 110701 (2012)
32. J.-H. Han et al., *Status of the PAL-XFEL Project, in Proceedings of the 2012 International Particle Accelerator Conference* (Louisiana, New Orleans, 2012)
33. E. Garwin, F. Meier, T. Pierce, K. Sattler, H.-C. Siegmann, A pulsed source of spin-polarized electrons by photoemission from EuO. Nucl. Instrum. Methods Phys. Res. **120**, 483 (1974)
34. D.T. Pierce, F. Meier, Photoemisison of spin-polarized electrons from GaS. Phys. Rev. B **13**, 5484 (1976)
35. C.K. Sinclair, R.H. Miller, A high current, short pulse, rf synchronized electron gun for the Stanford linear accelerator. IEEE Trans. Nuclear Sci. **NS-28**, 2649 (1981)
36. R.L. Sheffield, E.R. Gray, J.S. Fraser, The Los Alamos photoinjector program. N Nucl. Instrum. Methods Phys. Res. **A272**, 222 (1988)
37. M. Xie, *Design optimization for an x-ray free electron laser driven by the SLAC linac*, Proc. IEEE 1995 Particle Accelerator Conference, Vol. 183, IEEE Cat. No. 95CH35843 (1995)
38. K.-J. Kim, M. Xie, Self-amplified spontaneous emission for short wavelength coherent radiation. Nucl. Instrum, Methods Phys. Res. **A331**, 359 (1993)
39. H.P. Freund, W.H. Miner Jr., Efficiency enhancement in seeded and self-amplified spontaneous emission free-electron lasers by means of a tapered wiggler. J. Appl. Phys. **105**, 113106 (2009)
40. S. Krinsky, *On the definition of the number of temporal modes in the SASE output*, in Proceedings of the 27th International Conference on Free-Electron Lasers, (www.JACoW.org, 2005), p.94
41. H.P. Freund, Phase-matching segmented wigglers in free-electron lasers. Phys. Rev. E **70**, 015501(R) (2004)
42. N.A. Vinokurov, Multisegment wigglers for short wavelength FEL. Nucl. Instr. Meth. **A375**, 264 (1996)
43. H.P. Freund, P.J.M. van der Slot, D.L.A.G. Grimminck, I.D. Steya, P. Falgari, Three-dimensional, time-dependent simulation of free-electron lasers with planar, helical, and elliptical undulators. New J. Phys. **19**, 023020 (2017)
44. D. Ratner et al., *FEL Gain Length and Taper Measurements at LCLS*, SLAC-PUB-14194 (2010)

45. B.W.J. McNeil, G.R.M. Robb, M.W. Poole, N.R. Thompson, Harmonic lasing in a free-electron laser amplifier. Phys. Rev. Lett. **96**, 084801 (2006)
46. E.A. Schneidmiller, M. Yurkov, Harmonic lasing in x-ray free-electron lasers. Phys. Rev. ST-AB **15**, 080702 (2012)
47. H.P. Freund, N.A. Yampolsky, Q. Marksteiner, Enhanced harmonic generation in x-ray free-electron lasers. Phys. Rev. ST-AB **17**, 010702 (2014)
48. C. Emma, K. Fang, J. Wu, C. Pellegrini, High efficiency, multiterawatt x-ray free-electron lasers. Phys. Rev. Accel. Beams **19**, 020705 (2016)
49. H.P. Freund, P.J.M. van der Slot, Studies of a terawatt x-ray free-electron laser. New J. Phys. **20**, 073017 (2018)
50. T.J. Orzechowski, B.R. Anderson, J.C. Clark, W.M. Fawley, A.C. Paul, D. Prosnitz, E.T. Scharlemann, S.M. Yarema, D.B. Hopkins, A.M. Sessler, J.S. Wurtele, High-efficiency extraction of microwave radiation from a tapered-wiggler free-electron laser. Phys. Rev. Lett. **57**, 2172 (1986)
51. X.J. Wang, H.P. Freund, W.H. Miner Jr., J.B. Murphy, H. Qian, Y. Shen, X. Yang, Efficiency and spectral enhancement in a tapered free-electron laser amplifier. Phys. Rev. Lett. **103**, 154801 (2009)
52. Y. Jiao, J. Wu, Y. Cai, W.M. Fawley, J. Frisch, Z. Huang, H.-D. Nuhn, C. Pellegrini, S. Reiche, Modeling and multidimensional optimization of a tapered free-electron laser. Phys. Rev. ST-AB **15**, 050704 (2012)
53. W.M. Fawley, "Optical guiding" limits on extraction efficiencies of single-pass, tapered wiggler amplifiers. Nucl. Instrum. Methods Phys. Res. **A375**, 550 (1996)
54. K.L.F Bane, M. Sands, *The short-range resistive wall wakefields*, SLAC-PUB-95-7074 (1995)
55. K.L.F. Bane, G. Stupakov, *Resistive Wall Wakefield in the LCLS Undulator Beam Pipe* (SLAC-PUB-10707, 2004)
56. S.S. Baturin and A.D. Kanareykin, *New method of calculating the wakefields of a point charge in a* 51. K.L.F. Bane and G. Stupakov, *Roughness tolerances in the undulator vacuum chamber of LCLS-II*, in the Proceedings of the LINAC2014 Conference, Geneva, Switzerland, 2014, p. 708
57. A.W. Chao, *Physics of Collective Beam Instabilities in High Energy Accelerators* (Wiley, New York, 1993)

Chapter 13
Optical Klystrons and High-Gain Harmonic Generation

Optical klystrons have been in use for decades, and the first use was in an ultraviolet free-electron laser oscillator [1]. An optical klystron (OK) is fundamentally composed of a *modulator* wiggler that imposes a velocity modulation on the electrons followed by a magnetic dispersive section that enhances the modulation prior to injection into a *radiator* wiggler that takes the interaction with the modulation-enhanced electrons to saturation. The magnetic dispersive element is, typically, a three- or four-dipole chicane. We have discussed the use of strong chicanes to compress the electron bunch and weak chicanes to control the phase shift between the electrons and the optical field in the gaps between wigglers. The chicanes used in optical klystrons are intermediate between these other two applications. The *radiator* can be tuned to the fundamental or a harmonic of the *modulator*, in which case the interaction is referred to as high-gain harmonic generation (HGHG) [2, 3]. At the present time, an extreme ultraviolet through soft x-ray HGHG user facility is operational at FERMI-Elletra in Italy [4–7].

13.1 The Physical Concept

The fundamental interaction in an optical klystron is illustrated schematically in Fig. 13.1, where the electron trajectory through the *modulator*, the chicane, and followed through the *radiator* is indicated by the line showing the wiggler motion in the *modulator* and *radiator* as well as the displacement in the chicane. The phase space at the upper left in the figure represents the exit from the modulator and the entrance to the chicane. The modulation in the phase space here is due to the interaction in the *modulator*, which may arise due to a high-power seed in a short wiggler, or the modulation that arises in a longer wiggler due either to startup from noise or to a low power seed. As shown in the Appendix and in Eq. 12.22, the path length of an electron through the chicane scales inversely as the square of the energy. As a result, the higher energy electrons in the tail of the phase space will "catch up"

H. P. Freund, T. M. Antonsen, Jr., *Principles of Free Electron Lasers*,
https://doi.org/10.1007/978-3-031-40945-5_13

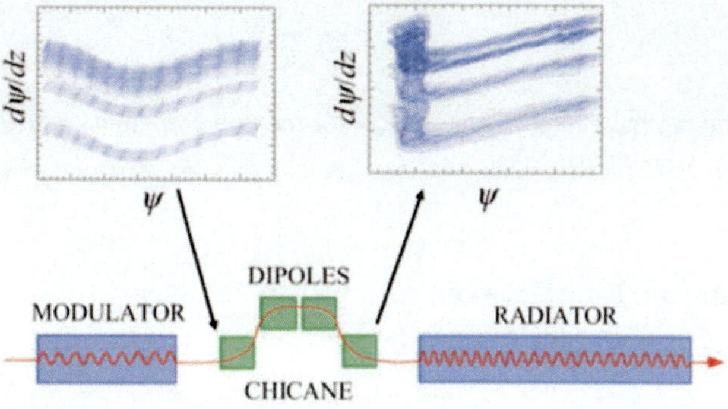

Fig. 13.1 Schematic illustration of the effect of the chicane in an optical klystron

with the lower energy electrons near the center or head of the bunch, as shown in the phase space in the upper right of the figure, which represents the phase space at the exit of the chicane and which then propagates to the entrance of the *radiator*. The enhanced bunching acts to pre-condition the electron beam to radiate strongly in the *radiator*.

The optical klystron has several advantages over a conventional long wiggler. In a conventional wiggler, the interaction between the electrons and the optical field results in a density and velocity modulation that increases exponentially as the interaction proceeds. The velocity modulation, in particular, is associated with an increasing energy spread over the course of the wiggler. This increasing energy spread ultimately results in a degradation of the resonant interaction and can limit the extraction efficiency. In contrast, the enhanced bunching in the chicane occurs with at most a modest increase in the energy spread, which can in some circumstances result in an efficiency that is enhanced relative to that in a conventional wiggler. In addition, the bunching process in the chicane is faster than the increasing modulation in a conventional wiggler, thereby shortening the overall interaction length. In principle, therefore, the OK can result in both a shorter interaction length and a higher extraction efficiency over a conventional long wiggler.

It should be recognized that the interaction in an optical klystron is sensitive to the energy spread [9]. The ratio of the path length, s, through a chicane of four dipoles of equal lengths and field strengths to the resonant wavelength, λ, is

$$\frac{s}{\lambda} = \frac{2}{3}\frac{L_d^3\Omega_0^2}{\gamma^2 c^2 \lambda}, \tag{13.1}$$

where L_d is the dipole length, Ω_0 is the cyclotron frequency corresponding to the dipole field strength, γ is the relativistic factor corresponding to the electron energy, and c is the speed of light in vacuo. As a result, the change in the path length with respect to variations in the energy is

$$\frac{\Delta s}{\lambda} = -2\frac{s}{\lambda}\frac{\Delta\gamma}{\gamma}. \tag{13.2}$$

However, we must require that $\Delta s < \lambda$ for the bunching in an optical klystron to be effective, which means that $\Delta\gamma/\gamma < \lambda/2\ s$.

13.2 Comparison Between an Optical Klystron and a Conventional Wiggler

The nonlinear formulation described in Chap. 6 is here used to compare the performance of an optical klystron with that of a conventional wiggler [8]. We assume that the electron beam has an energy of 82.25 MeV and a bunch charge of 650 pC. The pulse shape is assumed to be parabolic with a full width of 1.0 psec. The normalized emittance ranges over 10–15 mm-mrad and the rms energy spread over 0.1–0.3%. The radiator and modulator wigglers are assumed to provide weak focusing with parabolic-pole-faces, and the electron beam is injected with the matched-beam radius. The period of both the *radiator* and *modulator* is 3.0 cm, and the amplitude of the modulator is 5.0 kG over a length of 1.77 m. This represents a total of 59 periods, and the first and last periods describe and up- and down-taper for the field. Hence, the *modulator* is characterized by 57 periods with a uniform field strength. The *radiator* has an initial amplitude of 5.0 kG. The chicane is composed of four hard-edge dipoles each of which is 0.09 m in length and which is separated by gaps of 0.03 m. This results in an overall chicane length of 0.45 m, and the dipole fields are varied between 2 and 3 kG. The gaps between the chicane and the wigglers are also assumed to be 0.03 m, so that the overall gap between the *modulator* and *radiator* is 0.51 m long. The OK is tuned to a wavelength of 1.06 μm, and the seed laser is assumed to provide a parabolic pulse with a full width of 1.0 psec to match the electron bunch and with a peak power of 1.0 kW corresponding to a pulse energy of about 0.67 nJ. For the present parameters, therefore, we have that $|\Delta s/\lambda| \approx 475|\Delta\gamma/\gamma|$ for a 2.0 kG dipole field, which means that the energy spread must be on the order of about 0.2% or less. The conventional tapered wiggler amplifier uses the same beam and optical parameters, but the wiggler is lengthened.

The optimal dipole fields are dependent upon the modulation induced on the electron bunch due to the interaction in the *modulator*, and this, in turn, depends upon the beam parameters, the *modulator*, and the seed power. The evolution of the optical pulse for an emittance of 10 mm-mrad and energy spreads of 0.1%, 0.2%, and 0.3% is shown in Fig. 13.2 where the optimal choices of the dipole fields (2.4–2.5 kG) are indicated. It is evident that the optical field grows exponentially in the modulator indicating that velocity modulation is taking place. The growth of the optical field ceases in the gap/chicane between the wigglers. The enhanced bunching in the chicane pre-conditions the field for rapid growth in the radiator, and saturation is found over an additional length of about 1.8 m. The overall distance

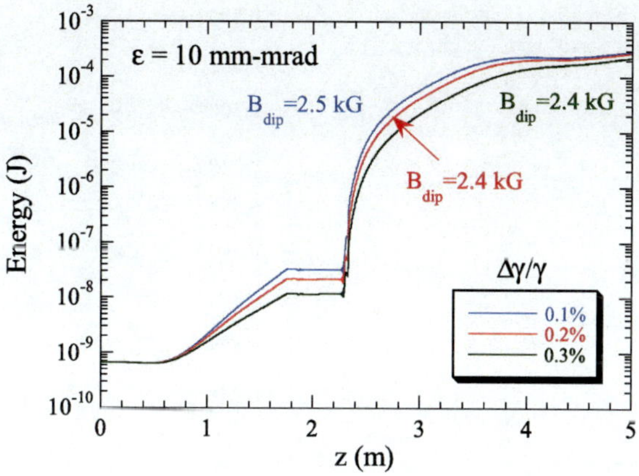

Fig. 13.2 Evolution of the optical pulse for an emittance of 10 mm-mrad and energy spreads of 0.1%, 0.2%, and 0.3% for the optimal dipole fields [8]

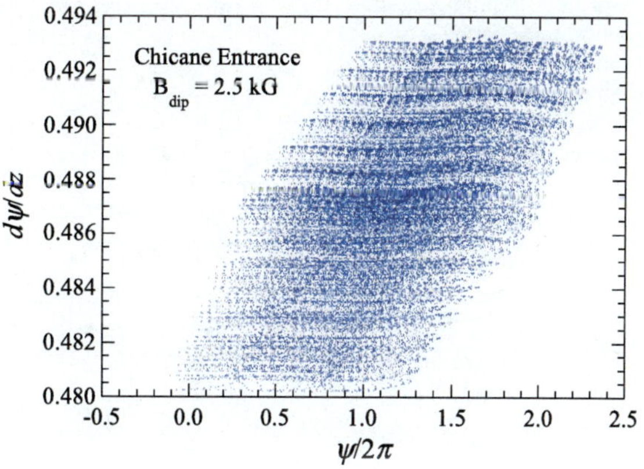

Fig. 13.3 Phase space at the entrance to the chicane [8]

to saturation is about 3.8 m. The pulse energy at saturation decreases with increasing energy spread as expected. The saturated pulse energy is 0.24 mJ at an energy spread of 0.1%, which decreases slightly to 0.21 mJ as the energy spread increases to 0.2%. As expected on the basis of the path length argument above, the saturated pulse energy decreases more dramatically to 0.15 mJ as the energy spread increases to 0.3%.

The phase space dynamics in the chicane are illustrated in Figs. 13.3 and 13.4, which show the phase space at the entrance to the chicane (Fig. 13.3) and at the exit from the chicane (Fig. 13.4) for an emittance of 10 mm-mrad, an energy spread of 0.1%, and the optimal dipole field of 2.5 kG. As shown in Fig. 13.3, the phase space for a *slice* that is one wavelength long exhibits some dispersion as well as a modulation due to the interaction in the modulator. The effect of the chicane is dramatic, as shown in Fig. 13.4, where (1) substantial dispersion is found as the original *slice* is now smeared over about $\delta\psi \approx 16\pi$ (i.e., 16 wavelengths) and (2) pronounced bunching is exhibited every wavelength. It is this bunching that drives the amplification in the radiator so strongly.

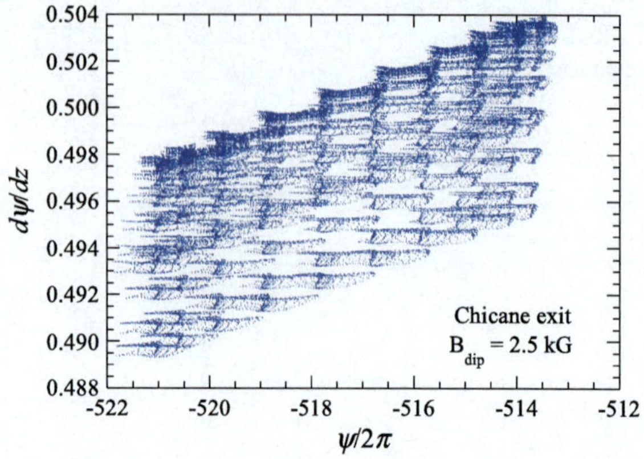

Fig. 13.4 Phase space at the exit from the chicane [8]

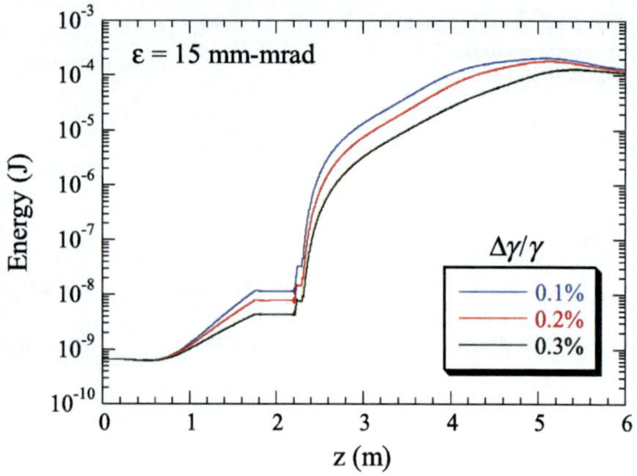

Fig. 13.5 Evolution of the optical pulse for an emittance of 15 mm-mrad and energy spreads of 0.1%, 0.2%, and 0.3% for the optimal dipole fields 8]

The interaction is degraded as the emittance increases but is still relatively strong for an emittance of 15 mm-mrad. The amplification of the seed pulse for an emittance of 15 mm-mrad and energy spreads of 0.1%, 0.2%, and 0.3% is shown in Fig. 13.5, where strong amplification is still found. However, the interaction is weaker than what was found at 10 mm-mrad as the overall saturation distance has increased to about 5.1 m and the saturated pulse energies have dropped to 0.21 mJ, 0.18 mJ, and 0.13 mJ for energy spreads of 0.1%, 0.2%, and 0.3%, respectively.

A comparison between the OK and a single wiggler is shown in Figs. 13.6, 13.7, and 13.8 for an emittance of 10 mm-mrad and energy spreads of 0.1%, 0.2%, and 0.3%, respectively. In the case of 0.1% energy spread, the two cases reach similar extraction efficiencies, but the OK reaches saturation over a shorter overall length by approximately 0.6 m. The advantage in saturation distance for the OK is retained as the energy spread increases to 0.2% (see Fig. 13.7). As discussed previously, coherent bunching in the chicane begins to fail as the energy spread increases to 0.3% and the performance of the OK is noticeably degraded. However, as shown in

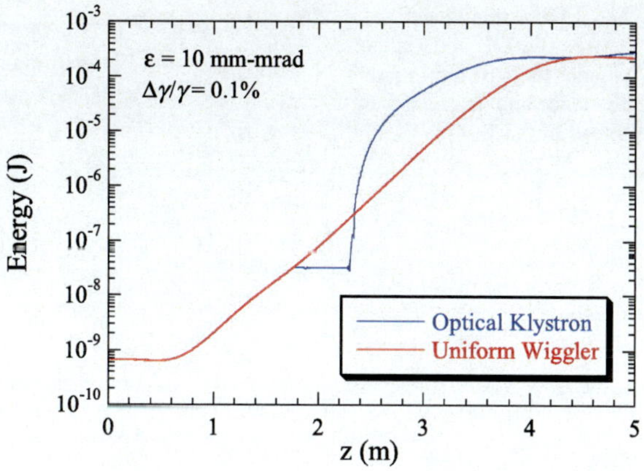

Fig. 13.6 Comparison between the OK and a single wiggler for a 10 mm-mrad emittance and an energy spread of 0.1% [8]

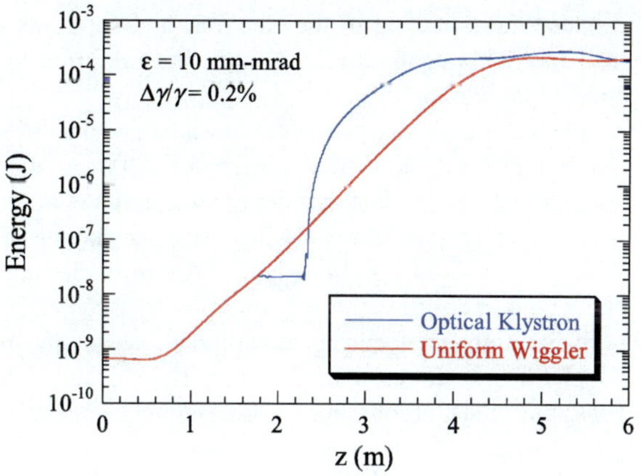

Fig. 13.7 Comparison between the OK and a single wiggler for a 10 mm-mrad emittance and an energy spread of 0.2% [8]

Fig. 13.8, the performance of the single wiggler is degraded even more than the OK, which now shows both a shorter saturation distance and a higher extraction efficiency.

13.3 The Multistage Optical Klystron

As described in the preceding chapter, many x-ray free-electron lasers rely on self-amplified spontaneous emission and are configured using multi-segment wigglers with quadrupoles inserted between the wigglers in order to provide for enhanced focusing of the electron beam. In addition, weak chicanes are also inserted in the gaps between the wigglers to control the relative phase shift between the electrons and the optical field. However, it has been proposed [9–13] that these chicanes may be configured, using dipole fields that are stronger than needed for a phase shifter, to

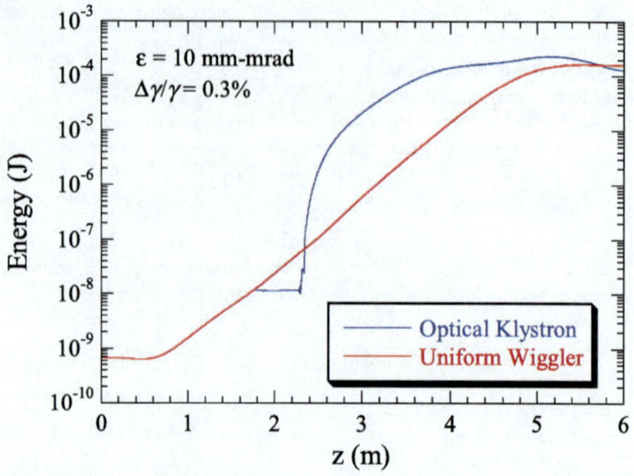

Fig. 13.8 Comparison between the OK and a single wiggler for a 10 mm-mrad emittance and an energy spread of 0.3% [8]

enhance the bunching of the electrons and so act as a multistage optical klystron (MSOK). This configuration is sometimes referred to as a *distributed optical klystron* in the literature.

The specific configuration that is under consideration here employs a 14.35 GeV electron beam with a peak current of 3400 A and a normalized emittance of 1.0–1.5 mm-mrad. For simplicity, we assume that each wiggler segment is a weak-focusing parabolic-pole-face wiggler having a period of 3.0 cm, a field strength of 13.2 kG, and a length of 4.8 m (including entry and exit transitions of one wiggler period) with a gap of 2.0 m between each segment, which obviates the need for a strong focusing quadrupole lattice. The matched beam radius in this wiggler line is approximately 88 μm. The chicanes are located in the centers of the gaps and consist of four dipoles of equal field strengths and have lengths of 0.4 m for a total length of 1.6 m. The resonant wavelength for these parameters is in the neighborhood of 1.5 Å.

The single-segment performance exhibits a minimum exponential gain length of 6.21 m for an energy spread of $\Delta\gamma/\gamma = 0\%$ and increases to 6.64 m for $\Delta\gamma/\gamma = 0.01\%$ but increases rapidly thereafter and rises to 8.04 m when $\Delta\gamma/\gamma = 0.02\%$. The saturated power shows a similar sensitivity to energy spread. It should be remarked that this is for the natural focusing in the PPF wiggler, while strong focusing with a more compact electron beam would result in a shorter gain length. However, the simplifications associated with the natural focusing in a parabolic-pole-face wiggler are suitable for the purpose of illustrating the both the advantages and disadvantages to the MSOK, and similar advantages are expected for a system with strong focusing.

The sensitivity of the MSOK to the beam energy spread is determined from Eqs. 13.1 and 13.2. For the parameters of interest here, $|\Delta L/\lambda| = 8.82 \times 10^3 \Delta\gamma/\gamma$. As a result, requiring that $|\Delta L/\lambda| < 1$ for coherent bunching implies that $\Delta\gamma/\gamma < 0.01\%$, which is similar to the sensitivity of a long, single-segment wiggler. Of course, the actual criterion is more complicated than this because the path length through the chicane also depends on the angle of entry into the chicane. Therefore, the requirement on beam quality is a complex function of both emittance and energy spread.

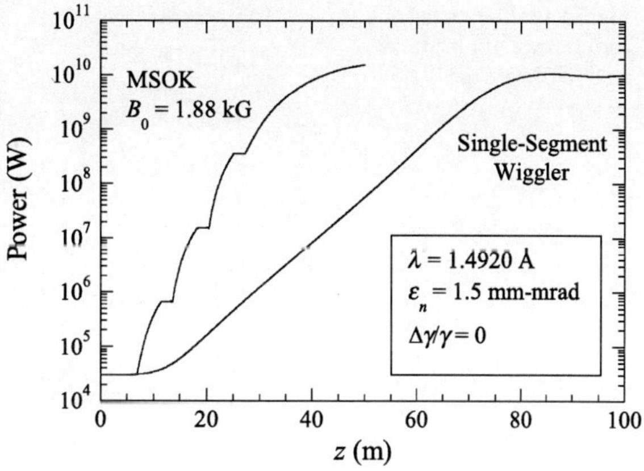

Fig. 13.9 Power growth for the MSOK and a single-segment wiggler [13]

While most of the x-ray free-electron lasers that might benefit from a MSOK rely on SASE since there are no seed lasers at these wavelengths, simulations using a high-power seed are adequate to illustrate the MSOK concept. As such, we perform single time-slice simulations of this configuration. The first case we consider is that of an emittance of 1.5 mm-mrad with an energy spread $\Delta\gamma/\gamma = 0$, and we compare the performance of an MSOK with a dipole field of 1.88 kG with that of a single long wiggler. Since the gap separating the wiggler segments is fixed, the wavelength that yields the optimal phase match between the electrons and the optical field is 1.492 Å. A comparison between the performance of the MSOK and the single, long wiggler is shown in Fig. 13.9 for an initial seed power of 30 kW. It is evident that the MSOK saturates at a higher power (15.5 GW) than the single-segment wiggler (10.7 GW). In addition, the MSOK saturates after only five wiggler segments and a length of 50 m, while the single-segment configuration saturates over a length of 85 m. The essential point, however, is that there is no region of exponential gain in the MSOK. Rather, the growth rates are faster than exponential in each wiggler segment.

The enhanced bunching in the chicane causes the growth in the MSOK to be faster than exponential. The growth rate in a wiggler segment after a chicane is initially faster than exponential but decreases to the expected exponential growth rate after a distance of approximately one gain length. However, the gain length in this design exceeds 6 m which is longer than the 4.8 m long wiggler segments. As a result, the beam and wave are extracted from each wiggler segment before the growth rate can roll over to the exponential rate. This process is repeated in each unit cell until the power approaches saturation. At some point approaching saturation, the effect of the enhanced bunching becomes ineffective. As a result, the final wiggler segment is longer than the preceding segments and is designed to bring the interaction to saturation.

The variation in the MSOK performance for five wiggler segments and 50 m total length is shown in Fig. 13.10 as a function of B0. The optimal B_0 of 1.88 kG is clearly indicated; however, there are large amplitude performance fluctuations with B_0. This is explained because the ratio of the path length through the chicane to the wavelength shown in Eq. 13.1 shows that

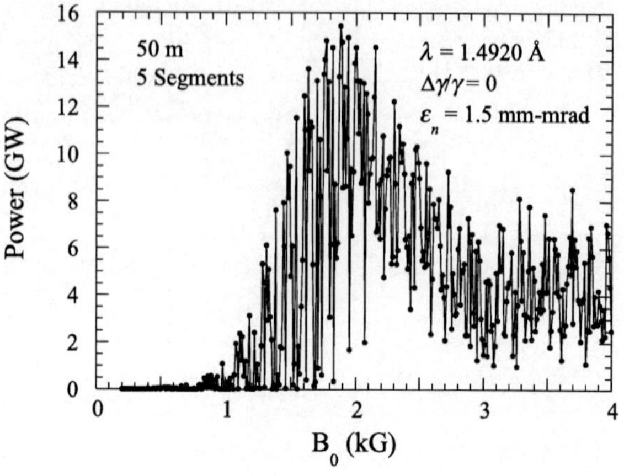

Fig. 13.10 The variation in performance of the MSOK with the dipole field [13]

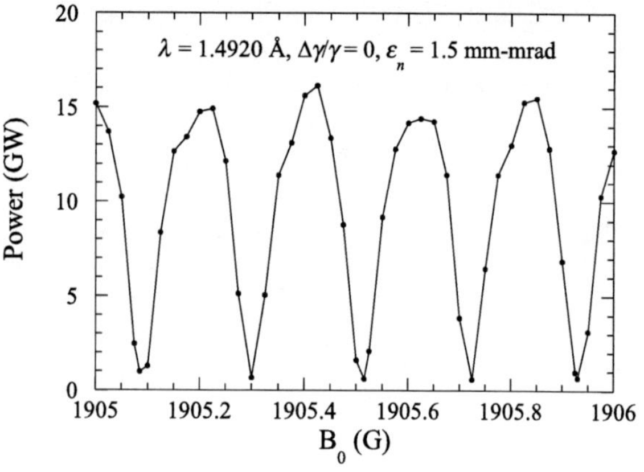

Fig. 13.11 Fine-scale variation of the MSOK performance with the dipole field strength

$$\frac{\Delta L}{\lambda} = \frac{4}{3} \frac{L_d^3 B_0}{\gamma^2 c^2 \lambda} \Delta B_0, \tag{13.3}$$

for an electron propagating paraxially. As a result, $\Delta L/\lambda \approx 4.9\Delta \text{B}_0$ for the parameters of interest; hence, the path length changes by one wavelength when B0 changes by about 0.20 G. The MSOK performance, therefore, is very sensitive to the dipole field strength and oscillates as B_0 changes by this amount. This periodicity is observed in simulation as shown in Fig. 13.11 where we expand the scale from Fig. 13.10 to show several such variations. As a result, the MSOK requires careful design and control of the fields in the chicanes.

This behavior is illustrated in Fig. 13.12 where we plot the evolution of the power with distance for a normalized emittance of 1.5 mm-mrad and for a single-segment wiggler and the MSOK with $\Delta\gamma/\gamma = 0$, 0.001%, and 0.002%. The evolution of the power for a single-segment wiggler is virtually independent of energy spread over

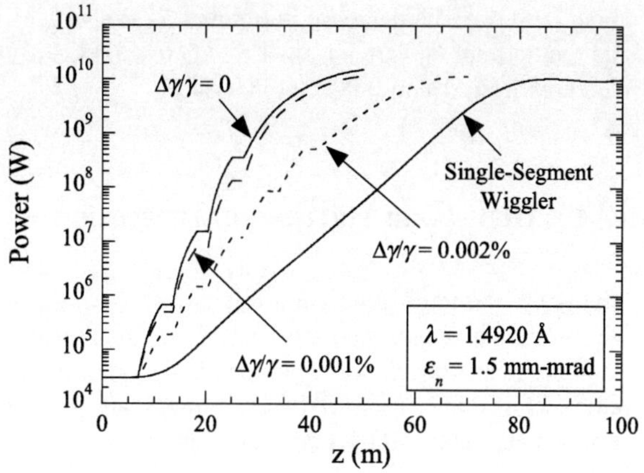

Fig. 13.12 Relative performance of the MSOK for an emittance of 1.5 mm-mrad [13]

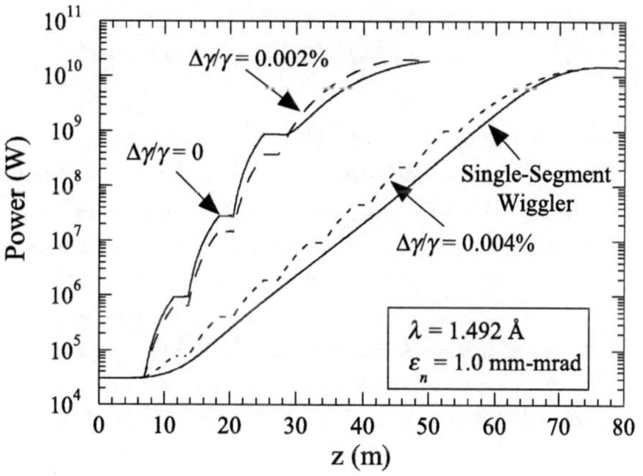

Fig. 13.13 Relative performance of the MSOK for an emittance of 1.0 mm-mrad [13]

this range. It is clear that there is little change in the performance of the MSOK as the energy spread increases to 0.001%. However, the performance degrades rapidly as the energy spread increases beyond this value, and seven wiggler segments and an overall length of 70 m are required to reach saturation in the MSOK for $\Delta\gamma/\gamma = 0.002\%$. This saturation length is only 17.6% shorter than that for the single-segment wiggler. The net benefit from the MSOK vanishes almost entirely when the energy spread increases to 0.0025%. It should also be noted that the optimal B_0 decreases from 1.88 kG to 1.82 kG to 1.47 kG as the energy spread increases from 0% to 0.001% to 0.002%.

The constraint on the energy spread is less severe for a lower emittance. This is illustrated in Fig. 13.13, where we plot the evolution of the power with distance for a normalized emittance of 1.0 mm-mrad and for a single-segment wiggler and the MSOK for $\Delta\gamma/\gamma = 0$, 0.002%, and 0.004%. It is clear that higher energy spreads can be tolerated for this lower emittance value, and the MSOK performance will substantially exceed that of the single segment wiggler for energy spreads below about 0.003%. The decrease in optimal B_0 is also seen for this choice of emittance, and the optimal B_0 varies from 1.82 kG to 1.63 kG to 0.70 kG as the energy spread increases from 0% to 0.002% to 0.004%.

13.4 High-Gain Harmonic Generation

High-gain harmonic generation refers to an optical klystron configuration in which the *radiator* is tuned to a harmonic of the fundamental resonance in the *modulator*. The nonlinear formulation described in Chap. 6, which was used above to simulate optical klystrons, self-consistently includes harmonic interactions and is used in this section to simulate HGHG configurations.

13.4.1 Second Harmonic Generation

We now consider an HGHG configuration in which the *modulator* is tuned to 10.6 μm and the *radiator* is tuned to 5.3 μm. Consider an electron beam with an energy of 40.7 MeV, a peak current of 120 A, a normalized emittance of 5.5 mm-mrad in both the *x*- and *y*-direction, and an rms energy spread of 0.05%. The *modulator* and *radiator* are assumed to be parabolic-pole-face designs with the electron beam matched into the *modulator*. The period and on-axis field strength of the *modulator* is 8.0 cm and 1.58 kG, respectively, with an overall length of 1.36 m (17 periods with entry and exit transitions of 4 periods in length). The *radiator* has a period of 3.3 cm, an on-axis field strength of 4.72 kG, and an overall length of 5.412 m (164 periods with an entry transition of 4 periods in length). The gap between the wigglers is 1.29 m in length. The chicane is composed of three dipoles in which the outer dipoles are oriented in the same direction with lengths of 7.5 cm. The center dipole is twice the length of the outer dipoles, and the field is oriented oppositely to that of the outer dipoles. This provides for an overall length of 30 cm. The first dipole is located a distance of 0.621 m from the exit of the *modulator*.

Steady-state, time-independent simulations are performed for a seed power of 700 kW at 10.6 μm. This is adequate to illustrate the basic properties of the HGHG mechanism. It is found in simulation that the optimal bunching in the chicane for this seed power is found for a dipole field strength of 2.05 kG. This is illustrated in Fig. 13.14 where we plot the electron phase space at (a) the entrance to the chicane and (b) at the exit from the chicane.

The growth of the second harmonic in the *radiator* is akin to the nonlinear harmonic generation mechanism in that it relies upon bunching that is due to the enhancement in the modulation imposed on the electron beam in the chicane. Thus, the second harmonic power rapidly grows from noise starting at the entrance to the *radiator*. This is shown in Fig. 13.15 where we plot the growth of the second

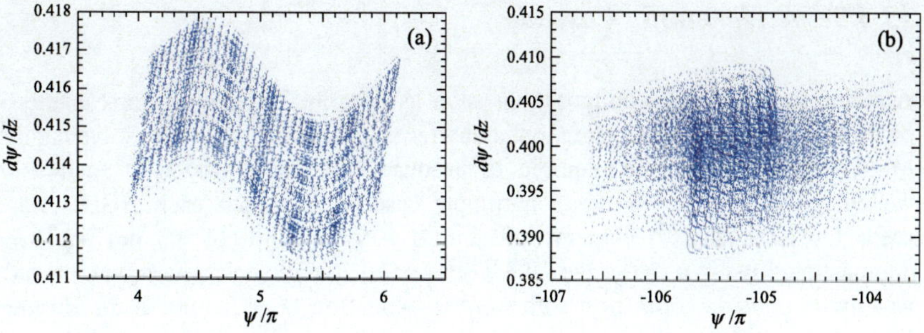

Fig. 13.14 The electron phase space (**a**) at the entrance to the chicane and (**b**) at the exit from the chicane

Fig. 13.15 growth of the second harmonic power in the *radiator*

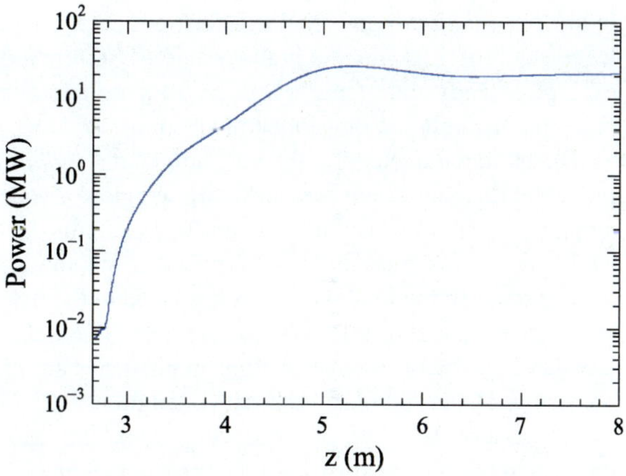

harmonic power in the *radiator*. Here, we observe that the power reaches a peak of about 26 MW at 2.3 m after the radiator entrance and then falls to an asymptotic level of about 20 MW.

The example that we have discussed is in rough correspondence to the HGHG experiment conducted at Brookhaven National Laboratory [2, 3]. The steady-state simulation is considered to be adequate to simulate this experiment because the electron bunch length of 6 psec is greater than the estimated slippage distance of 1 psec. The *radiator* used in the experiment was 2 m in length, and the measured output power at 5.3 μm was about 17 MW. The result shown in Fig. 13.15 shows a power of about 16.6 MW after 2 m of the *radiator*, which is close to the measured value.

13.4.2 A Harmonic Cascade

A single-stage HGHG configuration, such as described above, is not capable of producing fully coherent radiation at extreme ultraviolet or x-ray wavelengths. Multi- HGHG stages are required to produce coherent emission at such short wavelengths. This constitutes a harmonic cascade in which each HGHG stage seeds the next stage. Coherent emission at a wavelength of 4.3 nm has been demonstrated using a two-stage HGHG cascade [6], and such a harmonic cascade was under consideration for a soft x-ray free-electron laser facility at the Berliner Elektronenspeicherring-Gesellschaft für Synchrotronstrahlung (BESSY) [14]. Here, we discuss the simulations of one configuration that was under consideration in the BESSY technical design report that makes use of a cascade of four HGHG stages followed by a high-gain amplifier stage and that starts with a seed pulse at 279.5 nm in the first *modulator* and results in coherent output from the final amplifier stage at a wavelength of 1.24 nm. As in the previously discussed 10.6 μm → 5.3 μm HGHG example, steady-state (single time *slice*) simulations are discussed for this design using the formulation developed in Chap. 6.

The basic parameters for the *modulators* and *radiators* for these stages are shown in Table 13.1, in which we show the wiggler periods, lengths, and peak on-axis amplitudes, as well as the resonant wavelengths for each. The *modulators* and *radiators* are assumed to be plane-polarized parabolic-pole-face designs. Note that the *radiators* in the first two stages are tuned to the fifth harmonic of the *modulator*, while the third and fourth *radiators* are tuned to the third harmonic of the associated *modulators*. The dispersive sections in all stages are identical and contain chicanes consisting of four dipoles, each with a length of 0.25 m and a separation of 0.13 m. The separation between the preceding *modulator* and the first dipole and the last dipole and the succeeding *radiator* are both 0.13 m, yielding a total separation length between the *modulators* and *radiators* of 1.65 m. The optimal bunching in the chicanes is found in simulation by varying the dipole field strengths.

The electron beam we use has an energy of 2307 MeV, a current of 1750 A, a normalized emittance of 1.5 mm-mrad in both the x- and y-directions, and an rms energy spread of 0.01%. The procedure used in the simulation is to propagate the electrons and the optical field through the *modulator*, the chicane, and the radiator in each stage. The output from the *radiator* in each stage is taken to be the seed for the *modulator* of the next stage. The BESSY design calls for phase shifters between each

Table 13.1 Summary of the essential parameters for the harmonic cascade

	Modulator				Radiator			
Stage	λ_w (cm)	N_w	B_w (kG)	λ_{seed} (nm)	λ_w (cm)	N_w	B_w (kG)	λ_{rad} (nm)
1	12.2	18	11.886	279.5	9.2	40	7.9936	55.9
2	9.2	22	7.9936	55.9	7.0	86	5.0558	11.18
3	7.0	30	5.0558	11.18	5.0	180	4.3180	3.73
4	5.0	69	4.3180	3.73	2.85	225	4.6885	1.24
Amplifier				1.24	2.85	630	4.6775	1.24

Table 13.2 Separation distances between the different stages

Stage	Separation distance (m)
1 → 2	1.7450
2 → 3	2.0300
3 → 4	1.6500
4 → amplifier	1.5675

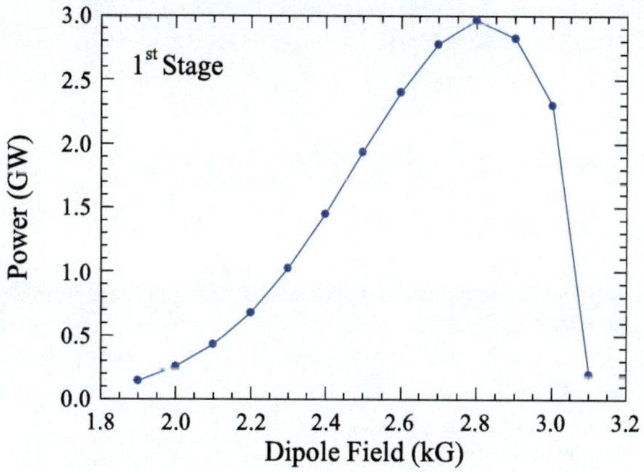

Fig. 13.16 Variation in the output power of the first stage with the dipole field strength

stage so that the optical field is synchronized with a different portion of the electron beam in each stage; hence, we assume that there is no loss of energy of growth in the emittance so that the optical field from the preceding *radiator* interacts with is *fresh electron bunch* in the subsequent *modulator*. As a result, we need to specify the separation distance between various HGHG stages, and these are summarized in Table 13.2

Simulation of the first stage requires the optimization of the dipole field in the chicane to maximize the output power from the *radiator*. The results of this scan in dipole field strength are shown in Fig. 13.16, where we find an output power of 3.34 GW using a dipole field of 2.80 kG.

A plot of the evolution of the power and spot size of the optical field (at 55.9 nm) through the first *radiator* and beyond to the start of the second *modulator* is shown in Fig. 13.17. The rapid increase in power at the start of the *radiator* is characteristic of the pre-bunched beam, and the output power is 3.34 GW. Note also that (1) the radiation has not saturated at the end of the *radiator*, and (2) the light propagating beyond the *radiator* to the start of the second *modulator* shows vacuum diffraction. Within the *radiator*, the light expands less rapidly than in vacuum indicating that there is some optical guiding but that the guiding is not perfect.

The optical field as determined in simulation of the first stage is used as a seed (at 55.9 nm) into the second *modulator*. It is again necessary to optimize the dipole field strength in the second stage. The optimization of the chicane for the second stage is shown in Fig. 13.18 where we plot the output from the *radiator* at a

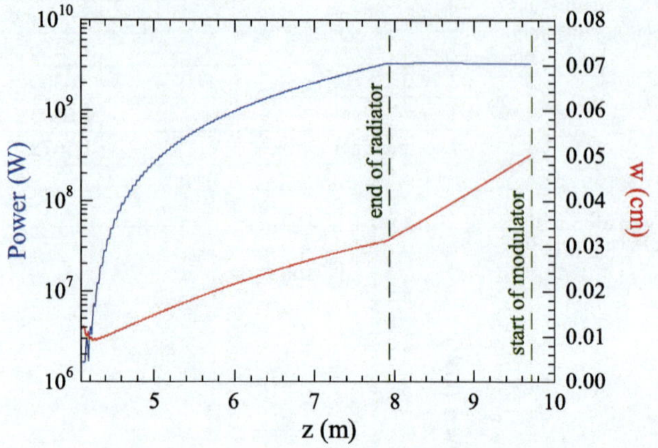

Fig. 13.17 The power and spot size of the optical field through the first radiator up the second modulator

Fig. 13.18 Variation in the output power of the second stage with the dipole field strength

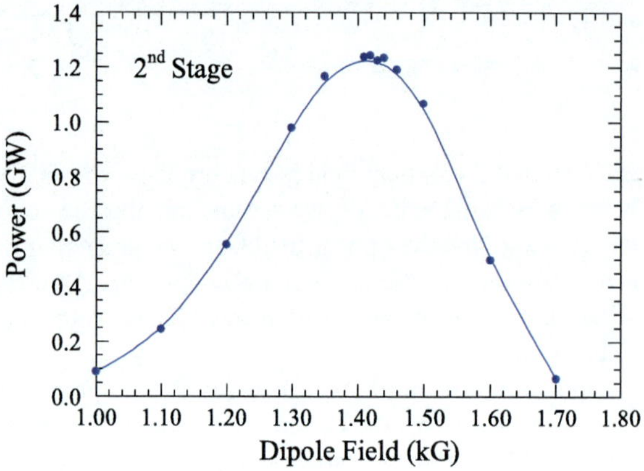

wavelength of 11.18 nm. The optimal dipole field strength in the second chicane is found in simulation to be 1.42 kG resulting in an output power of 1.24 GW from the second *radiator*.

The transverse mode structure of the harmonic radiation at the end of the second *radiator* is shown in Fig. 13.19. This field is then propagated to the entrance to the third *modulator* as an input seed for the third stage. Although the *x*- and *y*-axes are normalized to the waist size of the mode and comparisons with the transverse mode structure of the output of the first *radiator* must be made with care, it is correct to conclude that the spot size at the output of the second *radiator* is smaller than that of the first *radiator*. This is due, in part, to the fact that the shorter wavelength results in a longer Rayleigh range and less mode expansion in the *radiator*.

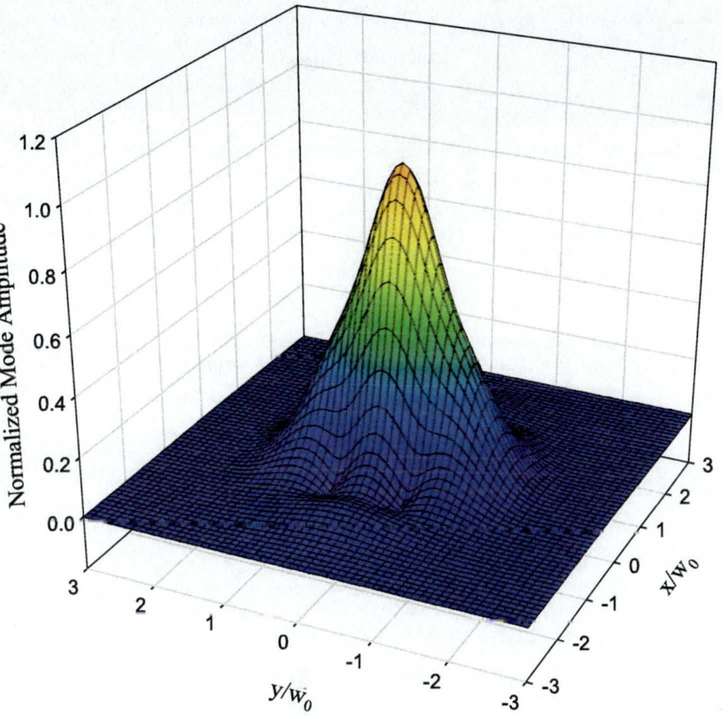

Fig. 13.19 The transverse mode pattern after the second *radiator*

Turing to the third stage, we plot the variation in the output power from the *radiator* versus the dipole field strength in Fig. 13.20. The optimal dipole field strength for the third stage is about 0.775 kG producing a power of 1.48 GW at a wavelength of 3.73 nm.

The transverse mode pattern at the exit of the third *radiator* is shown in Fig. 13.21, and it is safe to conclude that the spot size has decreased still further with the decrease in the wavelength and the increase in the Rayleigh range. Once again, this field is propagated to the entrance to the fourth *modulator* as an input seed.

The optimization of the dipole field in the chicane for the fourth stage showed that the presence of the chicane had a relatively small effect on the output power (at 1.24 nm) from the fourth *radiator*. This is shown in Fig. 13.22 where we plot the output power versus the dipole field strength. Recall that the output power for the preceding stages fell to near zero as the dipole field decreased toward zero. It is evident from Fig. 13.22, however, that the harmonic power does not fall to zero as the dipole field decreases. Indeed, while the optimal power of 0.25 GW is found using a dipole field of 0.20 kG, this is not substantially more power than we would predict using no chicane at all. This is reflected in the phase space after the fourth chicane, as shown in Fig. 13.23 for the optimal dipole field. Note that the degree of

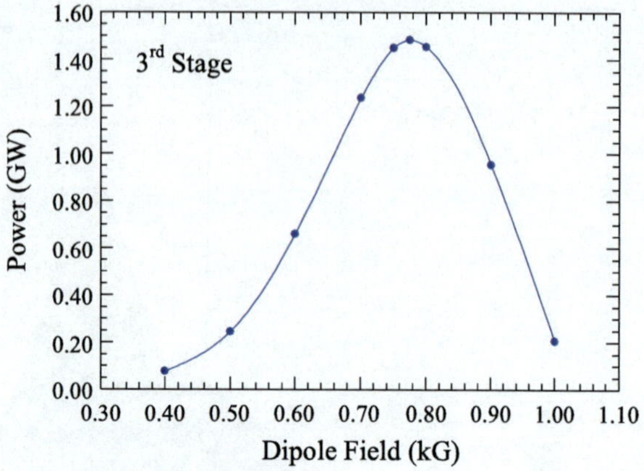

Fig. 13.20 Variation in the output power of the third stage with the dipole field strength

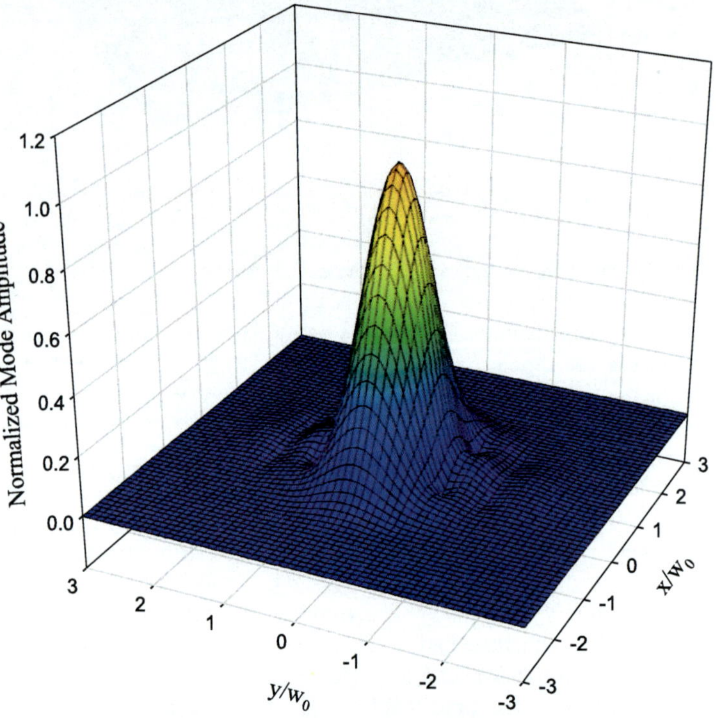

Fig. 13.21 The transverse mode pattern after the third *radiator*

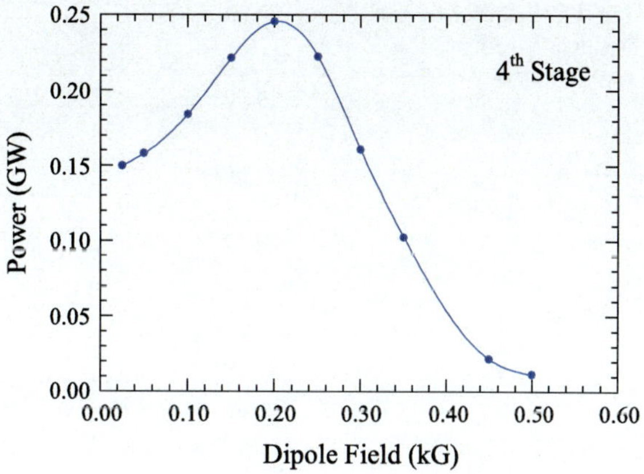

Fig. 13.22 Variation in the output power of the fourth stage with the dipole field strength

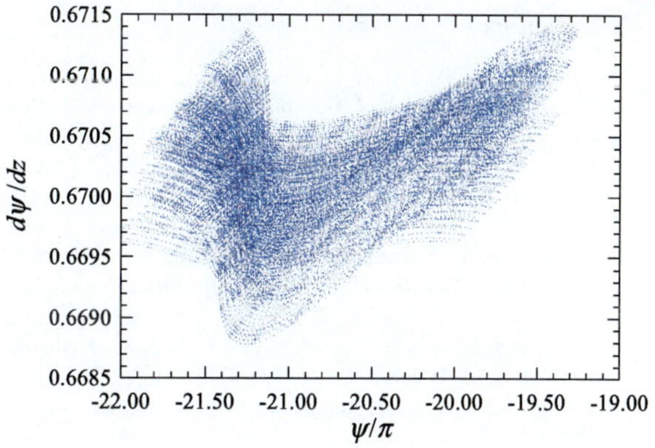

Fig. 13.23 Phase space after the chicane in the fourth stage

bunching achieved is substantially different from that found in the preceding stages, and we conclude that it is simply more difficult to bunch the beam at this short a wavelength. Therefore, it may be possible to operate without the dispersive section or with a shorter *modulator*.

The amplification in the final stage is shown in Fig. 13.24, where the output power of about 4.22 GW at a wavelength of 1.24 nm represents a gain of about 20 over the output power from the fourth HGHG stage. This represents the optimal gain found by scanning over a range of wiggler field strengths.

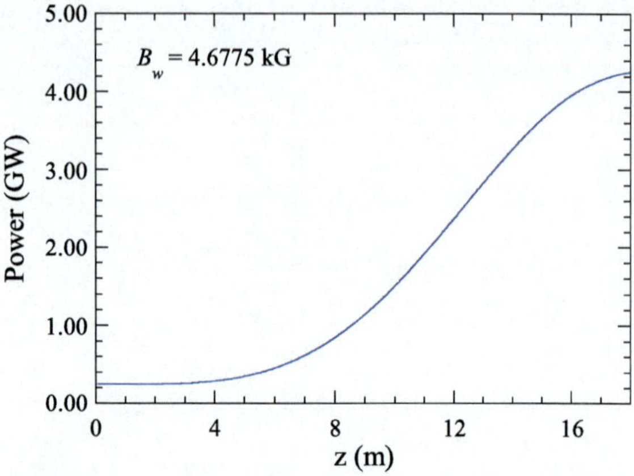

Fig. 13.24 Power versus position in the final amplifier stage

References

1. G.N. Kuliapanov, V.N. Litvinenko, I.V. Panaev, V.M. Popik, A.N. Skrinsky, A.S. Sokolov, N.A. Vinokurov, The VEPP-3 storage ring optical klystron: lasing in the visible and ultraviolet regions. Nucl. Instrum, Methods Phys. Res. **A296**, 1 (1990)
2. L.H. Yu, M. Babzien, I. Ben-Zvi, L.F. DiMauro, A. Doyuran, W. Graves, E. Johnson, S. Krinsky, R. Malone, I. Pogorelsky, J. Skaritka, G. Rakowsky, L. Solomon, X.J. Wang, M. Woodle, V. Yakimenko, S.G. Biedron, J.N. Galayda, E. Gluskin, J. Jagger, V. Sajaev, I. Vasserman, High-gain harmonic-generation free-electron laser. Science **289**, 932 (2000)
3. A. Doyuran, M. Babzien, T. Shaftan, L.H. Yu, L.F. Dimauro, I. Ben-Zvi, S.G. Biedron, W. Graves, E. Johnson, S. Krinsky, R. Malone, I. Pogorelsky, J. Skaritka, G. Rakowsky, X.J. Wang, M. Woodle, V. Yakimenko, J. Jagger, V. Sajaev, I. Vasserman, Characterization of a high-gain harmonic-generation free-electron laser at saturation. Phys. Rev. Lett. **86**, 5902 (2001)
4. E. Allaria et al., Highly coherent and stable pulses from the FERMI seeded free-electron laser in the extreme ultraviolet. Nat. Phot. **6**, 699 (2012)
5. E. Allaria et al., Two-stage seeded soft x-ray free-electron laser. Nat. Phot. **7**, 913 (2013)
6. D. Gauthier et al., Spectrotemporal shaping of seed free-electron laser pulses. Phys. Rev. Lett. **115**, 114801 (2015)
7. E. Roussel et al., Multicolor high-gain free-electron laser driven by seeded microbunching instability. Phys. Rev. Lett. **115**, 214801 (2015)
8. H.P. Freund, Comparison of free-electron laser amplifiers based on a step-tapered optical klystron and a conventional tapered wiggler. Phys. Rev. ST-AB **16**, 060701 (2013)
9. V.N. Litvinenko, High gain distributed optical klystron. Nucl. Instrum. Meth. **A304** (1991)
10. V.A. Bazylev, M.M. Pitatelev, Multisectional FELs with dispersion and undulator sections. Nucl. Instrum, Methods Phys. Res. **A358**, 64 (1995)
11. N.A. Vinokurov, Multisegment wigglers for short wavelength FEL. Nucl. Instrum, Methods Phys. Res. **A375**, 264 (1996)
12. H.P. Freund, G.R. Neil, Nonlinear harmonic generation in distributed optical klystrons. Nucl. Instrum, Methods Phys. Res. **A475**, 373 (2001)
13. G.R. Neil, H.P. Freund, Dispersively enhanced bunching in high-gain free-electron lasers. Nucl. Instrum, Methods Phys. Res. **A475**, 381 (2001)
14. Technical Design Report. The BESSY soft x-ray free electron laser. https://www.helmholtz-berlin.de/media/media/grossgeraete/beschleunigerphysik/fel/fel_tdr.pdf

Chapter 14
Electromagnetic-Wave Wigglers

The physical mechanism in the free-electron laser depends upon the propagation of an electron beam through a periodic magnetic field. Both incoherent and coherent radiation results from the undulatory motion of the electron beam in the external fields, which permits a wave-particle coupling to the output radiation. Coherent radiation depends upon the stimulated emission due to the ponderomotive wave formed by the beating of the radiation and wiggler fields. The wiggler field itself may be either magnetostatic or electromagnetic in nature. Although the bulk of experiments as of this time have relied upon magnetostatic wigglers with either helical or planar polarizations, the fundamental principle has also been demonstrated in the laboratory using a large-amplitude electromagnetic wave to induce the requisite undulatory motion in the electron beam [1, 2].

The basic difference between magnetostatic and electromagnetic-wave wigglers lies in the frequency of the output radiation, which depends upon both the wiggler period and the beam energy in both cases. In the case of a magnetostatic wiggler, the wavelength of the output radiation scales as $\lambda \approx \lambda_w/2\gamma_b^2$ where λ_w denotes the wiggler period and γ_b is the bulk relativistic factor of the beam. In contrast, the wavelength of the output radiation for an electromagnetic-wave wiggler scales as $\lambda \approx \lambda_w/4\gamma_b^2$. As a result, for fixed wiggler periods and beam energies, the electromagnetic wave wiggler will produce shorter output wavelengths. As a consequence, electromagnetic-wave wigglers become attractive alternatives to magnetostatic wigglers the production of short wavelengths when the electron beam energy is constrained.

Several different configurations have been proposed, and analyzed, to make use of electromagnetic-wave wigglers [3–19]. The earliest of these is the simplest and involves nothing more than the use of a large-amplitude radiation pulse from some convenient source that is launched in synchronism with the electron beam. An interesting variant on this concept makes use of an external radiation source of moderate to high intensities to *pump-up* a resonant cavity to extremely high intensities prior to the injection of an electron beam. In this concept, the electromagnetic wave constitutes a standing

H. P. Freund, T. M. Antonsen, Jr., *Principles of Free Electron Lasers*,
https://doi.org/10.1007/978-3-031-40945-5_14

wave of the cavity and contrasts with the earlier design, which employs a traveling the electromagnetic wave. One proposal for such a design would make use of a long pulse or CW Gyrotron oscillator [15, 17] as a source of radiation at wavelengths below approximately 1 cm to *pump-up* the resonant cavity. This would permit the generation of infrared radiation with wavelengths in the range of 300 μm with relatively modest beam energies of the order of 1 MeV. The most ambitious concept is the two-stage free-electron laser [10–12], which is a variant on an oscillator configuration. In this concept, an electron beam propagates through a magnetostatic wiggler located within a resonant cavity of some kind. The radiation generated by this means will itself act as an electromagnetic-wave wiggler to generate still shorter wavelength radiation on the electron beam. Indeed, this mechanism can operate within any long-pulse, high-intensity oscillator design. The principal difficulty with this concept as a source of short wavelength, high-intensity radiation is that if the interaction in either stage reaches sufficiently high efficiencies, then the electron beam quality can be degraded and can *quench* both stages of the interaction.

Whatever the source of the electromagnetic wave, however, the principal difficulty is the same as that found with short period magnetostatic wigglers. Specifically, in order for the oscillatory motion of the electron beam to reach the large amplitudes necessary to achieve high gains and efficiencies, a large amplitude signal must be generated. At the present time, it is still an open question as to which is the most advantageous configuration for this type of alternate wiggler design. In this chapter, therefore, we shall analyze a relatively simple configuration that consists of a uniform circularly polarized electromagnetic wave propagating antiparallel to the electron beam. Three basic issues will be addressed. First, the single-particle trajectories will be treated for a system that includes an axial solenoidal magnetic field [8]. The purpose for this is to provide for enhanced confinement of the electron beam. Note, in this regard, that the typical helical magnetostatic wiggler will act to confine the electron beam against defocusing due to self-field effects through the transverse gradients of the field. In addition, planar wigglers can also be designed with parabolic pole faces for enhanced confinement. However, it appears to be difficult to tailor a large-amplitude electromagnetic wave for such a purpose, and some additional means to ensure beam confinement will be necessary. Second, the small-signal gain for such a combined electromagnetic-wave wiggler/axial magnetic field configuration will be addressed [13]. Finally, while tapered magnetostatic wigglers are relatively easy to construct, tapered electromagnetic-wave wigglers present technical difficulties in design due to problems in mode control. For example, the coupling coefficient between modes in a tapered waveguide depends upon the slope of the taper. Difficulties may ensue, therefore, if the slope of the taper is comparable to the coupling coefficient mediating the free-electron laser interaction. Hence, an alternate efficiency enhancement scheme for electromagnetic-wave wiggler configurations is described, which employs a tapered axial magnetic field [14].

14.1 Single Particle Trajectories

The specific configuration to be investigated consists in the propagation of an electron beam counter to that of a circularly polarized electromagnetic wave. As such, the electromagnetic wave acts as a wiggler that induces an undulatory motion on the beam. In order to study the electron trajectories under the combined influence of an axial solenoidal magnetic field and a large-amplitude circularly polarized electromagnetic wave, we assume that the electromagnetic wave is approximately uniform in the transverse direction and write the electric and magnetic fields of the wave as

$$\mathrm{B}_w = B_w\left[\widehat{\boldsymbol{e}}_x \cos(k_w z + \omega_w t) + \widehat{\boldsymbol{e}}_y \sin(k_w z + \omega_w t)\right], \tag{14.1}$$

and

$$\mathrm{E}_w = -\frac{\omega_w}{ck_w} B_w\left[\widehat{\boldsymbol{e}}_x \sin(k_w z + \omega_w t) - \widehat{\boldsymbol{e}}_y \cos(k_w z + \omega_w t)\right], \tag{14.2}$$

where the subscript "w" is used throughout to denote quantities associated with the electromagnetic-wave wiggler, B_w denotes the amplitude of the magnetic field, and ω_w and k_w denote the angular frequency and wavenumber of the wiggler. Observe that the Poynting vector for this for these fields is

$$\mathrm{S}_w = -\frac{1}{4\pi}\frac{\omega_w}{ck_w} B_w^2 \widehat{\boldsymbol{e}}_z, \tag{14.3}$$

which demonstrates that the electromagnetic-wave wiggler describes a backward propagating wave for $\omega_w > 0$ and $k_w > 0$. It should be remarked that a magnetostatic wiggler is recovered in the limit in which ω_w vanishes.

The orbit equations for an electron in the combined fields are given by

$$\frac{d}{dt}\mathrm{v} = -\frac{e}{\gamma m_e}\left[\left(\mathrm{I} - \frac{1}{c^2}\mathrm{v}\mathrm{v}\right)\cdot \mathrm{E}_w + \frac{1}{c}\mathrm{v} \times (B_0 \widehat{\boldsymbol{e}}_z + \mathrm{B}_w)\right], \tag{14.4}$$

and

$$\frac{d}{dt}\gamma = -\frac{e}{m_e c^2}\mathrm{v}\cdot \mathrm{E}_w, \tag{14.5}$$

where **I** is the unit dyadic. For convenience, we transform to the rotating *wiggler-frame*

$$\widehat{\boldsymbol{e}}_1 \equiv \widehat{\boldsymbol{e}}_x \cos(k_w z + \omega_w t) + \widehat{\boldsymbol{e}}_y \sin(k_w z + \omega_w t), \tag{14.6}$$

$$\hat{e}_2 \equiv -\hat{e}_x \sin(k_w z + \omega_w t) + \hat{e}_y \cos(k_w z + \omega_w t), \tag{14.7}$$

$$\hat{e}_3 \equiv \hat{e}_z. \tag{14.8}$$

In this frame, the orbit equations take the form

$$\frac{d}{dt}v_1 = -\left[\Omega_0 - k_w\left(v_3 + v_p\right) - \Omega_w \frac{v_1 v_p}{c^2}\right] v_2, \tag{14.9}$$

$$\frac{d}{dt}v_2 = \left[\Omega_0 - k_w\left(v_3 + v_p\right)\right] v_1 - \Omega_w\left(v_3 + v_p\right) + \Omega_w \frac{v_2^2 v_p}{c^2}, \tag{14.10}$$

$$\frac{d}{dt}v_3 = \Omega_w\left(1 + \frac{v_3 v_p}{c^2}\right) v_2, \tag{14.11}$$

and

$$\frac{d}{dt}\gamma = -\gamma \frac{v_2 v_p}{c^2}\Omega_w, \tag{14.12}$$

where $\Omega_{0,w} = eB_{0,w}/\gamma m_e c$ and $v_p = \omega_w/k_w$ denotes the phase velocity of the electromagnetic-wave wiggler. Observe that these equations are analogous to Eqs. 2.10–2.12 for the magnetostatic helical wiggler.

The steady-state orbits for this configuration are found under the requirement that the total energy of the electron remains constant (i.e., $d\gamma/dt = 0$). This implies that $v_2 = 0$, which in turn means that v_1 and v_3 are constant as well. Denoting the constant energy and axial velocity by γ_0 and $v_\|$, we find that the constant transverse velocity is given by [8]

$$v_1 = v_w \equiv \frac{\Omega_w\left(v_\| + v_p\right)}{\Omega_0 - \left(\omega_w + k_w v_\|\right)}, \tag{14.13}$$

where now $\Omega_{0,w} = eB_{0,w}/\gamma_0 m_e c$. The assumption of an electromagnetic-wave wiggler that is uniform in the transverse direction requires that $v_w << v_\|$. Since the energy is a constant for the steady-state trajectories, v_w and $v_\|$ are related through

$$v_\|^2 + v_w^2 = \left(1 - \gamma_0^{-2}\right)c^2. \tag{14.14}$$

The dispersion relation between ω_w and k_w are determined in a self-consistent fashion by the dielectric properties of the medium. In this case, we assume that the beam is cold and uniform in the transverse direction. Hence, the dispersion equation that relates the frequency and wavenumber is

$$\omega_w^2 - c^2k_w^2 - \frac{\omega_b^2\left(\omega_w + k_w v_\parallel\right)}{\gamma_0\left(\omega_w - \Omega_0 + k_w v_\parallel\right)} = 0. \tag{14.15}$$

Equations 14.13–14.15 are sufficient to determine v_w, $v_\parallel$, and k_w for fixed values of B_0, B_w, γ_0, and ω_w.

The orbital stability of these steady-state trajectories is determined by a straightforward perturbation analysis. We write $v_1 = v_w + \delta v_1$, $v_2 = \delta v_2$, $v_3 = v_\parallel + \delta v_3$, and $\gamma = \gamma_0 + \delta\gamma$ and find that the orbit equations are

$$\frac{d}{dt}\delta v_1 = -\left[\Omega_0 - k_w\left(v_\parallel + v_p\right) - \Omega_w\frac{v_w v_p}{c^2}\right]\delta v_2, \tag{14.16}$$

$$\frac{d}{dt}\delta v_2 = \left[\Omega_0 - k_w\left(v_\parallel + v_p\right)\right]\delta v_1 - (\Omega_w + k_w v_w)\delta v_3 - \frac{v_w}{\gamma_0}k_w\left(v_\parallel + v_p\right)\delta\gamma, \tag{14.17}$$

$$\frac{d}{dt}\delta v_3 = \Omega_w\left(1 + \frac{v_\parallel v_p}{c^2}\right)\delta v_2, \tag{14.18}$$

and

$$\frac{d}{dt}\delta\gamma = -\gamma_0\frac{v_p}{c^2}\Omega_w\delta v_2, \tag{14.19}$$

correct to first order in the perturbations. If we take the derivative of Eq. 14.17 and substitute the values for the derivatives of δv_1, δv_3, and $\delta\gamma$ from Eqs. 14.16, 14.18, and 14.19, then we obtain

$$\left(\frac{d^2}{dt^2} + \Omega_r^2\right)\delta v_2 = 0, \tag{14.20}$$

and

$$\frac{d}{dt}\left(\frac{d^2}{dt^2} + \Omega_r^2\right)\begin{pmatrix}\delta v_1\\ \delta v_3\\ \delta\gamma\end{pmatrix} - 0, \tag{14.21}$$

where

$$\Omega_r^2 \equiv \left[\Omega_0 - \left(\omega_w + k_w v_\parallel\right)\right] \times \left[\Omega_0\left(1 + \frac{v_w^2}{c^2}\frac{c^2k_w^2 - \omega_w^2}{\left(\omega_w + k_w v_\parallel\right)^2}\right) - \left(\omega_w + k_w v_\parallel\right)\right]. \tag{14.22}$$

Orbital instability occurs whenever ${\Omega_r}^2 < 0$.

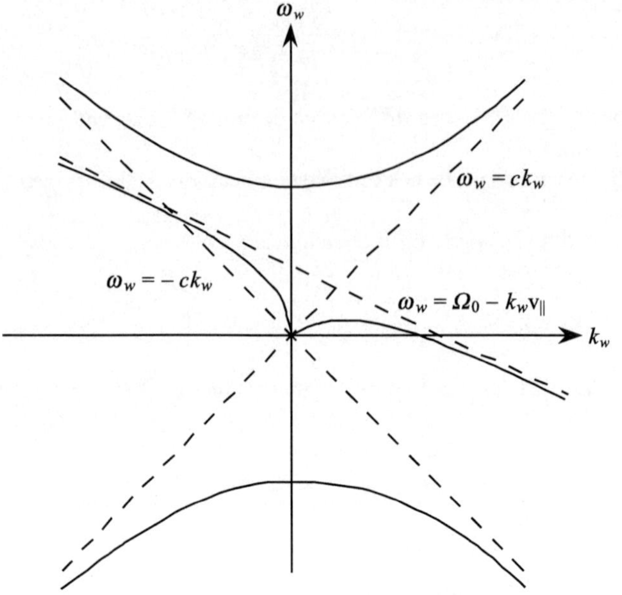

Fig. 14.1 Schematic illustration of the roots of the dispersion equation for the electromagnetic wave

Solution of Eqs. 14.13–14.15 must take cognizance of the properties of the dispersion Eq. 14.15, which exhibits three distinct branches. A schematic illustration of these branches is shown in Fig. 14.1. Two branches of interest exist corresponding to backward propagating waves in the first quadrant. These are (1) an electromagnetic electron cyclotron wave, which is found over a restricted range of wavenumbers given approximately $0 < k_w < \Omega_0/v_\parallel$, and (2) the electromagnetic escape mode, which is found for frequencies above the cutoff frequency $\omega_w \geq \omega_{co}$, where

$$\omega_{co} \equiv \frac{\Omega_0}{2}\left[1 + \sqrt{1 + \frac{4\omega_b^2}{\gamma_0 \Omega_0^2}}\right]. \tag{14.23}$$

The character of the steady-state orbits is dependent upon the choice of wave mode. As in the case of the magnetostatic helical wiggler, two classes of orbit are found. The generalization of Group I trajectories occurs when $\Omega_0 < \omega_w + k_w v_\parallel$ and corresponds to waves on the electromagnetic escape branch. In contrast to the Group I trajectories in a magnetostatic wiggler, however, these orbits are stable. In order to understand this, observe that the orbital stability criterion for these trajectories ($\Omega_r^2 > 0$) requires that

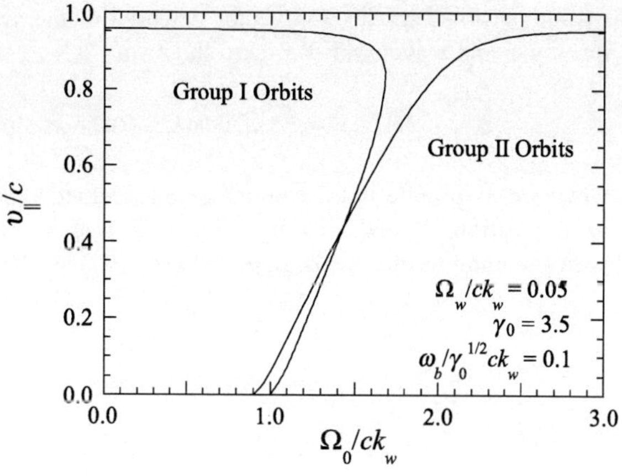

Fig. 14.2 Group I and Group II trajectories found in the electromagnetic-wave wiggler

$$\Omega_0\left(1+\frac{v_w^2}{c^2}\frac{c^2k_w^2-\omega_w^2}{\left(\omega_w+k_wv_\parallel\right)^2}\right)<\left(\omega_w+k_wv_\parallel\right). \tag{14.24}$$

This condition is trivially satisfied since waves on the escape branch are supraluminous (i.e., $\omega_w > ck_w$). The generalization of Group II trajectories of the magnetostatic wiggler occurs when $\Omega_0 > \omega_w + k_wv_\parallel$ and corresponds to waves on the electromagnetic electron cyclotron branch. In this case, the stability criterion is

$$\Omega_0\left(1+\frac{v_w^2}{c^2}\frac{c^2k_w^2-\omega_w^2}{\left(\omega_w+k_wv_\parallel\right)^2}\right)>\left(\omega_w+k_wv_\parallel\right). \tag{14.25}$$

These orbits are stable as well since the waves on this branch are subluminous. Hence, in contrast to the case of a helical magnetostatic wiggler, all orbits in a circularly polarized electromagnetic-wave wiggler are stable. An example of the variation of the axial velocity of each orbit group with the axial magnetic field is shown in Fig. 14.2 for $\gamma_0 = 3.5$, $\Omega_w/ck_w = 0.05$, and $\omega_b/\gamma_0^{1/2}ck_w = 0.1$. Group I trajectories for these parameters represent a multivalued function of the axial magnetic field and occur for $\Omega_0/ck_w \le 1.68$. In contrast, Group II trajectories are single-valued and occur for $\Omega_0/ck_w \ge 0.94$.

14.2 The Small-Signal Gain

The small-signal gain is due the beating of the wiggler and radiation fields and is governed by a nonlinear pendulum equation. In the present case, the interaction gives rise to the stimulated scattering of a right-hand circularly polarized wave propagating parallel to the electron beam. Within the context of the idealized one-dimensional model, the vector potential of the scattered wave of angular frequency ω and wavenumber k takes the form

$$\delta \mathbb{A}(z,t) = \delta A\left[\widehat{\boldsymbol{e}}_x \cos(kz-\omega t) - \widehat{\boldsymbol{e}}_y \sin(kz-\omega t)\right], \tag{14.26}$$

where the amplitude and wavenumber are assumed to be slowly varying functions of axial position. In the rotating wiggler frame, the electric and magnetic fields corresponding to this vector potential are

$$\delta \mathrm{E}(z,t) = -\frac{\omega}{c}\delta A[\widehat{\boldsymbol{e}}_1 \sin\psi(z,t) + \widehat{\boldsymbol{e}}_2 \cos\psi(z,t)], \tag{14.27}$$

and

$$\delta \mathrm{B}(z,t) = k\delta A[\widehat{\boldsymbol{e}}_1 \cos\psi(z,t) - \widehat{\boldsymbol{e}}_2 \sin\psi(z,t)], \tag{14.28}$$

where

$$\psi(z,t) = (k+k_w)z - (\omega-\omega_w)t, \tag{14.29}$$

denotes the generalized ponderomotive phase.

The nonlinear pendulum equation is obtained by a perturbation analysis of the orbit equations to first order in the radiation fields about the steady-state trajectories. Hence, writing $\upsilon_1 = \upsilon_w + \delta\upsilon_1$, $\upsilon_2 = \delta\upsilon_2$, $\upsilon_3 = \upsilon_\parallel + \delta\upsilon_3$, and $\gamma = \gamma_0 + \delta\gamma$, we obtain

$$\begin{aligned}\frac{d}{dt}\delta\upsilon_1 = &-\left[\Omega_0 - k_w\left(\upsilon_\parallel + \upsilon_p\right) - \Omega_w\frac{\upsilon_w\upsilon_p}{c^2}\right]\delta\upsilon_2 + \frac{e\delta A}{\gamma_0 m_e c}\\ &\times\left[\omega\left(1-\frac{\upsilon_w^2}{c^2}\right) - k\upsilon_\parallel\right]\sin\psi,\end{aligned} \tag{14.30}$$

$$\begin{aligned}\frac{d}{dt}\delta\upsilon_2 = &\left[\Omega_0 - k_w\left(\upsilon_\parallel + \upsilon_p\right)\right]\delta\upsilon_1 - (\Omega_w + k_w\upsilon_w)\delta\upsilon_3 - \frac{\upsilon_w}{\gamma_0}k_w\left(\upsilon_\parallel + \upsilon_p\right)\delta\gamma\\ &+\frac{e\delta A}{\gamma_0 m_e c}\left(\omega - k\upsilon_\parallel\right)\cos\psi,\end{aligned} \tag{14.31}$$

$$\frac{d}{dt}\delta\upsilon_3 = \Omega_w\left(1+\frac{\upsilon_\parallel\upsilon_p}{c^2}\right)\delta\upsilon_2 + \frac{e\delta A}{\gamma_0 m_e c}\frac{\upsilon_w^2}{c^2}\omega\sin\psi, \tag{14.32}$$

and

$$\frac{d}{dt}\delta\gamma = -\gamma_0\frac{\upsilon_p}{c^2}\Omega_w\delta\upsilon_2 + \frac{e\delta A}{\gamma_0 m_e c}\frac{\upsilon_w^2}{c^2}\omega\sin\psi. \tag{14.33}$$

Equations 14.31–14.34 are a straightforward generalization of Eqs. 14.16–14.19. Differentiating Eq. 14.32 with respect to time, we find that

$$\left(\frac{d^2}{dt^2}+\Omega_r^2\right)\delta v_2 = -\frac{e\delta A}{\gamma_0 m_e c}$$
$$\times\left[(\omega - kv_\parallel)(\omega_w - \Omega_0 - k_w v_\parallel) + \frac{v_w^2}{c^2}\frac{\Omega_0(\omega\omega_w + c^2 kk_w)}{\omega_w + k_w v_\parallel}\right]\sin\psi \qquad (14.34)$$

where terms in $d\psi/dt$ have been neglected. Hence,

$$\frac{d}{dt}\left(\frac{d^2}{dt^2}+\Omega_r^2\right)\delta v_3 = -\frac{e\delta A}{\gamma_0 m_e c}\frac{v_w}{v_\parallel}\frac{\omega-\omega_w}{\gamma_0\gamma_\parallel^2}\sin\psi$$
$$\times\left[\Omega_r^2 + \gamma_\parallel^2\Omega_0\left(\frac{v_w}{v_p + v_\parallel}\right)^2\left(1+\frac{v_\parallel v_p}{c^2}\right)^2(\omega - \Omega_0 - k_w v_\parallel)\right], \qquad (14.35)$$

where $\gamma_\parallel{}^2 = (1 - v_\parallel{}^2/c^2)^{-1}$. Under the assumption that the phase and the perturbed velocity vary more slowly than Ω_r, we obtain the pendulum equation

$$\frac{d^2}{dz^2}\psi \cong \delta a\frac{v_w}{v_\parallel}\frac{(\omega-\omega_w)^2 c}{\gamma_0\gamma_\parallel^2 v_\parallel^3}\Phi_{em}\sin\psi, \qquad (14.36)$$

where $\delta a = e\delta A/m_e c^2$,

$$\Phi_{em} \equiv 1 - \left(1+\frac{v_\parallel v_p}{c^2}\right)^2\frac{\gamma_\parallel^2\Omega_0\left(\frac{v_w}{v_p+v_\parallel}\right)^2}{\left[1+\left(\frac{v_w}{v_p+v_\parallel}\right)^2\left(1-\frac{v_p^2}{c^2}\right)^2\right]\Omega_0 - (\omega_w + k_w v_\parallel)}, \qquad (14.37)$$

and we have made use of the relation

$$\frac{d}{dt}\delta v_3 \cong \frac{v_\parallel^3}{\omega-\omega_w}\frac{d^2}{dz^2}\psi. \qquad (14.38)$$

Observe that Φ_{em} is the counterpart of Φ [Eq. 2.24] for the magnetostatic helical wiggler, and reduces to this function in the limit in which ω_w vanishes. Eq. 14.37 governs the axial bunching of the electron beam in the presence of the wiggler and radiation fields.

The behavior of Φ_{em} as a function of the axial magnetic field is shown in Fig. 14.3 for parameters corresponding to Group I and II orbits in Fig. 14.2. The essential character of this function is similar to that of Φ for the magnetostatic wiggler [see Fig. 14.3]. The principal differences arise due to the facts that (1) Group II orbits do not extend to the $\Omega_0 = 0$ limit and (2) there is no orbital instability for the electromagnetic-wave wiggler. The latter implies that, although Φ_{em} can become large, it displays no singularity.

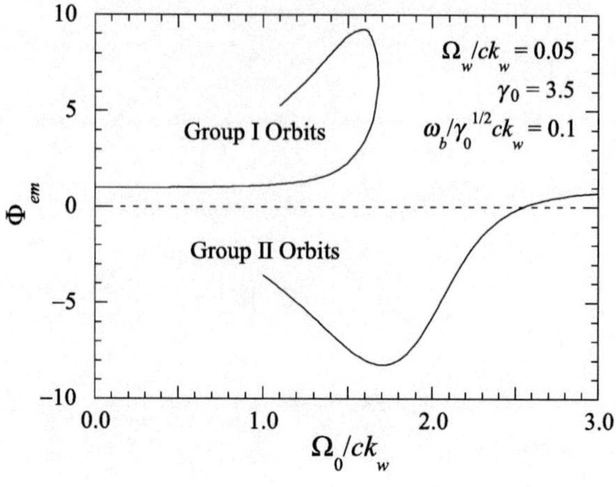

Fig. 14.3 Plot of Φ_{em} as a function of the axial magnetic field

The small-signal gain is determined from Maxwell's equations in the idealized one-dimensional limit. The source current is assumed to represent an ideal monoenergetic electron beam and can be expressed as

$$\delta \mathrm{J}(z,t) = -en_b v_0 \int_{-\infty}^{\infty} dt_0 \sigma(t_0) \frac{p(t,t_0)}{p_z(t,t_0)} \delta[t - \tau(t,t_0)], \tag{14.39}$$

where v_{z0} is the initial axial electron velocity, n_b is the ambient beam density, $\sigma(t_0)$ is the distribution in entry times (t_0) at which the electrons cross the $z = 0$ plane, $\mathbf{p}(t,t_0)$ is the electron momentum at time t for a particle which crossed the $z = 0$ plane at time t_0, and $\tau(z,t_0)$ is the Lagrangian time coordinate defined in Eq. 4.27. Substitution of the representations of the source current and vector potential into Eq. 4.28 yields two coupled equations for the evolution of the amplitude and wavenumber

$$\left[\frac{d^2}{dz^2} + \left(\frac{\omega^2}{c^2} - k^2\right)\right]\delta a \cong \frac{\omega_b^2}{c^2}\frac{v_w}{c} \int_{-\infty}^{\infty} dt_0 \sigma(t_0) \cos\psi \delta[t - \tau(z,t_0)], \tag{14.40}$$

and

$$2k^{1/2}\frac{d}{dz}\left(k^{1/2}\delta a\right) \cong -\frac{\omega_b^2}{c^2}\frac{v_w}{c} \int_{-\infty}^{\infty} dt_0 \sigma(t_0) \sin\psi \delta[t - \tau(z,t_0)], \tag{14.41}$$

where ω_b denotes the beam plasma frequency.

Under the quasi-static assumption for a large amplitude electromagnetic-wave wiggler, electrons that enter the interaction region within an integral number of beat-wave periods T [$= 2\pi/(\omega - \omega_w)$] execute identical trajectories. Hence, both $\tau(z,$

$t_0) = \tau(z,t_0 + 2\pi N/T)$ and $\psi(z,\tau(z,t_0)) = \psi[z,\tau(z,t_0 + 2\pi N/T)]$ for integer N. Making use of this symmetry, we find that upon averaging Eqs. 14.41 and 14.42 over a beat-wave period

$$\left[\frac{d^2}{dz^2} + \left(\frac{\omega^2}{c^2} - k^2\right)\right]\delta a \cong \frac{\omega_b^2}{c^2}\frac{v_w}{c}\langle \cos\psi\rangle, \tag{14.42}$$

and

$$k^{1/2}\frac{d}{dz}\left(k^{1/2}\delta a\right) \cong -\frac{\omega_b^2}{c^2}\frac{v_w}{c}\langle \sin\psi\rangle, \tag{14.43}$$

where the average

$$\langle(\cdots)\rangle \equiv \frac{1}{2\pi}\int_0^{2\pi} d\psi_0 \sigma(\psi_0)(\cdots), \tag{14.44}$$

is over the initial phase $\psi_0 = -\omega t_0$. The dynamical Eqs. 14.43 and 14.44 simplify under the assumptions that (1) the wave frequency is much greater than the beam plasma frequency and (2) that second-order derivatives of the amplitude and phase may be neglected. In this case, we obtain $\omega \approx ck$ and

$$2k\frac{d}{dz}\delta a \cong -\frac{\omega_b^2}{c^2}\frac{v_w}{c}\langle \sin\psi\rangle. \tag{14.45}$$

The gain in power over a length L is expressed as

$$G(L) \equiv 2\frac{\delta A(z=L) - \delta A(z=0)}{\delta A(z=0)}, \tag{14.46}$$

can be obtained by integration of Eq. 14.46. Under the assumption of low gain [i.e., $G(L) << 1$], we obtain

$$G(L) \cong \frac{\omega_b^2}{c^2}\frac{v_w}{c}\frac{1}{\delta a(z=0)}\int_0^L dz\langle \sin\psi\rangle. \tag{14.47}$$

The small-signal gain is obtained by integration of Eq. 14.48 under the assumption that the electrons are on untrapped trajectories. This has been discussed within the context of the low-gain limit for a realizable helical magnetostatic wiggler in Chap. 4. The untrapped trajectories in the electromagnetic-wave wiggler are of the form $\psi = \psi_0 + \Delta kz + \delta\psi$, where

$$\Delta k \equiv k + k_w - \frac{\omega - \omega_w}{\upsilon_\parallel}, \tag{14.48}$$

and the perturbed phase satisfies

$$\frac{d^2}{dz^2}\psi \cong \kappa^2 \sin(\psi_0 + \Delta kz), \tag{14.49}$$

where

$$\kappa^2 \equiv \frac{c(\omega - \omega_w)^2}{\gamma_0 \gamma_\parallel^2 \upsilon_\parallel^3} \frac{\upsilon_w}{\upsilon_\parallel} \delta a \Phi_{em}. \tag{14.50}$$

Therefore, the solution for the phase subject to the initial conditions $\delta\psi(z=0)=0$ and $d\delta\psi(z=0)/dz=0$ is

$$\psi \cong \psi_0 + \Delta kz + \frac{\kappa^2}{\Delta k^2}[\sin\psi_0 - \sin(\psi_0 + \Delta kz) + \Delta kz \cos\psi_0], \tag{14.51}$$

which in turn implies that

$$\langle \sin\psi \rangle \cong -\frac{\kappa^2}{\Delta k^2}[\sin\Delta kz - \Delta kz \cos\psi_0]. \tag{14.52}$$

As a result, the small-signal gain takes on a form [13], which is analogous to that found for the magnetostatic wiggler (4.59)

$$G(L) \cong -\frac{\omega_b^2 L^3 k}{8\gamma_0 \gamma_\parallel^2 \upsilon_\parallel^3} \frac{\upsilon_w^2}{\upsilon_\parallel^2} \left(\frac{\omega - \omega_w}{\omega}\right)^2 \Phi_{em} F(\Theta_{em}), \tag{14.53}$$

where $\Theta_{em} = \Delta kL/2$ and F is the spectral function defined in Eq. 4.60. It is a straightforward matter to demonstrate that this expression for the small-signal gain reduces to the result for a helical magnetostatic wiggler in the limit in which the ω_w vanishes (Fig. 14.4).

The extrema for the gain occur for $\Theta_{em} \approx \pm 1.3$ at which $F(\Theta_{em}) \approx \pm 0.54$, which corresponds to the frequencies

$$\omega \cong \left(1 - \frac{\upsilon_\parallel}{c}\right)\left[\omega_w + k_w \upsilon_\parallel \left(1 \mp \frac{2.6}{k_w L}\right)\right]. \tag{14.54}$$

Note that for $\Phi_{em} > 0$ (< 0) the maximum gain is found to correspond to $\Theta_{em} \approx 1.3$ (-1.3); consequently, the effect of the axial magnetic field is to cause a relative phase shift between Group I and II classes of trajectory. In either case, however, the maximum gain is given approximately by

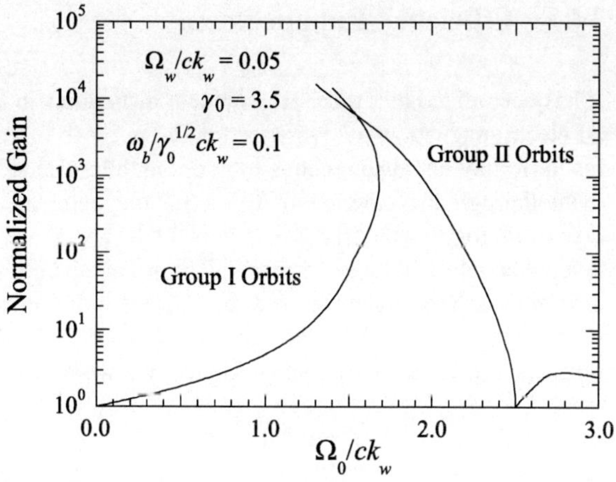

Fig. 14.4 Graph of the normalized value of the peak gain for Group I and Group II orbits corresponding to the orbits illustrated in Fig. 14.2

$$G_{\max}(L) \cong 0.034\left(1+\frac{\upsilon_{\|}}{c}\right)\frac{\omega_b^2}{\gamma_0 c^2 k_w^2}\frac{\upsilon_w^2}{\upsilon_{\|}^2}(k_w L)^3\frac{(\omega_w+ck_w)^2}{ck_w(\omega_w+k_w\upsilon_{\|})}|\Phi_{em}|. \tag{14.55}$$

The resonant enhancement in both υ_w and Φ_{em} due to the axial magnetic field can result in a substantial enhancement in the gain. As an example, observe that $\upsilon_w/c \approx 0.052$ and $\Phi_{em} = 1.0$ in the absence of an axial magnetic field for the parameters shown in Figs. 14.2 and 14.3. This results in a maximum gain of $G_{\max}(L) \approx 3.69 \times 10^{-6}$ $(k_wL)^3$. In contrast, $\upsilon_w/c \approx 0.53$ and $\Phi_{em} \approx 1.68$ when $\Omega_0/ck_w = 1.68$ near resonance. As a consequence, the maximum gain is enhanced by several orders of magnitude, and we find that $G_{\max}(L) \approx 2.48 \times 10^{-3}$ $(k_wL)^3$. A more detailed variation of the maximum gain as a function of the axial magnetic field is shown in Fig. 10.4 in which the maximum gain [normalized to the value of $G_{\max}(L)$ for $B_0 = 0$] is plotted versus Ω_0/ck_w.

It is also clear from Eq. 14.54 that the resonant frequency can be as high a twice that for a magnetostatic wiggler. Consider the limit in which ω and $\omega_w >> \omega_b$, so that $\omega_w \cong ck_w$. In the relativistic regime, $\beta_{\|} \approx 1$, and the resonant frequency becomes

$$\omega \cong 4\gamma_{\|}^2 k_w c\left(1 \mp \frac{1.3}{k_w L}\right). \tag{14.56}$$

This can also be seen by noting that the resonance is found when $d\psi/dz = (k + k_w)\upsilon_{\|} - (\omega - \omega_w) \approx 0$, which implies that $\omega \cong 4\gamma_{\|}^2 k_w c$ when $\omega_w \cong ck_w$ and $\beta_{\|} \cong 1$.

14.3 Efficiency Enhancement

It has been noted that there are practical difficulties in the control of the tapering of an electromagnetic-wave wiggler for the purpose of efficiency enhancement. As a result, it may be advantageous to adopt an alternate scheme of efficiency enhancement through the tapering of the axial magnetic field. This technique has been discussed for magnetostatic wigglers in Chap. 5, in which it was shown that a tapered axial field was an equivalent to a tapered wiggler for the purpose of efficiency enhancement. In order to formulate the problem for an electromagnetic-wave wiggler, it is assumed that the axial guide field is uniform for $z \leq z_0$ and displays a linear taper thereafter. Hence, we write [14]

$$B_0(z) = \begin{cases} B_0 & ; z \leq z_0 \\ B_0[1 + \kappa_0(z - z_0)] & ; z > z_0 \end{cases}, \tag{14.57}$$

where B_0 is the amplitude of the axial field in the uniform-field region and κ_0 [$= dB_0(z)/dz$] represents the normalized scale length for variation of the axial field.

The response of the electron beam to the tapered axial field can be determined by a first-order perturbation analysis about the steady-state trajectories in κ_0. The equations for the perturbations in δv_1, δv_3, and $\delta\gamma$ are unchanged and are given by Eqs. 14.31, 14.33, and 14.34. The effect of the tapered axial field is upon the equation for δv_2, which becomes

$$\begin{aligned} \frac{d}{dt}\delta v_2 = {} & \left[\Omega_0 - k_w\left(v_\parallel + v_p\right)\right]\delta v_1 - (\Omega_w + k_w v_w)\delta v_3 - \frac{v_w}{\gamma_0}k_w\left(v_\parallel + v_p\right)\delta\gamma \\ & + v_w\Omega_0\kappa_0(z - z_0) + \frac{e\delta A}{\gamma_0 m_e c}\left(\omega - kv_\parallel\right)\cos\psi. \end{aligned} \tag{14.58}$$

As a result of this, the nonlinear pendulum equation in the presence of the tapered axial magnetic field is

$$\frac{d^2}{dz^2}\psi \cong \delta a \frac{v_w}{v_\parallel}\frac{(\omega - \omega_w)^2 c}{\gamma_0 \gamma_\parallel^2 v_\parallel^3}\Phi_{em}(\sin\psi - \sin\psi_{res}), \tag{14.59}$$

where

$$\sin\psi_{\text{res}} \equiv \frac{\kappa_0}{k + k_w}\frac{\gamma_0\beta_\parallel}{\delta a}\frac{v_\parallel + v_p}{v_w}\frac{\Phi_{em} - 1}{\Phi_{em}}\left(1 + \frac{v_p v_\parallel}{c^2}\right), \tag{14.60}$$

and $\beta_\parallel = v_\parallel/c$. Eq. 14.59 describes the trapping of electrons in the ponderomotive potential formed by the electromagnetic-wave wiggler and the radiation field, and the term in $\sin\psi_{res}$ describes the bulk acceleration or deceleration due to the tapered axial magnetic field.

In order to determine the effect of the tapered field upon the efficiency of the interaction, the gain must be calculated using Eq. 14.48 under the assumption that the taper begins at the point at which the bulk of the electron beam has become trapped. Hence, the phase average is performed for $\psi \approx \psi_{res}$, and the enhancement in the gain over an additional length L becomes

$$\Delta G(L) \equiv 2\frac{\delta A(z=z_0+L)-\delta A(z=z_0)}{\delta A(z=z_0)}$$

$$\cong \kappa_0 L \frac{\omega_b^2}{c^2 k_w^2}\frac{\gamma_0 \beta_\parallel^2}{\delta a^2(z_0)}\frac{k_w(\omega_w+k_w v_\parallel)}{k(\omega-\omega_w)}\frac{1-\Phi_{em}}{\Phi_{em}}\left(1+\frac{v_p v_\parallel}{c^2}\right)^{-1}. \tag{14.61}$$

The efficiency enhancement associated with this enhancement in the gain is calculated from the ratio of the increase in the Poynting flux to the electron beam power flux. Since the Poynting flux increases by an amount $\Delta P = (\omega k/2\pi)\Delta G(L)\delta A^2(z_0)$, the efficiency enhancement is [14].

$$\Delta\eta(L) \cong \kappa_0 L \beta_\parallel \frac{k_w(\omega_w+k_w v_\parallel)}{k(\omega-\omega_w)}\frac{1-\Phi_{em}}{\Phi_{em}}\left(1+\frac{v_p v_\parallel}{c^2}\right)^{-1}. \tag{14.62}$$

In order to estimate the efficiency enhancements that are possible by this mechanism, we consider parameters consistent with those discussed previously for the uniform field. Specifically, $\gamma_0 = 3.5$, $\Omega_w/ck_w = 0.05$, $\omega_b/\gamma_0^{1/2}ck_w = 0.1$, and we assume that $\lambda_w = 1$ cm. We first consider Group I orbits and assume that $\Omega_0/ck_w = 0.5$, which corresponds to an axial magnetic field of 18.7 kG. For this choice of parameters, the orbit parameters are: $v_\parallel/c \approx 0.956$, $v_w/c \approx 0.067$, and $\Phi_{em} \approx 1.018$. In addition, the frequency of the electromagnetic-wave wiggler is 30.2 GHz, and the maximum gain for the interaction occurs at a resonant frequency $\omega/ck_w \approx 44.6$ and a wavelength of 220 μm. As a consequence, the efficiency enhancement per unit length $\Delta\eta(L)/L \approx -0.035\kappa_0$. Thus, if the axial magnetic field is tapered downward to zero, then it is possible to obtain an efficiency enhancement of 3.5% of the initial electron beam power. As the magnetic resonance is approached more closely, the efficiency enhancement can be increased. To see this, note that if we double the axial magnetic field to 37.5 kG, then $\omega/ck_w \approx 41.7$ (for $\lambda \approx 240$ μm), $v_\parallel/c \approx 0.953$, and $\Phi_{em} \approx 1.117$. Hence, the efficiency enhancement per unit length increases to $\Delta\eta(L)/L \approx -0.10\kappa_0$, and an efficiency enhancement of the order of 10% is possible if the field is tapered to zero. This efficiency enhancement can be increased still further as the axial field approaches the magnetic resonance more closely; however, the resonant wavelength of the interaction will also increase. In short, quite appreciable enhancements in the interaction efficiency for the electromagnetic-wave wiggler are possible by means of a tapered axial magnetic field.

References

1. V.L. Granatstein, P. Sprangle, R.K. Parker, J.A. Pasour, M. Herndon, S.P. Schlesinger, Realization of a relativistic mirror: Electromagnetic backscattering from the front of a magnetized relativistic electron beam. Phys. Rev. A **14**, 1194 (1976)
2. V.L. Granatstein, S.P. Schlesinger, M. Herndon, R.K. Parker, J.A. Pasour, Production of megawatt submillimeter pulses by stimulated magneto-Raman scattering. Appl. Phys. Lett. **30**, 384 (1977)
3. P. Sprangle, V.L. Granatstein, L. Baker, Stimulated scattering from a magnetized relativistic electron beam. Phys. Rev. A **12**, 1697 (1975)
4. P. Sprangle, A.T. Drobot, Stimulated backscattering from relativistic unmagnetized electron beams. J. Appl. Phys. **50**, 2652 (1979)
5. A.T. Lin, J.M. Dawson, Nonlinear saturation and thermal effects on the free-electron laser using an electromagnetic pump. Phys. Fluids **23**, 1224 (1980)
6. H.R. Hiddleston, S.B. Segall, Equations of motion for a free-electron laser with an electromagnetic pump field and an axial electrostatic field. IEEE J. Quantum Electron. **QE-17**, 1488 (1981)
7. H.R. Hiddleston, S.B. Segall, G.C. Catella, Gain-enhanced free-electron laser with an electromagnetic pump field, in *Physics of Quantum Electronics: Free-Electron Generators of Coherent Radiation*, ed. by S.F. Jacobs, G.T. Moore, H.S. Pilloff, M. Sargent, M.O. Scully, R. Spitzer, vol. 9, (Addison-Wesley, Reading, 1982), p. 849
8. H.P. Freund, R.A. Kehs, V.L. Granatstein, Electron orbits in combined electromagnetic wiggler and axial guide magnetic fields. IEEE J. Quantum Electron. **QE-21**, 1080 (1985)
9. A. Goldring, L. Friedland, Electromagnetically pumped free-electron laser with a guide magnetic field. Phys. Rev. A **32**, 2879 (1985)
10. J.A. Pasour, P. Sprangle, C.M. Tang, C.A. Kapetanakos, High-power two-stage free-electron laser oscillator operating in the trapped particle mode. Nucl. Instrum, Methods Phys. Res. **A237**, 154 (1985)
11. S.B. Segall, M.S. Curtin, S.A. Von Laven, Key issues in the design of a two-stage free-electron laser. Nucl. Instrum. Methods Phys. Res. **A250**, 316 (1986)
12. I. Kimel, L.R. Elias, G. Ramian, The University of California at Santa Barbara two-stage free-electron laser. Nucl. Instrum. Meth. **A250**, 320 (1986)
13. H.P. Freund, R.A. Kehs, V.L. Granatstein, Linear gain of a free-electron laser with an electromagnetic-wave wiggler and an axial guide magnetic field. Phys. Rev. A **34**, 2007 (1986)
14. H.P. Freund, Efficiency enhancement in free-electron lasers driven by electromagnetic-wave wigglers. IEEE J. Quantum Electron. **QE-23**, 1590 (1987)
15. B.G. Danly, G. Bekefi, R.C. Davidson, R.J. Temkin, T.M. Tran, J.S. Wurtele, Principles of gyrotron powered electromagnetic wigglers for free-electron lasers. IEEE J. Quantum Electron. **QE-23**, 103 (1987)
16. J. Gea-Banacloche, G.T. Moore, R.R. Schlicher, M.O. Scully, H. Walther, Soft X-ray free-electron laser with a laser undulator. IEEE J. Quantum Electron. **QE-23**, 1558 (1987)
17. T.M. Tran, B.G. Danly, J.S. Wurtele, Free-electron lasers with electromagnetic standing-wave wigglers. IEEE J. Quantum Electron. **QE-23**, 1578 (1987)
18. A. Sharma, V. Tripathi, A whistler pumped free-electron laser. Phys. Fluids **31**, 3375 (1988)
19. R. Hofland, D.C. Pridmore-Brown, Optically-pumped free-electron laser with electrostatic reacceleration. Nucl. Instrum. Methods Phys. Res. **A285**, 276 (1989)

Chapter 15
Chaos in Free-Electron Lasers

The past four decades has seen the emergence of a new field of science known as chaos or nonlinear dynamics.

Before going into a description of chaos and nonlinear dynamics as it applies to free-electron lasers, it is useful to introduce some basic concepts from that field. The discussion given here will be elementary and qualitative. A reader who is interested in the detailed points of nonlinear dynamics should read one of the several books that have been written on the subject [1–3]. First, what is chaos? A system that is chaotic exhibits extreme sensitivity to initial conditions. Small differences in the initial conditions for the system ultimately lead to large differences in the system's variables. For a chaotic system, the rate at which small differences become large differences is exponential with time. This implies that precise prediction of the future values of the system variables after a certain time is almost impossible. A small error in the initial conditions will ultimately lead to a large error in the predicted value of a system variable, and increasing the accuracy of the initial conditions by an order of magnitude results in only a modest increase in the length of time over which the solution is accurate. In chaotic systems, the sensitivity to initial conditions is measured by the Lyapunov exponent, which is the time average rate of exponential divergence of two solutions of the system equations whose initial conditions differ by an infinitesimal amount.

There is an important distinction between the phase space trajectories of a deterministic system that is chaotic and trajectories of a system which is driven externally by random noise. For deterministic systems, the trajectories may still be restricted to certain regions of phase space. For example, the motion of a particle in a two-dimensional time-independent potential is described as a trajectory in a four-dimensional phase space. Since the Hamiltonian is conserved, for a given initial energy, the particle trajectory is constrained to a three-dimensional volume in the four-dimensional phase space. The nature of the motion in this three-dimensional volume depends on whether or not the trajectory is chaotic. A nonchaotic trajectory is characterized by the existence of another constant of motion that further restricts the trajectory to lie on a two-dimensional surface. For example, if the potential is

H. P. Freund, T. M. Antonsen, Jr., *Principles of Free Electron Lasers*,
https://doi.org/10.1007/978-3-031-40945-5_15

circularly symmetric, then angular momentum is conserved. Such a trajectory is said to be integrable, as is the region of phase space filled with such trajectories. A chaotic trajectory in this system is not constrained to a surface but will eventually fill up a three-dimensional volume. This volume might be bounded by surfaces on which the trajectories are non-chaotic. Thus, the chaotic trajectory is constrained to stay within a certain subregion of the available phase space. For the two degrees of freedom Hamiltonian problem under discussion, the two-dimensional surfaces on which nonchaotic trajectories are constrained to lie are known as KAM (Kolmogorov, Arnold, and Moser) surfaces [1, 2]. The KAM theorem states that a small perturbation to the system will destroy only a small number of these surfaces. For example, if we start with a circularly symmetric potential, then angular momentum is conserved, and all trajectories lie on two-dimensional surfaces. Suppose we now add a small nonsymmetric perturbation to the potential. Angular momentum is no longer conserved, but this does not mean that every trajectory becomes chaotic. Most trajectories will still be nonchaotic with a constant of motion that differs slightly from the angular momentum. A relatively small fraction of the trajectories will have no such constant, and will chaotically visit a small three-dimensional region of phase space. As the size of the nonsymmetric perturbation is increased, the regions of phase space that are filled by chaotic trajectories increase, and correspondingly, the integrable regions will decrease in size.

A different type of constraint on trajectories in phase space occurs for non-Hamiltonian or dissipative systems. For such systems, the phase space volume is not conserved. Specifically, if one started out a group of initial conditions lying on a surface in phase space, then the volume of phase space inside that surface would not be constant in time as it is for Hamiltonian systems. For a purely dissipative system, the volume inside the surface will progressively shrink with time. As a consequence, there is a tendency for the solutions for different initial conditions to become attracted to the same trajectory. Such a trajectory is known as an *attractor*. A simple example is a damped pendulum. In the absence of any external forcing, all initial conditions ultimately lead to the state where the pendulum is motionless in the bottom of its potential well. The attractor is a point in the two-dimensional phase space of the pendulum. If the pendulum is driven by a weakly time-varying sinusoidal force, then the attractor will be a limit cycle corresponding to steady oscillation of the pendulum. In this case, the attractor is one-dimensional. If the amplitude of the driving force is increased, the motion becomes chaotic and nearby trajectories diverge exponentially in time at the same time that phase space volume contracts. This leads what are known as *strange attractors* [3], which have fractional dimension. The attracting trajectory of the chaotic system does not fill up all of the available phase space; in fact, it consists of a geometrically complex set of points which has measure zero.

In general, physical systems are characterized by a number of parameters. For example, a free-electron laser oscillator is characterized by the electron beam current, the wiggler field strength, the mirror reflectivity, etc. The behavior of the system can vary as the parameters of the system are changed. A system may exhibit chaotic behavior for some values of the parameters but not for others. What is frequently of interest is the way in which the system behavior changes as a parameter is varied. In particular, how does the transition from non-chaotic to chaotic behavior occur? This is referred to as the route to chaos, and it is believed that the route to chaos in most systems falls into one of a few different classes.

In the present chapter, we shall discuss two types of chaos. Specifically, Hamiltonian chaos arises due to self-electric and self-magnetic in the single-particle trajectories of electrons propagating through a helical magnetostatic wiggler and dissipative chaos in multimode free-electron laser oscillators. The study of chaos in free-electron lasers is relatively new, and these two aspects should not be thought of as the only sources of chaos in free-electron lasers. However, the onset of chaos in both cases acts to degrade the efficiency or spectral purity of the free-electron laser, and it appears likely that the study of chaos in coherent radiation sources is useful, primarily, as a guide to which operating regimes to avoid.

15.1 Chaos in Single-Particle Orbits

One example of chaos that has recently been discussed in the literature appears due to the effect of the self-electric and self-magnetic fields of the electron beam on the dynamics of the single-particle trajectories in free-electron lasers [4–6]. These analyses have treated helical wiggler configurations in the presence of an axial solenoidal magnetic field [see Sect. 2.1] and found that the effect of the self-fields can induce chaos in the single-particle trajectories in the vicinity of the gyroresonance. As we demonstrated in Chaps. 4 and 5, the growth rates and saturation efficiencies of the free-electron laser interaction are enhanced in the vicinity of the gyroresonance [where the Larmor period associated with the axial magnetic field is comparable to the wiggler period]. Hence, the ultimate impact of the orbital chaos induced by the self-fields can be to degrade the interaction.

However, caution must be used in the interpretation of this conclusion due to practical limitations imposed by the injection of the electron beam into the wiggler. As discussed in Sect. 5.1, it becomes progressively more difficult to inject the electron beam onto well behaved steady-state trajectories as the gyroresonance is approached [see Figs. 5.2 and 5.3]. Hence, as shown in Figs. 5.7 and 5.8, it is already evident that the gain and efficiency decrease in the vicinity of the gyroresonance. Therefore, the relative impact on the electron orbits of (1) the onset to chaos due to self-fields and (2) the injection process must be evaluated before a definitive conclusion can be reached on this issue.

The analysis we present here, unlike most treatments of classical chaos, is developed from the particle orbit equations rather than a Hamiltonian formalism. In addition, in the interests of simplicity and clarity, we limit the discussion to the idealized one-dimensional limit. The inclusion of wiggler inhomogeneities in the context of a fully three-dimensional analysis [4, 6] requires immediate recourse to numerical methods that obscures the essential physics of the interaction.

15.1.1 The Equilibrium Configuration

The external magnetostatic fields are given by the idealized wiggler field in conjunction in the idealized one-dimensional limit [Eq. 2.5] with an axial solenoidal field; hence,

$$\mathrm{B}_{ext} = B_0\widehat{e}_z + B_w\left(\widehat{e}_x \cos k_w z + \widehat{e}_y \sin k_w z\right), \tag{15.1}$$

which can be represented in terms of a vector potential of the form

$$\mathrm{A}_{ext} = B_0 x\widehat{e}_y - \frac{B_w}{k_w}\left(\widehat{e}_x \cos k_w z + \widehat{e}_y \sin k_w z\right). \tag{15.2}$$

The steady-state orbits in this idealized limit are of the form

$$\mathrm{v}_0 = v_w\left(\widehat{e}_x \cos k_w z + \widehat{e}_y \sin k_w z\right) + v_{\|}\widehat{e}_z, \tag{15.3}$$

where the magnitudes of the transverse and axial velocities are determined by solution of Eqs. 2.15 and 2.16. Recall that the idealized steady-state solutions are valid as long as $|v_w/v_{\|}| \ll 1$. More general solutions can be found in terms of the elliptic functions [see Eqs. 2.31, 2.37, and 2.38]. The steady-state trajectories exhibit a resonant enhancement in the transverse velocity when $\Omega_0 \approx kwv_{\|}$, and orbital instability is found in the vicinity of the gyroresonance for Group I trajectories in which $\Omega_0 < kwv_{\|}$.

We are interested in the effect of the self-electric and self-magnetic fields of the electron beam upon these trajectories. We shall assume for this purpose that the electron beam has a flat-top density profile of the form

$$n_b(r) = \begin{cases} n_0 & ; 0 \le r \le r_b \\ 0 & ; r > r_b \end{cases}, \tag{15.4}$$

as well as a flat-top current profile $\mathbf{J}_b(r) = -en_b(r)v_b\,\hat{e}_z$, where n_{b0} is the uniform ambient beam density and v_b denotes the bulk axial velocity of the beam. Observe that the axial beam velocity is assumed to be constant, and no attempt is made herein to obtain a self-consistent solution for axial electron velocity. In addition, the effect of the transverse wiggle-motion of the beam on the self-fields is neglected [as a higher order effect] in the idealized treatment.

Within the context of this beam configuration, the self-electric field within the beam (i.e., $r \le r_b$) can be shown to be [7]

$$\mathrm{E}_s = -\frac{m_e\omega_b^2}{2e}\left(x\widehat{e}_x + y\widehat{e}_y\right), \tag{15.5}$$

where $\omega_b{}^2 \equiv 4\pi e^2 n_{b0}/m_e$. This is associated with the scalar potential

$$\Phi_s = \frac{m_e \omega_b^2}{4e} r^2. \tag{15.6}$$

The self-magnetic field and associated vector potentials within the beam are

$$\mathbf{B}_s = \frac{m_e \omega_b^2}{2e} \beta_b \left(y\hat{\mathbf{e}}_x - x\hat{\mathbf{e}}_y \right), \tag{15.7}$$

and

$$\mathbf{A}_s = \frac{m_e \omega_b^2}{4e} \beta_b r^2 \hat{\mathbf{e}}_z, \tag{15.8}$$

where $\beta_b \equiv v_b/c$. The question of chaos in the single-particle trajectories is analyzed by solution of the orbit equations subject to the external magnetostatic fields and self-fields.

15.1.2 The Orbit Equations

The orbit equations for an electron in combined idealized helical wiggler and axial guide magnetic fields are given in Eqs. 2.10, 2.12, and 2.12 in the absence of self-electric and magnetic fields. The inclusion of self-fields results in equations of the form

$$\frac{d}{dt} p_x = -\frac{\Omega_0}{\gamma} p_y + \frac{\Omega_w}{\gamma} p_z \sin k_w z + m_e \frac{\omega_b^2 x}{2} \left(1 - \beta_b \beta_z \right), \tag{15.9}$$

$$\frac{d}{dt} p_y = \frac{\Omega_0}{\gamma} p_x - \frac{\Omega_w}{\gamma} p_z \cos k_w z + m_e \frac{\omega_b^2 y}{2} \left(1 - \beta_b \beta_z \right), \tag{15.10}$$

and

$$\frac{d}{dt} p_z = -\frac{\Omega_w}{\gamma} \left(p_x \sin k_w z - p_y \cos k_w z \right) + \frac{\omega_h^2}{2\gamma c} \beta_b \left(x p_x + y p_y \right), \tag{15.11}$$

in rectangular coordinates, where $\Omega_{0,w} \equiv eB_{0,w}/m_e c$ are the cyclotron frequencies associated with the axial and wiggler fields, $\gamma \equiv (1 + p^2/m_e^2 c^2)^{1/2}$, and $\beta_z \equiv v_z/c$. It follows from these equations that

$$\Gamma = \gamma - \frac{\omega_b^2 r^2}{4c^2}, \tag{15.12}$$

is a constant of the motion, which is related to the total energy. In addition, the canonical momenta in the presence of the self-fields are

$$P_x = p_x + m_e \frac{\Omega_w}{k_w} \cos k_w z, \tag{15.13}$$

$$P_y = p_y - m_e \Omega_0 x + m_e \frac{\Omega_w}{k_w} \sin k_w z, \tag{15.14}$$

and

$$P_z = p_z - m_e \frac{\omega_b^2}{4c} \beta_b r^2. \tag{15.15}$$

Observe that the self-fields appear explicitly only in the expression for P_z.

15.1.3 The Canonical Transformation

It will now prove to be convenient to make the canonical transformation [4] $(x,y,z,P_x, P_y,P_z) \to (\varphi,\psi,z',P_\varphi,P_\psi,P_{z'})$,

$$x = \sqrt{\frac{2P_\varphi}{m_e\Omega_0}} \sin(\varphi + k_w z') - \sqrt{\frac{2P_\psi}{m_e\Omega_0}} \cos(\psi - k_w z'), \tag{15.16}$$

$$y = -\sqrt{\frac{2P_\varphi}{m_e\Omega_0}} \cos(\varphi + k_w z') + \sqrt{\frac{2P_\psi}{m_e\Omega_0}} \sin(\psi - k_w z'), \tag{15.17}$$

$$z = z', \tag{15.18}$$

$$P_x = \sqrt{2m_e\Omega_0 P_\varphi} \cos(\varphi + k_w z'), \tag{15.19}$$

$$P_y = \sqrt{2m_e\Omega_0 P_\psi} \cos(\psi - k_w z'), \tag{15.20}$$

and

$$P_z = P_{z'} - k_w P_\varphi + k_w P_\psi. \tag{15.21}$$

Note that ψ as used in the canonical transformation should not be confused with the ponderomotive phase. In terms of these canonical coordinates, the orbit equations can be expressed as

$$\frac{d\varphi}{dt} = \left(\frac{\Omega_0}{\gamma} - k_w v_z\right) - \frac{\Omega_w}{\gamma k_w}\sqrt{\frac{m_e \Omega_0}{2P_\varphi}}\cos\varphi - \frac{\omega_b^2}{2\Omega_0}\left(1 - \beta_b \beta_z\right) \times \left[1 - \sqrt{\frac{P_\psi}{P_\varphi}}\sin(\varphi + \psi)\right], \tag{15.22}$$

$$\frac{d\psi}{dt} = k_w v_z - \frac{\omega_b^2}{2\Omega_0}\left(1 - \beta_b \beta_z\right)\left[1 - \sqrt{\frac{P_\varphi}{P_\psi}}\sin(\varphi + \psi)\right], \tag{15.23}$$

$$\frac{d}{dt}P_\varphi = -\frac{\Omega_w}{\gamma k_w}\sqrt{2m_e \Omega_0 P_\varphi}\sin\varphi - \frac{\omega_b^2}{\Omega_0}\sqrt{P_\varphi P_\psi}\left(1 - \beta_b \beta_z\right)\cos(\varphi + \psi), \tag{15.24}$$

$$\frac{d}{dt}P_\psi = -\frac{\omega_b^2}{\Omega_0}\sqrt{P_\varphi P_\psi}\left(1 - \beta_b \beta_z\right)\cos(\varphi + \psi), \tag{15.25}$$

and

$$\frac{d}{dt}P_{z'} = 0. \tag{15.26}$$

Hence, $P_{z'}$ is also a constant of the motion.

15.1.4 Integrable Trajectories

These equations of motion yield the steady-state trajectories discussed in Chap. 2 in the limit in which the self-fields can be neglected. In order to understand this, observe that when $\omega_b \neq 0$, P_ψ is a constant of the motion (denoted by $P_{\psi\,0}$) and that $d\psi/dt = k_w v_z$. If we require as well that $dP_\varphi/dt = 0$ for steady-state trajectories, then it follows that $P_\varphi = P_{\varphi 0}$ (constant), $v_z = v_\parallel$ (constant), and $\varphi = \varphi_0$ (constant) such that $\cos\varphi_0 = \pm 1$. Since φ is also constant for the steady-state trajectories, this requires from Eq. 15.22 that

$$\sqrt{\frac{2P_{\varphi 0}}{m\Omega_0}} = \pm \frac{\upsilon_w}{k_w \upsilon_\parallel}, \tag{15.27}$$

where υ_w denotes the wiggler-induced transverse velocity given in Eq. 2.15, and the "±" signs refer to the Group II and I trajectories, respectively, in order to ensure that $P_{\varphi 0}$ remains real. Note also that the energy conservation requirement implies that υ_w and $\upsilon_\parallel$ are related via Eq. 2.16. The values of $P_{\psi 0}$ and ψ are determined by the initial conditions on the steady-state orbit. In particular, $\psi = \psi_0 + k_w z$ where [we assume that $z(t = 0) = 0$]

$$x(t=0) = -\sqrt{\frac{2P_{\psi 0}}{m_e\Omega_0}} \cos\psi_0. \tag{15.28}$$

In addition, we must also have that

$$y(t=0) = -\sqrt{\frac{2P_{\varphi 0}}{m_e\Omega_0}} + \sqrt{\frac{2P_{\psi 0}}{m_e\Omega_0}} \sin\psi_0. \tag{15.29}$$

Together, Eqs. 15.27, 15.28, and 15.29 specify the values of $P_{\varphi 0}$, $P_{\psi 0}$, and ψ_0 corresponding to the steady-state trajectories in the absence of self-fields.

The characteristic electron phase space corresponding to the equations of motion in the limit in which the self-fields vanish is shown in Figs. 15.1 and 15.2 in both Group I and Group II regimes [4] for $\gamma = 3.0$ and $a_w = 0.2$ [recall that $a_w \equiv eB_w/m_ec^2k_w$]. Since both P_φ and P_ψ are constants, the phase space can be described by a normalized axial momentum

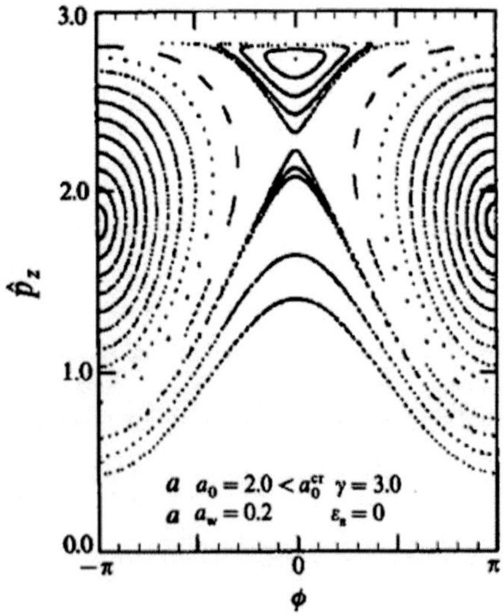

Fig. 15.1 Contour plots of the electron phase space corresponding to Group I trajectories in the integrable limit in which the self-fields are neglected [4]

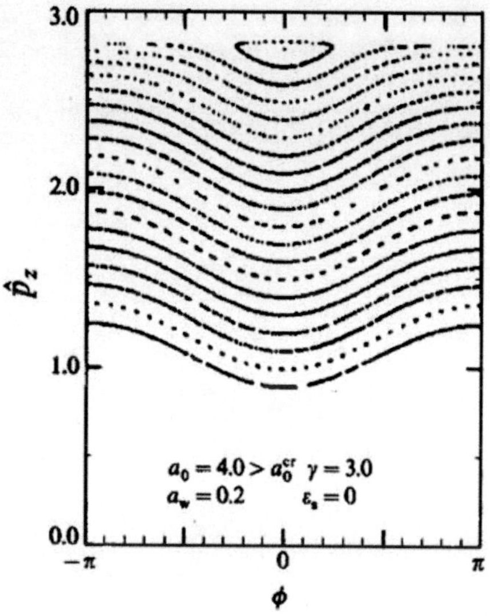

Fig. 15.2 Contour plots of the electron phase space corresponding to Group II trajectories in the integrable limit in which the self-fields are neglected [4]

$$\widehat{p}_z \equiv \frac{1}{m_e c}\left(P_{z'} - P_\varphi + P_\psi\right), \tag{15.30}$$

and φ. The case corresponding to Group I parameters for a weak axial magnetic field shown in Fig. 15.1 corresponds to a_0 [$\equiv eB_w/m_ec^2k_w$] = 2.0. The family of curves here describes the general trajectories, and the elliptic and hyperbolic fixed points evident in the figure correspond to the stable and unstable steady-state trajectories. The strong axial magnetic field (Group II) case is shown in Fig. 15.2, for which $a_0 = 4.0$. Observe that only an elliptic fixed point is evident for this case since, as discussed in Chap. 2, and there are no unstable Group II trajectories in the idealized one-dimensional limit. In the absence of the self-fields, no chaos is evident in either Group I or Group II trajectories (Fig. 15.3).

15.1.5 Chaotic Trajectories

Numerical integration of the complete equations of motion [Eqs. 15.22, 15.23, 15.24, 15.25, and 15.26] is required to demonstrate the onset of chaos. The first case under consideration corresponds to the Group I trajectories shown in Fig. 15.1 [4]. Here, it is assumed that $\Gamma = 3.0$, $a_w = 0.2$, $a_0 = 2.0$ and that $\omega_b^2/c^2k_w^2 = 0.16$. In addition, the initial value of P_ψ is $k_wP_\psi(t = 0)/m_ec = 0.0625$, while the initial axial momentum varies. The existence of chaotic orbits associated with the unstable Group I trajectories is evident in the figure.

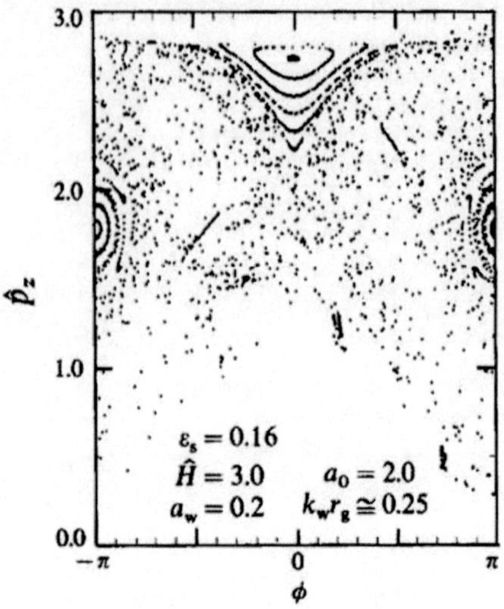

Fig. 15.3 Surface of section of the electron phase space at $\psi = 0$ corresponding to Group I trajectories in the chaotic regime [4]

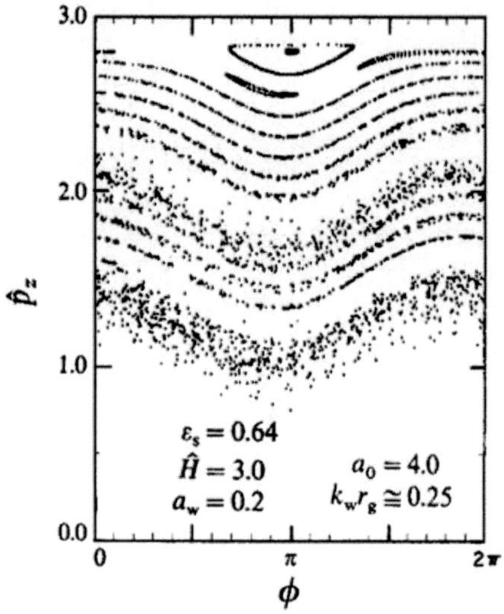

Fig. 15.4 Surface of section of the electron phase space at $\psi = 0$ corresponding to Group II trajectories in the chaotic regime [4]

The case corresponding to Group II parameters [1] is shown in Fig. 11.4 [this corresponds to the integrable limit shown in Fig. 11.2], where $\Gamma = 3.0$, $a_w = 0.2$, $a_0 = 4.0$, $\omega_b^2/c^2k_w^2 = 0.64$, and $k_wP_\psi(t = 0)/m_ec = 0.125$. The existence of chaotic orbits in this case is also evident, although higher values of self-fields are required since these orbits are stable in the absence of the self-fields. In this regard, it should be observed that unstable trajectories are found for Group II orbits in the absence of self-fields when three-dimensional effects are included [see Chap. 2] and that the conclusion that stronger levels of self-fields are required for the onset of chaos in Group II orbits does not hold in the three-dimensional case (Fig. 15.4).

The aforementioned analysis is restricted to the special case of an idealized helical wiggler in which the self-fields are described under the assumption that the axial velocity of the beam is constant and the transverse velocity can be neglected. The natural generalization of this model is to employ the self-consistent axial velocity in the expressions for the self-fields. The results of this generalization [5] yield qualitatively similar to those described for constant axial velocities as long as the densities are not too high.

While the self-fields are necessary in order to obtain chaotic trajectories for a helical wiggler, this is not the case for a planar wiggler. In the planar wiggler geometry, chaotic orbits have been obtained due merely to the presence of an axial solenoidal field [5].

15.2 Chaos in Free-Electron Laser Oscillators

The physics underlying the operation of a free-electron laser is classical, and one can expect that nonlinear dynamics and chaos theory might play an important role in many aspects of the study of these devices [8]. In this section, we discuss the nonlinear dynamics of electrons in the combined fields of the radiation and wiggler. This leads to the study of Hamiltonian motion with one or more degrees of freedom depending on the complexity of the model under consideration. If one regards the one degree of freedom description as the ideal, then the question that is raised is: does the inclusion of higher dimensional effects destroy the ideal picture? For example, are electrons expelled from the beam due to the spatial inhomogeneities of a realistic wiggler field [4, 5] or detrapped from the ponderomotive well by multifrequency fields [9]? While questions of beam confinement are important, the focus of nonlinear dynamics theory is often time asymptotic features of the electron motion, such as KAM surfaces, Lyapunov exponents, and Arnold diffusion. In practice, an individual electron spends only a limited time in the interaction region (except for storage ring free-electron lasers), and thus, the practical issue of beam confinement is not necessarily tied directly to the important issues of nonlinear dynamics.

A second type of nonlinear problem [10] concerns the self-consistent action and reaction of the electrons and the radiation in a free-electron laser oscillator. Here, the radiation bounces back and forth between mirrors, gaining power while interacting with the electron beam and losing power due to output coupling and losses. New electrons are continually injected (except in storage rings where they are re-accelerated) and lose energy (on average) to the field before exiting the interaction region. In any case, due to its open nature, the free-electron laser oscillator is a nonlinear dissipative system as opposed to a Hamiltonian one. The oscillator can, in principle, be run continuously and thus presents itself as a suitable system for the study of nonlinear dynamics and chaos. Indeed, conventional laser oscillators have already been studied extensively in this light.

15.2.1 Return Maps

The simplest example of a nonlinear dissipative system is the one-dimensional return map, and it is not hard to construct a model of a free-electron laser oscillator, which reduces to such a map. To do this, suppose that the electron beam is so energetic that we can assume that it moves at essentially the same speed as the radiation. Further, let us represent the time dependence of the radiation field entering the interaction region as

$$\delta\widehat{a} = \delta a_e(t)\exp(-i\omega t) + c.c., \tag{15.31}$$

where ω is the central frequency of the radiation and $\delta a_e(t)$ is a complex time-dependent amplitude. When the radiation enters the interaction region, it is met by a group of electrons that has yet to be affected by the radiation. Under the assumption of equal beam and radiation speeds, the radiation and electrons move together without slippage and interact resulting in an amplification of the radiation. The radiation that emerges from the interaction region will then depend on the radiation that entered and the beam current. We write this dependence in terms of a nonlinear gain function

$$\delta a_o(t) = [1 + I_b g(|\delta a_e(t - T_L)|)]\delta a_e(t - T_L), \tag{15.32}$$

where $\delta a_o(t)$ is the complex amplitude of the radiation emerging from the interaction region, I_b is a parameter representing the beam current, T_L is the time of flight through the interaction region, and the nonlinear (and possibly complex) gain function g depends only on the amplitude of the radiation due to the fact that the entering electrons are uncorrelated with the radiation phase.

The radiation that emerges from the interaction region is then reflected by a system of mirrors back to the entrance of the interaction region, during which time a fraction of power is removed as output and a fraction is lost due to dissipation. Let the reflection coefficient R denote these loss processes, and write the field at the entrance of the interaction region at time t in terms of the field at the exit at time $t - T_R$ where T_R is the time required for the radiation to return from the exit to the entrance of the interaction region, $\delta a_e(t) = R\delta a_o(t - T_R)$. Combining these equations, we thus arrive at a return map [10]

$$\delta a_e(t + T) = R[\delta a_e(t) + I_b g(|\delta a_e(t)|)]\delta a_e(t), \tag{15.33}$$

where $T = T_R + T_L$ is the total round-trip time for the radiation in the cavity. Such a map can of course be derived more rigorously. For example, (15.33) is precisely of the same form as the delay equation derived in Chap. 9 for the klystron model (9.55) in the absence of slippage. Further, as was discussed in Sect. 9.3, for the special case of negligibly small slippage, the radiation in an oscillator will evolve according to the form of Eq. 15.33.

The return map in Eq. 15.32 is slightly different from that which is traditionally studied, since it involves a continuous and complex amplitude $\delta a_e(t)$. The latter issue is easily resolved simply by taking the magnitude of both sides of Eq. 15.32. The result is a map for the magnitude $|\delta a_e(t)|$ in terms of the magnitude $|\delta a_e(t - T)|$. This map can be advanced independently of phase of $\delta a_e(t)$. By examining the phase of each side of Eq. 15.33, we obtain

$$\arg[\delta a_e(t)] = \arg[\delta a_e(t-T)] - \Omega|\delta a_e(t-T)|, \tag{15.34}$$

where $\Omega = -\arg[R(1 + I_b g)]$ from which the phase of $\delta a_e(t)$ is advanced. This phase does not affect the evolution of the amplitude in the present model.

The behavior of simple return maps of the form described is well known [11]. As the parameter I_b (i.e., the current) is changed, one observes the following behavior. For small I_b, the gain is insufficient to overcome losses, and the solution $\delta a_e(t) = 0$ corresponding to no oscillation is reached asymptotically in time. Above a threshold I_{st}, known as the start current, the $\delta a_e(t) = 0$ solution becomes unstable. In the simple case where g is real, the start current is given by $I_{st} = (1 - R)/(Rg(0))$. If the current exceeds the start current by a moderate amount, the solution asymptotes in time to one in which the amplitude of $\delta a_e(t)$ is constant, $|\delta a_e| = \delta a_o$, and the phase advances by an amount $-\Omega(a_o)$ on each iteration. In this equilibrium, the saturated nonlinear gain just balances losses at all times. In the special case of real g, the equilibrium $\delta a_e = \delta a_o$ is determined by the transcendental equation,

$$I_b = \frac{1-R}{Rg(\delta a_o)}. \tag{15.35}$$

At a current $I_b = I_2$, this equilibrium becomes unstable, and the sequence of a values oscillates between two values. For real g, the critical current I_2 is given by

$$RI_2\delta a_o \frac{\partial g}{\partial \delta a_o} = -2. \tag{15.36}$$

Higher values of current I_b above I_2 lead to the familiar sequence of period doublings until a critical value of current is reached I_∞ at which the sequence becomes chaotic. One can then construct bifurcation diagrams and calculate the *winding number* [8].

$$\omega = \lim_{N\to\infty} \frac{1}{N} \sum_{n+1}^{N} \Omega[\delta a_e(nT)]. \tag{15.37}$$

The bifurcation diagrams appear to be the same as one would obtain for the logistics map.

15.2.2 Electron Slippage

While the study of the return map is appealing because of its simplicity, it cannot be taken too seriously as a model of a free-electron laser. This is because the solutions of the return map eventually violate the assumptions under which the map was derived [10]. The map is a prescription for advancing a continuous time function $\delta a_e(t)$. One must initially specify $\delta a_e(t)$ and the interval $0 < t < T$ and from the map determine $\delta a_e(t)$ at a subsequent time. However, adjacent time slices do not affect each other. Thus, if our initial condition corresponded to a random complex function $\delta a_e(t)$ [as one would expect due to spontaneous noise], the time asymptotic solution for currents in the range $I_{st} < I_b < I_2$, is

$$\delta a_e(t) = \delta a_o \exp[-i\Omega t + i\phi(t)]. \tag{15.38}$$

where $\phi(t)$ is some arbitrary, random, periodic function of time, which depends on the initial conditions. That is, the radiation comes to an equilibrium where its amplitude is a constant, δa_o, but its phase is arbitrary. This effect is responsible for the initial, but erroneous, claim that the free-electron laser at the University of California at Santa Barbara was operating in a single mode [12]. There the constancy of the amplitude of the signal was argued to imply that only a single frequency was present. Subsequent theory [13, 14] demonstrated that the field could be of the form [Eq. 15.38], and owing to the time dependence of $\phi(t)$, many frequencies could still be present. A later measurement of the spectrum [15] confirmed this prediction.

A more serious problem occurs for currents $I_b > I_2$, in which case the sequence of $|\delta a_e(t - nT)|$ values might be, say, a period two sequence. If this is the case, it is impossible for $\delta a_e(t)$ to remain a continuous function of time. It will necessarily develop discontinuities at some sequence of times separated by T.

The assumption that becomes strained under these circumstances is that the beam and radiation move at the same speed. While electrons are in the interaction region, they continually slip behind the radiation, and by this process, information is communicated between different time slices of the radiation field. Observe that even if one slows down the radiation to match the speed of the beam, then dispersion will be introduced, and again the return map will not be valid. The electron slippage is characterized by the parameter

$$\varepsilon = \frac{L}{L_c}\left(\frac{c}{v_z} - 1\right), \tag{15.39}$$

where L and L_c are the length of the interaction region and twice the separation between the mirrors, c and v_z are the speeds of the radiation and the beam, and ε is typically small. In terms of our simple map, the result of including electron slippage is that the term representing the amplification of the radiation in Eq. 15.32 is not simply dependent on $\delta a_e(t)$, but depends in a complicated manner on all values of $\delta a_e(t')$ in the interval $t - \varepsilon T < t' < t$. This was illustrated by Fig. 9.2 in the discussion of slippage in Chap. 9.

A simple way of incorporating slippage into the model has been proposed [10]. It corresponds to consideration of an interaction region that consists of two parts: a prebunching region at the entrance and power extraction region at the exit. This is the optical klystron model that is discussed in Chap. 9. The simple result is that the return map becomes a two-time delay equation,

$$\delta a_e(t+T) = R[\delta a_e(t)] + I_b g(|\delta a_e(t-\varepsilon T)|\delta a_e(t-\varepsilon T)). \tag{15.40}$$

Introduction of slippage modifies the results of the return map in two important ways. First, for cases where the current I_b is only moderately above the start current, the final asymptotic state is one given by Eq. 15.37 but with $\phi(t)$ constant. That is, a single-frequency equilibrium is established. The time T_c to reach this coherent state depends on ε as [13]

$$T_c \approx \frac{T_d}{\varepsilon^2}, \tag{15.41}$$

where $T_d = -T/\ln R^2$ is the decay time of radiation in the empty cavity. As seen in Chap. 9, this formula applies to more realistic models of a free-electron laser as well.

Of greater concern from point of view of this discussion is that the presence of slippage leads to a new route to chaos. For a current I_b between three and four times the start current (depending on parameters), the single frequency equilibrium solution becomes unstable to the overbunch/sideband/spiking mode [16, 17] as shown in Chap. 9. The unstable solution is characterized by modulations of the amplitude $\delta a_e(t)$ with a period $2\varepsilon T$. The values of current for which the instability appears are those values for which the equilibrium amplitude is such that $d(\ln g)/d(\ln \delta a_o) \le -2$. In terms of our simple model, one can understand the onset of this instability in the following way [10]. If the dependence of the gain on the radiation amplitude becomes too strong, then the constant amplitude states becomes unstable to a perturbation, which changes sign every εT units of time. This perturbation constructively reinforces itself. Namely, a large value of the radiation field at time t produces a smaller amplification for the field at time $t + \varepsilon T$. This produces a smaller value of $\delta ae(t + \varepsilon T)$, which then results in a large amplification and a larger value for the field $\delta a_e(t + 2\varepsilon T)$, and so on.

Another feature of this instability is that it can occur even in low-gain oscillators. If we consider situations where the beam current is weak $I_b \ll 1$, then to develop sustained oscillations, one must have a reflection coefficient R close to unity. In such a case, the field is approximately periodic in time with period T. That is, if viewed over a relatively small number of periods $\delta a_e(t)$ would be periodic. Changes in $\delta a_e(t)$ occur only over a long time of the order $T_d = -T/\ln R^2$. In this case, the two-time delay map (15.39) can be transformed to a partial differential delay equation. Let us ascribe the nearly periodic behavior of $\delta a_e(t)$ to two-time variables, $\delta a_e(t) = \delta a_e(t_0, t_s)$, where $\delta a_e(t_0,t_s)$ is periodic in the variable t_0 with period T and changes slowly with t_s on the time scale T. This multiple time scale expansion is identical to that introduced in Chap. 9 where the round-trip index n was used as the slow time variable and the amplitude of the radiation was assumed to be periodic in t' with period T. Insertion of this form into the two-time delay equation gives

$$\delta a_e(t_0, t_s + T) = R[\delta a_e(t_0, t_s) + I_b g(|\delta a_e|)\delta a_e(t_0 - \varepsilon T, t_s - \varepsilon T)]. \qquad (15.42)$$

We now assume that R is close to unity and I_b is small so that a depends weakly on t_s. Thus, in the last term on the right-hand side, we neglect the slippage delay in the t_s dependence of δa_e. Further, we Taylor expand the t_s dependence of δa_e on the left-hand side to obtain,

$$T\frac{\partial}{\partial t_s}\delta a_e(t_0, t_s) = (R-1)\delta a_e(t_0, t_s) + I_b g(|\delta a_e|)\delta a_e(t_0 - \varepsilon T, t_s). \qquad (15.43)$$

Introducing the scaled time τ where $t_s = T\tau/[2(1-R)]$, we obtain an equation with a single parameter $I' = I_b/[2(1-R)]$

$$\frac{\partial}{\partial \tau}\delta a_e(t_0, \tau) = -\frac{1}{2}\delta a_e(t_0, \tau) + I' g(|\delta a_e|)\delta a_e(t_0 - \varepsilon T, t_s). \qquad (15.44)$$

The parameter I' reflects the fact that only the ratio of the current to the losses is of importance in the low gain limit.

The sequence of period doublings associated with the simple return map requires a current $I_b > I_2$, which is many times the start current in the limit $R \rightarrow 1$. This sequence is not permitted in the low gain model. On the other hand, the overbunch instability that arises due to finite slippage typically occurs for currents only several times the start current and is thus more likely to be observed and can be studied with the low-gain model.

Just above the threshold current for the overbunch instability, steady periodic modulations of the amplitude and phase of the radiation signal appear. The spectrum in this case consists of a discrete set of narrow lines spaced by an amount $\Delta f = 1/2\varepsilon T$ corresponding to the period of the overbunch instability. As current is increased, the temporal modulations develop into spikes. The spectrum in this case is still composed of discrete lines; however, the amplitudes of the components displaced from the fundamental increase. An example of such a spectrum appears in Fig. 15.5 where the magnitudes of the complex amplitudes of a set of modes are shown for a numerical simulation with a current about six times the start current. The system of equations, which were solved by computer, are the same as those which were described in Sect. 9.6.4 except a slippage parameter $\varepsilon = 0.25$ was used. Note that the spectrum consists of a set of modes that have a separation of four modes between each large mode. In the final state, the magnitudes of these modes are constant, and the phases are locked in such a way that the slow time frequency shift for each mode is proportional to mode number. As the current is raised, different types of behavior have been observed. In some numerical simulations, the amplitudes of the large modes became time dependent [10] in a periodic fashion, implying that the amplitude of the radiation was quasi-periodic with two frequencies. At higher currents, the

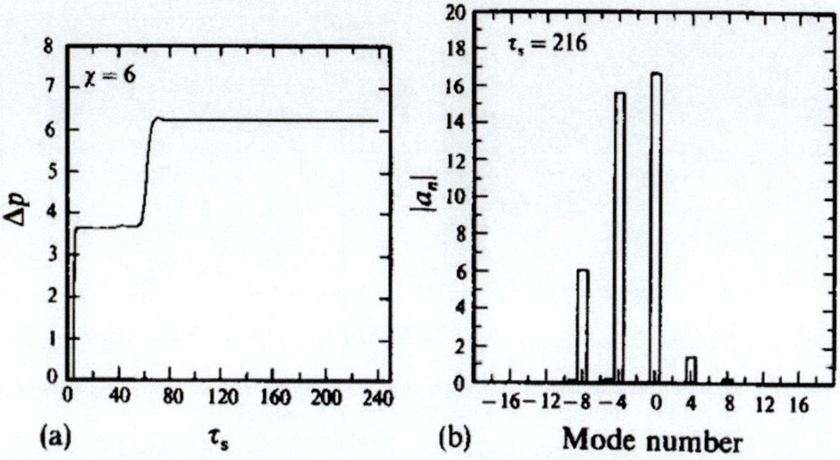

Fig. 15.5 The time history of (**a**) the normalized efficiency Δp and (**b**) the final spectrum from a simulation with 21 modes $\varepsilon = 0.25$, and a current six times the start current [18]

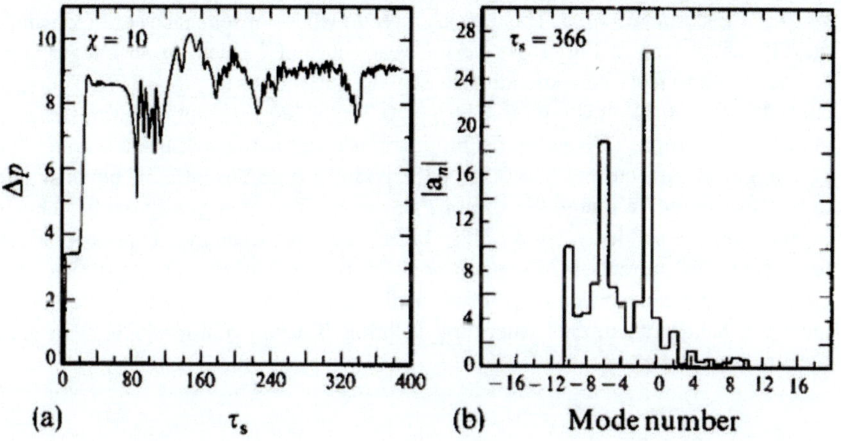

Fig. 15.6 The time history of (**a**) the normalized efficiency Δp and (**b**) the final spectrum from a simulation with 21 modes $\varepsilon = 0.25$ and a current ten times the start current [18]

quasi-periodic behavior makes a transition to chaos. This is illustrated in Fig. 15.6 where the time dependence of the efficiency and the final (nonstationary) spectrum are obtained from a computer run with a current ten times the start current [18]. However, the exact route is not yet known.

An alternate route [8] to chaos is predicted by the map Eq. 15.39. As the current is raised, the radiation goes through a sequence of transitions from (1) single frequency (constant amplitude) to (2) spiking mode (periodic amplitude) and (3) through a sequence of period doublings in which the spikes acquire structure on successively

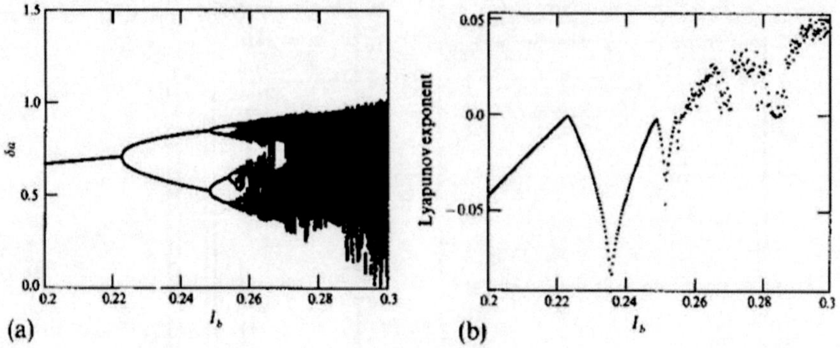

Fig. 15.7 Superimposed sequences of $\delta a(T)$ versus beam current from the map in Eq. 15.39 (**a**), and the corresponding values of the most positive Lyapunov exponents (**b**)

longer time scales given roughly by $T_m = 2^m \varepsilon T$. For higher currents, the periodic behavior gives way to quasi-periodic behavior and eventually chaos.

Figure 15.7 shows a bifurcation diagram for the map Eq. 15.39 with $\varepsilon = 1/8$, $g = 1 - |\delta a_e|^2$, and $R = 0.9$. The time asymptotic values of sequences of field values $a(m\varepsilon T)$ are plotted in Fig. 15.7a versus beam current I_b, with m an integer. The Lyapunov exponents corresponding to Fig. 1a are shown in Fig. 15.7b. The start current is $I_{st} = 0.11$. It is seen that at $I_b = 0.22$, the steady solution bifurcates to a period two solution. Two additional bifurcations occur at higher current producing a period 8 orbit. At a current $I_b = 0.2559$, the period 8 becomes unstable but is unable to become period 16 due to the size of ε. Instead, it becomes quasiperiodic. This is evident on an expanded version of Fig. 11.5b where the Lyapunov exponent is zero for a range of currents $0.2559 < I_b < 0.2566$. For currents above 0.2566, the Lyapunov exponent becomes positive, indicating chaos. At still higher currents, there are bands of currents where the solution is again quasiperiodic with zero Lyapunov exponent.

15.2.3 Pulsed Injection

Up to now, the beam current I_b has been assumed to be a time-independent parameter. However, free-electron lasers are often driven by accelerators such as rf linacs in which the beam consists of a stream of *micropulses* of duration T_p and separation T_a. The separation between *micropulses* T_p and the cavity round-trip time T are made to nearly (but not exactly) coincide. The difference in these times $T_\delta = T - T_a$ (often measured in microns assuming speed of light propagation) is called the cavity detuning.

There has been a great deal of study of free-electron lasers of this type [see Chap. 9] theoretically, numerically, and experimentally. It has been found that at sufficiently large amplitude δa_o, the radiation spontaneously develops spikes with a width determined by the slippage. However, the cavity detuning constitutes an extra parameter, which is effective in suppressing this modulational instability [19, 20]. Due to the extra parameter in this case, as well as the complication of the current pulse waveform, the specific route to chaos in such devices has yet to be determined.

The new effects of pulsed beams and cavity detuning can be incorporated into the derivation of the low-gain model by assuming that the radiation field is periodic in the fast time variable t_0 with period T_a as opposed to the cavity round-trip time T. The result is

$$\left(\frac{\partial}{\partial\tau}+\frac{T_\delta}{2(1-R)}\frac{\partial}{\partial t_0}+\frac{1}{2}\right)\delta a_e(t_0,\tau)$$
$$=I'(t_0-\varepsilon T)g(|\delta a_e(t_0-\varepsilon T,\tau)|)\delta a_e(t_0-\varepsilon T,\tau). \tag{15.45}$$

The field $\delta a_e(t_0,\tau_1)$ is now periodic with period T_a; it will be nonzero for times, which are coincident with the electron *micropulse*. Now assume that the current is peaked at $t_0=0$ with a width T_p characterizing the *micropulse*. If T_p is much smaller than T_a, then it is reasonable to rescale the fast time variable to T_p. That is, define $t_0'=t_0/T_p$, $\varepsilon'=\varepsilon\, T/T_p$ and $\delta=T_\delta/[2T_p\,(1-R)]$, and Eq. 15.44 becomes

$$\left(\frac{\partial}{\partial\tau}+\delta\frac{\partial}{\partial t_0'}+\frac{1}{2}\right)\delta a_e\left(t_0',\tau\right)=I_b\delta a_e g(|\delta a_e|)|_{t_0'-\varepsilon'} \tag{15.46}$$

where $\delta a_e\left(t_0',\tau\right)$ is periodic in t_0' with period T/T_p, which can be taken to be infinity. Preliminary investigations of this model show that it is relatively successful in predicting the onset of spiking in pulsed free-electron lasers. However, the accuracy of the differential time delay equation in the high current regime has not been checked.

15.2.4 Chaos in Storage Rings

It has recently been shown that time dependence of the radiation power in a storage ring can exhibit chaotic behavior when a parameter is varied periodically in time [21, 22]. Recall from the discussion of storage-ring free-electron lasers in Chap. 9 that the mutual interaction of the radiation and the beam gives rise to a weakly damped relaxation oscillation in which periodic bursts of radiation are excited. A large burst of radiation degrades the beam to the extent that the cavity losses exceed the gain. This effectively turns off the radiation. With the radiation off, the beam cools down until the net gain is again positive and a new burst appears. If the ring is stable, the relaxation oscillations eventually decay, and in the final state, the laser power is continuous in time. This relaxation oscillation has been successfully modeled with the following simple set of equations [23],

$$\frac{dI}{dt} = \frac{I(g-p)}{T}$$

$$\frac{d\sigma^2}{dt} = -2\nu_s\left(\sigma^2 - \sigma_0^2\right) + \alpha I \qquad (15.47)$$

$$g = g_0 \exp\left[-k\left(\sigma^2 - \sigma_0^2\right)\right],$$

where I is the intensity of the radiation in a *micropulse*, g and p are the gain and losses per time T, σ^2 is the normalized energy spread in the beam, σ_0^2 is the equilibrium energy spread in the absence of radiation, ν_s is the synchrotron radiation damping rate, and αI represents the rate of heating of the beam due to the interaction with the radiation. The final equation expresses the dependence of the gain on energy spread.

Equation 15.47 represents a two-dimensional, autonomous (time does not appear explicitly on the right-hand side), dissipative system. Such a system cannot be chaotic, and the solutions of this system eventually settle down to an equilibrium in which the energy spread and the radiation intensity are constants, a zero-dimensional attractor. In this respect, the system is rather like a damped nonlinear pendulum. In order to make the system chaotic, one of the parameters is varied periodically in time. This is analogous to a damped driven pendulum [24]. In the experiment, the gain is modulated periodically in time by periodically modulating the frequency of the rf accelerating field. This changes the arrival times of the beam *micropulses* and, hence, the effective cavity detuning. The result is that the gain is modulated at twice the frequency at which the rf field is modulated [this assumes that in the absence of modulation the cavity detuning has been optimized for maximum gain]. The modulation is accounted for in the model by multiplying the gain by a time dependent factor,

$$g = g_0 \exp\left[-k\left(\sigma^2 - \sigma_0^2\right)\right]\left(1 + a \sin^2\Omega t\right), \qquad (15.48)$$

The above system of equations was found to successfully model the results of a series of experiments performed on the Super-ACO storage ring. In particular, both the experiment and the model predicted period doubling in the bursting mode of the radiation. In addition, chaos in the temporal behavior of the radiation was observed for approximately the same parameters as in the experiment.

To summarize, as the current in a free-electron laser oscillator is raised, the radiation field progresses from being single frequency to broadband and chaotic. Simple model maps have been proposed, which predict how the transition to chaos occurs. Certain features of these maps agree with numerical simulation codes; however, detailed correspondence has yet to be verified. Experiments with pulsed beams are most likely to exhibit the behavior described here, but these experiments contain an additional parameter further complicating the picture. Finally, the practical question of whether chaos in these devices needs to be understood rather than simply avoided needs to be addressed.

References

1. A. Lichtenburg, M. Liberman, *Regular and Stochastic Motion* (Springer, New York, 1983)
2. R.Z. Sagdeev, D.A. Usikov, G.M. Zaslavky, *Nonlinear Physics: From the Pendulum to Turbulence and Chaos* (Harwood Academic Publishers, Chur, 1988)
3. E. Ott, Strange attractors and chaotic motions of dynamical systems. Rev. Mod. Phys. **53**, 655 (1981)
4. C. Chen, R.C. Davidson, Choatic particle dynamics in free-electron lasers. Phys. Rev. A **43**, 5541 (1991)
5. L. Michel, A. Bourdier, J.M. Buzzi, Chaos electron trajectories in a free-electron laser. Nucl. Instrum. Methods Phys. Res. **A304**, 465 (1991)
6. G. Spindler, G. Renz, Chaotic behavior of electron orbits in a free-electron laser near magnetoresonance. Nucl. Instrum. Methods Phys. Res. **A304**, 492 (1991)
7. R.C. Davidson, *Physics of Nonneutral Plasmas* (Addison-Wesley, Reading, 1990)
8. T.M. Antonsen Jr., Nonlinear dynamics of radiation in a free-electron laser, in *Nonlinear Dynamics and Particle Acceleration*, ed. by Y.H. Ichikawa, T. Tajima, (AIP Conference Proceedings No. 230, New York, 1991), p. 106
9. S. Riyopoulos, C.M. Tang, Chaotic electron motion caused by sidebands in free-electron lasers. Phys. Fluids **31**, 3387 (1988)
10. N.S. Ginzburg, M.I. Petelin, Multi-frequency generation in free-electron lasers with quasi-opticxal resonators. Int. J. Electron. **59**, 291 (1985)
11. M.J. Fiegenbaum, Quantitative universality for a class of nonlinear transformations. J. Stat. Phys. **19**, 25 (1978)
12. L.R. Elias, G.J. Ramian, J. Hu, A. Amir, Observation of single-mode operation in a free-electron laser. Phys. Rev. Lett. **75**, 424 (1986)
13. T.M. Antonsen, B. Levush, Mode competition and suppression in free-electron laser oscillators. Phys. Fluids B **1**, 1097 (1989)
14. T.M. Antonsen, B. Levush, Mode competition and control in free-electron laser oscillators. Phys. Rev. Lett. **62**, 1488 (1989)
15. B.G. Danly, S.G. Evangelides, T.S. Chu, R.J. Tempkin, G. Ramian, J. Hu, Direct spectral measurements of a quasi-cw free-electron laser. Phys. Rev. Lett. **65**, 2251 (1990)
16. Y.L. Bogomolov, V.L. Bratman, N.S. Ginzburg, M.I. Petelin, A.D. Yunakovsky, Nonstationary generation in free-electron lasers. Opt. Commun. **36**, 209 (1981)
17. W.B. Colson, R.A. Freedman, Synchrotron instability for long pulses in free-electron lasers. Opt. Commun. **46**, 37 (1983)
18. B. Levush, T.M. Antonsen Jr., Effect of nonlinear mode competition on the efficiency of low gain free-electron laser oscillator. SPIE Vol. **1061**, 2 (1989)

19. R.W. Warren, J.E. Sollid, D.W. Feldman, W.E. Stein, W.J. Johnson, A.H. Lumpkin, J.C. Goldstein, Near-ideal lasing with a uniform wiggler. Nucl. Instrum, Methods Phys. Res. **A285**, 1 (1989)
20. R.W. Warren, J.C. Goldstein, The generation and suppression of synchrotron sideband. Nucl. Instrum. Methods Phys. Res. **A272**, 155 (1988)
21. M. Billardon, Storage-ring free-electron laser and chaos. Phys. Rev. Lett. **65**, 713 (1990)
22. M. Billardon, Chaotic behavior of the storage-ring free-electron laser. Nucl. Instrum. Methods Phys. Res. **A304**, 37 (1991)
23. P. Ellaume, Macrotemporal structure of free-electron lasers. J. Phys. **45**, 997 (1984)
24. C. Grebogi, E. Ott, J.A. Yorke, Chaos, strange attractors, and fractal basin boundaries in nonlinear dynamics. Science **238**, 585 (1987)

Appendix

Electron Beam Optics

This appendix is intended to provide simple descriptions of basic beam optics and accelerator concepts that are important for free-electron lasers. It is not meant to be an exhaustive description of these topics, which are well studied and covered in detail in other sources that are listed in the references [1–4]. Rather, here, we present the minimal description needed to understand what is covered in the preceding chapters. The level of description is such that someone who has taken a course on classical mechanics at the advanced undergraduate or introductory graduate level should be comfortable with the presentation.

Equations of Motion

A main feature of interest of an electron beam produced by an accelerator and used in a free-electron laser is that all its electrons have essentially the same momentum value and that they be spatially localized. To achieve this, the beam must be confined transversely to its direction of propagation, which we take here to be the z-direction. The equations of motion for individual electrons are the relativistic generalization of Newton's law in the presence of Lorentz force,

$$\frac{d}{dt}\mathbf{p} = -e\left[\mathbf{E} + \frac{\mathbf{v}\times\mathbf{B}}{c}\right], \tag{A.1}$$

and

H. P. Freund, T. M. Antonsen, Jr., *Principles of Free Electron Lasers*,
https://doi.org/10.1007/978-3-031-40945-5

$$\frac{d}{dt}\mathbf{x} = \mathbf{v}, \tag{A.2}$$

where $\mathbf{p} = \gamma m_e \mathbf{v}$ is the mechanical momentum, m_e is the rest mass, and the relativistic factor is defined in two equivalent ways,

$$\gamma = 1/\sqrt{1 - |\mathbf{v}/c|^2} = \sqrt{1 + |\mathbf{p}/(m_e c)|^2}. \tag{A.3}$$

Here, we have expressed the electric and magnetic fields in electrostatic units (ESU) with c being the speed of light. The electric and magnetic fields are assumed to be macroscopic quantities. That is, they are smoothly varying in space and are determined by the average charge and current densities. The effects of the discreteness of the electron charge are ignored.

Hamiltonian Formulation

Equations (A.1, A.2, and A.3) can be cast in Hamiltonian form by introducing scalar and vector potentials (Φ,$\mathbf{A}$) to express the fields,

$$\mathbf{E} = -\nabla\Phi - \frac{1}{c}\frac{\partial}{\partial t}\mathbf{A}, \tag{A.4}$$

and

$$\mathbf{B} = \nabla \times \mathbf{A}, \tag{A.5}$$

and by defining a canonical momentum,

$$\mathbf{P} = \mathbf{p} - \frac{e}{c}\mathbf{A}. \tag{A.6}$$

The equations of motion then take the form,

$$\frac{d}{dt}\mathbf{P} = -\frac{\partial}{\partial \mathbf{x}}H(\mathbf{P}, \mathbf{x}, t), \tag{A.7}$$

and

$$\frac{d}{dt}\mathbf{x} = \frac{\partial}{\partial \mathbf{P}}H(\mathbf{P}, \mathbf{x}, t). \tag{A.8}$$

The Hamiltonian H is given by

$$H = m_e c^2 \gamma - e\Phi, \tag{A.9}$$

where the relativistic factor is now given by

$$\gamma = \left[1 + \frac{1}{m_e^2 c^2}\left|\mathbf{P} + \frac{e}{c}\mathbf{A}\right|^2\right]^{1/2}. \tag{A.10}$$

As mentioned, we take z to be the direction of propagation of the beam. Further, it is desirable to have a version of the equations that describes the evolution of electron coordinates as the electrons travel in z. We then convert Eqs. A.7, A.8, A.9, and A.10, which have time as the independent variable to a system in which z is the independent variable. We label the components of vectors transverse to z with a subscript $\perp$ and retain a subscript z for components parallel to the main direction of propagation. It will be useful in what follows to express an incremental change in Hamiltonian in terms of incremental changes in its arguments,

$$dH = d\mathbf{x}_\perp \cdot \frac{\partial H}{\partial \mathbf{x}_\perp} + d\mathbf{P}_\perp \cdot \frac{\partial H}{\partial \mathbf{P}_\perp} + dz\frac{\partial H}{\partial z} + dP_z\frac{\partial H}{\partial P_z} + dt\frac{\partial H}{\partial t}. \tag{A.11}$$

This will be used to re-express ratios of differentials occurring when we divide each of Hamilton's equations by *dz/dt* to make z the independent variable. For example,

$$\frac{d\mathbf{x}_\perp}{dz} = \frac{d\mathbf{x}_\perp/dt}{dz/dt} = \frac{\partial H/\partial \mathbf{P}_\perp}{\partial H/\partial P_z} = -\left.\frac{dP_z}{d\mathbf{P}_\perp}\right|_{dt = dH = dz = dx_\perp = 0} = -\frac{\partial P_z}{\partial \mathbf{P}_\perp}, \tag{A.12}$$

where the third equality in A.12 follows from A.11 by setting the increments dt, dH, dz, $d\mathbf{x}_\perp$ to zero. Repeating for the time derivative of each of the canonical coordinates recasts the differential equations of motion A.7 and A.8 as a Hamiltonian set with z as the independent variable and with the z component of canonical momentum serving as the Hamiltonian,

$$\frac{d\mathbf{x}_\perp}{dz} = -\frac{\partial P_z}{\partial \mathbf{P}_\perp}, \quad \frac{d\mathbf{P}_\perp}{dz} = \frac{\partial P_z}{\partial \mathbf{x}_\perp}, \tag{A.13}$$

and

$$\frac{dH}{dz} = -\frac{\partial P_z}{\partial t}, \quad \frac{dt}{dz} = \frac{\partial P_z}{\partial H}. \tag{A.14}$$

Here, the z component of canonical momentum is expressed as a function of the other variables,

$$P_z(\mathbf{x}_\perp, \mathbf{P}_\perp, H, z, t) = m_e c \left\{ \left(\frac{H + e\Phi}{m_e c^2} \right)^2 - \left[1 + \frac{1}{m_e^2 c^2} \left| \mathbf{P}_\perp + \frac{e}{c} \mathbf{A}_\perp \right|^2 \right] \right\}^{1/2} - \frac{e}{c} A_z. \tag{A.15}$$

Phase Space Density

Given the equations of motion, one can write an evolution equation for the distribution function giving the phase space density of electron coordinates. Here,

$$df = f(\mathbf{x}_\perp, \mathbf{P}_\perp, H, t; z) d^2x_\perp d^2P_\perp dH dt \tag{A.16}$$

is the number of particles occupying the small volume $d^2x_\perp d^2P_\perp dHdt$ centered at the point $\mathbf{x}_\perp$, $\mathbf{P}_\perp$, H, t in phase space. In writing the distribution function such that it depends on only six arguments and z, rather than six times the number of particles and z, assumes that particles respond only to the average fields. Thus, effects of discrete particle interactions are eliminated from the description.

Conservation of the number of particles then leads to a continuity equation for the distribution function, which in turn leads to the constancy of the distribution function along particle trajectories due to the incompressibility of the phase space flow inherent to Hamilton's equations,

$$\frac{df}{dz} = \frac{\partial f}{\partial z} + \frac{d\mathbf{x}_\perp}{dz} \cdot \frac{\partial f}{\partial \mathbf{x}_\perp} + \frac{d\mathbf{P}_\perp}{dz} \cdot \frac{\partial f}{\partial \mathbf{P}_\perp} + \frac{dt}{dz}\frac{\partial f}{\partial t} + \frac{dH}{dz}\frac{\partial f}{\partial H} = 0. \tag{A.17}$$

Phase Space Volume Conservation

The constancy of the distribution function along the phase space trajectory implies that the phase space volume occupied by a given number of particles is fixed as the particles travel in z. This fact in turn leads to the important role of the concept of beam emittance. Emittance is a measure of the volume of phase space occupied by the beam, and under the right conditions, suitably defined values of emittance are constant. For example, according to the description of the beam dynamics given by Eqs. A.13, A.14, A.15, A.16, and A.17, the total admittance representing the six-dimensional volume $d^2x_\perp d^2P_\perp dHdt$ is conserved. Lower dimensional emittances are also important. If the fields are time independent, then the last term in A.17 vanishes, and time can be integrated over in A.17 leading to a continuity equation preserving the four-dimensional emittance representing the volume

$d^2x_\perp d^2P_\perp$. Further, if the equations of motion in the two-dimensional transverse plane separate into two uncoupled systems, the two-dimensional emittances become relevant.

Paraxial Equations

We have yet to make the paraxial approximation, namely, focus on the case in which the transverse momenta and beam dimensions are small. The paraxial equations follow by expanding the axial canonical momentum assuming small potentials and small transverse momenta,

$$P_z(\mathbf{x}_\perp, \mathbf{P}_\perp, H, z, t) = P_{z0} - \frac{m_e\gamma_0}{P_{z0}} q\Phi - \frac{1}{2P_{z0}}\left|\mathbf{P}_\perp + \frac{e}{c}\mathbf{A}_\perp\right|^2 - \frac{e}{c}A_z, \tag{A.18}$$

where $P_{z0}(H) = c^{-1}\sqrt{H^2 - m_e^2c^4}$ and $\gamma_0(H) = H/(m_ec^2)$. The evolution of the energy H is to a good approximation uncoupled from the transverse motion. In particular, assuming the dominant time dependence is due to the axial component of the vector potential, which represents the accelerating electric field, Eq. A.14 becomes

$$\frac{dH}{dz} = -\frac{\partial P_z}{\partial t} = \frac{e}{c}\frac{\partial A_z}{\partial t}. \tag{A.19}$$

If the transverse extent of the beam is small compared with the spatial scale of the vector potential, all particles will experience the same acceleration and have the same energy.

The equations for the transverse canonical variables take the following form in this case,

$$\frac{d\mathbf{x}_\perp}{dz} = -\frac{\partial P_z}{\partial \mathbf{P}_\perp} = \frac{1}{P_{z0}}\left(\mathbf{P}_\perp + \frac{e}{c}\mathbf{A}_\perp\right), \tag{A.20}$$

and

$$\frac{d\mathbf{P}_\perp}{dz} = \frac{\partial P_z}{\partial \mathbf{x}_\perp} = \frac{m_e\gamma_0}{P_{z0}} e\frac{\partial \Phi}{\partial \mathbf{x}_\perp} - \frac{e}{cP_{z0}}\frac{\partial \mathbf{A}_\perp}{\partial \mathbf{x}_\perp}\cdot\left(\mathbf{P}_\perp + \frac{e}{c}\mathbf{A}_\perp\right) - \frac{e}{c}\frac{\partial A_z}{\partial \mathbf{x}_\perp}, \tag{A.21}$$

or in more familiar terms

$$\frac{dx_\perp}{dz} = \frac{\mathbf{v}_\perp}{{}_{z0}}, \tag{A.22}$$

and

$$\frac{d\mathbf{p}_\perp}{dz} = \frac{q}{v_{z0}}\left[-\frac{\partial\Phi}{\partial\mathbf{x}_\perp} + \frac{(\mathbf{v}\times\mathbf{B})_\perp}{c}\right], \tag{A.23}$$

where $v_{z0} = P_{z0}/\gamma_0 m_e$, $\mathbf{p}_\perp = \gamma_0 m_e \mathbf{v}_\perp$, and we have again assumed that the transverse components of the vector potential are time independent.

Focusing Magnetic Fields

Confinement of the beam in the transverse direction requires that the beam electrons experience an average force radially inward. This is achieved in practice by subjecting the electrons to a transverse Lorenz force that alternates in sign but has an average confining effect. As the beam current is low, we consider vacuum magnetic fields that are curl free as well as divergence free, $\mathbf{B} = -\nabla\chi(\mathbf{x}_\perp, z)$

$$\nabla_\perp^2\chi + \frac{\partial^2}{\partial z^2}\chi = 0. \tag{A.24}$$

The variation in the z-direction is weak in comparison to the transverse size of the beam. Consequently, we require $\nabla_\perp{}^2\chi = 0$. Expanding the scalar magnetic potential in a power series around the axis gives,

$$\chi = -B_D r\cos(\theta - \theta_D) - \frac{1}{2}B'_Q r^2 \sin[2(\theta - \theta_Q)], \tag{A.25}$$

where we have expressed transverse position in polar coordinates, ($x = r\cos\theta$, $y = r\sin\theta$). Here, B_D is the amplitude of a dipole magnetic field aligned at an angle θ_D with respect to the x-axis. The Cartesian components of this field are given by $B_x = B_D \cos\theta_D$ and $B_y = B_D \sin\theta_D$. Such a field would be used to deflect all the particles in the beam by an amount determined by their energy, as in a chicane that will be discussed subsequently. The quadrupole field is given in terms of the amplitude B'_Q and orientation angle θ_Q. When $\theta_Q = 0$, the Cartesian components of the magnetic field take the form, $B_x = B'_Q\, y$ and $B_y = B'_Q x$. Such a field is depicted in Fig. A.1. It would deflect electrons traveling in the +z-direction and on the x-axis away from the origin and deflect the electrons on the y-axis toward the origin. In principle, all the parameters in Eq. A.25 can be functions of z provided the scale length of their variation is sufficiently long compared with the distance to the axis.

We have retained in Eq. A.25 only the first two powers of radius in an expansion about the axis of propagation. As a consequence, the transverse dynamics will involve differential equations that are linear in electron displacement from the axis. This will greatly simplify the discussion here. It is important to note that higher-order terms leading to nonlinear dynamics are important, but will not be treated here. The interested reader is referred to the references at the end of this appendix.

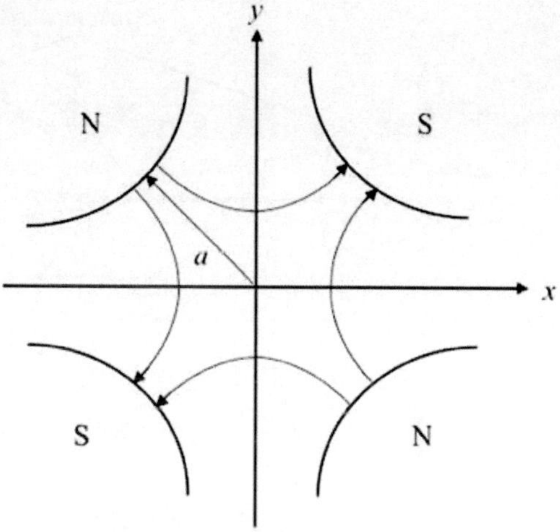

Fig. A.1 Illustration of a quadrupole field pattern

Periodically Varying Quadrupole Fields: FODO Lattice

We now consider the focusing ability of periodically varying quadrupole fields by examining solutions to the equations of motion in the transverse plane. We assume that there is no axial acceleration in this region, so the energy, H; relativistic factor, γ_0; and axial velocity, v_{z0}, are all constant. The magnetic field is taken to be a quadrupole field with $\theta_Q = 0$. The resulting equations for the transverse displacement are

$$\frac{d^2x(z)}{dz^2} = \frac{eB'_Q(z)}{m_e c \gamma_0 v_{z0}} x(z), \tag{A.26}$$

and

$$\frac{d^2y(z)}{dz^2} = -\frac{eB'_Q(z)}{m_e c \gamma_0 v_{z0}} y(z). \tag{A.27}$$

Equations A.26 and A.27 show that at a particular value of z, if the quadrupole field is focusing in one direction, then it is defocusing in the other. However, if B'_Q varies periodically in z, the net effect may be focusing in both directions. This follows from Floquet theory. If the coefficients in the above equations are periodic with period λ_Q, then solutions are of the form of periodic functions with period λ_Q multiplied by exponentials of the form exp(ikz). For some cases, k will be real implying that the trajectory is bounded (the electron is confined). In this case, the motion of the electron is two-wave number periodic with one period being the period of the magnetic fields and the second period being associated with the strength of the focusing. This second period is known as the betatron period.

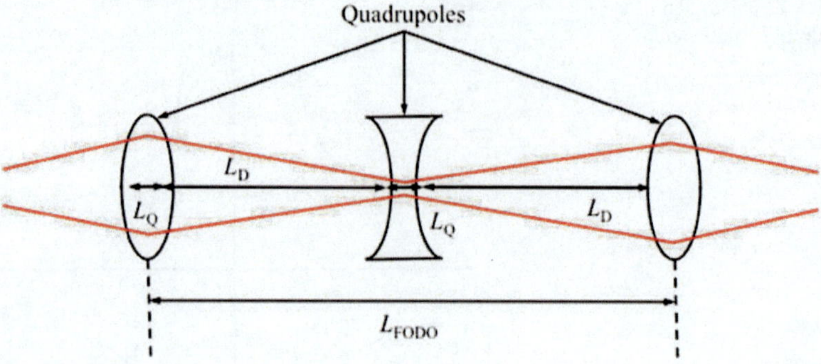

Fig. A.2 Schematic of a FODO lattice containing alternating focusing and defocusing quadrupoles, separated by drift region

As an example, we take the profile of the quadrupole field to consist of two regions of opposing field each of length L_Q, separated by field-free drift regions of length L_D,

$$\frac{eB'_Q(z)}{m_e c\gamma_0 v_{z0}} = \begin{cases} \kappa^2 & 0 < z \le L_Q \\ 0 & L_Q < z \le L_Q + L_D \\ -\kappa^2 & L_Q + L_D < z \le 2L_Q + L_D \\ 0 & 2L_Q + L_D < z \le 2L_Q + 2L_D \end{cases} \tag{A.28}$$

Such a profile is illustrated in Fig. A.2 and is termed a FODO lattice owing to its alternating focusing and defocusing magnetic field regions that are separated by drift space. We make a subsidiary approximation that L_Q is short so that the beam particles are deflected while passing through the quadrupole, but not significantly displaced. This is equivalent to the thin lens approximation in geometric optics and leads to the following map for the displacement of a particle from the axis and its derivative,

$$x'(L_Q) = x'(0) - \int_0^{L_q} dz\kappa^2 x(z) \simeq x'(0) - L_Q\kappa^2 x(0), \tag{A.29}$$

$$x(L_Q) = x(0) + \int_0^{L_Q} dz\ x'(z) \simeq x(0). \tag{A.30}$$

These relations can be expressed as a matrix multiplication,

$$\begin{pmatrix} x(L_Q) \\ x'(L_Q) \end{pmatrix} = \begin{bmatrix} 1 & 0 \\ -\kappa^2 L_Q & 1 \end{bmatrix} \begin{pmatrix} x(0) \\ x'(0) \end{pmatrix}, \tag{A.31}$$

where the column vector formed by the displacement and its deflection is known as the trace-space vector. The general case of an arbitrary axial profile of magnetic field also results in a matrix relation of the form of A.31, except that all elements of the matrix are nonzero. A general requirement on such a matrix is that its determinant must be united by virtue of the fact that the two-dimensional phase space area $dxdx' = dxdp_x/p_z$ is conserved in the case of linear quadrupole magnetic fields. Likewise, a similar matrix relation applies to the electron coordinates entering and leaving the field-free drift space,

$$\begin{pmatrix} x(L_Q + L_D) \\ x'(L_Q + L_D) \end{pmatrix} = \begin{bmatrix} 1 & L_D \\ 0 & 1 \end{bmatrix} \begin{pmatrix} x(L_Q) \\ x'(L_Q) \end{pmatrix}. \tag{A.32}$$

After traveling through one period of the lattice consisting of two drift spaces and two oppositely directed quadrupole field sections, the electron coordinates at the beginning and end of the period, $L_{FODO} = 2(L_Q + L_D)$, are given by a product matrix,

$$\begin{pmatrix} x(L_{FODO}) \\ x'(L_{FODO}) \end{pmatrix} = \begin{bmatrix} 1 & L_D \\ 0 & 1 \end{bmatrix} \begin{bmatrix} 1 & 0 \\ \kappa^2 L_Q & 1 \end{bmatrix} \begin{bmatrix} 1 & L_D \\ 0 & 1 \end{bmatrix} \begin{bmatrix} 1 & 0 \\ -\kappa^2 L_Q & 1 \end{bmatrix} \begin{pmatrix} x(0) \\ x'(0) \end{pmatrix}$$

or multiplying out the matrices

$$\begin{pmatrix} x(L_{FODO}) \\ x'(L_{FODO}) \end{pmatrix} = \begin{bmatrix} 1 - \zeta - \zeta^2 & L_D(2 + \zeta) \\ -L_D^{-1}\zeta^2 & 1 + \zeta \end{bmatrix} \begin{pmatrix} x(0) \\ x'(0) \end{pmatrix} \equiv \underline{\underline{M}} \cdot \underline{x}(0) \tag{A.33}$$

where

$$\zeta = L_D L_Q \kappa^2 = 5.9 \times 10^{-4} \frac{L_D L_Q[\mathrm{cm}^2] B'_Q[\mathrm{G/cm}]}{\beta\gamma_0}. \tag{A.34}$$

Matrix $\underline{\underline{M}}$ represents the one-period matrix, and $\underline{x}(z)$ represents the trace space column vector formed by the displacement and deflection. Note again the determinant of the matrix is unity. The trace-space vector can be represented as a superposition of the two eigenvectors $\underline{u}_{1,2}$ of the matrix $\underline{\underline{M}}$. After passage through N periods of the lattice, the coefficients of the two eigenvectors will be multiplied by factors $\lambda_{1,2}^N$ where $\lambda_{1,2}$ are the eigenvalues of the matrix associated with each eigenvector. The product of the eigenvalues is unity as a consequence of the fact that the matrix determinant is unity, which in turn follows from conservation of trace-space area. Thus, the eigenvalues are either complex conjugates of magnitude unity, or reciprocals with one eigenvalue greater than unity. In the former case, the trace-space coordinates oscillate with distance through the lattice corresponding to confinement. While in the latter case, the trace-space coordinates grow exponentially with distance indicating loss of confinement.

For the matrix in A.33, the eigenvalues are $\lambda_{1,2} = \cos\sigma \pm i\sin\sigma$ with $\cos\sigma = 1 - \zeta^2/2$ for $0 < \zeta^2 < 4$ corresponding to stable confined motion. One can think of the stability condition, $|\zeta| = |L_D L_Q \kappa^2| < 2$ as a restriction on the length of the drift space, or the strength of the quadrupole field. If this condition is not satisfied, the trajectories are overfocused in the sense that a trajectory with positive excursion and no deflection on entering one element of the quadrupole lattice will be deflected and cross the axis entering the next element with an even larger magnitude of displacement. On the other hand, if $|\zeta| = |L_D L_Q \kappa^2| \ll 1$, then the angle σ is small, and particles' trace-space coordinates change slightly from element to element. In this case, the effect of the lattice can be replicated by replacing the spatially varying quadrupole fields with an axially uniform, inwardly directed linear focusing force

$$\mathbf{F}_{\perp} = -m_e \gamma_0 v_{z0}^2 k_\beta^2 \mathbf{x}_{\perp}, \tag{A.35}$$

where $k_\beta = \sigma / L_{FODO}$ is the betatron wavenumber, L_{FODO} is the period of the lattice, and the angle σ is defined above in terms of the eigenvalues of the lattice matrix.

Twiss Parameters

In general, the single period matrix $\underline{\underline{M}}$ has three independent elements with the fourth determined by the requirement that the matrix's determinant is unity. The eigenvalues of the matrix, when the lattice is stable, are in the form of the complex conjugate pair, $\lambda_{1,2} = \exp(\pm i\sigma)$, where $\cos\sigma = \mathrm{Tr}\left[\underline{\underline{M}}\right]/2$, and Tr [] signifies the trace of the matrix, that is, the sum of the diagonal elements. The matrix can be characterized by the parameters $\widehat{\alpha}, \widehat{\beta}, \widehat{\gamma}$, known as the Twiss or Courant-Snyder parameters [1],

$$\underline{\underline{M}}(\sigma) = \left[\cos\sigma + \widehat{\alpha}\sin\sigma \widehat{\beta}\sin\sigma - \widehat{\gamma}\sin\sigma \cos\sigma - \widehat{\alpha}\sin\sigma \right]. \tag{A.36}$$

Here, the condition on the determinant being unity imposes the relation $\widehat{\beta}\widehat{\gamma} = 1 + \widehat{\alpha}^2$. For the example of the FODO lattice of Fig. A.2, the Twiss parameters are

$$\widehat{\alpha} = -\sqrt{\frac{(2+\zeta)}{(2-\zeta)}}, \tag{A.37a}$$

$$\hat{\beta} = \frac{2L_D}{\zeta}\sqrt{\frac{2+\zeta}{2-\zeta}}, \tag{A.37b}$$

and

$$\hat{\gamma} = \frac{2\zeta}{L_D\sqrt{(2+\zeta)(2-\zeta)}}. \tag{A.37c}$$

If one diagonalizes the matrix in A.36 using its eigenvalues and left and right eigenvectors, one has for the Nth power of the matrix,

$$\underline{\underline{M}}^N = \begin{bmatrix} 1 & 1 \\ -\frac{\hat{\alpha}-i}{\hat{\beta}} & -\frac{\hat{\alpha}+i}{\hat{\beta}} \end{bmatrix} \begin{bmatrix} e^{iN\sigma} & 0 \\ 0 & e^{-iN\sigma} \end{bmatrix} \begin{bmatrix} -i\frac{\hat{\alpha}+i}{2} & -i\frac{\hat{\beta}}{2} \\ i\frac{\hat{\alpha}-i}{2} & i\frac{\hat{\beta}}{2} \end{bmatrix} = \underline{\underline{M}}(N\sigma). \tag{A.38}$$

Here, the elements of the matrix on the left are the two column vectors that are the right-side eigenvectors of $\underline{\underline{M}}$. The elements of the matrix on the right are the two row vectors that are the left side eigenvectors of $\underline{\underline{M}}$. This form can thus be used to readily compute the trace-space trajectories through structures of arbitrary length.

Trace-Space Evolution

The conservation of phase space density and the evolution of second moments of the trace space coordinates are related in the case of linear focusing fields. The evolution of moments of the trace space density under the action of a matrix of the form (A.38) can be calculated as follows. Suppose that (x_0, x'_0) are the initial trace space coordinates, and under operation of a general, linear matrix

$$x = M_{11}x_0 + M_{12}x'_0, \tag{A.39a}$$

$$x' = M_{21}x_0 + M_{22}x'_0, \tag{A.39b}$$

are the coordinates some distance down the lattice. The following averages may then be found,

$$\langle x^2 \rangle = M_{11}^2\langle x_0^2 \rangle + M_{12}^2\left\langle x_0'^2 \right\rangle + 2M_{11}M_{12}\langle x_0 x'_0 \rangle \tag{A.40a}$$

$$\left\langle x'^2\right\rangle = M_{21}^2\left\langle x_0^2\right\rangle + M_{22}^2\left\langle x'^2_0\right\rangle + 2M_{22}M_{21}\left\langle x_0x'_0\right\rangle \tag{A.40b}$$

$$\left\langle xx'\right\rangle = M_{11}M_{21}\left\langle x_0^2\right\rangle + M_{22}M_{12}\left\langle x'^2_0\right\rangle + (M_{11}M_{21} + M_{22}M_{12})\left\langle x_0x'_0\right\rangle. \tag{A.40c}$$

Combining the moments as follows yields:

$$\varepsilon_{x,RMS}^2 \equiv \left\langle x^2\right\rangle\left\langle x'^2\right\rangle - \left\langle xx'\right\rangle^2 = \left\langle x_0^2\right\rangle\left\langle x'^2_0\right\rangle - \left\langle x_0x'_0\right\rangle^2, \tag{A.41}$$

where we have used the fact that the determinant of matrix $\underline{\underline{M}}$ is unity. Here, $\varepsilon_{x,RMS}$ is the root-mean-square trace space emittance and is a constant for one-dimensional motion in a linear focusing field.

The evolution with distance, more specifically with period number, of the trace space point (x,x'_N) is given by the matrix (A.38) multiplied by the initial condition. For a given initial condition, the trace space point can be expressed as a super position of the two eigenvectors of the one-period matrix, $\underline{\underline{M}}(\sigma)$. Thus, we can write

$$\begin{pmatrix} x_N \\ x'_N \end{pmatrix} = c_1 \exp[iN\sigma]\begin{pmatrix} 1 \\ -\dfrac{\widehat{\alpha} - i}{\widehat{\beta}} \end{pmatrix} + c_2 \exp[-iN\sigma]\begin{pmatrix} 1 \\ -\dfrac{\widehat{\alpha} + i}{\widehat{\beta}} \end{pmatrix}, \tag{A.42}$$

where the coefficients $c_{1,2}$ are determined by the initial conditions. Since the trace space vector must be real, we have $c_2^* = c_1 = |c_1|\exp[i\arg(c_1)]$. Consequently, the trace space point can be represented in terms of real functions,

$$\begin{pmatrix} x_N \\ x'_N \end{pmatrix} = 2|c_1|\begin{pmatrix} \cos\sigma_N \\ -\dfrac{\widehat{\alpha}}{\widehat{\beta}}\cos\sigma_N - \dfrac{1}{\widehat{\beta}}\sin\sigma_N \end{pmatrix}. \tag{A.43}$$

where $\sigma_N = N\sigma + \arg(c_1)$. Treating σ_N as a continuous variable, we see that the locus of points in trace space on which the trajectory falls is an ellipse in the $x - x'$ plane as shown in Fig. A.3. The area of the ellipse based on (A.43) is

$$A = 4\int_0^{2|c_1|} dx \frac{2|c_1|}{|\widehat{\beta}|}\sqrt{1 - \left(\frac{x}{2|c_1|}\right)^2} = \frac{4\pi}{|\widehat{\beta}|}|c_1|^2 \equiv \varepsilon_x \tag{A.44}$$

The RMS emittance is related to the trace space area in the following way. If the trace space distribution $f(x,x',z)$ is constant within an ellipsoidal region of phase space with area ε_x, then evaluation of (A.41) gives

$$\varepsilon_{x,RMS} = \frac{1}{4}\varepsilon_x. \tag{A.45}$$

Normalized Emittance

The discussion so far has focused on the trace-space area, which is conserved when the particle energy is constant. However, during the course of acceleration particle, energy increases, and trace space area changes. Under the specific case of adiabatic acceleration, the change in trace space area can be predicted. Assuming the accelerating field $\partial A_z/\partial t$ in A.19 to be constant across the beam, all particles will gain energy at the same rate. This will make the energy, H, in Eqs. A.20, A.21, A.22, and A.23 a prescribed function of axial distance. If energy changes slowly over the distance required for an electron to complete its betatron orbit, then the actions

$$I = \oint \mathbf{P}_\perp \cdot d\mathbf{x}_\perp = \oint \left(\mathbf{p}_\perp - \frac{e}{c}\mathbf{A}_\perp\right) \cdot d\mathbf{x}_\perp, \tag{A.46}$$

will be adiabatic invariants. If the magnetic fields have two planes of symmetry as in the case of the FODO lattice, there will be two such conserved actions. The contribution of the vector potential to A.46 represents a magnetic flux linking the betatron orbit that can be ignored. The action (A.46) can then be related to the trace space area via,

$$\oint \mathbf{p}_\perp \cdot d\mathbf{x}_\perp = m_e \gamma v_{z0} \oint \frac{d\mathbf{x}_\perp}{dz} \cdot d\mathbf{x}_\perp = m_e \gamma v_{z0} \varepsilon_x. \tag{A.47}$$

On the basis of Eq. A.47 and the many assumptions leading to it that are not always satisfied, a normalized emittance is introduced,

$$\varepsilon_n = \gamma \beta \varepsilon_x, \tag{A.48}$$

where $\beta = v_z/c$. If the normalized emittance is conserved, the trace space emittance decreases as electrons gain energy. For relativistic motion, the factor $\beta \approx 1$ and is often omitted from the definition of normalized emittance.

The Matched Beam Radius

The matched beam size is an important characteristic for determining the performance of a free-electron laser. In general, the rms beam extent is determined by the emittance, the beam energy, and the Twiss-β function and can be expressed as

$$x_{rms} = \sqrt{\frac{\varepsilon_n \hat{\beta}}{\gamma}}, \tag{A.49}$$

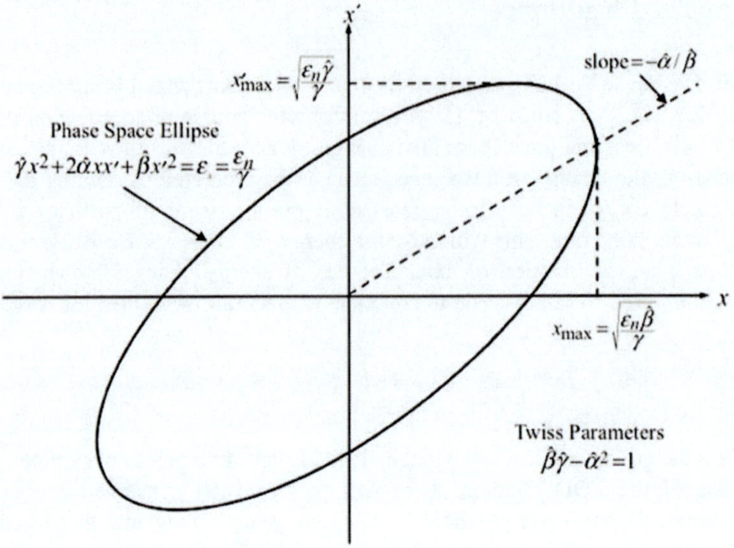

Fig. A.3 Locus of points in trace space that are mapped to each other by the matrix $\underline{\underline{M}}$ (A.37)

for any arbitrary transport line including the weak-focusing imposed by a wiggler or by the strong-focusing in a FODO lattice. The Twiss-β function for a weak-focusing wiggler corresponds to the betatron period discussed in Chap. 2.

In regard to a strong-focusing FODO lattice, as discussed previously, the ellipse in Fig. A.3 is a locus of points in trace space that map to each other after repeated applications of the FODO lattice map. All points within the ellipse remain within the ellipse as the beam passes through many quadrupole pairs. If the trace space distribution is initially uniform within the ellipse, it will remain uniform within the ellipse. As a consequence, at the end of each pair of quadrupoles, the x-extent of the beam will be the same, as indicated in the figure. This is taken to be the matched beam size. The optimal Twiss-β function for a FODO lattice is given by Eqs. A.34 and A.37b. Had the beam initially occupied a different ellipse, but with the same area, the radial extent of the beam would not be the same at the end of each element of the FODO lattice. Rather, it would oscillate from element to element. This is considered to be the case of an unmatched beam.

Bunch Compression in Chicanes

So far, we have discussed transverse dynamics of beam particles, primarily in time-independent fields. Here, we discuss the aspects of the dynamics associated with the temporal behavior of the beam. High-energy beams usually consist of pulses of current with durations that are much shorter than the period of the RF field that is responsible for the acceleration. The accelerating RF field is represented in our Hamiltonian description by the time-dependent axial component of the vector potential in Eq. A.19. If this term is oscillating sinusoidally in time with an angular frequency ω_0, then particles passing through the accelerating element at different times will be accelerated by different amounts. As it is desired that all beam particles have essentially the same energy, the duration of the beam pulse, T_p, must be short, $\omega_0 T_p \ll 1$. However, if the electric field changes during the duration of the pulse, there will be variations in beam energy, and these variations will be correlated with particle arrival time.

To model these correlations, we first consider particle dynamics in a phase space with coordinates arrival time at the accelerating gap and energy (t,H). We then transform these coordinates to new coordinates ($s = -ct$, $\gamma = H/m_e c^2$), where s is distance along the beam propagation path. We consider a beam pulse of duration T_p, length $l_{initial} = cT_p$ average energy H_0, and central arrival time $t_0 = -s_0/c$ at the accelerating gap. A plot of this beam in the scaled phase space $s - \gamma$ is depicted as the ellipse on the left in Fig. A.4.

Each time slice of the beam will receive a different increment of energy on passing through the gap. Let the vector potential in A.19 have the following form, $A_z(z,t) = A_0(z)\cos(\omega_0 t)$, where the amplitude $A_0(z)$ is localized in space near the accelerating gap at z_g. After passing through the gap, particles will have relativistic factor

$$\gamma = \gamma_0 + \Delta\gamma \sin[\omega_0(t - t_0) + \omega_0 t_0] \simeq \gamma_0 + \Delta\bar{\gamma} + (s_0 - s)\Delta\gamma', \tag{A.50}$$

where the parameters characterizing the relativistic factor are

$$\Delta\gamma = -\int dz \frac{e\omega_0}{m_e c^3} A_0(z), \tag{A.51a}$$

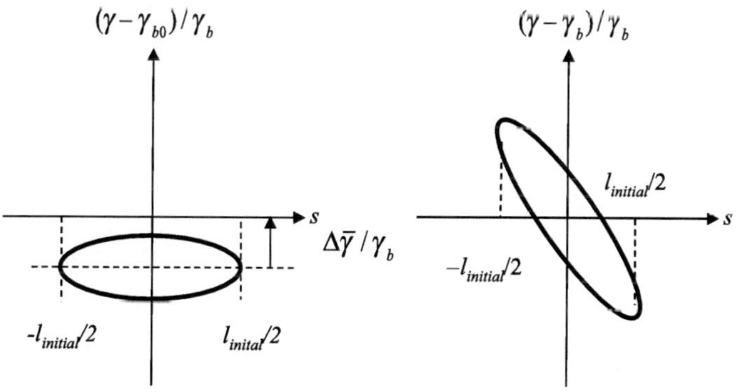

Fig. A.4 Energy – position phase space for a beam before and after an accelerating gap. The distribution after the gap show an energy chirp

$$\Delta\overline{\gamma} = \Delta\gamma \sin \omega_0 t_0, \tag{A.51b}$$

and

$$\Delta\gamma' = (\omega_0/c)\Delta\gamma \cos(\omega_0 t_0). \tag{A.51c}$$

The new distribution is shown as the right-hand ellipse in Fig. A.4. The systematic variation of mean energy with time (or equivalently $s = -ct$) is known as a chirp, and it can be used to compress the pulse in time when the beam is passed through a chicane.

The beams depicted in Fig. A.4 show a distribution of electron beam energies. As a consequence, one expects that the beam pulse will disperse in time due to the dispersion in propagation speeds. However, for ultra-relativistic beams with large relativistic factor γ, the propagation speed is close to the speed of light and relatively insensitive to energy, $v_z = c(1 - 1/\gamma^2)^{1/2}$. Thus, dispersion of the beam in time is a weak process. Dispersion in time can be enhanced by passing the beam through a chicane, which works by making the length of the travel path energy-dependent.

A schematic representation of a chicane is shown in Fig. 12.1. It consists of four regions of dipole magnetic field placed in the path of the electron beam. The sign of the magnetic field in the regions is selected so that the beam is first deflected from its axis, then deflected to a path parallel to the axis, and then deflected twice, returning the beam to its original axis. Because the deflections by the magnetic field are inversely proportional to the effective relativistic mass γm_e, the paths taken by beam electrons with different energy are different. Here, lower energy electrons in Fig. 12.1 have a longer travel path than higher energy electrons. What this means is

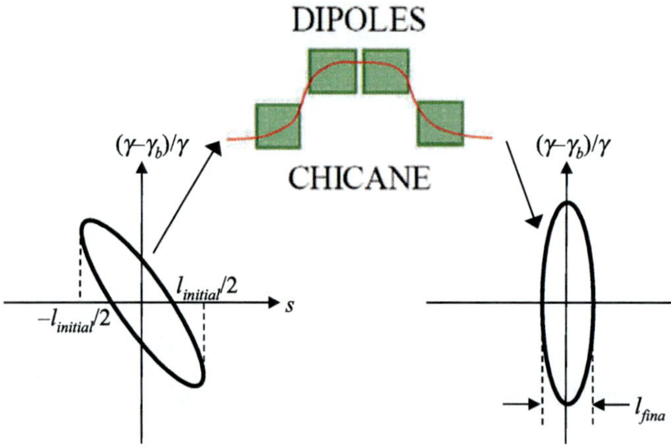

Fig. 12.1 Energy – time (or position) phase space for a single pulse of electrons, before and after passing through a dipole chicane

Fig. A.5 Dipole magnet showing deflection of an electron. A chicane consists of four of these

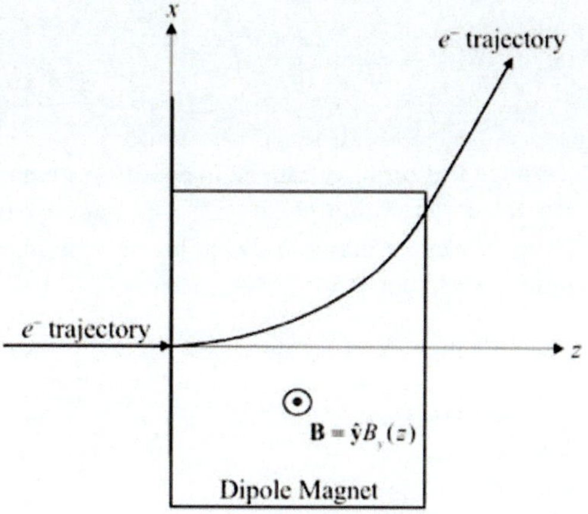

that a beam with a chirp such that in Figs. A.4 and 12.1, where the lower energy electrons come first, and the higher energy electrons come later, will compress and become shorter in time.

The degree of arrival time dispersion imparted by a chicane can be calculated from the paraxial equations. Considering the situation displayed in Fig. A.5, we can take the magnetic field $B_y(z)$ to be directed out of the page and generated from a vector potential,

$$A_x(z) = \int_{-\infty}^{z} dz' B_y(z'). \tag{A.52}$$

The Hamiltonian for this system (A.18) is now independent of spatial coordinates x and y, and consequently, the corresponding canonical momenta, P_x and P_y, are conserved quantities. We take these to be zero, assuming the entering beam is traveling parallel to the axis. This gives from A.18 and A.14,

$$t_f = t_i + \int_{z_i}^{z_f} dz \frac{\partial P_z}{\partial H} = \frac{\partial P_{z0}}{\partial H}\left[z_f - z_i + \frac{1}{2P_{z0}^2}\int_{z_i}^{z_f} dz \left|\frac{eA_z(z)}{c}\right|^2\right]. \tag{A.53}$$

It is the last term in A.53 that contributes the significant energy dependence to the travel time between the initial and final points. If the chicane magnetic field is modeled by regions of dipole field B_y of length L_B, with the first and second and third and fourth regions separated by distances L_D, the integral in (A.53) can be approximated as

$$\Delta t(\gamma) = \frac{1}{2P_{z0}} \frac{\partial P_{z0}}{\partial H} \int_{z_i}^{z_f} dz \left| \frac{eA_x}{c} \right|^2 = \frac{1}{c\beta^3\gamma^2} \left(L_D + \frac{2}{3} L_B \right) \left| \frac{eB_y L_B}{m_e c^2} \right|^2. \tag{A.54}$$

Through this term, variations in energy,γ, give rise to variations in travel time. Thus, if a relativistic beam pulse with mean energy factor $\gamma \gg 1$, a front to back energy chirp $\delta\gamma$, and a duration T_D, is to be compressed, one would pick parameters in (A.54) such that

$$\delta\gamma \frac{d\Delta t(\gamma)}{d\gamma} = -T_D. \tag{A.55}$$

Phase Space Maps

Equations A.50 and A.54 describe maps that take the coordinate pair (s_{in}, γ_{in}) before an element and map the pair to a new set of coordinates after the element (s_{out}, γ_{out}). In the case of the accelerating gap, (A.50), the mapping is

$$\gamma_{out} = \gamma_{in} + \Delta\gamma \sin[\omega_0 (s_0 - s_{in})/c + \omega_0 t_0], \tag{A.56a}$$

$$s_{out} = s_{in}. \tag{A.56b}$$

In the case of the chicane, (A.54), the mapping is

$$\gamma_{out} = \gamma_{in}, \tag{A.57a}$$

$$s_{out} = s_{in} - \frac{1}{(\beta\gamma)_{in}^2} \left(L_D + \frac{2}{3} L_B \right) \left| \frac{eB_y L_B}{m_e c^2} \right|^2. \tag{A.57b}$$

Both maps are area preserving in the sense that

$$\left| \frac{\partial s_{out}}{\partial s_{in}} \frac{\partial \gamma_{out}}{\partial \gamma_{in}} - \frac{\partial \gamma_{out}}{\partial s_{in}} \frac{\partial s_{out}}{\partial \gamma_{in}} \right| = 1. \tag{A.58}$$

Linearizations of these maps produce equations of the form of A.50. The coefficients in these equations are then labeled with subscripts based on which coordinates are being related. If we order the six-dimensional phase space coordinates $(x, P_x, y, P_y, s, \gamma)$ then s and γ are the fifth and sixth elements. The coefficient relating changes in s due to changes in γ is the denoted with subscripts 5 and 6,

$$\partial s_{out}/\partial \gamma_{in} = R_{56} = 1.1 \times 10^{-3} \frac{\left(L_B[\mathrm{cm}] B_y[G]\right)^2}{\gamma_{in}^3} \left(L_D[\mathrm{cm}] + \frac{2}{3} L_B[\mathrm{cm}]\right). \quad \text{(A.59)}$$

Higher-order derivatives are then indicated with repeated subscripts,

$$\frac{1}{2} \partial^2 s_{out}/\partial \gamma_{in}^2 = T_{566}. \quad \text{(A.60)}$$

Coherent Synchrotron Radiation (CSR)

Incoherent synchrotron radiation was addressed extensively in Chap. 3 for the case of electrons moving through undulators, that is, periodic magnetic fields. Because it is important to achieve short bunches with high peak currents in free-electron lasers driven by rf linacs, strong chicanes are used to compress the bunches. Coherent synchrotron radiation (CSR) occurs when the bunch length is shorter than the characteristic synchrotron radiation wavelength, and this is commonly found in the strong chicanes used to compress the electron bunches. The radiation from these particles has a characteristic spectral and angular distribution that can be calculated using the classical laws of electrodynamics as in Chap. 3. This is important because CSR in these chicanes leads to increases in both the emittance and energy spread, and this can negatively impact the radiation in the wiggler. Here, we first consider the radiation by particles moving with fixed speed. Radiation by single particles and an ensemble of particles will be addressed. Subsequently, we will consider the reaction of the radiation on the particles that modifies the particle trajectories and can cause unwanted bunching and energy spread.

The starting point is the distribution of radiated energy per unit frequency and per unit solid angle for a given current distribution as presented in the text by Jackson [5],

$$\frac{dI}{d\omega d\Omega} = \frac{\omega^2}{4\pi^2 c^3} \left| \int d^3x \, \mathbf{n} \times [\mathbf{n} \times \mathbf{J}(\omega, \mathbf{k} = \omega\mathbf{n}/c)] \right|^2. \quad \text{(A.61)}$$

Here, $\mathbf{J}(\omega,\mathbf{k})$ is the spatial and temporal Fourier transform of the current density, and the unit vector $\mathbf{n}$ points radially away from the current distribution, and its direction is labeled by the solid angle Ω. Representing the current density in terms of a set of discrete electrons gives

$$\mathbf{J}(\omega, \mathbf{k} = \omega\mathbf{n}/c) = -\sum_i e \int_{-\infty}^{\infty} dt \;\; \mathbf{v}_i(t) \exp[i\omega(t - \mathbf{x}_i(t) \cdot \mathbf{n}/c)]. \quad \text{(A.62)}$$

Generally, the contribution to the radiation is the strongest for times when the phase of the exponential in the integrand in A.62 is varying most slowly with time. This occurs when the instantaneous velocity is as close as possible to being parallel to the radiating direction **n**. Thus, the radiation is directed mainly along the tangent to the trajectory. If we consider the electrons to be moving along a planar trajectory ($x(z)$, $y = 0$) in the $x - z$ plane, and take the radiating direction to be in the $y - z$ plane, $\mathbf{n} = \widehat{\boldsymbol{e}}_y \sin\psi + \widehat{\boldsymbol{e}}_z \cos\psi$, then the main contributions to the integral in (A.61) come from z positions where there is a local maximum of $x(z)$ that is where the trajectory is closest to being parallel to the unit vector **n**. We label such a point (x_0,z_0) and parameterize the trajectory as a function of path length s from that point,

$$x(s) = x_0 - \frac{s^2}{2R}, \tag{A.63a}$$

and

$$z(s) = z_0 + s - \frac{s^3}{6R^2}, \tag{A.63b}$$

where R is the local radius of curvature of the trajectory, $R^{-1} = -d^2x(s)/ds^2$, and it can be verified that $(dx/ds)^2 + (dz/ds)^2 \cong 1$, to second order in s. In the limit in which the motion is ultra-relativistic, $\gamma \gg 1$, and correspondingly the angle of elevation of **n** is small, $\cos\psi \cong 1 - \psi^2/2$, the exponent then becomes,

$$\omega(t - \mathbf{x}_i(t) \cdot \mathbf{n}/c) \simeq \omega t_i - \frac{\omega z_0}{c} + \frac{\omega}{2}\left[(\gamma^{-2} + \psi^2)\tilde{t} + \frac{c^2\tilde{t}^3}{3R^2}\right], \tag{A.64}$$

where t_i is the time the ith electron arrived at z_0 and $\tilde{t} = t - t_i$. Under the same assumptions, the directional factor multiplying the exponential becomes

$$\mathbf{n} \times (\mathbf{n} \times \mathbf{v}) = (\widehat{\mathbf{y}} \sin\psi \cos\psi - \widehat{\mathbf{z}} \sin^2\psi)v_z - \widehat{\mathbf{x}} v_x \simeq \widehat{\mathbf{y}}\psi c + \widehat{\mathbf{x}}\frac{c^2\tilde{t}}{R}, \tag{A.65}$$

and the transformed current density becomes

$$\mathbf{n} \times (\mathbf{n} \times \overline{\mathbf{J}}) = -\sum_i ec \int_{-\infty}^{\infty} dt \left(\frac{c\tilde{t}}{R}\widehat{\mathbf{x}} + \psi\widehat{\mathbf{y}}\right) \exp\left[i\omega t_i - \frac{i\omega z_0}{c} + \frac{i\omega}{2}\left[\varepsilon\tilde{t} + \frac{c^2\tilde{t}^3}{3R^2}\right]\right], \tag{A.66}$$

where $\varepsilon = \gamma^{-2} + \psi^2$. This integral can be expressed in terms of Bessel functions of fractional order by changing the variable of integration to $\tau = \Omega_R \tilde{t} \varepsilon^{-1/2}$, $\Omega_R = c/R$ is the instantaneous rotation rate, and using

$$\int_{-\infty}^{\infty} d\tau \ \exp\left[i\frac{3}{2}\xi\left(\tau+\frac{1}{3}\tau^3\right)\right]=\frac{2}{3^{1/2}}K_{1/3}(\xi), \tag{A.67a}$$

and

$$\int_{-\infty}^{\infty} d\tau \ \tau\exp\left[i\frac{3}{2}\xi\left(\tau+\frac{1}{3}\tau^3\right)\right]=i\frac{2}{3^{1/2}}K_{2/3}(\xi), \tag{A.67b}$$

where

$$\xi=\frac{\omega}{3\Omega_R}\varepsilon^{3/2}. \tag{A.68}$$

This gives for the distribution of radiation intensity,

$$\frac{dI}{d\omega d\Omega}=3\frac{F(\omega)e^2}{\pi^2 c\varepsilon}\xi^2\left[K_{2/3}^2(\xi)+\left(\frac{\psi}{\varepsilon}\right)^2K_{1/3}^2(\xi)\right]. \tag{A.69}$$

The first term in A.69 is the intensity of radiation with electric field polarized in the plane of the motion, in this case in the x-direction, and the second term is the contribution from radiation with electric field polarized out of the plane of motion, in this case the y-direction.

The dependence of the radiation intensity on the number of electrons depends on the distribution of arrival times. This dependence is controlled by the double summation factor

$$F(\omega)=\sum_{i,i'}\exp[i\omega(t_i-t_{i'})]. \tag{A.70}$$

In the extreme case, all N electrons arrive at the same time $F=N^2$, and the radiation is the same as would be produced by a single particle of charge eN. If the electrons arrive uncorrelated over a period of time much longer than the inverse of the frequency, then the phases in the double sum are independent and uniformly distributed and $F=N$. If the electron arrival times are anticorrelated due to the fact that electron repel each other, then $F<N$ is possible. If the electrons form a bunch such that their arrival times are independent and identically distributed and given by the probability density function $f(t)$, which has a temporal dependence given by the bunch shape, then

$$\overline{F}(\omega)=\prod_j\int dt_jf(t_j)\sum_{i,i'}\exp[i\omega(t_i-t_{i'})]=N+N(N-1)\left|\int dt\ f(t)e^{i\omega t}\right|^2. \tag{A.71}$$

Thus, the factor $\overline{F}(\omega)$ makes a transition from N^2 at low frequencies below the inverse of the pulse duration, to N at high frequencies much above the inverse pulse duration. This transition will be smooth if the temporal profile $f(t)$ is smooth. If, however, the temporal profile acquires structure within the pulse duration, the radiation will be enhanced. Further, since there are typically a large number of electrons in a bunch, $N \approx 10^{10}$, the frequency must be substantially higher than the inverse bunch duration for the radiation to be incoherent.

The energy and angle dependence of the radiation are controlled by the factor $\varepsilon = \gamma^{-2} + \psi^2$ and through the argument of the Bessel functions ξ, defined in Eq. A.68. Integrating over frequency [5], the energy radiated per unit solid angle by a single electron is given by

$$\frac{dI}{d\Omega} = \frac{7}{16}\frac{e^2}{|R|}\varepsilon^{-5/2}\left[1 + \frac{5}{7}\frac{\psi^2}{\varepsilon}\right]. \tag{A.72}$$

Thus, for high-energy electrons, the radiation is contained within a cone of angle $\psi \sim \gamma^{-1}$.

The frequency and angular distributions of the radiation are determined primarily by the dependence on ξ in Eq. A.69. The radiation is peaked roughly for $\xi \cong 1$ through the dependence on the Bessel functions, and the amount of radiation becomes small quickly when $\xi > 1$. For a given frequency, the radiation is confined to angles such that $\varepsilon = \gamma^{-2} + \psi^2 < (3\Omega_R/\omega)^{2/3}$. Thus, for frequencies higher than $\omega_{\max} = 3\gamma^3\Omega_R$, the radiation becomes small.

We now illustrate the force of interaction between pairs of electrons moving in a bunch in the s-direction. If the electrons are moving in a straight line with the same speed, the force of interaction is easily calculated in a frame in which the electrons are stationary. In that frame, the field is derived from the electrostatic Coulomb potential, and the vector potential vanishes. On Lorentz transformation back to the lab frame, the potentials become

$$\phi = -\frac{e\gamma}{\left[s_\perp^2 + \gamma^2(s - vt)^2\right]^{1/2}}, \tag{A.73a}$$

and

$$A_s = -\frac{e\gamma\beta}{\left[s_\perp^2 + \gamma^2(s - vt)^2\right]^{1/2}}, \tag{A.73b}$$

where $\gamma = (1 - \beta^2)^{-1/2}$, and $\beta = v/c$. The resulting electric field $E_s = -\partial\phi/\partial s - c^{-1}\,\partial A_s/\partial t$ is given by

$$E_s = -\frac{e(s - vt)\gamma}{\left[s_\perp^2 + \gamma^2(s - vt)^2\right]^{3/2}}, \tag{A.74}$$

and the force is $F_s = -eE_s$. This field falls to zero according to an inverse square law, and it is reduced from the static value due to the fact that the electric field lines around a relativistic charge become concentrated in the transverse direction.

When the electrons follow a curved path, they accelerate, giving off radiation, and the radiation field modifies the Coulomb interaction force between beam particles. To evaluate this, we use the retarded solution of the wave equation in the Lorenz gauge,

$$[\phi, \mathbf{A}]_{x,t} = \int d^3x' dt' \frac{[\rho, \mathbf{j}/c]_{x',t'}}{|\mathbf{x} - \mathbf{x}'|} \delta(t' - t + |\mathbf{x} - \mathbf{x}'|/c), \qquad \text{(A.75)}$$

which gives the potentials at point $(\mathbf{x},t)$ due to source electrons at points $(\mathbf{x}',t')$. We take the charge and current density for the source electrons to be that of a single charge,

$$[\rho, \mathbf{j}]_{x',t'} = -e\delta(\mathbf{x}' - \mathbf{x}_p(t'))\left[1, \mathbf{v}_p(t')\right], \qquad \text{(A.76)}$$

where $\mathbf{v}_p(t) = d\mathbf{x}_p(t)/dt$ is the electron's trajectory. The particular curved trajectory of the source electron we will consider is illustrated in Fig. A.6.

The integrand in the four-dimensional integral in (A.75) is a product of four delta functions. Care must be taken in evaluation of the integral because the arguments of the delta functions are interrelated. To evaluate the integrals, we represent the four delta functions as the product of two delta functions whose arguments are the spatial coordinates transverse to the direction of travel of the electron at the retarded time, multiplied by a delta function whose argument is distance along the direction of travel, and finally multiplied by the delta function singling out the retarded time,

$$d^3x' dt' \delta(\mathbf{x}' - \mathbf{x}_p(t'))\delta(t' - t + d(\mathbf{x}', \mathbf{x})/c) = d^2s'_\perp ds' dt' \delta(\mathbf{s}'_\perp)\delta(\widehat{s})\delta(\widehat{t}), \qquad \text{(A.77)}$$

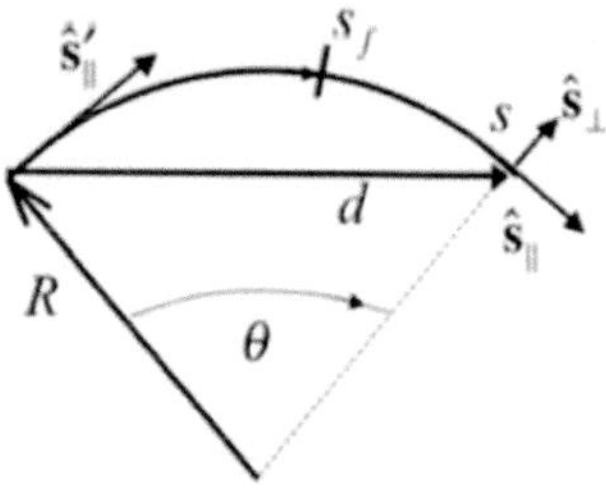

Fig. A.6 Curved field electron trajectory. An electron contributes to the potentials at the point labeled s at time t on the right. The contribution comes from an earlier time t' and when the electron is at the point labeled s' on the left. The radiation travels along the straight path of length d, while the field electron follows a curved path arriving at the point labeled s_f at time t

where $d(\mathbf{x}',\mathbf{x}) = |\mathbf{x}' - \mathbf{x}|$ is the distance between the source point and the observation point, $\hat{t} = t' - t + d(s',\mathbf{x})/c$ and $\hat{s} = s' - s_p(t')$. The integrals over time and distance along the direction of travel may be evaluated by changing variables, $dt'ds' = d\hat{t}d\hat{s}/J$, where the Jacobian J is given by

$$J = \left| \frac{\partial \hat{t}}{\partial t'} \frac{\partial \hat{s}}{\partial s'} - \frac{\partial \hat{t}}{\partial s'} \frac{\partial \hat{s}}{\partial t'} \right| = 1 - \beta \frac{\partial d}{\partial s'}. \tag{A.78}$$

As a result, we find for the potentials

$$\phi(\mathbf{x}, t) = -\frac{e}{dJ}, \tag{A.79a}$$

and

$$\mathbf{A}(\mathbf{x}, t) = -\frac{e\beta}{dJ} \hat{\mathbf{s}}', \tag{A.79b}$$

where $\hat{\mathbf{s}}'$ is a unit vector pointing along the direction of travel at the time of emission.

We are now in a position to calculate the electric field at the observation point. We specialize to the component of field in the direction of travel at the observation point $\hat{\mathbf{s}}$. We note from Fig. A.6 that if the emission and observation points are separated by an arc of length $s - s' \equiv \Delta s$ subtending an angle $\theta = \Delta s/R$, then

$$d(s, s') = 2R \left| \sin \frac{\Delta s}{2R} \right|. \tag{A.80}$$

This gives for the Jacobian (A.77)

$$J = 1 - \beta \frac{\partial d}{\partial s'} = 1 - \beta \cos \frac{\Delta s}{2R}. \tag{A.81}$$

Also, the location of the field electron at time t, which started at position s' at time t', is given by $s_p(t) = s' + u(t - t') = s - \Delta s + \beta d(\Delta s)$. Thus, we have

$$s - s_p(t) = \Delta s - 2\beta R \left| \sin \frac{\Delta s}{2R} \right|. \tag{A.82}$$

To construct the electric field, we need to take the derivative of the electrostatic potential with respect to distance s at constant time t and the derivative of the vector potential with respect to t at constant s. It can be seen from A.79, A.80, A.81, and A.82 that all quantities depend on s and i in the combination $s - s_p(t)$ through their dependence on Δs. Thus, we have, using A.79a and A.79b

$$E_s = -\frac{\partial \phi}{\partial s} - \frac{\hat{\mathbf{s}}}{c} \cdot \frac{\partial}{\partial t}\mathbf{A} = -\left(1 - \beta^2 \hat{\mathbf{s}} \cdot \hat{\mathbf{s}}'\right) \frac{\partial}{\partial s}\phi - \beta\phi\hat{\mathbf{s}} \cdot \frac{\partial}{c\partial t}\hat{\mathbf{s}}', \tag{A.83}$$

where $\phi = -e/(dJ)$ according to (A.79). In evaluating the derivative with respect to s in Eq. A.83, it is important to remember that it is taken with t held fixed. From (A.82) one sees,

$$\left.\frac{\partial}{\partial s}\right|_t = \frac{\partial \Delta s}{\partial s}\frac{\partial}{\partial \Delta s} = \frac{1}{1 - \beta \cos(\Delta s/2R)}\frac{\partial}{\partial \Delta s}. \tag{A.84}$$

Further, from Fig. A.6, one can derive,

$$\hat{\mathbf{s}} \cdot \frac{\partial}{\partial t}\hat{\mathbf{s}}' = \frac{v}{R}\sin\left(\frac{\Delta s}{2R}\right). \tag{A.85}$$

Considering the two terms in A.83, the first one is the modified Coulomb field, which exhibits an essentially inverse square law dependence on the separation between the source point and the observation point. The law is modified slightly due to the curvature of the trajectory. The second term, which is present only when the trajectory is curved, represents an additional electric force due to the radiated field. This force falls with distance only in the inverse power of the separation, as is expected as the radiated Poynting flux, which is quadratic in fields, falls as the inverse square of distance. The presentation here has been simple. However, a more general calculation can be found in textbooks on electromagnetism [5, 6], and the features of the interaction force found here are present in the more general calculations. The upshot is that beam bunches with temporal structure radiate more strongly than those with smooth temporal profiles in bends of the accelerator, and the radiated fields can influence neighboring electrons in the bunch and enhance the temporal structure.

The presentation of coherent synchrotron interaction here is quite idealized. In particular, the field equations have been solved for beam particles isolated from any structure. In practice, the beam propagates in a pipe, which is most often a conductor. This affects the fields, which can now be thought of as a superposition of waveguide modes, with each mode having a cutoff frequency, and a group velocity below the speed of light. Since beams are often highly relativistic, $\gamma \gg 1$, the reduction in the group velocity for a portion of the spectrum substantially changes the nature of the interaction force. As a consequence, calculation of coherent synchrotron interactions has become a computationally intensive effort [7–9] that is still an active field of research at this writing.

References

1. M. Reiser, *Theory and Design of Charged Particle Beams* (Wiley-VCH, Weinheim, 2008). isbn:978-3-527-40741-5
2. D.A. Edwards, M.J. Syphers, *An Introduction to the Physics of High Energy Accelerators* (Wiley-VCH, Weinheim, 2004). isbn:978-0-471-55163-8
3. A. Wolski, *Beam Dynamics Is High Energy Particle Accelerators* (Imperial College Press, 2014). isbn:978-1-78326-277-9
4. S. Bernal, *A Practical Introduction to Beam Physics and Particle Accelerators* (Morgan & Claypool Publishers, San Rafael, 2016). isbn:978-1-6817-4012-6
5. J.D. Jackson, *Classical Electrodynamics* (Wiley, 1999). isbn:0-471-30932-X
6. D. Griffiths, *Introduction to Electrodynamics* (Prentice Hall, 1999). isbn:0-13-805326-X
7. T. Agoh, K. Yokoya, Calculation of coherent synchrotron radiation using mesh. Phys Rev. STAB **7**, 054403 (2004)
8. D. Gillingham, T. Antonsen, Calculation of coherent synchrotron radiation in toroidal waveguides by paraxial wave equation. Phys. Rev. STAB **10**, 054402 (2007)
9. R. L. Warnock, D. L. Bizozero, An efficient computation of coherent synchrotron radiation in a rectangular chamber, applied to resistive wall heating, arXiv:1606.02671 (2016)

Index

H. P. Freund, T. M. Antonsen, Jr., *Principles of Free Electron Lasers*,
https://doi.org/10.1007/978-3-031-40945-5